Inside SolidWorks® *2003*

David Murray

THOMSON
★
DELMAR LEARNING

Australia   Canada   Mexico   Singapore   Spain   United Kingdom   United States

Inside SolidWorks 2003
David Murray

Business Unit Director:
Alar Elken

Executive Editor:
Sandy Clark

Acquisitions Editor:
James DeVoe

Development Editor:
John Fisher

Executive Marketing Manager:
Maura Theriault

Marketing Coordinator:
Sarena Douglass

Channel Manager:
Fair Huntoon

Executive Production Manager:
Mary Ellen Black

Technology Project Manager:
David Porush

Production Editor:
Betsy Hough

Editorial:
Carol Leyba, Daril Bentley

Cover Design:
Cammi Noah

Trademarks
SolidWorks is a registered trademark of SolidWorks Corporation. FeatureManager, PropertyManager, PhotoWorks, and FeaturePalette are trademarks of SolidWorks Corporation. Windows and the various Windows operating systems are registered trademarks of Microsoft Corporation. Parasolid is a registered trademark of Electronic Data Systems Corp.

Copyright © 2003 by Delmar Learning. a division of Thomson Learning, Inc. Thomson Learning is a trademark used herein under license.

Printed in Canada
4 5 6 XXXX 06 05 04 03

ALL RIGHTS RESERVED. No part of this work covered by the copyright hereon may be reproduced or used in any form or by any means—graphic, electronic or mechanical, including photocopying, recording, taping, Web distribution or information storage and retrieval systems—without the written permission of the publisher.

For more information, contact:
Delmar Learning
Executive Woods
5 Maxwell Drive
Clifton Park,
NY 12065-8007

Or find us on the World Wide Web at
http://www.delmarlearning.com

For permission to use material from this text or product, contact us by
Tel : 1-800-730-2214
Fax: 1-800-730-2215
www.thomsonrights.com

ISBN:1-4018-0510-8

NOTICE TO THE READER

Publisher does not warrant or guarantee any of the products described herein or perform any independent analysis in connection with any of the product information contained herein. Publisher does not assume, and expressly disclaims, any obligation to obtain and include information other than that provided to it by the manufacturer.

The reader is expressly warned to consider and adopt all safety precautions that might be indicated by the activities herein and to avoid all potential hazards. By following the instructions contained herein, the reader willingly assumes all risks in connection with such instructions.

The publisher makes no representation or warranties of any kind, including but not limited to, the warranties of fitness for particular purpose or merchantability, nor are any such representations implied with respect to the material set forth herein, and the publisher takes no responsibility with respect to such material. The publisher shall not be liable for any special, consequential, or exemplary damages resulting, in whole or part, from the reader's use of, or reliance upon, this material.

About the Author

David Murray is the Training Manager for CADimensions, Inc., located in East Syracuse, New York. His professional duties include supplying technical support and solutions for CAD and computer-related issues through CADimensions. He has been working with the SolidWorks software since its release in late 1995, and has since achieved the status of Certified SolidWorks Support Technician, Certified SolidWorks Instructor, and Certified SolidWorks Professional. *Inside SolidWorks 2003* is his fifth full-length work associated with the SolidWorks software. Previous versions of *Inside SolidWorks* have been made available to the international market in both Russian and Chinese.

Acknowledgments

Thank you to Daril Bentley and Carol Leyba for the effort they put forth in editing and layout. They really pulled out all the stops this time and helped put together a great book. Thanks to the people at SolidWorks Corporation for creating a fantastic product, and for their ongoing commitment. The guys and gals in technical support are of the highest caliber. Thanks to all the people who have shared tips over the years. It's little things that count, and you've helped me to help others. Last, thanks to Pete DiLaura and the staff at CADimensions, including Tara DiLaura, Mary Swedowski, Linda and Gary McHugh, Laurie and John Picinich, Tim Grove, Kevin Holbrook, Jim Francis, Tom Preville, and Steve Stojanovski (for his evolving programming expertise). You're the best group of people a guy could hope to work with.

This book is dedicated to my parents, Ron and Arlene Murray,
who taught me the difference between right and wrong.

Contents

Chapter 1: Computer-aided Design 1
Moving into the Digital Age 2
How SolidWorks Fits In 3
   Feature-based Modeling 3
   Parametric Modeling 4
   Solid Modeling 4
Design Intent 5
Boolean Modelers 7
Nonparametric Modelers 8
Parametric Modelers 8
Benefits of Solid Modeling 9
Downstream Benefits of Solid Modeling 11
SolidWorks and the Big Picture 12
SolidWorks API 13
Minimum System Requirements 13
Dual Processors? 14
Memory and Hard Drive Space 15
Graphics Cards 16
Monitor and Screen Size 16
Summary 17
Questions and Topics for Discussion 18
Optional Problem 18

Chapter 2: A SolidWorks Overview 19
Sketching Overview 19
Features Overview 20
Sketched Versus Applied Features 21
The SolidWorks Interface 22
   Beginning a New Part 22
   Toolbars 23
FeatureManager and the Work Area 25
   The Separator Bar 26
   PropertyManager 26

FeatureManager Pane Control 29
Renaming Objects 30
Using the Mouse Buttons 31
   Selecting Entities 31
   The Context-sensitive Menu 31
   Middle Mouse Button Functionality 32
Toggling Plane Display 32
Changing Views 35
   Using the Orientation Window 35
   Using the Standard Views Toolbar 36
   The Normal To Function 37
Using Templates 38
   Standard File Extensions 38
   Template File Extensions 38
   Modifying Document Properties and
      Saving Templates 39
Deciding on a Sketch Plane 41
Entering Sketch Mode 42
Sketch Color Codes 43
Understanding Error Symbology 44
   Overdefined Geometry 44
   What's Wrong? 45
The Origin Point 46
Sketch Basics 47
   Am I in a Sketch? 47
   System Feedback 48
   An Introduction to Constraints 49
Summary 52
Questions and Topics for Discussion 53

Chapter 3: Sketching Basics 55
Beginning a Sketch: A Quick Review 55
Sketching 56

Lines 56
Deleting Entities 58
Auto-transitioning to Arcs 60
Exiting Commands 61
Dragging Geometry 61
Command Persistence 62
Sketch Entities 63
Tangent Arc 64
3 Point Arc 64
Centerpoint Arc 65
Circles 65
Rectangles 66
Rectangles at an Angle 66
Parallelograms 66
Polygons 67
Ellipses 68
Centerpoint Ellipses 69
Parabolas 70
Points 70
Centerlines 71
Splines 71
Nonproportional Splines 72
Proportional Splines 73
Adding Control Points 73
Deleting Control Points 74
Simplifying a Spline 74
Spline Tangency Conditions 75
Moving Frames 76
Spline Minimum Radius and Inflection
  Points 77
Inspecting Sketch Curvature 77
Construction Entities 78
Sketch Guidelines 79
Contour Select Tool 81
Check Sketch For Feature 82
Keep It Simple 83
Fully Define Your Sketch 83
Units of Measurement 84
Grid/Snap Settings 84
Sketch Plane Indicators 85
Snap Behavior 86
Geometric Relations 86
Callouts 87
Deleting Relations 88

Dimensioning 89
Automatic Dimensioning 91
Pan and Zoom Commands 92
View Orientation 93
Additional Zoom Options 97
Zoom Limitations in Windows 98
Display Options 99
Wireframe 99
Hidden Lines Visible 99
Summary 104
Questions and Topics for Discussion 105
Optional Problem 106

Chapter 4: Castings 107
Getting Started: A Quick Review 107
Determine the Best Profile 108
Select the Appropriate Plane 108
Fully Define the Sketch 108
Creating Features 109
Base Features 109
Extrusion End Conditions 111
The Pivot Arm: An Overview 112
The Mid Plane End Condition 112
A Common Mistake 115
Editing Techniques 116
Editing a Sketch 116
Editing a Definition 118
Modifying Dimensions 119
Renaming Features 121
The Blind End Condition 121
The Up To Surface End Condition 121
The Through All End Condition 125
Cutting with an Open Profile 126
Mirroring Sketch Geometry 129
Mirroring Sketch Geometry Dynamically . . 130
Mirroring Existing Sketch Geometry 130
Multiple Bodies 134
Creating Planes 137
Starting a Plane 138
Summary 147
Questions and Topics for Discussion 148

Chapter 5: Turned Parts 149
Revolved Features 149
Begin with a Centerline 150

Rules Governing Revolved Sketch
    Geometry 151
Revolve Command Panel. 152
    Diameter Dimensions from Centerlines . . . 154
Dimension Properties. 155
    Modifying Properties via PropertyManager . 157
    Dimensioning to a Tangency Point 160
    Dual Dimensioning 162
Sketch Fillets 163
    Keeping Corners Constrained 164
    Adding Virtual Sharps Manually 165
Sketch Chamfers 165
Trimming 166
Extending 167
Converting Entities 168
Offsetting Entities 170
Summary 174
Questions and Topics for Discussion. 175

Chapter 6: Molded Parts 177
Thin-walled Parts. 178
    Thin-feature Parts 178
    Shelled Parts. 178
The Shell Command 178
    Shell Feature Panel. 179
    Multi-thickness Shell 181
Fillets and Rounds 182
    Fillets as Features 183
    Fillet Options 188
Selection Techniques 190
    Select Midpoint 190
    Select Chain. 191
    Select Loop. 191
    Select Tangency 191
    Select Other 191
Chamfers 193
    Angle-Distance. 193
    Distance-Distance 193
    Vertex Chamfer 194
The Hole Wizard 198
    Legacy Tab 200
    Holes on Nonplanar Faces 200
    Adding Favorites 200
    Simple Holes 201
Lofted Parts 202

Split Curve Command. 204
    Correct Selection of Loft Profiles 206
    Guide Curves 206
    Pierce Constraint 207
    Centerline Option 210
    Other Loft Options 211
    Tangency Conditions 212
Split Lines 215
The Silhouette Method. 217
Adding Draft 218
    Adding Draft with a Parting Line 218
    Adding Draft with a Neutral Plane 220
    Step Draft 221
    Face Propagation Options 222
Rib Tool. 223
    Linear Versus Natural 224
    Draft Reference for Ribs. 226
Draft Analysis 226
    Face Classification 227
    Draft Analysis Color Settings 229
Summary 232
Questions and Topics for Discussion. 234

Chapter 7: Patterns 237
Simple Pattern Alternatives 237
    Copying and Pasting Features 238
    The Control-Drag Technique 240
Linear Patterns. 241
    Pattern Seed Only 243
    Vary Sketch 243
    Deleting Instances from a Pattern 245
Circular Patterns 249
    Working with Circular Patterns 250
    Equal spacing Option. 251
Geometry Patterns. 252
Axes. 253
    Temporary Axes. 253
    User-defined Axes 254
Sketch-driven Patterns 258
Table-driven Patterns 260
    Coordinate Systems 261
    Modifying a Table-driven Pattern 263
Curve-driven Patterns 264
Mirroring Feature Geometry 267
    Mirroring and Patterning Faces and Bodies. 268

Symmetrical Parts. 269
Mirrored Parts, or "Left-Hand" Versions 269
External References 271
   Opening Referenced Files. 272
   Locking References. 272
   Breaking External References. 273
Sketch Step and Repeat Commands 273
   Linear Sketch Step and Repeat 274
   Circular Sketch Step and Repeat 277
Summary. 278
Questions and Topics for Discussion 280
Optional Problem. 281

Chapter 8: Sheet Metal 283
Thin-feature Parts 283
   Auto Fillet Option .285
   Revolving an Open Profile. 285
   Closed-profile Thin Features 286
   Cap Ends Option 287
Defining a Sheet Metal Part. 287
   Sheet-Metal1 288
   Base-Flange. 292
   Flat-Pattern1 293
Edge Flange. 295
Flange Position. 297
   Flange Position Offset Option 298
   Custom Relief Type and Custom Bend
      Allowance. 298
   Editing the Flange Profile. 299
Miter Flange. 300
Offset Miter Flange. .301
Tabs 302
Sketched Bends 303
Jogs 304
Hems. 306
Breaking Corners 307
Closing Corners 308
Lofted Bends. 310
Cutting Sheet Metal Parts 311
   Link to Thickness. .311
   Normal Cut. 312
   Working in the Flattened State 312
Forming Tools 315
   Using the Feature Palette 316
   Editing Palette Items 326

Modifying Part Color 328
Modify Sketch Command 329
   Rotating Using the Mouse 331
   Translating 331
   Scaling 332
   Flipping. 332
Converting to Sheet Metal 332
   Inserting Bends. 333
   Reordering Bends .336
   Rolled Bends. 336
   Legacy Sheet Metal Parts 337
Rip Features 338
Summary 340
Questions and Topics for Discussion 341
Optional Problem. 342

Chapter 9: Springs, Threads, and Curves . 343
Swept Features 343
   Valid Profiles 344
   Valid Paths 345
   Using Existing Edges as a Sweep Path 347
   Align with End Faces Option. 348
   Additional Sweep Options. 349
   Orientation and Twist Control 350
   Sweeping with Guide Curves. 353
Helical Curves 354
   Creating Helixes. .355
   Spirals 357
Springs 357
   Phase 1: Starting with a Circle. 358
   Phase 2: Defining the Helix 358
   Phase 3: Creating a Sketch Plane 358
   Phase 4: Sketching a Sweep Profile 359
   Phase 5: Exiting the Sketch and
      Creating the Sweep 359
Composite Curves 359
Threads. 364
   Creating Threads .364
   Rounding Threads 366
Springs and Threads: Final Comments 367
3D Sketcher 368
   3D Sketch Relations. 370
   Drawing Splines in 3D 371
Projected Curves 375
   Projecting a Sketch onto a Sketch. 375

Projecting a Sketch onto a Face......... 376
Curves Through Points 378
   Curve Through Reference Points 378
   Curve Through Free Points 378
   Curve Files 379
Section Properties 380
   How Section Properties Work 381
   Measurement Options Window Settings... 382
Mass Properties 382
   Changing the Coordinate System 384
   Mass Properties in Assemblies 384
   Placing a Point at the Centroid 385
Summary 385
Questions and Topics for Discussion....... 386
Optional Problem 387

Chapter 10: Part Configurations 389

Reasons for Configurations 389
   Assembly Performance................ 390
   Sheet Metal Forming Operations........ 390
   Part Families...................... 390
   Application-specific Requirements 390
   Design-specific Requirements 390
ConfigurationManager................. 391
   FeatureManager and
      ConfigurationManager Descriptions .. 392
   Properties for Newly Inserted Items....... 393
   Copying Configurations 394
Suppression States 395
   Parent/Child Relationships 395
   Other Methods of Changing Suppression .. 398
Dimensional Configurations 399
Other Configurable Objects 399
   Configuring End Conditions 400
   Configuring Geometric Relations 402
Nested Configurations................. 404
Summary 405
Questions and Topics for Discussion....... 406

Chapter 11: Design Drawings.......... 407

File Associativity 408
New Drawings 408
   Drawing Interface................... 410
   Drawing Sheet Formats 413
   Drawing Templates 416

Inserting Views 422
   Moving Views 424
   Modifying View Alignment............ 425
   Right-clicking Precautions............. 427
   Projected Views 427
   Auxiliary Views 428
   Named Views...................... 429
   Relative to Model View Method 430
   Section Views 431
   Aligned Sections 435
   Broken-out Section Views............. 441
   Detail Views....................... 442
   Broken Views 444
   Cropped Views..................... 446
   Empty Views and Creating Tables....... 447
View Appearance 448
   Hiding Individual Edges.............. 448
   Tangent Edges 449
   Changing a View's Scale 451
Dimensioning 452
   Moving Dimensions................. 457
   Dimension Favorites 459
   Reference Dimensions 461
   Extension Lines 462
Center Marks........................ 463
   Creating Center Marks 463
   Automatic Center Marks and Centerlines . 464
Layers.............................. 465
   Making a Layer Current.............. 467
   Deleting a Layer.................... 467
   Turning Layers On and Off 467
   Moving Objects Between Layers 468
Properties of Multiple Objects............ 469
Line Formatting...................... 469
Annotations......................... 470
   Notes 471
   Hole Callouts...................... 475
   Cosmetic Threads 476
   Weld Symbols 479
   Geometric Tolerancing 480
   Surface Finish Symbols 481
   Datum Features 481
   Datum Targets 482
   Multi-jog Leaders................... 482
   Dowel Pin Symbols 484

Blocks 484
    Creating and Editing Blocks............. 484
    Exploding Blocks 486
    Inserting Blocks 486
    Saving and Linking Block Files 487
    Imported Block Attributes 487
RapidDraft Files 488
    Creating RapidDraft Drawings........... 489
    Synchronizing Model Geometry 490
Tabulated Drawings 491
Summary................................. 492
Questions and Topics for Discussion 494
Optional Problem......................... 496

Chapter 12: Assemblies 497
Starting a New Assembly................... 498
Inserting Components 499
    Fixed or Floating?..................... 500
    FeatureManager Symbology 501
Moving and Rotating Components 502
Moving Components in Specific Directions .. 504
Mating Relationships 505
    Mate Options 507
    Finding Mates......................... 513
    Viewing Mates with PropertyManager..... 514
    Smart Mates 514
    Inserting Components with SmartMates ... 515
    Mate References 516
Palette Parts and Palette Assemblies 517
Component Patterns 519
    Creating Component Patterns 519
    Deleting Patterned Components 523
Hiding Components....................... 523
Component Suppression States 524
    Lightweight Components................ 524
    Resolving Components................. 527
Working with Subassemblies 528
    Component Subassemblies.............. 528
    Component Properties 529
Component Replacement 533
Reloading Components 535
Mate Troubleshooting and Repair........... 536
    Missing References 537
    Overdefined Mates 537
    Improper Alignment................... 538

Mate Diagnostics 538
Component Editing..................... 539
    Opening Component Files.............. 540
    Editing Components in Context 540
Interference Checking................... 544
Collision Detection 545
Dynamic Clearance 547
Physical Dynamics 548
Simulation 549
Assembly Features 551
    Feature Scope 551
    Assembly Feature Patterns 554
Assembly Drawings 554
    Assembly Section Views 555
    Hiding Components in Drawing Views.... 556
    Show Hidden Edges 557
Bill of Materials 558
    Bill of Materials Properties Window
        Options 559
    Editing a BOM 563
    Customizing a BOM Template 564
    BOM Properties 565
    Overriding BOM Part Number 566
Exploded Views 567
    Editing an Exploded View 569
    Expanding and Collapsing an Exploded
        View 570
    Copying Exploded Views 570
    Exploded Views in Drawings 571
Balloons 572
    Working with Balloons 572
    Balloons and Component Names 574
Summary 575
Questions and Topics for Discussion........ 576

Chapter 13: Cavities, Cores, and
Mold Making 579
The Cavity Command.................... 579
    Creating the Die...................... 580
    Creating the Cavity 581
Inserting Parts 582
    Referencing Configurations 585
    Creating the Mold 585
    Open-profile Cutting 586
Derived Parts 587

Splitting Parts 589
Creating a Core and Cavity............... 591
　Scaling 592
　Creating Shutoff Surfaces............. 593
　Radiated Surfaces..................... 596
　Lofted Surfaces....................... 598
　Creating the Mold Assembly............ 600
　Inserting a New Part into an Assembly 601
　InPlace Mates......................... 601
　Knitting Surfaces 603
　The Cavity Base Feature 605
　Completing the Core 607
　Final Touches......................... 607
Summary 608
Questions and Topics for Discussion....... 609

Chapter 14: Welded Assemblies 611
Adding Weld Beads...................... 611
　Weld Bead Mate Surfaces 615
　Weld Bead Special Situations 616
　Weld Bead Associativity 617
The Join Command...................... 619
　A Common Mistake When Joining....... 622
　Disassociating Joined Components 623
　Multiple Bodies in Joined Parts 623
Saving Assemblies as Part Files 623
Summary 626
Questions and Topics for Discussion....... 627

Chapter 15: Assembly Configurations 629
Adding Assembly Configurations 629
Assembly Configuration Properties......... 630
　Use configuration specific color Option ... 631
　Suppress features and mates Option...... 632
　Hide component models Option 632
　Suppress component models Option 632
　BOM-related Options 632
Controlling Component Configurations..... 633
　Using Assembly Configurations
　　Successfully...................... 635
　Showing Alternate Components......... 639
Controlling Configurations in Drawings..... 639
Summary 641
Questions and Topics for Discussion....... 642

Chapter 16: Design Tables............ 643
Linking Versus Embedding 644
Preparations for Design Table Creation 644
　Turning on Dimension Names 645
　Displaying All Model Dimensions........ 646
Creating a Design Table 646
　Adding a Blank Design Table 649
　Feature Suppression in Design Tables 650
　Design Table Options 653
　Adding Rows and Columns Automatically. 653
　Saving and Linking to Design Table Files.. 654
　ConfigurationManager Symbology 655
Design Table Editing..................... 656
Deleting a Configuration.................. 657
Assembly Design Tables 658
　Controlling Assembly Feature and
　　Mate Dimensions 659
　Controlling Component Suppression 660
　Controlling Assembly Feature and
　　Mate Suppression.................. 661
　Controlling Component Visibility 662
　Specifying a Component's Configuration .. 662
　Adding Notes and Controlling Part
　　Numbers.......................... 663
　Controlling Subassembly Expansion
　　in a BOM 663
A Pictorial Overview..................... 664
　Showing Design Tables in Drawings...... 666
　Design Table Summary and
　　Recommendations 667
Custom Properties 667
　Configuration-specific Properties 672
　Defining Custom Properties in Design
　　Tables............................ 672
　Associative Custom Properties 673
　Custom Properties and BOMs 675
Summary 675
Questions and Topics for Discussion........ 676
Optional Problem 677

Chapter 17: Equations................. 679
Preparatory Work 679
The Equation Creation Process........... 681
　Equations in Action.................... 683

Link Values 684	Solidifying Surfaces 754
Equations Window Buttons 688	Offset Surfaces 755
Summary. 692	Planar Surfaces. 756
Questions and Topics for Discussion 692	Manipulating Bodies 756
Optional Problem. 692	Moving, Copying, and Rotating Bodies 757

Chapter 18: Advanced Assembly Modeling 693

System Performance. 693
 Large Assembly Mode. 694
 Performance Options 696
Assembly Restructuring 700
Reorganize Components Command 702
Multi-user Environments and Copying Files. . 703
 Scenario 1: Working Independently 704
 Scenario 2: Working Over a Network 708
Envelopes 718
Envelope Selection Techniques. 720
Advanced Selection Techniques 721
Summary. 723
Questions and Topics for Discussion 724

Chapter 19: Advanced Feature Types 725

Domed Features 725
Shape Command 729
Advanced Filleting. 732
 Variable Radius Fillets 732
 Utilizing Movable Fillet Points 734
 Setback Fillets. 734
 Face Blends 736
 Full Round Fillets 739
 Continuous Curvature Fillets. 740
Face Curvature 740
Rolling Back FeatureManager 741
Reordering 743
Summary. 744
Questions and Topics for Discussion 745

Chapter 20: Working with Surfaces 747

Surfaces Versus Solids 748
Creating Surfaces 748
Trimming and Extending Surfaces. 750
 Deviation Analysis 751
 Untrimming Surfaces 751
 Trimming Surfaces 752
 Surface Lofting Techniques. 754

Cutting with Surfaces 759
Hiding and Showing Bodies. 760
Summary 761
Questions and Topics for Discussion 762

Chapter 21: Engraved and Embossed Text 763

Adding Text as a Feature. 763
 Aligning Text on a Curve 765
 Sketch Text Options 766
Sketch Text as a Feature 768
Text Features on Curved Surfaces. 771
Wrapping Text Around a Cylinder 773
Embossed Text on a Curved Surface 776
Summary 778
Questions and Topics for Discussion 779
Optional Problem. 780

Chapter 22: Importing and Exporting Files 781

Importing Files 781
 AutoCAD Drawing and DXF Files 784
 Parasolid Files 785
 IGES Files. 786
 STEP Files 788
 ACIS SAT Files 788
 Virtual Reality Markup Language 788
 Stereolithography Files. 789
 Proprietary Formats 790
 Dynamic Link Libraries 791
Exporting Files 791
 DWG and DXF Export Options. 793
 Creating a Map File 795
 Parasolid Export Options 798
 IGES Export Options 799
 STEP Export Options 801
 ACIS Export Options 801
 VDAFS Export Options 802
 VRML Export Options. 802
 Stereolithography Export Options 803
 Exporting Proprietary Formats 805
Saving Image Files 806

TIFF Files 806
TIFF Export Options 807
eDrawings. 810
Summary 811
Questions and Topics for Discussion. 811

Chapter 23: Customizing SolidWorks 813
Customize Window 814
Toolbars 814
Commands 815
Resetting Toolbars and Customized
   Settings. 816
Menus. 817
Keyboard. 820

Transferring Hot Keys 821
Macros 822
Feature Palette. 826
New SolidWorks Document Tabs. 828
Gradient or Image Backgrounds 829
Using Background Images 830
Manager Backgrounds and Schemes 831
Lighting . . . . : 832
Advanced Shading Properties 833
Light Sources 834
Summary 843
Questions and Topics for Discussion. 844

Index. 845

HowTos and Exercises

How-Tos

2-1: Starting a New SolidWorks Document . . . 23
2-2: Accessing an Item's Properties. 28
2-3: Changing the Names of FeatureManager Objects 30
2-4: Showing/Hiding Planes 33
2-5: Displaying Planes with Transparency and Adjusting Transparency and Color. 34
2-6: Enabling the Orientation Window 36
2-7: Modifying Document Properties 39
2-8: Saving a Template File. 40
2-9: Adding Relations 50
3-1: Sketching a Line Using the Click-Drag Method 57
3-2: Sketching a Line Using the Click-Click Method 57
3-3: Changing Command Persistence. 62
3-4: Adding Control Points to a Spline 74
3-5: Simplifying a Spline 75
3-6: Creating a Construction Entity 78
3-7: Using Check Sketch For Feature 82
3-8: Changing the Working Units 84
3-9: Adding Dimensions to a Sketch 89
3-10: Using the Autodimension Sketch Tool . . 91
3-11: Adding a User-defined View 94
3-12: Updating the Standard Views 95
3-13: Hidden Edge Display Style. 100
3-14: Modifying the Perspective Vanishing Point. 103
4-1: Reentering a Sketch. 115

4-2: Editing an Existing Sketch 116
4-3: Editing a Feature's Definition 119
4-4: Changing a Dimension Value 120
4-5: Dynamically Mirroring Sketch Geometry . 130
4-6: Mirroring Existing Sketch Geometry . . . 131
4-7: Adding a Symmetrical Relationship 131
4-8: Combining Multiple Bodies. 136
4-9: Creating a Plane Using Through Lines/Points. 138
4-10: Creating a Plane Using Parallel Plane at Point. 139
4-11: Creating a Plane Using At Angle 140
4-12: Creating a Plane Using Offset Distance 141
4-13: Creating a Plane Using Normal To Curve 142
4-14: Creating a Plane Using On Surface. . . . 144
5-1: Accessing a Dimension's Properties. 156
5-2: Modifying Tolerance Value Scale 158
5-3: Dimensioning to Tangency Points. 161
5-4: Turning on Dual Dimension Display. . . . 162
5-5: Creating a Sketch Fillet. 163
5-6: Creating a Sketch Chamfer 166
5-7: Using the Offset Entities Command 170
6-1: Creating a Shelled Part 179
6-2: Creating a Multi-thickness Shell. 181
6-3: Adding a Fillet as a Feature 184
6-4: Adding Multiple-radius Fillets. 186
6-5: Using Select Other 192

6-6: Creating a Hole with the Hole Wizard . . 198	8-21: Adding a Rip Feature 339
6-7: Using the Simple Hole Command 201	9-1: Creating a Sweep 346
6-8: Using the Split Curve Command 204	9-2: Creating a Helical Curve 356
6-9: Creating a Lofted Feature 205	9-3: Creating a Composite Curve 360
6-10: Creating a Split Line 216	9-4: Using the 3D Sketch Command 369
6-11: Adding Draft Using a Parting Line 218	9-5: Creating a Projected Curve 377
6-12: Adding Draft Using a Neutral Plane . . . 220	9-6: Creating a Curve Using Curve Through Free Points 379
6-13: Using the Rib Command 224	9-7: Defining the Density of a Part 383
6-14: Analyzing Draft 227	10-1: Adding a Configuration 391
7-1: Copying and Pasting 239	10-2: Accessing Parent/Child Data 396
7-2: Creating a Linear Pattern 242	10-3: Changing Feature Suppression States . . 397
7-3: Deleting Pattern Instances 245	10-4: Controlling End Conditions in Configurations 401
7-4: Creating a Circular Pattern 250	10-5: Controlling Geometric Relations in Configurations 402
7-5: Creating an Axis 254	11-1: Starting a New Drawing 409
7-6: Adding Ordinate Dimensions 258	11-2: Setting and Saving Toolbar Layouts . . . 410
7-7: Creating a Sketch-driven Pattern 259	11-3: Sketch Toolbar Auto-activation 412
7-8: Creating a Coordinate System 261	11-4: Editing a Drawing Sheet Format 414
7-9: Creating a Table-driven Pattern 263	11-5: Saving a Sheet Format 415
7-10: Creating a Curve-driven Pattern 265	11-6: Modifying a Sheet's Properties 418
7-11: Mirroring Feature Geometry 267	11-7: Adding Sheets to a Drawing 421
7-12: Performing a Linear Sketch Step and Repeat 274	11-8: Inserting the Three Standard Views . . . 422
7-13: Performing a Circular Sketch Step and Repeat 277	11-9: Hiding/Showing View Borders 424
8-1: Creating a Base Flange 288	11-10: Breaking View Alignment 425
8-2: Specifying a Bend Table 290	11-11: Horizontally or Vertically Aligning a View 426
8-3: Creating an Edge Flange 296	11-12: Creating a Projected View 427
8-4: Creating a Miter Flange 300	11-13: Creating an Auxiliary View 428
8-5: Creating a Tab 302	11-14: Creating a Named View 429
8-6: Creating a Sketched Bend 303	11-15: Creating a View Using Relative to Model 430
8-7: Creating a Jog 304	11-16: Creating a Section View 433
8-8: Creating a Hem 306	11-17: Creating an Aligned Section View . . . 436
8-9: Breaking Corners 308	11-18: Changing Part Crosshatch 437
8-10: Closing Corners 309	11-19: Changing View Crosshatch 438
8-11: Creating a Lofted Bend 310	11-20: Adding Area Hatch 440
8-12: Fold and Unfold Commands 314	11-21: Creating a Broken-out Section View . . 441
8-13: Inserting Features via the Feature Palette 317	11-22: Creating a Detail View 442
8-14: Creating a Library Feature 319	11-23: Creating a Broken View 445
8-15: Creating a Forming Tool 321	11-24: Establishing Break Gap and Break Line Extension 446
8-16: Modifying Folder Location Paths 325	11-25: Cropping a View 447
8-17: Controlling Dimension Access 326	11-26: Setting Default Display Characteristics . 449
8-18: Editing Colors 328	
8-19: Using the Modify Sketch Command . . . 330	
8-20: Inserting Sheet Metal Bends 333	

11-27: Modifying an Object's Line Style..... 450
11-28: Modifying View Scale 451
11-29: Inserting Dimensions 453
11-30: Hiding and Showing Dimensions 459
11-31: Adding Dimension Favorites 460
11-32: Turning Off Reference Dimension
       Parentheses...................... 461
11-33: Adding Center Marks 464
11-34: Creating Layers................... 465
11-35: Moving Objects Between Layers 468
11-36: Using the Line Format Toolbar 470
11-37: Adding Notes..................... 471
11-38: Inserting Cosmetic Threads........ 477
11-39: Adding a Multi-Jog Leader to an
       Annotation...................... 483
11-40: Defining a Block.................. 485
11-41: Converting to RapidDraft 489
12-1: Inserting Components Using the
      Insert Menu 500
12-2: Moving a Component 503
12-3: Rotating a Component 503
12-4: Adding Mate Relationships 507
12-5: Using Smart Mates................ 514
12-6: Creating a Component Pattern 520
12-7: Creating a Derived Pattern 522
12-8: Changing the Suppression State of
      Components..................... 525
12-9: Enabling Lightweight Components.... 526
12-10: Replacing Multiple Components..... 533
12-11: In-context Component Editing 541
12-12: Testing an Assembly for Interference . 544
12-13: Performing Collision Detection...... 546
12-14: Checking Dynamic Clearance....... 547
12-15: Editing the Feature Scope 552
12-16: Creating an Assembly Feature 553
12-17: Creating an Assembly Section View .. 555
12-18: Inserting a BOM.................. 559
12-19: Setting an Anchor Point 561
12-20: Adding Custom Properties......... 563
12-21: Exploding an Assembly 567
12-22: Copying an Exploded View 570
12-23: Showing an Exploded View in a
       Drawing 571
12-24: Adding Balloons.................. 573
13-1: Using the Cavity Command 581

13-2: Inserting a Part..................... 583
13-3: Derived Component Parts........... 588
13-4: Using the Split Command........... 589
13-5: Scaling a Part..................... 592
13-6: Using the Surface Fill Command 594
13-7: Creating a Radiated Surface 596
13-8: Creating a Lofted Surface 599
13-9: Creating In-context Parts 602
13-10: Creating a Knitted Surface 603
13-11: Creating the Cavity Component..... 606
14-1: Adding a Weld Bead 612
14-2: Joining Components 621
14-3: Saving an Assembly as a Part 624
15-1: Controlling a Component's
      Configuration.................... 634
15-2: Developing Assembly Configurations .. 635
15-3: Drawing View Configuration Selection. 640
16-1: Creating a Design Table............ 647
16-2: Editing a Design Table............. 656
16-3: Creating Custom Properties......... 668
16-4: Linking to Custom Properties 670
16-5: Modifying Custom Properties 671
17-1: Creating an Equation............... 681
17-2: Linking Dimension Values........... 685
18-1: Implementing Large Assembly Mode .. 695
18-2: Dissolving Subassemblies........... 700
18-3: Creating New Subassemblies........ 701
18-4: Copying a File with References...... 705
18-5: Copying Assemblies with External
      References 706
18-6: Reloading Components That
      Were Loaded Read-Only 712
18-7: Save Copies of Drawings with Models . 716
18-8: Inserting a Part as an Envelope 719
18-9: Inserting a New Envelope 720
18-10: Adding Selection Criteria 722
19-1: Creating a Domed Feature 726
19-2: Creating a Shape Feature........... 730
19-3: Creating a Variable Radius Fillet...... 733
19-4: Creating a Setback Fillet 735
19-5: Creating a Face Blend Fillet 736
19-6: Creating a Face Fillet with Hold Lines . 739
19-7: Reordering a Feature 743
20-1: Creating a Swept Surface........... 749
20-2: Extending a Surface 750

20-3: Trimming Surfaces with Surfaces 753	
20-4: Thickening Surfaces 755	
20-5: Moving or Copying Bodies........... 757	
20-6: Cutting with a Surface 760	
21-1: Creating Sketch Text 764	
21-2: Creating Engraved Text on a Curved Surface 771	
21-3: Creating Text on a Cylindrical Feature . 773	
21-4: Creating Embossed Text on a Curved Surface 777	
22-1: Importing a File into SolidWorks..... 781	
22-2: Exporting to Other File Formats 792	
23-1: Adding An Item to a Menu 817	
23-2: Deleting a Menu Item............... 819	
23-3: Renaming a Menu Item 820	
23-4: Adding a Keyboard Shortcut 821	
23-5: Recording a Macro 823	
23-6: Editing a Macro 824	
23-7: Running a Macro 825	
23-8: Assigning a Macro to an Icon........ 825	
23-9: Altering Background Colors......... 829	
23-10: Controlling Manager Backgrounds ... 831	
23-11: Adding and Editing a Light Source ... 835	

Exercises

4-1: Creating the Pivot Arm 113
4-2: Developing the Pivot Arm 122
4-3: Cutting Holes in the Pivot Arm 126
4-4: Adding Features to the Pivot Arm 132
5-1: Creating a Revolved Feature 153
5-2: Editing the Revolved Feature Sketch ... 172
6-1: Creating a Shaver Housing 195
6-2: Adding Features to the Shaver Housing . 229
7-1: Patterns on the Shaver Housing 246
7-2: Adding Features to the Valve Stem ... 255
9-1: Creating the Candle Holder 360
9-2: Creating a 3D Sketch 372
19-1: Using the Dome Command 727
21-1: Creating an Embossed Nameplate ... 769

Introduction

SOLIDWORKS IS PRODUCTION SOLID MODELING for the desktop in the Microsoft Windows environment. SolidWorks Corporation, the developer of the software, is focused on providing mechanical design solutions and depth of modeling for mechanical assemblies for desktop platforms. It is hoped that this book will allow you to learn and get the most out of the SolidWorks package.

Readership and Intent

Inside SolidWorks 2003 is written for the beginning to advanced SolidWorks user who wants to learn SolidWorks basics or expand existing knowledge of the software. This book takes you through beginner to increasingly more advanced SolidWorks functionality and thinking by describing and showing through examples and exercises how to apply the software to a multitude of widely divergent manufacturing processes.

The book also addresses the needs of the instructor. Students and those with little or no prior CAD experience will welcome the book's clear and easy-to-comprehend approach to solid modeling basics. In addition, each chapter ends with a "Summary" and a "Questions and Topics for Discussion" section. Some chapters are supplemented with an "Optional Problem" section. These sections are intended to reinforce the fundamentals and key points that will help new users reach higher levels of expertise and experience in a very short time.

In short, *Inside SolidWorks 2003* is intended to allow both the student and new user reach a highly productive level of SolidWorks program use. The book is also meant to guide current users of the software through specific commands and processes needed to model a wide range of parts from many manufacturing disciplines, and to serve as a reference for performing operations and solving problems.

Approach and Book Features

This book takes an easy-to-understand, step-by-step approach to SolidWorks functionality and its command procedures and combinations in a way that allows you to create real-world parts using a variety of manufacturing techniques. This approach is also intended to allow you to extrapolate what you have learned to new situations and new types of parts.

At the same time, the book imparts "how to" functionality. *Inside SolidWorks 2003* briefly describes the mind-set needed while working with the software; that is, the "why" and "when" of applying functionality to projects. The thinking involved and the mastery of functionality are reinforced through examples and exercises. Because of the wide range of processes covered, many SolidWorks users will find the book an invaluable resource and reference guide to SolidWorks command functionality.

Hands-on examples are consecutively numbered and are called How-Tos. These are functions and processes you need to know how to perform in operating the software. Consecutively numbered exercises bring several functions together to provide practice in building and manipulating parts and in performing more complicated processes. The text is also supplemented with consecutively numbered reference tables, with lists, and with notes, tips, and warnings. Notes highlight important information, tips provide time- and work-saving suggestions, and warnings steer you clear of actions that potentially cause problems.

Version Specificity

The SolidWorks software is always evolving to bring to the end user an ever more powerful and full-featured solid modeling program. SolidWorks Corporation has periodically released major upgrades to its very popular software. *Inside SolidWorks 2003* is based on SolidWorks 2003. This book contains all of the information you need to learn and productively operate the SolidWorks software, whatever version of the software you are using. However, future versions of SolidWorks may have enhancements that could not be foreseen or included in this edition.

Prerequisites to Using This Book

Readers of *Inside SolidWorks 2003* should be familiar with one of the following operating systems.

- Windows 95 or 98
- Windows ME (Millennium Edition)
- Windows NT 4.0
- Windows 2000
- Windows XP

> **NOTE:** *Not all of the listed operating systems are supported by the SolidWorks software. See Chapter 1 for system requirements.*

It would also benefit the reader to be somewhat familiar with Microsoft Excel. This only holds true if the reader wants to take advantage of SolidWorks Design Table (family-of-parts table) functionality or its automated bill of material (BOM) routine.

Prior experience with drafting or computer-aided drafting or design is helpful, but not a requirement. Prior experience with SolidWorks is not a requirement, as *Inside SolidWorks 2003* will start new users on the correct path from the beginning.

Note to the Reader

Although every effort has been made to make *Inside SolidWorks 2003* as seamless as possible, some of the screen shots or graphics may be slightly different than what you see on your computer screen. This is due to the nature of constantly evolving software and differences in operating system interfaces. Even so, we are highly confident this book will provide you with an in-depth and insightful educational experience with what has become one of the industry leaders in the solid modeling market. Happy modeling!

CHAPTER 1

Computer-aided Design

THERE ARE PROBABLY MORE ACRONYMS used in the computer industry than in any other (except perhaps governmental departments). The acronym CAD used to refer to computer-aided drafting. After all, drafting is where computers were most often used when they began to become more widely available at an affordable price for most people. Individuals did not start out by creating elaborate 3D solid models. The technology had not really been introduced yet. Most of the companies that were using CAD software were using it to create their 2D standard engineering layouts. These days, CAD is more often taken to mean computer-aided design. When both technologies are referred to, CADD is used (computer-aided design and drafting).

In the past, a typical scenario might happen like this. An individual might start out with a sketch on a napkin or something similarly crude. Afterward, the idea could get put down on paper. Then the Engineering Change Notices (ECNs) would follow, along with last-minute design changes, and so on. Perhaps even a prototype or two would be developed. Finally, the supervisor would accept and release the final version, copies of the prints would get sent to manufacturing, and the parts would be manufactured. The steps might vary from company to company, but generally speaking the processes would be similar.

In the SolidWorks environment, the process begins in much the same way. The person with the idea in his head starts with a sketch. Instead of using a napkin, however, he would use SolidWorks. The term *sketch* applies very well here. SolidWorks is a sketch-based modeler, and the sketch is a rough general outline of what the part might look like.

Moving into the Digital Age

Anyone who has been a drafter for any length of time is familiar with a drafting board. Broken pencil leads, eraser crumbs, and even an occasional coffee-ring stain were just a few of the things a draftsman had to live with. When computers emerged as a viable tool in the workplace, people discovered they did not have to put up with such common annoyances of the past.

Now there are a new set of annoyances; computer crashes and accidentally deleted files, to name a couple. Even so, most would agree that the electronic office is a big plus over the way things were. Sometimes you will hear it called the digital office, or sometimes a paperless office. These terms all mean essentially the same thing, and are used to describe the typical office environment in today's digital workplace.

Most offices still revolve around detailed drawings, but some offices truly are paperless, sending 3D models directly to rapid prototype machines from their computers. Stereolithography rapid prototype files (polygonal representations of a model) can be used to create prototypes, sometimes in a matter of hours, allowing designers and engineers to hold an actual part in their hands.

CAD is being used for much more than 2D layouts these days. Look around and you will see CAD software programs being used to design almost everything we use. 3D models of designs can be created, tested, and analyzed on the computer before ever actually being built, thereby reducing errors and shortening a product's time to market.

SolidWorks was the first of its kind to break into the mid-range price bracket for full-functionality, solid modeling, Windows-based CAD software. SolidWorks gives the CAD operator the tools to perform prototyping, assemblies, dynamic motion, mass properties, and so on. Additionally, there are many third-party applications that will run directly inside the SolidWorks interface. SolidWorks' term for the companies who produce these types of integrated add-ons is Gold Solution Partners.

Analysis programs can test a model for stress, strain, and thermal conditions; can find the resonant frequency of a part; can show air flow through or around a model; and much more. Assemblies can be assembled on the computer, and interference checks can be made. Mass properties can be obtained, and the amount of material for a specific part can be calculated, all automatically by the computer.

Animations can be created to show the dynamic movement of parts in an assembly, or to show assembly sequences for those putting the product together. Kinematics packages help people visualize dynamic motion and apply conditions such as gravity or spring effects. Photorealistic rendering utilities can produce beautiful images for use in advertising brochures. There are even document management tools that will automatically track ECNs or revisions for you. The list goes on and on, and it gets longer every week.

The benefits of CAD can easily be seen. Even so, there are many companies that continue to use drafting boards or use computers strictly for their 2D layouts. The reasons may differ. Perhaps the cost of computer hardware or software is a limiting factor. Technophobia (the fear of technology) may be what is keeping a company from taking the plunge into the digital age.

If your company is one that fits into one of these categories, you will probably find it necessary to make the jump to some sort of CAD design software in the near future. It may not happen tomorrow, it may not even happen next year, but sooner or later you will have to go digital. It used to be that CAD was a luxury. It is soon becoming a necessity, especially if you are a design engineer.

How can a non-digital-age design firm compete with another company employing CAD software, such as SolidWorks? The bottom line is that it cannot. This is the point at which CAD software ceases to become a luxury. To remain competitive in the computer age, one must learn to evolve with technology. It is not so much a question of whether your company should join the computer age, but when.

There are those that find the aspect of training very daunting or intimidating. Part of what makes a software program a great software program is its ease of use. Training should definitely be a consideration when making a purchase. It is a very important issue, but training does not have to be difficult. (Training is discussed in more detail later in this chapter.)

How SolidWorks Fits In

SolidWorks is much more than a simple CAD program. If it had to be summed up in one phrase, one could call SolidWorks a feature-based, parametric, solid-modeling design program. The following sections explain what this terminology is all about.

Feature-based Modeling

The term *feature-based* is something fairly new to the computer industry. It is used to describe the various component properties of a model. For example, a part can consist of various types of features, such as extrusions, holes, fillets, and chamfers. These features constitute the overall part, just as parts constitute an overall assembly. The mechanics behind some simple features are discussed in Chapter 3. Figure 1-1 shows typical features on a simple part.

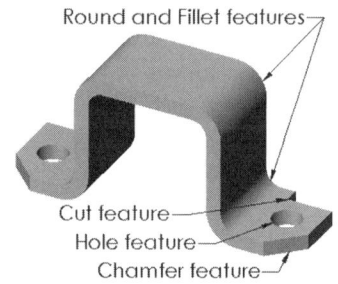

Fig. 1-1. Basic feature types on a simple part.

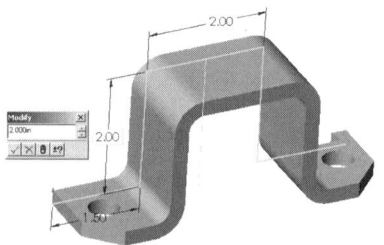

Fig. 1-2. Example of parametrics: modifying a dimension to alter the geometry.

Parametric Modeling

Parametric is a term used to describe a dimension's ability to change the shape of model geometry if the dimension value is modified. This is a different way of thinking for most people accustomed to nonparametric CAD software. For instance, when lines or arcs are being created in a 2D drafting layout, accuracy is extremely important. Machine shop personnel might be taking measurements from the drawing. In many CAD programs, if the geometry is not of the correct size, the dimensions themselves will not be accurate.

This is not the case with SolidWorks. Because SolidWorks is parametric, the dimensions drive the size and shape of the geometry, instead of the reverse. Figure 1-2 shows an example of what you might see when modifying a dimension (you will learn this technique in Chapter 3). Due to parametric capabilities, where dimensions are placed makes a big difference. This is known as *design intent,* which is discussed in material that follows. It is one of the most important fundamental aspects of SolidWorks.

Solid Modeling

A solid model implies that there is actually a 3D solid model, with density and mass, that you can hold in your hands. These properties are true of the computer-generated solid model except for the ability to hold it in your hands. However, for all intents and purposes, a computer model might as well be real. The model on the computer screen can be assigned a specific density, depending on what type of material it is to be made of. It has a center of gravity, otherwise known as its centroid. It also has weight and volume, at least as far as the computer is concerned.

You can rotate a solid model on the screen just as if it were actually sitting in the palm of your hand. You can measure it, and can extract a great deal of information from it (see figure 1-3) as easily and perhaps more easily than if it were the genuine article.

Design Intent

Fig. 1-3. Extracting mass property data from a solid model is only a few mouse clicks away.

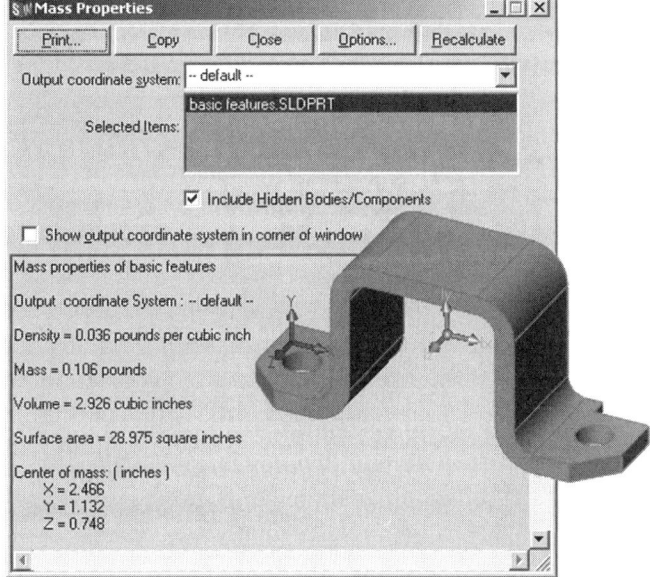

There are a number of ways in which we observe material objects in our everyday lives. Most of us see the world as a series of 3D objects we must act upon or interact with. These objects may be organic or inorganic, alive or inanimate. If we pick up a rock from the ground we know that it has weight and that it has solidity. We know these things and much more about the rock because our senses tell us so. The computer, on the other hand, does not have this luxury.

Understanding the question of just how SolidWorks interprets the data it is given can sometimes help you feel more comfortable with the software. In the case of SolidWorks, the software understands all the surfaces that make up the outer boundary conditions of the model. This is known as the model's *topology*. SolidWorks also understands what side of each face is the inside versus the outside. The various faces of the model are "knitted" by the software and in that way the software can understand that the model is "solid" geometry.

Design Intent

What may very well be more important than any other aspect of the software is the fact that SolidWorks is a design tool. There is more to this simple statement than meets the eye. The ability to go back to some earlier stage in the design process and make changes by editing a sketch or changing some dimensions is extremely important to a designer. This is a major advantage that SolidWorks has over its nonparametric competition, and even some of its parametric peers.

Design intent is how your model will react when dimension values are changed. This is very significant in a parametric modeling program because parametric dimensions control your model. Examine figure 1-4. The person that placed the dimensions on this sketch determined that the angle of the line on the right should be at an angle of 30 degrees. Additionally, the hole in the sketch should be located from the bottom right corner of the sketch. These parameters are what the designer deemed most important for this particular model.

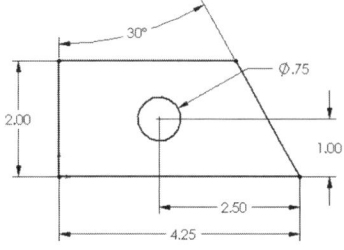

Fig. 1-4. Design intent, example 1.

Remember that the placement of dimensions is very important, because they are being used to drive the shape of the geometry. Take the horizontal 4.25-inch dimension in example 1. If this value is increased, the overall length of the part will be increased and the 30-degree angle will be maintained. The hole will maintain its position from the bottom right corner. If the locations of any of the dimensions are changed, how the geometry is driven will be changed.

In figure 1-5 (example 2), you see that what is important to the individual designing this part is the horizontal dimension at the top of the part, rather than the angle of the line on the right. In this second example, changing the 4.25-inch horizontal dimension will alter the length of the sketch, but the angle of the line will change. The line's angle will not be maintained, as it was in example 1.

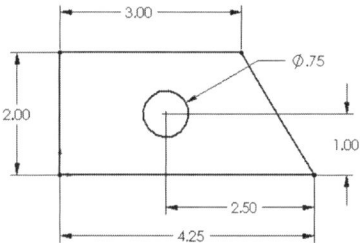

Fig. 1-5. Design intent, example 2.

- **NOTE:** *Keep in mind that the dimensioning scheme can be changed at any time. You are not locked into a specific design. You can also design without dimensions, rough out a sketch, and then later go back and fully define it.*

Take a look at one last example (example 3), shown in figure 1-6. Perhaps the design calls for the hole to be a particular distance from the top left corner, as opposed to the bottom right. In this case, the dimension arrangement would be best. But then again, it all depends on your design intent.

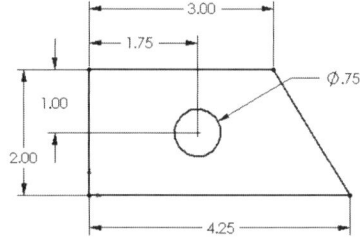

Fig. 1-6. Design intent, example 3.

It is obvious that there can be quite a few variations on dimension placement. If there are so many choices, how can you be sure you will make the right decisions? The answer is very simple. Add dimensions on any part of the sketch geometry you want to control, and do not worry if you make a mistake. The ability to go back and add, remove, and change dimensions is never more than a few mouse clicks away. Chapter 3 covers the commands for adding, deleting, and modifying dimensions.

> **NOTE:** *Do not be overly concerned with dimensioning to datum points or stacked tolerances in the part. These issues can be addressed in the drawing layout. Be more concerned with your design intent at this stage.*

Boolean Modelers

The word *Boolean* is normally associated with Boolean algebra. George Boole, shown in figure 1-7, lived from 1815 to 1864 and was the British mathematician and logician that invented Boolean algebra.

Fig. 1-7. George Boole.

Boolean algebra is still very important today to many mathematicians, and contributed significantly to the design of modern computers. So much so, in fact, that the word *Boolean* is still applied to operations performed by certain solid modeling software. Solid modelers that use Boolean operations to create solid models on a computer are sometimes referred to as Boolean modelers in the CAD industry.

A Boolean solid modeler works much differently than a parametric modeler such as SolidWorks, and merits further explanation. Imagine you were sitting at your computer CAD workstation and wanted to design a new part using a Boolean modeler. Perhaps you would start out with a general shape for the profile of that part and then extrude or revolve the profile. This would constitute the base feature.

Now assume you needed to create a cut in this part. A typical sequence of events would be to create another solid body that would resemble the cut. In other words, the material that needs to be removed to create the cut would have to be created as a solid object first. Next, this solid shape could be subtracted from the original part. Depending on the software program being used, this operation might be called a subtraction, or difference, command. Figure 1-8 shows what this scenario might look like prior to creating the cut.

It is fairly obvious that using any type of Boolean modeler is very inefficient. There are not many Boolean modelers left on the market today. However, there are some CAD programs that give the operator the ability to perform Boolean operations in addition to the other functionality inherent in the software.

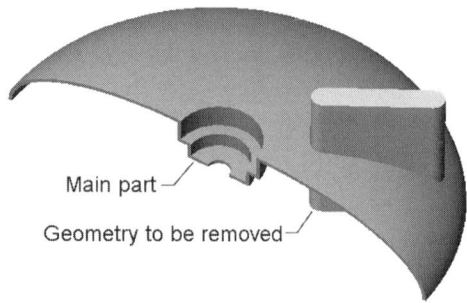

Fig. 1-8. Preparing to create a cut using a Boolean modeler.

Nonparametric Modelers

Standard solid modeling software that does not use any sort of parametric technology would probably be the next step up from a Boolean modeler. The scenario here would typically be one or two steps less than what would be expected with the Boolean modeler. In fact, a nonparametric modeler would probably function very similarly to SolidWorks, at least to begin with.

Using the same example as in the previous section, assume that once again a cut needed to be created through an existing model. Instead of having to create another solid body, now you simply create a sketch projected down through the model, thereby cutting away anything the sketch intersects. Figure 1-9 shows the same model used in the previous section, and a sketch that could be used to create the same cut.

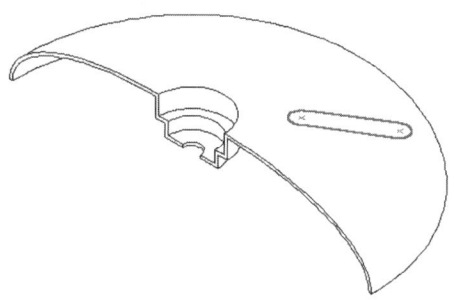

Fig. 1-9. Preparing to create a cut using sketched geometry.

The action of performing the cut would very likely be extremely similar to how a cut is created in SolidWorks. That is, the sketched geometry is projected through the part, perpendicular to its original sketch plane. The material that is removed is the same shape as the original sketch geometry. This begs the question of how parametric technology is different.

Parametric Modelers

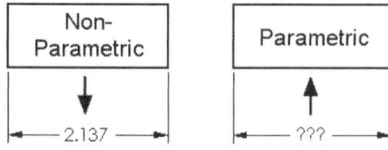

Fig. 1-10. Parametric dimensions drive geometry, as in the example at right.

The main difference between a nonparametric modeler and a parametric one can be summarized in one simple concept. In a nonparametric modeler, the geometry drives the dimension; in a parametric modeler, the dimension drives the geometry. This simple concept is illustrated in figure 1-10.

Benefits of Solid Modeling

It is extremely important to make certain everything in the model is accurate with a nonparametric modeler. If it is not, the detailed drawing will not be accurate and the model will probably get manufactured incorrectly. When working with a parametric modeler, it is possible to create the geometry in an almost carefree way, because the work gets done when the dimensions get placed on the model. For example, when a dimension is placed, SolidWorks will typically come back and ask you "what would you like the value of the dimension to be?" You would type in a value, and SolidWorks would make the geometry conform to that value, whether it be a radius of an arc, the length of a line, or something else.

The benefits of a parametric modeler are far-reaching. Dimension values can be changed at any time and the model will update accordingly. Instead of reworking the model, just tweak a few dimensions. This is demonstrated in figure 1-11.

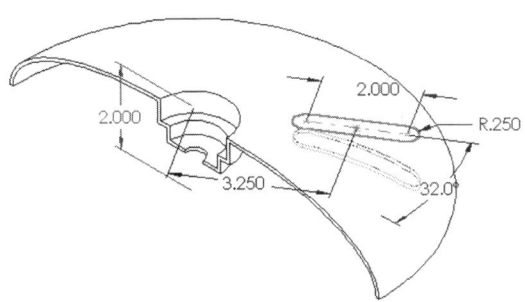

Fig. 1-11. A parametric feature being created in SolidWorks.

Once again we see the by now familiar model. This time, however, parametric dimensions are available because the model was created in SolidWorks. Any of the dimensions can be modified to change the shape of the underlying feature or sketch geometry. All of the techniques to accomplish these tasks, and many more, are covered in subsequent chapters. The model with the completed cut is shown in figure 1-12.

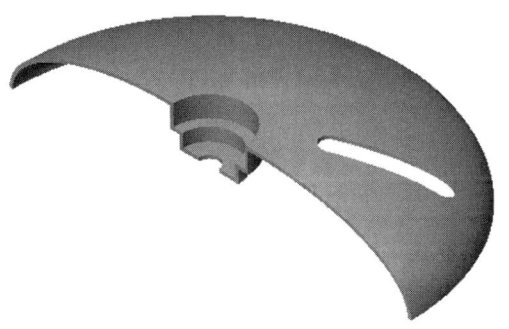

Fig. 1-12. Model with the completed cut feature.

Benefits of Solid Modeling

There are so many benefits to working with solid modeling software over strictly 2D drawings it would be difficult to talk about all of them in this book. However, there are some very simple reasons why solid modeling is often preferred over working with standard 2D geometry. The most apparent reason is ease of use. If you are a designer currently working in the 2D world, an immediate benefit would be the ability to see the part as a 3D image as it is being created.

Many mechanical parts contain features that are very difficult to visualize in 2D. Because it is possible to add features directly to a 3D model, it becomes possible to see the model grow and evolve into a creation you "built." As funny as it may sound, this ability to visualize the model on screen allows you to make your mistakes faster. There are always going to be design issues. Working in 3D helps to resolve those issues faster.

There are different degrees of 3D geometry. The most basic is wireframe geometry, which consists of lines and arcs (and sometimes free-form mathematical curves known as "splines") joined end to end to make up a 3D model. It is easy to see where the term *wireframe* originated; the model itself is akin to wires of varying shapes glued end to end. This is where 3D modeling got its start. Working with wireframe geometry can get very frustrating very quickly, mostly because trying to create a model out of "wires" can be extremely difficult. Where more complex models are concerned, it is downright impossible, in which case some sort of workaround must be found that does not really show the part as it is intended.

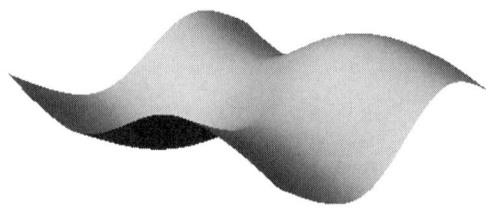

Fig. 1-13. An example of a surface.

Following this brief history of computer-aided modeling, we come to surface geometry. Surfaces go beyond simple wireframe geometry, and contain more mathematical information. Take, for example, the image shown in figure 1-13. The only way to create this type of surface in wireframe would be to approximate it with a mesh of lines. Approximations, obviously, are not accurate.

Not all surfaces are wavy shapes that look like the surface of the ocean. A surface can be a planar square face as well. If we proceed along this vein, imagine a number of surfaces all meeting at their edges. Perhaps the surfaces are six square faces that form a cube, or perhaps they are something much more complex.

Whatever the case, all of the surfaces come together nicely with no overlaps or gaps. In such a case, a solid modeling program can take such surfaces and turn them into a solid. The surfaces are known as the topological data. The software understands what side of each surface is the inside and which is the outside. SolidWorks understands details such as the model's mass, density, center of gravity, and certain other physical properties.

Due to the innate nature of solid modeling software, it is possible to create geometry that would be impossible to model with a wireframe CAD program. Figure 1-14 shows an example of this, including a portion of a part that contains three fillets. (For the sake of simplicity, the word *fillet* is used throughout this book when describing either fillets or rounds.)

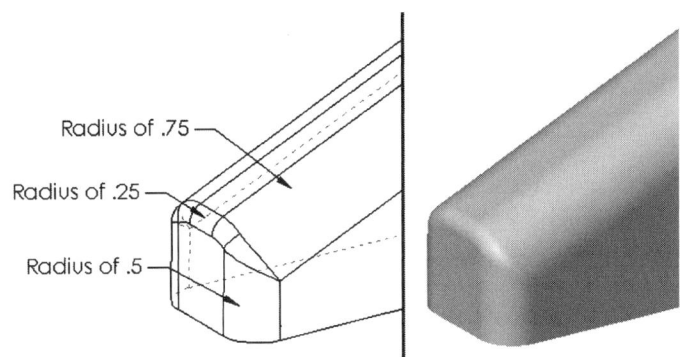

Fig. 1-14. What happens when three fillets meet.

In this illustration, the .75-inch fillet was added first, and then the .5-inch fillet. Finally, the .25-inch fillet was added, creating a blended surface that would have been very difficult to create in 3D wireframe. In SolidWorks, the task only took a few seconds. It is difficult to imagine how one might even go about creating this part with a 3D wireframe modeler.

It should be noted that even though numerous images in this book are shown in wireframe view, this is strictly for clarity. All parts created in SolidWorks are solid models. There are a number of view options explored in detail in Chapter 3.

Downstream Benefits of Solid Modeling

Another benefit of solid modeling is the ability to determine material properties, as previously mentioned. Because the computer is constantly keeping track of the model and its structure and characteristics, mass property data can be extracted in a matter of seconds.

Consider for a moment the area of industry related to the manufacture of bottles or containers. It is a fairly simple process of determining the exact volume for the inside of a part and knowing, for instance, how much liquid a container might hold.

Changing the density of the part is an option that can be performed at any time. Material properties can then be determined for, say, a specific alloy. Moments of inertia, section properties, and other data can be obtained at any stage of the design process.

All companies need to sell their products, and many have the need to create brochures or advertisements. Many CAD programs allow for some degree of shaded rendering. Because SolidWorks already keeps track of the topology information of the model, it requires no extra skills to render the model. Set the color, add a few lights if desired, and, when you are happy with the view, save the image. This can be done through a screen capture or by saving a TIFF (Tagged Image File Format) image directly from within SolidWorks. You can even specify your own background image to use with SolidWorks. If true photorealistic images are required, add-on programs such as PhotoWorks will accomplish this for you.

As mentioned earlier in this chapter, there are many other functions one can perform on a solid model. Finite element analysis (FEA), motion simulation, revision control, piping, harness and cabling, and kinematics are just a few. There are third-party programs that completely integrate with SolidWorks that let the user accomplish many tasks only dreamed of ten years ago. Questions regarding programs of this nature should be directed to your local value-added reseller or SolidWorks salesperson. They are there to answer your questions and help you find the best products to suit your needs.

SolidWorks and the Big Picture

There are a number of reasons SolidWorks is an attractive software. Of primary importance is that SolidWorks was created with ease of use and power at the top of the priority list. The Parasolids geometry engine employed by SolidWorks is a very powerful modeling kernel. It allows the user to perform functions that would make other programs crash and burn. Some of these powerful routines are explored later in the book.

SolidWorks is also very easy to use. The software was written exclusively to take advantage of the Windows operating system. SolidWorks will run only on a Windows operating system or its future incarnations.

The very fact that SolidWorks was developed on the Windows platform is a big plus, as the software did not have to be ported over from a DOS- or UNIX-based operating system. Porting from one platform to another almost always results in problems of one sort or another, a common trait being severe performance degradation. Modeling software ported from UNIX operating systems has traditionally required huge hardware upgrades. SolidWorks does not suffer these problems.

Because SolidWorks was written for Windows from the ground up, it has many of the benefits of Windows-based programs. For example, cut-and-paste functionality exists throughout SolidWorks. The ability to drag a feature from one model into another for the purpose of duplicating the feature instead of recreating it is another great advantage.

SolidWorks allows having numerous files open at one time. Toolbars can be resized and customized, much like toolbars in Microsoft Word or Excel. Macros can be recorded and played back using the SolidWorks Basic programming language. Excel-based spreadsheets can be used to create family part tables that drive part geometry in a SolidWorks part file. This is standard Windows Object Linking and Embedding (OLE, pronounced "o-lay") functionality.

Solid modeling software is what 3D wireframe CAD software always strove to be, but could not. However, times change, and the evolution of software and computer hardware is an ongoing force. SolidWorks Corporation has succeeded in creating solid modeling design software with unmatched functionality and ease of use, at a price point that puts it in reach of many companies and individual designers and engineers.

Solid modeling may not be for the architectural or mapping professions. However, for designers and engineers creating mechanical elements, whether toys or trains, solid modeling is the way to stay ahead of the competition and make the most productive use of your time.

SolidWorks API

There are a number of ways third-party developers or end users can create programs or interact with SolidWorks in a more intimate way. For the casual user, there is a macro recording function that allows for recording SolidWorks commands. These macros can be edited and analyzed by the user. SolidWorks macros have a specific *.swp* file extension and use an internal programming language known as SolidWorks Basic.

> **NOTE:** *Older versions of SolidWorks Macros used the* .swb *extension. When these macros are edited, SolidWorks automatically saves them with the new* .swp *extension.*

On a higher and more technical level, developers can interact with SolidWorks using the SolidWorks API (Application Programming Interface). SolidWorks supports Visual Basic and Visual C++ programming languages. There is help available in an API help file located in the SolidWorks directory structure. Look for the file *apihelp.chm* in the directory *SolidWorks\lang*. You should also be able to access this information using the SolidWorks Help menu.

Writing programs using the SolidWorks API or other means is outside the scope of this book. For more information regarding this topic, it is suggested you access the SolidWorks web site at *www.solidworks.com*.

Minimum System Requirements

If you are trying to decide on what type of computer to purchase to run SolidWorks, this section will tell you what you need to know. There are a number of factors to consider, including just what it is you intend on doing with SolidWorks. Keep in mind that *minimum requirements* is a term that should fit loosely. Your minimum requirements may be much more than those SolidWorks Corporation recommends. The minimum system requirements for running SolidWorks are as follows.

- A Pentium III class Intel or an Athlon processor, 733 MHz or better.
- 256 MB of RAM.
- CD-ROM drive (for installing the software).
- Graphics card that supports OpenGL acceleration. (Check SolidWorks web site at *www.solidworks.com* for the current list of supported video cards. Note that not all of the listed cards support hardware OpenGL acceleration, which is preferable.)
- A Windows-based operating system. Currently supported operating systems follow.
  — Windows 98 SE
  — Windows ME (Millennium Edition, not recommended)
  — Windows NT 4.0 (must have Service Pack 6 or better)
  — Windows 2000 (recommended)
  — Windows XP (recommended)
- 300 MB of hard drive space (more if add-ons have been purchased).
- Internet connection (not required, but highly recommended).
- Internet Explorer 5.0 or later.
- Microsoft Excel 97, Excel 2000, Excel 2002 (for BOM and design table functionality).

Now let's run through the previous list again to find out what you will really need. CAD programs benefit from a faster processor, obviously. This is mostly due to the mathematical computations that must be performed. A faster processor means less wait time. It is definitely a good idea to go with the fastest processor you can afford. Additionally, many SolidWorks users have found the AMD processors (such as the Thunderbird line) make great workstations. Consider AMD to be a viable alternative to Intel processors when making your purchase.

Dual Processors?

This is a very common question. Does SolidWorks benefit from having more than one CPU? Currently, yes, but only minimally. Some (but not all) SolidWorks graphics functions will take advantage of multiple processors. Is it worth the extra cost of a more expensive motherboard and an added processor? Probably not. However, is SolidWorks all you work with?

If all you will be doing is work with SolidWorks, purchasing multiple processors will not be cost effective. If, on the other hand, you frequently work with numerous programs at once, more than one processor will be to your advantage. If you perform some action in SolidWorks that is going to take some time, you can always switch to a different program and get some other tasks out of the way. A good example would be creating helical threads in SolidWorks. While waiting for the operation to complete, you could open up your e-mail program and respond to all that mail that has been piling up.

If you do decide to go the multiple processor route, make sure you use the Windows NT, 2000, or XP operating systems. Windows 98 or ME do not support multiple processors, and you will be throwing your money away.

Memory and Hard Drive Space

The amount of memory you purchase depends on what you want to accomplish with the SolidWorks software. Even if you are not creating complex parts, you may want to increase the amount of RAM you have to 512 MB, especially if using Windows 2000 or XP. These operating systems take up a lot of system resources. Assemblies with a few dozen components that are fairly simple in nature will do fine with 256 MB. Larger assemblies, of over 50 components, would benefit from more memory.

There are those that want to create assemblies of hundreds and even thousands of components. With assemblies like that, you will want to stock up on as much memory as you can afford, or that your motherboard can handle. Memory in the range of 512 MB or even 2 gig of RAM is not unheard of. It really depends on your specific situation. If you have any questions, contact your local SolidWorks reseller. The general rule of thumb, though, is that you can never have too much memory!

Hard drive space really is not an issue anymore. Hard drives have so much disk space that anything you purchase will more than likely have much more than enough disk space. If not, you can always pick up another 20 gig or so at the local computer store.

Do not feel pressured into buying an SCSI hard drive, either. ATA-100 hard drives perform very well, even when opening assemblies with many components. SCSI hard drives just do not seem to have enough increased speed to warrant the increase in cost. For those that want superior hard drive performance, consider a striped disk array (RAID level 0).

- **NOTE:** RAID arrays require a minimum of two hard drives. RAID 0 (better known as striping) will essentially double hard drive performance. However, striping has zero fault tolerance. In other words, if one drive is lost, all data is lost. Search the Internet for more information on RAID arrays.

Graphics Cards

One of the first questions most people ask is "What type of graphics card do you recommend?" There are too many graphics cards on the market and they change too quickly to give a good answer to that question. However, what you should look for in a graphics card is fairly straightforward.

Without getting into a rundown of the current technology, there are a few things that will make your shopping for a graphics card much easier. The main difference between graphics cards on the market today is that some are best for games and some are best for CAD. A 3D accelerator card is not necessarily going to make your CAD models fly around on the screen. Gaming cards have a high polygon fill rate, whereas CAD cards benefit more from geometry (polygon) setup engines.

Look for a graphics card that has a geometry setup engine built into the card. If the card in question can create millions of polygons locally, that is less work your main processor has to do. In addition, look for a card that will accelerate OpenGL (Open Graphics Language) in hardware as opposed to software. This will make a significant improvement on graphics performance.

There is some controversy over which is better, PCI- or AGP-based cards. Without getting into that controversy, suffice it to say that Solid-Works will operate just fine with whichever type of card you use. An advantage of AGP-based graphics cards is that they leave an extra PCI slot open. For this reason alone, it will benefit you to go with an AGP card.

Monitor and Screen Size

If you will be staring at a computer monitor a good portion of the day, you will want to see what it is you are doing without suffering eye fatigue. Purchasing the right size monitor is a major factor. Not much of a decision to be made here, really. When doing CAD work, make sure the monitor has a minimum of 17 inches of diagonal viewing area. A 19- or 21-inch monitor might be better, assuming you have the budget and available desk space.

Any monitor bought today will have the capability to run at high resolutions and refresh rates. The resolution is measured in pixels, and the lowest resolution SolidWorks should be run at is 1,024 by 768. If lower resolutions are used, seeing all icons on the toolbars and all dialog boxes becomes very difficult. It is possible to run at higher resolutions as long as your eyes can see the smaller icons and text. If you do purchase a large monitor, you will be able to run at very high resolutions and still be able to read text.

> **NOTE**: *The highest resolution your computer can go is limited only by your graphics card and monitor.*

If you are considering a flat panel display, know that they are typically locked into one particular resolution. This is the major downside of flat-panel displays (that, and their price). Even if you find a flat-panel display that will let you change the resolution, anything other than the recommended resolution of the display will have a very poor visual quality.

The refresh rate of a monitor is important because that is what controls how often the screen is repainted on the monitor. This refresh setting is established by your graphics card, but can easily be set through your Windows display settings. One item to look for when purchasing a graphics card is the RAMDAC (random access memory digital-to-analog converter) speed. A speed of 300 MHz or better is quite common. A higher frequency on the RAMDAC will enable larger monitors to maintain quicker refresh speeds at higher resolutions.

A typical default setting for the screen refresh frequency is 60 Hz. At 60 Hz, the refresh rate will cause most people to experience eye strain or even develop headaches. This is especially true under fluorescent lighting because of the 60-Hz flicker of the bulbs. Any recently manufactured monitor will handle much better than 60-Hz refresh speeds. Common settings are 75, 85, and even 100 Hz and higher. The higher the refresh rate, the better. Contact your system administrator or consult your Windows manual to discover how to increase the refresh rate.

> **NOTE:** *Flat-panel displays do not suffer the flicker of CRT monitors due to their nature.*

Summary

In this chapter, you have discovered that SolidWorks is a feature-based, parametric, solid modeling design tool. By gaining some insight in how a computer interprets CAD geometry, you should have a broader understanding for what goes on behind the scenes while building a solid model.

It should also be apparent that parametric dimensions within Solid-Works help to build intelligence into the models you create, thereby making design changes a simple task rather than a tedious chore. The fact that SolidWorks incorporates parametric dimensions makes it a true design tool. Boolean solid modelers cannot compete in this arena.

You have also learned in this chapter that there are a great many benefits from working with solid modeling software. These benefits include ease of use when creating a solid model, to the extent of being able to see and manipulate the shaded model on screen.

Other benefits of solid modeling include the wealth of information you can obtain from solid geometry (such as material properties, centroid data, measurements, and interference checks), all before a part is ever actually built. There are also the numerous third-party applications, such as kinematics and finite element analysis software, which let you extend the functionality of SolidWorks far beyond its initial scope.

Questions and Topics for Discussion

1. State what the following acronyms stand for: CAD, ECN, FEA, and API.
2. Optional: Discuss the meanings of the previous acronyms.
3. What is a stereolithography file?
4. Using personal reasons, list three benefits of CAD (electronic) drawings over paper drawings.
5. List four objects that could be considered features of a solid model.
6. What does the term *parametric* refer to?
7. Describe the meaning of design intent.
8. True or False: geometry drives dimension values in a parametric modeler.
9. State two advantages of working with a solid modeler over 3D wireframe geometry.
10. SolidWorks was originally written for which operating system: UNIX, DOS, Windows, Linux, or Irix 4.3?
11. Name three operating systems (or operating system versions) SolidWorks will run on.

Optional Problem

List the components you would use to build a computer workstation for SolidWorks. Include which operating system you would use. Go into as much detail as you like.

CHAPTER 2

A SolidWorks Overview

BEFORE YOU CAN WORK WITH SOLIDWORKS, you have to know how to think like SolidWorks. This is especially true if certain habits have been developed from working with other software programs. Certain programs are easier than others from which to make the transition. Those who have worked with other parametric solid modelers usually find the transition to be quite smooth. Those who have worked strictly in 2D usually have the most difficult time, because of the old habits and mind-set developed when working with such programs.

For the CAD population accustomed to working in a 2D environment, the jump to 3D solid modeling is not a major concern. That is, at least as far as SolidWorks is concerned. Other CAD systems are not nearly as easy to learn as SolidWorks. In addition, certain individuals find it difficult to "see" 3D images in their mind. If you fall into this category, do not despair. This book should be able to help, and SolidWorks can work wonders toward visualization.

Sketching Overview

When discussing the mind-set needed for working with SolidWorks, there are two topics that need to be expanded upon. The first topic is that of sketching. When speaking of a parametric modeler, the ability to edit dimensions and change the shape of a model is the most obvious aspect of parametrics. (The terms *parametric* and *nonparametric* are used here to make a distinction and do not necessarily fully describe SolidWorks.)

This functionality presents itself within the first sketch drawn. This capability is sometimes unexpected by those who are used to working in 3D wireframe, or even with 2D layout geometry. With nonparametric programs, it is imperative that geometry be created with the utmost accuracy. When a drawing is not accurate, dimensions can be interpreted incorrectly and parts built to incorrect specifications. This obviously results in many unhappy individuals.

When sketching in a modeler such as SolidWorks, it is not necessary to create geometry with 100-percent accuracy at this phase of the process. Lines, arcs, and additional geometry need not be created with exact dimensions in mind. To help you overcome this seemingly inappropriate behavior, take the word *sketch* literally. A sketch should be just that, a sketch. It need only be the approximate size and shape of the feature (or part) you are trying to create.

When dimensions are added to a sketch, the geometry will change size and shape, depending on the values of the dimensions. This is where the work gets done, and is the essence of parametric modeling. This way of thinking goes hand in hand with the notion of design intent, as mentioned in the previous chapter. Where dimensions are placed will make a difference as to how your model changes when a dimension is altered. It is very important to remember that dimensions do much of the work in SolidWorks.

> **NOTE:** *In summary, when sketching in SolidWorks, the sketch need only be the approximate size and shape of the part or feature being created. When dimensions are added, they will drive the size and shape of the geometry.*

Features Overview

There is another area of concern related to the SolidWorks way of thinking that has to do with SolidWorks being a feature-based modeler. Features have been discussed earlier. As you should recall, features can be fillets or chamfers, extrusions, cuts, holes, and so on. Features can add material or take it away. Even when removing material from the model, a feature is still being added. This is how the model should be viewed in your mind's eye. Take a simple hole as an example. Try not to think of it as just a hole, but as a feature. There is, of course, a reason for this way of thinking.

Wireframe geometry does not consist of features, rather of simple lines and arcs and other geometric entities. Likewise, Boolean-based modelers do not contain features, although the models created with such programs may contain solid geometry. The term *Boolean*, first mentioned in Chapter

Sketched Versus Applied Features

1, refers to the ability to add or subtract solid geometry to or from other solid geometry. The term *feature* implies that there is more to the specific parts of the model than meets the eye, and this is the reason for thinking in terms of features. Feature modelers have more built-in intelligence than Boolean modelers, as you will discover shortly.

➜ **NOTE:** *Features contain built-in intelligence and are fully editable components of a model.*

A feature has properties associated with it. It has dimensions and constraints that can be edited at any time. A feature also has intelligence built into it. The sketch used to create a feature can be edited, its dimensions can be modified, and its properties (color or name, for instance) can be changed.

Sketched Versus Applied Features

The feature types discussed in the next two chapters are sketched features. These feature types require a little more user intervention than applied features. In particular, they require the user to create a sketch. That sketch is then typically extruded or revolved to create a feature. Lofted or swept features are also possible, and are covered later in this book. Figure 2-1 shows a simple sketch with dimensions. Figure 2-2 shows an extruded feature created from the sketch. Figure 2-3 shows the same sketch as a revolved feature.

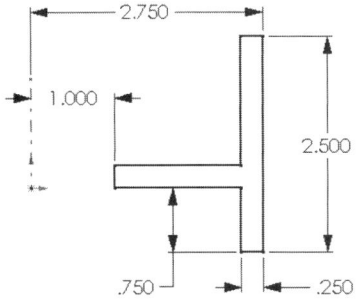
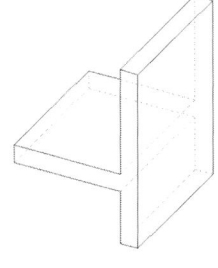
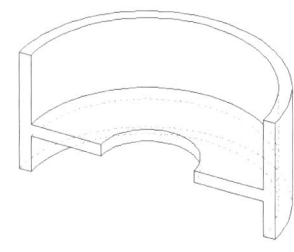

Fig. 2-1. A simple sketch prior to being turned into a feature.

Fig. 2-2. An extruded feature.

Fig. 2-3. A revolved feature created from the same sketch.

Because the techniques used when creating a sketched feature are very important, a large portion of this chapter is devoted to discussion of the various sketch tools and to certain things a SolidWorks user needs to be aware of when creating a sketch. Many of these topics are covered later in this chapter, and in Chapter 3 you will practice creating some actual sketch entities.

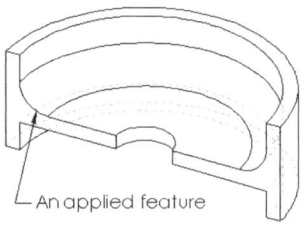

Fig. 2-4. A fillet is an applied feature.

Applied features do not require a sketch. They are simply applied directly to the model, with very little user intervention required. A very common applied feature would be a fillet. In the case of a fillet, an edge or edges are selected, a radius is specified, and SolidWorks does the rest. Figure 2-4 shows an example of an applied feature.

The SolidWorks Interface

As previously mentioned, SolidWorks is strictly a Windows-based program. As such, it uses Windows conventions. An example is the shortcut keys used for copying to or pasting from the Windows clipboard. Ctrl-C and Ctrl-V are the hotkey combinations used to copy and paste objects to and from the clipboard, respectively. It does not matter if you are using Microsoft Office, Netscape Composer, or JASCI's Paint Shop Pro, the same hotkeys will work. Considering that SolidWorks takes advantage of these standard Windows functions, typical Windows hotkeys will function in SolidWorks the same way you would expect them to in any other Windows program.

Many of the icons used in the SolidWorks toolbars are icons commonly used by many programs. The File New, File Open, and File Save icons are now the de facto standard icons found in many programs. This holds true for other icons. Of course, only solid modeling CAD software (and maybe a handful of other niche market software programs) would have icons for extrusions, shelling, and so on. SolidWorks, in this case, invented its own icons. You will probably find that the icons employed by SolidWorks are very easy to use and visually represent their functions very well.

Beginning a New Part

Because you will be exploring the SolidWorks interface in the material that follows, it would be best if you knew how to start a new part. This way, you can follow along on your computer. How-To 2-1 shows you how to start any new SolidWorks document, whether it is a part, assembly, or drawing.

HOW-TO 2-1: Starting a New SolidWorks Document

To begin a new SolidWorks document, perform the following steps.

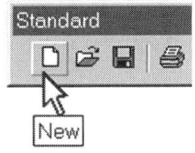

Fig. 2-5. The New icon.

Fig. 2-6. New SolidWorks Document window.

1. Select New from the File menu, or click on the New icon, shown in figure 2-5.

2. Select the type of new file (Part, Drawing, or Assembly) from the New SolidWorks Document window, shown in figure 2-6 (your tabs will look different).

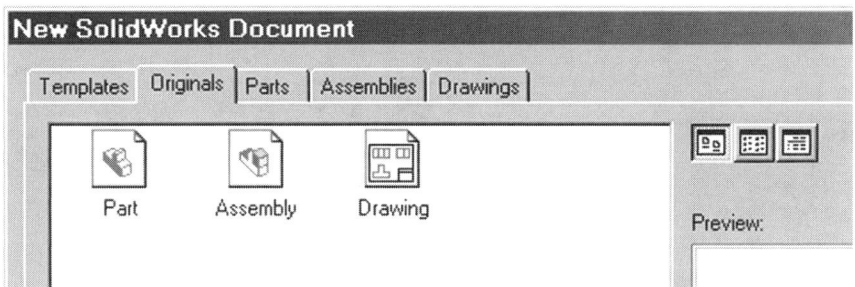

3. Click on OK to start the new file.

The icons present in the New SolidWorks Document window are actually templates. Later in this chapter you will learn more about SolidWorks template files. You will also learn how to customize the tabs found in the New SolidWorks Document window. For now, let's continue exploring the basic SolidWorks interface.

Toolbars

Under the View pull-down menu, you will find the Toolbars submenu that lists all of the toolbars available to the SolidWorks user. The various toolbars can be toggled on and off by clicking on the toolbar's name. The View > Toolbar submenu is shown in figure 2-7.

SolidWorks contains a number of standard and specialized toolbars. Included are toolbars you will not be able to do without, and those you may only occasionally use. Most of the available toolbars are customizable, and a few are not (such as the Font and Web toolbars). You should not worry about customizing the toolbars at this time. Become familiar with the program first. Once you can find your way around SolidWorks, then perhaps try customizing the toolbars.

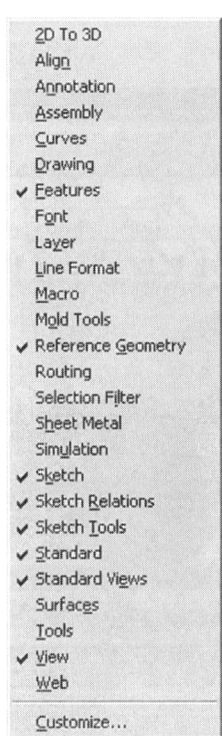

When the SolidWorks software goes through a new release, the toolbars sometimes change. This is due to functionality and enhancements being added to the program. Additionally, because most toolbars can be customized, the toolbar images pictured in this book may look slightly different from yours. This is nothing to be concerned about, but is something you should be aware of in case you notice discrepancies between what is on your screen and what you see in this book.

Some of the more commonly used toolbars are shown in the material that follows, along with short descriptions of the basic functions of each toolbar. Toolbars not mentioned here are covered later in this book. Figure 2-8 shows the Standard toolbar.

Fig. 2-8. Standard toolbar.

All of the icons normally associated with file operations or with the Windows clipboard can be found on the Standard toolbar. Users can also print or perform a print preview from this toolbar. Other commands include rebuilding a part, redrawing the screen's graphics, and accessing the Web toolbar or the Help function. If you do not understand the exact function of some of the icons mentioned here, rest assured that their meanings will be explained over the course of this book.

Fig. 2-7. The Toolbar submenu can be found under the View menu.

The View toolbar, shown in figure 2-9, performs all functions related to panning and zooming a model. It is also possible to control how the model is displayed, such as in wireframe or shaded mode. You will find that this toolbar is one you will use constantly throughout a SolidWorks editing session.

Fig. 2-9. View toolbar.

Sketching is another common function in SolidWorks. Likewise, you will often access many of the icons on the sketch-related toolbars. A good example is the Sketch icon itself (third from the left on the Sketch toolbar), which looks like a pencil drawing lines. It is used whenever you want to enter sketch mode. The Sketch toolbar is shown in figure 2-10.

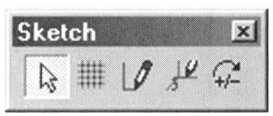

Fig. 2-10. Sketch toolbar.

FeatureManager and the Work Area

Sketch relations are sometimes referred to as dimensions or constraints. Constraints are often referred to as geometric relations. All types of relations can be added or deleted through the use of the Sketch Relations toolbar, shown in figure 2-11.

When you want to get down to the nitty-gritty of creating a SolidWorks sketch, the Sketch Tools toolbar is what you want. From here, circles, arcs, lines, and other entity types are created. This is also the toolbar to look to when mirroring, trimming, or extending sketch geometry, or when other sketch-related functions are required. The Sketch Tools toolbar, shown in figure 2-12, is one of the largest of the available toolbars.

Fig. 2-11. Sketch Relations toolbar.

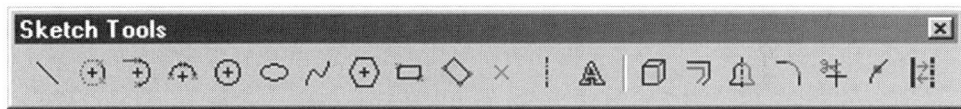

Fig. 2-12. Sketch Tools toolbar.

FeatureManager and the Work Area

FeatureManager is a part of SolidWorks you should become familiar with right away. It is the key to editing your model. Think of FeatureManager as a chronological history of events. The feature you created first will be near the top of FeatureManager, and the feature you created last will be at the bottom. FeatureManager is shown in figure 2-13.

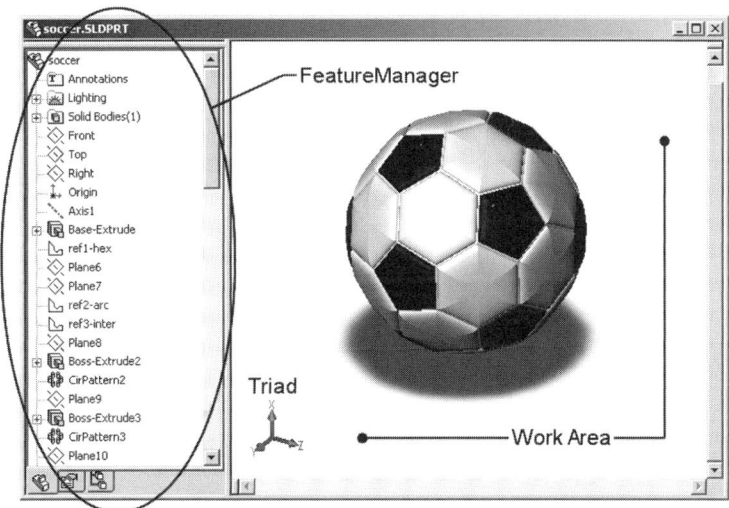

Fig. 2-13. SolidWorks' FeatureManager and work area.

The work area needs little explanation. It is where you will see the model you are working on. It is where you will sketch and where you will see your SolidWorks model come to life as you add features to it. FeatureManager is what lists those features you create. It is a timeline and a history tree of what you have done. Get accustomed to the term FeatureManager because it is one you will hear throughout the course of this book.

The triad, which can also be seen in figure 2-13, represents the X, Y, and Z axes. The triad rotates as the model is rotated, giving the user an idea as to where the coordinate system axes are in relation to the model.

The Separator Bar

There is a vertical bar that separates FeatureManager and the work area. This vertical bar can be repositioned by placing the cursor over the bar, holding down the left mouse button, and dragging the bar to a different position. When the cursor is over the bar, the cursor turns into a double-sided arrow. When this symbol appears, repositioning the bar is possible.

From time to time the separator bar needs to be repositioned so that the names of the features in FeatureManager can be seen in their entirety, or so that more of the work area can be used. If this vertical bar is repositioned and the file is saved, the next time the file is opened SolidWorks remembers where the bar was placed.

When repositioning the vertical bar, there is one position where the bar will want to "snap" into position. This is the ideal position for the vertical bar in most cases. It will allow you to see options in PropertyManager (see next section) in their entirety and still maintain a maximum work area.

✓ **TIP:** *Set the vertical bar to its ideal or preferred position prior to saving a template (more on templates later in this chapter).*

PropertyManager

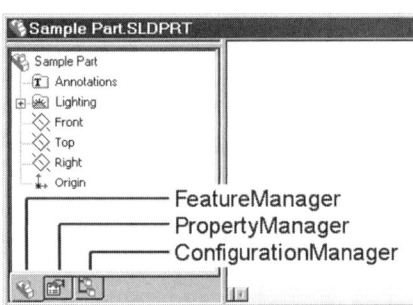

You may have noticed small tabs at the bottom of FeatureManager. There are a total of three tabs, shown in figure 2-14. Sometimes there are more tabs, depending on the third-party software you may have purchased in addition to SolidWorks.

Fig. 2-14. Tabs at the bottom of FeatureManager.

FeatureManager and the Work Area

By clicking on any of these tabs, you can navigate to the various "managers" within SolidWorks. FeatureManager you should already be somewhat familiar with. ConfigurationManager is discussed in more detail in Chapter 10. This particular section will familiarize you with PropertyManager.

PropertyManager's main function is to display the various options and input parameters when executing a command. To complete a command, use the buttons at the top of the PropertyManager display. These buttons are shown in figure 2-15, and are quite self-explanatory. Once the applicable parameters have been input into PropertyManager, click on OK to accept the parameters and carry out the command, or on Cancel to exit without completing the command. The Help button is context sensitive, and will display help for whatever command you are currently engaged in. Next and Back buttons will not always be present. They only appear when PropertyManager requires multiple steps to complete a command.

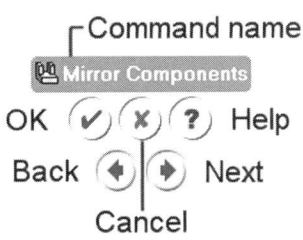

Fig. 2-15. PropertyManager command input buttons.

✓ **TIP:** *If it becomes necessary to access FeatureManager while in a command, click on the name of the command at the top of PropertyManager. This displays the "flyout" FeatureManager. Clicking anywhere on the PropertyManager background closes FeatureManager.*

PropertyManager has functions other than those that help you carry out commands. When sketching or dimensioning, PropertyManager can be set to automatically appear in place of FeatureManager. You will explore why you might want to use PropertyManager in a moment. First, you should be aware of an option that controls how PropertyManager functions. This option is found in the Tools > Options menu in the System Options tab. You will find this option titled Auto-show PropertyManager in the General section. The option is shown in figure 2-16.

For the time being, make sure the Auto-show PropertyManager option is checked. This will allow you to see how it functions when you begin sketching. Before beginning to sketch, however, let's look at an example of PropertyManager in use, shown in figure 2-17.

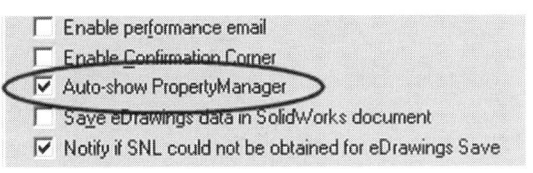

Fig. 2-16. Auto-show PropertyManager option.

28 Chapter 2: A SolidWorks Overview

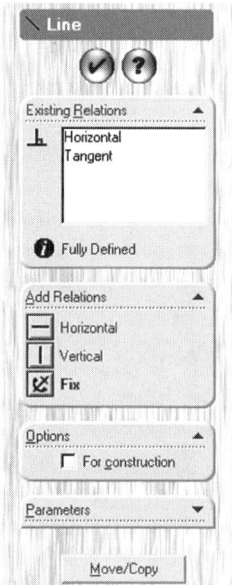

Fig. 2-17. An example of PropertyManager in use.

The example in the illustration is showing the properties of a line. Note the various sections available in PropertyManager. Existing geometric relations are shown, and directly below that, new relations can be added. Additionally, it is possible to turn a sketch entity into construction geometry. (More on this later.)

Another function of PropertyManager, shown in figure 2-18, is the ability to modify a number of aspects of the geometry, from its endpoint location to the overall length of the line. All entity types (i.e., arcs, circles, endpoints, and so on) have their own various parameters that can be modified by accessing PropertyManager. Because it is very easy to simply drag geometry anyway, and because it is the dimensions and geometrical relations that define and control a sketch, PropertyManager has limited significance with regard to sketch geometry.

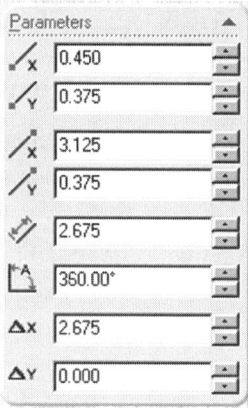

Fig. 2-18. Parameters section of PropertyManager.

With the Auto-show PropertyManager option enabled, PropertyManager flashes on screen in place of FeatureManager every time an entity is sketched or whenever a sketch entity is selected. Even without the Auto-show option turned on, it is still possible to access an item's properties. How-To 2-2 takes you through this process.

How-To 2-2: Accessing an Item's Properties

To access an item's Properties, perform the following steps.

1. Make sure you are not in a command by pressing the Esc key on the keyboard.

2. Right-click on a sketch entity (or other object) and select Properties from the menu.

FeatureManager and the Work Area **29**

By right-clicking on an object, PropertyManager will be displayed and you will have access to the same information that would be available if PropertyManager had been automatically shown. Why is this even important? Because almost all objects in SolidWorks have properties, not just sketch entities. This includes features, assembly components, drawing views, even SolidWorks document files themselves. It is as important that you know how to access the properties of sketch entities as knowing how to access the properties of any object. The right mouse button is the key, and this is very consistent throughout the software.

✓ **TIP:** *Nearly all objects in SolidWorks have properties that can be accessed simply by right-clicking on the object.*

Later in this book you will learn how to perform many different functions in SolidWorks, from changing the color of an assembly component to changing the appearance of a dimension. And guess what? It is all done via properties.

FeatureManager Pane Control

Another aspect of FeatureManager not quite as apparent as the vertical bar is a horizontal bar that allows for multiple panes. This horizontal bar is found at the very top of FeatureManager. As with the vertical bar, the cursor will change into a double-sided arrow when positioned over the horizontal bar. When this happens, drag the horizontal bar down to a new position, thereby creating a new pane in FeatureManager. An example of this is shown in figure 2-19.

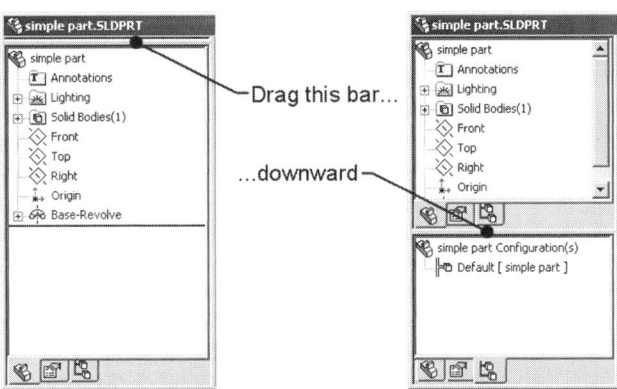

Fig. 2-19. Creating a new pane in FeatureManager.

Once FeatureManager is split into two panes, the tabs found at the bottom of FeatureManager can be used in any combination. As discussed earlier, these tabs are for PropertyManager and ConfigurationManager. To change back to having one pane, simply drag the horizontal bar back to its original position.

Renaming Objects

There are three default planes created by SolidWorks whenever a new part is begun. These planes are known as *Plane1*, *Plane2*, and *Plane3*. It is not necessary, but a good technique, to rename these planes to something more meaningful, such as Front, Top, and Right. You can change the names of these three planes prior to saving a file as a template. You will learn how to create and use templates shortly. For the time being, How-To 2-3 shows you how the names of the planes (or any item in FeatureManager) can be modified.

How-To 2-3: Changing the Names of FeatureManager Objects

To change the name of an object in FeatureManager, perform the following steps.

1. Click on the item in FeatureManager whose name you want to change, wait a moment, and then click again. This is known as a slow double click.
2. Type in the new name for the object.
3. Press Enter.

It is highly recommended you rename *Plane1*, *Plane2*, and *Plane3* to Front, Top, and Right, respectively. These changes will only affect your current part file. However, if this current file is saved as a template file, the changes can be propagated to future files. This eliminates the need to repeatedly adjust various settings, such as renaming the default planes.

It should be noted that the technique used for renaming objects in FeatureManager is the same technique used by the Windows operating system for renaming files. If you already know how to rename files in Windows, renaming objects in FeatureManager should be a snap. One additional point is that nearly anything in FeatureManager can be renamed using this technique. Sketches, features, planes, and many of the other objects you will learn about in this book can all be renamed.

You may wonder why you are renaming the planes Front, Top, and Right, in that particular order. The reason is due to how SolidWorks sets up its default views. First, take a look at these planes to get a better understanding of what you have to work with and how these three planes are oriented. Turning these planes on will require the right mouse button, which is explained in the next section.

Using the Mouse Buttons

SolidWorks does not require a three-button mouse, but it would definitely be to your advantage if your SolidWorks workstation had one. If SolidWorks is your primary tool on the computer, it would be beneficial to invest a little extra cash to obtain a three-button mouse. Even better would be a three-button mouse that has a wheel for the middle button. The uses of the three mouse buttons and middle wheel button are explained in detail in the material that follows.

Selecting Entities

The left mouse button's primary purpose is to select entities and commands. There are also many pick-and-drag operations used with this button that can be performed when sketching, as you will discover shortly. The left mouse button can also be used to edit dimensions through double clicking, or for drag-and-drop functionality. These functions will seem like second nature to you after the first four or five chapters of this book.

When an object is selected in the work area, it turns yellow or green (depending on the object). There are only a few exceptions to this rule, which will be pointed out when those exceptions make themselves apparent. It is also possible to select objects in FeatureManager. Whatever the case, one click on an object with the left mouse button will select that object, whether it be a model object, toolbar, or menu command.

- **NOTE**: *To deselect model items, simply click anywhere in a blank (empty) portion of the work area.*

The Context-sensitive Menu

The right mouse button is used for bringing up what is known as a context-sensitive menu. By clicking the right mouse button (i.e., right-clicking), you can bring up a menu containing various options. The options that appear are dependent on what is right-clicked on. Hence the term *context sensitive*.

Right-clicking while working in sketch mode may bring up a number of shortcuts to often-used commands. Right-clicking on a feature or sketch in

FeatureManager will result in a menu containing various editing options. As you begin to learn SolidWorks, you will see that the right mouse button menus are very predictable. The context-sensitive menus are not filled with needless commands, but instead have a very logical and well thought out selection of choices that directly pertain to the task at hand.

Middle Mouse Button Functionality

If you have access to a three-button mouse, you will find the middle mouse button very useful in rotating, zooming, and panning. By holding down the middle button and moving the mouse, the model on the screen can be rotated, identically to using the Rotate command. Rotating a model can be a little cumbersome to the uninitiated, but will soon become a function any SolidWorks user will not want to live without.

To zoom in or out with the middle mouse button, hold down the Shift key while depressing the middle mouse button and moving the mouse. To pan, use the Control key. Panning, zooming, and rotating model geometry are discussed in detail in Chapter 3. If you feel like experimenting, by all means open a sample part and try out the middle mouse button for size. More than likely you will like the fit.

Some three-button mouse drivers will program the middle mouse button for a special function, such as an automatic double click. This can be very useful in Windows or other programs, but will keep SolidWorks from assigning its own functionality to your mouse's middle button. If your middle button is not functioning properly in SolidWorks, remove any special functionality assigned to the middle button by your mouse driver software. Typically, this can be determined via the Windows control panel by double clicking on the Mouse icon. See your company's system administrator or your hardware vendor if you have further questions.

If the middle mouse button of your mouse is a wheel, zooming just got easier. Simply rotating the wheel will allow for zooming in and out of a model. All other middle mouse button functions described in the preceding paragraph will still function as advertised.

✓ **TIP:** *When zooming in to a model using the wheel, position the cursor over the area you would like to zoom in to.*

Toggling Plane Display

The right mouse button can be used to toggle planes on or off (in Solid-Works terminology *show* or *hide*, respectively). The Show and Hide commands are accessed via the right mouse button's context-sensitive menu,

Toggling Plane Display

and act as toggle switches. That is, a plane is either being shown or hidden at any one time. Show all three planes to see what they look like. How-To 2-4 provides practice in performing this operation. Figure 2-20 shows the Front plane being turned on.

Fig. 2-20. Showing the Front plane.

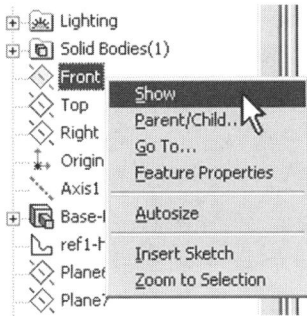

How-To 2-4: Showing/Hiding Planes

To show or hide planes, perform the following steps.

1. Right-click on any plane listed in FeatureManager and select Show from the menu.

2. To reverse the process, right-click on the same plane and select Hide.

✓ **TIP:** *If planes do not display in the work area after showing them, make sure they are globally enabled by checking the Planes option under the View menu.*

Planes being shown will appear as a gray outline or translucent (read on to discover the option that controls plane appearance). If the plane border appears with yellow "handles," it is because it is still selected. Deselect it by clicking in a blank. Be aware that it is possible to right-click on the plane itself in the work area in order to hide it. You must click directly on the plane's border for this to work. When a plane is hidden, its icon in FeatureManager appears ghosted.

It should be noted that planes actually extend infinitely in all directions, and that the gray border is there for display purposes only. When sketching on a plane, you are not limited to sketching within the gray borders of the plane.

A very nice display option of SolidWorks' is to display planes using a translucency effect that shows one side of a plane in one color and the opposite side of a plane in another color. This is quite a nice effect, and it is suggested you turn this option on. How-To 2-5 shows you how to display planes with transparency.

How-To 2-5: Displaying Planes with Transparency and Adjusting Transparency and Color

To display planes with transparency, and to adjust transparency and color, perform the following steps.

1. Select Options from the Tools menu.

2. Click on the System Options tab and select the Display/Selection section.

3. Check the option named *Display shaded planes*. You will find it near the bottom of the window.

You could quit here and the transparent shaded planes would be displayed instead of planes with simple gray borders. But while you are at it, you might as well learn how to adjust the transparency and color of the planes as well. Continue with the following steps.

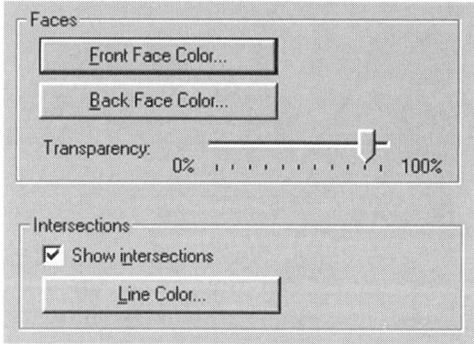

Fig. 2-21. Plane display options.

4. Select the Document Properties tab and select the Plane Display section. This will display the options shown in figure 2-21.

5. Make the desired changes to the front and back faces of planes, and change the transparency.

6. Click on OK when finished.

Changing Views

If running through this chapter for the first time, it probably is not prudent to change the color and transparency of the planes at this time. That information is only here for your future reference. Also for reference, the *Show intersections* option (shown in figure 2-21) will show the intersections of planes as dashed lines. The color of these lines defaults to black, but can be modified by the user. An example of turning on the *Display shaded planes* option is shown in figure 2-22.

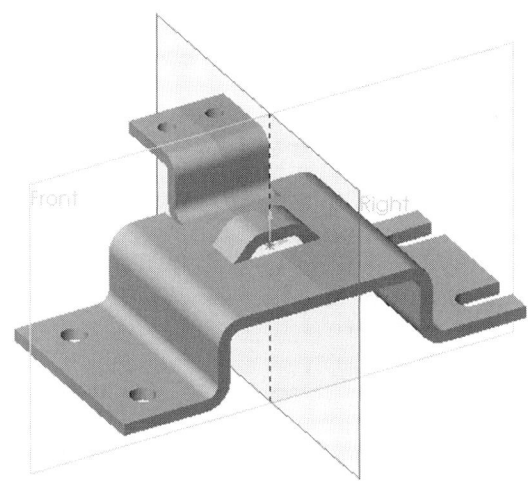

Fig. 2-22. Using the Display shaded planes option.

✓ **TIP:** *The triad can be turned on and off via the* Display reference triad *option, found below the* Display shaded planes *option.*

Changing Views

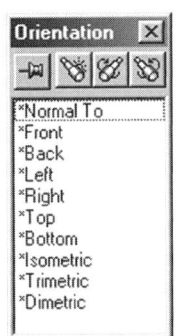

Fig. 2-23. Orientation window.

Now that you have learned how to hide and show planes, you can move on to the next stage and discover how to change the view orientation. You have already learned one method, which is to hold down the middle mouse button and move the mouse. The following section will show you other methods and options.

Using the Orientation Window

SolidWorks contains a number of preset views that can be called up at any time. These views are standard views, such as top, front, right, and isometric. They are accessed through the Orientation window (shown in figure 2-23), which is accessed by performing the steps outlined in How-To 2-6.

How-To 2-6: Enabling the Orientation Window

To enable the Orientation window, perform the following steps (there is more than one choice for step 1, any one of which will accomplish the same thing).

1. To open the Orientation window, perform one of the following.

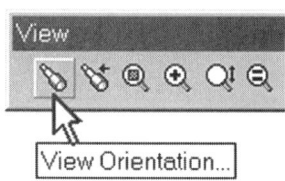

Fig. 2-24. View Orientation icon.

- Click on the View pull-down menu and select Orientation.
- Press the spacebar on your keyboard.
- Click on the View Orientation icon, shown in figure 2-24.

2. Using the left mouse button, double click on the desired view.

That is all there is to it. You might want to click on the "pushpin" button present on the Orientation window. This will keep this window open so that it is easily accessible. In other words, the window will be anchored to the desktop, and will remain on top of the SolidWorks window, even when you perform other functions.

There are other buttons on the Orientation window you have probably noticed. It is a good idea to try to stick to basics at this time. Therefore, discussions of the functions of these buttons appear with the design drawing topics covered in Chapter 11.

✥ **NOTE:** *Make certain you understand that the items listed in the Orientation window are not planes, but system views. Actual plane entities are listed in FeatureManager. It is important that you understand the difference between the two.*

Using the Standard Views Toolbar

Fig. 2-25. Standard Views toolbar.

Another alternative to using the Orientation window to change your view is to employ the Standard Views toolbar, shown in figure 2-25. This toolbar does not contain quite as many options as the Orientation window, but it is very easy to use. There is an icon for an Isometric view, but no Dimetric or Trimetric options, as with the Orientation window.

Changing Views

All it takes to change a view using the toolbar is one simple mouse click, as is the case when accessing any toolbar icon. It is usually a good idea to keep this handy toolbar available, because it is one you will probably use quite often. Find a good place to dock the toolbar where you can conveniently access it.

The reason the default planes were named Front, Top, and Right, in that particular order, should start to become apparent. SolidWorks sets up its views so that they are oriented in a particular logical fashion. Naming the planes the way you did in How-To 2-3 matched up the planes with the system views SolidWorks creates by default. If you now go to the Front view, you see a plan view of the Front plane. If you double click on the Right view, you see a plan view of the Right plane, and so on.

The term *plan* view may not be familiar to everyone. A plan view is simply looking at a particular plane straight on. In other words, the plane being viewed is parallel to the computer screen. Any planar face on a model or any plane in the model file can be shown in a plan view. This is discussed in the next section.

The Normal To Function

While checking out the Standard Views toolbar, you may have discovered a view called Normal To. This is nothing more than a plan view of a selected face. The key word here is *selected*. Usually, Normal To is used when there is a particular face or plane in the model you want to see a plan view of.

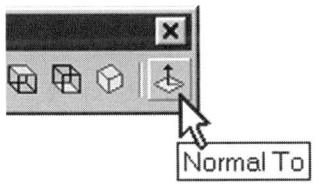

Fig. 2-26. Normal To icon on the Standard Views toolbar.

It is necessary to have a way of viewing a particular face or plane in a plan view. This undoubtedly aids the user, especially when sketching. Sometimes faces of the model cannot be seen "flat on" in any of the existing system views. By selecting the face or plane and double clicking on Normal To in the Orientation window, you can get the desired plan view of the selected face or plane. The actual view displayed will completely depend on the face or plane selected. By the way, you can also click on the Normal To icon on the Standard Views toolbar (shown in figure 2-26) to get the same result.

✓ **TIP:** *When using Normal To, if you Ctrl-select a second planar face or plane, SolidWorks will attempt to point that second face toward the top of the screen.*

Using Templates

Now that the three planes have been renamed, this would be an excellent time to save this file as a template. Before saving the template, you need to know a little about templates first.

A document template is a way of saving a SolidWorks part, assembly, or drawing file with specific parameters set up by the user. There is no limit to the variations of templates that can be created. Templates can be created for different drawing sheet sizes, for part files created in English or metric units, or for different standards, such as ANSI or ISO dimensioning standards. These are the most common reasons for saving a template. You may have others.

Standard File Extensions

Template files have a different file extension than the regular three SolidWorks document types. What are file extensions? They are the characters at the end of a file name the Windows operating system uses to identify a file type. This has more to do with Windows than it does with SolidWorks. For your reference, the following are the file extensions used by the standard three SolidWorks document types.

- Part files: *.sldprt*
- Assembly files: *.sldasm*
- Drawing files: *.slddrw*

Template File Extensions

The file extensions used by templates differ from those used by the standard document types so that SolidWorks (and Windows) can tell them apart. The naming convention used for the template file extensions is very logical, as are most things in SolidWorks. The following are the file extensions used by SolidWorks templates.

- Part templates: *.prtdot*
- Assembly templates: *.asmdot*
- Drawing templates: *.drwdot*

If you are fairly new to computers and do not fully understand the reasons for file extensions or what their significance is, it is not that important. The computer adds file extensions to your file names automatically,

Using Templates

so it is nothing the computer user has to worry about. When the time comes, *Inside SolidWorks* will tell you what you need to know regarding file extensions.

Modifying Document Properties and Saving Templates

Once you have a part file (or any SolidWorks document) set up the way you like, that file can be saved as a template. There are many optional parameters that can be adjusted. It would be a little overwhelming to go through all of these various parameters now. Many of these parameters will be discussed throughout the book. However, it is a good time to discuss where these parameters are located, so that you can explore them to some extent.

The parameters cited previously are known as document properties. They consist of many things, such as grid, unit, and dimensioning standard settings. Be careful altering too many settings prior to saving a template, because you may change something you should not. Basically, stick with what you know, such as modifying the Units setting, if desired. How-To 2-7 shows you how to modify a document's properties.

HOW-TO 2-7: Modifying Document Properties

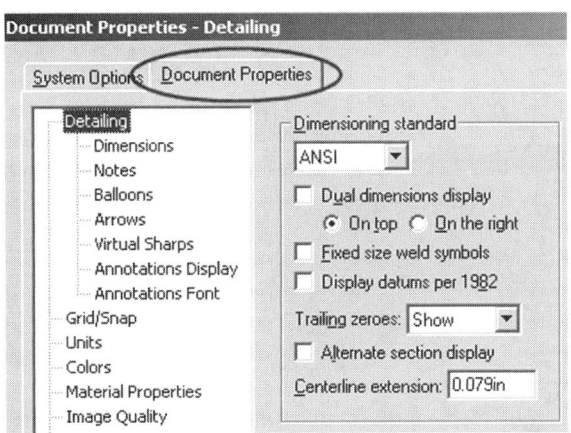

Fig. 2-27. Modifying a document's properties.

To modify a document's properties prior to saving the document as a template, perform the following steps.

1. Select Options from the Tools menu.

2. Click on the Document Properties tab. Anything changed within this tab will be saved with the document.

3. Select the desired category from the pane on the left (see figure 2-27).

4. Make the desired changes.

5. Click on OK when finished.

> **NOTE:** *Changes made via Document Properties only affect the current document. Changes made via System Options globally affect every SolidWorks document.*

The Document Properties window is shown in figure 2-27. In the illustration, the Detailing section is shown, though any of the options in any of the sections in the Document Properties tab can easily be altered. Once the properties are set the way you like, it is time to save the file as a template. How-To 2-8 takes you through this process.

HOW-TO 2-8: Saving a Template File

To save a file as a template, perform the following steps.

1. Select Save As from the File menu.

2. From the *Save as type* drop-down list (see figure 2-28), select Part Templates as the type of file to be saved.

3. Navigate to the appropriate directory where you would like the file to be saved. By default, this is the *Templates* directory. It is suggested you save the template there for now.

4. Type in a name for the file. Use a logical name that makes sense to you.

5. Click on the Save button.

Fig. 2-28. Save as type drop-down list.

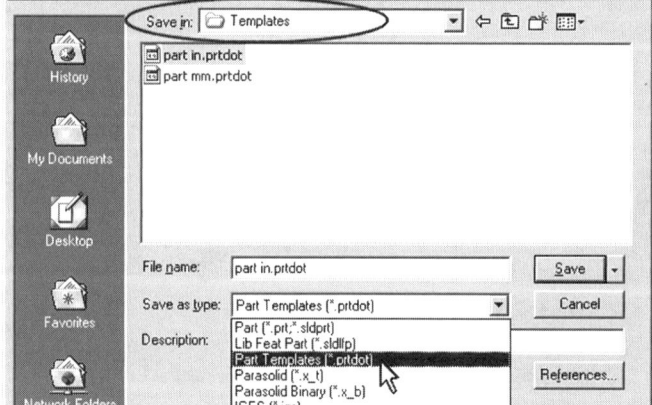

Now you can use the newly saved template whenever a new SolidWorks part is begun. You learned how to start a new SolidWorks document in How-To 2-1. All that is left is to specify the appropriate template when starting the new document. It is just a matter of clicking on the template you wish to use. If the entire process of creating, saving, and using a template seems easy, that is because it is. Now you are ready to learn more of the basics.

Deciding on a Sketch Plane

One of the first decisions you need to make is which plane to begin sketching on. It is not mandatory to use one of the default planes, although that is a typical starting point. It is possible to create other planes on which to sketch. (Plane creation is discussed in Chapter 4.)

It is important to remember to visualize the part you are about to create before actually beginning a sketch. For example, what is going to be the top of the part, what will be the front, and so on. It is possible to change a model's orientation after the design process has been started, but it is much easier if you make up your mind first and get the part's orientation right to begin with. Once an initial sketch plane has been decided on, the first sketch can be created and the first feature can be built.

The first feature built is considered the base feature. The sketch used to create the base feature should describe the basic overall shape of the part, but should not contain every bit of information needed to describe every feature. It is usually best to keep things simple. The general rule of thumb is to keep sketch geometry simple and create a part with a large number of features, rather than with elaborate sketch geometry and fewer features. Following this rule will give you more control over the part and make editing the part an easier task. Keeping features simple also aids in troubleshooting if and when things go wrong.

Take a look at a sample part, which may make it easier to understand the importance of starting off on the correct plane. A good example is the camera model because it has a front face that is recognizable, as shown in figure 2-29. In other words, what should be the front side of the camera is obvious.

In the case of the camera, what face should be designated the front is very cut and dry. However, not all parts have such clearly defined faces. If that is the case, it is simply a matter of choice. Keep in mind that when the drawing layout is created, whatever face of the part is considered the front face will be the face displayed as the Front view. This holds true for the Top and Right views as well.

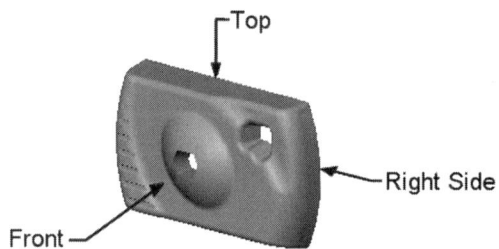

Fig. 2-29. Determining what face is the front side of the camera.

A question that often gets asked at this point is "Can the orientation of the part be changed after it has been created?" The answer is yes, it can, but this is not the time to discuss that topic. You will discover a method used to reorient a model in the next chapter (see "Updating Views").

Entering Sketch Mode

Once a plane has been chosen, it is time to enter sketch mode. This is done by clicking on the Sketch icon, shown in figure 2-30. The order is not important, except for the first sketch. In other words, you can click on the Sketch icon first and then select a plane. When starting the first sketch, however, SolidWorks will automatically place you on the Front plane, which may not be what you want.

Fig. 2-30. Entering a sketch.

❧ **NOTE:** *If the Sketch icon is selected without selecting a plane first, Solid-Works will place you on the Front plane. This only occurs when starting the very first sketch of a new model.*

There will be a few telltale clues that will point out the fact that sketch mode has been entered. First, the sketch icon will appear as though it has been depressed. Second, you should notice that a new item has been added to FeatureManager, shown in figure 2-31.

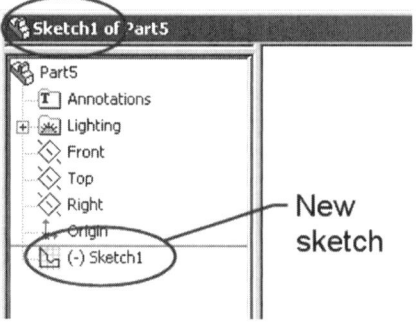

Fig. 2-31. Sketch1 added to FeatureManager.

It should be pointed out that there is a small minus sign (–) to the left of the *Sketch1* object. This tells the user that the particular sketch in question is *underdefined*. What this term means is that the geometry in the sketch has not been completely dimensioned or constrained. The rules that govern the sketch have not been fully imposed on the sketch. Once dimensions or constraints have been added to the sketch, and the sketch completely defined, the small minus symbol to the left of the *Sketch1* object will disappear. This change takes effect only after the feature has been created and the part is rebuilt. It also takes us directly into the next topic, sketch color codes.

Sketch Color Codes

There are a number of colors SolidWorks uses to tell you that certain conditions exist in a sketch. The most common color codes are blue, black, and red. Green or yellow simply means that something has been selected, and is not considered a sketch color code. SolidWorks allows you to work with underdefined geometry, unlike more constricting CAD programs. If sketch geometry is underdefined, it will be color-coded blue. This is the way sketch geometry usually appears when first drawn.

As more and more constraints or dimensions are added to a sketch, some of the geometry will begin to turn black. This means that the geometry has become fully defined. When the entire sketch is black, the sketch has become completely defined. Adding any other dimensions at this point would serve to overdefine the sketch. Overdefined sketch geometry turns red, and should always be cleared up as soon as possible. If you choose not to correct overdefined geometry, you will likely create more problems for yourself down the road. Table 2-1 outlines the color codes used by SolidWorks.

Table 2-1: SolidWorks Color Codes

Color	Significance
Black	Fully defined
Red	Overdefined
Blue	Underdefined
Pink	No solution found
Yellow	Invalid solution found
Brown	Dangling relation

Occasionally there are situations in which a valid solution simply cannot exist within a particular sketch. This happens because dimensions or constraints have resulted in a sketch that cannot be drawn because of its geometric nature. This turns the sketch geometry pink, and the sketch geometry is not solved. Another rare occurrence is that the sketch is actually solved but has invalid geometry, such as a zero-length line, in which case the offending geometry will turn yellow. This, however, seldom happens.

The last color code, brown, means that there is either a dangling dimension or constraint in the sketch. An example of a dangling dimension would be if a circle were dimensioned to another feature and at some point in the future the feature were deleted. SolidWorks would not delete the dimension. Instead, it would assign the dangling attribute to it. It would then be up to you to edit the sketch and fix the dangling dimension, which you could easily spot due to its brown color. You will learn more about dimensioning in Chapter 3.

Understanding Error Symbology

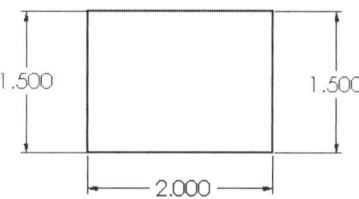

Fig. 2-32. An overdefined sketch.

An error situation can arise when too many dimensions or relations have been added to a sketch. This overdefines the sketch and is an undesirable situation. In figure 2-32, it is clear why the sketch has been overdefined. Not all such sketches will be as easily diagnosed. This is one reason it is good practice to keep sketch geometry simple. However, as you gain experience with the software, more complex sketch geometry can be created without fear of losing mastery over the sketch.

Overdefined Geometry

In figure 2-32, there are two dimensions that are clearly conflicting. Both vertical 1.500-inch dimensions are driving the height of the rectangle. Because either dimension can have a different value, this causes a conflict, even though they currently have an equal value. In this particular example, there are two options. Either delete one of the vertical dimensions, or delete the horizontal relation on one of the horizontal lines. You will learn how to do this shortly.

When sketch geometry has been overdefined, SolidWorks alerts you to this condition a number of ways. One way the software warns the user is to turn the color of all conflicting entities the color red, as discussed previously. You may also notice error symbols in FeatureManager telling you there is a problem and also where the problem lies, as shown in figure 2-33. An overdefined sketch is also indicated by a plus (+) sign. These error symbols only appear after a rebuild (or similarly, when exiting a sketch). Rebuilds occur, logically enough, when "building" features.

Understanding Error Symbology

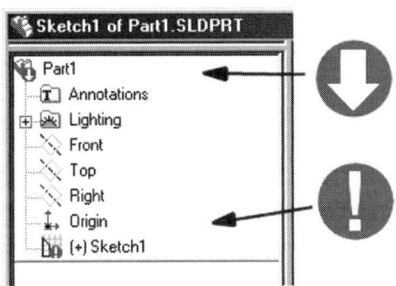

Fig. 2-33. Warnings in FeatureManager.

You typically will not see error symbols attached to a sketch in FeatureManager (at least not resulting from overdefined sketch geometry), as long as you make it a point to repair any overdefined conditions prior to building a feature or exiting a sketch. Red sketch geometry will clue you in to the problem before it ever gets to the point of FeatureManager error symbols appearing.

What's Wrong?

The What's Wrong function goes hand in hand with the error symbols shown in figure 2-33. A red circle with a downward-pointing arrow indicates a problem exists someplace in FeatureManager. Sometimes there are many features and you will literally have to scroll through a list of features to find the error. A red circle with a white exclamation point pinpoints where a problem is. By right-clicking on the problem area in FeatureManager, you receive an option named What's Wrong?

If the What's Wrong function is activated, a window with an explanation of the problem is shown. An example of this is shown in figure 2-34. While learning the SolidWorks software, you will invariably experience error messages from time to time. This is fine, and is all part of the learning process. Do not be afraid to experiment, because errors can be an educational experience.

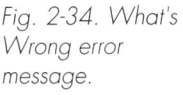

Fig. 2-34. What's Wrong error message.

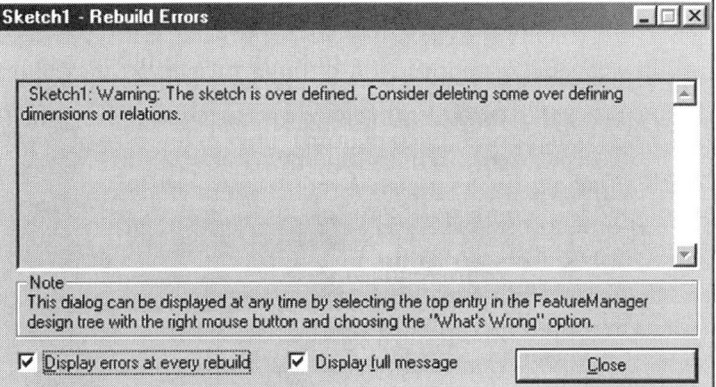

Whenever a problem arises, whether from overdefined sketch geometry or something else, every effort should be made to correct the situation. If the problem is not remedied, more problems are likely to occur further down the road. At various stages throughout the book, you will look at common errors, how they occur, and how to fix them.

The Origin Point

Fig. 2-35. Origin point.

The origin point is indicated by two gray arrows in the center of the SolidWorks screen. Think of the origin (as it is commonly referred to) as the world 0,0,0 reference point. 0,0,0 refers to the x,y,z coordinates of the Cartesian coordinate system. The origin is also the intersection of the default three planes listed in FeatureManager. The small gray arrow represents the x axis, and the long gray arrow represents the y axis. The origin is shown in figure 2-35.

If there is no origin, it may be because the display of origins has been turned off. To make sure origins are being displayed, click on the View pull-down menu and make sure there is a check in front of the Origins option. If there is not, select it to turn Origin display on.

When entering sketch mode, you can tell you are in a sketch because the red arrows of the sketch origin will appear. The red sketch origin arrows are not always in the same position as the world origin point. If in a sketch, look for the red arrows. You may have to alter your view to see them, but they will definitely be there. The red sketch origin arrows are always aligned with whatever plane you happen to be sketching on.

✓ **TIP:** *If the view is rotated, the origin will also rotate. This helps to give an indication of your current sketch plane and can aid in establishing your bearings.*

The fact that the origin acts as a 0,0,0 reference point, or that the sketch origin arrows represent the x and y axes, is not all that important. What is extremely important, however, is that the sketch origin acts as an anchor point. Do not let this seemingly trivial fact escape you!

➥ **NOTE:** *The sketch origin acts as an anchor point.*

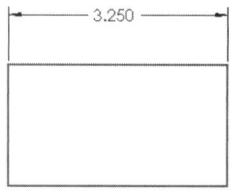

Fig. 2-36. What side moves if the dimension is changed?

Why is this so important? The importance may be summed up in two words: predictable behavior. If a rectangle were sketched in space and not anchored in position, and a horizontal dimension were added to the rectangle (see figure 2-36), which side of the rectangle would move if the dimension were changed? Try to answer this question prior to moving on to the next paragraph.

Did you come up with an answer to the question? Admittedly, this was something of a trick question. The answer should have been "it is impossible to tell." The rectangle does not exhibit predictable behavior. If, as a designer, you cannot predict how your model is going to behave, you are certainly not going to be able to incorporate your design intent into the model.

Using the same scenario, if the rectangle's bottom left corner were anchored at the sketch origin, you could say beyond a shadow of a doubt that it would be the right side of the rectangle that moves if the dimension is changed. In this case, predictable behavior is exhibited by the geometry, and your design intent can be maintained.

Sketch Basics

Sketching is really where the fun starts. This is typically the point at which you hunker down and start getting your hands dirty. It is also the place that will determine the difference between whether the part being built is a well-constructed part or a sloppy part. A well-constructed part makes good use of existing geometry, fully defined sketch geometry, and well-placed dimensions. A well-constructed part will also incorporate your design intent, so that a few modified dimensions can be made without fear of introducing rebuild errors in the model. This section is very important to becoming a proficient SolidWorks modeler. Learn it well.

Am I in a Sketch?

You should be able to answer that question without hesitation. At any one point in time it is imperative that you understand what is being edited. In other words, are you editing a sketch or editing the part? You may also think of this as being in a sketch or being outside a sketch. Knowing what you are editing becomes more important in assemblies because there is more to keep track of. For instance, it is possible to edit a sketch, component, or assembly (explored in Chapter 12).

So how can one tell if they are in a sketch? Some methods you have already learned. This is a good time to summarize. There are four basic ways to tell if you are currently editing a sketch. They are as follows.

- The Sketch icon will appear depressed (pushed in).

- The red sketch origin arrows will be visible (though the view may need to be changed).

- A new *Sketch* object will appear at the bottom of FeatureManager.

- The title bar will list what is being edited.

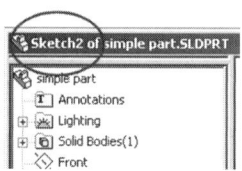

Fig. 2-37. An indicator that a sketch is being edited.

As shown in figure 2-37, the fourth indicator that a sketch is being edited is to look at the Windows title bar. If a sketch is not being edited, only the name of the model will be shown. If you are editing a sketch, it will say so. What could be easier?

✐ **NOTE:** *To exit out of a sketch, simply click on the Sketch icon.*

Do not be concerned with the number following the sketch. SolidWorks automatically increments the number of every new sketch. If the current sketch is exited and another sketch begun, it would be named *Sketch2*. This is SolidWorks' way of keeping track of things. This holds true for a number of objects in SolidWorks. Part file names are another perfect example. When a new part is started, its default name is *Part1*. Subsequent new parts would be named *Part2*, *Part3*, and so on. Obviously, you can rename the part files anything you want.

✓ **TIP:** *If you exit a sketch by accident, right-click on the sketch in FeatureManager and select the Edit Sketch option.*

System Feedback

Sketching can be very easy and user friendly, as long as some simple rules are adhered to. One such rule has to do with staying focused on the cursor. SolidWorks will tell you just about everything you need to know, as long as you watch the cursor. This information is known as system feedback.

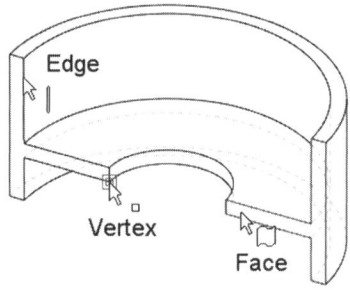

Fig. 2-38. System feedback should be used when selecting entities.

During the sketch process, SolidWorks will give you a steady stream of information. This information is dependent on where the cursor happens to be positioned at any given time. You will find that symbols are attached to the cursor. These symbols change to show you what the cursor is pointing at, or indicate to you that something is about to happen. System feedback, as indicated in figure 2-38, should be used to identify objects about to be selected.

Sketch geometry can be created quickly and easily with geometric relations added on the fly by SolidWorks. These relations are added automatically, but it lets you know what relations are being added through system feedback. This is a desirable situation. The cursor informs you exactly what constraints are about to be added. As long as you are paying attention and keeping your eye on the cursor display, very little can go wrong. If a constraint is accidentally added, it is possible to remove relations at any time.

Sketch Basics

If for some reason you want to sketch without the benefit of constraints being added automatically, it is possible to turn off this feature. It is not recommended that you turn Automatic Relations off, but you should be familiar with the option. Under the Tools menu, select Sketch Settings and look for the Automatic Relations option. It should be checked. Obviously, unchecking it will turn Automatic Relations off.

You may notice that the terms *constraint* and *geometric relations* appear to mean the same thing. In the context of SolidWorks, the two terms are interchangeable. Throughout this book, both terms may be used, but their meanings are identical.

✎ **NOTE:** *It cannot be stressed enough how important system feedback is. Watch the cursor!*

An Introduction to Constraints

Because constraints are added automatically, it is important to discuss what they are before sketching. When you do begin sketching for the first time in SolidWorks, you will see the system feedback informing you of what constraints are about to be added. A constraint is a rule or condition set on an object or between objects. For instance, a line may be constrained vertically or horizontally. Once the line is drawn horizontally, it will remain horizontal unless the relation is removed. The line can be repositioned or resized, but must remain horizontal in order to satisfy the condition placed upon it.

There are many other constraints that can be placed on sketch objects, all of which have cursor symbols associated with them that are displayed as the sketch is created. If a line is about to be drawn parallel to another line, the cursor displays the symbol for parallelism. If an arc is about to be drawn tangent to a line, the tangent symbol is displayed. All in all, there are over a dozen types of constraints. Not all can be added automatically as the sketch is being created. Some constraints must be added manually by the user. The automatic relations added by SolidWorks are horizontal, vertical, tangent, parallel, perpendicular, midpoint, and coincident.

Not all of the terms used to describe SolidWorks constraints are known by users using a solid modeler for the first time. Most are self-explanatory, but for the sake of thoroughness, a table describing the various constraints available to the SolidWorks user is provided later in this chapter.

The Add Relation icon, shown in figure 2-39, is located on the Sketch Relations toolbar, right where you would expect it. The icon appears as a perpendicularity symbol. Clicking on this icon activates PropertyManager

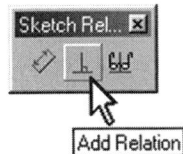

Fig. 2-39. Add Relation icon.

Chapter 2: A SolidWorks Overview

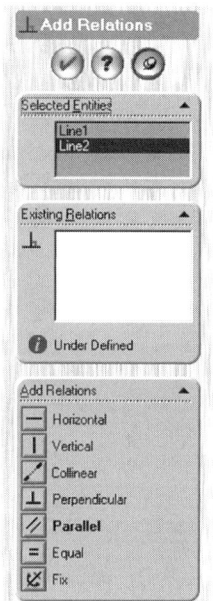

and allows for adding geometric relations, depending on the entities selected.

The Add Relations panel allows for adding specific relations that depend entirely on what is selected. For instance, it would be impossible to constrain two lines to be tangent to each other. A line and arc, on the other hand, would not be a problem. Only relations that can physically be added will be shown. Figure 2-40 shows an example of PropertyManager after clicking on the Add Relation icon.

PropertyManager will appear when sketching an entity (such as a line). This assumes that Auto-show is enabled, discussed previously. Geometric relations can be added without even clicking on the Add Relation icon. Because this is the case, it would serve well to understand what PropertyManager is displaying regarding geometric relations. This topic is explored in more detail in the following chapter. For now, try How-To 2-9, which takes you through one method of adding geometric relations.

Fig. 2-40. PropertyManager after clicking on the Add Relation icon.

How-To 2-9: Adding Relations

To add a geometric relation, perform the following steps.

1. Click on the Add Relation icon.
2. Select the entity (or entities) to add a relation to.
3. In PropertyManager, click on the icon to the left of the relation you want to add. Be aware that the relation is added as soon as you click on the icon.
4. Repeat steps 2 and 3 as necessary.
5. Click on the green check when finished.

•◦ **NOTE:** *It is not actually necessary to click on the Add Relation icon. Simply selecting an entity (or entities) will suffice. All the Add Relation icon really does is to keep the Add Relations panel visible so that more than one relation can be added at a time.*

Sketch Basics

Sketch geometry will move position or change shape to accommodate the relations being added. Use Table 2-2, which follows, to better understand what the various constraints will accomplish. The term *arc* in this table refers to circles or arcs. The term *point* refers to point entities, endpoints of lines and arcs, centerpoints of arcs and ellipses, and so on. This table is meant to give a wide range of examples, but does not show every combination of possibilities.

Table 2-2: Geometric Relations and Examples of Use

Relation	Sketch Entities Used	Changes Sketch Entities Undergo
Horizontal or Vertical	One or more lines, or two or more points	Lines become horizontal or vertical; points are aligned horizontally or vertically.
Parallel	Two or more lines	Lines become parallel.
Perpendicular	Two lines	Lines become perpendicular.
Collinear	Two or more lines	Lines lie on the same theoretically infinite line.
Coincident	A point and a line, arc, parabola, or ellipse	Point lies on the line, arc, parabola, or ellipse.
Midpoint	A point and a line	Point remains at the midpoint of the line.
Intersection	Two lines and one point	Point remains at the intersection of the lines. This can be a projected intersection.
Coradial	Two or more arcs	Items share the same centerpoint and have an equal radius.
Tangent	An arc, ellipse, or spline, and a line, arc, parabola, ellipse, or spline	Items remain tangent.
Concentric	Two or more arcs, or a point and an arc	Arcs share the same centerpoint, or a point remains at the arc's centerpoint.
Equal	Two or more lines, or two or more arcs	Line lengths or arc radii remain equal.
Symmetric	A centerline and two points, lines, arcs, ellipses, splines, or parabolas	Items remain symmetrical about the centerline.
Pierce	A sketch point and an axis, edge, line, arc, or spline	Sketch point is coincident to where the axis, edge, line, arc, or spline pierces the sketch plane.
Merge Points	Two sketch points (including endpoint)	Points are merged into a single point (similar to coincident).
Fix	Any item	Item's size and location are locked. Points (such as a line's endpoints) and entities (such as a line itself) can be fixed independently of each other.

It is not intended that you memorize this table. Its purpose is for reference. Adding relations is normally a fairly straightforward procedure. The descriptions of the various relations are meant to help the reader who is unfamiliar with some of the terminology being used by SolidWorks.

Sometimes sketch geometry may look as though it is already constrained, when in reality it is not. Take a very simple example in which one line appears as though it is perpendicular to another. If you are not sure the geometry is constrained, simply try moving it around. This is done by placing the cursor over an entity, holding down the left mouse button, and moving the mouse. This is known as "dragging" geometry, and is an excellent technique for determining what relations may already be applied to the sketch geometry, or what relations need to be added.

Creating sketch geometry and then dragging that geometry around to visualize its constraints is one of the first things you will do in the next chapter. Sketch techniques and guidelines will be explored, along with what it takes to create various sketch entities. You will also see how to place dimensions on sketch geometry, and the simple guidelines you should follow when sketching.

Summary

This chapter has introduced you to the mind-set needed when working with the SolidWorks program. The transition from 2D to 3D CAD software can sometimes be intimidating. Learning to think in terms of features and sketch geometry is a major step in understanding how SolidWorks functions.

The SolidWorks interface makes use of typical standard Windows functionality, such as pull-down menus and toolbars. This familiar Windows look and feel, combined with the logically organized menu structure, makes for an easy-to-learn software program. SolidWorks also makes intelligent use of the mouse. The left mouse button is primarily used for selecting entities or commands, and the right mouse button brings up a context-sensitive menu. Use the middle mouse button, if available, for panning, zooming, or rotating the model. The SolidWorks FeatureManager shows a chronological history of the various features within your model.

Templates are necessary when starting a new SolidWorks document. A few important parameters, or document properties, can be set up beforehand, and then the file can be saved as a template file. This makes starting a new file easy because the properties can be set once, then forgotten.

Selecting the correct plane to sketch on is an important first step in sketching. It is important to understand what side of a new model is going to be the top, the front, and so on. Use the right mouse button to click on planes in FeatureManager to toggle plane display on and off. Click on the Sketch icon to enter sketch mode.

There is a great deal of system feedback SolidWorks is constantly showing you during the sketch process or while selecting entities. It is extremely important to pay close attention to the cursor, as SolidWorks will add constraints to sketch geometry during the sketch process. Which constraints are added will be displayed by the cursor. This system feedback is an integral part of SolidWorks.

Fully defined sketch geometry will be color-coded black. Underdefined geometry will be blue. It is okay to work with underdefined geometry, but it is best to fully define all sketch geometry. If geometry is fully defined, its behavior can be predicted if dimensions are changed. Red geometry is overdefined and should be corrected as soon as possible.

Questions and Topics for Discussion

1. Why is it not necessary to be 100-percent accurate when sketching in SolidWorks?
2. What is the difference between sketched and applied features?
3. What are the default plane names when starting a new SolidWorks part?
4. Is it possible to rename planes? If so, what would you name them, and why?
5. What are four signs that you are in sketch mode?
6. What is meant by the term *context sensitive* when referring to the right mouse button?
7. What color will sketch geometry be if it is underdefined? Overdefined? Fully defined?
8. What is the significance of the origin point?
9. Why is it very important to pay attention to the cursor while sketching?
10. What are constraints?
11. How many constraints can you name without referring to SolidWorks or this book?

12. Describe the method used for saving a template.

13. How can the Auto-show PropertyManager option be turned on?

14. What icon could you click on to establish a plan view of any selected plane?

15. What is the difference between Document Properties and System Options?

CHAPTER 3

Sketching Basics

NOW THAT YOU HAVE LEARNED ABOUT most of the fundamental aspects of SolidWorks, it is time to begin sketching. Sketching must occur before a feature can be created. SolidWorks is in fact a sketch-based modeler. A sketch is the 2D geometry that precedes the 3D solid model. As with any foundation, whether it be for a house or for CAD software, the way in which the foundation is laid is extremely important. This may very well be the most important chapter in this book because of this fact.

There are some guidelines that should always be followed when creating sketch geometry. These guidelines are explored following an examination of the various types of sketch entities. Bear in mind during the upcoming examples that it is imperative you watch the cursor as you sketch. You should be watching for the system feedback SolidWorks displays. If this topic is hazy in your mind, it might be a good idea to review the section on system feedback in Chapter 2. Remember, constraints will be added automatically as you sketch.

Beginning a Sketch: A Quick Review

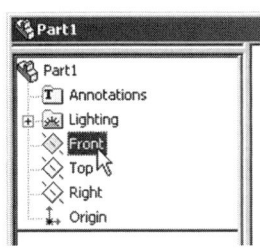

Fig. 3-1. Selecting the Front plane.

You may recall from the section "Entering Sketch Mode" in Chapter 2 that a plane must be selected in order to create a sketch. This is accomplished by simply clicking on the desired plane in FeatureManager. If the plane is currently being shown in the sketch area, you could also just as easily select it from there. Let's start by selecting the Front plane to sketch on, as shown in figure 3-1.

Now enter sketch mode. This will begin a brand new sketch on the plane you have selected. Enter sketch mode by clicking on the Sketch icon, shown in figure 3-2.

56 **Chapter 3: Sketching Basics**

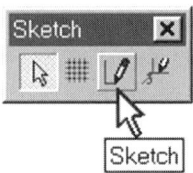

Fig. 3-2. Entering sketch mode.

Keep in mind that SolidWorks places the user on the Front plane by default, so it was not really necessary to select the Front plane prior to entering sketch mode. However, the only time the Front plane is used by default is when starting the first sketch. Subsequently, you will always be required to select a plane.

✧ **NOTE:** *Plane selection and entering sketch mode can take place in any order after the first sketch.*

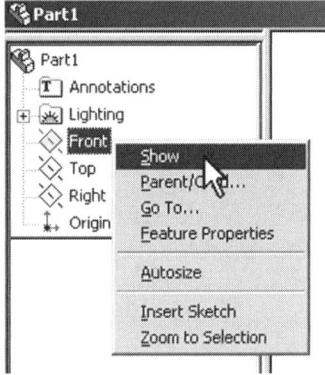

Fig. 3-3. Showing the sketch plane.

It is not a requirement, but it is also good technique to show the plane being sketched on. This generally helps new users orient themselves. It also serves as a reminder as to what plane is being sketched on. You learned how to perform this operation in the last chapter. Right-click on the sketch plane in FeatureManager, shown in figure 3-3, and select Show.

Note that you will be able to change the size of a plane's border by dragging the small green boxes on its border. Selecting a plane will display its handles. It is also possible to reposition a plane by dragging its border from some point other than a green handle. Resizing planes is for cosmetic value only. The plane actually extends infinitely in all directions. Sketching need not be done within the visible borders of the plane.

Sketching

There are two methods that can be employed when sketching any type of sketch entity. These methods are known as click-drag and click-click. Each has its benefits, which are explored in the following section. Sketching lines is the most basic, so let's start there. Both methods of sketching can be used with other entity types (such as circles, arcs, and so on), but for the sake of describing both sketch methods the Line command will be used.

Lines

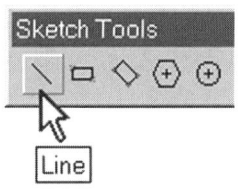

Fig. 3-4. Line icon.

To begin sketching a line, click on the Line icon, shown in figure 3-4, or use the pull-down menus (Tools > Sketch Entity), whichever is easiest for you. It is assumed you have entered sketch mode at this time. How-To 3-1 takes you through the process of sketching a line using the click-drag method.

Sketching **57**

How-To 3-1: **Sketching a Line Using the Click-Drag Method**

To sketch a line using the click-drag method, perform the following steps.

1. Click on the Line icon, or select Tools > Sketch Entity > Line.

2. Hold down the left mouse button where you want the line to start.

3. Move the mouse, and then let go of the mouse button to position the second endpoint.

The first thing you will notice when sketching is how the cursor changes appearance. Where once there was an arrow, there is now what appears to be a marker. Along with the marker there is a symbol of a line. If you had been sketching an arc, there would have been an arc symbol. If you had been sketching a circle, a circle symbol, and so on.

Pay attention to the system feedback as you sketch. If you did not notice any symbols associated with the cursor the first time around, sketch more lines. Try to make the cursor display the letters H or V, for example. These letters represent horizontal and vertical constraints being placed on the line if you were to end the line while the symbol was visible.

Inferencing lines are another feature of system feedback. These are blue, dashed lines that help you to align sketch entities with other endpoints or to center points within the sketch. Inferencing lines do not add relations; they are there simply to aid in the sketch process. Now let's see how the second method of sketching differs from the first. How-To 3-2 takes you through the process of sketching a line using the click-click method.

How-To 3-2: **Sketching a Line Using the Click-Click Method**

To sketch a line using the click-click method, perform the following steps.

1. Click on the Line icon, or select Tools > Sketch Entity > Line.

2. Click once where you want the line to start. (Do not hold down the left mouse button!)

3. Click once where you want the line to end.

4. Click to establish subsequent endpoints, and continue adding line segments.

5. To break off the command, press the Escape (Esc) key.

Note that to create a single line segment the click-drag method is most fitting. To create more than one segment, use the click-click method. Either method can be applied to all types of sketch entities. Which method is used depends on what you are trying to accomplish and on personal preference.

During the sketch process, notice the information supplied by the cursor. This is the system feedback explored in Chapter 2. For instance, a numerical readout divulging a line's length will appear while creating lines. Use the numerical readout to sketch objects approximately of the size you want the object to be. But remember, this is only a sketch. The size of a sketch entity does not have to be exact, because the dimensions will control, or drive, the entity's size. (Dimensions are examined after discussion of the various sketch entity types.)

Another aspect of sketching is how SolidWorks automatically selects the very last entity you have sketched. This will come in handy for specific tasks. For instance, to draw a line parallel to another line, sketch a diagonal line first, and then sketch another line approximately an inch away from the first. Note that SolidWorks attaches a parallel symbol to the cursor when you are parallel to the first line. Another example would be if an object were sketched and you decided you did not like how it turned out. Because the object is already selected, you need only press the Delete key and try again.

In regard to the topic of deleting objects, you should practice drawing lines until you feel comfortable with both methods of sketching. Pay special attention to the system feedback as you sketch, and become familiar with the symbols.

> **NOTE:** *There is one system feedback symbol whose meaning may not be obvious the first time you see it. This is the coincident symbol, which looks like a light bulb with rays of light emanating from it. Coincident means "on," which would translate to a line's endpoint being "on" (or coincident with) another object.*

Deleting Entities

If the sketch area is starting to fill up with entities, you probably want to delete some of them to give yourself some more room to experiment.

Sketching

There are a number of ways in which entities can be deleted from a sketch, but first you must select those entities. Table 3-1 provides guidelines on the various entity selection options.

Table 3-1: Entity Selection Options

Function	Operation
To select one entity	Click on the entity with the left mouse button.
To select more than one entity	Hold down the Control (Ctrl) key while selecting entities with the left mouse button.
To select multiple entities at once	Click in an empty portion of the sketch area and drag a rectangle around the entities to be selected. Only entities completely enclosed by the rectangle will be selected.
To remove entities from the selection set	Hold down the Control (Ctrl) key and select the entity to be deselected.
To clear the entire selection set	Select in a blank area of the screen.

When a sketch entity is selected, it is green; and if it is not green, it is not selected. With that in mind, and knowing how to select sketched objects, you can do a little housekeeping and get rid of some of that unwanted clutter. The following are three ways in which you can delete entities once they are selected.

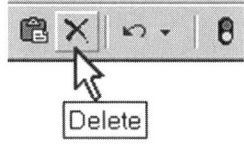

Fig. 3-5. Delete icon.

- Select Delete from the Edit pull-down menu.
- Click on the Delete icon, shown in figure 3-5.
- Press the Delete key.

The Delete key may be quicker for some people, but the choice comes down to user preference. It should be noted also that the Delete key works for deleting anything in SolidWorks. All you have to do is select the object to be deleted and press the Delete key. Features, drawing views, parts in an assembly, and just about anything else can be deleted in this fashion. For significant items, such as features, you will get a confirmation window warning you that something is about to be deleted if you proceed.

Auto-transitioning to Arcs

When sketching lines, it is possible to automatically transition to sketching arcs with little effort. This is accomplished in a number of very simple ways. The only requirement is to use the click-click method of sketching. You are encouraged to experiment and try this on your own. To begin, sketch a line using the click-click method. Then position the cursor as if you were about to establish the endpoint of a second line segment. You will see the typical thin line, which acts as a preview. Next, perform one of the following steps.

- Press the A key.

- Move the cursor back to the endpoint of the first line and then away again.

At this point, an arc preview should be visible in place of the line initially present. Clicking will position the endpoint of the arc, and then you can proceed from there, either drawing another line or auto-transitioning to another arc. In this fashion, a series of lines and arcs can easily be created, all of which will be tangent to the previous entity.

✓ **TIP:** *If you auto-transition to an arc and then change your mind, press the A key to transition back to a line.*

By moving the cursor back to the previously sketched object, then away again, it is possible to sketch an arc in one of four directions. The arc will either be tangent to the line or tangent to a perpendicular vector of the line. In this way, up to eight different tangent arcs can be created. All eight possible arcs are shown in figure 3-6.

Fig. 3-6. Possible arcs created by auto-transitioning.

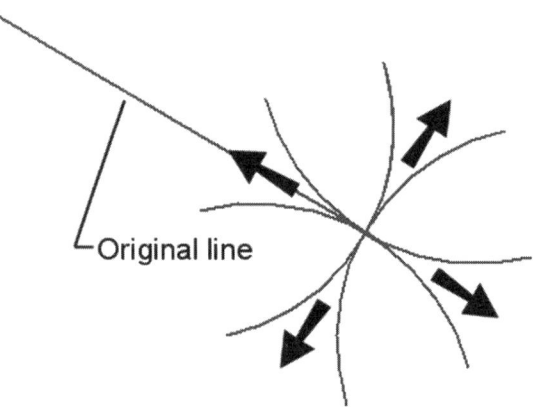

Exiting Commands

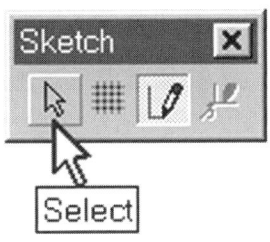

Fig. 3-7. Select icon.

When not in a command, you are in what is known as select mode. This just means that it is possible to select things. The cursor looks like an arrow when in select mode. You can enter select mode a number of ways. Use any of the following options to exit a command and enter select mode. The Select icon is shown in figure 3-7.

- Deselect the icon currently in use (i.e., deselect the Line icon if currently sketching lines).
- Click on the Select icon.
- Right-click in an empty area of the sketch and click on Select.
- Click on Select in the Tools pull-down menu.
- Press the Escape (Esc) key.

Pressing the Escape key is easiest for most people, but this is strictly a matter of personal preference. Do whatever works best for you. In addition, if you know what command you are going to use next, do not bother to exit the existing command; simply click on the next icon you need to use.

Dragging Geometry

Once in select mode, you can drag geometry to reshape and resize it. Any geometric relations placed on sketch geometry must be maintained during the drag process. This can allow you to quickly and easily see constraints added to the geometry. Dragging geometry is also useful in determining what geometric relations or dimensions must be added to the geometry. This is very valuable information to a SolidWorks user.

What part of a sketch is dragged will make a difference as to how the sketch will react. This can most easily be demonstrated using a simple line. Draw a line at an angle, exit the command, and then place the cursor over the line and drag the geometry to a new position by holding down the left mouse button. Note that the entire line moves but remains parallel to its original location.

Next, try dragging a single endpoint of the same line. Note that it is possible to reposition just that one endpoint being dragged. You will notice that other geometry reacts differently as well. A circle will drag differently from its center point versus its perimeter, and so on.

As more dimensions and relations are added to a sketch, less and less geometry remains blue. Sometimes it is difficult to figure out why geometry is not fully defined. This is precisely why it is so beneficial to be able to drag geometry. When the question "why isn't the sketch defined" arises, drag any remaining blue geometry and you will almost always find the answer you are looking for.

Command Persistence

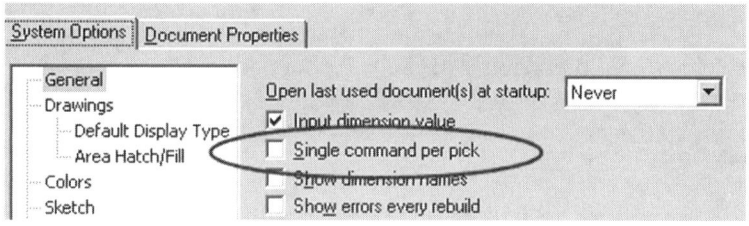

Fig. 3-8. Single command per pick option.

Whether or not you remain in a command after carrying out that command depends on an option named *Single command per pick*, shown in figure 3-8. If this option is enabled, the icons will work a little differently. Specifically, clicking on an icon will allow you to use that command once, with the command then exiting automatically. If *Single command per pick* is not checked, commands persist until canceled. In other words, you will remain in the command until you decide to exit the command or until a new command is entered. How-To 3-3 takes you through the process of changing command persistence.

How-To 3-3: Changing Command Persistence

To modify the *Single command per pick* option, thereby altering whether or not a command persists when used, perform the following steps.

1. Select Options from the Tools pull-down menu.
2. Select the General section in the System Options tab.
3. Modify the *Single command per pick* option as desired.

It should be noted that even with *Single command per pick* enabled, double clicking on an icon will allow for repeated use of that icon. If *Single command per pick* is not checked, one click on an icon is all it takes to use that command indefinitely. This is another of those options that depend on the preference of the user. This book will assume you do not have *Single command per pick* enabled, so you may want to leave it unchecked for now.

Sketch Entities

Well, you are obviously going to need to sketch objects other than lines and tangent arcs. This section introduces you to all of the different types of sketch entities and how to create them. Any special properties individual sketch entities may have are noted, along with any quirks or noteworthy behavior they may exhibit.

> **NOTE:** *It is very likely that not every icon mentioned in this section will be present on your Sketch Tools toolbar. It is the default nature of SolidWorks to not include all icons on all toolbars. SolidWorks toolbars can be customized to include any missing icons, or to remove those not frequently used. Customization is discussed in Chapter 23. If an icon is not present on your toolbar, it is suggested you access the command via the Tools > Sketch Entity pull-down menu.*

It would be very redundant to list the steps for creating each type of sketch entity using both the click-drag and click-click methods. This book lists the steps involved in creating a particular entity. It will be up to the reader to decide which method, click-drag or click-click, is used. Steps involved specify the "pick points." In the case of a line, for instance, you would see the following.

1. Pick to establish the line's start point.
2. Pick to establish the line's endpoint.

This explanation works whether you decide to drag the cursor between the start point and endpoint or whether you decide to simply click to establish these points. In this way, the creation process of each sketch entity type can be described in a very concise manner, which should serve as an excellent reference guide when you need it.

With that said, let's dive into it. It is recommended you try creating the various sketch entities on your own. This will give you a chance to get your hands dirty, and hopefully will allow you to feel more comfortable with the software. Feel free to experiment, and remember to watch the cursor for any system feedback SolidWorks is relaying to you.

> **NOTE:** *In each of the sections that follow it is assumed you have already clicked on the appropriate sketch entity icon.*

Tangent Arc

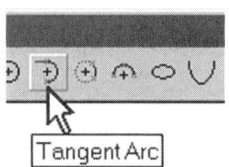

Fig. 3-9. Tangent Arc icon.

You basically already know how to create tangent arcs using auto-transitioning, but there is an icon for creating them as well. Sketching a tangent arc is quite easy, but there is one stipulation. You must select an existing entity endpoint before sketching the arc. The Tangent Arc icon is shown in figure 3-9. The following steps take you through the process of creating a tangent arc.

1. Pick an existing endpoint where the arc is to start.
2. Pick to establish the tangent arc's endpoint.

Tangent arcs can be drawn tangent to lines or other arcs. The system feedback must be used so you will know when you are over an endpoint. The cursor will change to display a symbol that looks like a small square. That is when you know it is safe to create the arc. If you do not begin the tangent arc on an endpoint, you will receive an error message stating so.

✑ **NOTE:** *Arcs can be drawn tangent to splines also, but the approach is slightly different. This topic is covered in the upcoming section on splines.*

When a tangent arc has been created, a tangent constraint is added between the arc and the tangent entity. This relationship will remain if any of the associated geometry is moved or dragged to a new position.

3 Point Arc

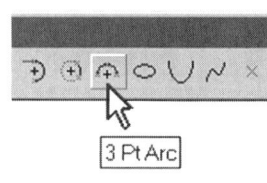

Fig. 3-10. 3 Point Arc icon.

The 3 Point Arc option, the icon for which is shown in figure 3-10, is commonly used when the locations for the arc's endpoints are known. For example, you want to place an arc between two existing entities. This, however, is not a requirement. Three-point arcs can be created anyplace. To create a three-point arc, perform the following steps.

1. Pick to establish the start point of the arc.
2. Pick to establish the endpoint of the arc.
3. Pick to establish the arc's radius.

Bear in mind that when performing step 3 the pick point establishes where the arc will pass through, which in turn defines the arc's radius. Watch the system feedback prior to completing step 3 and it will give an indication of what the included angle and radius of the arc are going to be. This is shown in figure 3-11.

Sketch Entities

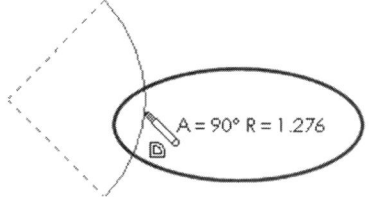

Fig. 3-11. System feedback when creating a three-point arc.

Centerpoint Arc

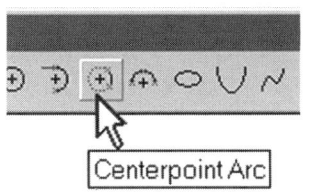

Fig. 3-12. Centerpoint Arc icon.

The Centerpoint Arc icon, shown in figure 3-12, is the trickiest of the three arc creation commands. To create a centerpoint arc, perform the following steps.

1. Pick to establish the arc's centerpoint.

2. Pick to establish where the arc will start.

3. Pick to establish where the arc will end.

As you may have noticed already, the second pick point will also define the radius of the arc. Once again, as is always the case, let the system feedback give you an "in-the-ballpark" estimate of the arc's radius and included angle. In the end, dimensions or geometric relations will control the geometry.

Circles

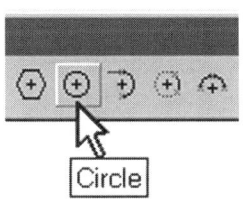

Fig. 3-13. Circle icon.

Creating circles requires very little effort from the user. It is a very simple operation to sketch a circle. The Circle icon is shown in figure 3-13. To create a circle, perform the following steps.

1. Pick to establish the circle's center point.

2. Pick to establish the circle's radius.

Dragging the completed circle will modify the circle in different ways, depending on the point being dragged. Dragging the circle's perimeter will allow you to resize the circle. Dragging the circle's centerpoint will allow you to reposition it.

Rectangles

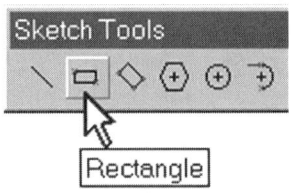

Fig. 3-14. Rectangle icon.

Rectangles are constrained to be rectangles, meaning that the top and bottom lines of the rectangle are horizontal and its sides are vertical. The Rectangle icon is shown in figure 3-14. To create a basic rectangle, perform the following steps.

1. Pick to establish one corner of the rectangle.

2. Pick to establish the opposite corner of the rectangle.

Rectangles at an Angle

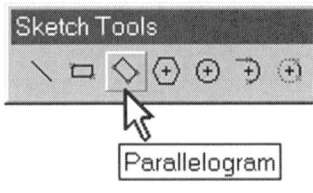

Fig. 3-15. Parallelogram icon.

A command other than the Rectangle command is used to create a rectangle at an angle. This may seem odd at first, but makes more sense once you understand how an angled rectangle is constrained. Because none of the sides of an angled rectangle are horizontal or vertical, it was easier for SolidWorks to combine the ability to create an angled rectangle with a different function. Creating an angled rectangle actually involves the Parallelogram icon, shown in figure 3-15. To create a rectangle at an angle, perform the following steps.

1. Pick to establish one corner of the rectangle.

2. Pick to establish the length of the first side of the rectangle.

3. Pick to establish the length of the adjacent side of the rectangle.

As noted previously, make certain you click on the Parallelogram icon prior to creating an angled rectangle. If you click on the Rectangle icon by accident, the process will not work correctly. This is usually about the time most people wonder how one goes about creating a parallelogram, especially because the icon is already being used to create an angled rectangle. Read on for your answer.

Parallelograms

The Parallelogram icon, shown in figure 3-15, has a dual identity. It can be used to create both a rectangle at an angle and a parallelogram. Different relations are automatically added when using the Parallelogram icon, depending on whether or not you are creating a parallelogram or an angled rectangle. An angled rectangle will have two perpendicular relations and

Sketch Entities

one parallel relation associated with it. In the case of a parallelogram, the opposite sides of the parallelogram are geometrically constrained to be parallel, just as you would imagine them to be.

If you want to create a parallelogram, the steps are exactly the same as for creating an angled rectangle, except that you need to hold down the Ctrl key while establishing the length of the adjacent side of the parallelogram. To create a parallelogram, perform the following steps.

1. Pick to establish one corner of the parallelogram.
2. Pick to establish the length of the first side of the parallelogram.
3. Holding the Ctrl key down, pick to establish the length of the adjacent side of the parallelogram.

Polygons

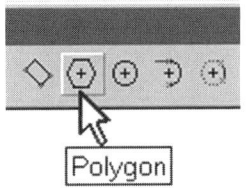

Fig. 3-16. Polygon icon.

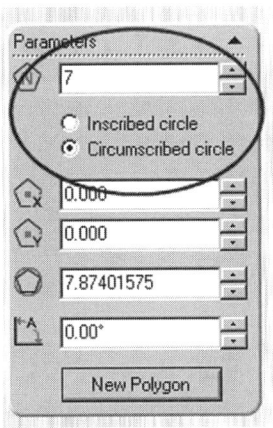

Fig. 3-17. Modifying the properties of a polygon.

The easiest way to create polygons is through the use of the Polygon icon, shown in figure 3-16. What is occurring behind the scenes when a polygon is sketched is that SolidWorks is busy patterning one side of the polygon. There is also a construction circle that helps control the polygon geometry. One side of the polygon is made tangent to the construction circle prior to patterning. This is all done automatically by SolidWorks, and is transparent to the user. To create a polygon, perform the following steps. Note that the first two steps will require making use of PropertyManager.

1. Specify the number of sides the polygon should have.
2. Specify whether the polygon should be inscribed or circumscribed.
3. Pick to define the center of the polygon.
4. Pick to define the diameter of the polygon.

When creating a polygon, it is necessary to take advantage of the sketch PropertyManager, an example of which is shown in figure 3-17. This is because it is necessary to specify how many sides the polygon should have. Also decide on whether or not the polygon will be inscribed or circumscribed about the construction circle. As for the rest of the options in the polygons PropertyManager, they really do not serve much purpose, so they are not discussed here. You will want to define the polygon through the use of dimensions and constraints anyway, which the settings in PropertyManager cannot accomplish.

✓ **TIP:** *Specifying the number of sides and the Inscribed/Circumscribed option can be performed immediately after a polygon has been created, as well as prior to its creation.*

One option for defining the polygon is to dimension the circle's diameter. One side of the polygon could also be dimensioned. The choice depends on your design intent. You might also want to consider adding a vertical or horizontal relationship to one of the sides of the polygon to firmly establish its orientation. This will keep the polygon from rotating.

To modify the number of sides a polygon has, simply right-click on one side of the polygon and select Edit Polygon. This will open the sketch PropertyManager and allow you to change the number of sides the polygon has, along with the other associated options. Additionally, there is a menu item named Edit Polygon in the Tools > Sketch Tools menu. Right-clicking on the polygon is easier, though.

Once the PropertyManager is open, just change the number of sides as necessary. Either use the spin box arrows or type in a value. The choice is yours. Because the number of sides the polygon has is not considered a dimension by SolidWorks, it will be necessary to edit the sketch containing the polygon to modify the number of sides it contains. Editing sketch geometry is discussed further later in this chapter.

Ellipses

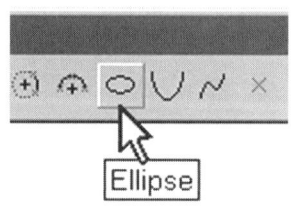

Fig. 3-18. Ellipse icon.

An ellipse is normally defined by the length of each of its two axes. The major axis is the long axis; the minor axis is the shorter of the two. The Ellipse icon is shown in figure 3-18. To create an ellipse, perform the following steps.

1. Pick to establish the center of the ellipse.

2. Pick to establish the first axis length.

3. Pick to establish the second axis length.

It makes no difference whether the first axis is the major or minor axis. Once the ellipse has been created, you can drag its shape or rotate it by dragging its vertex points. Reposition the ellipse by dragging the ellipse itself (from some point other than a vertex point). An ellipse will have four vertex points that can be dimensioned, as shown in figure 3-19. This is normally an appropriate way of controlling the size and shape of an

Sketch Entities

ellipse. It would also be common to add construction lines that can be dimensioned in order to control the rotation angle of the ellipse. Adding dimensions is covered later in this chapter.

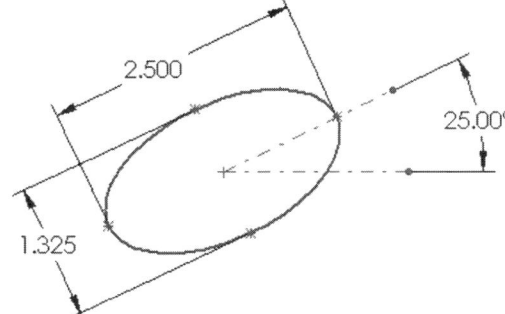

Fig. 3-19. Controlling an ellipse through dimensions.

Centerpoint Ellipses

Centerpoint ellipse is another term for an elliptical arc. The command was given the name Centerpoint Ellipse because it functions in a manner similar to that of the Centerpoint Arc command. It is one of the most involved of all sketch entity commands. Creating a centerpoint ellipse begins exactly the same as using the Ellipse command. However, it involves an additional operation that determines the arc's length. There is no icon for this command, so you must use the pull-down menus to access it (Tools > Sketch Entity). To create an elliptical arc using the Centerpoint Ellipse command, perform the following steps.

1. Pick to establish the elliptical arc's centerpoint.
2. Pick to establish the first axis length.
3. Pick to establish the second axis length (this pick point will also determine where the elliptical arc begins).
4. Pick to establish the elliptical arc's endpoint.

The Centerpoint Ellipse command can be a little tricky until you have gone through it a few times. Do not worry about being exact. You can always drag the sketch geometry after the arc is completed. Note that an elliptical arc has four vertex points, just like would be seen with a complete ellipse. If you have not drawn one of these arcs on your screen, figure 3-20 shows an example of what an elliptical arc might look like. The arc's endpoints differ from the quadrant points and are noted in the figure.

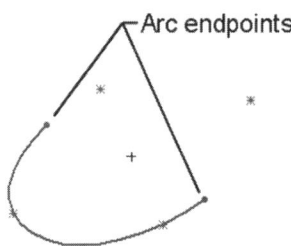

Fig. 3-20. An example of an elliptical arc.

Elliptical arcs can be difficult to manage. Dimensioning the arc is fairly easy. Attach dimensions to the quadrant points, just like with a complete ellipse. Where the difficulty comes in has to do with the endpoints of the arc and how they are related to the quadrant points. The arc's endpoints may or may not be coincident with the quadrant points, so coincident relations may need to be added or removed (deleting relations is covered in an upcoming section).

The included angle of the arc may need to be defined with a dimension as well. Typically, the arc will be attached to other geometry, which would help to define the arc. If tangent relationships are required, they can be added between the arc and some other adjacent object.

Parabolas

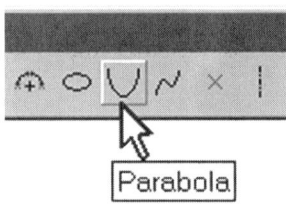

Fig. 3-21. Parabola icon.

A parabola is one of those sketch entities most designers do not need very often, unless they are in a particular manufacturing discipline. Perhaps you fall into this category. Like the Ellipse icon, the Parabola icon (shown in figure 3-21) may not be present on your Sketch Tools toolbar. If that is the case, use the Tools > Sketch Entity pull-down menu.

A parabola consists of a focal point, a directrix, and a parabolic curve. The curve's endpoints can be positioned to the user's liking. The steps that follow take you through the process of creating a parabola. If you are unfamiliar with the terms used in the steps, use figure 3-22 for guidance.

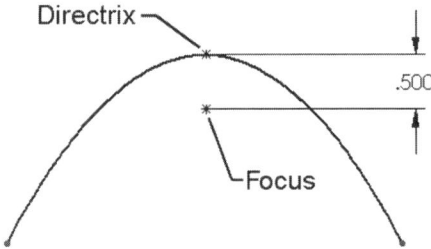

Fig. 3-22. Terms used to define a parabola.

1. Pick to establish the focus.
2. Pick to establish the length of the directrix.
3. Pick to establish where the parabolic arc will begin.
4. Pick to establish where the parabolic arc will end.

Points

Some items that can be sketched do not actually contribute to feature geometry. This is the case with sketch points and centerlines. For instance, if you were to create a shape out of centerlines and attempt to extrude it, no solid geometry would be created. As a matter of fact, SolidWorks would inform you that no contour geometry was present and would refuse to create the feature.

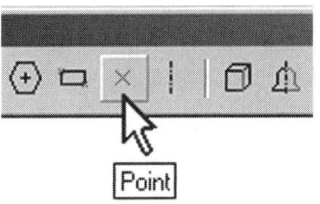

Fig. 3-23. Point icon.

Points can be created by clicking on the Point icon (shown in figure 3-23) and picking where you want to place the point entity. Sometimes points can be beneficial when trying to dimension or constrain other entities. It would be difficult to show you some meaningful examples at this stage without further elaborating on other SolidWorks functionality. Later in the book, you will see some examples of points used for specific functions. No steps are shown for creating a point because there is only the one step anyway: pick to create the sketch point.

Centerlines

Centerlines are also known as construction lines. There is no difference between the two as far as SolidWorks is concerned. Centerlines have a special function in SolidWorks. Specifically, you will need a centerline when performing any of the following functions.

- Mirroring sketch geometry
- Adding symmetrical constraints
- Creating revolved features
- Creating reference or construction geometry

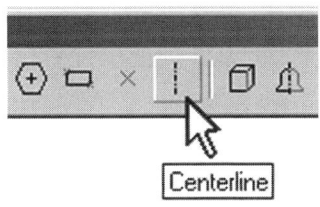

Fig. 3-24. Centerline icon.

As far as adding symmetrical constraints and mirroring sketch geometry are concerned, the end result is exactly the same. You will learn how to mirror sketch geometry in Chapter 4, and creating revolved features is covered in Chapter 5.

Very often centerlines are used for the sake of dimensional references or as construction geometry. Because they do not contribute to feature geometry, they are perfect for this task. Centerlines are also very easy to create. You use the Centerline icon, shown in figure 3-24. The process used to create a centerline is identical to that used to create lines.

Splines

Spline entity types are best used for creating free-form curves. Splines take the place of French curve drawing guides in the CAD world. Because splines have many peculiarities associated with them, and because there are many things you need to be aware of when working with spline curves, this section has been devoted completely to spline entities.

SolidWorks employs two types of splines: proportional and nonproportional. Nonproportional splines are created by default if you click on the Spline icon (shown in figure 3-25) and sketch a spline. Both spline types can be reshaped if you drag their control points. Proportional (fixed shape) splines retain their shape if their endpoints are moved. The steps that follow will take you through the process of creating a spline, and are the same whether creating a proportional or nonproportional spline.

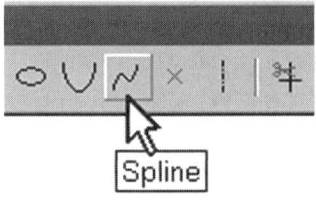

Fig. 3-25. Spline icon.

1. Pick to establish the start point of the spline.
2. Pick to establish subsequent control points of the spline.
3. Double click to establish the endpoint of the spline.

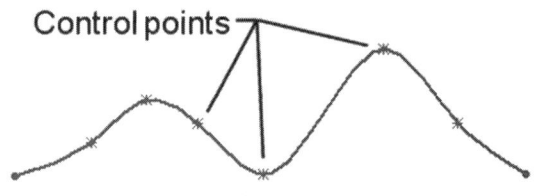

Fig. 3-26. An example of a spline.

In place of step 3, it is also possible to press the Esc key after establishing the endpoint of the spline. This has the effect of ending the spline and simultaneously exiting the Spline command. If double clicking to end the spline, you will remain in the Spline command, at which point you could either exit the command or proceed directly to another command. An example of a completed spline is shown in figure 3-26. Some of the spline's control points are noted for reference.

Nonproportional Splines

As previously stated, nonproportional splines behave somewhat differently than their relatives, proportional splines. A nonproportional spline is what you have been creating if you are performing the exercises. Note that splines contain control points, which can be used to change the shape of the spline. If dragging a spline by a control point, you will change the shape of the spline. Dragging the endpoints of a spline will also change its shape if the spline is a nonproportional spline. Dragging on the spline itself (some location other than a control point) will allow you to reposition the spline.

Splines

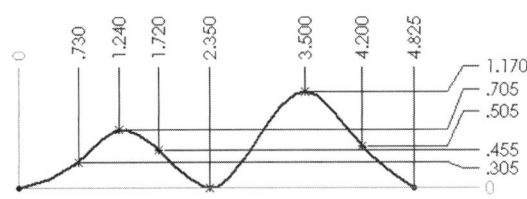

Fig. 3-27. Dimensioning a nonproportional spline.

Because nonproportional splines have a larger degree of freedom of movement and are not as tightly regulated as proportional splines, their control points can easily be dimensioned. Figure 3-27 shows an example of a dimensioned nonproportional spline. Ordinate dimensions were used in the example, a dimension type explored later in the chapter.

Proportional Splines

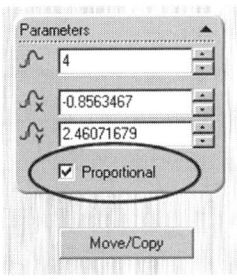

Fig. 3-28. Making a spline proportional.

Proportional splines start out as nonproportional splines. What makes the spline proportional is an option in PropertyManager, shown in figure 3-28. You may decide a proportional spline would serve you better because its size can be scaled up or down proportionally. If this is the case, simply select the spline and check the Proportional option.

Of course, proportional splines look just like nonproportional splines. They just react differently when their endpoints are moved. Try it for yourself and you will see the difference. You will, in essence, be able to scale a proportional spline by dragging one of its endpoints. By their very nature, proportional splines do not require as many dimensions to fully define them.

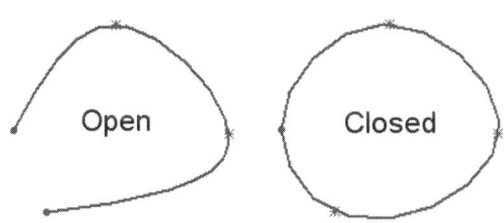

Fig. 3-29. Examples of open and closed splines.

Splines can be open or closed. To create a closed spline, place the spline's endpoint on its start point. This has to be done during the initial creation of the spline and cannot be accomplished after ending the spline command. For instance, dragging the endpoint of a previously created spline to its start point will result in an error message. Figure 3-29 shows simple examples of both open and closed splines.

Adding Control Points

Once a spline has been created, it may be necessary to add additional control points, allowing for a greater degree of control over the spline's shape. Such is the case when it is discovered that there are not enough control points to achieve the desired contour. More control points would remedy this situation, and it is definitely easier to add control points than to delete

74   *Chapter 3: Sketching Basics*

the spline and start from scratch. The newly inserted control points can then be used to further define (or perhaps "refine") the shape of the spline. How-To 3-4 takes you through the process of adding control points to a spline.

HOW-TO 3-4: Adding Control Points to a Spline

To add control points to a spline, perform the following steps.

1. Right-click on the spline and select Insert Spline Point.

2. Using the left mouse button, click on the spline to add a control point at the desired location. Repeat as necessary.

3. Press the Esc key when finished.

Once the new control points have been added, they can be dragged to reshape the spline, or dimensioned as required. Deleting control points is even easier, and is covered in the next section.

Deleting Control Points

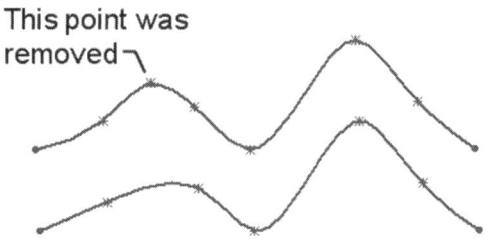

Deleting spline control points is even easier than adding them. Just click on the spline's control point you want to delete and press the Delete key. SolidWorks will recalculate the new spline for the remaining control points. See figure 3-30 for a before-and-after depiction of a spline with a control point removed.

Fig. 3-30. Removing a control point from a spline.

Simplifying a Spline

There is another way of removing spline control points, and that is to simplify the spline. This method is performed by specifying a lesser tolerance level for the spline. This is sometimes referred to as a more loose tolerance. By specifying a lesser tolerance level, fewer control points are required, thereby reducing the number of control points on the spline.

The difference between deleting control points and simplifying a spline is that during the simplification process SolidWorks attempts to retain the original shape of the spline as much as possible. However, with

Splines

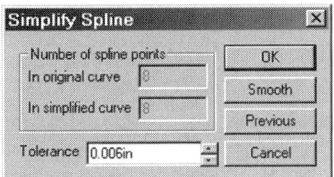

Fig. 3-31. Simplify Spline window.

a looser tolerance setting, the spline becomes straighter. It loses its ability to retain its original curvature due to the lower tolerance setting. How-To 3-5 takes you through the process of simplifying a spline, rather than simply deleting control points. The Simplify Spline window is shown in figure 3-31.

How-To 3-5: Simplifying a Spline

To simplify a spline, perform the following steps.

1. Right-click on the spline and select Simplify Spline.
2. Enter a new tolerance for the spline.
3. Click on the OK button.

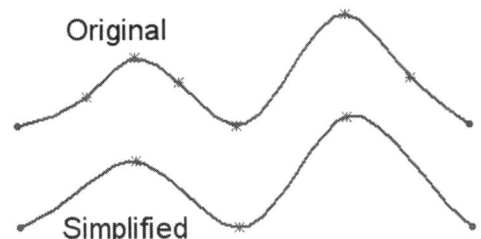

Fig. 3-32. A spline before and after it has been simplified.

Figure 3-32 shows a spline that has been simplified. Note that the simplified version retains much of the original's basic shape. The simplified spline just has fewer control points. In the spline pictured, a tolerance value of .4 inch was used, which resulted in the spline using five control points, as opposed to the original eight.

Another method of simplifying the spline would be to use the Smooth button found on the Simplify Spline window. This loosens the tolerance in incremental steps, thereby reducing the number of control points. The Previous button reverses this effect, similar to an undo function.

✓ **TIP:** *When entering a new tolerance, press Enter to see an updated preview prior to clicking on the OK button.*

Spline Tangency Conditions

Lines and arcs can be made tangent to a spline, but there is a specific procedure that must be used to accomplish this task. For example, simply creating a tangent arc at the end of a spline is not enough. Even though

SolidWorks will let you create the arc, it will not be tangent to the spline. You must physically add the tangent geometric relation between the spline and arc.

Splines can be made tangent to lines, centerlines, arcs, elliptical arcs, and parabolas. If the entity the spline is tangent to moves, the spline will reform itself to accommodate the tangent relationship. Adding geometric relations were first introduced in the previous chapter. Read the section "Adding Relations" in Chapter 2 to understand how to add geometric relations (such as a tangent relation between a line and spline).

Moving Frames

Fig. 3-33. An example of a moving frame.

A moving frame, shown in figure 3-33, is another way of manipulating a spline. To add a moving frame, right-click anywhere on the spline and select Moving Frame. Once the moving frame is visible, drag the arrows of the frame to change the tangency conditions at that location of the spline. The frame can be moved by dragging the point at the center of the frame.

When a control point of a spline is selected, a frame-like object appears. This, however, is not the same as a moving frame. When a moving frame is added to a spline, it is similar to adding a control point. However, a moving frame can be locked, whereas a control point cannot. When locked, a moving frame is locked to that precise point on the spline. To lock a moving frame, right-click on the moving frame and select Lock Moving Frame on Curve. The moving frame can still be rotated, but it will no longer slide along the spline as before. It is suggested that you experiment with moving frames yourself to best see how the spline is affected.

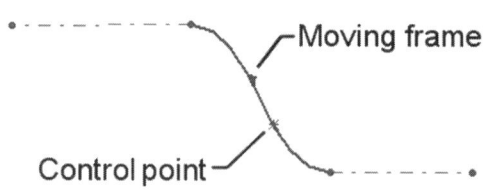

Fig. 3-34. Control points and moving frame points look different.

Once a moving frame is locked, it can also be unlocked. To reverse the process and unlock a moving frame, right-click on the moving frame and select Unlock Frame. Moving frame arrows look identical to the arrows displayed when a control point is selected. However, moving frame points and control points can easily be distinguished from each other. Moving frames are displayed as small triangular-shaped points on the spline, whereas control points are displayed as asterisks. These are shown in figure 3-34.

Spline Minimum Radius and Inflection Points

Two diagnostic utilities available for splines are the Show Minimum Radius and Show Inflection Points options. Both options can be accessed via the context-sensitive menu when right-clicking on a spline. Both options are toggle switches and can be turned on or off by simply checking or unchecking the appropriate option.

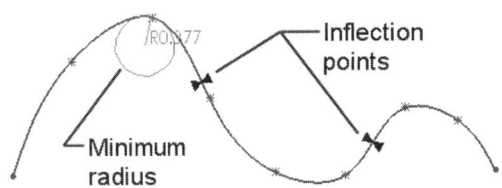

Fig. 3-35. Displaying a spline's minimum radius and inflection points.

Minimum radius is self-explanatory. There would only be one portion of a spline that is determined to have a minimum radius. Inflection points are the points along a spline that momentarily become flat. In other words, curvature is zero at the inflection points. A spline does not necessarily have to have inflection points, but if it does, the inflection points are represented as small "bow ties." Figure 3-35 shows an example of a spline whose minimum radius and inflection points are being displayed.

Inspecting Sketch Curvature

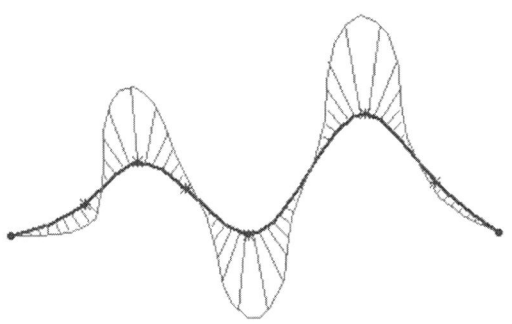

Fig. 3-36. Evaluating the curvature of a spline.

To inspect the curvature of a spline, right-click on the spline and select Show Curvature. Figure 3-36 shows the curvature information for a spline. The curvature information appears as a blue, highlighted area known as a comb. The comb will update dynamically as the spline is reshaped. Try dragging a control point with the curvature information displayed and you will discover this for yourself. To turn off the curvature data, right-click on the spline and uncheck Show Curvature.

It should be noted that any sketch entity that has curvature can have its curvature information displayed. This includes arcs, circles, and parabolas. Entities such as lines and rectangles do not contain curvature, so obviously there will be no option for displaying curvature information.

Construction Entities

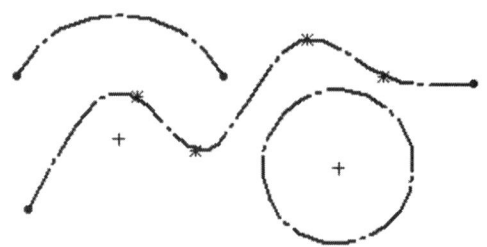

Fig. 3-37. Construction entities.

As was noted earlier, construction lines (also known as centerlines) do not contribute to solid geometry. For this reason, construction lines can be used for a variety of tasks, including being used as reference objects or to aid in adding constraints or dimensions. It is possible to create other types of construction entities (see figure 3-37) in SolidWorks besides construction lines. To do this, create any sketch entity just like you normally would, and then change it into a construction entity. How-To 3-6 takes you through this process.

How-To 3-6: Creating a Construction Entity

To create a construction entity, perform the following steps.

Fig. 3-38. Construction Geometry icon.

1. Select the sketch entity to be changed into a construction entity.

2. Click on the Construction Geometry icon, shown in figure 3-38.

These same steps can be used to change a construction entity back into its original format. As a matter of fact, if you perform these same steps on a centerline, you will see that it, too, can be changed into a standard sketch line.

✏ **NOTE:** *To change more than one entity at a time into construction geometry, hold the Ctrl key down when selecting the entities, and then right-click on any one of the selected entities to access its properties.*

The Construction Geometry icon can have an alternate way of functioning that might prove useful. If nothing is selected prior to clicking on the Construction Geometry icon, it enters what might best be described as "toggle mode." When in this mode of operation, sketch objects can be

Sketch Guidelines

selected one after the other. Any sketch object with a continuous line type will convert to construction geometry, and vice versa.

There is another way to change items into construction entities, or vice versa. This second method requires accessing an object's properties using PropertyManager. After selecting a sketch entity, one of the options in PropertyManager will be an item titled *For construction*. Checking or unchecking this option will turn an object into a construction entity or back to having a regular solid line type. Whatever method you use has the same result, so choose the method you find most convenient.

✓ **TIP:** *Turning a circle into a construction circle is the ideal way of creating a bolt hole circle.*

Sketch Guidelines

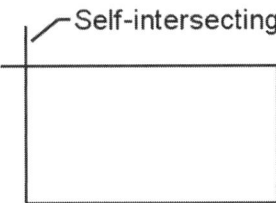

Fig. 3-39. Self-intersecting geometry.

There are a few guidelines you should keep in mind when creating a SolidWorks sketch that will eventually become a feature. These are simple rules to follow, but must be obeyed. SolidWorks allows for not having to worry about exact size and shape when sketching, as dimensions and geometric relations will drive the geometry. However, it would not be wise to be sloppy. For instance, sketch profiles should be non self-intersecting. Figure 3-39 shows an example of self-intersecting sketch geometry. This rule must not be broken under any circumstances or an error message will result.

- Sketch rule 1: Avoid self-intersecting geometry.

If a sketch contains self-intersecting geometry, clean up the corners and get rid of any dead-end geometry. It is easy to create a sketch whose corners all meet perfectly in the first place. Just make it a point to pay attention to the system feedback.

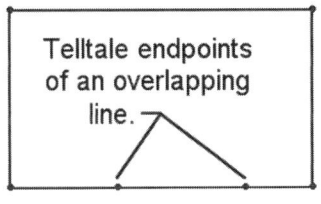

Fig. 3-40. Overlapping geometry.

Another type of problem is overlapping geometry. In figure 3-40, there is a line overlying the top of an existing line. These errors are sometimes very difficult to find because the overlapping geometry is very difficult to see. In the figure, you may be able to make out two of the overlapping line's endpoints. The endpoints of the extra line are visible due to an option that makes this possible.

- Sketch rule 2: Avoid overlapping geometry.

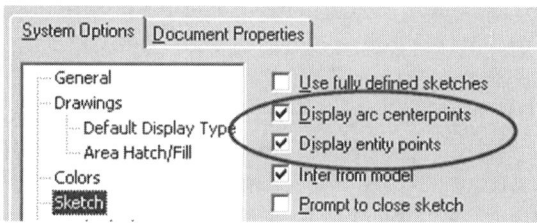

Fig. 3-41. Activating two important system options.

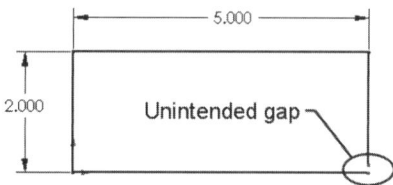

Fig. 3-42. Note the small gap in the profile.

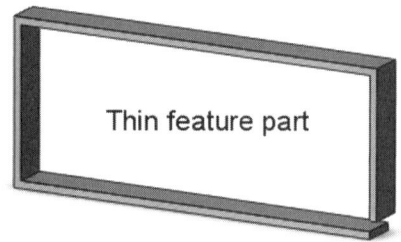

Fig. 3-43. The unintended result of extruding an open profile.

There are two options in the System Options window you will almost certainly want to enable. These options are titled *Display arc centerpoints in part/assembly sketches* and *Display entity points in part/assembly sketches*. The first option allows for seeing the centerpoints of arcs and circles. The second option displays endpoints as small dots. This second option will aid you when creating sketch geometry, and will aid in troubleshooting. The option, found in the Sketch section of the System Options window, is shown in figure 3-41. Make sure you have these options checked.

One last rule to follow would have to do with closing a profile. In this case, the "rule" is not really a steadfast rule, but more of a guideline for certain feature types. Leaving gaps in sketch geometry will result in a feature type that is probably not intended. Open profiles are legal for certain functions and for thin features, but not when trying to create typical solid geometry. As shown in the following figures, the SolidWorks user was attempting to create a simple extruded part. However, he did not realize that a small portion of the sketch was open (figure 3-42). Figure 3-43 shows the result of extruding an open profile, and the reason for the result.

Sometimes you might intend to sketch an open profile in order to create what is known as a thin feature. You may also use open profiles for more advanced feature types. Swept features, for instance, typically use open profile curves as a sweep path, or trajectory. All of these feature types are covered in upcoming chapters.

One last sketch guideline has to do with nested profiles, or sketch profiles within other sketch profiles. Profiles within profiles are acceptable to a point, but keep in mind that your editing options will be limited. Consider the scenario depicted in figure 3-44. Here is a case in which the circles are

Sketch Guidelines

nested within the larger profile of the rectangle. Because the circles and rectangle are part of the same sketch, they will have the same feature definition. What exactly does this mean to the user?

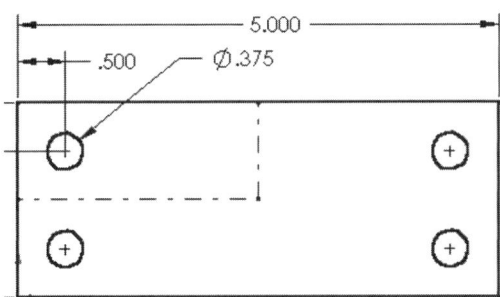

Fig. 3-44. Nested profiles.

If the holes in the plate are to always go all the way through the plate, the sketch with the nested profiles would suffice. However, if at some point in the future the designer changes her mind, a fair amount of editing would be required. If the circles were part of a separate sketch, they would have a separate definition as well. The depth of the holes could be independently changed with respect to the depth of the plate.

Sketches should be separated into editable chunks of geometry. Do not try to cram everything into a few complex sketches. It makes for a part that is difficult to manage, difficult to troubleshoot, and very inflexible.

Contour Select Tool

Occasionally you may find it convenient to create a sketch containing more geometry than SolidWorks legally allows for creating a feature. Reasons vary, but the point is that it is not necessary to use an entire sketch for a feature. Rather, you can choose specific regions, or contours, with which to "grow" the desired feature. Contours can be open or closed, and can consist of any geometry in an otherwise invalid sketch.

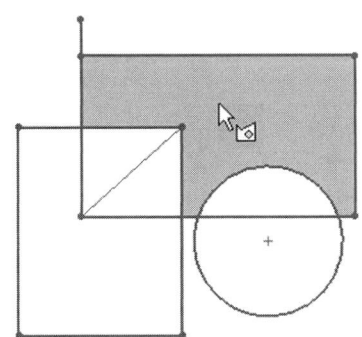

Fig. 3-45. Selecting a contour.

In essence, the Contour Select tool makes it possible to use what SolidWorks would otherwise consider an illegal sketch. Examine figure 3-45, and note how the Contour Select tool highlights a region consisting of the surrounding self-intersecting sketch geometry. The highlighted region could then be extruded, as an example.

Multiple contours can be used from the same sketch. Additionally, contours can be selected whether or not you are editing a sketch. Contours can even be selected while defining a feature. In short, the Contour Select tool adds a great amount of flexibility to how sketch geometry can be created and put to use for basic feature creation.

To access the Contour Select tool, right-click in the work area or on a sketch in FeatureManager and select Contour Select Tool. Moving the cursor over the sketch geometry will cause various contours to dynamically highlight. Select the desired contour and create the feature as usual. Feature creation is discussed in much more detail throughout this book.

✏ **NOTE:** *Although the Contour Select tool makes it possible to work with otherwise invalid sketch geometry, do not use this tool as a crutch. Contours are limited in what they can be used for, and it is important that you understand the basic sketch guidelines discussed in the previous section.*

Check Sketch For Feature

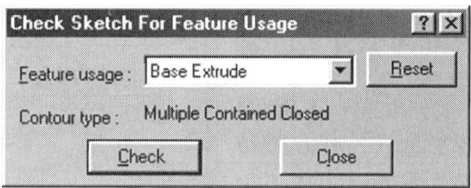

Fig. 3-46. Check Sketch For Feature Usage window.

The Check Sketch For Feature function is very useful for anyone learning the SolidWorks software. You have the ability to make SolidWorks check a sketch and tell you if it is a valid sketch for any particular feature type. This function is found in the Tools > Sketch Tools menu. It is a very straightforward and easy-to-use command. The window for this function is shown in figure 3-46. How-To 3-7 takes you through the process of using the Check Sketch For Feature command.

How-To 3-7: Using Check Sketch For Feature

To use the Check Sketch For Feature command, perform the following steps.

1. Select Check Sketch For Feature from the bottom of the Tools > Sketch Tools menu.

2. Using the drop-down list, select the feature type you want to create.

3. Click on the Check button.

4. Click on Close when finished checking your sketch.

SolidWorks will display a message informing you as to whether or not you have a valid sketch for the feature type you chose. If there is a problem with your sketch, SolidWorks will tell you what is wrong. If there are no problems with the sketch, a message will inform you of this as well.

The Check Sketch For Feature command is commonly used when troubleshooting existing geometry. When an existing sketch goes bad due to some sort of design change, Check Sketch For Feature will tell you what is wrong. When used for troubleshooting a sketch that has already been turned into a feature, the correct feature will automatically be selected from the Feature Usage drop-down list. All the user has to do after calling up the command is to click on the Check button (previous step 3).

Keep It Simple

Keep in mind that constraints and dimension placement are important factors in creating a good sketch. Place dimensions on the geometry you want to be able to control. Dimensioning can be very difficult with large and elaborate sketch geometry. As previously noted, it is better to create simplified sketch geometry and a greater number of features than complex sketch geometry and fewer features.

Do not try to incorporate every last detail of the model into a sketch. This will make editing the part much easier farther down the road if design changes are needed. It will also make error recognition and troubleshooting the model a great deal easier. Mistakes will happen and errors will occur, even once you become proficient with the software. If you do make a mistake, hunting for it through a maze of sketch geometry and interrelated dimensions and constraints can be difficult. If the sketch geometry is simple, troubleshooting is much easier.

Fully Define Your Sketch

SolidWorks does not require you to completely define sketch geometry. This means that if you did not want to add any constraints or dimensions, you would not have to. This flexibility can be nice, but it is not necessarily good practice. It is good practice to fully dimension and constrain all of your sketch geometry. This allows you greater control over your sketches because the sketch must follow the set of conditions you have placed on it. If you do not have full control over a sketch, the sketch will not have predictable behavior.

Use the origin point to your advantage when creating the first sketch of a part. Remember that the origin acts as an anchor point. This means that it will lock your sketch in place, thereby allowing it to be fully defined. Sketch geometry created later can be anchored or dimensioned to features that come before them. This creates a flexible model, one that is not prone to errors if design changes are needed later on.

☛ **NOTE:** *Always fully define your sketch geometry.*

Units of Measurement

One of the options usually set up before beginning a new part is the working units. Typical working units are inches and millimeters, but other options are available. How-To 3-8 takes you through the process of changing the working units of measurement. Note that the units can be changed at any time, even after a model is completed.

How-To 3-8: Changing the Working Units

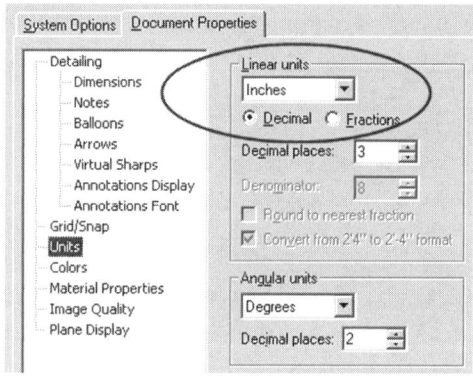

Fig. 3-47. Changing the units of the model.

To change working units, perform the following steps.

1. Select Options from the Tools menu.
2. Select the Units section from the Document Properties tab.
3. Select a linear unit from the drop-down list (see figure 3-47).
4. Select an angular unit, or modify any of the other optional parameters in the Units section.
5. Click on OK when finished.

When the units of a part are modified, the change will only affect the current document. To have the units of measurement already set when starting a new part (or assembly or drawing), the document must be saved as a template with the desired units already specified. That way, when a new document is begun, the template with the proper units can be selected. See the section "Using Templates" in Chapter 2 for more information on this matter.

Grid/Snap Settings

The Document Properties section Grid/Snap is shown in figure 3-48. Placing a check in the *Display grid* setting will turn on the grid. The grid is strictly a user preference. If you find it useful, by all means turn the grid on. Some find the look of a clean, unencumbered screen more desirable. Others like the grid turned on because it helps orient the user as to what plane they are sketching on. If *Display grid* is turned on, the grid will only be displayed while in an active sketch.

Grid/Snap Settings

If you decide to use the grid, the other optional settings in the Grid/Snap section will allow you to customize the grid's appearance. Major grid spacing allows for changing the spacing between the main grid lines. *Minor-lines per major* sets how many secondary grid lines are placed between the major grid lines. In other words, if the *Major grid spacing* option is set to 1 inch and *Minor-lines per major* is set to 4, there will be a grid line every quarter inch.

Incidentally, the Grid icon, shown in figure 3-49, is a shortcut to the Grid/Snap section of the Document Properties window. However, this does not mean that you can only adjust the Grid or Snap settings once the window is open. Feel free to change any of the System Options or Document Properties necessary in the Grid/Snap section.

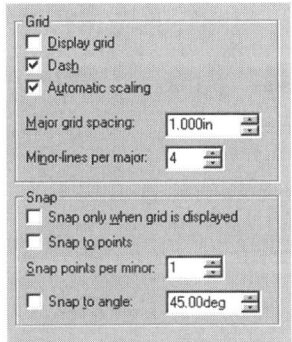

Fig. 3-48. Grid and Snap options.

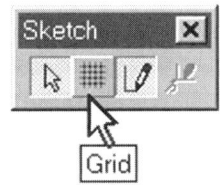

Fig. 3-49. Grid icon.

Sketch Plane Indicators

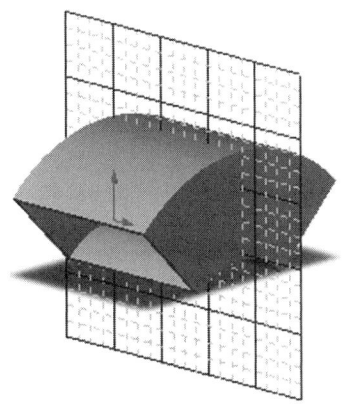

Fig. 3-50. Display grid.

As mentioned previously, the display grid can be useful because it serves as an indicator as to what plane is currently being sketched on. If this is the reason you find the grid useful, you should be aware of other options that serve as sketch plane indicators. Figure 3-50 shows an example of what the display grid might look like.

The grid only displays when in a sketch, so it can also be useful to new users in that it serves to indicate when a sketch is being edited. The grid also helps those not accustomed to reading system feedback. For example, the grid can help gauge the length of a line being sketched. It should be noted that system feedback also accomplishes this task, and system feedback is what you should be in the habit of using.

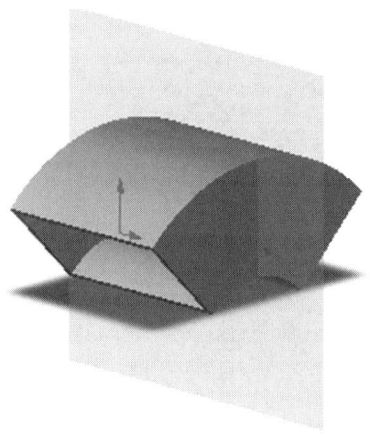

Fig. 3-51. Sketch display plane.

If the grid is not your cup of tea, another alternative is to use the *Display plane when shaded* option. An example of the display plane is shown in figure 3-51. You will find the *Display plane when shaded* option in the Sketch section of the System Options (Tools > Options). Similar to the display grid, the display plane only appears when editing a sketch. Because the display plane is transparent, it has a tendency to slow down graphics performance on some older computers.

Some users prefer to have full control over what planes are shown, and when. This is typical of those who have more experience with the software. Chapter 2 discussed how to manually display a plane. There is something to be said for working with an uncluttered screen without grids and shaded planes appearing every time a sketch is started. However, these options are there to help, and you should use them if they benefit you.

Snap Behavior

SolidWorks does have an option for turning on a snap grid function, but it is not recommended by this author. Turning on snap behavior can actually impede the user's ability to sketch. Consider the fact that SolidWorks is a parametric program. This means that dimensions and constraints are controlling the sketch. Why would one need to snap a line to a precise length when a dimension is going to be driving the length of the line anyway? You would not need a snap function, and that is precisely the point.

If on the off chance you do decide to turn on the snap behavior, the *Snap points per minor* setting will control how many snap points there are between the minor grid lines. A good setting for this would probably be 1, but then again it is not advised that the snap behavior be turned on anyway. Feel free to experiment, however, because that is the best way to learn.

Geometric Relations

Geometric relations (constraints) were touched upon in Chapter 2. Here you will learn about them in more depth. Throughout the rest of the book, you will employ them in practice.

Chapter 2 discussed adding geometric relations. You learned that constraints can be added in primarily two different ways. One method would be while sketching. This is why paying attention to system feedback is so

Geometric Relations

important. If you do not watch for the feedback displayed by the cursor, you have no idea what constraints were just added.

Another method of adding constraints is to use the Add Relations icon. This allows for manually adding constraints after sketch geometry has already been created. The steps for adding geometric relations were covered in the last chapter.

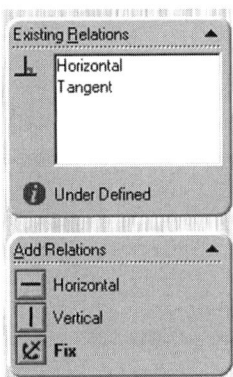

PropertyManager displays relations that have already been added to a selected object. It will also show relations between objects if more than one object is selected. This information is displayed in the area labeled Existing Relations. In the case of the example shown in figure 3-52, a sketched line has been selected. What PropertyManager is telling you is that there is already a horizontal and tangent relationship associated with the line. Below that you see the words *Under Defined*. This tells you the current status of the sketch.

There is another area (also shown in figure 3-52) titled Add Relations. These are the relations that are physically possible to add to the selected object. PropertyManager will not show relations that are not applicable to the object (or combination of objects) selected. Note that there is one relation, Fix, displayed in bold text. The relation displayed in bold is the relation SolidWorks thinks you want to add. You will find that more than 50% of the time SolidWorks guesses correctly. This time, however, it did not.

Fig. 3-52. Relation information in PropertyManager.

When the software guesses correctly, it makes finding the relation you want a little easier. Other than that, the bold text does not serve any purpose. Be forewarned that the Fix constraint is one you will typically want to stay away from. It locks whatever is selected in position and allows for absolutely zero flexibility. The Fix relation has its occasional uses, but it is highly recommended you avoid it when defining sketch geometry.

Callouts

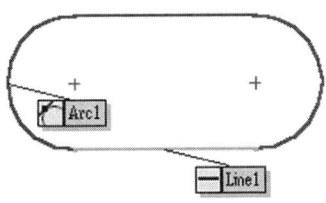

Fig. 3-53. Examples of callouts.

A callout is a small tag attached to an object that helps you understand something about that object. For example, callouts are used to show what relationships are associated with a particular sketch entity. An example of callouts is shown in figure 3-53.

In figure 3-53, the callouts were displayed by double clicking on the line at the bottom of the sketch. You can see that there is a tangent relation between the line and the arc on the left. There is also a horizontal relation associated with the line. Apparently there is no tangent relationship associated with the arc on the right, even though it looks as though they are tangent.

Callouts will also be displayed when an existing relation is selected from PropertyManager. For example, selecting a constraint from the Existing Relations list box will display a callout for that constraint. This makes it easy to delete a particular constraint, especially when there is more than one constraint of the same name associated with a particular entity. This could occur if, for instance, a line were tangent to two arcs, one on either side of the line. If it were necessary to delete one of the tangent relations but not the other, you would want to know which one to delete. The callouts make this task easy. The following section discusses deleting geometric relations.

Deleting Relations

The easiest way to delete a geometric relation is to delete its callout. Clicking on a callout will select it and turn the callout yellow. Once that has been done, pressing the Delete key is all it takes to remove a geometric relation from sketch geometry.

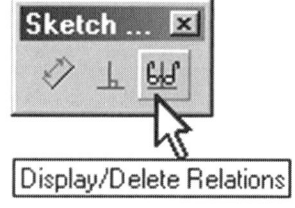

Fig. 3-54. Display/Delete Relations icon.

Another method of deleting constraints is to use PropertyManager. By selecting a relation from the Existing Relations list box, it then becomes possible to delete that relation by pressing the Delete key. So as you can see, deleting geometric relations is a very easy task to perform.

One last method used for displaying or deleting geometric relations is via the Display/Delete Relations command, whose icon is shown in figure 3-54. It really is not necessary to use the Display/Delete Relations command to show or remove relations from just a few sketch entities. Where this command becomes beneficial is when it is necessary to see all relations associated with an entire sketch. The command is also helpful when troubleshooting.

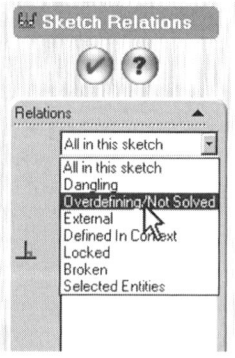

Fig. 3-55. Using the Display/Delete Relations command.

The best part of the Display/Delete Relations command is the drop-down list shown in figure 3-55. Note the highlighted selection Overdefining/Not Solved in the figure. If this option were selected, only overdefining or unsolved relations would be displayed. There are other options as well, such as locked or broken relations, discussed later in the book. For now, it is enough to understand that relations can in fact be filtered out. Only certain relations that have a particular characteristic, such as overdefining, are displayed.

Dimensioning

You probably will not need the Display/Delete Relations command on a daily basis, but it is good to know it is available. If your sketch turns red because it is overdefined, you may find this command very useful.

Dimensioning

Placing dimensions on a sketch is a very straightforward process. Most of the dimensions placed on a sketch (and on a part in general) transfer to the 2D drawing later on. This is another reason it is important to fully define sketch geometry. If all sketches are fully defined, theoretically, the 2D drawing will have all required dimensions for the manufacturing department to build the part. Do not worry about 2D drawings just yet, as they are covered in Chapter 11.

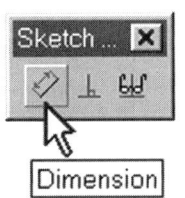

Fig. 3-56. The Dimension icon.

There is only one dimension tool in SolidWorks, aptly named Dimension, whose icon is shown in figure 3-56. This one icon does it all, from aligned dimensions to ordinate dimensions and everything in between. The dimension type created largely depends on what is selected to place the dimension on. If a circle is selected, a diameter dimension is created. If a line is selected, a linear dimension is created, and so on.

There are some dimensioning techniques that require a more elaborate explanation, but this section will get you started with the basic dimensioning skills needed to get by. In short, dimensioning requires picking the objects to be dimensioned, and then picking a location to place the dimension. It would be redundant to cover every possible dimension type here in depth. Many specific dimension tips and tricks are covered throughout this book. For now, How-To 3-9 will take you through the process of adding generic dimensions to a sketch.

How-To 3-9: Adding Dimensions to a Sketch

To add a dimension to a sketch, perform the following steps.

1. Select the Dimension icon.
2. Click on the item you wish to dimension.
3. Click to position the dimension.
4. Type in a value for the dimension if prompted.
5. Press Enter or click on the green check to accept the dimension value.

Chapter 3: Sketching Basics

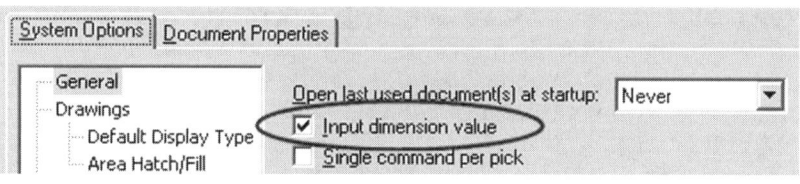

Fig. 3-57. Input dimension value *option*.

Step 4 states that you should type in a value for the dimension if prompted. Whether or not you are prompted for a value depends on a particular setting in the System Options setting (select Options from the Tools menu). This optional setting is known as *Input dimension value* and is shown in figure 3-57. Make sure the option is checked. This will make SolidWorks ask for a value each time a dimension is added to the sketch.

How-To 3-9 shows how to add a dimension in the most basic and simplified way possible. However, sometimes it is necessary to add a dimension between two objects, such as when dimensioning the angle between two lines. What throws most new users off is the dimension's preview. The point to remember is "don't stop short"! Pick the second line and the preview will automatically update to show the correct dimension. The following figures should help clarify this. In figure 3-58, the angled line at the top of the triangle was selected while in dimension mode. Note the preview, which will change according to the position of the cursor.

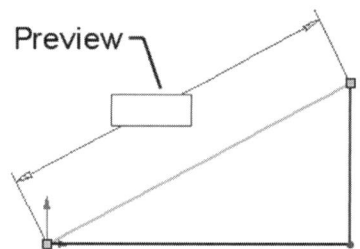

Fig. 3-58. *Initial preview after selecting the angled line.*

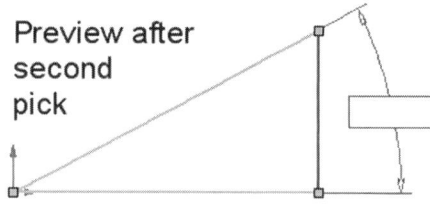

Fig. 3-59. *Preview after selecting the second line.*

Again, the main point here is to not stop short. In other words, continue picking whatever it is you wish to dimension. If you want to dimension the angle between two lines, two lines are what you must pick. After selecting the horizontal line, the preview shown in figure 3-59 will be displayed.

The rest is a no-brainer. Pick where you would like to see the dimension value be positioned and the Modify window, shown in figure 3-60, will be displayed. The easiest way to finish up this process of adding a dimension is to simply type in a number. Do not click in the blue area first, where you see the current value displayed. Just type. Once you have typed in a value, press Enter and move to the next task.

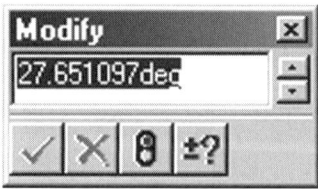

Fig. 3-60. *Modify window.*

Users from other CAD programs may have a tendency to pick endpoints as opposed to lines when adding a linear dimension. This is not necessary.

Dimensioning

Just select the line. It is one click less and will do the job just as well. The following are a couple of other pointers that will set you on your way.

- To delete a dimension, select the dimension with the left mouse button and press the Delete key.

- To modify a dimension value, double click on the dimension with the left mouse button.

Automatic Dimensioning

When positioning sketch geometry dimensionally is more important than positioning with geometric relations, automatic dimensioning may be the tool for you. Automatic dimensioning also works wonders for imported 2D geometry, which is discussed further in Chapter 22.

Automatic dimensioning, otherwise known as autodimensioning in SolidWorks lingo, can save time in certain situations. Let's look at an example, such as that shown in figure 3-61. This is a simple stamped metal plate that contains a fair amount of sketch geometry. Considering the nature of the part, and that all cuts will always go completely through the model, it is easier to place all of the sketch geometry in one sketch than it is to extrude the entire sketch as one feature.

Only a few constraints have been added to the sketch shown in figure 3-61. Only dimensions need to be added at this point, whereupon the feature can be created. Autodimensioning makes adding the dimensions easy. Using dimensioning schemes (such as baseline, chain, or ordinate), we can tell SolidWorks what types of dimensions should be added. How-To 3-10 takes you through the process of using the Autodimension function.

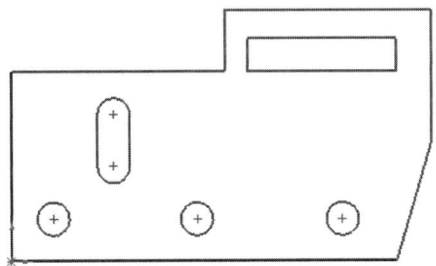

Fig. 3-61. Sketch for a stamped metal plate.

HOW-TO 3-10: Using the Autodimension Sketch Tool

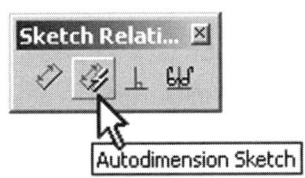

Fig. 3-62. Autodimension Sketch icon.

To add dimensions automatically to a sketch, perform the following steps.

1. Select Autodimension Sketch from the Tools > Dimension menu, or click on the Autodimension Sketch icon (shown in figure 3-62).

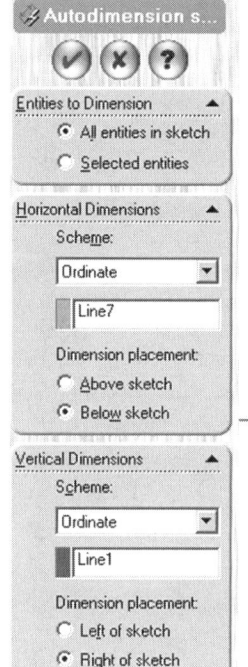

Fig. 3-63.
Autodimension Sketch panel.

2. In the Autodimension Sketch panel, shown in figure 3-63, specify the dimensioning scheme to use for both horizontal and vertical dimensions.

3. Specify a point or line where the dimensioning should start. By default, this is the leftmost point or line for horizontal dimensions and the lowermost point or line for vertical dimensions.

4. Specify where the dimensions should be placed, such as above or below the sketch.

5. Click on OK to create the dimensions.

To see how well autodimensioning works, examine figure 3-64. Solid-Works added 17 dimensions in the blink of an eye. Not only that, it knew enough to add the proper ANSI standard text 3X before the hole diameter. This is because the three holes were constrained to be equal (see "Adding Relations" in Chapter 2) prior to autodimensioning.

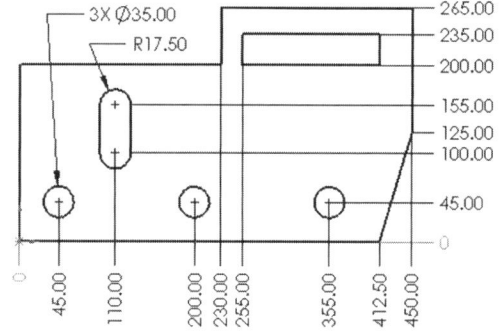

Fig. 3-64. After autodimensioning the sketch.

Autodimensioning is not for every sketch. It is better suited, for example, to occasions in which a series of holes needs to be drilled through a plate and those holes do not conform to a pattern, or for situations involving imported geometry. With simple sketch geometry involving a few dimensions, you are better off adding them one at a time, making sure the dimensions are placed correctly, thereby imparting your design intent.

Pan and Zoom Commands

Probably the last topic that should be covered prior to actually building your first part has to do with being able to move the part around on screen.

Pan and Zoom Commands

The ability to zoom in close, pan the model from side to side, or even rotate the model is critical in order to get any work done. This section will teach you how to accomplish all of those functions and more.

The zoom command icons are located on the View toolbar and are fairly easy to figure out on your own. For the sake of being thorough, they are discussed here. Figure 3-65 shows only the left-hand side of the View toolbar. The right-hand side of the toolbar contains display options, which are discussed in material to follow.

On the View toolbar there are a variety of icons that allow you to pan or zoom the view, or change it in a number of ways. They are all easy to learn, with the possible exception of the Rotate icon, which sometimes takes a little practice. In the section that follows, the full selection of zoom icons is described, and the corresponding hot keys, if any, are listed. You are encouraged to experiment with these various options to gain familiarity with them.

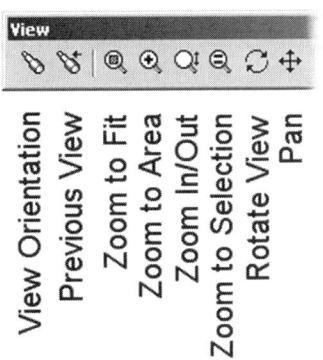

Fig. 3-65. Pan and zoom icons of the View toolbar.

✧ **NOTE:** *It may be helpful to open an existing sample part in order to experiment with the following commands.*

View Orientation

View Orientation hot key: Spacebar

This icon opens the Orientation window, which you learned the basics of in the last chapter. Once the Orientation window is open, double click on any of the listed views to display that view. Use the pushpin to keep the Orientation window on top of the work area. What was not discussed in the last chapter were the icons above the Orientation window, shown in figure 3-66. These icons allow for saving user-defined views or resetting the system views. This functionality is explained in the following material.

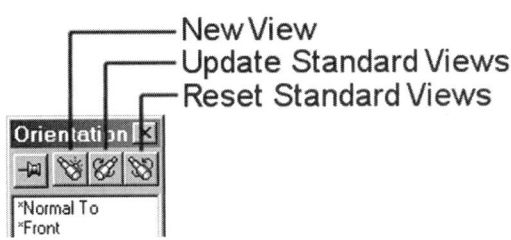

Fig. 3-66. Orientation window icons.

System Views

Any view with an asterisk before it is a system view created by SolidWorks. System views cannot be deleted. Parts and assemblies will have a number of system views. 2D drawings will only have one system view (Full Sheet). Drawings are covered in Chapter 11.

New Views

The New View icon is for adding your own views. This can be done in parts, assemblies, or drawings. User-defined views are saved with the file, so they will be available next time the document is opened. How-To 3-11 outlines how to add a user-defined view.

How-To 3-11: Adding a User-defined View

To add a user-defined view to the listing in the Orientation window, perform the following steps.

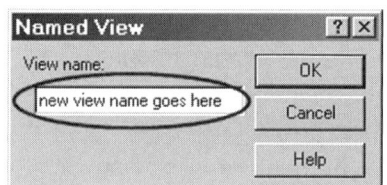

Fig. 3-67. Typing in a name for the new view.

1. Establish the desired view using the Pan, Zoom, and Rotate commands.
2. In the Orientation window, click on the New View icon.
3. Type in a name for the new view, as shown in figure 3-67.
4. Click on OK.

The new view name will be added to the list, and will stay there forever unless deleted. User-defined views will not have an asterisk before them. It is possible to delete user-defined views by selecting the view and pressing the Delete key.

Not only is the orientation of the part remembered when saving a user-defined view, but so is its size and positioning on screen. Make sure you have sufficiently zoomed or panned the object on screen before saving your view. If you make a mistake, or do not like a view you created, you will have to delete it and try again. This is not a handicap because creating views is so easy anyway.

Updating Views

Update Standard Views and Reset Standard Views go hand in hand. Update Standard Views is used for changing the default orientation of the views. For instance, if you want the Right view to actually be the Front view, Update Standard Views would allow you to change that. Think of updating the views as redefining the views, as that is a good way of describing the function. How-To 3-12 takes you through the process of using the Update Standard Views icon.

How-To 3-12: Updating the Standard Views

To update (redefine) the standard system views, perform the following steps.

1. Change to a view you would like to redefine, using any method you wish.

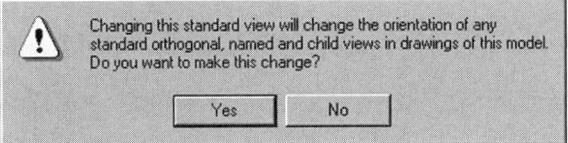

Fig. 3-68. Heed the warning message!

2. Select (one click only) the view in the Orientation window you would like to set the current view to.

3. Click on the Update Standard Views icon. A warning message, shown in figure 3-68, will appear.

4. Click on Yes to update the standard system views, or click on No to keep the default system view arrangement.

The reason for the warning message, and what it actually means, is that any drawing of the model whose views you are updating will be affected. For example, imagine a top/front/right-side view of a part in a 2D drawing. These views are directly linked to the solid model. If the system views of the model are redefined, the drawing is directly affected. Specifically, the drawing views will update to accommodate the new view definitions. If you have not yet created a drawing of the model, the warning can be safely ignored.

One reason users often update the system views is because the model was not created on the proper plane to begin with. It is always best to try to begin sketching on the correct plane in the first place. That is, if you are creating the first sketch for the part and it is a profile of the part as seen from the top, you should be sketching on the Top plane. This is always the best practice, because the part will be oriented correctly when finished, and the system views can stay as they are.

Resetting Views

Where the Reset Standard Views icon comes into play is when you change your mind after updating the standard views. It is also possible to accidentally mess up the system views if not paying attention to what you are doing. This is not unheard of among new users.

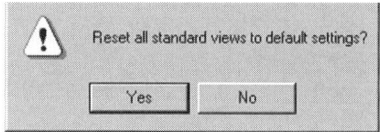

Fig. 3-69. Resetting the views to their default settings.

If you click on the Reset Standard Views icon, you will see a query message asking if you are sure you want to reset the views (shown in figure 3-69). Simply click on Yes and you will be back to the default view arrangement, with no harm done. Any changes made using the Update Standard Views icon will be negated.

Previous View

The Previous View icon reverts back through the last ten previous views. You will probably never need to go back through more than ten views, so that number should suffice.

Zoom To Fit

Zoom To Fit hot key: F

Zoom To Fit is analogous to an auto scale function. The model will be resized to fit on screen, with a little room left to maneuver around the outside of the part. This is the easiest way to get an overall view of the part.

Zoom To Area

Zoom To Area requires a small amount of user intervention. If Zoom To Area is selected, pick and drag opposite corners of a window that define the area you want to zoom in on. The smaller the window, the closer you will zoom in.

Zoom In/Out

Zoom In/Out hot keys: Z, Shift + Z

Zoom In/Out also requires some mouse work on the part of the user. Once the icon has been selected, hold the left mouse button down and move the mouse either forward or backward. Moving the mouse forward is like walking toward the screen, with the object on screen increasing in size. Moving the mouse backward is like moving away from the screen.

If the mouse you are using has a middle mouse button wheel, scrolling the wheel will zoom the model in and out. Using the mouse wheel behaves slightly differently than using the Zoom In/Out icon, however. In the case of the wheel, zooming in will be accomplished with respect to where the cursor happens to be positioned. Using the icon zooms in or out with respect to the screen's center and does not depend on cursor location.

One last option is to hold the middle mouse button and Shift key down simultaneously while moving the mouse. This can be quicker than clicking on the icon if your hand is already near the keyboard, and is preferred by some people. Note that this technique is identical to using the icon, and not to using the wheel.

Zoom To Selection

Zoom To Selection allows you to quickly and effortlessly zoom to a particular entity. Usually a face is selected that you want to zoom in on, and then the Zoom To Selection icon is clicked. This will zoom in on the selected entity until it fills the screen.

Rotate View

Rotate View is probably the most difficult of all zoom commands to get accustomed to. New users sometimes find it somewhat awkward. However, with a little practice, using Rotate View will seem like second nature to you. Just hold the left mouse button down while gently moving the mouse to get a feel for how this command works.

It is not necessary to click on the Rotate View icon if your mouse has a middle mouse button or wheel. Simply hold the middle button or wheel down while moving the mouse.

Rotate View has another capability if you desire a higher degree of control while rotating the part. Specifically, it is possible to rotate about an edge, about a selected position on a plane, or about a vertex point. All you need to do is select an edge, plane, or vertex point while in the Rotate View command. This helps to precisely control the point you want to rotate about, making it easier to get the exact view desired. This option is only possible in conjunction with the Rotate View icon and not the middle mouse button.

Pan

Pan hot keys:
Ctrl + Arrow keys

Pan is the last of the zoom icons and can be used to pan the model right, left, up, or down. Panning will always be parallel to the screen. After clicking on the Pan icon, hold the left mouse button down and move the mouse to slide the model about the screen. This function is quite easy to perform and does not require much practice, if any.

Another pan option involves the middle mouse button again. Hold down the middle mouse button (or wheel) and the Ctrl key simultaneously while moving the mouse. This function is identical to using the Pan icon.

Additional Zoom Options

In addition to the previously listed zoom and pan commands, there are a few extra hot key combinations that do not have any counterpart icons. These additional hot keys, along with those previously mentioned, are listed in Table 3-2 for your convenience. Additionally, middle mouse button shortcuts are listed in Table 3-3 for easy reference.

Table 3-2: Zoom, Pan, and Rotate Hot Keys

Function	Hot Key(s)
Rotate view	Arrow keys
Rotate view in 90-degree increments	Shift + Arrow keys
Rotate model parallel to the screen	Alt + Left or Right Arrow keys
Pan the view	Ctrl + Arrow keys
Zoom in	Shift + Z
Zoom out	Z
Zoom to Fit	F
Front view	Ctrl + 1
Back view	Ctrl + 2
Left view	Ctrl + 3
Right view	Ctrl + 4
Top view	Ctrl + 5
Bottom view	Ctrl + 6
Isometric view	Ctrl + 7

Table 3-3: Middle Mouse Button Shortcuts

Procedure	Function
Middle mouse button	Similar to Rotate View command (see Rotate View)
Middle mouse wheel (scrolling)	Similar to Zoom In/Out (see Zoom In/Out)
Middle mouse button + Shift	Same as Zoom In/Out
Middle mouse button + Ctrl	Same as Pan

Zoom Limitations in Windows

Due to some limitations of Windows 98 and ME, SolidWorks has a limit to how far you can zoom in on the model. There is a toggle switch in the System Options window that overcomes this limitation. The downside is that you may experience some degradation of performance while working in non-shaded display modes. A recommendation would be to not override the zoom limitation unless needed. If you notice performance problems, you may want to turn off the zoom limitation override.

You can find the zoom limitation toggle switch in the Performance tab of the System Options window (Tools > Options). The option is labeled

Enable clipping for zoom limitation in Windows 98\ME. Checking this option will allow you to zoom in closer to the model than if the option were turned off.

It should be noted that Windows NT, 2000, and XP do not suffer from this problem. This option will be grayed out if SolidWorks is being run on any of these Windows operating systems. Of course there would have to be some sort of limit as to how far a SolidWorks user could zoom in or out in a model anyway. There would also have to be limitations on the physical size a computer-generated model can be in SolidWorks. (The technical reasons are complex and beyond the scope of this book.)

To put it simply, a SolidWorks user will be limited to creating parts no smaller than .1 micron (.0001 mm) and no larger than 1 kilometer. It should also be noted that approaching these limitations, even though not hitting them, may cause unexpected results in modeling and graphics display.

Display Options

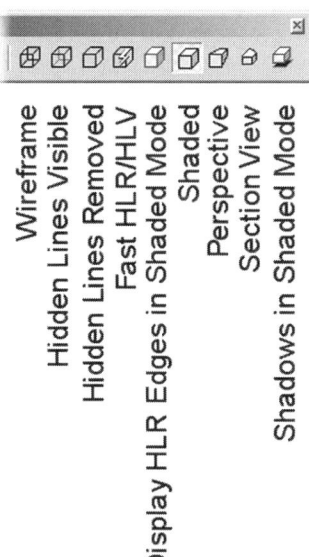

Fig. 3-70. View toolbar display icons.

Most of the main display options are fairly self-explanatory. However, there are a lot of little tweaks and adjustments that control how a model can be displayed in conjunction with the display options. This section should help clarify what all of your options are. The main display icons are located on the right-hand side of the View toolbar, shown in figure 3-70.

Wireframe

In wireframe mode, all model edges are displayed, which is not recommended for most tasks. Keep in mind that the term *wireframe* in the context of display options has to do with display only. The model itself is still a solid model, with all of the characteristics of a solid model, no matter how it is displayed.

Hidden Lines Visible

All edges hidden behind the model will be displayed as dashed or gray lines. Whether gray or dashed depends on a setting in the System Options. How-To 3-13, which follows, takes you through the process of setting whether hidden lines are displayed as solid light gray or dashed gray lines when employing the Hidden Lines Visible display option. Figure 3-71 shows the setting for changing this option.

100   **Chapter 3: Sketching Basics**

Fig. 3-71. Setting how hidden edges should appear.

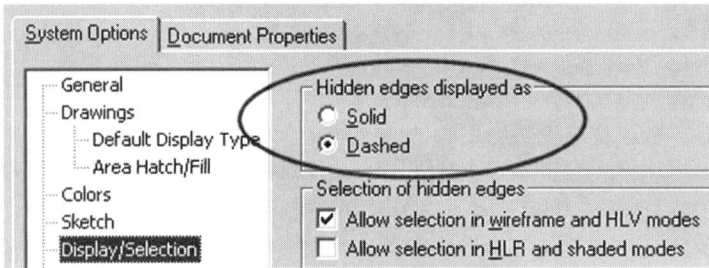

HOW-TO 3-13: Hidden Edge Display Style

To set the style of hidden edges when using the Hidden in Gray display option, perform the following steps.

1. Select Options from the Tools menu.

2. In the System Options tab, select the Display/Selection section.

3. In the *Hidden edges displayed as* section, select either Solid or Dashed.

4. Click on OK.

Selecting Hidden Edges

Whether or not hidden edges can be selected is controlled by another option. The option, also shown in figure 3-71, is titled *Selection of hidden edges*. You can specify whether or not it is possible to select hidden edges when using the Wireframe and Hidden Lines Visible options, and whether or not it is possible to select hidden edges when using the Hidden Lines Removed and Shaded display options. These are both independent toggle switches, and really depend on the user's preference.

✓ **TIP:** *Check only the first option in the* Selection of hidden edges *section. This way, your ability to select hidden edges activates when switching to Hidden Lines Visible display mode.*

Hidden Lines Removed

As the name implies, any model edges hidden behind the model will not be displayed.

Fast HLR/HLV

What this partial acronym stands for is "Fast Hidden Lines Removed/Hidden Lines Visible." What this means in English requires further elaboration. This option is a toggle switch that can be turned on or off. It affects Wireframe, Hidden Lines Visible, and Hidden Lines Removed display modes. When turned on, the model is treated like a shaded model, even though it is being displayed in a wireframe style display mode. This is also known as a facetted display mode, and typically will result in faster graphics display, but at the cost of a less accurate graphics display.

When rotating a model that is being displayed in either Hidden Lines Removed or Hidden Lines Visible display mode, SolidWorks typically degrades the display while the model is being rotated. For example, the model will be temporarily displayed in wireframe mode, and any horizon lines (such as those along the length of a cylinder) will not be displayed at all. When the model stops rotating, the hidden lines are calculated and the model is correctly displayed on screen. One benefit of the Fast HLR/HLV option is that SolidWorks displays the model in the correct state during the rotation process. In other words, you do not have to wait to let go of the mouse button before the model's hidden lines are displayed correctly.

Image Quality settings go hand in hand with the Fast HLR/HLV toggle switch. They control the quality of the model being displayed on screen. These Image Quality settings are explained in the next section.

Image Quality

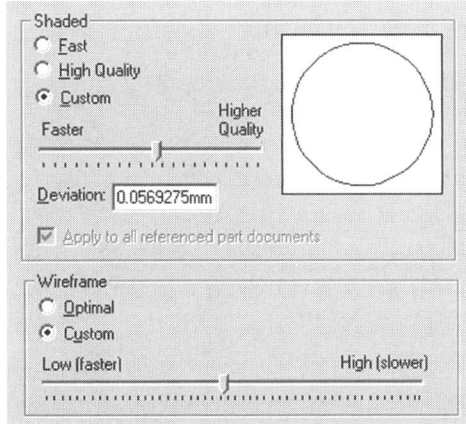

Fig. 3-72. Image Quality settings.

The quality of the model on screen can be controlled for both shaded and wireframe display modes independently. To put it simply, the higher you set the display quality for the model, the more facets it will have (with regard to shaded image quality). The image quality itself is controlled through slider bars, shown in figure 3-72.

To make adjustments to the image quality, access the section titled Image Quality in Document Properties (Tools > Options menu). By clicking on the Custom option, you can gain control of the slider bars. Moving the slider to the left decreases quality, and moving it to the right increases quality. The other settings (such as Fast, High Quality, and Optimal) are just presets. They do not allow for moving the slider bars.

> **NOTE:** Increasing image quality will degrade performance. Find a setting that is a good compromise between quality and performance.

When the Fast HLR/HLV option is turned on, the image quality is controlled via the slider bar in the Shaded section (see figure 3-68). When turned off, the Wireframe slider bar controls the quality of the display for wireframe display modes. Note that when Shaded display mode is used, the model is always facetted and the Wireframe slider bar has no effect.

Increasing the Shaded quality will increase the number of facets used to display the model. (Actually, it decreases the chordal deviation, but increased facet number is a close enough approximation.) This results in curved surfaces appearing smoother. The Wireframe slider increases the accuracy of the wireframe display, as opposed to the number of facets used, but the end result is much the same. In short, wireframe display will look better when Fast HLR/HLV is turned off.

Shaded

Using the Shaded option, the model will be displayed in shaded mode. As described in the previous section, the slider bar found in the Image Quality section of Document Properties will control the appearance of the shaded model. Other options that only affect shaded display mode are discussed in the following material.

✓ **TIP:** *Models saved while shaded make for better previews when opening files.*

Display HLR Edges in Shaded Mode

This option is another toggle switch and only works for shaded mode. When turned on, the edges of the model are highlighted. This includes tangent edges. Any edges hidden behind the model are not shown.

Shadows in Shaded Mode

When toggled on, shadows are displayed as if a light were shining onto the model from above the Top plane. The lowermost physical point of the model determines the shadow plane, but the shadow plane is always parallel to the Top plane. This is a nice effect, but it decreases graphics performance.

Perspective

When perspective is toggled on, a vanishing point will be used to display the model. This gives the model a more lifelike appearance. If Perspective view is enabled, it is possible to alter the vanishing point distance. In other words, the relative distance the model is from the user's eye can be

Display Options

increased or decreased. How-To 3-14 takes you through the process of modifying the degree of perspective.

HOW-TO 3-14: Modifying the Perspective Vanishing Point

To modify the vanishing point distance when in Perspective view, perform the following steps.

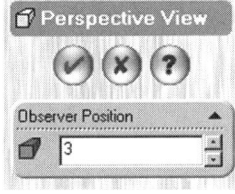

Fig. 3-73. Modifying perspective.

1. Select Modify from the View pull-down menu.
2. Select Perspective.
3. Specify a value for Observer Position, as shown in figure 3-73.
4. Click on OK.

When changing the perspective Observer Position value, a smaller value increases the degree of perspective. The default value is 3. It is possible to type in a value less than 1. This greatly increases the degree of perspective. Increasing the value of the setting has a much more subtle effect.

With regard to enabling the Perspective option, it is not always a good idea to edit the part with Perspective enabled. Depending on the degree of perspective, certain objects may look odd, such as planes or notes and dimensions. Usually, perspective views are used as an enhancement when saving snapshots of the model or creating rendered images.

Section View

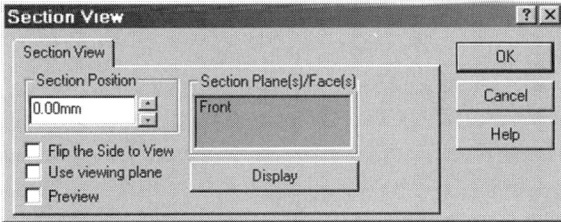

Fig. 3-74. Section View window.

Clicking on the Section View icon opens the Section View window, shown in figure 3-74. You must select a plane that will be used to section the model. By default, this is the Front plane, but you can delete the Front plane from the list box and choose a different plane. Use the Section Position setting to offset the selected plane through a particular location in the part. By clicking on the Display button, you can see what the section view will look like. The Flip the Side to View option creates a section view on the opposite side of the section plane.

The *Use viewing plane* option allows you to use your computer screen as the section plane. The Preview option will automatically update the display as adjustments are made to the Section View window. Do not check Preview if you want manual control (by clicking on the Display button) when the section view is updated.

It should be mentioned that more than one plane can be used when creating a section view. Planar faces can also be used. If more than one plane or planar face is selected, it is even possible to specify an offset distance for each plane or face individually.

✓ **TIP:** *Select a plane prior to entering the Section View command and the selected plane will be used instead of the Front plane.*

Once a section view has been established and you click on the OK button to close the window, the section view remains. To turn off the section view, simply click on the icon again or uncheck the Section View option in the pull-down menus (View > Display > Section View). To modify an existing section view, select Section View from the View > Modify menu.

Do not confuse the Section View function with creating section views in a drawing. The procedure for creating design drawing section views is much different and is covered later in the book. If you need to read up on creating drawing section views at this time, see Chapter 11.

How you decide to view the model as you work is strictly a user preference. It is recommended you use whatever is easier for you. Most people will probably find switching between the Shaded and Hidden Lines Visible options the most beneficial.

Summary

You have learned a great deal in this chapter. To begin with, you should be feeling more comfortable with all of the sketch entity types and how to sketch them. Delete unwanted sketch geometry by selecting it and pressing the Delete key. Drag blue geometry to help determine what constraints have been added to the sketch, or to determine what constraints need to be added.

You have also seen firsthand how to add and remove constraints and dimensions. Double click on sketch entities to view the constraint callouts. Use the SolidWorks color codes to determine if your sketch is underdefined or fully defined. Fully constrained geometry will appear black, and underdefined will appear blue. It is always good practice to fully define sketch geometry because it gives tighter control over what is happening in the sketch. If a sketch is not fully defined, its behavior is not predictable and your design intent has not been incorporated into the sketch.

It is okay to use centerlines and points as reference geometry, if the need arises. Centerlines and construction lines are one in the same as far as SolidWorks is concerned. Construction geometry does not contribute to solid geometry when creating features.

Keep sketch geometry simple. The sketch will be easier to maintain and control that way. It is better to have a larger number of features and less complex sketch geometry. This makes for a flexible model that is more editable. Due to the nature of geometric relationships, an overly complex sketch can be very difficult to maintain. It can also make troubleshooting a difficult task if something goes wrong while making design changes.

Use hot keys and mouse shortcuts to be more productive. For example, the F key will fit the model to the screen, and the middle mouse button will rotate the model. Work in any view you find convenient. Save user-defined views using the Orientation window to make it easy to return to a particular view.

Use a display mode that suits your needs, yet makes it easy to select geometry being edited. Hidden Lines Visible allows for seeing the part clearly and makes it easy to pick hidden edges on the other side of the model. Shaded mode is a good choice as well, and by adjusting the image quality it is possible to find a good setting that allows for reasonable graphics performance while maintaining a decent image quality on screen.

Questions and Topics for Discussion

1. What key do you hold down when you want to select more than one entity? Do you need to hold down this key to select multiple objects while in a command?

2. Name the three icons used to create arcs.

3. What is meant by the term *auto-transition*?

4. What are callouts, and how can they be accessed?

5. What is the significance of the *Input dimension value* option?

6. How would you turn a circle into a construction circle?

7. Why is it good practice to fully define sketch geometry?

8. Describe what happens if you do not select a plane prior to clicking on the Sketch icon when creating your first sketch?

9. What hot key would you use to scale the model to the screen?
10. What function(s) does the middle mouse button accomplish?
11. Describe how to get to the option for controlling image quality.
12. How can a dimension preview be locked in when positioning a dimension?
13. Describe the difference between click-drag and click-click sketch modes.

Optional Problem

Create a sketch that contains one each of every object in the Sketch Entity menu (except for the Route Lines and Text options, which have not been discussed).

Chapter 4

Castings

CAST PARTS CAN BE CREATED EASILY using the SolidWorks Extrude command. The Extrude command also happens to be one of the most basic SolidWorks feature types, so cast parts make an excellent place to begin working with SolidWorks.

In the last chapter, you learned about sketching, adding and deleting constraints, and the basics of adding dimensions. Throughout this book, you will be applying what you have learned in previous chapters and expanding on that information. Portions of this book are dedicated to applying the knowledge you have gained directly to parts you will create in exercises within each chapter.

The layout of this book is designed with the student in mind. Much more can be absorbed and retained when the student is trying firsthand the topics covered in the chapter. People have a tendency to wool gather or even fall asleep when an instructor does nothing but lecture. Any person who has attended college knows this to be true, unless the instructor is speaking on an interesting topic or is simply an outgoing and energetic person. Lab exercises are much more interesting. Therefore, let's start with a quick review, examine extruded base features briefly, and then move on to your first exercise.

Getting Started: A Quick Review

Before getting into the creation of cast parts, it would be beneficial to reiterate a few of the important topics covered in the previous chapters with regard to sketching. Most of these topics apply to all sketched features, not just extrusions, but now is still a good time to reinforce these important issues.

Determine the Best Profile

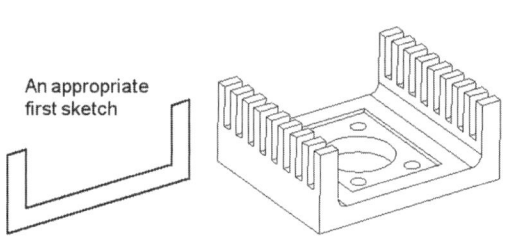

Fig. 4-1. A logical first sketch for the heat sink part.

Determining the best profile goes hand in hand with selecting the appropriate starting plane. As a general rule, select a profile that will describe the overall shape of the part without being overly complex. When it comes to sketch geometry, keep it simple. Simple sketch geometry is easier to manage and will provide more control when performing edits later on, because there will be fewer relationships between sketch geometry that can conflict with each other.

Examine figure 4-1. Note that even though there are a number of features that make up this part, the sketch used to create the base feature is actually quite simple. Incidentally, the sketch for the base feature was created on the Right plane. This makes perfect sense if the final orientation of the heat sink should appear as it does in the isometric view shown.

Select the Appropriate Plane

Before creating a sketch for the base feature, decide which side of the part would be best suited to represent the part's top, front, and right sides. Keep this in mind when selecting the initial sketch plane. This will aid you in the creation process later on, because you will have already determined in your mind's eye the correct orientation of the part. The part will also be oriented correctly when it comes time to generate a 2D drawing of the model, which is probably the most important reason for starting off on the proper plane.

Fully Define the Sketch

SolidWorks does not require that you add dimensions and constraints, but it is extremely prudent to do so. Assume for a moment that you are creating a part and are unsure of what the dimensions should be. All you know is the approximate size and shape, so you leave the dimensions off. Two days later, you work out what the final dimensions should be. Now you must edit all of your sketches and add the dimensions.

In another scenario, assume you add the dimensions and use arbitrary dimension values. In this situation, you no longer need to edit the sketch to add the dimensions. All that is necessary is to access the dimensions with a simple double click. The point is that dimensions cannot be accessed if they have not been added in the first place, and fully defining a sketch is impossible without them. Usually it is best to add dimensions when creating a sketch and be done with it, even if you are unsure of the value. Without geometric relations, it is impossible to predict how a sketch will behave.

✓ **TIP:** *Use dimensions and constraints to your advantage, and incorporate your design intent into the model.*

The easiest way to see what constraints are being added while you sketch is to pay attention to the system feedback. Additionally, use the left mouse button to try to drag underdefined geometry. It is amazing how much information you can deduce from dragging entity endpoints.

✓ **TIP:** *SolidWorks communicates through system feedback. Always watch the cursor!*

Creating Features

There are four basic types of sketched features. Features can be created by extruding a sketch, revolving about a centerline, sweeping a sketch along a path, or lofting (blending) two or more closed profile sketches. The names for these commands are Extrude, Revolve, Sweep, and Loft, respectively. Any one of these sketched features can be a boss or a cut; that is, they can either add or remove material.

All features fall into two categories: sketched and applied. Sketched features require a sketch. This is the feature type for which you will need the most practice. Applied features are applied directly to the model and do not require a sketch. Therefore, there is typically much less manual labor involved in creating an applied feature.

Base Features

The very first feature created in a part file is called the base feature, and nearly every SolidWorks solid model starts with one. The base feature should represent the general overall shape of the part. With regard to sketched features, there are five steps you will have to perform over and over in order to create a feature and build up the model. These five steps are summarized in the sections that follow. Remember them well, because you will be following them often.

Select a Plane or Planar Face

You must always have a plane on which to sketch. This can be one of SolidWorks' default planes or one that you create. It can also be a planar (flat) face of existing geometry.

Enter Sketch Mode

Enter sketch mode by simply clicking on the Sketch icon, shown in figure 4-2. One click with the left mouse button is all it takes.

Fig. 4-2. Sketch icon.

Create the Sketch

Most of the mechanics involved in creating a sketch were covered in the previous chapter. If you need help creating sketch geometry, review Chapter 3.

Add Dimensions and Constraints

Adding dimensions and constraints is not a requirement, but is good practice. Adding constraints helps achieve your design intent on the part. Adding the dimensions is good technique simply because you cannot fully control the model without them. Additionally, a sketch that is not fully defined may inadvertently change, producing unwanted results.

Create the Feature

Creating features is what you will be spending a lot of time on from this point forward. There are many options, and many features that can be created in SolidWorks. This book will show you a large majority of those feature types and how to create them.

These five steps to creating a sketched feature are a generality. You would not perform all of these steps when creating, for example, an applied feature, because no sketch is required. This should become self-evident to you over time.

After the first sketch has been created and you are ready to turn that sketch into a solid feature, you will notice that the Insert menu contains a submenu named Base. This is where you will look to create the first base feature. Once the first base feature has been created, this menu item changes to the word *Boss* and remains Boss from that point forward.

Once the first base feature has been created, building up the rest of the model is primarily a matter of adding more features to the model. These features may be sketched or applied, and may add material or remove material. The earlier chapters in this book deal primarily with extrusions and revolved features. Sweeps and lofts are covered in more detail later in the book, as they are generally more complex feature types.

A detail you may notice regarding the menu structure is that only certain menu items are available at a given time. This is because SolidWorks

Creating Features

knows what tasks you can complete with the geometry contained within a sketch at any given time. For example, revolved features require a centerline in the sketch. If there is no centerline present, SolidWorks will not make the Revolve menu item available. Menu items will be grayed out if appropriate conditions have not been met.

When creating your first base feature, normally the sketch you just created is open (active). In other words, you are editing the sketch. When this is the case, SolidWorks assumes the open sketch is the one you want to use to create the feature. If a sketch is not currently open, you must select the sketch before SolidWorks will allow you to extrude or revolve it. This is done by selecting the sketch from FeatureManager.

Extrusion End Conditions

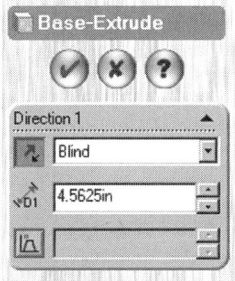

Fig. 4-3. PropertyManager parameters for an extrusion.

When creating an extruded feature, it is important to specify how you want to terminate the extrusion. In other words, how you want to specify the conditions that will determine how long the extrusion will be. In its simplest form, you would specify a distance for the extrusion and the sketch will be extruded that distance. This is known as a Blind end condition. All end conditions are specified via PropertyManager, shown in figure 4-3.

There are numerous end condition types, not all of which are always available. Examples of the various end condition types are shown in figure 4-4, which gives an idea of what each end condition type accomplishes. The image is just to show some quick examples. All end condition types will be explored throughout this book, so do not feel it is necessary to commit this information to memory just yet.

Fig. 4-4. Various end condition types and their results.

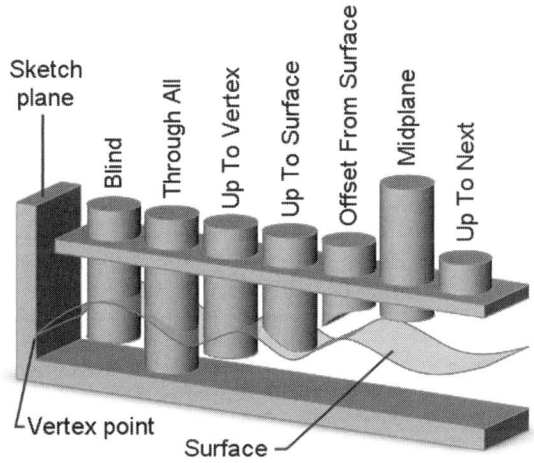

The Pivot Arm: An Overview

Fig. 4-5. Pivot arm part.

Most cast parts incorporate draft, and that will be one of the main points of focus in this chapter, along with basic feature creation. Other topics you will be introduced to include cut extrusions, renaming features, the Hole wizard, mirroring, and trimming and extending sketch geometry. Various end condition types in the Extrude panel will also be explored. At the end of the exercise you will also see how easy it is to make design changes on a completed part. This will include modifying dimension values and editing definitions of features. The pivot arm you will create in this chapter is shown in figure 4-5.

This part will start like any other. A new file must be created and the working units should be established. The pivot arm will be created using inches as the unit of measurement. If you have forgotten how to do this, the exercise that follows will guide you through this process. As always with a base feature, you must begin by selecting the sketch plane, entering sketch mode, and creating the sketch that will be used to create the base feature.

The pivot arm's base feature and a likely plane on which to create its sketch need to be determined. Looking at this part, it is fairly safe to say that the main body of the part would be a good candidate for the base feature. Remember to keep the sketch geometry simple. Try to break the part up into its individual feature components.

✏️ **NOTE:** *Boss or cut extrusion directions are always perpendicular to the profile plane, without exception.*

The Mid Plane End Condition

Cast parts are created (cast) from a mold. The mold itself can be of a wide variety of types. The mold might be of the basic metal variety, or it could be a wax mold, a sand mold, or any of the numerous techniques used for creating cast parts. The point is that there are many ways to create a cast part, but the specific casting technique or casting process used typically does not matter when it comes to creating a cast part in SolidWorks.

Many cast parts will contain draft so that they will come out of the mold more easily. If you are not familiar with the term *draft*, the following simple example should help. In figure 4-6, one part contains draft and the other does not. Which do you think would come out of the mold more easily?

The Pivot Arm: An Overview

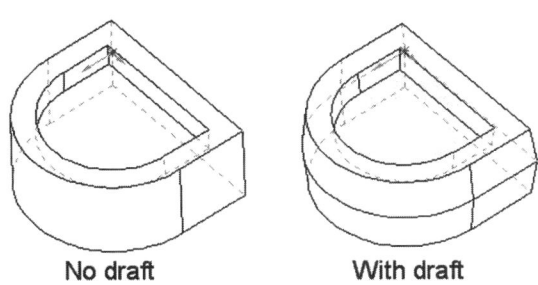

Fig. 4-6. Two parts, one with draft and one without.

The line where the draft begins is known as the parting line. This is where the mold splits in two and pulls apart. Creating the actual molds is covered in detail in Chapter 13.

The Mid Plane end condition type allows a sketch to be extruded in opposite equidistant directions simultaneously. Combine this with draft during the extrusion process to get a good start on the creation of a cast part. Using the Mid Plane end condition and adding draft simultaneously creates the parting line for you.

Exercise 4-1 shows you how to create the pivot arm part that incorporates the topics previously discussed. The Mid Plane end condition and draft will be added simultaneously in this exercise.

EXERCISE 4-1: Creating the Pivot Arm

This exercise steps you through the process of creating the pivot arm. You will begin by starting a new part and selecting the appropriate units to work with. You will then create the sketch and extrude it using the Mid Plane end condition type while adding draft to create the base feature.

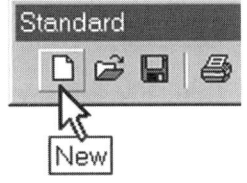

Fig. 4-7. The New icon.

1. Select File > New, or click on the New icon (see figure 4-7), to begin a new SolidWorks document.
2. Select the Part template and click on OK.
3. Right-click on the Front plane and select Show.
4. Click on the Sketch icon.
5. Create the sketch shown in figure 4-8. Add dimensions and constraints as needed to fully define the sketch geometry.

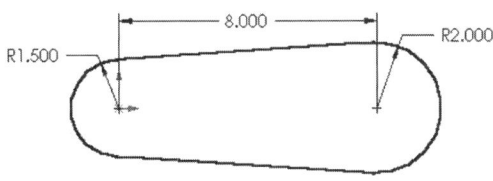

Fig. 4-8. Sketch profile for the base feature.

It is not necessary to show the sketch plane in order to sketch on it, but sometimes it helps to maintain your bearing. If you do not feel as though you need to see the plane you are sketching on, do not feel obligated to show it. It was also not necessary to select the Front plane prior to sketching on that plane. This is because SolidWorks will default to the Front plane if you do not pick one. A reminder: this default action only occurs when creating the first sketch. Afterward, you will always be required to select a plane or planar face either before or after clicking on the Sketch icon.

It will be necessary to attach the sketch to the origin point before it becomes fully defined. You may have to drag the center of the left arc to the origin. This will establish a coincident relation to the origin, thereby anchoring the sketch. If the sketch is black (fully defined), you are ready to create the base feature. The sketch you just completed should still be active, meaning that the Sketch icon should still appear depressed. There should also be only one sketch listed in FeatureManager. If this is not the case, start a new part and try again. Otherwise, continue with the following steps.

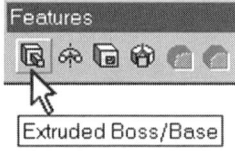

Fig. 4-9. Extruded Boss/Base icon.

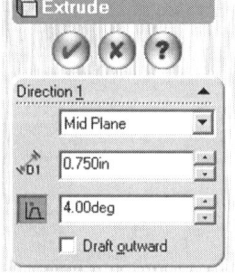

Fig. 4-10. PropertyManager settings for the base feature.

6. Select Insert > Base > Extrude, or click on the Extruded Boss/Base icon, shown in figure 4-9.

7. In PropertyManager, change the end condition from Blind to Mid Plane.

8. Set the depth to .75 inches.

9. Click on the Draft button and set the value to 4.00 degrees. Make sure *Draft outward* is not checked. Use figure 4-10 for guidance.

10. Click on OK (the green check). The part should look as it appears in figure 4-11.

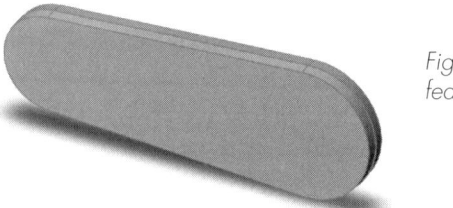

Fig. 4-11. Pivot arm base feature.

11. Make sure to save your work at this point, as the pivot arm will be used throughout this chapter. Select File > Save, or click on the Save icon.

The Pivot Arm: An Overview

12. Navigate to the directory of your choice and type *Pivot Arm* as the name for the file.

13. Click on the Save button to save your work.

A Common Mistake

NOTE: *Skip this section if you have successfully completed the first portion of the pivot arm exercise.*

There are some common mistakes new users have a tendency to make. One such mistake is the tendency to accidentally exit out of the sketch before the feature is created. To complicate things, a second mistake is to accidentally start a new sketch when in reality it is the original sketch the user should be editing. Read on, and chances are you will understand why you are having difficulty completing the pivot arm base feature. The topic of editing a sketch is covered in greater detail immediately following this section.

First, if your model does not look the same as the model pictured, it could be that you are simply not seeing the same view. This is not really important to the model, it just alters your perspective of the model. Change to a trimetric view using the Orientation window (discussed in Chapter 2).

Next, ask yourself the following questions. Is some or all of the sketch geometry gray instead of blue or black? Do you see more than one sketch in FeatureManager? If so, perform the steps outlined in How-To 4-1.

How-To 4-1: Reentering a Sketch

To reenter a sketch, perform the following steps.

1. Click on the Sketch icon to exit out of the current sketch, if you have not already done so. If the Sketch icon does not look "pushed in," you are not currently in a sketch.

2. You need to edit the sketch you were originally in (typically the first sketch listed in FeatureManager). Right-click on the sketch in FeatureManager and select Edit Sketch. Now the sketch is active and you can edit it.

Did you remember to add draft? Did you use Mid Plane as the end condition? Would you like to double check the dimensions you placed in the sketch? If so, the section that follows will help you do all of this and more.

Editing Techniques

The following material is critical to being a productive SolidWorks user. One of SolidWorks' main strong points is the ease with which you can edit existing material. This section shows you how to implement a few of SolidWorks' most basic editing commands. Make it a point to learn this material well.

Editing a Sketch

There are times when you will find it necessary to get back to the sketch used to create a feature in the first place. Sketch geometry is hidden and absorbed by the feature it was used to create. The reasons for editing a sketch will vary, but the process is always the same. The right mouse button comes into play for this procedure, as it does quite often while working with SolidWorks. To edit an existing sketch, perform the steps outlined in How-To 4-2.

How-To 4-2: Editing an Existing Sketch

To edit an existing sketch, perform the following steps.

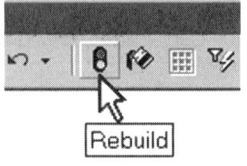

Fig. 4-12. Rebuild icon.

1. In FeatureManager, right-click on the sketch you want to edit.
2. Select Edit Sketch.
3. Make the desired changes.
4. Select Edit > Rebuild, or click on the Rebuild icon (shown in figure 4-12).

Be aware that if you exit out of a sketch, whether on purpose or by accident, the sketch geometry will turn gray. When preparing to create an extruded or revolved feature, it is not necessary to exit out of the sketch. When the feature is created, the sketch geometry is absorbed by the feature and the sketch is exited automatically. If you see gray sketch geome-

Editing Techniques

try, it could mean you have accidentally exited the sketch and you should edit it in order to reenter that same sketch.

Hot key = Ctrl + B

Rebuilding a Model

Anytime a dimensional change is made to your SolidWorks model, or if a sketch has been edited, the model must be rebuilt. Rebuilding the part is how you tell SolidWorks to take into consideration the changes you have made. This does not necessarily mean you have to rebuild after every change. For instance, it is perfectly acceptable to modify multiple dimensions before rebuilding the part.

If you are in the process of editing a sketch, there are a number of ways in which you can exit that sketch and rebuild the part simultaneously. Each of the following is a method of exiting a sketch.

- Click on the Sketch icon.
- Right-click in an empty area of the work area and select Exit Sketch.
- Uncheck Sketch from the Insert menu.
- Select Rebuild from the Edit pull-down menu.
- Press Ctrl + B (the hot key combination for Rebuild).
- Click on the Rebuild icon.

Of the previous options for exiting a sketch, the last is one of the most commonly used. However, use whatever method is easiest for you. Any of these options will rebuild the part and exit you out of a sketch (assuming you are currently editing one). If you are not currently editing a sketch, rebuilding the part will do simply that. If you have made any changes whatsoever, they will be taken into account when the part is rebuilt.

What does rebuilding the part actually accomplish? Technically speaking, the modified portion of the part file's database is recalculated. The feature data is read from memory and processed by your computer using the part's current dimensional values and attributes. In simpler terms, SolidWorks recreates the part using the latest dimensional values you gave it.

Whenever a dimension is modified, sooner or later you should rebuild the part. This includes dimensions associated with a feature, sketch, or plane. It is acceptable to postpone rebuilds if there is more than one dimensional change to make. When you are finished making those changes, however, rebuild the model.

If you make the mistake of forgetting to rebuild the model after making dimensional alterations, SolidWorks will eventually prompt you to rebuild when you attempt to save the file. This warning prompt, shown in figure 4-13, should be taken seriously. Not rebuilding the model before saving is asking for trouble. If your modifications caused problems, you will not know it until you open the part at a later date. If someone else is opening your model, they may be in for a nasty surprise. Upon opening, the model will be rebuilt, and any errors made in your previous alterations will become apparent.

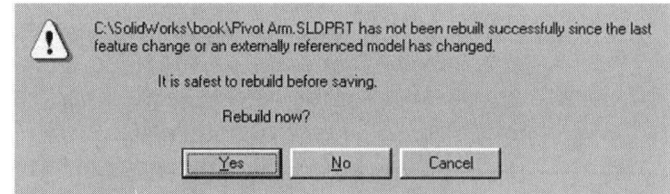

Fig. 4-13. A warning to rebuild the part before saving.

Certain operations will automatically rebuild the part, such as when editing a feature's definition. There is also an option to regenerate a model, as opposed to rebuild. Editing a feature's definition is covered in material to follow. For now, read on to understand the difference between a rebuild and a full regeneration.

Hot Key = Ctrl + Q

Regenerating a Model

When performing one of the previously mentioned steps to rebuild a part, the model is not really completely rebuilt. In other words, the part file's entire database is not completely recalculated. Only the features marked as needing a rebuild by the software are recalculated. In contrast, it is possible to fully regenerate the entire part database by forcing a full regeneration. A full regeneration takes longer than a rebuild, so be forewarned: if the model is complex, you will have to wait a little longer.

It is usually not necessary to do a full regeneration. Sometimes, however, it can serve a useful purpose. If for some reason your model does not seem to rebuild correctly, try performing a regeneration. This happens rarely and is really nothing to be concerned with at this time, but is worth mentioning.

Editing a Definition

The mechanics of editing a definition are exactly the same as for editing a sketch. All it requires is the right mouse button. Why would you want to edit a definition? The definition defines what the sketch is doing. If editing the sketch is not giving you enough options, chances are that editing the definition will.

Editing Techniques

The definition of a feature defines what was done to the sketch in order to create the feature. In the case of an extrusion, that definition includes information regarding the extrusion's end condition, such as the depth of the extrusion. The definition also includes data on whether or not the extrusion contains draft, and if so, how much and in what direction. Any or all of the feature definition parameters can be edited.

All feature types (whether extrusions or revolved, swept, or lofted features) contain a definition. For that matter, both sketched and applied features contain definitions. This means that features such as fillets and chamfers also have definitions.

No matter what the feature type, the mechanics of editing a feature's definition are exactly the same. This is one of the advantages of the software. Consistency in menu structure and in functionality makes SolidWorks much easier to learn and use. How-To 4-3 takes you through the process of editing a feature's definition.

HOW-TO 4-3: Editing a Feature's Definition

To edit a feature's definition, perform the following steps.

1. In FeatureManager, right-click on the feature to be edited.
2. Select Edit Definition.
3. Make the desired changes.
4. Click on OK to accept the changes.

Of course, you can always click on Cancel in PropertyManager if you change your mind about making any changes. This holds true for nearly all SolidWorks functions, not just when editing a feature's definition.

When you click on OK to accept the changes you made to a feature's definition, the part is automatically rebuilt. This is a safe assumption on SolidWorks' part, because if you did not want to make any changes you would not have edited the definition in the first place. SolidWorks must rebuild the part if you are to see the changes take place.

Modifying Dimensions

This is the easiest to implement of the editing options previously mentioned. If you do not need to make any modifications to sketch geometry, and you do not need to change the definition in any way, it is a simple matter to modify the dimensions of a part. Perform the steps outlined in

120 Chapter 4: Castings

How-To 4-4 to change the value of any dimension associated with any object in SolidWorks.

How-To 4-4: Changing a Dimension Value

To make dimensional changes, perform the following steps.

1. Using the left mouse button, double click on the desired object in FeatureManager to access its dimensions. The object can be a sketch, feature, or anything else.

2. In the work area, double click on the dimension to be modified.

3. Specify a new value for the dimension, and press Enter or click on the green check.

4. Rebuild the model to see the changes you have made.

✓ **TIP:** *You can modify more than one dimension on more than one sketch or feature before rebuilding the model. This can save time waiting on rebuilds.*

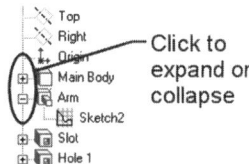

Fig. 4-14. The feature tree can be expanded and collapsed.

Click on the plus sign (+) located to the left of any sketched feature to see the sketch that belongs to that feature (see figure 4-14). This is known as expanding the feature tree. The same can be done with folders (directories) in any Windows operating system to see the files that folder contains. Collapsing the feature tree is done the same way, except that there will be a minus sign (–) instead of a plus sign.

If you double click on a sketch, you will see the dimensions associated with that sketch. Double clicking on a feature shows the sketch dimensions as well as the dimensions associated with the feature definition. Feature dimensions are added by SolidWorks, not the user. Take an extruded sketch as an example. If a sketch were extruded 3 inches with 5 degrees of draft, those 3-inch and 5-degree dimensions would be added to the model automatically.

The reason this is being pointed out to you is to make it easier to find a particular dimension when the need arises. Dimensions added by you during the sketch process will be color coded black. Any dimensions added by SolidWorks automatically during the feature creation process will be color coded blue. This should not be confused with the color codes associated with underdefined or fully defined sketch geometry. Dimension colors are just a convenience to make it easier to find certain dimensions when modifications need to be made. Do not read any more into it than that.

Renaming Features

Renaming features is a simple process that was discussed in Chapter 2. As a reminder, nearly anything in FeatureManager can be renamed using a slow double click. If you are already familiar with renaming files in Microsoft Windows, you will know how to rename objects in SolidWorks.

When the list of features in FeatureManager grows long, as it has been known to do, it is sometimes very difficult to find the feature you are looking for. Renaming features makes it that much easier to find them later. This makes life easier for you, as well as others who may be editing the same part at a later date.

Renaming features is certainly not a requirement, but it is a good idea, especially if a model is going to contain many features. Even if you only rename a few of the more critical features in the part, it will help you six months from now to remember where things are and how you went about building the part. Common practice is to not worry about renaming every item, such as fillets and sketch geometry, in FeatureManager.

In certain cases, it is desirable to rename sketches, and even dimensions. The reasoning behind this has to do with how dimension names can be used in an Excel spreadsheet or in equations. This is material covered in chapters 16 and 17, which deal with the creation of design tables and equations. What it all really boils down to, though, is that renaming FeatureManager objects (or even dimensions) is a way to document how a model was built or what is being done inside a model file. It makes it easier to find what you need to edit.

The Blind End Condition

The Blind end condition is one of the more common and easier to understand of the end condition types. It means that the sketch you are extruding will continue blindly in one direction for whatever distance you specify. You will see this end condition used in many of the exercises of this book.

The Up To Surface End Condition

There are a number of end condition types that allow you to specify an extrusion distance by relating that distance to a surface or plane. The surface does not necessarily have to be planar (flat). In other words, the surface can be curved or some sort of wavy, free-flowing shape.

The Up To Surface end condition allows you to extrude a boss or cut up to a specified surface. This is convenient in many cases, especially if it incorporates the design intent you are looking for. In the case of the pivot arm, the arm containing the slot should always be flush with the outer surface of the main body on one side. The following material explores why Up To Surface would accomplish this task.

The base feature of the pivot arm is currently .750 inches wide. The slotted arm that protrudes at an angle from the base extrusion is only half as wide as the base extrusion. If the sketch for the slotted arm is made on the same plane as the base extrusion, it could be extruded to half the distance (.375) for the desired results. However, what if the main feature's .750-inch depth is altered? The slotted arm would still be at .375 inches, which is not the intent. The slotted arm is flush with the back of the pivot arm part, and this is the way it should stay, regardless of the base feature depth.

This is where Up To Surface would be beneficial. By sketching the slotted arm on the same plane as the base feature, the sketch can be extruded up to the surface of the back of the pivot arm. This condition will be maintained by SolidWorks, no matter what the thickness of the base feature.

In Exercise 4-2, you will continue development of the pivot arm to see how all of this comes together. You will also edit some definitions to see how easy it is to make design changes in SolidWorks.

EXERCISE 4-2: Developing the Pivot Arm

To further develop the pivot arm, perform the following.

1. Open the pivot arm model created in the previous exercise. Perform a slow double click (with the left mouse button) over the base-extrude feature while in FeatureManager and rename the feature *Main Body*. Press Enter to accept the new name for the feature.

Now it is time to create the next feature. The slotted arm must begin near the center of the main body and be flush with the rest of the model on the back. The pivot arm is shown in figure 4-15.

To complete the next feature, you will first need to select a plane to sketch on, and then create the sketch for the slotted arm. To complete this process, continue with the following steps.

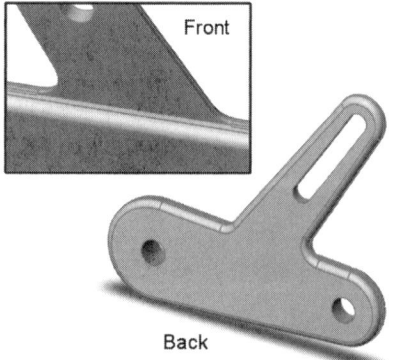

Fig. 4-15. Note how the arm attaches to the main body.

2. Select the Front plane.
3. Click on the Sketch icon.
4. Change to a view option under Front.
5. Complete the sketch as shown in figure 4-16.

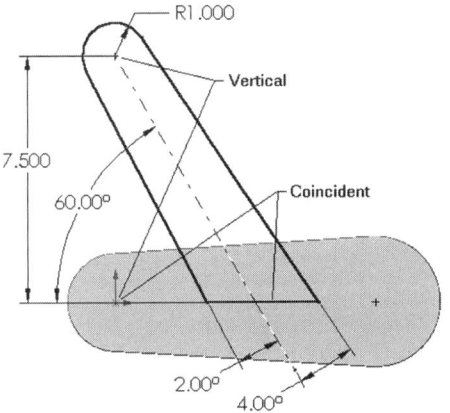

Fig. 4-16. Sketch for the arm feature.

Figure 4-16 shows almost everything you need to know to complete the sketch. You may have to add some relations on your own. Two of these relations are pointed out, along with the entities the relations should apply to. Be careful when selecting entities, as it is possible to select either a line or its endpoints, and it makes a difference.

Once you have the sketch completed, you are ready to create the boss extrusion. Make sure your sketch geometry is fully defined. If your sketch still contains geometry that is blue, fully define the sketch before proceeding. To create the arm feature, continue with the following steps.

6. Select Insert > Boss > Extrude, or click on the Extruded Boss/Base icon.
7. If necessary, change to a view such as Isometric so that you can see the feature preview better.
8. Specify a Blind end condition and a Depth of .375 inches (you will change this later).
9. Click on OK to accept the settings.

The Blind end condition works fine, but you will now make some design changes to see how a poor choice made while defining a feature (specifically, the end condition) can adversely affect the model. Continue with the following steps.

10. Rotate the part so that you can see it from the back.
11. Double click on the Main Body feature option in FeatureManager to access its dimensions.
12. Double click on the .750 dimension, indicated in figure 4-17.

Chapter 4: Castings

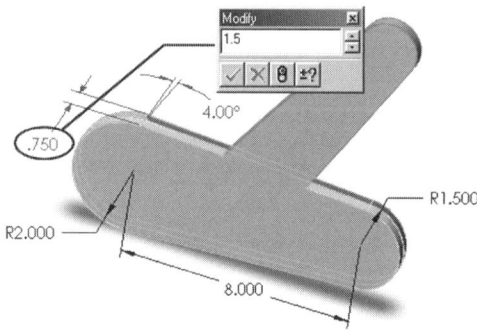

Fig. 4-17. Modifying a dimension's value.

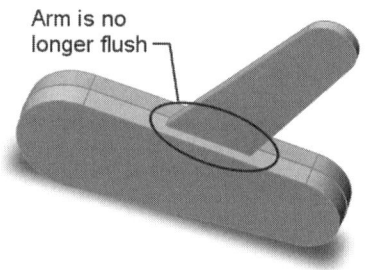

Fig. 4-18. Pivot arm after a rebuild.

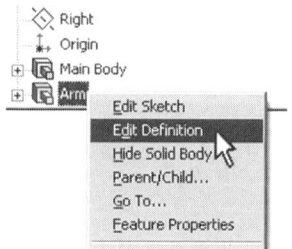

Fig. 4-19. Editing the definition of the arm feature.

13. In the Modify box that appears (see figure 4-17), type in the value *1.5*.

14. Click on the green check or press Enter to accept the change.

15. Click on the Rebuild icon.

The rebuilt pivot arm is shown in figure 4-18. You will next modify the end condition of the boss extrude to use the Up To Surface end condition. When using this end condition, a surface to extrude up to must be selected. You will also add some draft to the arm feature. The part will be rebuilt automatically upon accepting the modifications. Begin by renaming the boss-extrude feature.

16. Rename *Boss-Extrude1* as *Arm*.

17. Right-click on the arm feature and select Edit Definition (see figure 4-19).

18. Change the end condition from Blind to Up To Surface.

19. Select the surface shown in figure 4-20.

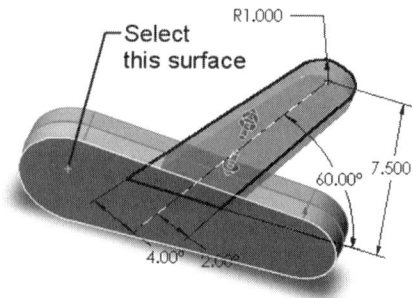

Fig. 4-20. Selecting the surface to extrude up to.

20. Click on the Draft button and specify 4 degrees, as shown in figure 4-21.

Editing Techniques

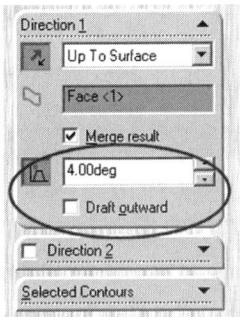

Fig. 4-21. Enabling draft.

21. Make sure Draft Outward is not checked.

22. Click on OK to accept the changes.

23. Double click on the main body feature. (You are going to reset the extrusion depth of the Main Body feature option back to its original value to see if the back face remains flush, per the design intent.)

24. Double click on the 1.500-inch dimension and change it to .750 inches. (Do not accidentally change the radial dimension.)

25. Press Enter to accept the change.

26. Rebuild the part.

27. Save your work.

Was the design intent maintained this time around? The answer to that question should be yes. The back of the pivot arm should once again be flush with the main body of the part. Now you have seen firsthand how easy it is to modify dimensions or edit the definition of a feature within SolidWorks. In the material that follows, you will continue building the pivot arm toward completion.

The Through All End Condition

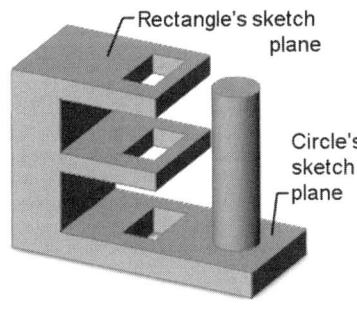

Fig. 4-22. Using the Through All end condition.

Another often-used end condition type is Through All. No surface selection is required when performing an extrusion using Through All. Cuts are typically performed when using this option, but bosses can also take advantage of the Through All end condition.

Technically, when using the Through All end condition, the extrusion continues until it makes its way up to the last surface in the model. The surface does not have to intersect the extrusion. This may sound somewhat contradictory. However, figure 4-22 should make it clear.

In figure 4-22, a rectangle was sketched on the top surface. An extruded cut operation was then used to cut the rectangle down through the part. A circle was sketched on the bottom surface, as shown in the figure. Note that the circle was created in a separate sketch. (This was necessary, of course, as the second sketch is on a different planar face.) An extruded boss operation was used on the circle to extrude it upward the entire length of the part. In both cases, the Through All end condition was used.

Cutting with an Open Profile

It is possible to use an open profile to cut a part. However, the only end condition available in such a situation is Through All. Think of sawing a piece of wood in half with a hand saw. If you were to saw only halfway through, you would still have one piece of wood. Likewise, stopping halfway through a solid model would confuse SolidWorks. You must slice through all in order to get two pieces.

When using an open profile to create a cut, one side of the part must be discarded. Although it is possible to have multiple solid bodies in a part file, the Split command would be required if you wanted to keep both halves of the part after the cut was completed. This is explored in more detail in Chapter 13. In Exercise 4-3 you will continue building the pivot arm. The Through All end condition will be used to create the slot and one of the holes. You will also get some good practice with adding geometric relations. Open the pivot arm model and follow along.

EXERCISE 4-3: Cutting Holes in the Pivot Arm

Begin this exercise by opening the pivot arm part file, if you have not already done so. Rotate the part so that you can see the back flat face of the part. This is the face you should select for the next sketch plane.

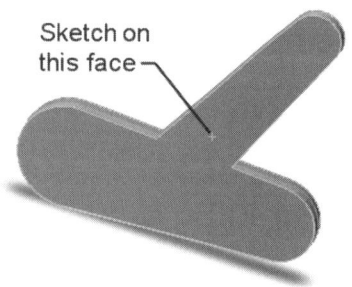

Sketch on this face

Fig. 4-23. Selecting a planar face to sketch on.

In this exercise, you will first create the slot on the arm, and then one of the holes on the main body. Use the steps that follow if you need guidance accomplishing this task.

1. Select the face on which to begin sketching. This is shown in figure 4-23.
2. Click on the Sketch icon.
3. Create the sketch for the slot. Use figure 4-24 for guidance.

Editing Techniques 127

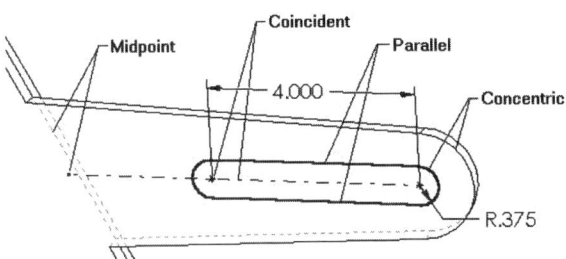

Fig. 4-24. Sketching the slot.

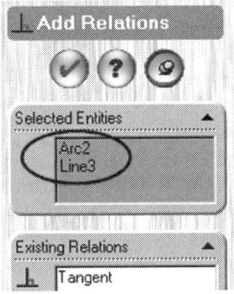

Fig. 4-25. The Selected Entities list box will list what has been selected.

Some of the constraints in this sketch may be tricky for new users. Pay special attention to what is selected before adding constraints. If you do not have the correct items selected, the correct constraint will not be available. The Add Relations display panel in PropertyManager will show you what is selected. Pay close attention to the Selected Entities portion of this panel, shown in figure 4-25.

A Note on the Midpoint Constraint: One of the constraints that causes new users a particular amount of grief is Midpoint. When adding a midpoint relation between a line and a point, select the line, not the line's midpoint. SolidWorks will automatically find the midpoint for you.

With regard to the midpoint constraint you will be adding in the exercise (see figure 4-25), you will need to select the endpoint of the centerline and the edge hidden behind the arm feature. This is a perfectly acceptable maneuver. Constraining your sketch geometry to an object that is not on the same plane as the sketch will cause the object to be projected to the sketch plane. In other words, SolidWorks sees the objects in a plan view and will constrain them as if they are on the same plane. This holds true for many relationships, including coincident, concentric, collinear, and of course, midpoint.

When you do find it necessary to constrain a point to a midpoint of a line or edge, keep in mind that the point can be a point entity or an endpoint of a line. It could be the center of a circle, too, for that matter. Typically, however, it winds up being a line's endpoint that gets constrained to another line or edge's midpoint.

Once the slot sketch is finished, you are ready to create the slot feature. Continue with the following steps to create this slot.

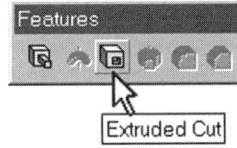

4. Select Insert > Cut > Extrude, or click on the Extruded Cut icon, shown in figure 4-26.

Fig. 4-26. Extruded Cut icon.

5. Select Through All as the end condition.

6. Click on the Reverse Direction icon if necessary. This is the icon to the left of the end condition listing, where you just selected Through All.

7. Click on OK to create the slot.

Over the course of these last few exercises, you may have noticed an arrow in the work area whenever performing an extrusion. This arrow is a handle that can be dragged with the left mouse button to dynamically alter the extrusion distance. It can be a useful design tool, but in the long run it is better to type in a precise value if using a blind or midplane end condition.

No draft was added to the slot feature because the slot will get machined in after the pivot arm is cast. This is also true of the hole you are about to add. Before you add the next feature, rename the last feature created *Slot*. Your FeatureManager should now look similar to that shown in figure 4-27. To begin adding the next feature, continue with the following steps.

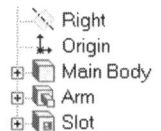

Fig. 4-27. Your FeatureManager should look similar to this.

8. Select the face that was used to sketch the slot.

9. Click on the Sketch icon.

10. Sketch a circle and add a dimension, as shown in figure 4-28. (Hint: If you draw the circle at the origin point, it will be constrained there automatically.)

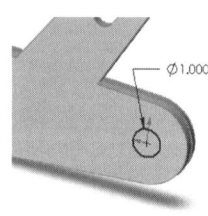

Fig. 4-28. Sketching the circle for the hole feature.

Waking Up Centerpoints: When creating the base feature of the pivot arm, the location of the origin point on your model may be different than the model displayed in this book. If that is the case, the circle for the hole you are about to create will have to be constrained using a different technique. One method of doing this is to add a concentric relationship between the circle and an existing arc edge on the model. Another method that would essentially accomplish the same thing would be to "wake up" the centerpoint of the arc edge so that it could be inferred to.

Waking up a centerpoint of an edge can be accomplished by pausing over any arc or circular edge with the cursor. Note that you must be in a sketch command (e.g., Line or Circle) and not in select mode. The centerpoint of the arc or circular edge will then be projected onto the current sketch plane. At this point, it would be possible to place the center of the new circle over the newly displayed centerpoint of the arc edge, thereby inferring to it. This adds a coincident relation between the arc's centerpoint and the center of the circle, in essence making them concentric. Use this method if necessary to complete the sketch, and then continue with the following steps.

11. Select Insert > Cut > Extrude, or click on the Extruded Cut icon.

12. Click in the work area to establish the extrude direction.

13. Set the end condition to Through All.

14. Click on OK. The pivot arm should appear as it does in figure 4-29.

15. Rename the new feature *Hole 1* (it is okay to use spaces).

16. Click on the Save icon (shown in figure 4-30) to save your work.

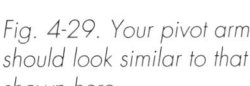

Fig. 4-29. Your pivot arm should look similar to that shown here.

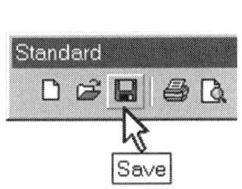

Fig. 4-30. Save icon.

Mirroring Sketch Geometry

Symmetrical relationships can be added in a number of ways. Sketch geometry can be mirrored, in which case SolidWorks adds symmetrical constraints to the sketch geometry. Symmetrical relationships can also be added with the Add Relations panel. This method is not as commonly used, because normally when a designer knows she wants symmetrical geometry she will sketch it that way.

There are three methods of creating symmetrical sketch geometry, and you will explore them all here. Sketch geometry can be mirrored dynamically as you sketch or manually after the geometry has been sketched. You can also achieve symmetrical sketch geometry by adding a symmetric relationship to the geometry after it has been created. Whichever method you use, symmetrical geometry always requires a centerline.

Mirroring Sketch Geometry Dynamically

Once a symmetrical relationship has been set up, either by adding symmetry as a geometric constraint or through mirroring, the centerline must not be deleted. Doing so would delete the symmetrical relationship. The steps for mirroring sketch geometry dynamically are outlined in How-To 4-5.

How-To 4-5: **Dynamically Mirroring Sketch Geometry**

It is assumed in the following series of steps that you are already in sketch mode. No other preparations are necessary. To dynamically mirror sketch geometry, perform the following steps.

1. Sketch a centerline.

2. Make sure the centerline is selected (which it should be if you just sketched it).

3. Select Tools > Sketch Tools > Mirror, or click on the Mirror icon, shown in figure 4-31.

4. Create the desired sketch geometry. Stay on one side of the mirror line or the other (shown in figure 4-32) and the entities will be dynamically mirrored to the opposite side as you sketch them.

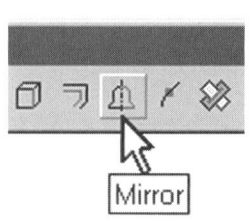

Fig. 4-31. Mirror icon.

Fig. 4-32. A centerline being used as a dynamic mirror line.

Note that the centerline contains symbols indicating that Dynamic Mirroring is turned on. These symbols are the small hash marks on either end of the centerline. To turn off dynamic mirroring, click on the Mirror icon again. The hash marks should disappear, and dynamic mirroring will be disabled.

Dynamic mirroring is certainly the easiest way in which to create symmetrical sketch geometry. One point you should be aware of, however, is to not cross over the centerline while sketching. This has a tendency to create overlapping sketch geometry, which can cause problems when the feature is created.

Mirroring Existing Sketch Geometry

It is also possible to mirror sketch geometry after it has been created. You cannot mirror centerlines, but the sketch can have more than one centerline present. If there is more than one centerline, make sure you select

Mirroring Sketch Geometry

only the centerline you want to mirror about. How-To 4-6 takes you through the process of mirroring existing sketch geometry.

How-To 4-6: Mirroring Existing Sketch Geometry

To mirror existing sketch geometry, perform the following steps.

1. Select the geometry to be mirrored.
2. Select a centerline to use as the mirror line.
3. Click on the Mirror icon.

Note that you cannot mirror sketch geometry unless you are actively editing the sketch that contains the geometry to be mirrored. In addition, do not forget to hold down the Ctrl (Control) key when selecting more than one entity. By the way, it is only necessary to Ctrl-select objects when not in a command.

As mentioned previously, SolidWorks adds symmetrical relationships automatically when entities are mirrored. However, you can add this constraint on your own if you wish. You still need a centerline, and the results will be the same as mirroring when all is said and done. To add a symmetrical relationship, perform the steps outlined in How-To 4-7.

How-To 4-7: Adding a Symmetrical Relationship

To add a symmetrical relationship, perform the following steps.

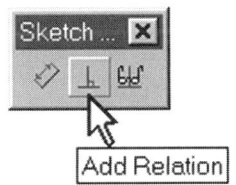

Fig. 4-33. Add Relation icon.

1. Click on the Add Relation icon, shown in figure 4-33.
2. Select the two entities to be made symmetric. These would typically be two similar objects, one on either side of a centerline.
3. Select the centerline the entities are to be symmetrical about.
4. Click on the Symmetric button in the Add Relations panel.
5. Click on OK when finished.

Be aware that a relation is added as soon as its respective icon is clicked in the Add Relation panel. Clicking on OK only serves to exit the command. It is possible to add symmetry between points, lines, arcs, or any other entities. Try dragging the geometry on one side of the mirror line by picking and dragging vertex points. Any entity that is symmetrical will respond on the opposite side of the centerline when its counterpart is dragged to a new position.

✦ **NOTE**: *If symmetrical objects are trimmed or extended (discussed in the next chapter), the symmetrical relationship will be deleted.*

The next features to be added to the pivot arm are a symmetrical recessed area on the back of the main body feature and an octagon-shaped extrusion on the front of the part. Exercise 4-4 takes you through these operations.

EXERCISE 4-4: Adding Features to the Pivot Arm

In this portion of the ongoing pivot arm exercise, a recessed area is added to the back of the model. You will also create an octagonal extrusion used to implement easier maintenance on the assembly the pivot arm belongs to. This is a fictitious assembly that will not be examined in this book, although assemblies and related topics are covered in subsequent chapters.

Make sure the pivot arm file is open. Make use of the following steps if you feel you need them. Some of the more confident readers may decide to use the illustrations and just skim over the steps. Do not do this unless you feel you are ready.

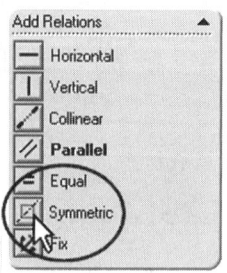

Fig. 4-34. Sketch for the recess feature.

1. Change to a Back view of the pivot arm.

2. Start a new sketch on the back face of the part.

3. Create the sketch shown in figure 4-34.

   Draw the horizontal centerline first and turn on dynamic mirroring. This will make creating the sketch much easier.

4. Use the Centerpoint Arc command to create the 1.5-inch arc and its smaller concentric counterpart. You will only have to worry about half the sketch, in that the other half will be created for you.

5. Use the Tangent Arc command to add the rounded ends, and then add the centerlines to control the included angle of the slot. You can turn off dynamic mirroring before putting the centerlines in. Use figure 4-35 as guidance.

Mirroring Sketch Geometry

Fig. 4-35. Creating the sketch for the recess feature.

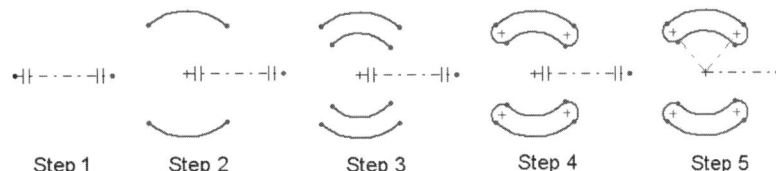

Step 1   Step 2   Step 3   Step 4   Step 5

Once the basic sketch geometry is in place, adding the dimensions is easy. There will more than likely be a few tangent relations that will have to be added manually as well. This is material that has already been covered, so the details will not be covered again here. Finish the recess feature by continuing with the following steps.

Fig. 4-36. Completed recess feature.

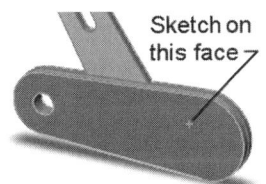

Fig. 4-37. Selecting a face to sketch on.

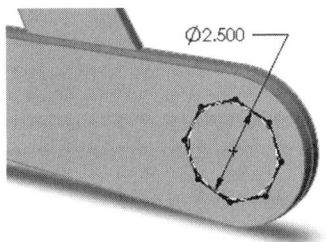

Fig. 4-38. The completed octagon sketch.

6. Cut the sketch into the part at a depth of .25 inch. Use 1 degree of draft.

7. Rename the feature *Recess*. The completed recess feature is shown in figure 4-36.

8. Change the view so that you are looking at the front of the part.

9. Start a sketch on the face shown in figure 4-37.

10. Create an octagon that is centered on the large side of the Main Body feature, as shown in figure 4-38. Use the Inscribed circle option found in the PropertyManager after clicking on the Polygon icon. (Hint: wake up the centerpoint of the arc edge before positioning the octagon's centerpoint.)

11. Add a 2.5-inch dimension to the construction circle used to constrain the octagon.

12. Add a horizontal relation between two opposite points on the octagon. This is to orient the octagon and keep it from spinning.

There are a few different techniques that could be used to keep the octagon from turning at its centerpoint. Of course, it will not really turn unless you force it to, but the point is that the sketch should be fully defined. The technique used in step 12 is to add a horizontal relation between two opposite points. It would have been just as easy to make one edge of the octagon horizontal or vertical. It strictly depends on the designer's intent.

With regard to step 10 and using the Inscribed circle option, this is only important in the respect that you may want to dimension the polygon a certain way. For example, if the polygon is inscribed within a circle, the diameter of the circle will reflect the measurement from corner to corner of the polygon. If the polygon is circumscribed about the circle, the diameter of the circle will define the measurement between flats.

13. Extrude the sketch .75 inch, with 1 degree of draft.

14. Rename the new feature Octagon.

15. Save your work. The finished octagon feature is shown in figure 4-39.

In the next chapter, you will continue to use the pivot arm in additional exercises, so make sure you save your work. These exercises will show you how to add fillets and how to use the Hole wizard, among other things. You will also learn how to create revolved or turned parts.

Fig. 4-39. Pivot arm after adding the octagon feature.

Multiple Bodies

Part files can contain separate chunks of material. They do not have to touch. These separate chunks are considered bodies. The ability to have multiple bodies in a part file expands the flexibility a user has when designing parts. Finished parts should typically consist of one body, however, and you should not confuse having separate bodies in a part file with components in assemblies.

When creating a feature other than the first base feature, an option named *Merge result* is available in PropertyManager. This option is shown in figure 4-40. *Merge result* is always turned on by default, but there are times when you may need to turn it off. Those occasions will be self-evident once we examine how multiple bodies can be used.

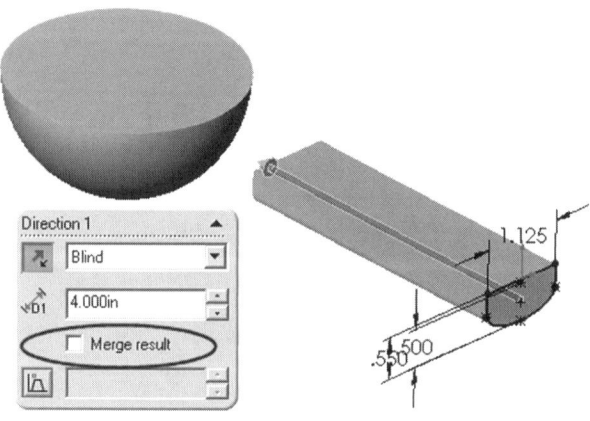

Fig. 4-40. Merge result option.

Multiple Bodies

The object being modeled in figure 4-40 is an ice cream scoop. Note that the Merge result option has been turned off. This is because the handle feature is not connected to the scoop. The handle was created in this fashion to make it easier to create a feature that blends between the scoop and the handle.

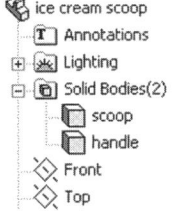

When multiple bodies are created, a *Solid Bodies* folder is automatically established in FeatureManager, which lists those bodies. An example of the *Solid Bodies* folder is shown in figure 4-41. Bodies take the name of the last feature created where the Merge result option was turned on. When the results are not merged, the list grows.

Fig. 4-41. Solid Bodies folder in FeatureManager.

When there are multiple solid bodies, and a new feature is created, the new feature can be merged with specific bodies. This further adds to the extreme flexibility presented by SolidWorks. In the case of the ice cream scoop, a feature will be added that blends the handle with the scoop. This is shown in figure 4-42. The feature being added utilizes the Loft command, discussed in Chapter 6.

Fig. 4-42. Adding a feature between solid bodies.

When creating this third feature, the Merge results option is left on, which in turn gains access to the Feature Scope panel in PropertyManager. The Feature Scope panel, shown in figure 4-43, allows for specifying which solid bodies are merged when the feature is created. In the case of the ice cream scoop, the new feature is being merged with the scoop only, and not the handle. Incidentally, the Auto-select option merges only those bodies that touch. This option is turned on by default.

Once the new feature has been created, the new body can be shelled out separately without affecting the handle. (The Shell command is discussed in Chapter 6.) Finally, any multiple bodies can be combined into one using the Combine command. The final results are shown in figure 4-44.

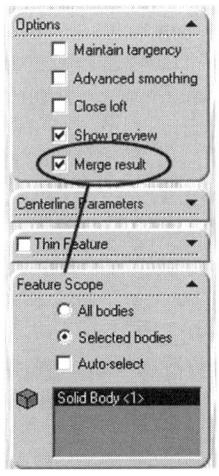

Fig. 4-43. Feature Scope panel.

Fig. 4-44. The ice cream scoop's basic shape has been modeled.

The Combine command is a simple command well-suited to working with multiple bodies. How-To 4-8 takes you through the process of using the Combine command.

HOW-TO 4-8: Combining Multiple Bodies

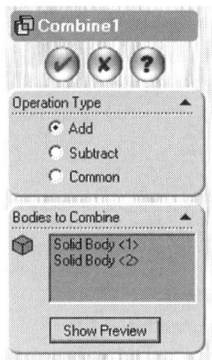

Fig. 4-45. Combine panel.

The Combine command can be used to add multiple bodies together (Add operation), subtract bodies from another body (Subtract operation), or find the common intersecting area shared between bodies (Common operation). The steps that follow describe the process of adding bodies together. The remaining two operations (Subtract and Common) are similar to Add and will not be described in separate How-Tos.

1. Select Combine from the Insert > Features menu.
2. Select the operation to perform from the Combine panel, shown in figure 4-45. Add is used for this example.
3. Select the bodies to be added together.
4. Click on OK.

When bodies are added together, the resultant feature will be listed in the *Solid Bodies* folder in FeatureManager. The bodies that were added together will no longer be listed in the *Solid Bodies* folder.

If a Subtract operation is performed, you will be asked to select a Main Body that other bodies will be subtracted from. If the Common operation is performed, only the area common to all selected bodies will result. It is not possible to combine bodies that do not intersect, as this would result in an "empty" body.

Not all models are good candidates for multiple bodies. For example, the pivot arm would not benefit at all from this functionality. Using multiple bodies when modeling is more of a convenience feature than anything else. As a matter of fact, SolidWorks did not contain multiple body support until SolidWorks 2003!

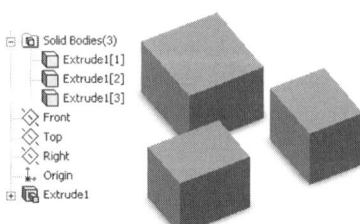

Be aware that multiple bodies do not necessarily have to be physically separated. This fact was reinforced with the ice cream scoop example just prior to combining the bodies. Furthermore, one feature can create more than one body. This would be the case if multiple profiles were extruded simultaneously, for instance. Such a situation is shown in figure 4-46.

Fig. 4-46. Multiple bodies from one feature.

Multiple bodies are a core aspect of how SolidWorks functions, and their presence is felt throughout the software. However, making use of multiple bodies is not a requirement. The ability to define user-defined planes, on the other hand, is a necessity. In light of this, let's finish this chapter by examining how planes can be defined in a number of ways.

Creating Planes

Often you will need to create a new plane for one reason or another. A common reason is simply because you need a plane to sketch on, and the Front, Top, or Right planes are not satisfactory for the purpose. Creating planes is easy, but the method used is dependent on the geometry you have to work with.

For some of us, it has been a while since geometry class in high school. However, most of the plane creation options are very logical, and often a matter of common sense. For instance, it is possible to define a plane with three non-collinear points. This is very basic geometry, and is an example of one of the options you have at your disposal for defining reference planes in SolidWorks.

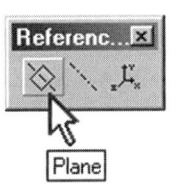

Fig. 4-47. Reference Geometry toolbar.

As most SolidWorks commands go, the command for creating reference planes can be accomplished via the menu structure or a toolbar. The toolbar in this case is the Reference Geometry toolbar, shown in figure 4-47. There is also a Reference Geometry menu under the main Insert menu. The toolbar is very handy, and is one you will probably want to keep available, mostly for the sake of the Plane icon. Planes are just one of those objects that get created quite frequently, so the toolbar will save you time.

The steps used to create a plane are provided in material to follow. First, however, take a look at the various methods available. There a six methods in all, and each method uses a different combination of geometry to define a plane. Use Table 4-1 to choose the method best suited to your needs. Once again, which method is used depends on what geometry you have to work with.

✺ **NOTE:** *In regard to Table 4-1, the term* edge *refers to a model edge. Point can mean sketch point, endpoint, or midpoint, whereas* vertex *is an endpoint on a model edge. Line refers to a sketch line. The term* curve *means any sketch line, centerline, arc, spline, parabola, axis, or edge.*

Table 4-1: Plane Creation Methods

Method	Description	Required Geometry
Through Lines/Points	Plane passes through three points or a line and point	A combination of three points or vertices, or an axis, edge, or line and a point or vertex
Parallel Plane at Point	Parallel to a plane and passing through a specified point	Plane or face and point or vertex
At Angle	At an angle from another plane and passing through a line	Plane or face and axis, edge, or line
Offset Distance	Offset from another plane	Plane or face
Normal to Curve	Perpendicular to a curve and passing through a specified point	Any curve and a point or vertex
On Surface	Tangent to a surface, typically cylindrical or conical, where an axis, edge, line, or plane intersects said surface	Surface and an axis, edge, line, or plane

Starting a Plane

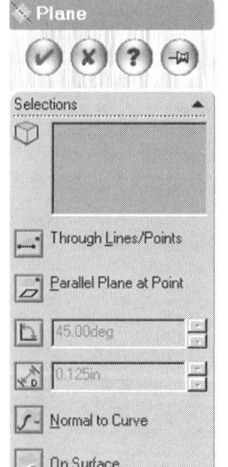

Fig. 4-48. PropertyManager options when creating a reference plane.

Whatever method you decide to use, creating a reference plane always starts in the same fashion. That is to say, you click on the Plane icon (shown in figure 4-47) or perform an Insert > Reference Geometry > Plane selection in the pull-down menus. Once that has been done, PropertyManager will display the panel shown in figure 4-48.

Some plane creation methods are easier than others. There are even a few shortcuts that can be used to create planes. One such shortcut creates an offset plane. Another creates a plane normal to a curve. These shortcuts are explored in the material to follow, and are clearly marked for easy reference.

The order in which the plane creation methods are presented in this section follows the order in which the options appear in the Plane panel in PropertyManager. To create a plane using any of the methods available in SolidWorks, complete the steps in one of the How-To sections that follow. Let's begin with creating a plane using Through Lines/Points, outlined in How-To 4-9.

How-To 4-9: Creating a Plane Using Through Lines/Points

This method allows for defining a plane by specifying three (3) points through which the new plane will pass. A single point and a line can also be used. Perform the following steps to create a plane using the Through

Creating Planes

Lines/Points option (method). Figure 4-49 shows an example of a plane created using three points.

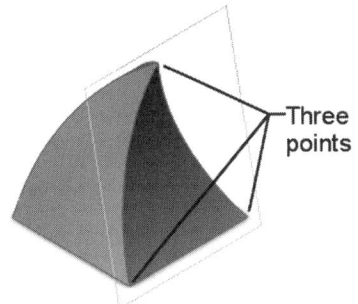

Fig. 4-49. An example of a plane created via three points.

1. Click on the Through Lines/Points button.

2. Select any combination of three (3) sketch points, midpoints, or vertex points. Alternatively, select any combination of a single sketch point, midpoint, or vertex point and a single line, linear edge, or axis.

3. Click on OK when finished.

✓ **TIP:** *In any of the plane creation methods used, it is not necessary to click on the appropriate button first. For example, in this How-To, skipping step 1 and simply selecting three points would have sufficed.*

Note in figure 4-49 that the horizontal edge at the bottom of the part could have been used in place of the bottom two points to create the same plane. It would make no difference.

A Word on Selecting Midpoints

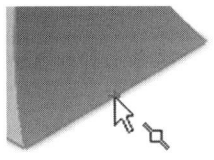

Fig. 4-50. Selecting a midpoint.

You will notice that one of the options under How-To 4-9 is to select midpoints. If you move along a linear edge with your cursor, you will eventually see a midpoint symbol appear. This midpoint symbol is displayed as system feedback, shown enlarged in figure 4-50. Alternatively, right-clicking on an edge will give access to the Select Midpoint option. This will force SolidWorks to select the midpoint for you, which is often easier. Let's continue with the methods of plane creation by examining the Parallel Plane at Point method, the process for which is outlined in How-To 4-10.

How-To 4-10: Creating a Plane Using Parallel Plane at Point

This method allows for defining a plane by specifying a plane or planar face the new plane is to be parallel to, and a point through which the new plane will pass. The point can be a sketch point, endpoint of a sketched object, or endpoint of a model edge. Perform the following steps to create a plane using the Parallel Plane at Point option (method). An example of a plane created using the Parallel Plane at Point method is shown in figure 4-51.

140   Chapter 4: Castings

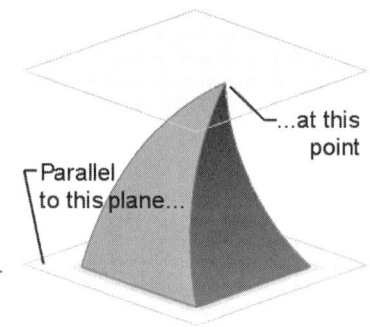

1. Click on the Parallel Plane at Point button.
2. Select a plane or planar face.
3. Select a sketch point, endpoint of a sketched object, or endpoint of a model edge.
4. Click on OK when finished.

Fig. 4-51. An example of a plane created using Parallel Plane at Point.

As you can see, the Parallel Plane at Point method is quite simple. This holds true for the At Angle method as well, described in How-To 4-11.

How-To 4-11: Creating a Plane Using At Angle

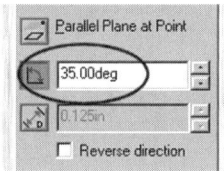

Fig. 4-52. Entering an angle for the angled plane.

This method allows for creating an angled plane. Occasionally it is necessary to create some construction geometry prior to using this option, in order to get the new plane positioned correctly. Of course, this depends on the situation and model requirements. Any sketch line or construction line can be used as a basis for determining the "hinge" location for the angled plane. Alternatively, any linear model edge can be used as well. A plane or planar face must also be specified, which will be the plane from which the angle is measured. Perform the following steps to create an angled plane.

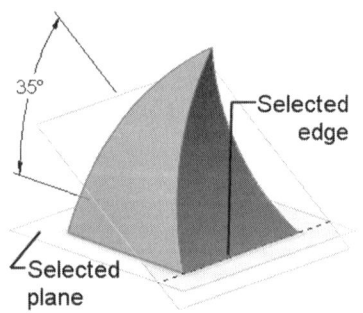

Fig. 4-53. An angled plane.

1. Click on the At Angle button.
2. Select a plane or planar face.
3. Select a sketch line or linear model edge.
4. Enter an angle for the new angled plane, such as that shown in figure 4-52. Use the Reverse direction option if necessary, also shown in the figure.
5. Click on OK when finished. An example of an angled plane is shown in figure 4-53.

The plane's angular dimension is shown in figure 4-53 to help illustrate how the angle value is being applied, but this should be self-evident. As is the case with any driving dimension, the angle value can be altered after the plane has been created.

Another method of creating a plane is to use the Offset Distance option. How-To 4-12 takes you through this process.

How-To 4-12: Creating a Plane Using Offset Distance

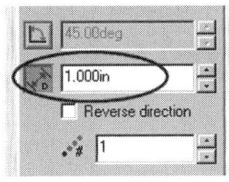

Fig. 4-54. Creating an Offset Distance plane.

This method allows for creating a plane offset from an existing plane or planar face. An offset distance must be supplied, similar to supplying an angle value for an angled plane. Figure 4-54 shows the parameters that will appear if the Offset Distance plane creation button is selected. To create an Offset Distance plane, perform the following steps. Figure 4-55 shows an example of an Offset Distance plane.

1. Click on the Offset Distance button.

2. Select a plane or planar face the new plane will be offset from.

3. Specify an offset distance for the new plane (see figure 4-54).

4. Select the Reverse direction option if necessary to reverse the offset direction. Optionally, specify the number of offset planes to be created. This defaults to a value of 1. Increasing the number of planes creates a series of planes similar to a linear pattern of planes with the spacing determined by the offset distance specified in step 3.

5. Click on OK when finished.

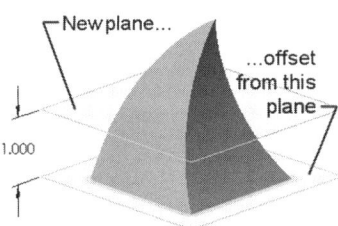

Fig. 4-55. An example of an Offset Distance plane.

Offset planes are extremely common. You will probably find yourself creating them on a regular basis. Step 4 of How-To 4-12 allows for creating a series of planes, all parallel to one another, by specifying some value greater than 1. For instance, if you were to type in a value of 3 when creating an offset plane, and a value of 1 inch for the offset distance, the result would be a series of four planes (the original plus three new planes) 1 inch apart. This is convenient if it is necessary to create cross sections when building a model. This option is available when creating angled planes as well.

A Shortcut for Offset Distance Planes

There is an alternative to creating an Offset Distance plane that is very convenient. It is a shortcut that involves holding down the Ctrl (Control) key. This shortcut works only if you want to create a plane offset from another plane, not a planar face. Nonetheless, it is still a very nice shortcut.

To create an offset plane using the shortcut method, hold down the Ctrl key, and then hold the left mouse button down while the cursor is positioned over the plane's border. Drag the cursor away from the plane and you will see a new plane spring into view. You will also see that PropertyManager is now displaying the Plane panel. Release the mouse button, and then the Ctrl key, and type in a value for the offset distance as you normally would when creating an offset plane. Click on OK or press the Enter key and you are finished. Let's continue with the plane creation methods by examining the Normal To Curve option.

How-To 4-13: Creating a Plane Using Normal To Curve

The Normal To Curve plane creation method is sometimes referred to as the Perpendicular To Curve method. Both terms mean exactly the same thing. A normal in CAD lingo refers to a perpendicular vector. If the term *normal* is unfamiliar to you, just replace it with the word *perpendicular* anytime you see it used by the SolidWorks software.

To create a plane using the Normal To Curve method, a curve must be selected, along with a point the new plane will pass through. The resulting plane will be perpendicular to the selected curve and will pass through the selected point. This may sound confusing, but hopefully it will not after you see an example.

Be aware that a curve can be any number of things. For example, a sketch entity such as a line, arc, or spline is considered a curve. Model edges are considered curves as well. Even axes can be selected when using this plane creation method. Points can be sketch points, endpoints, or vertex points. The following steps take you through the process of creating a Normal To Curve plane. Figure 4-56 shows an example of a plane created using Normal To Curve.

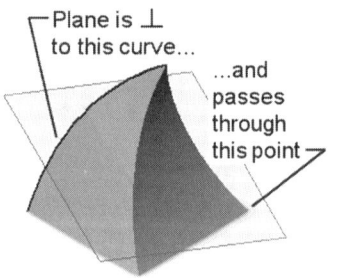

Fig. 4-56. An example of a Normal To Curve plane.

1. Click on the Normal To Curve button.
2. Select the curve the new plane is to be perpendicular to.
3. Select a point the new plane will pass through.
4. Click on OK when finished.

If you examine figure 4-56, you might think there should be numerous solutions for the positioning of the new plane, given the selected geometry. However, consider that the selected edge does not have a consistent radius. It is spline-like in nature. Even if it were an arc with a consistent radius, the selected point (a vertex point) would not be the centerpoint of the arc. In light of these facts, there can be only one solution for the placement of the new plane.

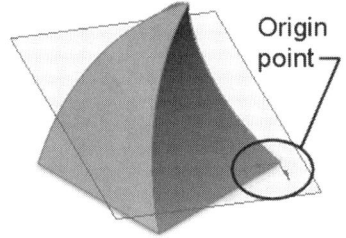

Fig. 4-57. Set origin on curve option not checked.

When creating a Normal To Curve plane, there is an option called *Set origin on curve*. This option is irrelevant if the point being selected to define the plane is already set on the selected curve. Such would be the case if a spline were used to define the new plane, along with one of its control points. In the case of the example shown in figure 4-56, the vertex point exists at some distance from the curve.

The *Set origin on curve* option is sometimes significant when it comes down to creating a sketch on the new Normal To Curve plane. In figure 4-57, the *Set origin on curve* option has not been checked. As you can see, the origin point is at the same location as the point selected to define the plane.

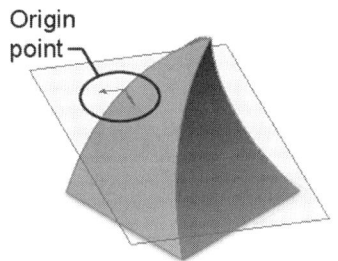

Fig. 4-58. Set origin on curve option has been turned on.

In figure 4-58, the *Set origin on curve* option has been checked. When a sketch is begun on the plane, the origin point appears attached to the curve instead of the point used in defining the plane. This often does not make any difference when sketching on the Normal To Curve plane unless it is necessary to use the origin for adding geometric relations. For most work, you can safely ignore this option.

A Shortcut for Normal To Curve Planes

When it is necessary to create a plane normal to a sketch curve (a sketched entity, in other words), the manual method of creating the plane described in How-To 4-13 must be used. However, if the new plane must be normal to a model edge, there is an excellent shortcut that will save you some time. All that is needed is to select near the endpoint of the model edge where the new plane should be, and then to click on the Sketch icon. A new plane will be created normal to the selected edge and passing through the edge's endpoint.

You really should try this to see how it works. It is as easy as it sounds. Typically a plane or planar face is selected prior to sketching. When implementing this shortcut, do not worry about selecting a plane. Solid-Works creates the plane for you. Open up any existing SolidWorks model and give this shortcut a try.

✓ **TIP:** *This shortcut works on curve types other than model edges. For example,* helical curves and 3D sketch splines, to name two, will also accept this shortcut. These curve types are discussed later in the book.

Let's continue with the examination of plane creation methods by working through the process of creating a plane via the On Surface option, outlined in How-to 4-14.

How-To 4-14: Creating a Plane Using On Surface

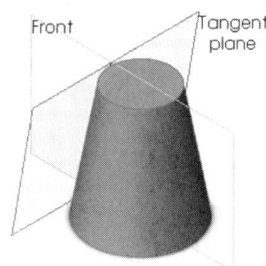

Fig. 4-59. On Surface plane creation method, example 1.

Of all the plane creation methods, creating a plane on a surface is often the trickiest to implement. The On Surface plane creation method has a variety of options you will be exploring in this section. In this first example, shown in figure 4-59, the only two items selected were the conical surface of the cone and the Front plane. As can be seen from the illustration, the new tangent plane is tangent to the conical face where the Front plane intersects the face.

The following steps will vary slightly in the examples, depending on exactly how the On Surface plane is created. In regard to the first example, the process is fairly simple.

1. Click on the On Surface button.
2. Select a cylindrical or conical surface.
3. Select a plane that passes through the first surface.
4. Use the Other Solutions button if necessary to flip the new plane's tangency from one side of the cylindrical or conical surface to the other.
5. Click on OK when finished.

It should be noted that this first example assumes that the reference plane in step 3 passes through the center of the cylindrical or conical surface. If this is not the case, a different scenario is necessary. Let's assume there is a plane offset some distance from the center of the cone. In this second example, the resultant plane will pass through the center of the cone and not be tangent to the conical surface at all. Furthermore, there will be an option for adjusting the angle of this new plane with respect to the reference plane (which would be the offset plane in this example).

Figure 4-60 shows what you might see using the scenario described in this second example. Superimposed on the image is a view of PropertyManager that shows the setting for adjusting the new plane's angle.

Creating Planes

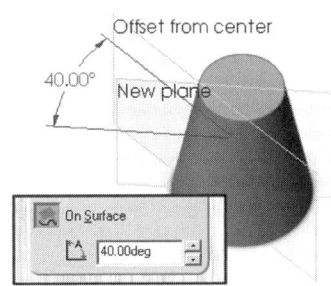

Fig. 4-60. On Surface plane creation method, example 2.

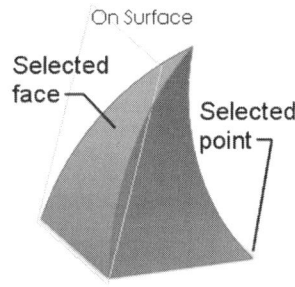

Fig. 4-61. On Surface plane creation method, example 3.

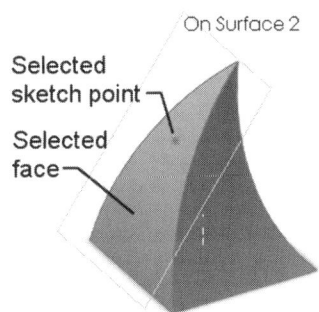

Fig. 4-62. On Surface plane creation method, example 4.

To this point, you have been using faces and planes in the creation of On Surface planes. However, points can be used as well. In figure 4-61, the On Surface plane creation method is implemented in one of its most basic ways. Note that the items selected consist of a nonplanar face and a vertex point.

The combination of selected objects works in example 3 because of the normal vector described by the point and the face. If a line were drawn from the selected point perpendicular to the selected face, the line would represent the normal used to define the new On Surface plane. The new plane is tangent to the face where the normal vector is defined. Considering that an imaginary line drawn from the point to the face is perpendicular to the face at only one position, this combination of entities works for defining the On Surface plane.

For this third example, simply substitute a point for the plane specified in step 3 of How-To 4-14 and you have the required process. If there is only one solution available for the entities selected, there will be no Other Solutions button displayed in PropertyManager.

Sticking with the theme of selecting a point, let's use a sketch point instead of a vertex point. In example 4, shown in figure 4-62, a point was sketched on one of the nonplanar faces of the model. This was done using the 3D Sketcher, discussed in Chapter 9. Suffice it to say for now that the sketched point can be constrained to a particular face of the model. By controlling the position of the point on the face, the tangency point of the new On Surface plane is also controlled.

Creating a plane tangent to a nonplanar face at a particular location was difficult prior to availability of the 3D Sketcher. SolidWorks now makes it very easy. Continuing with this theme, you discover that simple 2D sketch points can also be used to define an On Surface plane. Sometimes the sketch point does not reside on the surface where the new plane must be tangent. This is often the case with a standard 2D sketch point. The point resides in space somewhere, sketched on a plane or planar face.

Because the sketch point lies at some location away from the nonplanar face where the On Surface plane must be tangent, there is a choice to be made as to how the point is projected to the nonplanar face. It can be projected perpendicular to its sketch plane or simply to the closest location found on the surface. Figure 4-63 should help you visualize this.

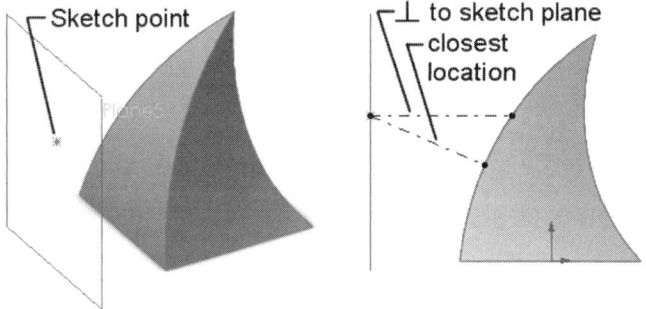

Fig. 4-63. A sketch point can be projected in two ways.

Note the sketch point in figure 4-56. This point was sketched on a plane some distance from the model. When seen from the side, it is easy to see how the point can be projected in two directions, either perpendicular to the sketch plane or closest to the surface, whichever you choose. The options used to control this functionality will appear in PropertyManager, and are shown in figure 4-64. The Reverse option can be used if the projection direction needs to point the opposite way. This option is only available if *Project onto surface along sketch normal* is used.

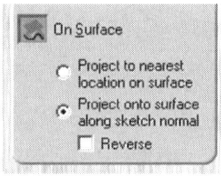

Fig. 4-64. Specifying how the point should be projected.

There is still one other combination of entities that can be used for defining an On Surface plane. This combination consists of a surface, an edge, and a plane. This is one of those cases for which a picture speaks a thousand words, so let's start there. Figure 4-65 shows two views of a vase. The view on the left shows the curve highlighted for purposes of reference. Note the point of tangency, which defines the location of the new On Surface plane. But what defines the point of tangency?

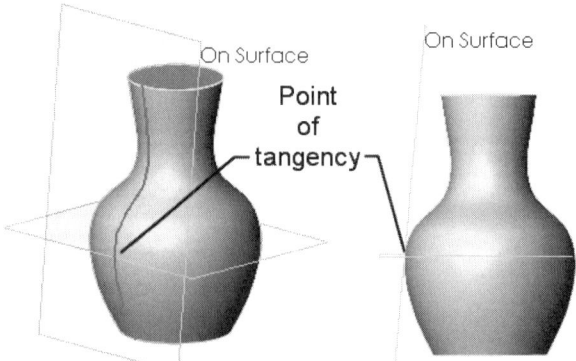

Fig. 4-65. Using a plane, surface, and curve to define a point of tangency.

The point of tangency is defined by selecting the surface on which the new plane is to be created, along with a curve and a plane. The curve should reside on the surface, but should not lie on the plane. Where the curve intersects the plane defines the point of tangency.

Sometimes the On Surface method does not seem to work properly. This is usually a case of operator error. You just have to know what to select. Where you can get into trouble is when selecting the curve. For example, sketch curves will not work. Neither will projected curves. On the other hand, split lines and model edges work just fine. (Chapter 9 deals with curves in depth.)

This section has taught you how to create planes using a variety of methods. The main reason for creating planes is so that you will have something to sketch on. With the wide variety of plane creation techniques at your disposal, you should be able to place a plane just about anywhere you desire. In addition, with your knowledge of sketching and editing tools, you are well on your way to becoming proficient with the SolidWorks software.

Summary

In this chapter you have learned a great deal about creating extruded features and the various end conditions that can be used to specify how the extrusion should terminate. You have also seen how easy it can be to add draft to a part during the extrusion process. The Mid Plane end condition type lends itself very nicely to drafted parts because the draft can be added simultaneously for both extrusion directions. Additionally, adding draft during a mid-plane extrusion creates a parting line automatically by its very nature.

Very important editing commands were covered in this chapter. Specifically, you learned how to edit a sketch or edit a feature's definition by clicking with the right mouse button over the sketch or feature in FeatureManager. You also learned how to modify dimensions by double clicking on a feature or sketch in FeatureManager to access its dimensions, and then double clicking on the dimension value.

You should also now be more familiar with mirroring sketch geometry, and the various methods used to create mirrored geometry. Sketch geometry can either be mirrored after it has been sketched, or dynamically as it is being sketched. You should also have greater facility and confidence in adding relations.

You have also learned some good SolidWorks work habits in this chapter. One such habit is to rename features as you create them, to help make those features more easily recognizable. This is helpful if you need to go back at some later time and make design changes, or if somebody unfamiliar with the part is trying to understand the model.

The final section on creating planes has given you the ability to position planes anyplace you might find it necessary. Sometimes a little additional construction geometry may be needed to define a plane, such as sketch points, but this is perfectly acceptable. Positioning of planes is important because a plane's position will determine the placement of a sketch and the extrusion direction of the feature. Being able to create a variety of planes for sketching will also greatly aid you in creating more complex feature types, covered in subsequent chapters.

Questions and Topics for Discussion

1. What are the five basic steps involved in creating a sketched feature?
2. Name at least four extrusion end condition types.
3. Describe the procedure for renaming a sketch or feature.
4. Describe the procedure for editing a feature's definition.
5. How can you tell if you have exited out of a sketch (whether on purpose or accidentally)?
6. What is the procedure for entering a previously existing sketch?
7. Can you use the Through All end condition during a Boss/Extrude operation? Explain your answer.
8. When creating a base feature for a cast part, what end condition and/or options will automatically create a parting line?
9. How do you access a feature's dimensions if you do not need to edit its sketch or definition?
10. Describe two methods of mirroring sketch geometry.
11. What happens to mirrored geometry when the centerline used for mirroring is deleted?
12. What does it mean when you see gray sketch geometry in the model?
13. Describe four methods of creating planes.
14. Why is it so important to pay attention to the cursor?
15. Describe exactly what type of geometry you would select to constrain the end of one line to the midpoint of another.

Chapter 5

Turned Parts

As far as SolidWorks is concerned, the main difference between cast parts and turned parts is the base feature used to create such parts. Turned parts are generally created by a lathe or similar manufacturing process. This chapter deals with parts that contain radial symmetry, at least at some point in the design process.

Many chapters of this book describe how parts manufactured by a specific process can be modeled in SolidWorks. That does not mean to say, however, that a part with radial symmetry cannot be a cast part. It may very well be that you have a cast part that could most easily be modeled with the Revolve command, or a turned part with the Extrude command.

The point is that this book covers a number of manufacturing processes and the most common means of modeling those types of parts with SolidWorks. However, keep in mind that the first and foremost deciding factor of how you actually model the part is the basic shape of the part, not how the part will be manufactured.

Revolved Features

Turned parts are typically very well suited for manipulation with the Revolve command. The Revolve command parameters, which you will be working with in material to follow, are very simple. There are not as many options associated with the Revolve command as with the Extrude command. Usually when a sketch is being revolved, you must decide the number of degrees to revolve the sketch, and in which direction, clockwise or counterclockwise.

There are a few simple guidelines that must be followed when creating a revolved feature. These guidelines are easy to remember and should not pose a problem. First, a revolved feature must have something to define the axis of revolution. This is where centerlines come into play once again.

Begin with a Centerline

You must have at least one centerline in your sketch when creating a revolved feature. There is no way around it. Therefore, it is usually easiest to draw the centerline and be done with it. There is a reason you would want to draw the centerline before dimensioning, which is explained in material to follow.

The Strategy

When creating a revolved base feature, it is often best to draw the centerline horizontally or vertically out from the origin point. There is, of course, a reason for this. Revolved parts are typically symmetrical. By drawing the centerline from the origin point, the part will be symmetrical about the origin. Because the three default planes pass through the origin, they will also pass directly through the center of the revolved part. This can sometimes benefit you in the long run, such as when creating cuts or other revolved features on the model.

It is not necessary to draw the centerlines horizontally or vertically. This just happens to be the most common approach when beginning a revolved part. There may be situations in which you must create a revolved feature at some odd angle from the main body of the part, in which case the centerline would be at some user-defined angle. It all depends on the part.

Using More Than One Centerline

There is nothing wrong with having more than one centerline in your sketch. You can have a hundred centerlines without suffering any repercussions, although there would probably never be a justifiable reason to have that many centerlines in one sketch.

If there is more than one centerline in your sketch, make sure you select one of them first to represent the axis of revolution. Do this before creating the revolved feature. If you do not select one of the centerlines, SolidWorks has no idea which one you want to revolve about, and you will receive an error message. SolidWorks is intuitive, but it cannot read

Rules Governing Revolved Sketch Geometry

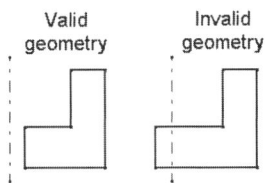

Fig. 5-1. Valid and invalid sketch geometry for a revolved feature.

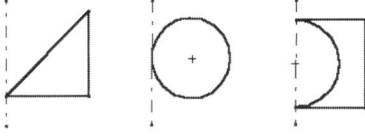

Fig. 5-2. Touching the centerline at an isolated point: examples of invalid sketch geometry.

Fig. 5-3. Examples of valid sketch geometry for a revolved feature.

your mind. To select the appropriate centerline, click on it with the left mouse button.

Probably the most straightforward rule is to not cross the centerline. When you are creating a sketch for a revolved feature, under no circumstances should you allow any sketch geometry to cross the centerline. Figure 5-1 shows valid and invalid geometry for a revolved feature.

Another steadfast rule to follow when creating a revolved feature is to never let the sketch geometry touch the centerline at an isolated point. During the revolve process, zero-thickness geometry would be created, and that is not allowed. Solid modeling software has a difficult time mathematically defining such geometry. Figure 5-2 shows examples of geometry not allowed with the Revolve command.

You may think that the second and third examples in figure 5-2 are legal operations, but they are not. In the second operation, the circle is tangent to the centerline. Mathematically speaking, the point of tangency is a single point. In the last example, the sketch touches the centerline at two points. However, each point can be defined as an isolated point, which rules that scenario out as well.

Allowing the sketch to touch the centerline is perfectly acceptable, as long as the sketch follows the previous rules. Any of the examples shown in figure 5-3 would make for an acceptable revolved feature sketch.

Note in figure 5-3 that a sketch does not have to touch the centerline. In that case, you would wind up with a hole in the center of the model, assuming you revolved it a full 360 degrees. Of course, you can revolve the sketch to fill any included angle that is necessary. (The second example in figure 5-3 would create a sphere.)

The following summarize the basic rules governing a sketch used to create a revolved feature.

- The sketch must contain a centerline.

- First select a centerline to revolve about if there is more than one centerline in the sketch.

- Never let sketch geometry cross the centerline.
- Never touch the centerline at an isolated point.

The guidelines governing basic sketch geometry in general still apply (such as when creating a sketch for an extruded feature). If you need to refresh your memory as to what those guidelines are, see the "Sketch Guidelines" section in Chapter 3.

Revolve Command Panel

Fig. 5-4. Revolve panel.

When initiating the Revolve command, the Revolve command panel is displayed in PropertyManager. This panel is shown in figure 5-4. As long as the sketch you create for a revolved feature follows the previous guidelines, the mechanics behind creating the feature itself are very straightforward. There are only a few options you need to be concerned with. As stated previously, these options require you to specify the number of degrees to revolve the sketch and in what direction, clockwise or counterclockwise (CW or CCW).

As you can see in figure 5-4, there are not many parameters needed to define the revolved feature being created. The drop-down list for the revolve types consists of the options One-Direction, Mid-Plane, and Two-Direction. In other words, do you want to revolve the sketch in one direction only (CW or CCW), in both directions equal amounts, or in both directions different amounts? The choice is yours. More often than not, use of the One-Direction option is all that is necessary.

If it becomes necessary to use the Two-Direction option, you will need to plug in angular values for each revolve direction. For instance, it would be possible to enter 75 degrees for the CW direction, and 40 degrees for the CCW direction. The Selected Contours panel is available if you wish to select contours from the sketch with which to create the revolved feature.

Exercise 5-1 takes you through the creation of a revolved feature. You will more than likely find the process quite simple. During the exercise, concentrate on your sketching technique. Pay close attention to the cursor and system feedback, and take care to connect lines from endpoint to endpoint.

EXERCISE 5-1: Creating a Revolved Feature

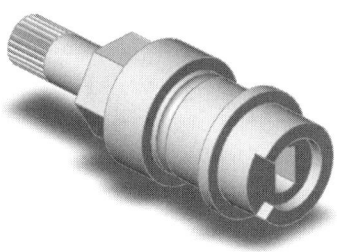

Fig. 5-5. The completed valve stem model.

Begin by starting a new part. Make sure your units are set to inches, with precision set to three decimal places. Save the part right away and give it the name Valve Stem. Figure 5-5 shows the valve stem when completed, though the valve stem will not actually be completed in its entirety in this exercise.

You will start by sketching the profile of the valve stem on the Right plane and revolving it 360 degrees. You will begin the valve stem in this chapter and complete it in subsequent chapters. To create the valve stem base feature, perform the following steps.

1. Right-click on the Right plane and select Show.

2. Make sure the Right plane is still selected, and then click on the Sketch icon.

3. Change to a Right view using one of the techniques learned previously.

4. Create the sketch for the base feature, as shown in figure 5-6.

Fig. 5-6. Creating the sketch for the valve stem.

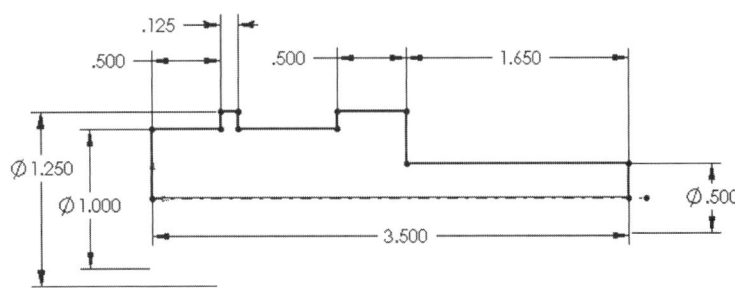

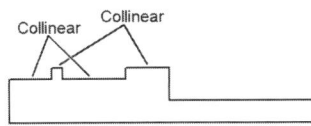

Fig. 5-7. Adding collinear constraints.

5. Add collinear constraints, if needed, to the lines shown in figure 5-7 (for clarity, dimensions are not shown).

Make it a point to fully define the sketch by adding any necessary constraints. Remember, your goal should be to turn all sketch geometry black. When a sketch is black, the geometry is fully defined and it will exhibit predictable behavior. This is always in your best interest.

Diameter Dimensions from Centerlines

Eventually, a model such as the valve stem is going to get placed in a detailed drawing, complete with dimensions and various annotations. When you begin creating drawings, in Chapter 11, you will see that model dimensions can be used in drawings. This eliminates the need to recreate them. This being the case, it is sometimes convenient to make sure the dimensions look right in the model. Such is the case with the diameter dimensions shown in figure 5-6.

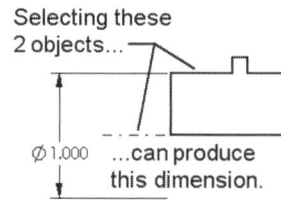

Fig. 5-8. Creating a diameter dimension using a centerline.

Note that there are a total of three (3) diameter dimensions in figure 5-6. These diameter dimensions can be created quite easily, including diameter symbols, but you have to know the proper technique. The secret in creating these dimensions lies in which entities are selected. To add a diameter dimension, make sure the centerline is one of the two entities selected. SolidWorks assumes you will be creating a revolved feature about the centerline, and allows for creating diameter dimensions. Figure 5-8 should help explain this task.

The most common mistake made when trying to add a diameter dimension is to select the endpoint of the centerline, rather than the centerline itself. You must select the centerline, or the diameter dimension will not be created. Additionally, make sure to move the cursor to the opposite side of the centerline in order to see the correct diameter dimension preview.

✓ **TIP:** *Keep in mind that right-clicking locks in a dimension preview.*

Do not worry if the dimensions do not look exactly as they do in this book. In material to follow, you will learn how to change some of the properties of dimensions. However, before tackling dimension properties, let's complete the creation of the revolved feature. To do this, perform the following steps.

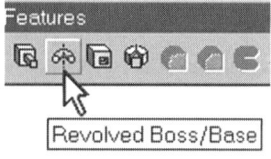

Fig. 5-9. Revolved Boss/Base icon.

1. Select Revolve from the Insert > Boss/Base menu, or click on the Revolved Boss/Base icon, shown in figure 5-9.

2. Make sure the Angle option is set to 360 degrees.

3. Click on OK. Figure 5-10 shows the valve stem at this stage in the process.

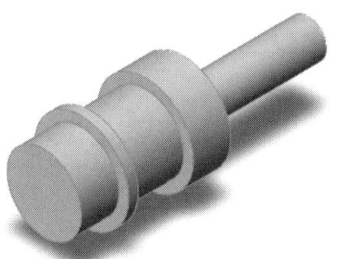

Save your work, and read on to see how you can change the appearance of dimension arrows, which side the leader is on, and other aspects of a dimension's appearance.

Fig. 5-10. Valve stem so far.

Dimension Properties

Almost every object in SolidWorks has what are known as properties. What types of properties an object has depend on the type of object it is. Dimensions, for instance, have properties that determine whether arrows are inside or outside the extension lines of the dimension, whether the dimension is a diameter or radial dimension, the dimension's precision, dual dimensioning settings, tolerance values, and a host of other options.

In the following material you will take a look at how you can modify many of these properties, and what some of your options are. Consider the fact that the dimensions you add to a sketch will be used later on, in the 2D detail drawing. In SolidWorks, there is no sense in recreating dimensions if you do not have to, especially if you have already added them.

Now consider how you might want those dimensions to look in the drawing. In many cases, the fine-tuning of a dimension's appearance is done in the drawing and not in the part. A perfect example of this is controlling where the ends of the extension lines are positioned. As a matter of fact, controlling that particular aspect of a dimension must be done in the drawing, and not in a part. However, some aspects of a dimension's appearance can be modified most conveniently while creating them in the 3D part file.

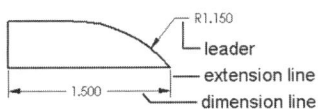

Fig. 5-11. Dimension terminology.

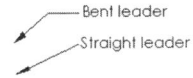

Fig. 5-12. Leader lines.

Figure 5-11 is for those not familiar with the terminology used in regard to dimensions. The extension lines shown in the illustration are sometimes referred to as witness lines. This book will always refer to them as extension lines.

SolidWorks refers to dimension lines as leader lines. However, when this book refers to a leader, it is referring to the dimension leader line typically associated with radial dimension, notes, or balloons, to name a few. Leaders, incidentally, can either be bent or straight, such as those shown in figure 5-12.

Most of a dimension's basic properties can be controlled via PropertyManager, first discussed in Chapter 2. However, the full range of a dimension's properties is tucked away in a different location. To access a dimension's properties, perform the steps outlined in How-To 5-1.

How-To 5-1: Accessing a Dimension's Properties

To access a dimension's full range of properties, perform the following steps.

1. Right-click on the desired dimension.

2. Select Properties from the context-sensitive menu.

Alternatively, perform the following steps.

3. Select a dimension.

4. Click on the More Properties button found at the very bottom of PropertyManager.

At this point, the Dimension Properties window will appear, which will contain all properties for the dimension. Different dimension types will have somewhat different properties. Therefore, some of the items found in the Dimension Properties window will not always be exactly the same.

A wide range of options can be modified, including but not limited to what units are being used, the decimal place precision, arrow style, dimension value font, tolerance display settings, and on and on. Most of these properties are very straightforward in nature and are therefore not described in detail here.

Probably one of the more important aspects to keep in mind when modifying a dimension's properties is that the changes only affect that one dimension. To change settings to dimension styles globally, the document's properties must be changed (see "Modifying Document Properties and Saving Templates" in Chapter 2). Specifically, the Detailing section and its various subsections under the Document Properties tab (shown in figure 5-13) will control the appearance of all dimensions and annotations in a drawing on a global level.

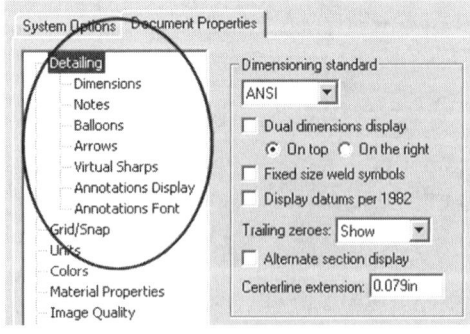

Fig. 5-13. Document Properties tab showing Detailing section and subsections.

Dimension Properties

✓ **TIP:** *Properties for multiple annotations can be modified at one time. By selecting multiple annotations and accessing the properties of any one of the selected annotations, the properties for all of them are shown, and common properties can be modified simultaneously.*

Modifying Properties via PropertyManager

It is often not necessary to access a dimension's properties when all that really needs to be done is to tweak a couple of basic dimension characteristics. Most basic elements of a dimension's appearance can be changed much more efficiently in PropertyManager. A portion of the Dimension PropertyManager is shown in figure 5-14. This is part of what you would see after doing nothing more than selecting a dimension. If PropertyManager is not visible after selecting a dimension, you must enable it (see Chapter 2).

The following can be changed in PropertyManager. If what needs to be changed is not among the following, in all likelihood it will be necessary to access the dimension's full properties.

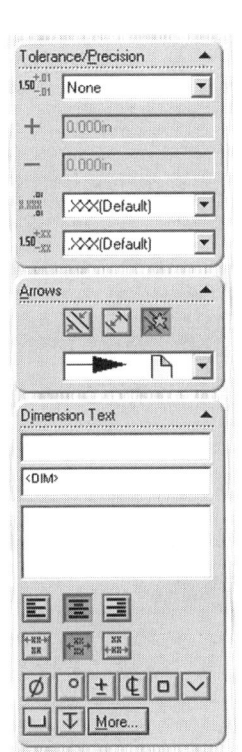

Fig. 5-14. Elements of a dimension's appearance can be changed in PropertyManager.

- Utilize dimension favorites
- Add tolerances such as basic, bilateral, and symmetric
- Modify tolerance values, such as in the case of bilateral or limit tolerance types
- Change the decimal place precision
- Change the dimension line's arrow position
- Modify the arrow type
- Add text to a dimension value
- Add various symbols to the dimension value

The list is fairly extensive, and most of the items do not require further explanation. For example, if you need to add a tolerance to a particular dimension, simply select the tolerance type from the drop-down list. Nothing to it. However, there are a few details that will prove valuable for a few of the properties, which are discussed in the material that follows.

∽ **NOTE:** *Dimension favorites are not discussed in the following material. The topic is better suited to Chapter 11, which discusses detailed design drawings. See the section "Dimension Favorites" in that chapter for further information.*

Dimension and Tolerance Settings

As should be evident from figure 5-14, adding tolerance values or changing dimension or tolerance value precision can be done in PropertyManager. It is just a matter of choosing the desired settings for the selected dimension or dimensions. However, to change the scale of tolerance values with regard to the dimension value, you must access a dimension's properties. How-To 5-2 takes you through the process of changing tolerance scale.

How-To 5-2: Modifying Tolerance Value Scale

In addition to setting the precision of tolerance values, the text size of tolerance values can also be altered. The following steps take you through this process.

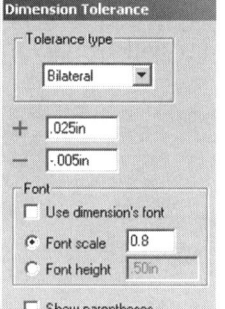

Fig. 5-15. Modifying tolerance parameters.

1. Click on the Tolerance button found at the bottom of the Dimension Properties window. A portion of the window is shown in figure 5-15.

2. Uncheck the *Use dimension's font* option.

3. Specify a scale (relative to the dimension font) or height for the tolerance values, as required.

4. Click on OK.

5. Click on OK again to accept the changes and exit the Dimension Properties window.

Using the steps outlined in How-To 5-2, it is possible to modify all tolerance parameters, but basic settings are more easily controlled via PropertyManager. When changing the tolerance size, using the Scale option is easiest. To modify the tolerance text size via the Height option, you must know what the current height of the dimension font is ahead of time, or the tolerance values will not look right. There is a preview area (not visible in figure 5-15) that can help.

✓ **TIP:** *To modify the precision of all dimensions in the model, use the Precision button in the Dimensions section of the Document Properties window. Note that changing the primary unit's precision has the same effect as changing the number of decimal places in the Units section.*

Dimension Arrows

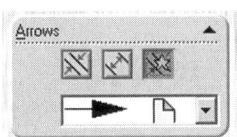

Fig. 5-16. Dimension arrow properties.

There are three options listed in PropertyManager associated with how dimension arrows will appear. These options are Outside, Inside, and Smart. The Smart option is the third button from the left, shown in figure 5-16. Using the Smart option, if there is insufficient room to position the arrows between the extension lines, the arrows will be placed outside.

The drop-down list below the buttons allows for changing the dimension arrow type. This is straightforward, but one item worth mentioning is the sheet of paper (or "document") pictured to the right of the arrow. This signifies that the arrow style will follow whatever is set as the default in the Document Properties window. This is typically what you would want.

✓ **TIP:** *Arrow positioning and style can be set for the entire document in the Dimensions section of the Document Properties window.*

Flipping Arrows: An Alternative Method

An alternative method of modifying a dimension's arrow placement via PropertyManager is to use the green points present on the dimension arrow when the dimension is selected. First select the dimension by clicking on the dimension value. Next, one click on one of the green dots is enough to flip the arrow (or arrows) to the opposite side. Needless to say, this is much more direct.

Flipping Leaders

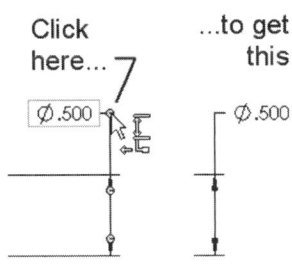

Fig. 5-17. Flipping dimension leaders.

On a related note, some dimensions have leaders that end in a small, horizontal line, an example of which is shown in figure 5-17. When this is the case, clicking on the green dot at the end of the horizontal line will flip the leader and value to the other side. Of course, the dimension must be selected first for the green dot to appear.

Adding Text to Dimension Values

Anytime text, symbols, or tolerances are added to a dimension value, that information is carried over to the drawing when the dimensions are inserted. This can be useful, especially when two different individuals are doing the designing and the detailed drawings. If the data is already present in the model, the drafter does not have to call up the designer and ask "What do you want the thread callout to be for this hole?"

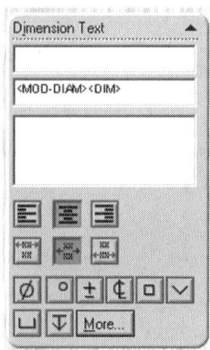

Fig. 5-18. Adding text to a dimension value.

To add text to a dimension value, click in the applicable field (see figure 5-18). There are three fields, with the middle field showing the code for the dimension's value. The code might read something like <MOD-DIAM><DIM>. You should not touch the less-than or greater-than symbols, or anything between them. This is the code that translates data to produce dimension symbols, as well as the actual dimension value shown on the drawing. Mess with the code, and you break the link.

Adding text to the field above the code will place text above the dimension value. Text can also be added in-line with the code (resulting in text in-line with the dimension value), or in the field below the code. More than one line of text can be added in the lower field. Symbols can also be added by clicking on the applicable buttons, also shown in figure 5-18. The More button will open up another window, which contains an expanded list of symbols. Use the additional buttons on the Dimension Text panel for justifying all of the text and symbols vertically or horizontally.

Dimensioning to a Tangency Point

Normally when dimensioning to an arc or circle, the extension line is attached to the arc's centerpoint. This default behavior is not always desirable, such as in the case of dimensioning an obround slot. Figure 5-19 shows what is meant by dimensioning to the tangency points of arcs. Note where the extension lines are attached to the sketch geometry.

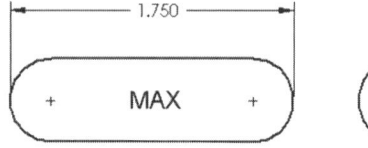
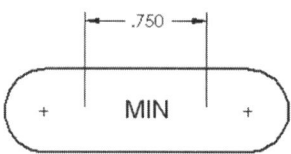

Fig. 5-19. Dimensioning to tangency points.

In figure 5-19, the minimum and maximum arc conditions are used as extension line attachment points, rather than the arc's centerpoint. Arc condition is SolidWorks' terminology for the minimum and maximum tangency points. As you can see in figure 5-19, SolidWorks can infer where the arc's minimum tangency condition is, even though no geometry is there. Of course, in this particular example you would not want to add a dimension to the minimum arc tangency points, but in other situations it may be necessary.

The trick to dimensioning to tangency points is to dimension to the arc itself, not the arc's centerpoint. What usually throws new users off is that the dimension will not attach itself to the tangency point automatically. It is up to you to modify the properties of the dimension and specify the minimum or maximum arc condition after the dimension has been

Dimension Properties   **161**

added. To see how this is done, perform the steps outlined in How-To 5-3. You may need to create a sample sketch so that you can experiment with this setting. Try sketching an obround slot similar to that shown in figure 5-19.

How-To 5-3: Dimensioning to Tangency Points

To establish dimension properties and an arc's (or circle's) tangency point, perform the following steps.

1. Click on the Dimension icon.

2. Add the required dimension, but make it a point to select the arc or circle, not its centerpoint. Note that the extension line will still go to the arc's or circle's centerpoint, but just ignore this for now.

3. Click to place the dimension as you normally would.

4. Right-click on the new dimension and select Properties.

5. Near the bottom right of the Dimension Properties window, select the applicable arc condition (see figure 5-20).

6. Click on OK to accept the changes.

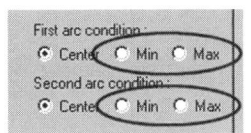

Fig. 5-20. Setting the arc condition.

If the arc condition options do not show up in the dimension's properties, it is almost certainly because you selected the arc's or circle's centerpoint when adding the dimension. Go back and try again, but this time avoid selecting the centerpoints! If the wrong arc condition is chosen, which can sometimes happen, just repeat the process and choose the proper arc condition.

When there is more than one arc or circle in the sketch, such as with a rounded slot, there will be options for both the first and second arc conditions. Both conditions will have the same three options of Center, Min, and Max. Center is the centerpoint of the arc, obviously. If there is only one arc or circle, only the First arc condition options will be shown.

Tangency Point Dimensions: An Alternative Method

If you have performed steps 1 through 3 in How-To 5-3 properly, there is another method you can use to attach the extension lines to the tangency points. If the dimension is selected, you should notice small green dots appear at the end of each extension line. Drag the green dot to the tangency point of the arc or circle. Repeat this process for the other extension line if necessary.

Under certain conditions this shortcut does not work. If this is the case, use the dimension's properties to change the arc condition, as outlined previously. Additionally, the extension line previews do not always look perfect when dragging the extension lines. Do not worry about this, as the dimension should look fine as soon as you let go of the left mouse button.

✓ **TIP:** *Arc length dimensions can be created by picking both endpoints of an arc and the arc itself.*

Dual Dimensioning

Dual dimensioning refers to dimensioning in both English and metric units. Usually the extra dimension is placed in brackets. A typical example of a dual dimension value follows: 1.00 [25.4].

Dual dimensions will not be used in this book. However, your company or business may require you to use them. Normally you would probably want dual dimension display to be a global setting so that all dimensions in the drawing will be affected. To turn on dual dimensioning, perform the steps outlined in How-To 5-4.

How-To 5-4: Turning on Dual Dimension Display

To turn on dual dimension display for the entire drawing, perform the following steps.

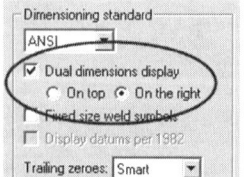

Fig. 5-21. Dual dimension display settings.

1. Select Options from the Tools menu.

2. Select the Detailing section in the Document Properties tab.

3. Place a check in front of the *Dual dimensions display* option, shown in the figure 5-21.

4. Specify whether you want the dual dimension value to be *On top* or *On the right* of the primary dimension value.

5. Click on OK when finished to accept your changes.

This setting is retroactive and will affect dimensions already present in the model. If there are only a select few dimensions you want to show dual dimensioning on, access the properties of those specific dimensions. There you will see an option titled *Display as dual dimension*.

Sketch Fillets

It should be noted that dual dimension display can be turned on in a 2D detail drawing quite easily. Therefore, it really is not necessary to enable this option in a part model. This is especially true considering that dual dimension display will clutter things up quite a bit.

There are quite a few options when it comes to making changes to the way a dimension looks, as you can see. Chapter 11 will cover many of the other options available. Many of these options are more important with regard to detail drawings, but do not play as much of a role when modeling a part.

The following section takes you through some new sketch topics, including adding sketch chamfers and fillets, trimming and extending sketch entities, extracting sketch geometry from model edges, and offsetting sketch entities or model edges. As mentioned earlier in the book, it is very important to develop good sketch practices and skills in SolidWorks, as that is the backbone of becoming a good modeler. Practice these skills well.

Sketch Fillets

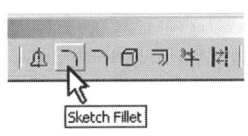

Fig. 5-22. Sketch Fillet icon.

If you want to add fillets directly into your sketch, you can use the Sketch Fillet icon, shown in figure 5-22. When you click on the Sketch Fillet icon, PropertyManager displays the appropriate panel, shown in figure 5-23. How-To 5-5 takes you through the process of creating a fillet in a sketch.

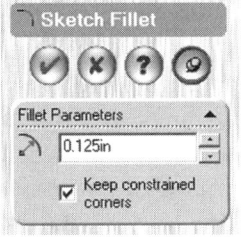

Fig. 5-23. PropertyManager's Sketch Fillet panel.

How-To 5-5: Creating a Sketch Fillet

It is suggested that you create a rectangle or two on your computer at this point, adding dimensions as well. This will give you something to experiment with when performing the steps that follow for creating a sketched fillet.

1. Click on the Sketch Fillet icon.

2. Specify a radius for the fillet.

3. Select two sketch entities to be filleted. These can be any combination of lines, arcs, splines, or even parabolas.

4. Click on OK when finished.

Be aware that sketch entities being filleted do not have to form a perfect corner. If there is a gap between the two entities, or if the entities overlap, SolidWorks will extend or trim them as necessary. If the entities overlap, click on the portion of the entities you want to keep when the fillet is added. If the two entities do happen to form a perfect corner, it is possible to select the corner to be filleted. This is one less click than having to select both entities.

Keeping Corners Constrained

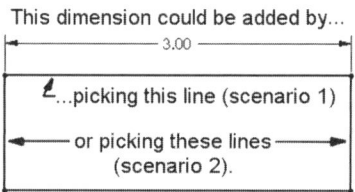

Fig. 5-24. The same dimension can be added multiple ways.

There is an option for keeping corners constrained. This option is necessary when there are existing dimensions (and sometimes geometric relations) on the items being filleted. If *Keep constrained corners* is not checked, any dimensions previously placed on the lines being filleted will usually be deleted, though this depends on how the dimension was added. Figure 5-24 and the material that follows explore this further.

Imagine a rectangle with a horizontal dimension along the top. To add this dimension, either the top horizontal line could be selected, or the two vertical sides of the rectangle could be selected. Both scenarios would result in the same dimension. However, the extension lines are related to different geometry. In the second case, the dimension would be retained if fillets were added to the rectangle. In the first case, the dimension would be lost. This is because the extension lines of the first dimension are related to the endpoints of the horizontal line. If you add the fillets and trim back the line's endpoints, the extension lines no longer have a termination point.

As a general rule, leave *Keep constrained corners* checked all the time. Rarely, if ever, will you need to turn it off. If you turn it off by accident and add a fillet to a line that contains a dimension, a warning message will appear, which is shown in figure 5-25.

Fig. 5-25. A warning that dimensions will be deleted.

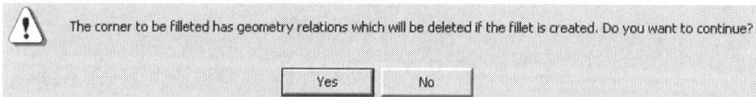

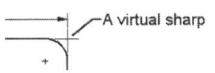

Fig. 5-26. A virtual sharp.

Assuming you take the smart way out and just leave *Keep constrained corners* checked, SolidWorks will add what is known as a virtual sharp wherever it needs to in order to maintain any previously created dimensions. Virtual sharps are just construction entities that give the extension lines someplace to reference. An example of a virtual sharp is shown in figure 5-26.

Adding Virtual Sharps Manually

You certainly do not have to use the Sketch Fillet command if all you want is to create your own virtual sharps manually. Rather, use the Point sketch tool (see "Points," Chapter 3). The Point icon has an alter ego, you might say. Typically, adding points to a sketch is done by clicking on the Point icon, and then clicking to position the points in the sketch.

The key to creating a virtual sharp is to select the lines or model edges whose intersection the virtual sharp will be added to. Do this first, and then click on the Point icon. The virtual sharp should appear at the proper intersection.

✓ **TIP:** *To control the appearance of virtual sharps, access the Virtual Sharps section of the Document Properties window.*

Sketch Chamfers

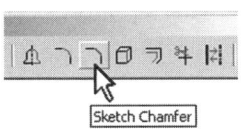

Fig. 5-27. Sketch Chamfer icon.

Chamfers can be added to a sketch in much the same way that fillets can. The process is almost identical. After clicking on the Sketch Chamfer icon (shown in figure 5-30), PropertyManager will display the corresponding panel of the same name. This panel is shown in figure 5-28. How-To 5-6 takes you through the process of creating a sketch chamfer.

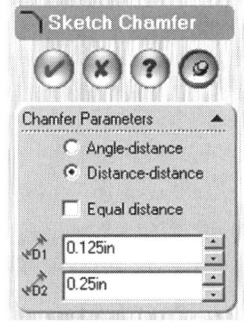

Fig. 5-28. Sketch Chamfer panel.

How-To 5-6: Creating a Sketch Chamfer

To create a chamfer in a sketch, perform the following steps.

1. Click on the Sketch Chamfer icon.
2. Specify how you would like to define the chamfer's dimensions by selecting either Angle-distance or Distance-distance.
3. Type in the applicable dimensional values.
4. Select the two sketch entities where the chamfer is to be applied.
5. Continue as required and click on OK when finished.

There is an *Equal distance* option that can be applied if the Distance-distance method of defining the chamfer is used. This option will not be available if the Angle-distance method is selected.

Similar to creating a sketch fillet, sketch chamfers can be created by selecting a corner instead of the two lines being chamfered. This is not recommended, because you have no control over which line has the first distance applied to it, and which line has the second distance. When selecting lines, as opposed to the corner, the first distance is applied to the first line, and so on. Of course, this is irrelevant if the distances are equal.

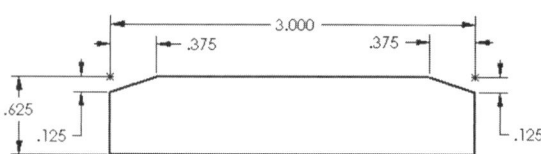

Fig. 5-29. A sketch with two chamfers.

For whatever reason, there is no *Keep constrained corners* option, as there is with the Sketch Fillet command. Corners are automatically constrained with points. Virtual sharps are not used as they are with the Sketch Fillet command. The reason for this is a mystery. An example of a rectangle with two sketch chamfers added is shown in figure 5-29.

Trimming

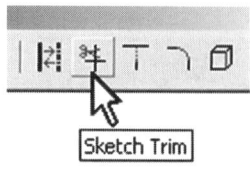

Fig. 5-30. Sketch Trim icon.

Because SolidWorks requires a non self-intersecting sketch, it is sometimes necessary to trim away some unwanted sketch geometry. Trimming is quite easy to accomplish in its simplest form. Using the Sketch Trim icon, shown in figure 5-30, is just a matter of clicking on what you want to get rid of. The Sketch Trim icon can be found on the Sketch Tools toolbar.

Extending

A very convenient aspect of the Sketch Trim function is its ability to delete standalone entity segments. What this means is that if you have an entity you want to delete, you can click on it while in the Sketch Trim command and delete it right then and there. There is no need to exit the Sketch Trim command, select the entity, and press the Delete key.

When using the Sketch Trim command, it may be necessary to trim out one particular section of a line, or it may be necessary to trim an object back to a particular entity. It may even be necessary to trim to an object that does not physically intersect the entity being trimmed. To trim to a specific entity, place the cursor over the portion of the entity you want to get rid of, hold the left mouse button down, and then drag to the entity you want to trim to. This option will work whether or not the entity being trimmed physically intersects the entity being trimmed to.

Try the Sketch Trim function on your own until you feel comfortable with how it functions. You will notice a preview that highlights what will be discarded as you position the cursor over the objects being trimmed. When you feel comfortable with trimming, move on to the next section, which involves extending entities.

Extending

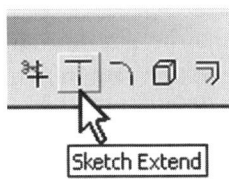

Fig. 5-31. The Sketch Extend icon.

You probably could have guessed that if there is a Sketch Trim command there is an Sketch Extend command as well. If you were able to get the hang of trimming, using Sketch Extend should pose no difficulty. The two commands are very similar. The Sketch Extend icon is shown in figure 5-31.

Like the Sketch Trim command, Sketch Extend will give you a preview of what the extended entity will look like before you actually click the left mouse button. When extending an object, it makes a difference as to which side of the entity you click on to extend. For example, clicking on the wrong side of a midpoint of a line will extend the line in the opposite direction.

When extending an entity, SolidWorks extends it to the very next sketch entity it encounters. Additionally, as with the Sketch Trim command, it is possible to drag the mouse to a specific entity to extend to, rather than have SolidWorks simply extend the entity to the next one it encounters. This will work whether or not the entity being extended physically intersects the entity being extended to. This is another way in which the Sketch Extend command is similar to the Sketch Trim command.

✓ **TIP:** *You can use the Sketch Trim icon as an extend command. Use the same drag technique used when extending to a specific entity. You will find that even though you are in the Sketch Trim command, SolidWorks lets you extend with it. (In the early days of SolidWorks, there was no Sketch Extend command.*

Sketch Trim did everything. Even though Sketch Extend is now a separate command, Sketch Trim still retains its old functionality.)

Practice extending some sketch objects (such as lines or arcs) until you feel comfortable with the command. Do not worry about creating a legitimate sketch you would use to create a feature. Just concentrate on practicing the commands, so that you understand how they work.

✏ **NOTE:** *If the Sketch Extend icon, or other icons, are not present on your Sketch Tools toolbar, you can add them using the customize capabilities of SolidWorks discussed in Chapter 23. If you do not feel comfortable customizing toolbars, the commands mentioned throughout this section are available through the Tools > Sketch Tools menu.*

Converting Entities

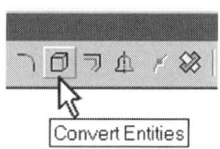

Fig. 5-32. Convert Entities icon.

The ability to convert existing edges of a feature is an extremely important function. There is more to this than meets the eye; therefore, the following material takes an in-depth look at this command. The Convert Entities icon is shown in figure 5-32. When using either the Convert Entities command or the Offset Entities command (discussed in the next section), you must be in a sketch.

It has been mentioned repeatedly that you must have a sketch in order to create a sketched feature, such as an extrusion or revolved feature. In that sketch geometry is a requirement of creating sketched features, it would be beneficial to directly convert existing edges of feature geometry into a sketch. This capability does exist, and it is basically just a shortcut for obtaining the necessary sketch geometry. The Convert Entities command essentially "extracts" sketch data from the existing model edges.

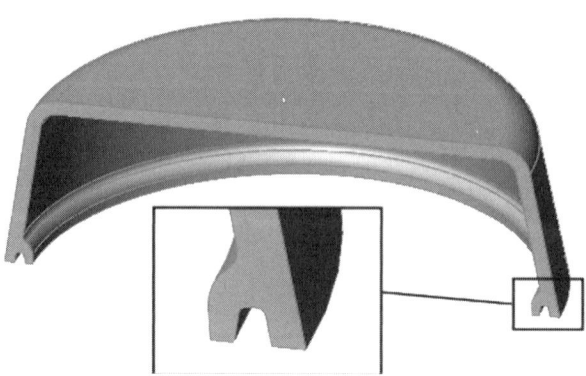

Let's consider a very simple example in order to better understand this principle. Figure 5-33 shows a battery pack cover to a remote control unit. The model is not finished yet. The first few features consist of revolved geometry and some fillets. The finished product needs to extend an additional 4 inches or so, and will have the same cross section as that shown in the figure. The

Fig. 5-33. Battery pack cover needs to be lengthened.

Converting Entities

finished model will eventually slide and snap into place on the remote control unit.

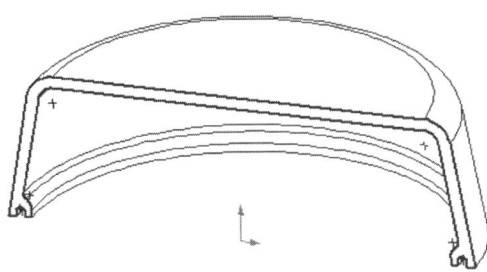

Fig. 5-34. Battery pack after extracting sketch geometry.

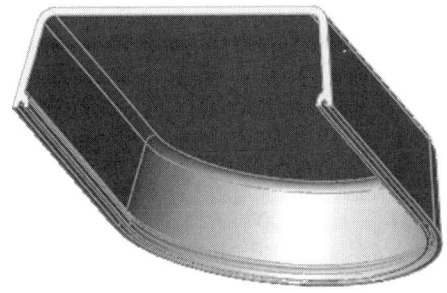

Fig. 5-35. Battery pack cover after adding the extrusion.

A blow-up of the grooved area shows the geometry in more detail. It is not all that complex, but trying to create a sketch around the perimeter of all this geometry would be a real chore. With the Convert Entities command, you need not give it a second thought. Simply start a new sketch on the end face of the model and click on the Convert Entities icon. Figure 5-34 shows what the sketch would look like if you were to actually carry out this task.

It took three clicks to complete the sketch shown in figure 5-34. The first click was on the front face of the model to inform Solid-Works where the sketch plane was. The second click was on the Sketch icon. While the face was still selected, a third click on the Convert Entities icon was all it took to convert all edges on the perimeter of the face into sketch geometry. The next task would be to extrude the sketch. The model of the completed feature is shown in figure 5-35. The view is from the underside, after adding the extrusion.

The steps for using Convert Entities are very simple. Select the model edge or edges to be converted, and then click on the Convert Entities icon. Because there are other choices, or situations that may arise when using the Convert Entities function, Table 5-1 is included to outline other scenarios you may encounter. Use this table as a reference in terms of how the Convert Entities command functions.

Table 5-1: Convert Entities Functionality

Selection	Result of Clicking on Convert Entities Icon
Model edges on the current sketch plane	Sketch geometry is extracted from the model edges.
Model edges not on the current sketch plane	Sketch geometry is extracted from the projection of the model edges perpendicular to the current sketch plane.

Selection	Result of Clicking on Convert Entities Icon
A model face	All edges of the face are extracted. If the face is not on the current sketch plane, edges are projected perpendicular to the sketch plane.
Sketch geometry from an existing sketch	Sketch geometry is extracted from the existing sketch. You must show the sketch being converted and select the geometry from the work area.

One very important aspect of the Convert Entities command has to do with relationships. Any sketch geometry that has been converted from a model edge (or from another sketch) has a relationship to the originating geometry. This means that if the original geometry changes size or shape, the sketch containing the converted geometry will change also. In fact, there is a direct dependency leading back to the original model edges.

The relationship created when converting geometry is known as an on edge relationship. When a model edge is converted into sketch geometry, its endpoints will appear to be fully defined. This is intended behavior. This behavior is convenient because you do not have to worry about fully defining converted sketch geometry as it is already fully defined. However, dragging the endpoints of a converted entity will "loosen" the endpoints, thereby rendering them underdefined. It would then be possible to dimension the endpoints or length of a line, for instance, without overdefining it.

Offsetting Entities

Offset entities display the same characteristics as converted entities. They are both dependent on the underlying geometry. Both converted and offset sketch geometry will appear black because it is fully constrained. Endpoints of offset entities will appear fully defined unless they a dragged. Like converted entities, offset entities will change shape if the dimensions of the underlying geometry are modified. The only difference is that offset entities allow you to supply an offset distance from the original edge.

The Offset Entities command has a bit more to it than the Convert Entities command. This being the case, How-To 5-7 takes you through the process of using the Offset Entities command.

How-To 5-7: Using the Offset Entities Command

You would implement the Offset Entities function just like you would the Convert Entities function. Select the edges you want to offset, and then click on the Offset Entities icon, shown in figure 5-36. However, the two

Offsetting Entities

commands diverge at that point. To use the Offset Entities command, perform the following steps (listed in their entirety).

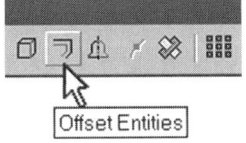

Fig. 5-36. The Offset Entities icon.

1. Select the model edges (or sketch geometry) to be offset.
2. Click on the Offset Entities icon located on the Sketch Tools toolbar. A preview will appear at this point.
3. Enter an offset distance in the Offset Entities panel displayed in PropertyManager (shown in figure 5-37).
4. Select the Reverse option to reverse the offset if necessary.
5. Click on OK when finished.

Fig. 5-37. Offset Entities panel in PropertyManager.

In the case of converting entities, you must select the entities first. That is not the case here. Select the entities to be offset either before or after clicking on the Offset Entities icon.

Offset entities will have an associated dimension that is added automatically by SolidWorks. This offset dimension is fully parametric and can be changed just like any other driving dimension. The Bi-directional option, like the name implies, offsets in two directions at the same time.

The Select chain option only applies to offsetting sketch geometry. Any sketch geometry can be offset, including geometry in the current sketch. If Select chain is checked and a sketch entity is selected, any sketch geometry connected to the selected segment will also be offset. This makes selecting sketch geometry much easier.

✓ **TIP:** *Instead of supplying a dimension value when offsetting entities, hold the left mouse button down and drag the preview to dictate the offset distance and direction.*

In Exercise 5-2 you will edit the sketch used to create the revolved feature and modify some of the properties of the dimensions. You will also create a groove feature that will later be used to create a circular pattern. This pattern will represent the portion of the valve stem used to grip the handle when the faucet is assembled. Some other processes you will get some practice with in this exercise are working with sketched chamfers, converting entities, and trimming.

EXERCISE 5-2: Editing the Revolved Feature Sketch

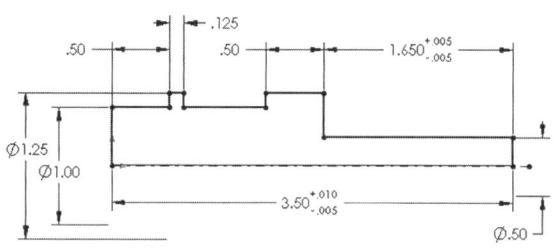

Fig. 5-38. Adding tolerance values and modifying dimension properties.

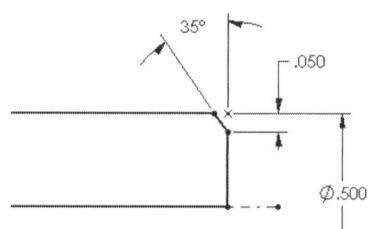

Fig. 5-39. Adding a sketched chamfer.

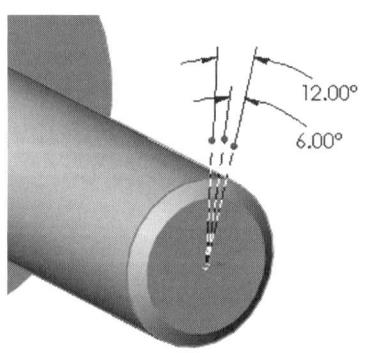

Fig. 5-40. Adding construction lines and dimensions.

Begin this exercise by opening the valve stem part you created in Exercise 5-1. Use some of the techniques learned in the Dimension Properties section to modify the way the dimensions appear in the sketch for the revolved feature.

1. Right-click on the Base-Revolve feature in FeatureManager and select Edit Sketch.

2. By changing the properties of some dimensions, try to make your sketch appear like that shown in figure 5-38. Add tolerances as required.

3. When you have finished modifying the dimensions, add a chamfer to the sketch, such as that shown in figure 5-39 (image enhanced for clarity).

4. Click on the Rebuild icon when finished modifying the sketch.

5. Start a sketch on the small end of the stem where the chamfer was just added.

Next, you will be creating the sketch profile used to cut the groove into the small end of the stem. This will most easily be accomplished by starting out with some construction geometry. You will also use Convert Entities, the Trim function, and the Construction Geometry command to finish up the sketch.

6. Add some construction geometry (centerlines), as shown in figure 5-40, to help control the sketch used to cut the groove. Add angular dimensions as well.

7. Convert the outer edge of the chamfer into sketch geometry. Add two lines, as shown in figure 5-41. Pay attention to system feedback!

8. Trim back the portion of the converted circular edge that is not needed, as well as the ragged ends of the construction lines (just for the sake of neatness).

Offsetting Entities

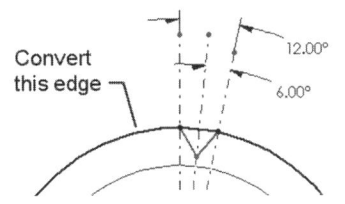

Fig. 5-41. Converting an edge and adding two lines.

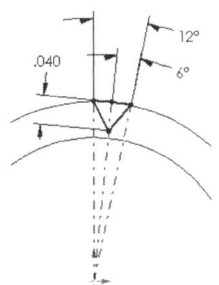

Fig. 5-42. The finished sketch.

9. Add a .040-inch dimension to control the depth of the groove.

10. Modify the properties of any dimensions as desired. Figure 5-42 shows the finished sketch. Note that the .040-inch dimension is an aligned dimension, not a vertical.

11. Create an extruded cut .50 inch deep.

12. Rename the new feature *Groove*. The complete groove feature is shown in figure 5-43.

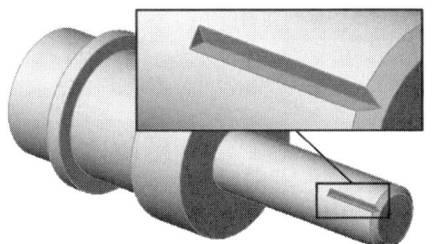

Fig. 5-43. The finished groove feature.

Next you will add a revolved cut, which will allow for the placement of an O-ring washer. First, let's resize the Right plane so that it more closely matches the size of the part.

13. Show the Right plane. (Hint: use your right mouse button in FeatureManager.)

14. Change to a Right view.

15. Right-click on the Right plane in FeatureManager and select Autosize. The plane should shrink to match the size of the part mode accurately.

16. Start a sketch on the Right plane.

17. Create the sketch shown in figure 5-44. You will need to add a coincident relation between the center of the circle and the silhouette edge of the valve stem.

18. Make sure you add a centerline along the axis of the model for the revolved feature.

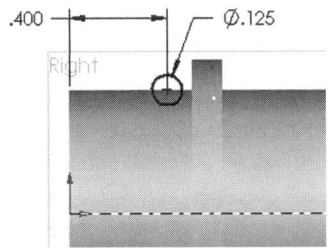

Fig. 5-44. Sketch for the O-ring groove.

19. Create a revolved cut 360 degrees around the model.

20. Rename the feature *O-ring groove*.

21. Save your work.

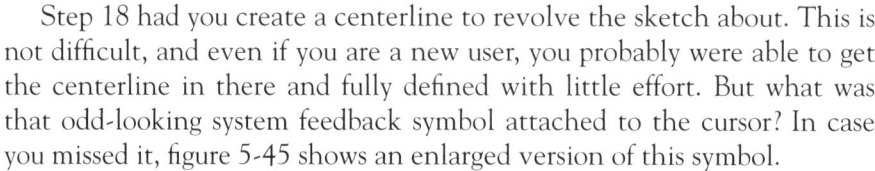

Fig. 5-45. System feedback: coincident with an axis.

Step 18 had you create a centerline to revolve the sketch about. This is not difficult, and even if you are a new user, you probably were able to get the centerline in there and fully defined with little effort. But what was that odd-looking system feedback symbol attached to the cursor? In case you missed it, figure 5-45 shows an enlarged version of this symbol.

The object above the cursor arrow is a light bulb with rays of light coming out of it. You have seen this before by itself. It is the system feedback you see when a coincident relation is about to be added: the light bulb is "on," meaning that the cursor is "on" an object (i.e., coincident). To the left of the arrow is what appears to be a centerline, but in actuality this represents an axis. To make a long story short, this symbol means that whatever you are sketching will be coincident with an axis.

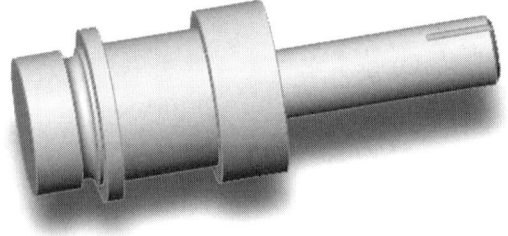

Fig. 5-46. The valve stem with a few new features.

Another system feedback you should have seen is the symbol for a silhouette edge. This symbol looks like a cylinder with one dark line traveling its length. If you managed to get through the exercise correctly, the valve stem should look similar to that shown in figure 5-46. You will finish this part in subsequent chapters, so make sure you tuck it away for safe keeping.

Summary

Now that you have finished Chapter 5, you can add another feature type to your list of SolidWorks tools. Revolved features can be used to create turned or lathed parts, or any part that is radially symmetrical. You must have a centerline in the sketch in order to create a revolved feature.

When creating a sketch for a revolved feature, the sketch must not cross the centerline or touch the centerline at an isolated point. With regard to dimensioning, it is possible to dimension to a centerline and thereby create the appearance of a diameter dimension. When brought into a 2D layout, the diameter dimension will conform to typical drawing standards.

Modifying the properties of a dimension allows you to change many aspects of the dimension. These properties include changing the direction of the arrows, modifying decimal precision, changing between radial or diameter dimension types, extending a linear dimension to the tangency point of an arc or circle, and so on. Quite often, these dimensional properties are fine-tuned in the layout. Sometimes, however, you may decide to make changes directly in the part file. Use the green handles associated with dimension lines to flip the arrows from one side to the other.

Sketch chamfers and fillets can be created quickly and easily from within a sketch. If sketch geometry forms a perfect corner, clicking on the corner is enough to add a fillet. SolidWorks will add a virtual sharp if there are dimensions attached to the corner being filleted. Virtual sharps can be added manually to a sketch through the use of the Point command.

Other important sketch tools were also covered in this chapter. You learned the importance of the Convert and Offset Entities commands, and that any converted or offset sketch entity is dependent on and associated with the edge it was converted or offset from. The only major difference between the two commands is that Offset Entities gives you the added ability to specify an offset distance from the edge it was converted from.

You also learned how to use the Sketch Trim and Sketch Extend icons. SolidWorks will give you a preview of the entity to be trimmed or extended. Clicking on an entity will trim or extend that entity to the next object it encounters. If you want to be more precise, you can pick an entity and drag it with the left mouse button to the specific entity you want to trim or extend to.

Questions and Topics for Discussion

1. Name at least two rules to follow governing sketch geometry when creating a revolved feature.

2. Can you have more than one centerline in a sketch for a revolved feature? Explain your answer.

3. What is the difference between a Boss/Revolve and a Cut/Revolve operation?

4. Explain the procedure for changing a linear dimension into a diameter dimension. Additionally, describe what to select when adding such a dimension in order for this capability to exist.

5. What procedure would you use to dimension to a tangency point of an arc?

6. If you wanted every dimension in the part to have a precision of three decimal places, what setting would you change? How about if you wanted to change the precision of one dimension only?

7. When extending a line or arc, does it matter which side of the midpoint you click on? Why?

8. What happens if you try to trim an entity that does not intersect with anything?

9. What is the main similarity between Convert Entities and Offset Entities? Additionally, what relation is added when either of these commands are used?

10. What is the main difference between Convert Entities and Offset Entities?

11. Explain how to change the size of the font used for a tolerance value.

12. What relation is being added when the light bulb system feedback symbol appears?

13. Can the properties of more than one dimension be modified at the same time? Explain your answer.

Chapter 6

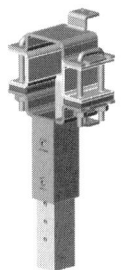

Molded Parts

IT WOULD BE IMPOSSIBLE TO COVER EVERY TYPE of molded part in a single book, much less in one chapter. Therefore, the type of component this book deals with is primarily the thin-walled type of molded part, usually plastic, that you might find in many places in your own household.

There are probably a dozen molded plastic parts within sight of where you are sitting. If you are reading this book at the beach on a deserted island, you are probably going to prove this author wrong. However, you get the idea. Plastic molded parts are very common.

The types of parts this chapter deals with are typical of many household appliances. For example, the casing of your telephone or answering machine is probably a plastic molded part. How about your electric shaver, or even your disposable razor handle? If you are working near your computer (assuming the reader of this book would own one), take a look at your mouse, or maybe the faceplate of the computer itself.

Many of these components would be considered thin-walled parts in SolidWorks terminology. The parts generally have a uniform wall thickness, probably with a few ribs added here and there for structural support. They usually also have a number of built-up plastic bosses that can be drilled through. These bosses would be used for putting screws through, in most cases, to attach the part to another component in the assembly.

This should give you a pretty good idea of where this chapter is heading. The main topics covered are the Shell feature and the Hole wizard. You will also learn how to add fillets, and how to use the Rib command to easily create ribs. Last, you will take a closer look at creating lofted features.

Thin-walled Parts

SolidWorks has two feature types that could be considered thin-walled parts. Therefore, it is necessary that you understand the terminology used by SolidWorks and in this book. The following sections explain the terminology and the concepts behind it.

Thin-feature Parts

The term *thin feature* is used to describe the feature type usually created with an open sketch profile. You can force SolidWorks to create a thin feature from a closed profile as well. The general concept behind thin features is to create the feature as you would any other feature type, such as an extrusion or revolved feature, but additionally specify a wall thickness. This is done directly though the Extrude or Revolve panel of PropertyManager, discussed in the previous two chapters. Chapter 8 covers how to create a thin-feature part, including steps that guide you through the process, along with an exercise.

Shelled Parts

When a part is shelled, a large portion of material is removed, leaving behind a specified wall thickness. The end result may be somewhat similar to a thin-feature part, but the technique used to carry out a shell command is quite different. Shelling is the first operation covered in this chapter, and is explained in detail in the following section.

The Shell Command

In its simplest form, the Shell feature is very easy to implement. In theory, the inside of the part is removed, leaving behind a thin wall. You have the ability to define the thickness of this wall. Typically at least one face is selected to be removed during the shell process. This might be akin to an opening in a part, through which excess material is poured out prior to its solidification in the mold. In SolidWorks, however, face selection is not required. If no face is selected, you wind up with a hollow part.

The Shell Command

An option available to you is to add a wall thickness to the outside of a part. This is something like dropping the part into a vat of molten wax and letting the wax cool and harden. The outer layer of wax would be all that is left, and the original part would be removed (assuming that this feat could actually be accomplished).

Shell Feature Panel

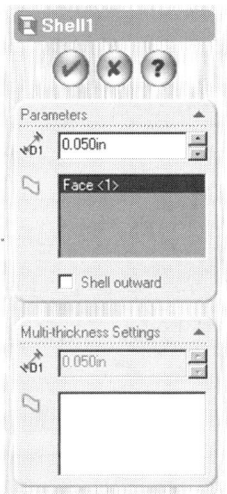

Fig. 6-1. Shell panel in PropertyManager.

The Shell feature panel is shown in figure 6-1. In particular, note the Parameter area. It is this area where a wall thickness must be specified and faces can be selected to be removed during the shell process. The *Shell outward* checkbox option allows you to create the type of shell analogous to the "hot wax" scenario previously described. How-To 6-1 takes you through the process of creating a simple shelled part.

HOW-TO 6-1: Creating a Shelled Part

To create a shelled part with a consistent wall thickness, perform the following steps.

1. Click on the Shell icon (shown in figure 6-2), or select Shell from the Insert > Features menu.
2. Specify a wall thickness for the shell.
3. Select the face or faces to be removed during the shell operation.
4. Specify *Shell outward* if necessary.
5. Click on OK to complete the shell.

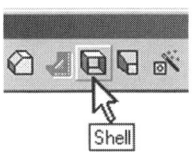

Fig. 6-2. Shell icon.

If you take a look at figure 6-3, you will see a simple example of the shell operation. On the left is the part before the shell was completed. On the right is the same part after the Shell command. In this example, a shell thickness of .050 inch was used, and the top face of the part was specified as the face to be removed.

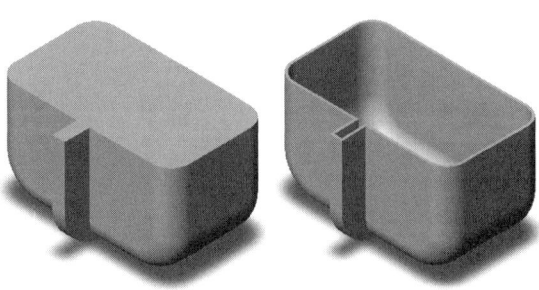

Fig. 6-3. Before and after performing a shell operation.

Fig. 6-4. The shell operation ignores the rib.

An important aspect of shelling is the built-in intelligence of the command. For instance, note the rib on the front side of the part in figure 6-4. This rib happens to be exactly .250 inch wide. Keeping this in mind, consider what might happen if the shell wall thickness were increased beyond an eighth of an inch. Would you expect the shell operation to fail? SolidWorks knows enough to ignore the rib and performs the shell anyway, as shown in the figure.

If we were to continue along these same lines, you would find that the same sort of built-in intelligence would ignore a rounded edge if the shell wall thickness were to increase beyond the radius of the round. The software would ignore the rounded edges and create sharp inside corners. The main point to be learned from this is that the shell operation continues to work, even in the face of geometric obstacles.

From a geometrical standpoint, the shell operation is a very difficult operation to perform. You must use some common sense when creating a shell. SolidWorks can only assume so much. Do not try to perform a shell if it is not physically possible. For example, do not try to perform a shell on a part with many small, intricate features.

There are some guidelines you might want to consider when performing a shell operation. One is that you should perform a shell early in the design process. Due to the Shell command's innate complexity, it is easier to perform a shell before you add features such as cosmetic fillets or threads.

Another consideration is the faces you are trying to remove during the shell process. For instance, you would typically not want to remove faces tangent to other faces. This would often result in some fairly odd-looking geometry, and earlier releases of Solid-Works would not perform a shell under such conditions. However, as the software matures, it becomes more powerful, and what once was not possible can now be done. Figure 6-5 shows the same sample part with an additional tangent face removed during the shell operation.

Fig. 6-5. Tangent face removed while shelling.

The Shell Command

When creating a shell feature, it is not mandatory to select a face to be removed. In other words, it is possible to create a completely hollow part. To perform this operation, simply execute the Shell command and do not select a face. You should still, however, specify a value for the desired thickness.

Multi-thickness Shell

One last function of the Shell command that should be covered here is its ability to perform what is known as a multi-thickness shell. What this means is that during the shell operation it is possible to specify varying degrees of thickness for specific faces.

In the Shell panel, you will see an area titled Multi-thickness Settings, shown in figure 6-6. If you click inside the list box within this area, it is possible to select faces on the model that are to have varying degrees of thickness. You can specify whatever thickness you want for individual faces, within reason. The same rules apply with multi-thickness shells as with standard shell features. The shell feature has to be physically (geometrically) possible or SolidWorks will refuse to process it.

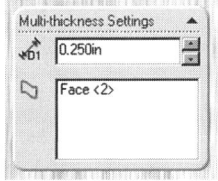

Fig. 6-6. Multi-thickness settings.

An example of a multi-thickness shell is shown in figure 6-7. In this example, the front face of the rib was selected. Note how SolidWorks displays the thickness value on screen, over the selected face. The end result is visible in the same image. A thickness of .050 inch was used for the overall thickness, but .25 inch was specified for the front face of the rib. How-To 6-2 takes you through the process of creating a multi-thickness shell.

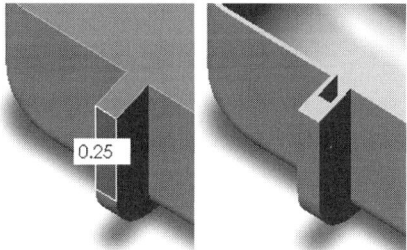

Fig. 6-7. Performing a multi-thickness shell and the end results.

How-To 6-2: Creating a Multi-thickness Shell

To create a multi-thickness shell, perform the following steps. Note that the first three steps are exactly the same for creating a shell of constant wall thickness. The same option for shelling outward is still available, but is not listed in the following steps, as it is not pertinent.

1. Click on the Shell icon, or select Shell from the Insert > Features menu.
2. Specify a wall thickness for the shell.
3. Select the face or faces to be removed during the shell operation.
4. Click in the list box area of the Multi-thickness Settings section to turn it a salmon color, and then select any faces that are to have a different thickness.
5. Select each face in the *Multi-thickness faces* list box and assign a thickness value to each. Make sure to assign a thickness value to every face listed if there is more than one. Values can be different if required.
6. Click on OK to complete the shell.

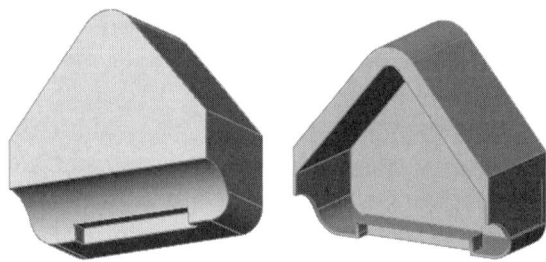

Fig. 6-8. Another multi-thickness shell example.

Figure 6-8 shows another example of what a multi-thickness shell might look like. Two faces at the front of the part, one planar and one curved, were selected for removal during the shell process. The three tangent faces that make up the "peak" of the model were selected as multi-thickness faces. These last three faces were given a larger wall thickness value than the rest of the model.

When creating a multi-thickness shell feature, it is not usually good practice to select faces that are tangent to other faces. The results can be unpredictable and potentially undesirable. If one face is selected that does have tangent faces, the tangent faces usually need to be selected as well. Otherwise, the shell operation may fail. This obviously depends on the type of part being created. You may just have to experiment to see what works and what does not.

Fillets and Rounds

Fillets and rounds can be added simultaneously without difficulty. Solid-Works does not really differentiate between the two. From a terminology

Fillets and Rounds

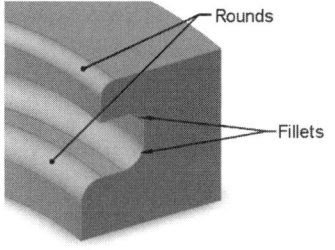

Fig. 6-9. Fillets and rounds.

standpoint, fillets are applied to inside edges and rounds to outside edges. Some simple examples are shown in figure 6-9.

NOTE: *For the sake of simplicity, both fillets and rounds are referred to as fillets throughout this book.*

Adding fillets is a very simple process, and in its simplest form you need only specify a radius for the edge being filleted. Whether you add a fillet as part of a sketch or as a feature is really up to you; the end result will be the same for simple fillets. More complex fillets need to be added as features, because the command parameters are much more flexible. In addition, feature fillets can propagate along tangent faces, whereas sketch fillets cannot.

As a general rule of thumb, it is usually best to add cosmetic fillets near the end of the design process. Where fillets affect the form or function of the model, they should be added earlier in the design process, rather than toward the end. Cosmetic fillets complicate the model and can slow down the graphics display due to their nature. More polygons are needed to render curved surfaces than flat surfaces. There is no sense in overly complicating the model if there is still design work to be done.

Fillets as Features

Instead of adding fillets in your sketch, you may very well decide to add them as a feature. There is not much difference in the end result, but the process is different. For example, two lines can be selected when adding a sketch fillet. But when applying a fillet to a model, the edge being filleted must be selected.

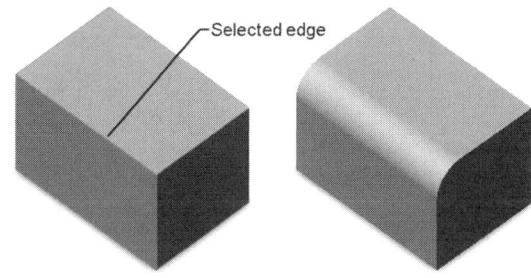

Fig. 6-10. Adding a fillet to a model edge.

When adding a fillet as a feature, you can select objects other than edges to be filleted. You can also select an entire face of a part, in which case every edge on that face would be filleted. Features can also be selected from FeatureManager, in which case every edge on the feature is filleted. What you select depends entirely on your design intent. Figure 6-10 shows an example of what would happen if an edge were selected on a particular part and a fillet applied. Keep in mind that you can select entities before or after the Fillet command is initiated.

Control-selecting: A Reminder

It is normally preferable to select objects after initiating a command, because you do not have to hold down the Control (Ctrl) key. As a reminder, hold the Ctrl key down anytime you want to select more than one entity prior to entering a command. This is standard operating procedure throughout SolidWorks.

Fig. 6-11. Wrapping a large-radius fillet around a small-radius fillet.

Large fillets are usually applied to a model first. It is generally desirable to have smaller fillets wrap around the tangent faces of a larger fillet. However, it is possible in SolidWorks to make a larger-radius fillet wrap around the faces of a smaller-radius fillet. This may not be a desirable outcome, however. Take a look at figure 6-11 for a better understanding of what is actually being accomplished in this type of situation. The edges are shown in black for clarity.

In figure 6-11, a small .125-inch-radius fillet was added first. The larger .5-inch-radius fillet was then added later. As you can see, the larger-radius fillet propagates around the tangent faces created by the first fillet. To add a fillet as a feature (not in a sketch), perform the steps outlined in How-To 6-3.

How-To 6-3: Adding a Fillet as a Feature

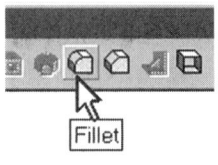

Fig. 6-12. Fillet icon.

First, bear in mind that these steps show how to apply a fillet directly to the model as a feature. The process is very simple, but you must not be editing a sketch when performing these steps. Note that the Fillet icon (shown in figure 6-12) can be found on the Features toolbar. To add a fillet as a feature, perform the following steps.

1. Click on the Fillet icon, or select Fillet/Round from the Insert > Features menu.

2. Specify a radius for the fillets in the Items to Fillet section, shown in figure 6-13.

3. Select the items to be filleted (typically edges).

4. Click on OK to create the fillets.

Fillets and Rounds

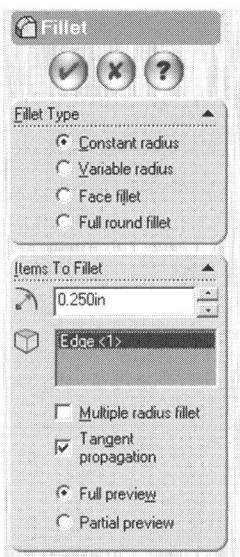

Fig. 6-13. Fillet panel in PropertyManager.

The previous steps show how to add a fillet, in its simplest form, using the Fillet command. Obviously, there is much more to the Fillet command than How-To 6-3 describes. Variable radius, face blend, and full round fillets are discussed in depth in Chapter 19. For now, though, some of the more important basic options you should be aware of are described in the material that follows.

Tangent Propagation

By default, SolidWorks will continue filleting any edges tangent to the original edge selected for the fillet. This is known as the *Tangent propagation* option, which can be turned off if desired. By default, *Tangent propagation* is always turned on, even if it was turned off the last time the Fillet command was used. When this option is turned off, SolidWorks miters any corners instead of flowing around tangent faces. Figure 6-14 shows two examples of a fillet, the same edge of which was selected in both cases. The first example shows tangent propagation turned off, whereas the last example shows the same part with *Tangent propagation* turned on.

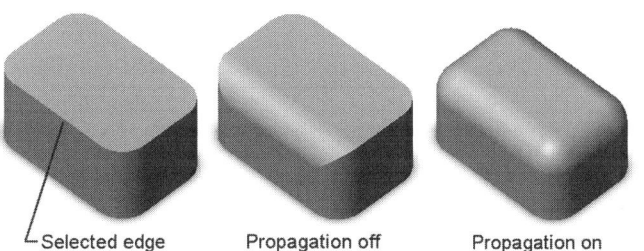

Fig. 6-14. Various effects of tangent propagation.

Multiple Radius Fillets

The *Multiple radius fillet* option is used when it is desirable to fillet more than one edge using different radial values in the same fillet command. Most of the time it is usually best to create similar radius fillets as the same feature. This just makes good sense, and also makes it easy to incorporate your design intent into the model. The *Multiple radius fillet* option is located above the *Tangent propagation* option in the Fillet panel.

It is difficult to come up with an example of where it would be better to create multiple-radius fillets as a single feature, as far as model geometry is concerned. The use of this option is more of a convenience issue than anything else. How-To 6-4 takes you through the process of adding multiple-radius fillets.

HOW-TO 6-4: Adding Multiple-radius Fillets

To create multiple-radius fillets as a single feature, perform the following steps.

1. Click on the Fillet icon, or select Fillet-Round from the Insert > Features menu.

2. Select the *Multiple radius fillet* option.

3. Specify a radius for the fillets (this can be changed later).

4. Select the edges to be filleted (faces can be selected, but they must not have common edges).

5. Select any object from the *Items to fillet* list box and specify the desired radius for that particular object. Use the Ctrl key to select more than one object at a time.

6. Click on OK to create the fillets.

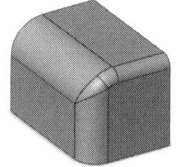

Fig. 6-15. Three fillets created simultaneously.

An example of fillets created with the *Multiple radius fillet* option is shown in figure 6-15. The fillets use radial values of .25, .50, and .75 inch. Note in particular the corner at which the fillets meet. It would be possible to achieve this same effect without the *Multiple radius fillet* option, but it would be more time consuming and would require three separate fillet features. To duplicate this same look, one would have to add the fillets independently, with the *Tangent propagation* option turned off. This could certainly be accomplished, but would not be nearly as efficient.

Fillet Previews

Fillet previews come in two varieties: full and partial. If the *Partial preview* option is used, only one edge will show a preview, even if a face is selected. The *Full preview* option shows previews for all edges that wind up being filleted, regardless of what was selected for filleting. Unless your computer is an older model and has a difficult time crunching data, you may as well use the *Full preview* option.

Filleting Faces

As mentioned previously, it is possible to fillet every edge on an entire face at once. You simply select the face to be filleted and SolidWorks adds fillets to every edge on that face. Figure 6-16 shows an example of this.

Fillets and Rounds

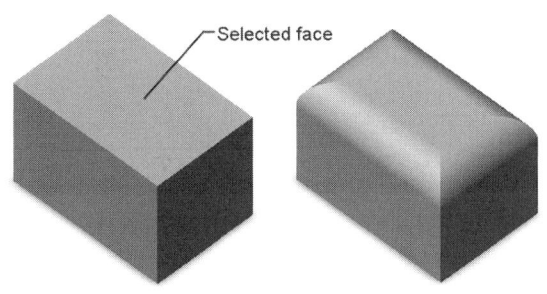

Fig. 6-16. Adding a fillet to every edge of a face.

Note that the corners of the face shown in figure 6-16 look as though they have been mitered. This is because there was no other way for SolidWorks to accomplish the fillet. If the outside corners had been filleted first, the fillet would have wrapped smoothly around the top edges of the part. This is just another example of how the order in which fillets are added can be very important.

Use care when filleting faces. Quite often you get more than you bargained for. Sometimes a face has more edges than was readily apparent. It is also likely that faces may contain outside and inside edges that produce undesirable results when filleted together as part of the same feature.

Filleting Features

We have been discussing adding fillets as features. But features themselves, such as bosses and cuts, can be selected when adding fillets. Features consist of faces and model edges. Because of this, selecting a feature to be filleted must be done carefully. Otherwise, a face or edge may be filleted instead of the entire feature.

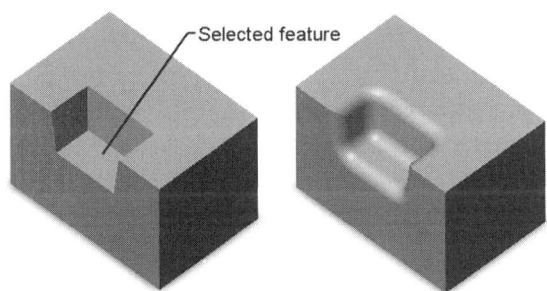

Fig. 6-17. The result of filleting a cut feature.

To select the feature correctly, do not simply click on the feature in the work area. Rather, right-click on the feature and use the *Select feature* option found in the menu while in the Fillet command. Another option would be to use the fly-out FeatureManager, mentioned in Chapter 2 (click on the command name in Property-Manager). Figure 6-17 shows an example of what happens when a feature is selected for filleting. In this first example, the feature is a cut.

Bosses can be filleted the same way as cuts. When features are selected for filleting, there is another option that plays an important role. This option is known as *Omit attach edges*, shown in figure 6-18. This option is only available if a feature has been selected. It is in the section of the Fil-

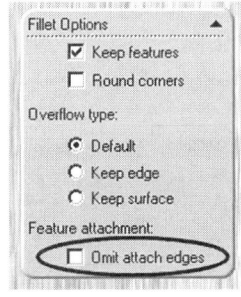

Fig. 6-18. Omit attach edges *option.*

188   Chapter 6: Molded Parts

let command panel labeled Fillet Options. (Material to follow deals further with the remainder of the fillet options.)

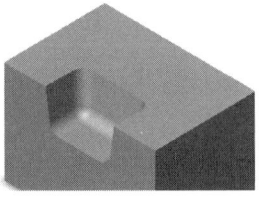

Fig. 6-19. Filleting features with the Omit attach edges option checked.

When Omit attach edges is checked, none of the edges that form a boundary between the existing model and the selected feature are filleted. Figure 6-19 shows two examples of this. The first example shows the same cut pictured earlier, only this time *Omit attach edges* has been checked. There is also a boss extrusion with the same option checked. Keep in mind that the entire features were selected for filleting, rather than edges or faces, which greatly simplifies the selection process.

Fillet Options

There is a section of the Fillet command panel titled Fillet Options, shown in figure 6-18. It contains a number of optional settings you should be aware of. These options are explored in the sections that follow.

Round Corners

Fig. 6-20. Note the difference with Round corners enabled (at right).

Only certain types of geometric situations can take advantage of the *Round corners* option. There are no special instructions required for using this option. Simply enable it by checking the option. Figure 6-20 shows an example of a part with *Round corners* disabled (left), and with the same option enabled (right).

Keep Features

Simply put, when the *Keep features* option is checked, features affected by the fillet are retained. This refers to features that are not being filleted themselves but are positioned on faces being directly affected by the fillet. Figure 6-21 shows this option in action better than can be explained.

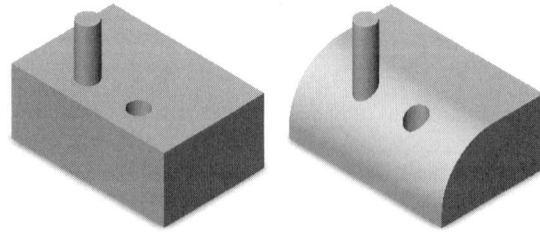

Fig. 6-21. Features remain after fillet is applied.

The *Keep features* option is checked by default, which is more than likely the way you will want to leave it. Traditionally, a SolidWorks user should have a good idea

Fillets and Rounds

of the order in which the features are to be added to the model. As many readers are aware, however, this is often not the case! Models change and revisions are almost inevitable.

Large fillets that affect the form or function of the model should be added prior to adding smaller features, which will in turn interact with the face of the fillet. Because foresight is not always as accurate as we might like, the *Keep features* option gives you a little more flexibility in creating the model. The option can be turned off, but in doing so any fillet created will absorb features affected by the fillet. In other words, the features will disappear from the model.

> **NOTE:** *Features absorbed by a fillet are not deleted from the model, they just disappear. This is due to a fillet's definition and geometric necessity. The features are still present, and if the fillet is deleted, the affected features will reappear.*

Overflow Control

When a fillet extends beyond an area that can physically accommodate it, something has to give. This is known as "overflow" in SolidWorks terms, and how the geometry surrounding the fillet is affected depends on the overflow setting. To make a long story short, either the surface of the fillet can change or the surrounding geometry can change. The choice is yours.

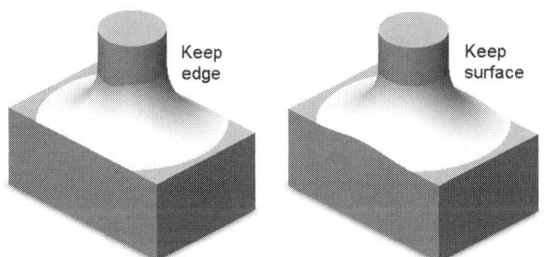

This is an option easily understood, as there are only two settings. Those settings are *Keep edge* and *Keep surface*. There is also a setting of Default, but that just means that SolidWorks will use the *Keep edge* setting. An example of each setting is shown in figure 6-22. A description of these settings follows the illustration.

Fig. 6-22. Overflow control for fillets.

Keep Edge

When *Keep edge* is used, the edge where the fillet overflows the geometry is kept the same. In other words, the edge does not change to accommodate the fillet. Instead, the fillet changes to accommodate the edge. In figure 6-22, note that the top of the block remains straight, whereas the top of the fillet dips down. The radius of the fillet is not changing. Rather, picture a rolling ball positioned between the block and cylinder. It must move away from the axis of the cylinder so that the edge of the block can remain consistent. This accounts for the dip you see where the fillet meets the cylinder.

Keep Surface

When *Keep surface* is used, the surface of the fillet remains constant, and the edges where the overflow takes place alter to accommodate the fillet. Note in figure 6-22 that the fillet appears very consistent, with the top of the fillet appearing as a perfect circle around the cylinder. The top edge of the block, on the other hand, is pulling upward. As with *Keep edge*, the radius of the fillet never changes. The center of the fillet remains at the correct distance from the cylinder, whereas the surrounding edges of the tapered pad adjust to conform to the surface of the fillet.

Selection Techniques

In Chapter 2 you learned that the left mouse button was used to select objects, and in Chapter 3 you discovered additional methods for selecting objects (see Table 3-1: Entity Selection Options). Now let's examine the remaining selection techniques that can be employed to select sketch entities, model edges, and features. Some of these techniques will prove invaluable when it is necessary to select model edges for filleting. These selection techniques will also prove very useful elsewhere, so remember them well.

Select Midpoint

This topic was originally discussed in Chapter 4, so we need not spend much time on it. The following rules apply in selecting midpoints.

- If in a sketch command (such as sketching lines), moving the cursor over another object will display the midpoint system feedback, allowing you to constrain to that midpoint.

- If it is necessary to select a midpoint (for whatever reason) and no system feedback is offered, right-click on the object and use the Select Midpoint option.

- If in the Add Relations command, selecting a midpoint is not possible, even by right-clicking on an object.

Why is it not possible to force SolidWorks to select a midpoint when in the Add Relations command? The reason has to do with how a midpoint relation can be added. It is a matter of semantics. Consider this: It is possible to add a midpoint relation between a point and a line. It is possible to add a coincident relation between a point and a point. It is not possible to add a midpoint relation between a point and a point (a midpoint is considered a point).

Select Chain

A chain is nothing more than a series of sketch entities connected end to end. When right-clicking on a sketch entity, the Select Chain option is presented, allowing for the selection of all entities in the chain.

Select Loop

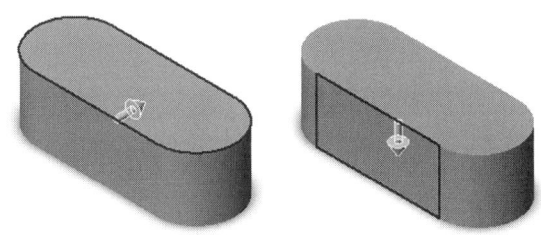

Fig. 6-23. Using Select Loop sometimes offers two choices.

Loops are similar to chains, but typically refer to model edges as opposed to sketch entities. Loops usually, but not always, have one or more edges common to a pair of loops. To put this another way, right-clicking on a model edge allows for selecting all edges that wrap around either of the edges' adjacent faces. Examine figure 6-23 to get a clearer picture of this.

When the Select Loop option presents you with two choices, using the left mouse button, click on the arrow to toggle to the other loop. Once the arrow is pointing in the proper direction, simply continue with whatever command is required to keep working.

Select Tangency

Like the Select Loop option, this option is available when right-clicking on model edges. As the name implies, all edges tangent to the original edge will be selected. Occasionally this option is necessary, even for commands such as the Fillet command. Even though the Fillet command contains an automatic tangent propagation function, using Select Tangent often gives SolidWorks an extra little boost in difficult situations.

Select Other

This is one of the most useful of the various selection options. It is also the trickiest to learn. Use Select Other to select edges or faces that prove difficult or impossible to pick any other way. Also use it to pick faces that are on the other side of the model, facing away from you, or that are located deep within components of an assembly. It is often much more convenient to use Select Other than to rotate a model to get at a particular model face. How-To 6-5 takes you through the process of the Select Other technique.

How-To 6-5: Using Select Other

When model faces are on the side of the model facing away from you, they are impossible to select using solely the left mouse button. Use the Select Other selection technique, outlined in the steps that follow, to select such faces without rotating the model.

1. Right-click over the face to be selected. Cursor position is critical! The cursor must be positioned directly over the face you are trying to select. This can obviously be difficult if working in Shaded display mode, because the face will not be visible. Change display modes if necessary.

2. Select Select Other from the menu.

3. Continue right-clicking until the proper face is highlighted.

4. Left-click to select the currently highlighted face.

After right-clicking in step 1, it does not matter at all where the cursor is positioned. Cursor positioning is only critical when initially right-clicking to implement the Select Other option. Note that the cursor has a small mouse displayed as system feedback. There is a small Y on the left button, and an N on the right. Once into the Select Other option, it is as if you are telling SolidWorks "No, that's not the face I want, try the next one" every time you click with the right mouse button. When you left-click, that tells SolidWorks "Yes, that's the one I want." SolidWorks selects the face, and you can continue your work.

Chamfers

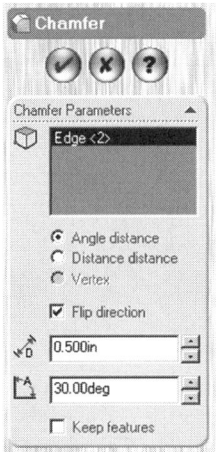

Fig. 6-24. Chamfer panel.

Chamfers are easier to add to a model than fillets because there are fewer options associated with chamfers. Basically, though, chamfers are applied the same way. Start the command, pick the objects to be chamfered, plug in some values to establish the size, and click on OK. There are a few other options that should be expanded upon, though. The sections that follow explore the three main options for creating chamfers.

The Chamfer panel is shown in figure 6-24. Like fillets, it is possible to select either an edge or a face when adding chamfers. Just remember that if a face is selected, chamfers will be applied to every edge on that face, so use discretion. Adding sketch chamfers was discussed in Chapter 5. In the material that follows, chamfers are applied directly to the model.

Angle-Distance

Angle-distance chamfers require you to specify a distance and an angle to define the chamfer dimensions. The preview arrow points in the direction the distance will be applied. Depending on the display mode, a cross-section representation of the chamfer should be visible as well. There will also be a callout with the current dimensions displayed. These graphical cues can all be seen in figure 6-25.

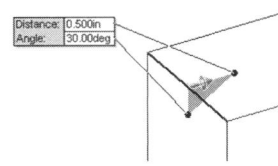

Fig. 6-25. Distance and angle parameters can be used to define a chamfer.

When the angle-distance chamfer parameters are used, there is an option that allows for flipping the direction of the preview arrow. The option is appropriately named *Flip direction*. Likewise, clicking on the arrow (which indicates the direction the distance value will be applied) will also flip the chamfer. The preview will update accordingly.

Distance-Distance

When defining a chamfer using the distance-distance parameters, there is no preview arrow or option to "flip" the chamfer. Use the callout and cross-section preview to tell how the distance values will be applied. Figure 6-26 shows what a user might see when adding a distance-distance chamfer.

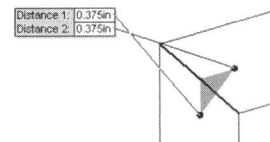

Fig. 6-26. Using two distances to define a chamfer.

If creating a chamfer using the Distance-Distance method, you will have the option *Equal distance*. This disables the setting for the second distance parameter and automatically sets that distance to whatever the first distance happens to be.

✓ **TIP:** *Be aware that parameters can be changed via the callouts. This holds true for any command that utilizes callouts, not just for chamfers. Click on the dimension to be changed, type in the desired value, and press Enter.*

Vertex Chamfer

Vertex chamfers are a little different than their counterparts. Only vertices can be selected for this type of chamfer, and three dimension values must be specified. There will be a cross-section preview, which is critical in determining which dimension values are being applied to each of the three edges that meet at the selected vertex point. Figure 6-27 shows an example of a vertex chamfer and preview.

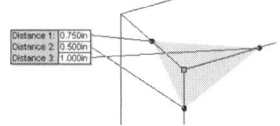

Fig. 6-27. Creating a vertex chamfer.

If this chamfer type is used, there will be an *Equal distance* option, equivalent to the *Equal distance* option in the Distance-Distance chamfer type. In the case of the vertex chamfer, equal distances are automatically assigned to all distance parameters. There is a *Keep features* option in the Chamfer panel as well, no matter which type of chamfer is being created. This option has the same meaning with regard to chamfers as it did in the Fillet command. (See the foregoing "Fillet Options" section for reference.)

By the way, vertex chamfers can only be created on corners with three edges, though the adjacent faces certainly do not need to be orthogonal. Geometry containing vertices with more than three edges is not all that common anyway.

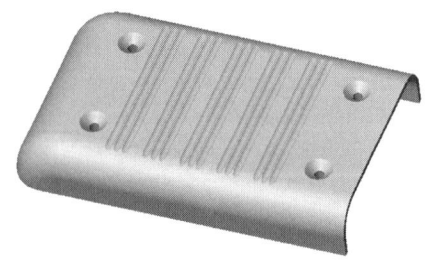

Fig. 6-28. Completed shaver housing.

Now it is time for you to put to practice some of the topics you have learned so far in this chapter, along with some of the other knowledge you have gained. In Exercise 6-1, which follows, you will begin creating a shaver housing, shown in figure 6-28. It is a fairly simple plastic part that makes up part of an electric razor. The features you will use to begin this part are features you learned previously. These features are an extrusion and a revolved feature. Afterward, you will add some fillets and a shell feature.

⇢ **NOTE:** *You will begin to notice that fewer and fewer steps are spelled out for you as you continue through this book. Some things you will be expected to remember from previous chapters and/or material. The exercises will get slightly more advanced without offering quite as much attention to detail. If you get stumped, return to earlier portions of the book so that you can refresh your memory on topics as necessary.*

Chamfers

EXERCISE 6-1: Creating a Shaver Housing

Begin a new part and save it as *Shaver Housing*. Start by sketching on the Front plane. You will begin the shaver housing here, by performing the following steps, and finish it in Chapter 7.

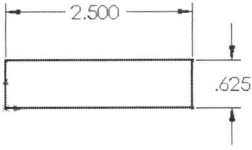

Fig. 6-29. Complete this sketch on the Front plane.

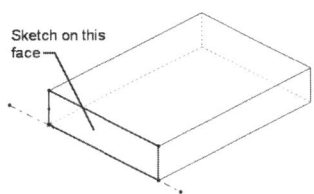

Fig. 6-30. Preparing to create a revolved feature.

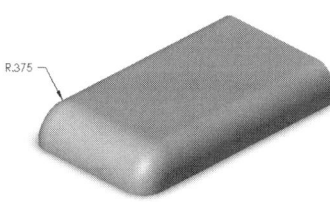

Fig. 6-31. After adding the fillets.

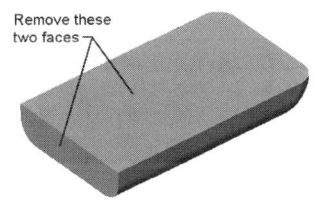

1. Create the sketch shown in figure 6-29, with dimensions.
2. Perform a Base Extrusion.
3. Specify a Blind end condition of 3.5 inches. Do not add any draft.
4. Click on OK to complete the extrusion.
5. Change to an Isometric view.
6. Start a new sketch on the front end of the part, as shown in figure 6-30.
7. Use Convert Entities to extract a new sketch, and then add a centerline (see figure 6-30).
8. Create a revolved feature that revolves 90 degrees. Flip the direction of revolution if required. As guidance, examine the preview SolidWorks presents.
9. Add .375-inch fillets to either side of the housing, as shown in figure 6-31. You will need to select one edge on either side of the part.

Next you will perform a shell operation. Figure 6-32 shows the correct two faces to select for removal during shelling. The model has been flipped upside down for illustration purposes. If your part does not look like that shown in the figure, try repeating the exercise and figure out where you went wrong. Otherwise, continue with the following steps.

10. Create a shelled part by selecting Shell from the Insert > Features menu.
11. Select the two faces shown in figure 6-32.

Fig. 6-32. Selecting faces to be removed during the shell operation.

12. Type in a wall thickness of 1/16 (inches) for the shell feature. Try actually typing in the text *1/16*. You will find that it converts automatically to .0625 inch (rounded to .063 if using three-place precision).

13. Click on OK to complete the shell.

14. Save your work.

It is possible to type in mathematical operators when being asked for dimensions. This essentially works as a built-in simplified calculator. It is possible to use mathematical operators such as the plus sign (+), minus sign (–), asterisk (*), and forward slash (/). These symbols function as plus, minus, multiply, and divide operations, respectively. You can also use parentheses, the carat symbol (^) for exponents, and trig functions. Table 6-1 outlines mathematical operators used for dimension values.

Table 6-1: Mathematical Operators Used for Dimension Values

Symbol	Meaning	Example	Outcome
+	Add value	.0625+1.375	1.4375
–	Subtract value	.325-.03125	.29375
*	Multiply values	3*.625	1.875
/	Divide values	1/16	.0625
^	Raise to the power of	2^.25	1.1892...
( )	Parentheses	(1/16)+(.125*2)	.3125
sin	Sine of (value)	sin(1)	.8415
cos	Cosine of (value)	cos(1)	.5403
tan	Tangent of (value)	tan(1)	1.557

If you need to use the trig functions, note that SolidWorks likes to work in radians, not degrees. If you are not used to working in radians, you can work in degrees and convert to radians. There are 2π radians in 360 degrees. Therefore, to convert to radians, divide the number of degrees by 180 and then multiply that value by π. (Trigonometric functions are best left to a mathematics class and are beyond the scope of this book.)

The shaver housing should now look like that shown in figure 6-33. The next portion of this exercise will have you create a boss. This boss will create material that can be drilled through for connecting one side of the housing to the other. Later on, these features will be patterned, which you

Chamfers

will learn how to do in Chapter 7. For now, finish Exercise 6-1 with the following remaining steps.

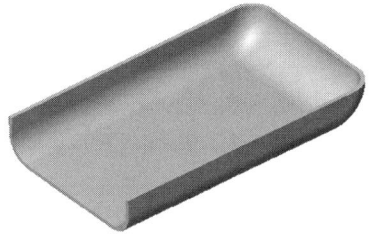

Fig. 6-33. After completing the shell feature.

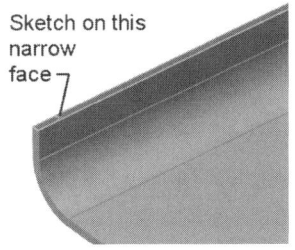

Fig. 6-34. Beginning a new sketch.

15. Start a new sketch on the narrow face created on the bottom of the part during the shell operation. Use figure 6-34 for reference. Zoom in if necessary!

16. Change to a plan view (using the Normal To View option).

17. Create the sketch shown in figure 6-35. Optionally, create a virtual sharp to dimension to.

18. Create a boss extrusion.

19. Use the Up To Next end condition.

20. Add 3 degrees of draft, and make sure draft is applied outward.

21. Click on OK to complete the operation. Figure 6-36 shows the boss added with draft, both before and after completing the extrusion.

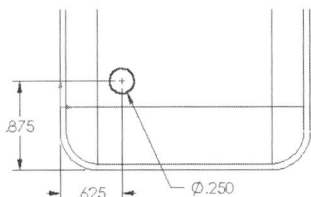

Fig. 6-35. Creating a sketch for the boss.

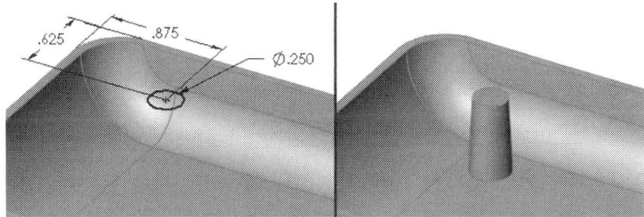

Fig. 6-36. Creating the extruded boss from a circle.

22. Rename the boss-extrude feature *Boss1*.

23. Save your work. You will need this part later.

The Hole Wizard

The Hole wizard is an easy and convenient means of creating a wide variety of hole types. Holes can be counterbored or countersunk, drilled or tapered, and many combinations of these options. There are sets of steps, or windows, that walk you through creating a hole. Creating a hole using the Hole wizard is not difficult, but there are a lot of parameters that need to be set carefully to obtain the proper hole size.

When you create a hole using the Hole wizard, SolidWorks remembers the hole callout data. Later, when a drawing layout is created, you have the opportunity to use that callout information in the drawing. You will explore hole callouts in Chapter 11, along with other annotations.

SolidWorks has a special way of handling holes created with the Hole wizard. Specifically, two sets of sketch geometry are created. The first sketch consists of nothing more than point entities, also known as sketch points. There may be a single point, or there may be many. The point is a locational point you use to place the center of the hole via constraints or dimensions.

The second sketch is used to create the hole itself. The hole is created from a revolved feature. A sketch profile of half the hole's center cross section is automatically created and then revolved about a centerline, with the centerline passing through the locational point. Even though there is a sketch (or rather, two sketches), a hole created with the Hole wizard is considered an applied feature because you do not have to manually create the sketch on your own.

All of this is transparent to the person creating the hole. All you do is fill in the desired dimensions and parameters in the Hole Wizard window, specify some locations, and SolidWorks does the rest. To create a hole using the Hole wizard, perform the steps outlined in How-To 6-6.

How-To 6-6: Creating a Hole with the Hole Wizard

To create a hole with the Hole wizard, perform the following steps.

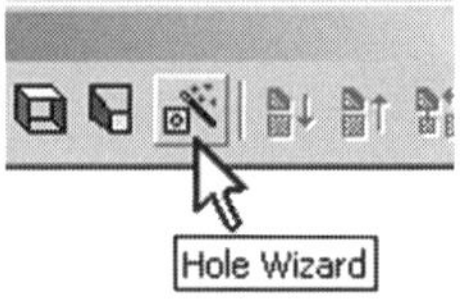

Fig. 6-37. Hole Wizard icon.

1. Select a face on the part on which to place the hole. You should not be in a sketch when you do this!

2. Select Insert > Features > Hole > Wizard, or click on the Hole Wizard icon found on the Features toolbar and shown in figure 6-37.

The Hole Wizard

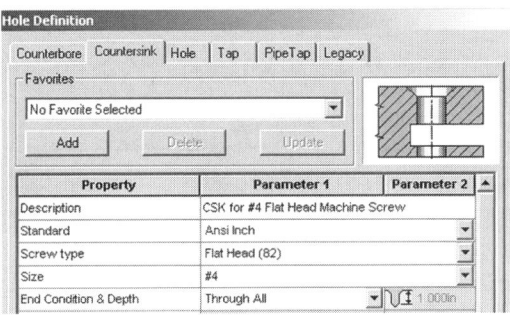

Fig. 6-38. The Hole wizard's Hole Definition window.

3. Select the appropriate tab for the type of hole you want to create. The Hole Definition window is shown in figure 6-38.

4. Specify the parameters to define the size of the hole. You may have to scroll through the list.

5. Make sure to adjust the end condition parameter as required; for example, using Through All, Blind, and so on.

6. Click on the Next button.

At this point you will be in the Point command. This often confuses those new to SolidWorks. Be aware that you do not have to stay in the Point command. Your choices are to either add points where copies of the hole will go, or add dimensions or relations to constrain the points already there. Actually, both of these tasks should be performed, but if you only want the one point (and subsequently the one hole), exit the Point command and constrain the single point to the proper location.

7. Add more points wherever a copy of the hole should be located.

8. Add dimensions or geometric relations to locate any points you created.

9. Click on Finish.

Depending on the hole type you decide to create, there will be different dimensions to supply values for. You may also need to select certain types of geometry to define the end condition. For instance, if the end condition were Up To Surface, it would be necessary to select a surface. Most of the time the end condition will be Blind or Through All, in which case you will not have to worry about selecting an additional surface.

When finished creating the hole, you will notice a new feature in FeatureManager. If you expand this feature (click on the little plus sign to the left of the feature's icon), you will see the two sketches used to create the hole. Either sketch can be edited. The sketch containing the locating points can have points added or removed from it, thereby changing the number of holes.

Legacy Tab

Just what is the Legacy tab in the Hole Wizard window? It is something you will probably want to avoid unless editing older SolidWorks models. The Legacy tab is what the old Hole wizard used to look like. It remains in the new Hole wizard for those times when an older model needs editing. For anything created in a newer version of the software, the other tabs in the Hole wizard present a much wider array of options.

Holes on Nonplanar Faces

The Hole wizard will indeed allow for placing holes on nonplanar faces. Faces can be cylindrical, conical, spherical, or any type of free-form shape. SolidWorks handles this by creating the locating points for the holes in what is known as a 3D sketch, explored in Chapter 9.

What is important to understand from a user standpoint is the order in which a face is selected and the Hole wizard command initiated. If a planar face is selected prior to the command being initiated, a 2D sketch will be used for the hole-locating points. This is more efficient for the computer to process. Additionally, it is much easier to locate points in a 2D sketch than in a 3D sketch.

If the command is started prior to picking a face to place the hole (or holes) on, a 3D sketch will be used for the locating points. This is true even if the selected face is planar! You want to be careful with this, because 3D sketches are processed differently. For example, placing points on different faces of the model, even though the faces are coplanar, may result in the Hole wizard failing to create all the holes.

✓ **TIP:** *Although it is not required, you should always select a face prior to starting the Hole wizard! This allows SolidWorks to create the proper locating sketch (2D or 3D) according to the face selected.*

Adding Favorites

Each tab, with the exception of the Legacy tab, has a section called Favorites. This Favorites section is used to save Hole wizard parameter settings for hole types you use a lot. You can add as many favorites as you wish. Each tab will have its own Favorites listing, so do not worry that they will get mixed up between tabs or hole types.

To add a favorite hole size, simply click on the Add button. All of the parameters currently set will be saved for the hole. You will be asked what to name your favorite, and you can accept the default name, add to it, or type in something completely different. This makes recalling your preferred settings for particular hole types very easy. Figure 6-39 shows an example of what you might see after clicking on the Add button.

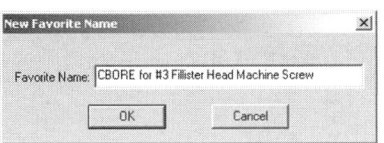

Fig. 6-39. Naming favorite hole parameters in the Hole wizard.

The Update and Delete buttons are fairly self-explanatory. Clicking on the Delete button will delete the currently displayed favorite from the drop-down list. The Update button will only be active if a change to one of your favorites is made. You can then decide on whether or not to update the current settings so that the favorite will always reflect those changes. The choice is yours.

Simple Holes

When it comes to creating simple holes that do not contain special geometry, such as countersunk or counterbored holes, you basically have two options. These options are explained in the next two sections.

Using the Hole Wizard for a Simple Hole

It is possible to use the Hole wizard to create a simple hole by specifying the Hole tab and plugging in the desired parameters. One of the benefits of using the Hole wizard for such basic holes is that the parameters for specific hole sizes are already spelled out. Hole sizes range from a #97 hole (.0059 inch) all the way up to 1 inch, though that does not stop you from using whatever diameter you want to type in.

Other benefits would be the ability to enter near- or far-side countersink diameters and angles, to enter drill point angles (assuming blind end conditions), and to add holes simultaneously. There is another option for creating simple holes that is not quite as flexible. This option is explained in the following material.

Using the Simple Hole Command

In the same pull-down menu that contains the Hole wizard is the option Simple. This is not to be confused with the Hole tab mentioned in the previous section. The interface for the Simple Hole routine is much different and very bare bones. How-To 6-7 takes you through the process of creating a simple hole using this method.

How-To 6-7: Using the Simple Hole Command

To create a hole using the Simple Hole command, perform the following steps.

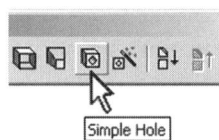

Fig. 6-40. Simple Hole icon.

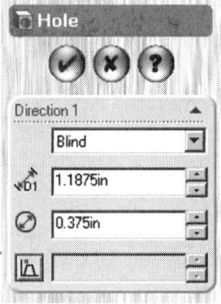

Fig. 6-41. Creating a simple hole.

1. Select a planar face on which you want to place the simple hole (this is required).

2. Select Simple from the Insert > Features > Hole menu, or click on the Simple Hole icon, shown in figure 6-40.

3. Enter values for end condition, diameter, and so on (see figure 6-41).

4. Click on OK to create the hole.

As you may have noticed, the Hole panel (shown in figure 6-41) is essentially a stripped-down version of the panel displayed when creating an extruded cut. You can specify an end condition, add draft, and do nearly anything you would normally be able to do when performing a basic extrusion. The only additional option is for the diameter of the hole.

On the downside, the user never gets the opportunity to tell Solid-Works just where it is the hole should be located. None of the sketch tools are available, and the opportunity to add points or position the hole when using the Hole wizard is missing in action when creating a simple hole. To position the hole, you must edit the sketch automatically created by Solid-Works when using the Simple Hole command.

Actually, it is possible to drag the center of the sketched circle to a specific location and drop it into place. However, this requires the use of system feedback and assumes you already have some sketch point or vertex point in position. This is rarely the case. Therefore, some may decide that using the Simple Hole command does not really have as many benefits as one might initially think. For simple circular holes, you may very well decide to simply sketch a circle and cut it through the model the old-fashioned way!

Lofted Parts

Many plastic parts found in the world today are molded parts. Many of these are blow molded or injection molded, or one of many other similar processes. Parts such as these often exhibit free-form shapes and surfaces.

Certain types of plastic parts (such as bottles for shampoo, detergent, or a host of other compounds) can best be created in SolidWorks using certain feature types. Sometimes it is possible to create a sweep using guide curves to obtain complex free-form shapes. Lofted features, described here, can also use guide curves. Guide curves are explained in this section and

Lofted Parts

will aid you in creating shapes unobtainable in simple extruded or revolved features.

Lofted features are one of the four main feature types possible in Solid-Works. So far you have discovered extruded and revolved features. In this chapter you will explore the Loft command. In Chapter 9 you will take a close look at swept features.

Lofted features are different from swept features in that the lofted feature requires at least two sketches that are closed profiles. (There are a few exceptions to this rule, which are explored later in this chapter.) The Loft command will take two closed profiles and blend them, creating material between the two profiles to form a boss, or removing material to create a cut.

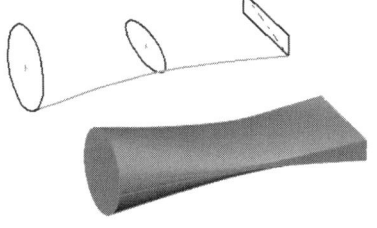

Fig. 6-42. A lofted feature.

You are not limited to two closed profiles. You can have three, four, or a dozen or more profiles. Neither do the profiles need to be on parallel planes. They can exist on many different angled planes, if necessary. Lofted profiles do not have to have the same number of segments, either. For example, you can loft between a square and a circle. Figure 6-42 shows such a feature, and the profiles used for the feature.

It is usually best to make sure all profiles have the same number of segments. It is not a requirement, but it is good technique. Otherwise, undesirable blending might occur. When the sketch profiles used in the loft have a different number of segments, they will also invariably have a different number of vertex points. In such a case, SolidWorks does not know what to do with the extra vertices.

Examining figure 6-42, you see that there are a total of three profiles: a circle, an ellipse, and a rectangle. The rectangle has four segments, and the ellipse and circle have one segment each. In this case, the blending works out fairly well, due to the ellipse having only one segment. Because the ellipse has no vertex points, the rectangle blends into it without a problem. Where problems start is if you were to try to blend, let's say, a rectangle into a triangle. SolidWorks would create the feature, but it would look odd.

In situations in which there is no getting around the fact that the profile sketches will have a different number of segments, a little finesse will usually solve the problem. For example, you might split a line in half so that it is actually two line segments, or break a circle up so that it consists of four arc segments instead of one circle. There is a sketch tool that is very well suited to this task, which is described in the section that follows.

Split Curve Command

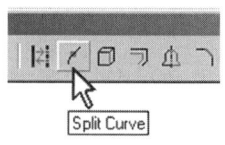

Fig. 6-43. Split Curve icon.

You can think of the Split Curve command as a "break" command. Sketch entities can be broken at any location. For example, a line can be broken in two so that it is two line segments instead of one. Any sketch objects can be split, including parabolas, circles, and splines.

To use the Split Curve command, perform the following steps. The Split Curve icon is shown in figure 6-43. The icon may be missing from your Sketch Tools toolbar, but you can add it using the procedure outlined in Chapter 23. Additionally, you may opt to simply use the pull-down menus instead. You must be editing a sketch, however, to work in this mode. How-To 6-8 takes you through the process of using the Split Curve command.

How-To 6-8: Using the Split Curve Command

To use the Split Curve command, perform the following steps.

1. Click on the Split Curve icon on the Sketch Tools toolbar, or select Split Curve from the Tools > Sketch Tools menu.

2. Click on a sketch entity where you would like a break (or split) to occur.

That is all there is to it. If you want the split to be at a particular spot, add any additional relations or dimensions required. When a line is split, the resultant segments are collinear. When a circle is split, the resultant arcs are co-radial. If a spline is split, the remaining spline segments will be tangent to each other.

Using the same series of profiles as shown in figure 6-42, let's see how Split Curve might benefit us. We can split the circle and the elliptical profiles so that each contains the same number of segments as the rectangle. Figure 6-44 shows these profiles side by side after applying the Split Curve command to them. The black dots are the split points. Dimensions and relations have also been added. Horizontal and vertical relations were used on the split points so that only one dimension would be required per profile.

✓ **TIP:** *To add an angular dimension (such as those shown in figure 6-44), pick three points, with the second pick being the centerpoint.*

Lofted Parts

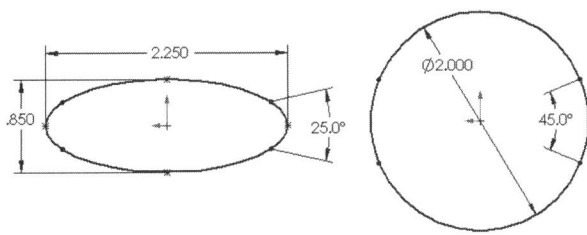

Fig. 6-44. After splitting the circle and ellipse profiles.

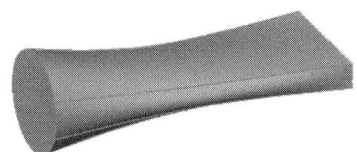

Fig. 6-45. The new loft feature using profiles with an equal number of segments.

If the loft is recreated with the new profiles, the blend lines can be directly manipulated by adjusting the angular dimensions on the profiles. This makes for having a much higher degree of control over the final lofted feature. Figure 6-45 shows the new loft feature. Compare this with figure 6-42 and see if you can tell the difference.

How-To 6-9 takes you through the process of creating a lofted boss. If you are creating a lofted cut, the process is the same, except that you use the Cut menu in step 3, in place of the Boss menu. Also note that when creating a base feature the Boss menu will be named Base. Hopefully you are pretty used to this behavior by now, as it is standard operating procedure for the SolidWorks software.

How-To 6-9: Creating a Lofted Feature

To create a lofted boss (otherwise known as a "boss-loft"), perform the following steps. The details for creating a sketch have been left out, as this topic has been covered in previous chapters.

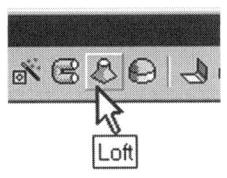

Fig. 6-46. Loft icon.

1. Create the sketch profiles for the loft.

2. Make sure to exit the last sketch when finished.

3. Select Loft from the Insert > Boss menu, or click on the Loft icon, shown in figure 6-46. The Loft panel will appear in PropertyManager (a portion of which is shown in figure 6-47).

4. Select the sketch profiles from the work area in the order you want the blend (loft) to occur.

5. If the preview looks good, and the correct vertex points are in alignment, click on the OK button to create the lofted feature.

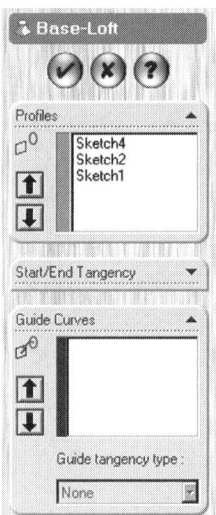

Fig. 6-47. Base-Loft panel.

If the loft you are attempting to create does not quite look the way it should, it might be due to your selection technique. Read the following section for some important information regarding this matter.

Correct Selection of Loft Profiles

It is important to understand which technique works best when selecting profiles to be lofted. The decision is quite simple. Make sure you select the profiles from the work area, not from FeatureManager. This allows you to select the profiles near the vertex points that should be matched from one profile to the next.

The preview line will tell you if you have correctly lined up the vertex points. The objective is to eliminate twisting during the loft. You will almost certainly notice the twisting effect in the preview if you are paying attention. The preview in the work area should show which vertices are being matched, at the very least. If the Show preview option is checked, an entire preview of the lofted feature will be displayed. In short, you are going to know if you selected the profiles incorrectly.

One other point to remember is to select the profiles in the order in which you want the loft to occur. This is not important when you have less than three profiles. However, when you have three or more, it becomes significant. Once again, the preview will indicate if the loft is not as it should be.

The order of the profiles in the Loft Profiles list box can be altered. It is relatively easy to select the profiles in the correct order, but on the off chance you did not select them in the correct order, use the Up and Down buttons. Select the profile that is out of sequence, and then click on the Up or the Down icon as required. You will see the selected sketch move up or down in the list box. The preview line will also update accordingly.

Guide Curves

Guide curves can be employed when it is necessary to have a greater degree of control over the lofted feature. A guide curve is essentially an extra sketch that directs the path of the loft. You can use one or more guide curves simultaneously. Guide curves are typically open profile sketches, but do not have to be.

As an example of where a guide curve or two can play a very important role, let's use the *Wedding Vase* model. To begin with, it would be beneficial to see what the underlying sketch geometry might look like. Figure 6-48 shows a total of seven sketches. The image can almost be thought of

Lofted Parts

as wireframe geometry, which is what CAD modelers sometimes used before solid modeling programs came to be.

In figure 6-48, note the five closed profiles. These are the loft profiles, and consist of a set of ellipses. Several sketch planes had to be set up ahead of time, sometimes using a little extra construction geometry. Such was the case with the two angled planes the top two ellipses were sketched on. The guide curves are the open profiles, which essentially connect the five ellipses.

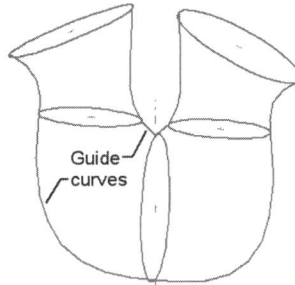

Fig. 6-48. Preliminary sketch geometry prior to creating a lofted part.

If a loft is attempted without the use of the guide curves, the outcome is not desirable. SolidWorks completes the loft, blending the elliptical profiles, but the resultant feature certainly does not resemble a vase (or at least not one you would want to give as a wedding present). The feature is shown in figure 6-49.

As soon as the guide curves are added, the shape of the model changes dramatically. A blend occurs between the profiles, which must follow the path of the curves. Figure 6-50 shows the resultant model. Add a few features and this vase might even be worthy of sitting in a department store display case somewhere.

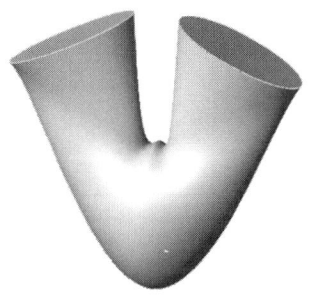

Fig. 6-49. A loft with no guide curves.

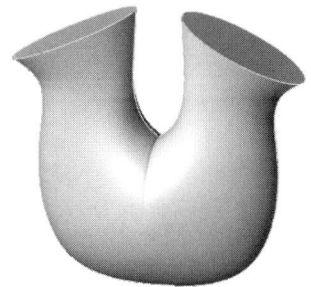

Fig. 6-50. The same loft with two guide curves.

Pierce Constraint

There is a special constraint in SolidWorks known as the "pierce" constraint. It is similar to a coincident constraint in that it typically constrains a point to some other object. The pierce constraint is a bit more restrictive than a coincident relation, however. Pierce is very often used with regard to guide curves when creating lofted or swept features.

Technically, the pierce constraint might be explained as follows. The sketch point being pierced is coincident at the point where another object pierces the sketch plane. This description is a little difficult to understand without some illustrations. Let's try to visualize the pierce constraint in comparison to a coincident constraint. Examine figure 6-51.

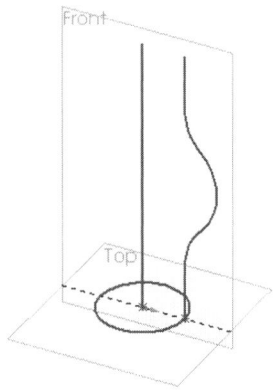

Fig. 6-51. Visualizing a pierce constraint.

Figure 6-51 shows three sketches. Think of the straight line as the path, the circle as the profile, and the third sketch as the guide curve. Imagine that the guide curve is a sharpened piece of stiff wire. It has been sharpened at the bottom, where it intersects the circle. It should be noted that a sketch point was added to the circle for the express reason of adding the pierce constraint. Now imagine that as the circle moves up the path the pierced point, as it were, must move to conform to the shape of the wire. Because the center of the circle will be locked onto the path, the pierce constraint causes the circle to change diameter.

The analogy being used is more accurately depicting a swept feature employing a guide curve, but it is easier to understand the pierce relation in this light. Swept features are explored more fully in Chapter 9, but because pierce relationships go hand in hand with guide curves this is a good time to talk about them.

How would the pierce constraint be different than a coincident constraint? Consider what would happen if a coincident relation were added between the guide curve and point on the circle. The guide curve (or at least a portion of it) would be projected perpendicular to the sketch plane of the circle. This projection would actually be a line, and that is what the point on the circle would be coincident with. From previous experience, we know that coincident does not necessarily mean "touching," but rather "on." In other words, the point would be free to move anywhere along an infinite line, as defined by the projection of the guide curve to the circle's sketch plane.

This can all be rather confusing. It might help to know that a coincident relation to some other object in the model means that the object, if not on the current sketch plane, will be projected to the sketch plane. With the pierce relation, there is no projection. The point being pierced must intersect the object piercing it.

With that said, a point being pierced can be any point associated with a sketch, such as an endpoint, center of a circle, or sketched point. The object doing the piercing can be an axis, edge, line, arc, spline, or any number of other curves we have not yet explored. Chapter 9 explores more of the available curve types found in SolidWorks.

The pierce constraint is added the same way any other constraint is added. Use the Add Relation icon. The entire scenario of creating a lofted feature with a guide curve might take place as follows.

- Start a new sketch and create a guide curve.

Lofted Parts   209

- Add some sketch planes if necessary, such as at either end of the guide curve.
- Create a sketch profile at one end of the guide curve. Use Add Relation to pierce the profile sketch at some point.
- Create another sketch at the other end of the guide curve, again using the pierce constraint.
- Exit the final sketch and perform the loft.

It is important to note that the guide curves are created first. If they are not present when the profiles are created, you cannot pierce the profiles with them. This sounds simple, but it is amazing how many Solid-Works users ignore this important fact. Let's go back to the wedding vase example and see how it was built.

Figure 6-52 shows the guide curve sketches. The guide curves in this case are both on the same plane, but it is important to note that they are separate sketches. Otherwise, they could not be classified as individual guide curves.

The guide curves in the case of the wedding vase are splines. The control points of the splines were dimensioned in order to fully define the splines and give the designer accurate control over spline shape. At some point in time, some planes would have to be added to serve as sketch planes for the loft profiles. In this particular case, the planes were added prior to the guide curves, and coincident relations were added between the spline endpoints and angled planes. This is solely due to the nature of this particular part.

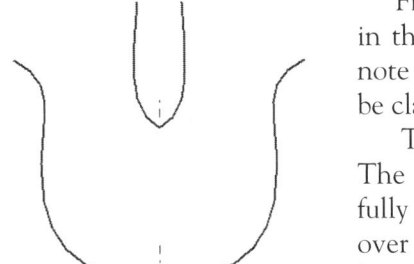

Fig. 6-52. Guide curves for the wedding vase.

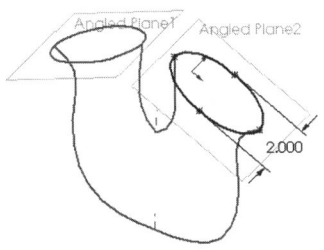

Fig. 6-53. Adding some sketch profiles for the loft.

Once the sketch planes and guide curves are in position, start adding some loft sketch profiles. Figure 6-53 shows the second ellipse being added. Only one dimension is needed to define one axis length of the ellipse. The major axis is defined through the use of the pierce constraint, with each major axis endpoint being pierced by either guide curve.

➥ **NOTE:** *It is not always a requirement to use the pierce constraint with guide curves when lofting. It depends on the situation. Good technique is to use pierce, as it will normally give you a higher degree of control.*

That should hopefully give you a better understanding of what the pierce relation is all about. It should also give you an indication of the general sequence of events involved in creating a lofted feature with guide curves. Sometimes it can be a little confusing. Quite frankly, a better option when lofting is to use the Centerline option, described in the following section.

Centerline Option

When it comes down to controlling the direction of a lofted feature, there is another option you have at your disposal that is more user friendly than employing guide curves and pierce relationships. This other option is known as the Centerline option. There is not a huge amount of difference between how centerlines and guide curves are implemented, but there is a subtle difference as to how they affect a lofted feature.

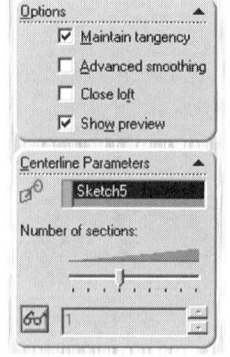

The Centerline option, shown in figure 6-54, is accessible via the Loft panel. Centerlines are used when your primary objective is to create a loft while gently nudging the lofted feature profiles in a particular direction. The centerline can be thought of as the "path" the loft will take.

How could one summarize the difference between using a guide curve and the Centerline option? Probably the best explanation would be that guide curves are used to control the shape of a profile and/or the direction of the loft, whereas the Centerline option is used to influence only the direction the loft takes. The difference is sometimes subtle, depending on the circumstances.

Fig. 6-54. Using the Centerline option.

To use the Centerline option, click in the area labeled Centerline and then pick the sketch to be used for the centerline. It should be stated that the sketch should not literally consist of centerlines. Rather, the sketch should consist of regular lines, arcs, or other sketch entities. So why is it called the Centerline option? The name is probably derived from the requirement that the centerline must intersect anywhere within the inside (or center) of the loft profiles, or at the very least intersect the profile itself.

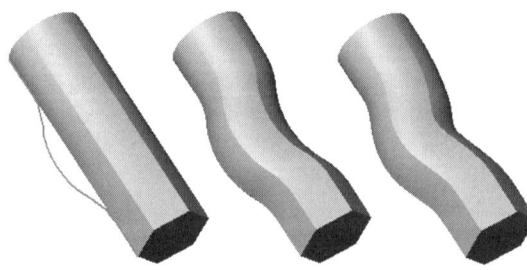

Let's take a look at one more simple example to hopefully clear up any remaining questions you might have about the Centerline option versus guide curves. Figure 6-55 shows a loft occurring between a circle and a hexagon. In the first example (far left), note the unused sketch. The sketch consists of two lines and three arcs, and runs the length of the loft, although this is difficult to discern in the image.

Fig. 6-55. Three examples of lofting, all using the same two profiles.

The middle example in figure 6-55 uses the sketch as a guide curve. It should be noted that the loft cross sections are all perpendicular to the loft direction. In the last example, the sketch is used as a centerline. Note that in this last example the loft cross sections are perpendicular to the centerline, rather than to the loft direction.

In the end, you will more than likely use whatever method is needed to create the model with the form and function you require. However, if it is possible to use the Centerline option, do so, as that is the easy way out. No pierce relation is required, which also means that you do not have to create the centerline prior to the loft profiles. Guide curves do have one big advantage over centerlines, though: you can use more than one guide curve at a time. You cannot, however, use guide curves and centerlines at the same time.

Other Loft Options

There are still a few other options that might prove useful for specific applications when using the Loft command. These are explained in the sections that follow.

Maintain Tangency

The Maintain Tangency option will attempt to maintain tangency between entities in the profiles during the loft if they are tangent to begin with. If you are lofting from a profile with tangent entities (such as a rounded slot) to a profile that does not contain tangent entities (such as a rectangle), the profile with the tangent entities will obviously not be able to remain tangent throughout the entire loft.

Advanced Smoothing

The Advanced Smoothing option works only if the loft profiles use circular or elliptical arcs. Often you may not see any difference. If you are creating complex lofted features and desire a smoother surface, try checking this option and see if it makes a difference.

Close Loft

If you are creating a loft that uses three or more profiles, and you want the last profile to connect back up with the first, check the Close Loft option. This will create, in effect, a closed-loop loft. You must have a minimum of three closed profiles you are lofting between for this option to work. Realistically, you would probably use more than three profiles. A good example might be four profiles, 90 degrees apart, used to create the shape of a ring.

Show Preview

Checking the Show Preview option will show a nice shaded preview, but sometimes at the expense of time. Lofts can take a while if complex in nature. If that is the case, you may want to uncheck this option and nix the preview.

An alternative preview function, at least when using centerlines, is to show the cross sections of the loft prior to creating it. In the Centerline Parameters panel (shown in figure 6-54), there is a slider bar that can be used to adjust the number of cross-section profiles generated for the preview. Sliding this bar to the right a short distance and then clicking on the Show Sections icon results in the sections being generated. It is then possible to use the increment/decrement arrows to cycle through the cross sections, thereby getting some idea of what the finished loft is going to look like.

Merge Result

The Merge Result option is only available after the first base feature has been created. When this option is checked, the new feature will become part of the existing model. Otherwise, multiple bodies are created. (See the section "Multiple Bodies" in Chapter 4 for more information regarding this topic.)

Tangency Conditions

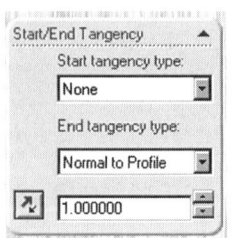

Fig. 6-56. Start/End Tangency panel.

The start and end tangency conditions for a lofted feature can be set in the Start/End Tangency panel, shown in figure 6-56. Tangency conditions allow for setting how the first and last profile in the lofted feature relate to other faces adjacent to the loft, or to the starting or ending profile planes.

A loft in its simplest form will blend from one profile to another with absolutely no regard for the orientation of the profile planes. Relate this to extrusion for a better understanding. When extruding a sketch, the extrusion direction is always perpendicular to the profile plane. Because this is not the case when lofting, it is sometimes advantageous to have some way of controlling (besides centerlines or guide curves) the direction of how the loft starts or ends.

To put it simply, tangency conditions control how the loft profiles are propagated from the sketch plane. If the tangency condition is set to Normal To Profile, the profile is lofted perpendicular (normal) to the profile sketch plane before it begins to conform to the next profile in the loft. This concept is depicted in the illustrations that follow. Figure 6-57 shows

Lofted Parts

two simple rectangular profiles that are orthogonal. The planes have been shown for clarity.

Figure 6-58 shows the lofted feature. Absolutely no tangency conditions have been used. This is the result of a loft created in its most basic form.

Figure 6-59 shows two examples. The image on the left shows the same loft, but this time with the Normal To Profile tangency condition enabled. It should be noted that the "start" profile in this case is the bottom rectangle, although this only really makes a difference as far as the tangency conditions go.

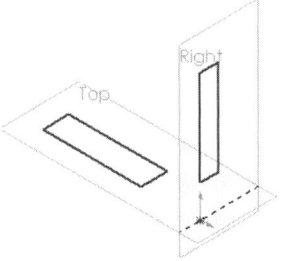

Fig. 6-57. Two profiles about to be lofted.

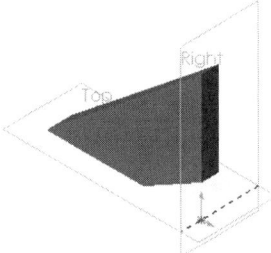

Fig. 6-58. After completing a simple loft.

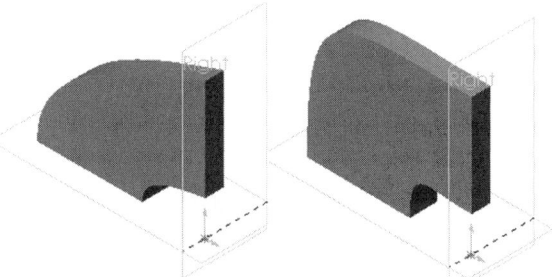

Fig. 6-59. Turning on tangency conditions.

The example on the right shows the loft with Normal To Profile enabled for both the start and end profiles. Additionally, the amount of force exerted by the tangency conditions has been increased, and therefore the tangency now exhibits more "pull" on the model. All in all there are three possible tangency conditions. These are summarized in the following material.

Normal To Profile

In the case of Normal To Profile, the loft will begin as though it were extruding perpendicular to the profile plane. When this option is used, the tangency can be perpendicular to the profile plane in either direction, so care should be taken. Large red preview arrows appear in the work area to indicate the direction in which the tangency is applied. These arrows can be reversed by clicking on the Reverse Tangent Direction icon found in the Start/End Tangency panel.

214   *Chapter 6: Molded Parts*

If Normal To Profile is used, adjust the tangency force as necessary. The force is a value from .1 to 10, but be sure not to overdo it or the loft may fail. Likewise, do not overdo it if the tangency arrows are pointing in the wrong direction.

Direction Vector

The Direction Vector option allows you to specify any sketch line or model edge in order to indicate what the tangency direction should be. This makes it possible to establish tangency in literally any direction, not just perpendicular to the profile plane. A sketch line can be created in a sketch near the profile plane and then used to establish the tangency direction, for instance. When this option is used, the tangency direction can be reversed, and the tangency force can be altered, identical to the Normal To Profile option.

All Faces

This is perhaps the better of the three possible tangency options, because it accomplishes a truly impressive task. When the All Faces option is used, any faces adjacent to the loft profile will be automatically selected by the SolidWorks software, and the newly lofted feature will attempt to make the faces of the new feature tangent to the existing faces of the model. Let's examine a plastic molded suitcase handle (or at least the beginnings of one), shown in figure 6-60, as an example.

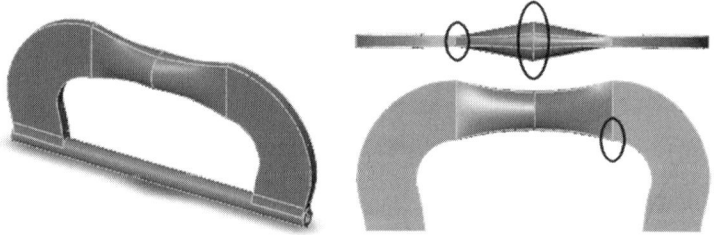

Fig. 6-60. Suitcase handle.

Figure 6-60 shows the suitcase handle from an angle, and from the front and top. Front and top views are shown in an incomplete state for the sake of clarity. Note that there are areas that do not blend smoothly, such as in the very center of the handle and other areas. These areas have been circled in the image. They represent the problem areas for which it would be better to have smooth transitions between features, rather than sharp transitions with hard edges.

The hard edges in the suitcase handle can be remedied by making use of the Normal To Profile and All Faces tangency control options. If we backtrack for a moment to see how this part was built, it will be clear why this is so. Figure 6-61 shows the first two features, both of which are lofts. The second feature is in progress of being made, and it is the previews and callouts that are shown in the image. Once the second feature was finished, the part was mirrored onto itself (which you will learn how to do in the next chapter).

If the start tangency condition of this second feature is set to All Faces, the second feature will blend very smoothly with the feature preceding it. Additionally, the first feature's definition can also benefit from being edited. In the case of the first feature, both start and end conditions will be set to Normal To Profile. This will do two things: it will make for a smooth transition when the part is mirrored, and it will help shape the part where it meets the second loft.

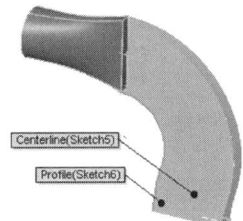

Fig. 6-61. Examining how the suitcase handle was made.

Figure 6-62 shows how the sharp edges have been eliminated. Although it is difficult to convey computer images through black-and-white printed media, it is still possible to see the rounded transitions between features.

Fig. 6-62. After enabling some tangency options.

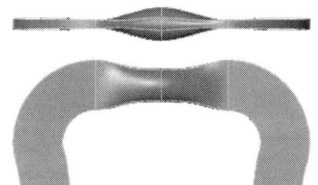

Split Lines

One command frequently associated with molded parts is the Split Line command. As a matter of fact, the Split Line command could have very easily been called the "Parting Line" command. Technically, though, a split line physically splits a face or set of faces, which is why it carries the name it does.

The Split Line command is often used for creating parting lines. In a case where a parting line is being created, a sketch (often just a simple line) is projected toward the faces to be split. Any sketch profile can be used to split a face, even a closed profile.

Creating a split line accomplishes a number of things. It allows you to see where the parting line of the part is. This parting line can then be used for adding draft, or later for mold making. Adding draft using a parting line is discussed in material to follow. To illustrate what the Split Line command can do for you, let's take a look at a molded plastic faceplate of a children's toy, shown in figure 6-63.

216   *Chapter 6: Molded Parts*

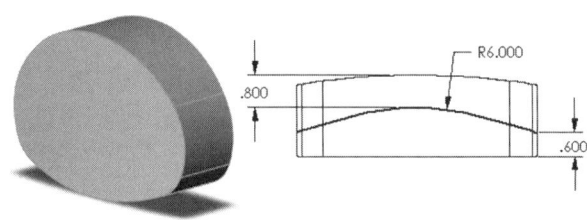

Fig. 6-63. Molded faceplate in need of a parting line.

Figure 6-63 shows the faceplate in an isometric view. It also shows the side of the faceplate. The dark line represents where the design specifications require the parting line to be. The parting line is represented by a sketch and is dimensioned, typical of any sketch. One aspect of this sketch worth mentioning is that its endpoints have been constrained to the silhouette edges of the model. This has been done in case the part changes size. If the part does change size, the parting line will shrink or grow with the model.

Next, the Split Line command will be used to project the sketch onto the faces containing the parting line. The steps involved in this process are provided in How-To 6-10. The faceplate is assessed after How-To 6-10 to see how it fared. You will also discover how to add draft to the same model using the newly created parting line. The steps outlined in How-To 6-10 can also be used as a precursor to many other feature types in SolidWorks, not just parting lines. The Split Line command is extremely versatile, especially with advanced fillets.

How-To 6-10: Creating a Split Line

To utilize the Split Line command, perform the following steps.

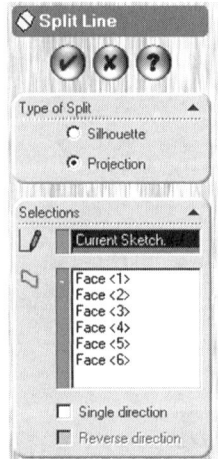

Fig. 6-64. Split Line panel.

1. Create a sketch that will be used to define the split line. The sketch must extend up to the edges of the faces to be split, or beyond, but must not fall short.

2. Select Split Line from the Insert > Curve menu. The Split Line panel will appear, which is shown in figure 6-64.

3. Select Projection as the method for defining the split line. (Silhouette is discussed in material to follow.)

4. Click in the Faces to Split list box and select the faces to be split.

5. Click on OK to complete the process and create the split line.

The Silhouette Method

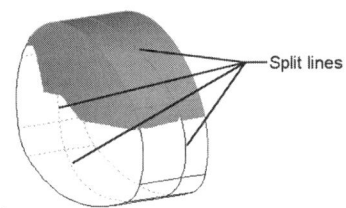

Fig. 6-65. After creating the split line.

Figure 6-65 shows a partially shaded view (for clarity's sake) of the faceplate after the split line has been created. It wraps all the way around the part, and for all intents and purposes can be considered the parting line. This example shows that the parting line does not have to be straight, if that is what your design requirements call for.

When working through the Split Line command, it is not necessary to select a sketch to project, as long as the sketch was being edited when the Split Line command was initiated. Otherwise, it would have been necessary to select a sketch. The *Single direction* option is useful if it is not necessary to project the sketch in both directions at once. There is no sense in making SolidWorks work any harder than it has to. As a general rule, use *Single direction* if feasible to increase efficiency of the part file. Obviously, the *Reverse direction* option can be used to flip which direction the sketch is projected in if *Single direction* is enabled.

The Silhouette Method

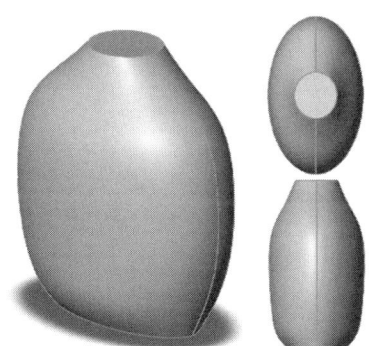

Fig. 6-66. Split line created with the silhouette method.

The silhouette split type mentioned earlier is used for creating split lines on parts without the need for a sketch. Instead of projecting a sketch, the user must supply a pull direction. The pull direction is the direction the mold would be pulled away from the part. This can be supplied by selecting an edge, axis, two vertex points, or a planar face. In the case of the face, the pull direction will be perpendicular to the face.

The silhouette method of adding a split line is very convenient and can save a lot of time. However, it is only meant to be used on particular parts. The faceplate would not have been a good candidate for the silhouette method because all of the faces that would have required the parting line were completely perpendicular to the mold's "pull" direction. Figure 6-66 shows a model that would be a good candidate. It has a free-form body with flat faces at the top of the neck and at the base only.

In the case of the example shown in figure 6-66, the front plane was used to define the pull direction, and the surface of the model was selected on which to define the parting line. Note that the parting line flows perfectly up either side of the model. The Silhouette option, in essence, extracts the silhouette edges from the model relative to the planar face you select.

Now that you know how to add a parting line to a part, you can take it one step further and apply draft using that parting line. The following section shows you how to do just that.

Adding Draft

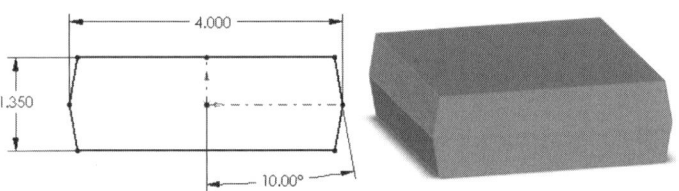

Fig. 6-67. Incorporating draft into a sketch.

To this point, you have leaned how to add draft using one of two methods. One method is to add draft during the extrusion process (Chapter 4). Another method, which has not been specifically stated, is to simply incorporate the draft directly into a sketch. This second method was not laid directly out for you, but you should have been able to interpolate this process from existing material already covered. Figure 6-67 serves as an example.

If a parting line were used to create the draft, it would be possible to place draft on all of the necessary faces in one move. In addition, adding draft as a feature is not limited to using only parting lines. Draft can be specified using a neutral plane as well. First, however, let's examine how to add draft with a parting line.

Adding Draft with a Parting Line

How-To 6-11 takes you through the process of adding draft as a feature. Keep in mind that because the draft is being added as an applied feature you will not be required to first create a sketch. The draft will be applied directly to the model, and because a split line was already added to the faceplate it will serve as an excellent example. The illustrations that follow will help guide you in the proper direction, and aid you in selecting the correct geometry.

How-To 6-11: Adding Draft Using a Parting Line

To add draft as a feature using the parting line method, perform the following steps.

Adding Draft

Fig. 6-68. Draft icon.

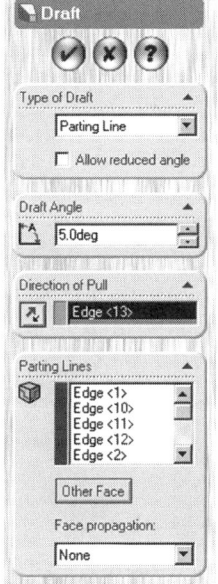

Fig. 6-69. Draft panel in PropertyManager.

1. Select Draft from the Insert > Features menu, or click on the Draft icon, found on the Features toolbar and shown in figure 6-68.

2. Select the Type of draft; in this case, Parting Line. Figure 6-69 shows the Draft command panel.

3. Specify the Draft angle.

4. Click in the Direction of Pull list box and select a linear edge or planar face to indicate the direction of pull. This will be the direction the mold would be pulled away from the model. If a planar face is selected, the direction of pull will be perpendicular to the face.

5. Click on the *Reverse direction* icon to flip the direction of pull if necessary.

There are a number of things going on at this point you should be aware of. The direction of pull will be represented with a callout and a large arrow (see figure 6-70). Clicking on this arrow is enough to flip the direction of pull. The next step is to select the parting lines. There will almost certainly be more than just a couple, and you will need to select them all. Use any of the various selection methods at your disposal. Once the parting lines are selected (step 6), many more arrows will appear. These smaller arrows point toward the faces being drafted. This can be reversed, but typically you will want all of the smaller arrows pointing in the same direction.

6. Click in the Parting Lines list box and select the parting lines.

7. Click on OK to complete the draft.

Figure 6-71 shows what the faceplate looks like after the draft has been added to both sides of the parting line. The drafting process was run through twice to achieve this effect. Because the direction of pull is different for each side of the parting line, all perimeter faces cannot be drafted at the same time.

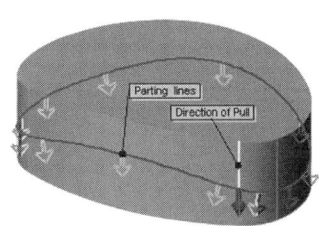

Fig. 6-70. In the process of adding draft.

Fig. 6-71. Faceplate with draft.

220   *Chapter 6: Molded Parts*

If for whatever reason you wish to reverse one of the smaller arrows, which point to the face being drafted on either side of the parting line, select the appropriate edge from the Parting Line list box and click on the Other Face button. You will rarely want to do this, however, because it will often cause the draft to fail due to geometric conditions. Use discretion with this option.

Adding Draft with a Neutral Plane

Whether you decide to add draft using a parting line or a neutral plane is really up to you, but this also depends on the model on which draft is being applied. Obviously, if the Parting Line method is used, the part must contain a parting line. That is not the case if the Neutral Plane method is used.

What exactly is a neutral plane? The neutral plane dictates where the draft will be measured from. The neutral plane does not necessarily have to be adjacent to any of the faces being drafted. However, it may make face selection easier (see the section "Face Propagation Options" in material to follow). How-To 6-12 takes you through the process of adding draft using a neutral plane.

How-To 6-12: Adding Draft Using a Neutral Plane

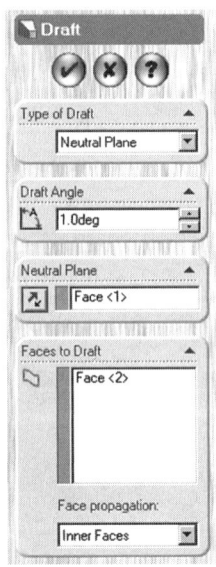

The following steps take you through the process of adding draft using a neutral plane. The process is similar to adding draft using a parting line, but without quite as many bells and whistles. Therefore, it is a bit more straightforward.

1. Select Draft from the Insert > Features menu, or click on the Draft icon.

2. Select Neutral Plane as the Type of Draft (shown in figure 6-72).

3. Specify the Draft Angle.

4. Click in the Neutral Plane list box and select a plane or planar face to indicate the Neutral Plane. This will determine where the draft angle is measured from.

Fig. 6-72. Adding neutral plane draft.

Adding Draft 221

5. Click on the Reverse Direction button (if necessary) to flip which side of the neutral plane the mold will be pulled away from. The preview arrow will indicate the direction of pull.

6. Click in the Faces to Draft list box and select the faces to which draft should be applied.

7. Click on OK to add the draft.

As is always the case, bear in mind which list box is salmon colored, as this indicates which list box is currently active. This is important when working with commands such as Draft, which contain more than one list box.

Step Draft

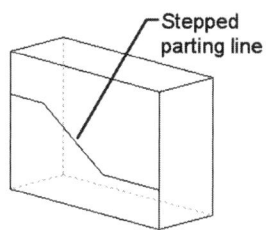

Fig. 6-73. A stepped parting line on the model prior to adding draft.

Step draft is used in a situation in which the parting line does not consist of a simple line or tangent entities. Instead, the parting line consists of line segments. This is more easily described visually than verbally. Take the simple example shown in figure 6-73. Note the stepped parting line created on the front face of the part. (This example will use a split line projected only onto the front face, to clearly illustrate the effects of step draft, rather than confusing the issue and wrapping the parting line around the entire part.)

Next, let's add parting line draft to the model and see what results are achieved. If draft were added to this part without using the Step Draft option, the results would be as shown in figure 6-74. SolidWorks adds the draft and places the middle face at an angle in order to accommodate the stepped parting line. A large draft value of 25 degrees was used to exaggerate the effect. The bottom face of the model was used to define the direction of pull.

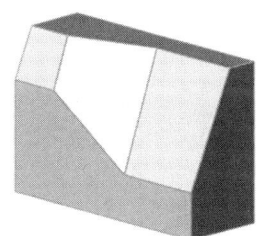

Fig. 6-74. Model with parting line draft.

When step draft is used, the results vary dramatically. When using step draft, the direction of pull must be specified in a distinct way. Specifically, a plane or planar face must dictate the direction of pull, whereas a model edge will not suffice. Direction of pull will be perpendicular to the plane, as usual. In figure 6-75, step draft was used, and once again the bottom face of the model was used to define the direction of pull.

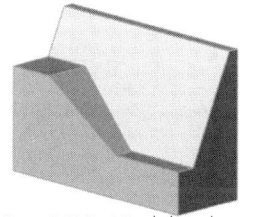

Fig. 6-75. Model with step draft.

There is one last option that should be mentioned. An option for setting the step-draft "steps" to either tapered or perpendicular will change the appearance of the draft. It should be noted that the step draft depicted in figure 6-75 is using the *Perpendicular steps* option. Figure 6-76 shows use of the *Tapered steps* option.

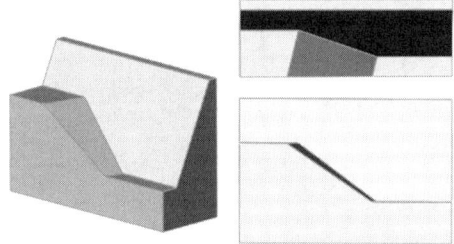

Fig. 6-76. Step draft with tapered steps.

Note in figure 6-76 that the middle step in particular appears in the top and front views. The middle step is not perpendicular to the original face, which was drafted as it would have been if the Perpendicular steps option were used. It is a fairly subtle difference.

Face Propagation Options

Face Propagation is no more than a convenient option that makes selecting parting lines or faces to be drafted a little easier. For example, if you are adding draft to a part that has perimeter faces that are all tangent, selecting *Along tangent* from the drop-down list will enable you to only have to select one parting line segment or one face, depending on the type of draft being created. Solid-Works will select all other segments of the parting line for you, or all of the perimeter tangent faces.

✓ **TIP:** *The* Along tangent *option is much easier to use with parting line draft. If you attempt to use this with neutral plane draft, the selected face must share a common edge with other faces to be drafted, and the edge must reside on a neutral or base plane. Results can be unexpected.*

When using the Neutral Plane method of defining draft, there are other options added to the Face Propagation drop-down list in addition to *Along tangent*. Table 6-2 outlines the various options associated with the Face Propagation feature. Figure 6-77 shows examples of some of the face propagation options.

Table 6-2: Face Propagation Options

Face Propagation Option	Description
None	This setting is the default. No face propagation will be used for the selection process. Faces must be individually selected.
Along tangent	All edges tangent to the selected parting line segment will be drafted.
All faces	All faces residing next to the neutral plane will be drafted.
Inner faces	All faces residing within the boundaries of the neutral plane will be drafted.
Outer faces	All faces residing on the perimeter of the neutral plane will be drafted.

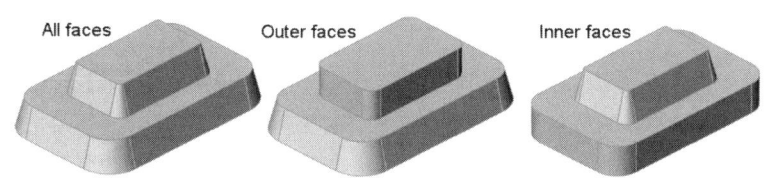

Fig. 6-77. Examples of Face Propagation options.

Sometimes it is difficult to tell what qualifies as inner or outer faces. Solid-Works' description of what constitutes these faces is a little vague. The recommendation is to just try one of the options and see how it looks. The draft definition is easy enough to change if you do not like the outcome. Besides, using any of the last three options means no face selection is necessary! You do not need any faces listed in the Faces to Draft list box. The last three options in the table are only available when using neutral plane draft.

Rib Tool

The Rib tool allows you to add ribs to a model with minimal effort. A sketch is required, but the sketch can be very simple in nature. This is the major benefit of the Rib tool. It allows for quickly and easily placing ribs in a model. That is not to say only simple ribs can be created. Quite the opposite, as very elaborate webbing can also be created with the Rib tool.

If a simple line is used as an example, we can examine what the Rib tool will accomplish. The length of the line is, for the most part, irrelevant. SolidWorks will project the line up to the next face it encounters. Good practice would be to sketch the line within the boundaries of where the rib should be positioned. Second, the line is "thickened," in a manner similar to that of a thin feature. This wall thickness is determined by the user when defining the rib. Third, the rib is extruded up to the next face it encounters. Let's look at an example.

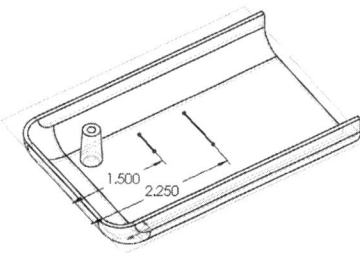

Fig. 6-78. Preparing a sketch for a rib.

In figure 6-78, a plane was created that is situated .200 inches below the top edge of the mouse cover. The two lines were then sketched on this plane. Note that the length common to the two lines has not been established. It does not need to be. This is the beauty of the Rib command. Plug in just enough information to establish your design intent, and then forget the rest.

In any other situation in which an attempt were made to extrude two open profiles, it would result in errors. However, with the Rib command, the results are two new ribs added to the model. Draft can even be added during the

224 Chapter 6: Molded Parts

creation process. Using the sketch shown in figure 6-78, the end result might look like that shown in figure 6-79.

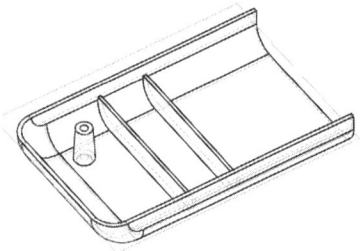

The Rib command itself is fairly straightforward. How-To 6-13 takes you through the process of creating ribs with the Rib command. It is assumed that an appropriate sketch has already been created. What constitutes an appropriate sketch can be just about anything. This includes closed or open profiles or a combination of the two. Splines can be used as well. The important thing to remember is this: any open profile's projection must be bounded by a face of the model. In other words, if a line (for example) is extended, the line must run into a wall.

Fig. 6-79. After creating the ribs.

How-To 6-13: Using the Rib Command

To create a rib using the Rib command, perform the following steps.

Fig. 6-80. Rib icon, found on the Features toolbar.

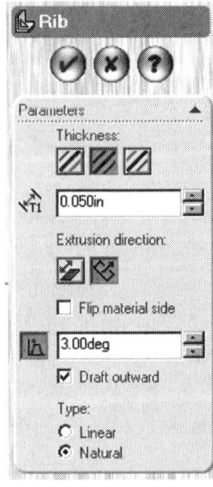

Fig. 6-81. Rib panel.

1. Click on the Rib icon (shown in figure 6-80), or select Rib from the Insert > Features menu.

2. In the Rib panel, shown in figure 6-81, enter a value for the thickness of the rib.

3. Establish to which side of the model the rib wall should be applied with respect to the sketch geometry. This will be one side or the other, or midplane (both sides), which is the default.

4. Using the Parallel to Sketch or Normal to Sketch icons and the *Flip material side* option, specify a direction for the rib's extrusion. Use the preview arrow as a guide.

5. Add draft if required.

6. Click on OK to create the rib feature.

Linear Versus Natural

An interesting aspect of the Rib command is its ability to extrude either parallel or perpendicular to the original sketch plane. This makes the command very flexible. Another interesting option gives you the ability to control how certain objects, such as arcs, are projected to the walls of

Rib Tool

the model. These are, specifically, the Linear and Natural options located at the bottom of the Rib panel.

An arc can be extended by projecting tangent lines from either endpoint (linear), or extended naturally by projecting the ends of the arc using the radius of the original arc. This is true of parabolic and elliptical arcs as well. The mathematical definition of the original parabolic or elliptical arc will be used to elongate the arc until it runs into a wall of the part. In the case of circular or elliptical arcs, the original arc segment will form a circle or ellipse if no boundary face is encountered.

✎ **NOTE:** *Splines are not affected by the Linear or Natural option.*

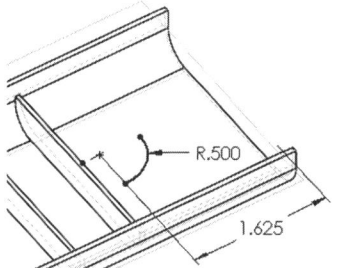

Figure 6-82 shows a simple sketched arc that will be used as an example for illustrating the Linear and Natural options. The endpoints of the arc have been constrained to control their locations, as you would want them to be if the Linear option were used. The reason for this will become clear in a moment.

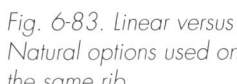

Fig. 6-82. A sketched arc that will be turned into a rib.

In figure 6-83, there are two examples. The example on the left shows the arc extending linearly until it connects with the walls of the model. The example on the right shows the same arc, only this time it has been extended naturally. The arc continues on its original path until it encounters either itself or a wall or walls of the model.

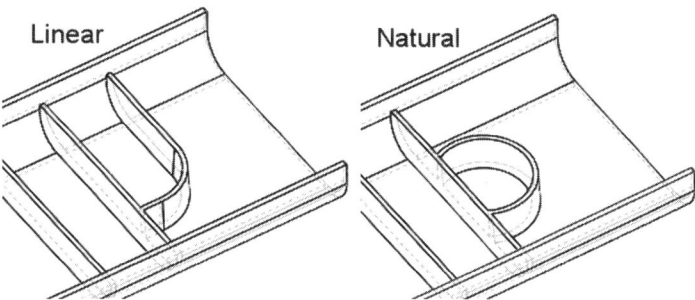

Fig. 6-83. Linear versus Natural options used on the same rib.

Obviously, if the arc is extended linearly, it will make a big difference as to the included angle and positioning of the arc's endpoints. Care must be taken to add dimensions or relations to the arc. Otherwise, the arc will extend in what will probably be an undesirable direction.

Draft Reference for Ribs

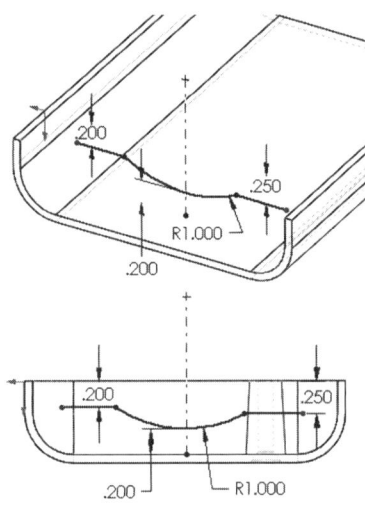

Fig. 6-84. Where would the draft be measured from?

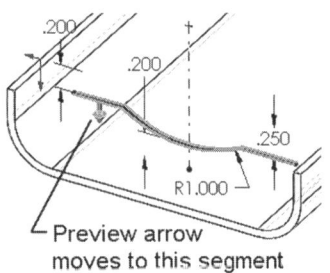

Not all ribs are created equal. Ribs are not all linear on top. Some are stepped, some have cutouts, and some are "barbecued" (but that is a topic for another book). Take, for example, the sketch shown in figure 6-84 (ribs created earlier are not shown for sake of clarity). Note the arc in the center, and the fact that the lines on either side of the arc are at different heights. Under such conditions, how would draft on the rib be defined, assuming it were added to the resultant rib?

The answer to the previous question is that you have a choice. When a sketch for a rib has multiple segments, the user can choose where the draft is measured (referenced) from. This is done via a button titled Next Reference. The preview arrow used to distinguish the direction of the extrusion also distinguishes what segment the draft will reference. Clicking on the Next Reference button serves to move the arrow from one sketch entity to the next.

Because the thickness of the rib to be created in this example is going to be .050 inch at the uppermost line segment, the draft should reference that specific line. Figure 6-85 shows the preview prior to creating the rib.

Fig. 6-85. Referencing the upper line segment for draft.

Draft Analysis

An important tool that goes hand in hand with molded parts is the ability to analyze the draft of a model. The Draft Analysis tool, found in the Tools menu, allows for checking if faces of the model exhibit positive or negative draft, among other things. Draft analysis does not take much effort from a user standpoint. In a nutshell, select a face to indicate direction of pull, enter a draft value, and click on a button. How-To 6-14 takes you through this process.

HOW-TO 6-14: Analyzing Draft

To analyze the draft of a model, perform the following steps. Note that the model should be displayed in shaded mode in order to use this function.

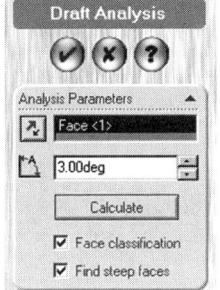

Fig. 6-86. Draft Analysis panel.

1. Select Draft Analysis from the Tools menu. The Draft Analysis panel will appear, a portion of which is shown in figure 6-86.

2. Select a face or plane that will determine the direction of pull. Direction of pull refers to the direction the mold will be pulled away from the model.

3. Click on the Reverse Direction icon if necessary to reverse the direction of pull.

4. Enter a value for the draft angle.

5. Click on the Calculate button.

Once the Calculate button is depressed, the model will be displayed using the colors designated in the bottom half of the Draft Analysis panel (not shown). If the *Face classification* option is *not* used, the model is checked for positive and negative draft, regardless of faces. What this means is that one face may contain one area (for example) with positive draft, but another portion of the same face may require draft. The colors are displayed accurately. In contrast to this, let's assume the *Face classification* option is used. With faces being classified as having positive or negative draft, entire faces are either one color or another. There are no color gradations within individual faces.

With *Face classification* turned off, an option for using a gradual color transition becomes available. This is the most accurate way of checking the amount of draft over the entire model, but may be more than most people require. Once the analysis calculation is complete, moving the cursor over the part will result in the draft for that location being displayed on the cursor. This "cursor readout" functionality will work whether or not the gradual color transition is used.

Face Classification

The *Face classification* option does have some benefits. First, the graphics usually appear cleaner. That is to say, the shaded part is a little easier on the eyes. Second, the model can be saved with the colors used to display the draft analysis. This could be useful if extensive rework is needed. (Read on for a warning regarding this functionality.)

There are some terms that might not be familiar to all readers with regard to how SolidWorks classifies certain faces. For this reason, Table 6-3 is provided, which outlines face classification designations. Note that the "straddle" and "steep" face designations are used only when the *Face classification* option is enabled. These designations are not otherwise necessary. The term *neutral plane* can be thought of as where the draft is measured from. It is typically where the mold would split. The term *draft specification* would refer to the value typed in by the user in the Draft Analysis panel.

Table 6-3: Face Classification Designations

Face Classification	Description
Positive draft	Faces have draft equal to or greater than the current draft specification on the side of the neutral plane that signifies the current direction of pull.
Requires draft	Faces on either side of the neutral plane that have less draft than required by the current draft specification.
Negative draft	Faces have draft equal to or greater than the current draft specification on the side of the neutral plane opposite the current direction of pull.
Straddle faces	Faces which have both positive and negative draft. Only curved faces will exhibit this condition, as planar faces will fall into either one category or the other.
Positive steep faces	Faces that have both positive draft and areas that will require draft. Only curved faces can exhibit this condition.
Negative steep faces	Faces that have both negative draft and areas that will require draft. Only curved faces can exhibit this condition.

When *Face classification* is used and you are done with the draft analysis, clicking on the OK button will result in a message being displayed. This message asks "Do you want to keep face colors?" Be forewarned: clicking on OK will change the colors of all faces on the model, and there is no easy way to revert to the original part color! The Undo command will function for a time, but once that option is gone, the only way to change each face back to its original color is to pick every individual face and remove the color from it using the Edit Color command (discussed in Chapter 8).

Draft Analysis Color Settings

The Color Settings section of the Draft Analysis panel, a portion of which is shown in figure 6-87, allows for changing the colors used to indicate the various face classifications. Feel free to change these colors as you see fit, but understand that any color alterations that are made will be saved as system settings and not saved with the part. Color changes will take effect every time you use the Draft Analysis command. There is no easy way to reset them to their default values.

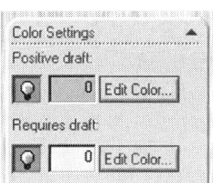

Fig. 6-87. Color Settings options in the Draft Analysis panel.

The small light bulb buttons will hide all faces with that particular face classification. This is good for seeing into a part to further examine other interior or difficult-to-see faces. The number present on the swatch of color is the number of faces that exhibit that face classification. For example, a number 14 in the yellow swatch indicates that there are 14 faces that require draft.

In Exercise 6-2, the *Shaver Housing* model is modified and a number of features are added. These features include a hole created with the Hole wizard, and some rib features. In Chapter 7, you will pattern some of the features created in this exercise.

EXERCISE 6-2: Adding Features to the Shaver Housing

Open the *Shaver Housing* model started earlier in this chapter. You should have *Boss1* created, per Exercise 6-1. Next, you will need to add a hole to *Boss1* so that a screw can be inserted through the hole and connect with the other half of the housing. Note that this is an exercise, not a How-To description. Steps you should be expected to know at this point in the book will not be spelled out in detail.

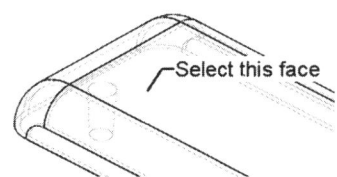

Fig. 6-88. Selecting a face for the new hole.

To create the countersunk hole, you will use the Hole wizard. This is the most economical method of creating a hole of this type. Remember that it is best to first select a face on which to place the hole. This will ensure that the Hole wizard creates a 2D sketch for the locating points. That is where you will begin. Selecting a face is depicted in figure 6-88. To create the hole through the boss, perform the following steps.

1. Select a face on which to place the hole, as shown in figure 6-88.

2. Access the Hole Wizard command.

3. Select the Countersink tab, and then plug in the following parameters.
   — ANSI inch, flat head (82), size #4
   — End Condition & Depth = Through All
   — Hole Fit & Diameter = Normal
   — Head Clearance & Type = Added C'bore at .020 inch

4. Leave all other parameters at their default values, and click on Next.

5. Note that you are in the Point command. Click on the Add Relations icon, and then add a coincident relation between the locating point and the end of *Boss1*. This will center the hole properly.

6. Click on Finish to create the hole.

Figure 6-89 shows the newly created countersunk hole as seen from two different vantage points. Next, you will add some ribs to strengthen the model. If you added ribs by following along with the How-To sections earlier, delete them now to make room for the new ribs.

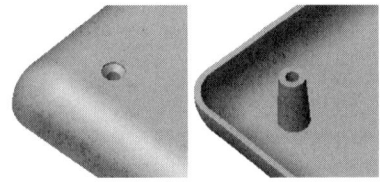

Fig. 6-89. After adding the countersunk hole.

7. Create a new plane that is .180 inch above the inside face of the shaver housing, as shown in figure 6-90. Rename this new plane *Rib Plane*.

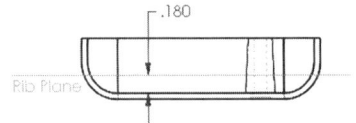

Fig. 6-90. Creating the rib plane.

8. Create the sketch shown in figure 6-91. This sketch will be used to create the ribs.

9. Using the Rib command, create ribs with a wall thickness of .050 inch using the Midplane (or "both sides") option and 3 degrees of draft. The extrusion direction should be obvious, as there is only one possible choice.

10. Add one more rib, down the middle of the part lengthwise. Use the same parameters as those for step 9. The final result is shown in figure 6-92.

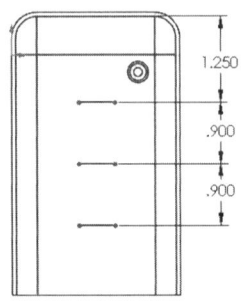

Fig. 6-91. Sketch for the ribs.

Draft Analysis

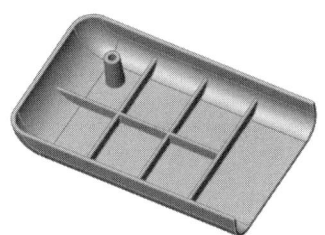

Fig. 6-92. The completed ribs.

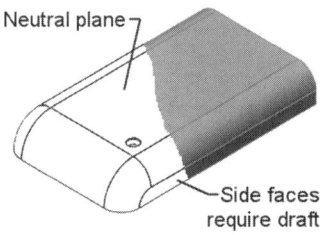

Fig. 6-93. Showing an Isometric view (and additional information).

If step 10 is causing you grief, try adding midpoint relations between the endpoints of the sketch line and the rounded end of the shaver housing and the rib on the opposite end of the part. Experiment if you have to, as that is one of the best ways to learn.

Performing a draft analysis shows that the part requires draft if the entire part is to have a draft of 3 degrees minimum. This last portion of the exercise will involve performing a draft analysis, and then editing the base feature sketch to change the draft on the sides of the part. Continue with the following steps.

11. Change to an Isometric view. If you have been following along correctly, the view on your screen should look like that shown in figure 6-93.

12. Select Draft Analysis from the Tools menu.

13. Using the neutral plane, shown in figure 6-93, analyze the draft using a 3-degree draft angle. Use the *Face classification* option.

14. Close out of the Draft Analysis command and do not keep the face colors when prompted. (You will only be prompted to keep the face colors if you click on OK. It is fine to click on the Cancel button.)

At this point you should be able to determine faces of the model that do not meet the current design requirements of 3 degrees of draft. It is necessary to alter the part in some way to establish 3 degrees of draft on the sides of the model. This can be accomplished by editing the base sketch of the part. Do not be afraid of altering the base sketch, as long as you do not change anything in a drastic way.

15. Edit the sketch for the base feature and change to a Front view.

16. Remove the vertical relationships on each of the lines on the ends of the rectangle.

17. Remove both dimensions. Because these are horizontal and vertical dimensions, they control the sketch in a way that is unsatisfactory.

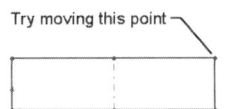

Fig. 6-94. Modifying the base feature sketch.

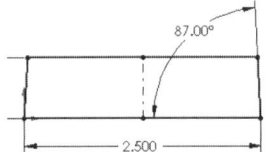

Fig. 6-95. Adding dimensions after dragging the sketch point.

18. Add a centerline between the midpoints of each horizontal line, as shown in figure 6-94. Make sure the centerline is constrained to be vertical.

19. Try dragging the top left or right points of the rectangle. Get a feel for how the sketch reacts. Attempt to move the point inward.

20. Add a dimension to control the draft via the sketch, as shown in figure 6-95.

21. Add dimensions to fully define the sketch.

22. Rebuild the model.

23. Perform another Draft Analysis using the same parameters as in step 13.

24. Save and close the *Shaver Housing* project.

After performing the final draft analysis, it is found that the inner faces of the countersunk hole show up as requiring draft. This can be ignored, as the hole will be machined. Additionally, the small thin edge at the end of the part shows up as requiring draft, but this can also be safely ignored.

This concludes the exercise on the shaver housing in this chapter. You will pick it up again in Chapter 7, so keep the shaver housing file handy.

Summary

In this chapter, you first learned about thin-walled parts. You also discovered that a thin feature is one usually created with an open profile, which is covered in detail in Chapter 8. The term *thin-walled* refers to shelled parts, whereas *thin feature* is the term reserved for extruded or revolved features given a wall thickness by the user.

The Shell command can be useful when creating a thin-walled feature because it allows you to remove, or hog out, the inside of a part and leave a wall behind whose thickness you specify. Alternatively, it is possible to add a wall thickness to the outside of the part, akin to dipping the part in a vat of hot wax. This essentially removes the entire original geometry, leaving behind the newly created shell. Multi-thickness shelled parts can be created with one command.

Summary

A fillet or round can be created on a part, thereby rounding the sharp corner of an outside edge or an inside edge. *Fillet* is the term used to describe a rounded inside edge, but is generally used to describe a fillet or a round. It is possible to fillet an entire face at once, if you want to fillet every edge on that face. Fillets and rounds can be added simultaneously, all within the same feature.

When fillets are added, the *Tangent propagation* option is selected by default. This function can be turned off if desired. When selecting items to be filleted, it is possible to select edges using a variety of options, including selecting chains, loops, or tangencies. This selection process works for other commands as well. When filleting, features can also be selected, in which case every edge associated with that feature is filleted.

Use the Select Other option from the right mouse button menu when trying to select a face on the opposite side of the part. The critical aspect of this command to remember is to position the cursor over the face you are trying to select while accessing the Select Other option.

The Hole wizard allows the creation of a wide variety of hole types, including countersunk, counterbored, and tapered. It is good technique to first select the face on which to place the hole. Holes can be placed on nonplanar faces using the Hole wizard. You should not be in a sketch when using this function. The Hole Wizard function allows you to position the hole's centerpoint before completing the command. Multiple holes can be added in one operation.

Lofted features can be used to create a wide variety of shapes. A lofted feature requires a minimum of two closed sketch profiles. The sketch planes do not need to be parallel. Quite the contrary, the sketch planes can be at any angle or position. The loft will blend the profiles to create a solid. The sketch geometry used to create a lofted feature should have the same number of segments per sketch. The Split Curve sketch tool is an excellent tool for accomplishing this.

When lofting, pay special attention to where you select the geometry to be lofted. One click per sketch in the work area is all it takes to select the geometry, but that click must be near the vertex points you are attempting to match during the loft operation.

Guide curves can be used while lofting to control the shape of a profile. The Pierce command should be used whenever a guide curve is used. Good technique is to create the guide curve first, and then create the profiles to be lofted. The profiles can then be pierced by the guide curve. The Centerline option can be used while lofting to guide the direction the loft takes. No pierce relation is needed when using the Centerline option. In addition, the centerline can be any sketch entity, and should not literally be a centerline. Guide curves and the Centerline option cannot be used simultaneously.

Tangency options when lofting allow for controlling how the loft propagates off the start and end profile planes. Both the start and end tangency conditions can be controlled. The loft can be made tangent to the profile plane's normal (perpendicular vector), a specified vector (which can be a sketch line or existing edge), or tangent to existing faces.

Other important topics learned in this chapter were how to create a parting line using the Split Line command, and how to add draft as a feature. Draft can be added either by defining a neutral plane or through the use of parting lines. The neutral plane defines where the draft angle is measured from. If a stepped parting line is used, it is possible to create step draft, where the drafted face pulls away from the existing geometry. Ribs can easily be added to a model with the Rib tool. Benefits of the Rib command include simplicity of sketch geometry and extrusion direction flexibility. Last, you learned how to analyze the faces of a model to see if those faces require draft.

Questions and Topics for Discussion

1. Describe the difference in the selection process when adding a fillet in a sketch as opposed to adding a fillet as a feature.
2. What happens to selected faces during a basic (same-thickness) shell operation?
3. Describe how to use the Select Other function.
4. When adding a hole with the Hole wizard, what happens if you neglect to select a face prior to starting the command?
5. What is a vertex chamfer?
6. How many sketches are created when using the Hole wizard, and what purpose do they serve?
7. What is the Legacy tab of the Hole wizard used for?
8. Is it possible to loft between a triangle and a circle?
9. How would you go about preventing twisting when creating a loft?
10. How does the pierce relation differ from coincident?

Questions and Topics for Discussion

11. How does the Loft command's Centerline option differ in function from using a guide curve?

12. When creating a loft, do the profile planes have to be parallel?

13. When adding draft as a feature, what is meant by the term *direction of pull*?

14. When adding draft using a parting line, there are small preview arrows that appear at every parting line segment. Describe what these arrows signify.

15. What is the difference between the Split Line and Split Curve commands?

16. Is it possible to use a closed profile when performing the Split Line command?

17. When creating a sketch for use with the Rib command, should you fully define the geometry? Explain your answer.

18. When using the Rib command, how many possible extrusion directions are there?

19. When performing a draft analysis, what face classification is typically determined by the color yellow?

Chapter 7

Patterns

MECHANICAL PARTS OFTEN CONTAIN PATTERNS. Many items you use in your everyday life also contain patterns of one sort or another. The peanut butter jar in your cupboard may have some sort of ridge pattern around its outside perimeter near the lid. The razor you use to shave with in the morning probably has some type of rib or groove pattern so that it is easier to hold onto.

If you were to walk around your house and look at things the way a designer or manufacturer would, you would see everything in a different light. Instead of picking up a knick-knack and thinking how it might look on the shelf in your dining room, you might instead wonder how you would build it. Take a look around your living room. Do you own a television with a remote? Chances are the remote control unit has buttons that form a pattern.

The point of this discussion is that patterns are a very common aspect of the design process. This chapter is devoted to patterns, both feature patterns and sketch patterns. A much simpler form of patterning is to mirror feature geometry, a process also covered in this chapter.

Simple Pattern Alternatives

Once you have taken the time to create a feature, it is often much more convenient to pattern that feature rather than to recreate the feature at another location. Depending on the situation and how many copies of a feature you require, it may be easier to perform a simple copy and paste.

In addition to linear and circular patterning, you will take a look at some simple alternative methods of copying feature geometry. Because copying and pasting feature geometry is such a simple task, it is discussed first.

Copying and Pasting Features

When it is only one or two additional features that are required, or if additional features do not conform to any type of pattern, copying and pasting features can be the best method to use. This functionality makes use of the Windows clipboard, and can be used to copy a feature from one location to another, or even from one part to another.

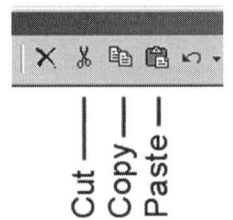

When copying and pasting features, the Edit menu can be used, or the "hot key" combination specifically designed for the task. There are also icons on the Standard toolbar can be used, as shown in figure 7-1. No matter what Windows program you are using, Ctrl-C will copy the selected object to the clipboard, and Ctrl-V will paste the content of the clipboard into your document. This is standard operating procedure in any Windows program, and not limited to SolidWorks.

Dangling Relationships

Fig. 7-1. Cut, Copy, and Paste icons.

When copying and pasting features, there are normally locating dimensions or constraints attached to the original feature. If this is the case, you will see a window asking you how you want to handle these relationships. You can either delete the locating dimensions or constraints, or you can leave them dangling.

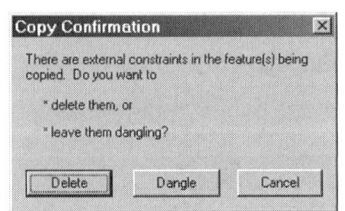

Fig. 7-2. Deciding what to do with a dangling relationship.

A dangling dimension might be a dimension that was attached to a deleted object. In the case of copying and pasting features, a dimension or constraint is considered dangling if it originally located the feature but is now left "hanging" because the copied feature is now in a different location. Because of this, SolidWorks gives you the ability to delete the now-meaningless relationships in the form of the window shown in figure 7-2.

If you leave a relation dangling (whether it is a dimension or geometric constraint), you must reattach the relation to other geometry. This can be done by clicking on the object containing the dangling relation and then dragging the red "handle" to a new location. Dangling dimensions will appear brown, and dangling constraints will make the associated sketch geometry appear brown.

Simple Pattern Alternatives

Bear in mind that you must be editing the sketch of the newly copied feature in order to add or repair relations. It is really quite simple to repair dangling relations, so do not let any of the red-circle error symbols in FeatureManager scare you. Just be aware that you need to repair the problems before continuing. You may decide it is easier to simply delete the relations and add new ones. It is really a personal preference that will not affect the model one way or the other.

It should be noted that if there are no locational relationships attached to the feature being copied, you will not even see the Copy Confirmation window (shown in figure 7-2). The feature will be directly pasted in wherever you tell it to be. In summary, if relations are deleted, new ones should be added. If they are left dangling, repair them prior to continuing.

Performing the Copy and Paste

To copy and paste a feature from one location to another (or from one part to another), perform the steps outlined in How-To 7-1. Make certain you are not editing a sketch, or copying and pasting will not work.

How-To 7-1: Copying and Pasting

To copy and paste a feature, perform the following steps. Note that any method (not just the Ctrl-C or Ctrl-V hot keys) can be used for copying or pasting, such as accessing the Edit menu or utilizing the Standard toolbar.

1. Select the feature from FeatureManager you want to copy to the clipboard.
2. Copy the item to the clipboard (Ctrl-C).
3. Select the face on the part to which you want to copy the feature.
4. Paste the item from the clipboard (Ctrl-V).
5. If there are dangling relationships, specify whether you want to delete them or leave them dangling.

At this point, the feature will be pasted at the specified location, which is the cursor location when the "receiving" face was selected in step 3. It is normally desirable to edit the sketch for the newly pasted-in feature in order to add locating dimensions or constraints at this time. Likewise, if relations were left dangling, edit the sketch to repair them. You should already be familiar with editing sketch geometry, which was discussed in detail in Chapter 4.

Copying and pasting has its limitations. Copying more than one feature at a time, for instance, often does not work. Do not attempt to copy multiple features that exist on different planes. This simply will not be allowed. When pasting a feature onto a part, make sure a planar face is selected. Nonplanar faces cannot be used for pasting features; nor can construction planes.

For those not familiar with "cutting" objects to the clipboard, it works exactly the same as copying. The only difference is that the object being cut is deleted from the model, and then pasted in at a different location. Obviously, you will need to be careful with this option.

The Control-Drag Technique

Another technique that can be used to add copies of features to a part is the control-drag technique. The result is exactly the same as if you were to copy and paste features. This is just a different way of accomplishing the same thing.

Control-dragging amounts to nothing more than holding down the Control (Ctrl) key and dragging a feature from FeatureManager to a planar face on the model in the work area. It should be noted that control-dragging a feature from the work area also works. Whether you use copy-and-paste techniques or the control-drag method is up to you. Use the method that works easiest for your situation.

> **NOTE:** *If control-dragging a copy of a feature from the work area to another location, the cursor must be positioned over a face on that feature for the procedure to work.*

If using the control-drag technique, make sure you are not actively editing a sketch. Otherwise, this method does not work. As the feature to be copied is being dragged, a preview will be displayed. The preview will appear to align itself with whatever planar face the cursor is over. Just release the mouse button, and the feature will be created at that position.

Because there is no way to precisely locate the feature while using the control-drag technique, the sketch must be edited afterward in order to position the sketch geometry at a particular location. Additionally, the Copy Confirmation window may appear if the feature being copied has locating relations. This is the same situation faced when using the copy and paste technique, and is resolved the same way.

If you hold down the Shift key while dragging feature geometry, this will cause the feature to be moved, not copied. This can be useful in certain situations, such as when trying to visualize where a particular feature should be placed. Dragging the feature around the model allows seeing the feature at various locations before finally deciding to place the feature at some specific point. It also makes editing the position of features extremely easy. What could be easier than simply dragging a feature to a new place!

✏ **NOTE:** *When using either the copy-and-paste or the control-drag technique to copy a feature, there is no association back to the original feature. The copies are independent features.*

Linear Patterns

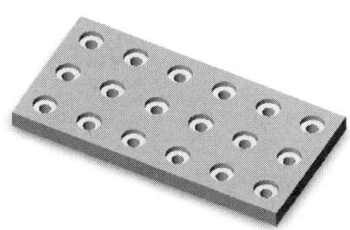

Fig. 7-3. A simple linear pattern.

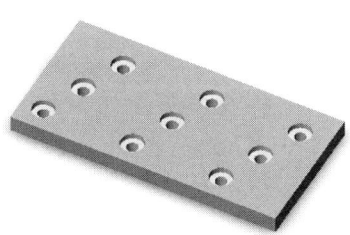

Fig. 7-4. Patterning in two non-orthogonal directions.

When you need to make a number of copies of one or more features, patterning is often the best method to use. Linear patterns are covered here. Other pattern types and the Mirror command (for mirroring features) are covered later in this chapter. Patterns in general can be employed more broadly than may be apparent. Do not underestimate the power of patterns in SolidWorks, as they may surprise you. A simple linear pattern is shown in figure 7-3.

You have the option of creating a linear pattern in either one or two directions. A two-directional pattern will create a grid-type pattern. Often, the pattern directions are orthogonal, but that is certainly not a requirement, as indicated in figure 7-4. The options associated with pattern commands will be something you learn about in this chapter.

One requirement when creating a pattern is to specify a direction for the pattern. This means you must select either an edge or a dimension that will signify the direction in which the patterned copies are created. If selecting a linear edge, that edge determines the pattern direction. If selecting a dimension, it must be a linear dimension, and the dimension line's arrows will point in the direction the pattern will take place.

An important aspect of patterning is that if the original patterned feature is modified, the changes will propagate to the other features in the pattern. Because of this, it is very easy to maintain control over a large number of features on the part with very little effort. In actuality, the pattern feature is considered a single feature within itself, even if it contains a thousand instances.

It probably goes without saying that parent/child relationships must be taken into consideration when patterning. For example, if you try to pattern a drill hole in a boss without patterning the boss, it should be quite obvious that the pattern will not work. The drill hole must have the boss to exist. Use common sense when patterning. How-to 7-2 takes you through the process of creating a linear pattern.

How-To 7-2: Creating a Linear Pattern

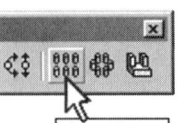

Fig. 7-5. Linear Pattern icon.

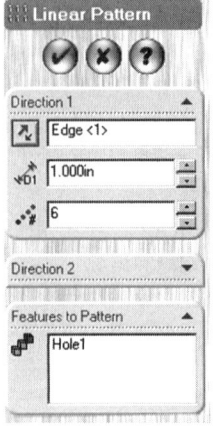

Fig. 7-6. A portion of the Linear Pattern panel.

To create a linear pattern, perform the following steps. The Linear Pattern icon, found on the Features toolbar, is shown in figure 7-5.

1. Select Insert > Pattern/Mirror > Linear Pattern, or click on the Linear Pattern icon. This will open the Linear Pattern panel, a portion of which is shown in figure 7-6.

2. Click in the Features to Pattern list box and select a feature or features to be patterned.

3. Click in the pattern direction list box for Direction 1, and select a linear edge or dimension to define the pattern direction.

At this point you should see a preview take shape on screen. For this reason alone it is best to specify the objects to be patterned, and a pattern direction, prior to plugging in any of the other parameters. Technically, though, these steps can be done in nearly any order.

4. Click on the Reverse Direction icon if necessary to flip the direction preview arrow.

5. Specify the Spacing (distance) from one instance to the next.

6. Specify the Number of Instances to create in the pattern. This number includes the original.

When typing in values for spacing or number of instances, press the Enter key to update the preview. Using the spin box arrows updates the preview automatically.

7. If patterning in two directions, repeat steps 3 through 6 for the Direction 2 parameters. Otherwise, click on OK to accept the settings and create the pattern.

There are a number of options available when creating patterns in general. Some of these options are common to all patterns, and some are specific to linear patterns. Let's look at some of these options.

Pattern Seed Only

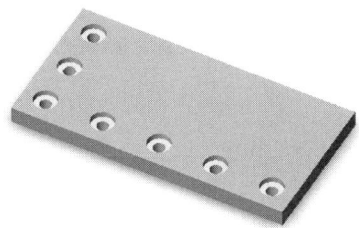

Fig. 7-7. The results of using the Pattern Seed Only option.

The Pattern Seed Only option is very easy to understand once it is seen in action. The term *seed* simply means the objects being patterned. The option is only available when patterning in a second direction. Typically when patterning in two directions, the second direction patterns the seed and every instance created from patterning in the first direction. If only the seed is patterned, however, none of the instances created from the first direction is patterned in the second direction.

Vary Sketch

The Vary Sketch option can be a very powerful feature when used correctly. This option is active only if you are working with sketch geometry that contains entities either converted or offset from existing feature edges, or that is constrained to existing geometry. The sketch must also be fully defined. One last criterion requires that a dimension be selected, as opposed to an edge, to establish the pattern direction. If all of these conditions are met, the Vary Sketch option will be enabled.

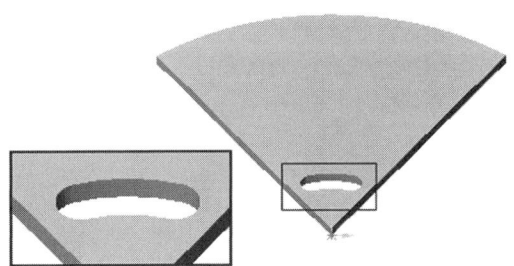

Fig. 7-8. A slot cut into the plate.

What does Vary Sketch accomplish? This is easiest to explain through example. In figure 7-8, a simple metal plate has been created that contains a single curved slot. The curved slot is rounded on both ends. The metal plate is pie-shaped.

Using this example, it would be convenient if the rounded slot could be patterned outward to create more slots. The dilemma occurs when creating the pattern. The design intent of the proposed slot pattern is that the distance between each rounded end of the slots and the edges of the pie-shaped plate be maintained during the pattern. However, during a typical pattern, these distances would not be maintained. The distance would be maintained relative to one edge of the pie shape, but not the other. This situation is depicted in figure 7-9.

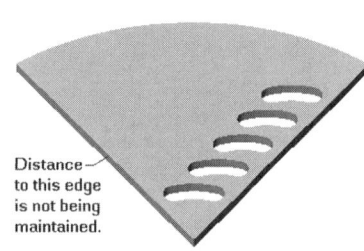

Fig. 7-9. A typical linear pattern of the slot.

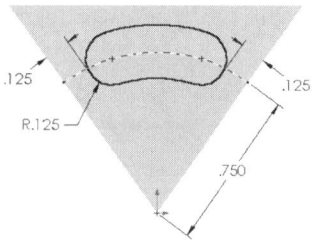

Fig. 7-10. This sketch would allow for using the Vary Sketch option.

What would be ideal is if the slot could be made to somehow maintain a constant distance from the left edge of the plate to the left side of the slot, just as the right side of the slot is doing. It is possible to accomplish this feat using Vary Sketch.

To take advantage of the Vary Sketch option, some forethought must be used when creating the sketch for the feature to be patterned. The sketch must be fully defined and must contain some sort of constraint or dimension to existing feature edges. Converted or offset entities would fall into this category, but these are not a requirement. Figure 7-10 shows the sketch used for the slot feature. Such a sketch would allow the Vary Sketch option to become enabled.

In this sketch, the .125-inch dimensions control the distance between the arcs and the plate edges. A dimensional property known as Arc Condition is used to move the dimension extension lines to the tangency points of the arcs. This property was explained in Chapter 5 (see How-To 5-3: Dimensioning to Tangency Points).

By dimensioning and constraining the sketch the way it is, the slot can be moved upward and away from the origin by altering the .750-inch dimension, and the rounded ends on either side of the slot will remain at constant distances from each edge of the pie-shaped plate. The .750-inch dimension plays an important role in the linear pattern. It specifies the direction of the pattern, and it is required to enable the Vary Sketch option. Vary Sketch will become active once the .750-inch dimension is selected to indicate the pattern direction.

> **NOTE**: *Using a dimension to indicate the pattern direction is required for enabling the Vary Sketch option.*

Fig. 7-11. Patterning the slot using the Vary Sketch option.

You saw what the patterned slot looked like earlier (figure 7-9) without the Vary Sketch option selected. Figure 7-11 shows the same pattern with the Vary Sketch option enabled. The difference is obvious.

When the Vary Sketch option is enabled, the relationships present in the sketch that constrain it to existing feature edges are taken into consideration. This allows the sketch to change shape as it is being patterned, thereby maintaining relationships to the rest of the part. It is not a bad idea to try this on your own. Make up some dimensions for the pie shape, and then model the slot. Try creating a linear pattern using Vary Sketch. Later you will perform a circular pattern on the entire model to create a round cover plate for a fan housing.

Deleting Instances from a Pattern

It is possible to remove instances from any type of pattern, but linear and circular patterns have a special option titled Instances to Skip. This gives the ability to choose which instances to keep and which to get rid of. You can also change your mind and quite easily get back any instance previously deleted, without recreating the entire pattern. This can be done at any time because the deleted instance data is saved with the part file. How-To 7-3 takes you through the process of deleting pattern instances.

HOW-TO 7-3: Deleting Pattern Instances

To delete an instance from a pattern, whether it is a linear or a circular pattern, perform the following steps. Note that these steps can be performed either during the pattern creation process or after the pattern has already been created. To delete instances after a pattern has been created, edit the pattern's definition.

> **NOTE:** *Use these same steps for recovering skipped instances.*

1. When defining the pattern (How-To 7-2) or while editing a pattern's definition (How-To 4-3), click inside the list box titled Instances to Skip.

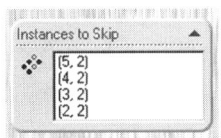

2. A series of dots will appear on screen. Click on a yellow dot to skip an instance, or on a red dot to add the instance back to the pattern. Any skipped instances will appear in the Instances to Skip list box, shown in figure 7-12.

Fig. 7-12. Skipping instances in a pattern.

3. Click on OK to complete the pattern.

An example of a linear pattern with skipped instances is shown in figure 7-13. Any instance can be skipped, as long as it is not the original feature the pattern was based on. Once an instance has been skipped, its instance position identifier is listed in the Instances to Skip list box. If a linear pattern, the identifier might read (5,2). To the user, this would read "patterned instance in the fifth position along the first direction and second position along the second direction." If an instance from a circular pattern, the number will read (x), where *x* is the number of the instance in the circular pattern, with the original being 1. Count clockwise or counterclockwise, depending on the circular pattern's direction.

To recover a skipped instance, the definition of the pattern must be edited. However, there is a shortcut for deleting pattern instances. If a face of a pattern instance is selected, pressing the Delete key will display the Pattern Deletion window, shown in figure 7-14. This is nothing more than a confirmation window, asking whether or not you want to delete just the selected instance or the entire pattern.

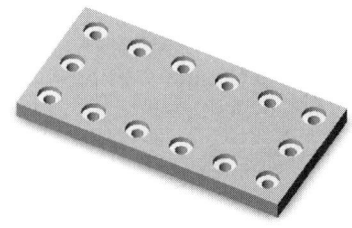

Fig. 7-13. A linear pattern with skipped instances.

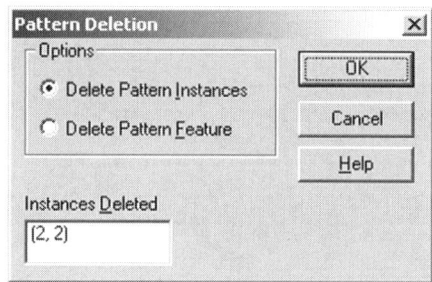

Fig. 7-14. Pattern Deletion window.

It is essential to select a face, not an edge, in order to use this shortcut. Additionally, more than one instance can be deleted simultaneously by holding down the Ctrl key and selecting faces from multiple instances. SolidWorks will even let you delete multiple instances from more than one pattern at a time. In this case, you will be presented with the Pattern Deletion window more than once.

You will next put some of the knowledge you have gained regarding linear patterns to the test. The shaver housing you started in Chapter 6 will be used in Exercise 7-1.

EXERCISE 7-1: Patterns on the Shaver Housing

To begin this exercise, open the *Shaver Housing* part you started in Chapter 6. Currently, there is only one boss containing a countersunk hole. There should be four such features. The boss and hole will be patterned simultaneously. To create the additional features, perform the following steps.

1. Click on the Linear Pattern icon.

Linear Patterns 247

2. Click in the Features to Pattern list box and select *Boss1* and the Hole Wizard feature from FeatureManager. (Hint: use the flyout FeatureManager by clicking on the name of the command, Linear Pattern, from the top of PropertyManager.)

3. Click in the Direction 1 list box and establish a pattern direction. The .875-inch dimension can be used. Click on the Reverse Direction button if necessary.

4. Specify two (2) instances for Direction 1.

5. Specify 2.75 inches for the spacing.

6. Click in the Direction 2 list box and establish a pattern direction. This time, use the .625-inch dimension. Once again, click on the Reverse Direction button if necessary.

7. Specify two (2) instances for Direction 2.

8. Specify 1.25 inches for the spacing.

9. Click on OK to accept the data and create the pattern.

Fig. 7-15. After creating the first pattern.

Fig. 7-16. Changing a plane's size.

When finished with this first pattern, the shaver housing should look as it does in figure 7-15. Next, an additional feature will be added, and then patterned. In particular, it would be beneficial to the design of the shaver housing to add grooves that would make gripping the shaver easier. In the next portion of this exercise, a simple cut will be created first, followed by a pattern of the cut. This will create a washboard effect that will keep the shaver from slipping in the hand of the person using it.

For this next feature, you will be sketching on the Right plane. This being the case, it is suggested that you show the Right plane. Resize the plane if you desire, and position the part as indicated in figure 7-16. Remember that you can resize a plane's border by dragging the green handles. This does not affect the model and is for aesthetic purposes only.

✓ **TIP:** *You can "autosize" a plane by right-clicking on its border and selecting Autosize. This sizes the plane to the part or the geometry used to define it. The plane will resize if the underlying geometry changes size. Manually resizing a plane turns off the Autosize function.*

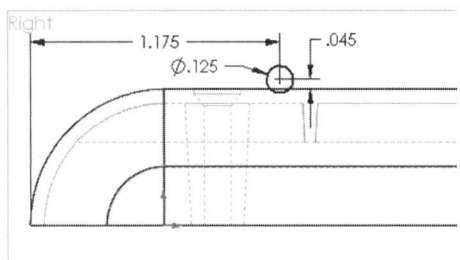

Fig. 7-17. Creating the sketch for the traction groove.

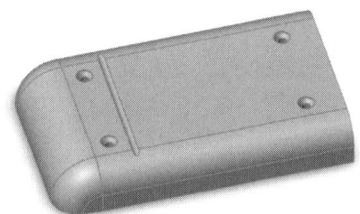

Fig. 7-18. Shaver with the traction groove.

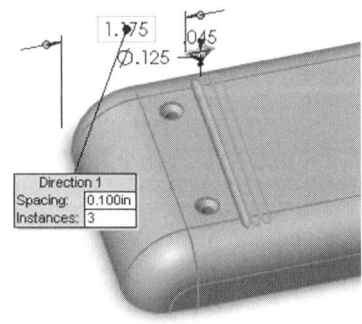

Fig. 7-19. Creating the first groove pattern.

The sketch that needs to be created is shown in figure 7-17. This will be used to create the first traction groove. The feature will then be patterned. This should be fairly easy for you this time around. Afterward, the pattern itself will be patterned again.

10. Create the sketch shown in figure 7-17.

11. Perform a Cut-Extrude. Use the Through All end condition, and click on the Reverse Direction button if necessary.

12. Rename the cut-extrude *Traction Groove*. The feature is shown in figure 7-18.

13. Click on the Linear Pattern icon.

14. Select the *Traction Groove* feature to be patterned.

15. Specify a pattern direction. In figure 7-19, the groove's 1.175-inch dimension was used. Reverse the direction if necessary.

16. Set the number of instances to 3.

17. Set the spacing to .100 inch.

18. Click on OK to finish the pattern.

Next, you will pattern the pattern. Nothing unique or different needs to be accomplished in regard to this task. However, as a side note it is worth mentioning that when the pattern is patterned the original feature need not be selected, as it is automatically included as being part of the first pattern. Continue with the following steps.

19. Click on the Linear Pattern icon again to create another pattern feature.

Circular Patterns

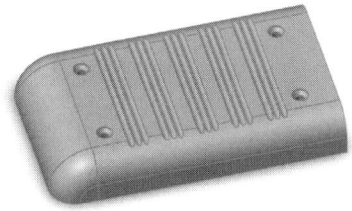

Fig. 7-20. Final Traction Pattern feature.

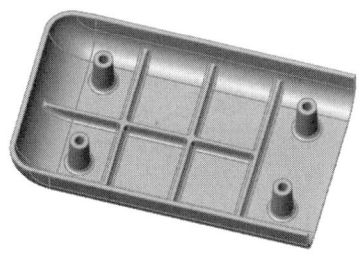

Fig. 7-21. After adding fillets to the shaver housing.

20. Select the last pattern you created as the feature to be patterned.

21. Specify a pattern direction. This time, use a linear edge that runs along the length of the shaver housing. Reverse the direction if necessary.

22. Set the number of instances to 5.

23. Set the spacing to be .4875 inch.

24. Click on OK to finish the pattern.

25. Rename this last pattern *Traction Pattern*.

26. Save your work.

The new groove patterns on the shaver housing should look like those shown in figure 7-20. Optionally, add some .050-inch fillets to the ribs and bosses on the inside of the shaver housing.

In the example shown in figure 7-21, fillets were added by selecting the edges at each rib junction and the bottoms of the bosses first, and then picking edges where the ribs connect to the inside surface of the part. In this way, all ribs were added in two operations.

Circular Patterns

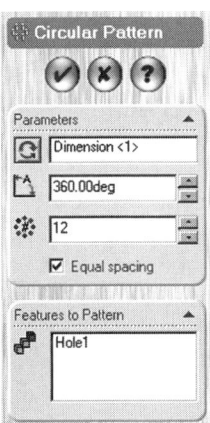

The Circular Pattern panel, shown in figure 7-22, is very similar to that used for linear patterns. The only major difference between defining linear and circular patterns would be the requirements for what dictates the pattern direction. With regard to circular patterns, an axis is the object most frequently used to define a circular pattern's direction of rotation.

Just for the record, it is possible to use an axis for linear patterns, though this is rare. If you find yourself in a situation in which an axis might be useful for a linear pattern, keep in mind that the axis will determine the pattern direction, just as an edge would. You will read more about axes, both temporary and user-defined, later in this chapter.

Fig. 7-22. A portion of the Circular Pattern panel.

Working with Circular Patterns

Another way to determine circular pattern direction is to use a dimension. For a linear pattern, you would select a linear dimension. For a circular pattern, an angular dimension is required. In the case of an angular dimension, the point where the extension lines for the angular dimension would theoretically intersect dictates the center of rotation for the pattern. It would be necessary to use a dimension to dictate circular pattern direction, for example, on a part in which an axis was not readily available.

When working with cylindrical parts, it is almost always more convenient to use an axis or temporary axis for the center of rotation. In other cases, such as with noncylindrical parts, it might be tedious to create an axis. In such a situation, it would be easy to add some sort of angular dimension in the sketch of the feature being patterned. Even a reference dimension would suffice. Such is the case in the example shown in figure 7-23.

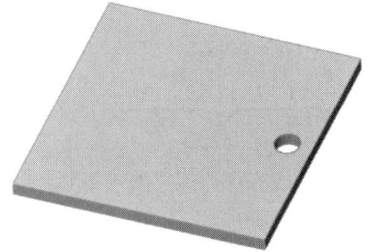

Fig. 7-23. How to pattern this hole in a circular fashion?

Normally when dimensioning the circle in the original sketch for this feature, linear dimensions would be used. Figure 7-24 shows the dimensions typically added, in the form of 10-mm and 50-mm linear dimensions and a diameter dimension. The image also shows additional centerlines and an angular reference dimension. It is the angular dimension that will dictate the center of rotation for the circular pattern. The centerlines are only there for the sake of the dimension.

How-To 7-4 takes you through the process of creating a circular feature pattern. Be aware that when it is necessary to select a dimension to define the pattern direction, whether it is a linear or circular pattern, it is helpful to first select the feature to be patterned. SolidWorks will then display the dimensions associated with that feature. (There are other ways of displaying dimensions, which are addressed in later chapters.)

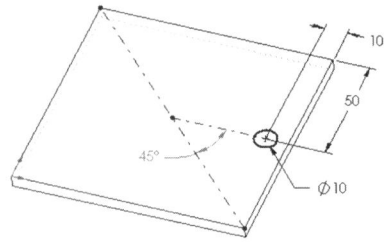

Fig. 7-24. This sketch makes a circular pattern easy.

How-To 7-4: Creating a Circular Pattern

To create the circular pattern, perform the following steps. A dimension will be used in this How-To, but an axis could be substituted for the dimension if one were available. The section titled "Axes" in material to follow explains how to create axes for a variety of uses.

Circular Patterns

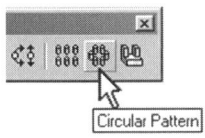

Fig. 7-25. Circular Pattern icon.

1. Select Insert > Pattern/Mirror > Circular Pattern, or click on the Circular Pattern icon, shown in figure 7-25.
2. Click in the Features to Pattern list box and select a feature or features to be patterned.
3. Click in the Pattern Axis list box and select an axis, linear edge, or angular dimension to represent the axis of rotation.
4. Specify the angle between instances.
5. Specify the number of instances.
6. Click on the Reverse Direction icon if necessary.
7. Click on OK to create the circular pattern.

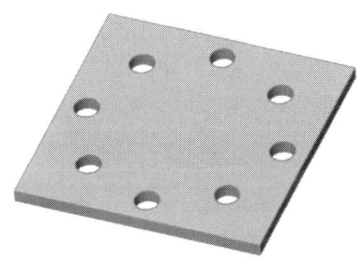

Fig. 7-26. Completed circular pattern.

Figure 7-26 shows an example of what a circular pattern might look like. With regard to deleting patterned instances, linear and circular patterns follow the same set of rules. An Equal spacing option in the Circular Pattern panel acts as a built-in equation (explored in the following section).

✓ **TIP:** *When creating patterns, specifying the objects to be patterned and defining an object to indicate the pattern direction first will result in a preview that will update as the number of instances or spacing is changed.*

Equal spacing Option

The *Equal spacing* option is only available with circular patterns. This option acts as a sort of built-in calculator. When checked, whatever value is typed into the Angle parameter is automatically divided by whatever value is specified for the Number of Instances parameter.

If Equal spacing is checked, the Angle parameter will be regarded as the Total Angle option and will automatically change to 360 degrees. The user can override the 360-degree setting, but to do so would be counterproductive. There is typically no reason to set Total Angle to anything but 360 degrees if you want equal spacing. For example, if three small holes are "equally spaced" over an included angle of 90 degrees, it would be just as easy to set the Angle setting to 30 degrees and leave *Equal spacing* disabled. There is, however, one exception.

With *Equal spacing* enabled, a design change can be made quite easily regarding the number of instances in a circular pattern. The number of instances can easily be altered, and the spacing between the instances is automatically updated to fill the desired included angle. If the designated included angle happens to be less than 360 degrees, then so be it. *Equal spacing* still works.

✓ **TIP:** *The number of instances in a linear or circular pattern is treated as a dimension. When double clicking on a pattern in FeatureManager, the number of instances can be changed via a double click, just as any other parametric dimension.*

Geometry Patterns

An option available for all feature patterns is the *Geometry pattern* option. The *Geometry pattern* option is also available when mirroring feature geometry. A geometry pattern, simply put, disregards the definition of the original feature. It takes only the geometric information of the feature (i.e., size and shape) and patterns it. There are no special requirements when performing a geometry pattern. Simply check the *Geometry pattern* option and complete the pattern as you normally would.

Because creating a geometry pattern means the software will have to do slightly less math (no end condition calculations), it would be reasonable to think the file size would be smaller. This is often the case, but to such a small extent that it would not be beneficial to create a geometry pattern strictly for the sake of reducing file size. Neither does a geometry pattern save time, even when creating large patterns. In short, use geometry patterns for reasons of a geometric nature.

To see what a geometry pattern will look like in comparison to a traditional "non-geometry" pattern, examine figure 7-27. The image shows a plate and a boss positioned in one corner. The boss was extruded using the Up To Surface end condition. The angled plane was used as the terminating surface, which is the reason the boss has an angled top.

Figure 7-28 shows the same model with the boss patterned in two directions using a standard linear pattern. Note the change in height of the patterned instances. The reason for this has to

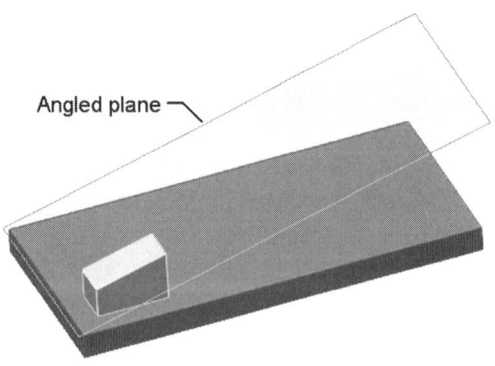

Fig. 7-27. Original feature before patterning.

do with the end condition being taken into account during the patterning process.

When the Geometry pattern option is turned on, SolidWorks ignores the end condition and patterns the original geometry. The shape of the original is maintained because of this. In the example shown in figure 7-29, note that the patterned instances all look exactly the same as the original. This is a direct result of creating a geometry pattern.

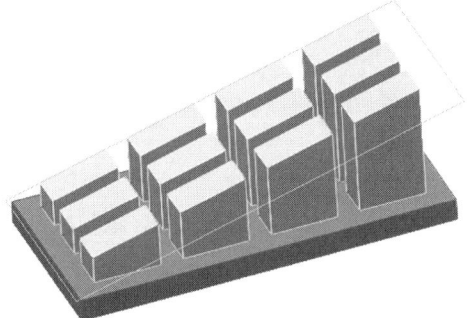

Fig. 7-28. After creating a typical linear pattern.

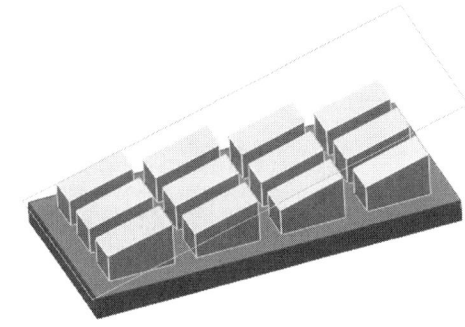

Fig. 7-29. After creating a geometry pattern.

Axes

In that circular patterns can benefit from having an axis present in the part, this is a good time to talk about axes. The first fact you should know about axes in SolidWorks is that there are two types: those created by SolidWorks and those created by the user. Axes created by SolidWorks are known as temporary axes, which are explored first because they are the most basic of the two.

Temporary Axes

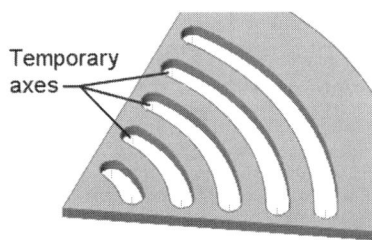

Anytime a cylindrical or conical face is created as a by-product of creating a feature, a temporary axis is created along with the feature. Temporary axes can be used for creating circular patterns, mating components in assemblies, and other functions. Figure 7-30 points out some temporary axes on a part used for demonstration purposes earlier in this chapter.

Fig. 7-30. Temporary axes.

A temporary axis is called "temporary" because theoretically another feature could completely remove the cylindrical or conical face being used by SolidWorks to define the axis. If the face is removed, so is the related axis, and hence the term *temporary axis*. From a cosmetic standpoint, there is not much you can do with temporary axes. You cannot rename them and cannot lengthen or shorten them; nor can you decide which temporary axes to display on an individual basis.

Displaying the temporary axes within a SolidWorks file is simple enough. Under the View menu, look for the Temporary Axes menu item, and make sure it has a check mark in front of it. Clicking on the Temporary Axes menu item will serve to toggle the display of temporary axes on and off.

NOTE: *Certain faces, such as those created by adding a constant-radius fillet as a feature, do not display temporary axes.*

User-defined Axes

Creating an axis is much the same as creating a plane. How you create the axis depends on the geometry you have to work with. You can define an axis by two vertex points, an existing edge, the intersection of two planes, and so on.

A user-defined axis can be renamed and will appear in FeatureManager as a feature. You can change the length of a user-defined axis (referred to simply as an axis from this point on) to make it appear more relevant to the part it belongs to, or simply for aesthetics. You can also control which axes you want to hide or show, just as you can for planes. How-To 7-5 takes you through the process of creating an axis.

How-To 7-5: Creating an Axis

To create a user-defined axis, perform the following steps.

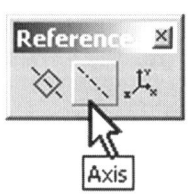

Fig. 7-31. Axis icon.

1. Select Insert > Reference Geometry > Axis, or click on the Axis icon, shown in figure 7-31. The icon is located on the Reference Geometry toolbar.

2. In the Reference Axis window that appears (see figure 7-32), select the method with which to define the axis. This will depend on the geometry you have to work with.

Axes

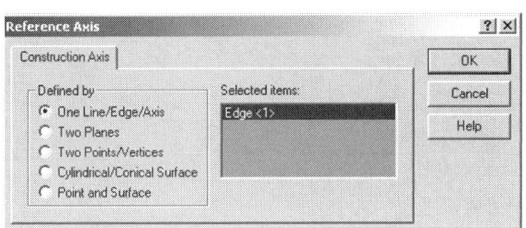

Fig. 7-32. Reference Axis window.

3. Select the applicable geometry for defining the axis.

4. Click on OK.

Figure 7-33 shows a user-defined axis. It is possible to right-click on the axis either in the work area or in FeatureManager to hide it or show it, using the same procedure you would for hiding or showing a plane.

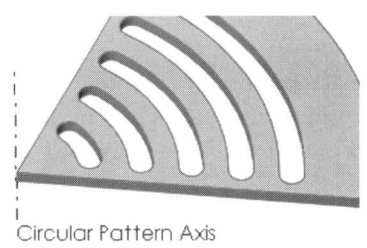

Fig. 7-33. A user-defined axis.

✏ **NOTE:** *To globally toggle the display of all user-defined axes, make sure Axes is checked in the View menu. This is important! None of the axes you create will be visible if the ability to view them is not enabled.*

One of the parts created in an exercise in Chapter 5 was the *Valve Stem*. That same part is used in Exercise 7-2. Now that feature patterns have been discussed in detail, finishing up the valve stem should not prove any difficulty. A number of other features will be added in addition to a pattern, so this exercise will give you some good practice.

EXERCISE 7-2:

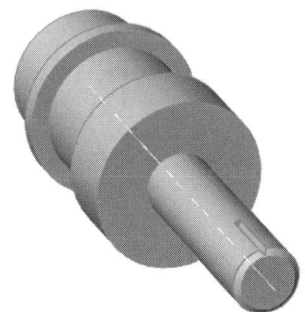

Fig. 7-34. Valve stem with temporary axes being shown.

Adding Features to the Valve Stem

In this exercise, you will pattern the groove on the *Valve Stem* part originally created in Chapter 5. This feature will represent the area the faucet handle will grip onto. Some additional features, including a countersunk hole, will be added to the stem. Begin by opening the *Valve Stem* file. Use the illustrations that follow as guidance. Figure 7-34 shows the valve stem. Note that the temporary axes are being shown.

1. Turn on the display of temporary axes by selecting View > Temporary Axes. You will use a temporary axis for the pattern.

2. Click on the Circular Pattern icon, or select Insert > Pattern/Mirror > Circular Pattern.

3. Select the temporary axis that runs through the center of the part for the Pattern Axis.

4. Select the Groove feature from FeatureManager as the Feature to Pattern. You should see a preview at this point.

5. Check the *Equal spacing* option.

6. Specify 24 for the Number of Instances. The Total Angle need not be supplied because it will default to 360 degrees automatically.

7. Click on OK to accept the settings and build the pattern.

8. Rename the pattern *Groove Pattern*.

9. Turn off the display of temporary axes, in that they are necessary at this point.

Figure 7-35 shows what the groove pattern should look like. Next, let's add a countersunk hole to the end of the stem. This should be old hat for you, in that the Hole wizard was covered in the last chapter. Continue with the following steps.

10. Select the face at the end of the stem on which to place the hole.

11. Click on the Hole Wizard icon.

12. Select the Countersink tab.

13. Specify an ANSI Inch, Flat Head (82), #4 screw. Use a blind end condition of .750 inch. Leave all other parameters at their default settings.

14. Click on Next.

15. Locate the single locating point at the center of the stem by adding a concentric relationship to the circular edge.

16. Click on Finish.

Figure 7-36 shows the valve stem with the countersunk hole. The last few features are simple features that will serve as good practice. Try to finish the valve stem by adding these last few features using the skills you have already learned. The sketch geometry and dimensions will be shown, but you are on your own regarding adding the proper relations to fully define the sketches. Continue with the following steps.

Fig. 7-35. After creating the circular groove pattern.

Fig. 7-36. Valve stem with a countersunk hole.

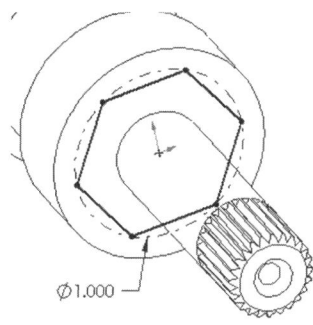

17. Add the six-sided extrusion to the valve stem using the dimensions shown in figure 7-37. Rename the feature *Hex Nut*. The extrusion depth is .500 inch.

18. Add a cut at the opposite end of the stem that is .150 inch deep. The sketch for the cut is shown in figure 7-38, along with the completed feature.

Fig. 7-37. Adding the Hex Nut *feature.*

Fig. 7-38. Creating the Recess *feature.*

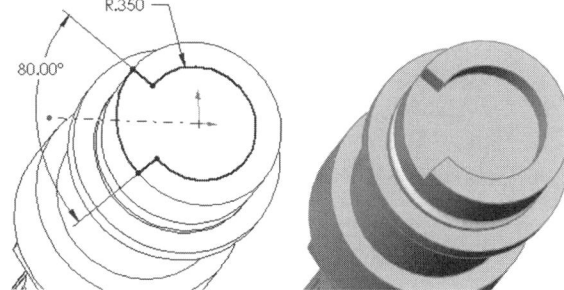

19. Rename the feature *Recess*.

20. Create one final sketch at the bottom of the *Recess* feature. The sketch is shown in figure 7-39.

21. Cut the sketch to a depth of 1 inch.

22. Save your work.

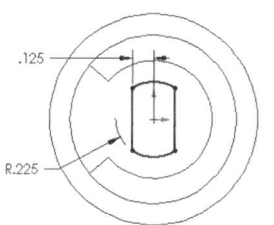

Fig. 7-39. Creating a sketch for one last cut.

Fig. 7-40. Completed valve stem part.

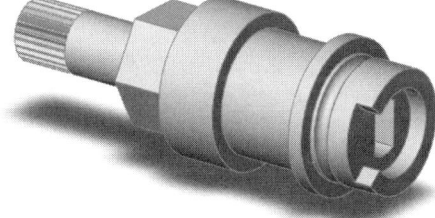

The valve stem could probably use a few more features before it is fully functional, but for the sake of brevity the exercise will end here. In addition, the valve stem should have ideally been created as an assembly, with at least a few components that turn to open and close the valve. However, if this is a purchased part that will be going into a higher-level assembly, consider that it probably only needs to be shown in an assembly drawing and does not necessarily have to function like an assembly itself. The completed part should look as it does in figure 7-40.

Sketch-driven Patterns

Not every pattern is a perfect linear or circular pattern. Sometimes it is desirable to specify points where instances should be placed, as if telling SolidWorks to "drop a copy here." Sketch-driven patterns are very flexible, convenient, and easy to edit because they allow for placement of patterned instances wherever there is a sketch point. All that is required is a sketch containing points.

The benefits of a sketch-driven pattern are that the pattern can be irregular or chaotic in nature. It is also extremely easy to create and modify. Because the pattern is based on sketch points, little effort is required on the user's part. Figure 7-41 shows a part with a set of features, each feature of which needs to be copied to various locations within the part.

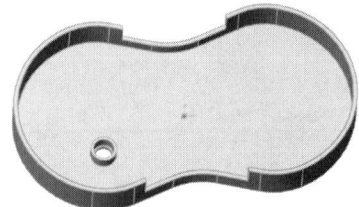

Fig. 7-41. A feature in need of patterning.

To prepare for creating the pattern, all that must be done is to create a sketch on the same plane as the feature to be patterned, and then to add sketch points wherever a copy of the original is to be placed. Such is the case in figure 7-42. First the points were placed in their approximate positions, and then the dimensions were added to fully constrain them. In some cases, geometric relations were used, such as horizontal or vertical. The sketch shown in the figure uses ordinate dimensions, but you can use any dimension type.

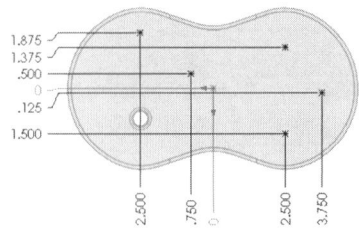

Fig. 7-42. Adding sketch points.

Because we have not explored ordinate dimensions yet, let's first learn how they can be added to a sketch. How-To 7-6 takes you through the process of adding ordinate dimensions.

How-To 7-6: Adding Ordinate Dimensions

To add ordinate dimensions to a sketch, perform the following steps.

1. Click on the Dimension icon.
2. Right-click anywhere in the work area and select Ordinate Dimension, Vertical Ordinate, or Horizontal Ordinate.
3. Select an object to serve as the 0 datum point.
4. Pick to position the 0 datum point value.

Sketch-driven Patterns

5. Select additional points to add further dimensions.

The Ordinate Dimension option allows for creating ordinate dimensions at an angle. Because the dimension lines are not forced to be either horizontal or vertical, pick a linear edge when establishing the 0 baseline location.

To add a new ordinate dimension value to an already existing series, make sure you are not in any other command, and then right-click on one of the ordinate dimension values. You should see a submenu titled Display Options. That is where you will find the Add To Ordinate option. There will also be some other nice functions, such as Jog and Re-Jog. These options, along with the Break Alignment option (also found in the right mouse button menu), give you quite a bit of flexibility in regard to ordinate dimensioning.

Now that you can add ordinate dimensions, let's proceed to creating a sketch-driven pattern. How-To 7-7 takes you through this process. Because a sketch is a requirement with this type of pattern, for your reference, that portion of the process is included in the steps.

How-To 7-7: Creating a Sketch-driven Pattern

To create a sketch-driven pattern, perform the following steps.

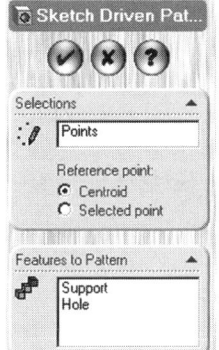

Fig. 7-43. Sketch Driven Pattern panel.

1. Create a sketch containing sketch points. Use the Point command to establish the points. Add a point wherever copies of the original feature being patterned will be placed.

2. Add dimensions or geometric relations as required to locate the points.

3. Exit the sketch.

4. Select Insert > Pattern/Mirror > Sketch Driven Pattern (no icon is available). This accesses the Sketch Driven Pattern panel, shown in figure 7-43.

5. Click in the Reference Sketch list area and select the sketch containing the points.

6. Click in the Features to Pattern list box and select the feature or features to be patterned.

7. Select whether to pattern by the selected object's centroid or some other reference point.

8. Click on OK to create the pattern.

The *Reference point* option allows for specifying a reference location on the parent feature, which will dictate where the patterned copies get positioned. By default, the centroid of the object being patterned is used. If there is more than one feature being patterned at a time, the centroid of the largest feature is used. If a reference point is used, select a sketch point, entity endpoint, or vertex point somewhere on the model. When patterning features such as holes, using the centroid is fine, but when patterning groups of features or features with irregular shapes you will probably need to use a reference point.

> **NOTE:** *A centroid is an object's center of gravity.*

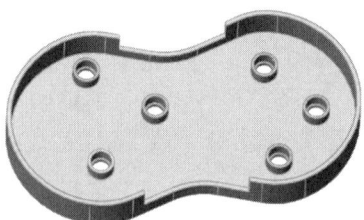

Fig. 7-44. After completing the sketch-driven pattern.

If the *Reference point* option still is not making sense, just think of it as the "handle" where the feature gets picked up. When it gets put back down on the sketch points, the "handle" is positioned on those sketch points. Figure 7-44 shows an example of a sketch-driven pattern.

If an instance in the pattern needs to be repositioned, modify the dimensions associated with that instance's respective sketch point. This is assuming the sketch points in the reference sketch were dimensioned in the first place, which they certainly should have been.

To take modifications one step further, edit the sketch that contains the sketch points. This would be known as the reference sketch. To delete an instance in the pattern, delete its respective sketch point. To add an instance, add a sketch point. It could not be any easier. The patterned instances are associated with the original feature, and therefore modifying the original will automatically update the copies.

Table-driven Patterns

Table-driven patterns are different from sketch-driven patterns in one major fundamental way. Instead of the need to create sketch points that define the patterned instance locations, a table of points is used to describe those locations. In other words, a set of *x*- and *y*-axis coordinates determines where the patterned instances get positioned.

One additional item required when creating a table-driven pattern is a reference entity known as a coordinate system. Because of this, coordinate systems are discussed in the following section. Make sure to read the next section before attempting to create a table-driven pattern.

Coordinate Systems

A coordinate system consists of a set of *x*, *y*, and *z* axes, and a point that defines the coordinate's 0 reference point. Think of the SolidWorks origin point as a part or assembly's world coordinate system. It is the intersection of the three default planes and is the Cartesian coordinate system's zero reference point.

Certain functions of the SolidWorks software can alternatively employ user-defined coordinate systems (as opposed to the part's origin point). A perfect example of this would be when obtaining the mass properties of a part (discussed in more detail in an upcoming chapter). When obtaining the principal moments of inertia, for example, it might be preferable to reference a location other than the part's origin. In other cases, a coordinate system is simply a requirement, such as in the case of a table-driven pattern.

Creating a user-defined coordinate system is a simple enough process. An existing point must be selected. This can be a vertex point or a sketch point of some sort. A sketch point can be an actual sketch entity, endpoint, centerpoint, and so on. If the need arises, the user can control the direction in which the *x*, *y*, or *z* axis points. If this is the case, a line or edge must be selected as well, which will dictate the direction of the desired axis of the new coordinate system.

Additionally, a plane or planar face can be used in place of a line or edge, in which case the associated axis will be perpendicular to the plane or face. How-To 7-8 takes you through the process of creating a user-defined coordinate system.

How-To 7-8: Creating a Coordinate System

To create a user-defined coordinate system, perform the following steps.

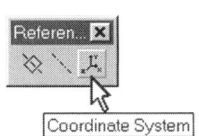

Fig. 7-45.
Coordinate System icon.

1. Select Insert > Reference Geometry > Coordinate System, or click on the Coordinate System icon, shown in figure 7-45, located on the Reference Geometry toolbar.

2. Select a point for the new coordinate system.

3. Optionally, click in the X Axis, Y Axis, or Z Axis box of the Coordinate System window (see figure 7-46) and select a plane, planar face, edge, or line to define the direction of that axis.

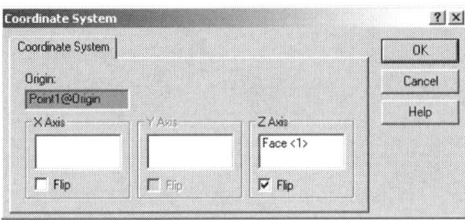

4. Repeat step 3 for an additional axis if required, and flip the axes if you wish.

5. Click on OK when finished.

Fig. 7-46. Coordinate System window.

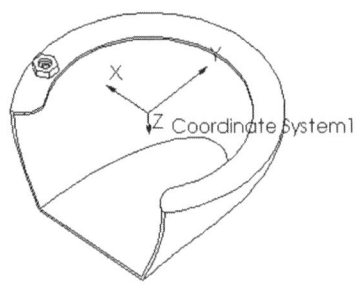

Fig. 7-47. A user-defined coordinate system.

After creating a user-defined coordinate system, it will appear in FeatureManager as *Coordinate System1*. You will find it possible to rename the coordinate system, just like any other feature in FeatureManager. If the coordinate system is not visible in the work area, turn on coordinate system visibility by selecting Coordinate Systems from the View menu. A user-defined coordinate system is shown in figure 7-47. The origin point was used in defining this coordinate system, which may seem redundant, but it was nonetheless necessary due to the requirements of a table-driven pattern.

✓ **TIP:** *Coordinate systems can be hidden or shown, just as planes can be, by right-clicking on the coordinate system in FeatureManager.*

Now that you understand user-defined coordinate systems, you can move on to table-driven patterns. It should be mentioned that a coordinate system is needed for a table-driven pattern because that is what the x-y coordinates in the table will reference. Without the coordinate system, the table coordinates have no 0 reference point or reference direction.

In addition to the coordinate system, a table-driven pattern also requires that x-y coordinates be specified, which will dictate the placement of the patterned instances. This can be accomplished in one of two ways, as follows.

• Create x-y coordinates on the fly by typing them in.

• Import a previously existing table of x-y coordinates.

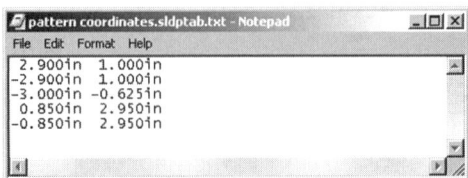

Fig. 7-48. A set of coordinates for a table-driven pattern.

If a table is to be used, the coordinate table file must have either a *.txt* extension or SolidWorks' own *.sldptab* extension. Either is fine, and in either case the file itself is actually only a simple text file. Figure 7-48 shows an example of what one of these coordinate table files might look like. Windows Notepad was used to create the table shown.

Table-driven Patterns

How-To 7-9 takes you through the process of creating a table-driven pattern.

How-To 7-9: Creating a Table-driven Pattern

The following steps show you how to create a table-driven pattern. It is assumed that the feature or features to be patterned already exist.

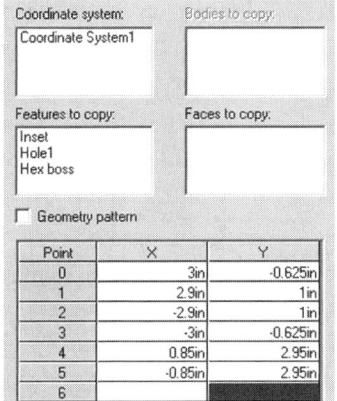

Fig. 7-49. A portion of the Table Driven Pattern window.

1. Select Insert > Pattern Mirror > Table Driven Pattern.

2. In the Table Driven Pattern window (a portion of which is shown in figure 7-49), click in the *Features to copy* list box and select the features to be patterned.

3. Click in the *Coordinate system* list box and select the previously defined coordinate system. The part's origin point will not suffice for this.

4. Specify the *x-y* locations for the patterned instances by double clicking on the appropriate cells in the Point listing, or optionally browse for a file that contains the point coordinates.

5. Click on OK to create the pattern.

The *Reference point* option serves the exact same function it did with a sketch-driven pattern. A table-driven pattern was performed on the part shown in figure 7-50.

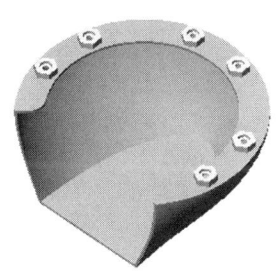

Fig. 7-50. A completed table-driven pattern.

Modifying a Table-driven Pattern

Modifying a table-driven pattern is slightly different than its sketch-driven counterpart. First, deleting an instance in a table-driven pattern can be done by selecting a face on any of the instances to be deleted. This is akin to deleting an instance in a standard linear or circular pattern. As a matter of fact, the Pattern Deletion confirmation window that appears is the same one that appears when deleting an instance in a linear or circular pattern.

Another way to delete an instance in a table-driven pattern is to edit the definition of the pattern, select one of the instance rows from the table by clicking on its corresponding Point number, and press the Delete key.

An interesting aspect of table-driven patterns is the ability to edit any of the table coordinate dimensions associated with the pattern. By double clicking on the pattern feature in FeatureManager, access to the table coordinate dimensions is gained. Double click on any of the dimensions to modify them, just as you typically would with any feature in FeatureManager. Depending on the number of features in the pattern, you may wind up with a lot of dimensions on the screen. If this is the case, just edit the definition of the pattern instead. Regarding 2D design drawings, the coordinate dimensions will appear on the drawing as they do on the part.

The other buttons found in the Table Driven Pattern window allow for saving the points as a SolidWorks SLDPTAB text file (Save and Save As). This assumes that the coordinates were manually entered in the first place. The Browse button, obviously, is for hunting down a text file containing the coordinates.

Curve-driven Patterns

If you thought we were done with patterns, the answer is no, not quite yet. SolidWorks provides yet another method of patterning features: along a curve. The curve can be a model edge or sketch geometry. As a matter of fact, the curve can be a nonplanar curve. (Various methods of creating curves are discussed in Chapter 9.)

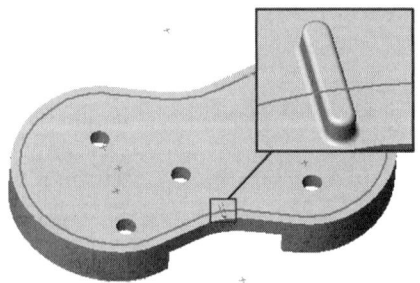

Fig. 7-51. Preparing for a curve-driven pattern.

The objects that constitute a valid curve are not unlimited, but there are really only a few stipulations. Only one curve can be used, and all segments in the curve must be tangent. This means that if you are using a sketch to drive the pattern, all sketch entities must be tangent. The easiest way to accomplish this task is to simply add tangent relations, which was first discussed in Chapter 2.

Because there is no option for propagating along tangent edges built into the Pattern command, creating a sketch is common. This way, the entire sketch can be used to drive the pattern. As is the case with any feature pattern, the sketch must be exited prior to attempting the pattern. Figure 7-51 shows a sketch created preparatory to creating a curve-driven pattern. The image also shows a blowup of the small feature being patterned.

Curve-driven Patterns

The sketch shown in figure 7-51 was created using the Offset Entities sketch tool, which made creating the curve extremely easy. How-To 7-10 takes you through the process of creating a curve-driven pattern.

How-To 7-10: Creating a Curve-driven Pattern

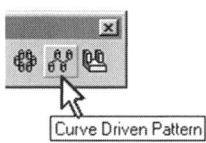

Fig. 7-52. Curve Driven Pattern icon.

To create a curve-driven pattern, perform the following steps. If using a sketch to drive the pattern, make it a point to create the sketch first. The steps that follow assume that a sketch or curve of some sort has already been created. If using a sketch, make sure to exit the sketch first.

1. Select Insert > Pattern/Mirror > Curve Driven Pattern, or click on the Curve Driven Pattern icon, shown in figure 7-52, located on the Features toolbar.

2. Click in the Pattern Direction list box and select a sketch, model edge, or some other curve to drive the pattern for Direction 1. A portion of the Curve Driven Pattern panel (accessed in step 1) is shown in figure 7-53.

3. Optionally, check Direction 2.

4. Optionally, click in the Pattern Direction list box and select a sketch, model edge, or some other curve to drive the pattern for Direction 2.

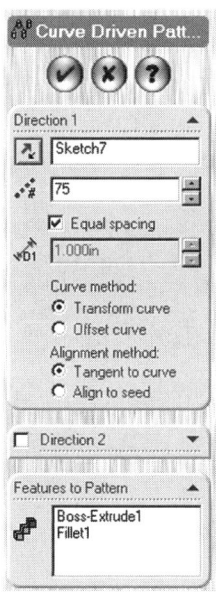

Fig. 7-53. Portion of the Curve Driven Pattern panel.

5. Click in the Features to Pattern list box and select the features to be patterned.

6. Specify the number of instances to be included in the pattern. This is the total number of instances, as is typical for other pattern types.

7. Check the *Equal spacing* option if desired.

8. If the *Equal spacing* option is not used, enter a value for the spacing.

9. Click on OK to create the pattern.

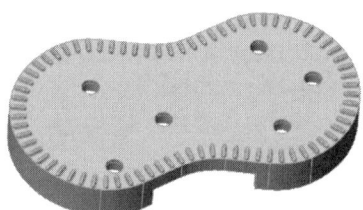

Fig. 7-54. Example of a curve-driven pattern.

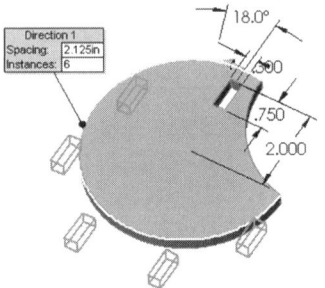

Fig. 7-55. Transform curve *and* Align to seed *options.*

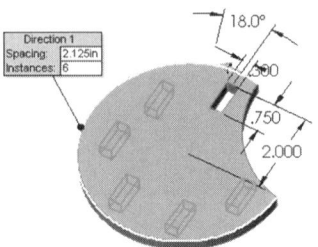

Fig. 7-56. Offset curve *and* Align to seed *options.*

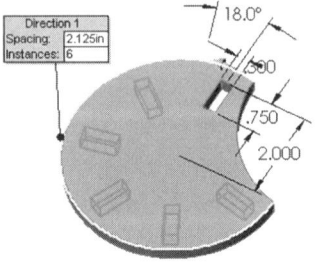

Figure 7-54 shows an example of a curve-driven pattern. Note that the original feature maintains the proper orientation even as it continues around the curved perimeter of the part. Quite a few copies were created in this particular pattern, which can sometimes take a while, especially on a computer with older hardware. (But having to wait a few minutes for a large number of instances to pattern is always a good excuse for a coffee break.)

There are two options that play a role in the outcome of the pattern. These options are Curve method and Alignment method. They are both easy to understand once seen in action. In figure 7-55, the rectangular cut in the upper right-hand portion of the part is being patterned. The large outer arc edge is used for Direction 1. In this first example, curve and alignment methods are set to *Transform curve* and *Align to seed*, respectively.

As is evident from the preview being displayed in figure 7-55, using the current combination of options would position certain instances completely off the part. This will not do. In particular, it is the *Transform curve* option that is getting us into trouble. The pattern follows the shape of the curve precisely. If *Transform curve* is changed to *Offset curve* instead, the centroid of the feature being patterned stays at a precise distance from the curve. This distance is determined by whatever the original feature's distance is from the curve. Figure 7-56 shows the same model with *Offset curve* turned on.

In another permutation of the Curve method and Alignment method options, *Align to seed* can be set to *Tangent to curve*. In figure 7-57, that is exactly what has been done. Because the edge being used to drive the pattern is an arc, the results look similar to a circular pattern. It should be noted that the *Tangent to curve* alignment method was used in the example shown in figure 7-54.

If you performed How-To 7-10, you are aware that a second curve can be specified for Direction 2. Things can get fairly interesting when driving a pattern along two curves. To be frank, it is a complex maneuver that does not always work as expected. It really depends on what is being patterned and the nature of the curves. The example shown in

Fig. 7-57. Offset curve *and* Tangent to curve *options.*

figure 7-58 is a completed pattern using two curves. The arc edge on the right side of the part was used for Direction 2.

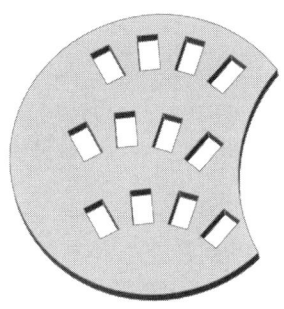

Note in figure 7-58 that the rectangular cuts follow the larger arc quite nicely. The cuts also follow the second curve (smaller arc), but the second direction is using the *Transform curve* option, even though *Offset curve* has been selected. This serves to illustrate that the Curve method settings do not necessarily apply to Direction 2.

✓ **TIP:** *If Direction 2 is used, a curve does not have to be selected, in which case Direction 2 will pattern in a linear fashion.*

Fig. 7-58. Patterning along two curves.

Mirroring Feature Geometry

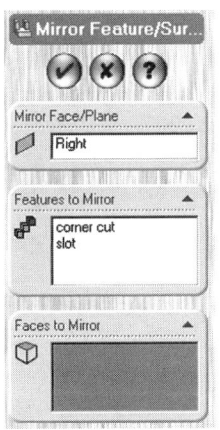

Fig. 7-59. Mirror Feature panel.

When sketch geometry is mirrored in a sketch, a centerline is needed. However, when feature geometry is mirrored, a plane must be used. The Mirror Feature panel, shown in figure 7-59, is extremely easy to use. To mirror a feature, simply select a plane or planar face to be the mirror plane, and select the items you want to mirror. You will see a preview of the mirrored objects. Click on OK to accept the new feature. This is all there is to it.

Parent/child relationships play a role when mirroring features (just as they do when patterning). For instance, you would not be able to mirror a hole in a boss without mirroring the boss also. In other words, features that are children cannot be mirrored without their parents. In addition, if a plane is used as the mirror plane, you cannot delete the plane without also losing the mirrored features. In this case, the plane would be a parent of the mirrored features.

The Geometry Pattern option means the same thing here as it does when creating patterns. There is absolutely no difference. If Geometry Pattern is checked, no end condition information is taken into account. How-To 7-11 takes you through the process of mirroring feature geometry.

How-To 7-11: Mirroring Feature Geometry

To mirror features, perform the following steps.

1. Select Insert > Pattern/Mirror > Mirror, or click on the Mirror icon, shown in figure 7-60, located on the Features toolbar.

268 Chapter 7: Patterns

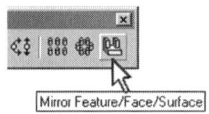

Fig. 7-60. Mirror icon.

2. Select the plane or planar face to use as the mirror plane.

3. Select the items to be mirrored.

4. Click on OK to mirror the features.

Figure 7-61 shows a before-and-after image of a part that had two features mirrored. Sometimes it is necessary to create a plane first that will serve as a mirror plane. In the case of the model shown in the figure, the Right plane was convenient for this purpose because the original base feature sketch had been centered on the origin. Centering the first sketch on the origin point is common practice when creating symmetrical parts.

Fig. 7-61. Mirroring feature geometry.

Mirroring and Patterning Faces and Bodies

Fig. 7-62. Mirror PropertyManager.

Throughout the previous sections dedicated to various patterning commands and mirroring, you may have noticed areas in each of the command interfaces that allow for selecting faces or bodies. These areas can be seen in figure 7-62, and in this case happen to be from the Mirror PropertyManager. Incidentally, bodies must be patterned or mirrored separately from faces or features.

The ability to pattern or mirror faces is extremely important to those working with imported geometry. Because imported models rarely contain any feature data, there are no features that can be selected. In such a case, the ability to select faces becomes critical. Faces can be mirrored or patterned with the same result as if features had been mirrored or patterned.

There are limitations to patterning or mirroring faces. If the faces being mirrored (for example) do not form a closed boundary condition, Solid-Works will fail to perform the operation. This is not necessarily true when mirroring faces for the sake of cutting away geometry, but such is the case in figure 7-63. Three faces on the left side of the part are being mirrored forward. This can be seen in the preview. However, the operation fails because the mirrored faces are essentially hanging out in space by themselves. This does not constitute valid geometry.

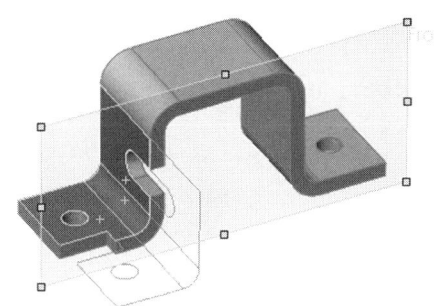

Fig. 7-63. These faces do not form a closed boundary.

On a somewhat similar note, surfaces can be patterned and mirrored, and no closed boundary condition is required. Although the terms *face* and *surface* seem to describe similar objects, they are generally used to describe different types of data. Faces are thought of as the portions of a solid model that make up a closed boundary. Surfaces can exist by themselves and do not necessarily have to be used to form a solid. (Read more about surfaces in Chapter 20.)

Bodies can also be patterned or mirrored. Simply use the same commands you would to pattern or mirror features. Although patterning or mirroring bodies is not as common a task, it does add to the flexibility available when creating a model.

Symmetrical Parts

Sometimes, as in the case of symmetrical parts, it is easier to create half of a part and then mirror that half to create the whole. This can be done with the same Mirror command discussed in the previous section. The main difference would be to click inside the Bodies to Mirror list box and select the model from the work area. Assuming the model is a single body, it can then be mirrored about the planar face of your choosing.

Because multiple bodies can exist in the same part file, it is possible to mirror about planes. This could theoretically result in two (or more) detached pieces of material. Similarly, even if using a planar face on the model to mirror about, unchecking the *Merge solids* option would also result in separate pieces of geometry. Unless you have ulterior motives, leave *Merge solids* checked.

Mirrored Parts, or "Left-Hand" Versions

Another option open to you is the ability to create a mirror image of a part. A new part is created that is a mirror image of the original. The new part will be a separate SolidWorks part file. Additionally, a relationship (or "link") is created to the original part. Therefore, if the original part is modified, the changes will appear in the mirrored part as well.

Essentially what is happening when a part is mirrored is that Solid-Works uses the original part as a base feature for the mirrored part. As an example, the sample part used in the "Mirroring Feature Geometry" section will be used again to illustrate how the Mirror Part command works.

270 **Chapter 7: Patterns**

Figure 7-64 shows the part to be mirrored. The face being used for the purpose of creating a mirrored part is based solely on the fact that it is easy to select. The selected face will have a bearing on the new part's orientation, but this really is not significant.

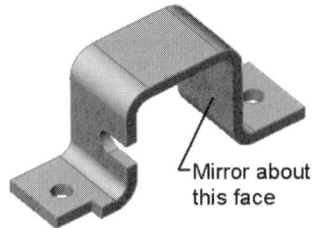

Fig. 7-64. Part to be mirrored.

The Mirror Part command is found in the Insert menu. A plane or planar face must be selected prior to accessing the command. Once the command has been accessed, PropertyManager displays the Insert Part panel, shown in figure 7-65.

The reason for the panel is to allow for importing any additional reference geometry that might be relevant. Of significant note is the Cosmetic Thread option. If creating a left-hand version of a large part with numerous cosmetic threads, checking the Cosmetic Thread option keeps you from having to recreate all of those cosmetic threads in the mirrored part. Likewise, reference plane, axes, and surfaces can also be brought over from the original part file into the mirrored part.

SolidWorks will create a new part using the original as the base feature. Note in figure 7-66 that the two cut features are now on the opposite side. In addition, note the content of FeatureManager. The new part's base feature will use the name of the original part, with the text _Mirrored appended to it. This additional text will disappear once the file is saved.

Fig. 7-65. Panel displayed in PropertyManager.

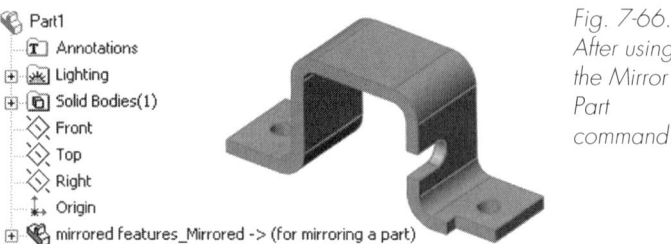

Fig. 7-66. After using the Mirror Part command.

Also note that an external reference has been created, which is symbolized by the arrow after the feature name (and also after the name of the part at the top of FeatureManager once saved). This confirms the fact that this newly created part is indeed dependent on the original part. Any changes made to the original will be propagated to the mirrored part. Likewise, if the originating part file is deleted or renamed, you will not be able to open the mirrored part. This is a unidirectional dependency only. Features added to the left-hand version will not propagate to the original.

One last addition to FeatureManager is the name found in parentheses after the external reference arrow symbol. The wording that can be seen in the image reads "for mirroring a part." This is the name of the part configuration being referenced by the base feature. Configurations are a topic unto themselves and are explored in their entirety in Chapter 10. External references are very important and can be explained somewhat more easily. Thus, they are explored in the section that follows.

External References

Anytime a part references another, an external reference is created. This is exactly what happened in the previous section. One part (call it part A) was used to create a second part (part B). Part B is now dependent on part A. If a change is made to part A, the change is propagated to part B. This is a one-way street only. Any changes made to part B will not propagate to part A.

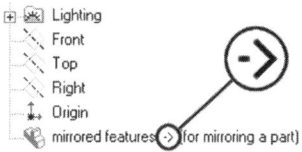

Fig. 7-67. External reference symbols.

External references are denoted by a small arrow symbol after the name of the feature or part with the external reference. This is shown in figure 7-67. The "symbol" is actually a dash followed by a greater than character (->), but it serves the purpose. Any part containing an external reference will have a symbol attached directly to the part's name. This is to let you know there is an external reference somewhere in the part. Additionally, a symbol will appear on whatever feature actually contains the external reference.

If at any time you want to view the data associated with an external reference, you may do so. By right-clicking on any feature containing an external reference, it is possible to view those references. Selecting the List External Refs option will open the External References window, partially shown in figure 7-68.

Fig. 7-68. Listing external references.

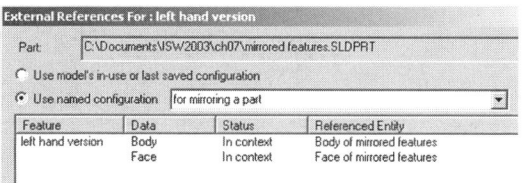

The path data at the top of the window will list the path to the part or assembly the feature is referencing. Other data is given as well, such as what type of entity is being referenced, and whether or not the reference is *In context* or *Out of context*. If a reference is out of context, it means that the referenced geometry is unavailable because the file being referenced is not open at that time. Opening the referenced file will cause the status of the reference to be changed to *In context*.

Opening Referenced Files

It cannot be stressed enough that it is extremely important to realize when an external reference is out of context or not. *In context* means in association with. The meaning in SolidWorks is identical to that associated with speech. For instance, words can be taken out of context, which means their original meaning is lost. When the same words are read in context to the original phrase, their true meaning is discovered. The same holds true with features that are out of context or in context.

Features that are out of context will appear with a question mark character (->?). Be aware that when externally referenced files are not loaded into memory (opened, in other words), they are *out of context*. What exactly does this mean to you? It means you may not be viewing the true nature of the model being displayed on screen. Geometry may have changed. If the referenced file were altered in some way, you would not know it because it has not been opened yet to allow the referencing file to update.

To load the externally referenced file, simply right-click on the feature containing the reference and select Edit In Context. This will open the referenced file automatically. Performing a rebuild on the file with the external reference will cause the question mark to disappear. When that happens, you can rest assured the geometry being displayed on screen is up to date.

✓ **TIP:** *The Load Referenced Documents option found in the External References section of the System Options (Tools > Options) offers the ability to always load externally referenced files automatically. Selecting the All option enables this functionality.*

Locking References

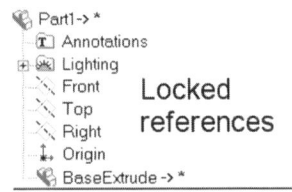

Fig. 7-69. Locked external references.

When references are locked, it means that any changes made to the referenced file will not show up in the file containing the reference. This can be useful if it is desirable to put a temporary "hold" on a model. If at a future time it becomes necessary to incorporate the changes made to the referenced file, the external references can be unlocked and the changes made to the referenced file will become incorporated in the model containing the reference. Figure 7-69 shows that an asterisk symbol (*) is used when external references have been locked. To lock (or unlock) external references, use the Lock All or Unlock All buttons found in the External References window.

Breaking External References

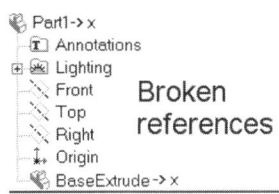

Fig. 7-70. Broken external references.

Fig. 7-71. Breaking references is permanent!

It has always been possible to break external references on sketch geometry by editing a sketch and using the Display/Delete Relations function to remove any unwanted external references. In the situation of a mirrored part, however, the external references are associated with a feature, not sketch objects. In this case, the external references can be broken using a different technique. Note that any external references can be broken; not just those created by mirroring a part. The symbol used to denote broken references is an x, shown in figure 7-70.

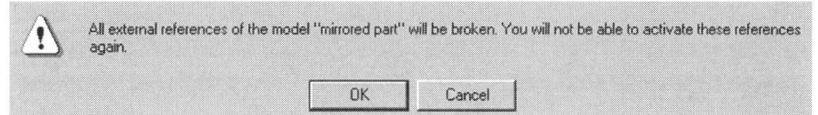

Once references have been broken, they cannot be reestablished, as indicated in figure 7-71. Take this as a word of warning. Make sure it is what you want.

✗   **WARNING:** *When clicking on the Break All button (External References window) to break the external references, a small warning message appears (shown in figure 7-71). If this process is continued, the external references will be broken and the box listing those references will clear.*

You will come into contact with external references again further into this book. External references are created when using the Base Part command (Chapter 13), and when creating features in the context of an assembly (Chapter 18).

Sketch Step and Repeat Commands

There are two versions of the step-and-repeat command. There is the Linear Sketch Step and Repeat command and the Circular Sketch Step and Repeat. They are both very similar to creating feature patterns. Patterning feature geometry is recommended over patterning sketch geometry. Patterning features is more flexible. However, there still may be times when the ability to pattern sketch geometry proves useful. Step-and-repeat commands are performed strictly on sketch geometry, never features, and are therefore considered sketch tools.

Linear Sketch Step and Repeat

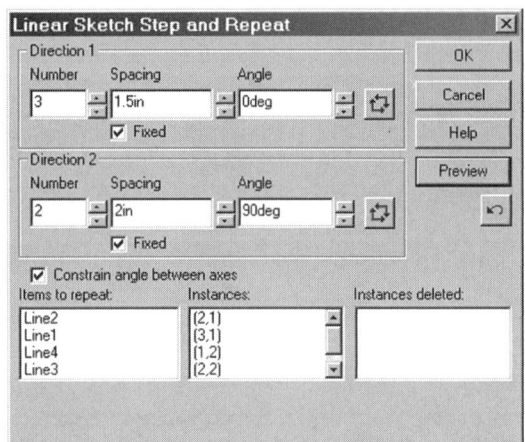

Fig. 7-72. Linear Sketch Step and Repeat window.

The Linear Sketch Step and Repeat window is shown in figure 7-72. As the name suggests, this window will allow you to create copies of selected sketch geometry in a linear fashion. It is possible to create the linear pattern in a single direction, or two directions at the same time.

The Process

One notable difference the linear step-and-repeat function has over feature patterning is that there is no requirement to select an object that dictates the pattern direction. Instead, step-and-repeat functions reference the sketch origin. The *x* axis of the origin, represented by the smaller of the two red arrows, would correlate to an angle of 0. The *y* axis would represent 90, and so on. This is exactly how a Cartesian coordinate system is arranged. You will want to bear this important fact in mind when performing a step and repeat. How-To 7-12 takes you through the process of using the Linear Sketch Step and Repeat command.

How-To 7-12: Performing a Linear Sketch Step and Repeat

To perform a Linear Sketch Step and Repeat, perform the following steps. Note that the software gives some interesting system feedback when selecting objects while in step-and-repeat commands. This is an oddity, but should not affect your ability to carry out the command.

> **NOTE:** *If the number of instances for a particular direction is set to 1, the Spacing and Angle parameters will be grayed out. Do not let this fool you. The Spacing and Angle will automatically become available as soon as the Number parameter is greater than 1.*

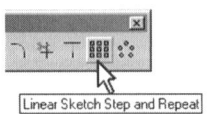

Fig. 7-73. Linear Sketch Step and Repeat icon.

1. Click on the Linear Sketch Step and Repeat icon (shown in figure 7-73), or select Linear Step and Repeat from the Tools > Sketch Tools menu.

2. Select the entities to be repeated. This can be one or more sketch entities.

Sketch Step and Repeat Commands 275

3. Specify the Number of Instances (total) for Direction 1.

4. Specify the Spacing between instances.

5. Specify the Angle for Direction 1, where an angle of 0 is along the sketch origin's *x* axis.

6. Repeat steps 3 through 5 for Direction 2, if required.

7. Click on OK to create the pattern.

These steps cover the basics for creating a step-and-repeat pattern, but there are a few other options available in the window. For instance, select Fixed for either direction if you want a dimension added automatically for the spacing distance. Additionally, select *Constrain angle between axes* if you want SolidWorks to add a dimension to control the angle between pattern directions. Both of these options are recommended. Do not confuse the Fixed option with the Fix geometric relation option. They are not related to each other in any way.

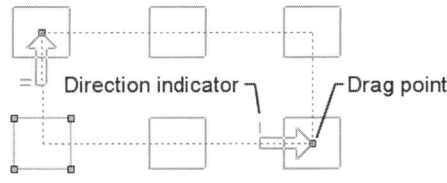

Fig. 7-74. Step-and-repeat preview.

If there are any instances you want to have removed from the pattern, select those instances from the Instances list box and press the Delete key. The instances will be displayed in the *Instances deleted* list box. Deleted instances can be retrieved (added back in) by removing their instance numbers from the *Instances deleted* list box. It is possible to tell the instances apart because of the direction indicators in the preview, shown in figure 7-74.

Obviously, the arrow labeled 1 is indicative of Direction 1, and so on. Instance numbers follow a grid pattern. That is, an instance number of (3,2) in the Instances list box refers to an instance positioned over 3 in direction 1 and across 2 in direction 2.

The dots at the arrow points can be dragged, thereby changing the spacing dynamically. The Reverse Direction buttons will flip the pattern direction from one side to the other. In actuality, all that is happening when the Reverse Direction buttons are used is that SolidWorks adds or subtracts 180 degrees from the Angle value. You could do this manually almost as easily.

Figure 7-75 shows a completed sketch pattern. The dimensions were added by SolidWorks automatically, as were the construction lines. The linear dimensions were added because the Fixed option was checked, along with the *Constrain angle between axes* option. The 90-degree dimension is a result of the latter. Other dimensions and relations should still be added to the original sketch geometry to fully define it before using the sketch for a feature. As mentioned previously, fully defining your sketch geometry is always good practice.

A typical question at this point is whether to create a pattern from sketch geometry using the step-and-repeat function or to create a feature pattern. Feature patterns contain dimensions that can be parametrically modified at any time and do not require as much user intervention or manual labor. You will probably find feature patterns to be more flexible. Read the next section on how to edit a step-and-repeat pattern and you can make a more intelligent decision for yourself.

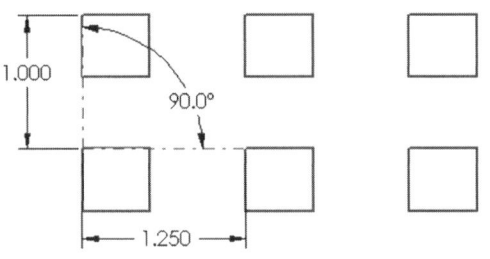

Fig. 7-75. After completing a Linear Step and Repeat.

Editing a Sketch Step and Repeat Pattern

Once a step and repeat has been performed, the pattern parameters can still be edited. To modify the spacing between instances, alter the dimension value associated with that distance. The dimension may be added automatically, or it can be one that was placed by the user. Whatever the case, double click on the dimension to change its value. This holds true whether the dimension is for the spacing or any other value related to the step-and-repeat geometry.

To change the number of instances within the pattern, right-click on any of the sketch entities in any of the pattern instances. You will see an option for Edit Linear Step and Repeat. This holds true for the Circular Step and Repeat command as well. Selecting the Edit Linear Step and Repeat option will bring the original window back up, and the number of instances can be adjusted as required.

A quirk of the step-and-repeat commands occurs when editing the linear or circular step and repeat. The Spacing and Angle options will be grayed out. This is because the proper way of adjusting these values is to modify the dimensions associated with the geometry via double clicking. Nevertheless, the spin box arrows still seem to work. Clicking on the spin box arrows will even make the preview change. However, clicking on OK closes the window and none of the changes will take place. So again, to change the Spacing or Angle values, just double click on the associated dimensions.

Circular Sketch Step and Repeat

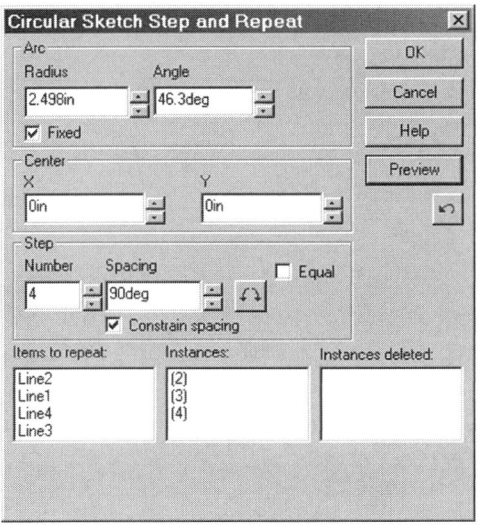

The circular step-and-repeat function is the counterpart to the linear step-and-repeat function. Most of the options are identical, but there are a few that will require further explanation. A sample of the Circular Sketch Step and Repeat window is shown in figure 7-76.

Try running through a circular step and repeat so that you can get the hang of it. The trickiest part will be understanding how the Angle and Total angle options work. The Angle option helps to dictate where the center of the arc defining the pattern will be located. How-To 7-13 takes you through the process of performing a circular step and repeat.

Fig. 7-76. Circular Sketch Step and Repeat window.

How-To 7-13: Performing a Circular Sketch Step and Repeat

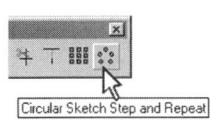

Fig. 7-77. Circular Sketch Step and Repeat icon.

To create a Circular Sketch Step and Repeat pattern, perform the following steps. The Circular Sketch Step and Repeat icon is shown in figure 7-77.

1. Click on the Circular Sketch Step and Repeat icon, or select Circular Step and Repeat from the Tools > Sketch Tools menu.

2. Select the sketch entities to be repeated.

3. Specify the Radius of the arc used to define the circular pattern.

4. Specify the Angle used to determine the pattern's centerpoint. The Angle value uses the direction of the origin's x axis as the 0 angle reference. If a line were drawn from the center of the original sketch objects being patterned outward in the direction of the x axis, the angle between that line and the one shown in the preview will equal the Angle setting. (A sample preview is shown in figure 7-78.)

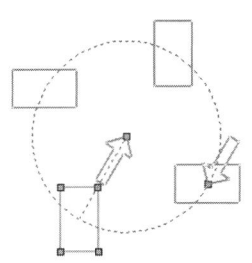

Fig. 7-78. A sample Circular Step and Repeat preview.

5. Alternatively, change the X and Y values, which determine the center of the pattern. Note that these values reference the sketch origin, and changing them will in turn alter the Radius and Angle parameters.

6. Specify the total Number of instances in the pattern.

7. If using the Equal option, specify the Total angle. If not, specify the Spacing.

8. Click on OK to complete the pattern.

If the entire circular step-and-repeat process seems cumbersome, that is because it probably is. After you have been through it a couple of times, it gets easier. The fact of the matter is that it is simply more user friendly to create a feature pattern. When it comes down to it, though, the method you use is up to you, and will not make all that much difference as far as the part is concerned.

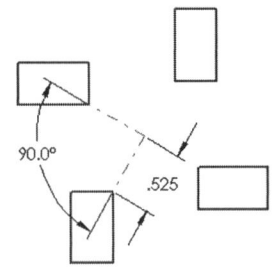

Fig. 7-79. A completed Circular Step and Repeat pattern.

The options found in the Circular Step and Repeat command have the same meaning as their counterparts in the Linear Step and Repeat command. The Fixed option will force SolidWorks to add a dimension that dictates the distance between the original objects being patterned and the pattern center, such as the .525-inch dimension shown in figure 7-79. The 90-degree dimension is representative of the Constrain spacing option, only available when the Equal option is not being used.

During the creation process, dragging the points at the tips of the arrows results in the parameters updating. The arrow point on the perimeter of the pattern will adjust the *Total angle* or Spacing option. The arrow point in the center will alter the Radius, Angle, and Center X and Y values. Modifying a circular step and repeat works the same way as the linear step and repeat (see the previous section "Editing a Sketch Step and Repeat Pattern").

Summary

In this chapter you learned about patterns and how they can be useful, as well as the various forms of mirroring geometry. Patterns can be linear or circular in nature. In either case, an axis, dimension, or edge must be selected that determines the pattern direction or center of rotation. There are also patterns that can be defined via sketch points, *x-y* coordinates, or curves. These are sketch-driven, table-driven, and curve-driven patterns, respectively.

Like most anything else in SolidWorks, patterns are parametric. This includes the number of instances, which can be accessed by double clicking on a linear or circular pattern feature in FeatureManager. Instances in a pattern can be deleted by selecting a surface of that instance and pressing the Delete key. They can be brought back by editing the definition of the pattern feature. Linear patterns can be in one or two directions, and need not be orthogonal.

Once a direction has been specified for a pattern, and the objects to be patterned have been selected, a preview will be displayed. For circular patterns, the preview arrow points along the center of rotation for the pattern. Axes are the most common way of specifying the center of rotation for a circular pattern. Temporary axes are created automatically by SolidWorks. User-defined axes can be independently hidden or shown, and lengthened or shortened. User-defined axes also appear in FeatureManager.

Use copy and paste when only a few copies of the features are required but those instances do not necessarily constitute a pattern. The hot key combinations of Ctrl-C and Ctrl-V can be used to copy and paste, respectively. The hot keys are a Windows standard, and will work for any Windows program. Make sure to select a planar face on which to place the copied feature before pasting it in. Also bear in mind that this functionality exists only when not editing a sketch.

Patterned features will contain the original feature definition. If the Geometry Pattern option is enabled, feature definition information is not used. The patterned items will be duplicates of the original feature's geometry only. As is the case with any pattern, geometry pattern or otherwise, modifying the original feature will update all instances in the pattern. This is not the case when performing a simple copy and paste, as there is no link to the original.

Geometry can be mirrored by using a mirror plane or planar surface. It is also possible to mirror all of the geometry in an entire part. Use the Mirror Body list box in the Mirror command and select the model from the work area to accomplish this task. If a mirror image of a part is needed, you can employ the Mirror Part command, in which case a new part will be created that is an exact mirror image of the original. This latter function also creates an external reference to the original part file. If the original part is modified, the changes will propagate to the mirrored part. This effect is unidirectional only, and therefore if the mirror image part is modified, it will not affect the original.

Sketch-driven patterns are very useful when a pattern is going to be somewhat irregular in nature. A sketch is required that contains sketch points. Wherever a point is located, a pattern instance will appear in the completed pattern. Deleting a point in the reference sketch deletes its associated pattern instance.

Sketch step-and-repeat functions are another way of creating patterns. Step-and-repeat functions can only be used in a sketch to pattern sketch geometry. Feature patterns are typically easier to work with and provide more flexibility.

Questions and Topics for Discussion

1. Name three objects that can be used for specifying a linear or circular pattern's direction.

2. What type of information is not taken into account when the Geometry Pattern option is enabled?

3. Is it possible to pattern a pattern?

4. Describe two methods of deleting a linear or circular pattern instance.

5. Fill in the blank: Sketch mirror is to centerline as feature mirror is to _____.

6. Describe a process used to copy and paste a feature.

7. What is meant by the term *dangling dimension*?

8. How would you go about fixing a dangling dimension?

9. What is meant by the symbol shown here (->) when it is found in FeatureManager?

10. What is meant when this same symbol contains a question mark (->?) immediately following it?

11. What user-defined object is required by a table-driven pattern?

12. What advantages does a user-defined axis have over a temporary axis?

13. What file extensions are acceptable when using a file to define the coordinates of a table-driven pattern?

14. How is a pattern instance deleted in a sketch-driven pattern? How would you get this instance back? Explain.

15. How would an ordinate dimension value be added to a string of previously created ordinate dimensions?

16. How can one edit the number of instances in a sketch step-and-repeat pattern?

Optional Problem

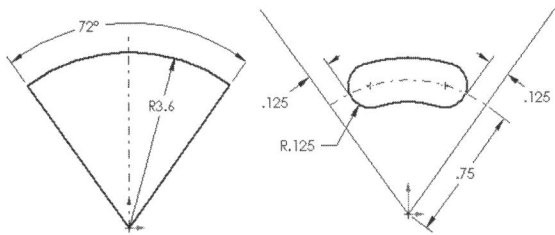

Fig. 7-80. Sketch for the base feature and slot.

Fig. 7-81. Nearly completed round cover plate.

Fig. 7-82. Finished round cover plate.

Create the round cover plate that was used earlier in this chapter for various examples. The model was never fully completed. If you were working along with the book and already have the part partially finished, use what you have. Otherwise, the dimensions you will need are provided here. The base feature sketch and the sketch for the rounded slot (discussed in the "Vary Sketch" section) are shown in figure 7-80. Extrude the second sketch through the part as a cut.

Next, pattern the slot using a linear pattern with the Vary Sketch option. Use a spacing of .500 inch and set Total instances to 5. Select the linear dimension to indicate the pattern direction. Refer to the "Vary Sketch" section if you need help completing this feature. Figure 7-81 shows the part to this point.

Create an axis at the point of the pie-shaped part. A temporary axis could also be used, but a user-defined axis is sometimes nice because of its added capabilities. Rename the axis *Pattern Axis*. Create a circular pattern of the entire part using the newly defined axis. Include the base feature, the slot, and the slot pattern as the items to be copied. Specify a total of five (5) copies spaced 72 degrees apart. The finished part is shown in figure 7-82. Alternatively, feel free to use the *Equal spacing* option. If *Equal spacing* is used, the *Total angle* should be set to 360 degrees.

You might think it would have been easier to simply use the short edge at the point of the wedge instead of bothering with creating an axis. It is true that an edge can be used for circular patterning. However, in this case the edge would not work, because the nature of the pattern actually eliminates the edge used to create it. Once this happens, SolidWorks can no longer define the pattern's center of rotation. Using the user-defined axis relieves this problem because the definition of the axis has been established prior to the pattern's definition. The order or sequence of events can be very important in SolidWorks.

CHAPTER 8

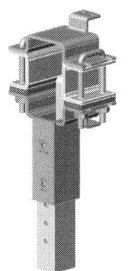

Sheet Metal

SHEET METAL PARTS COME IN A LARGE ARRAY of sizes and shapes, but they all have at least a few characteristics in common. They almost always have bends, there may be some sort of punched-out or cut-out shapes or holes, and often some sort of relief cuts may be needed so as to not stress the metal.

In this chapter you will learn how to create thin-feature parts, define sheet metal parts, add bends, and perform numerous functions related to working with a sheet metal part. This includes showing the part in its flattened state and working on the part in either its flattened or formed state.

From a SolidWorks perspective, sheet metal is defined when extruding a base flange feature. Bends can also be added to a model, either created natively in SolidWorks or imported, in order to define it as a sheet metal part. When a model is defined as a sheet metal part in the SolidWorks software, it exhibits special properties, and various additional functionality is available for editing and developing the model. First, however, let's explore thin-feature parts and how they are different from sheet metal parts.

Thin-feature Parts

In Chapter 6 you explored thin-walled parts, which are created using the Shell command. Thin-feature parts differ in that they do not require the Shell command and are usually created with an open profile sketch.

It should be noted that thin-feature parts are not sheet metal parts. They can be turned into sheet metal parts by employing the Bends command. You will see how to do this in the second half of this chapter.

NOTE: *Just because thin features are called thin does not mean that they are not solid model parts. Make no mistake: parts created from thin features are still considered solid geometry. They have material properties just like any other solid part, and volume, mass, and centroid data can be extracted from such models.*

Let's examine how a thin-feature part could be created. What would happen if an open-profile sketch were extruded? Figure 8-1 shows a very simple open profile sketch.

Even though the profile is open, the sketch can be extruded, as you learned in Chapter 4. If you have SolidWorks available to you, feel free to create a sketch and try this out for yourself. When the Extrude panel is displayed, note that the Thin Feature option is checked, as shown in figure 8-2.

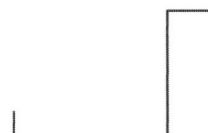

Fig. 8-1. An open profile sketch.

Because the sketch is an open profile, SolidWorks forces you to create a thin-feature part. You will see an option for specifying how thick you want the material to be. This wall thickness can be added to one side of the sketch or the other by clicking on the Reverse Direction button. This is typically dependent on which side of the material the dimensions are critical for, either the inside or outside. Make it a point to keep an eye on the preview so that the wall thickness is applied in the proper direction.

Fig. 8-2. Thin Feature option is enabled.

The Type drop-down list contains three selections for specifying how the wall thickness will be applied. These options are One-Direction, Mid-Plane, and Two-Direction. The Mid-Plane selection will apply a wall thickness equally in both directions (the total of which is user-defined). Two-Direction allows for entering precise values for the wall thickness on either side of the profile sketch. Usually there is only one area available for typing in the desired wall thickness, but if Two-Direction is selected there will be two settings for thickness.

When creating a thin feature, there are still the end-condition parameters that must be entered. This aspect of the Extrude panel has not changed, and thus you specify the end condition type and distance as you always have. It is even possible to add draft while extruding a thin feature. Figure 8-3 shows the possible outcome of extruding the sketch shown in figure 8-2.

Thin-feature Parts

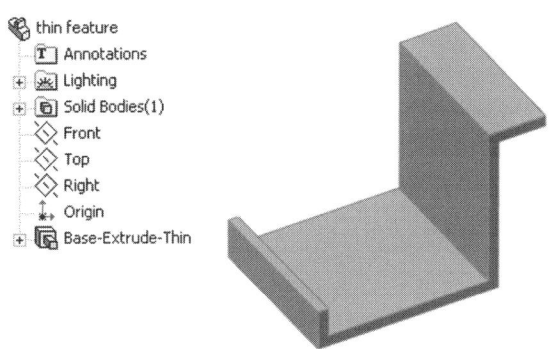

The word *Thin* has been appended to the Base-Extrude feature name in Feature-Manager as an added indicator of what type of feature it is. However, you can still give the feature any name you want. The Auto Fillet option is another aspect of creating thin features and is described in the following section.

Auto Fillet Option

Fig. 8-3. The beginnings of a thin-feature part.

If the part you are creating is fairly simple, or you just want to add some fillets to the thin-feature part, the *Auto-fillet corners* option is one possibility. There are certain conditions under which you would not want to use auto-filleting, however. For example, the user has no control over individual bends when using *Auto-fillet corners*. The fillet radius parameter is applied to the inside of all sharp corners, so it is an "all or nothing" setting. A part for which *Auto-fillet corners* has been enabled is shown in figure 8-4.

In figure 8-4, the sharp corners are replaced by "fillets," though they are not really fillets in the true sense of the word. The fillets more closely resemble bends, but do not confuse these with sheet metal bends, because there are very few similarities other than appearance.

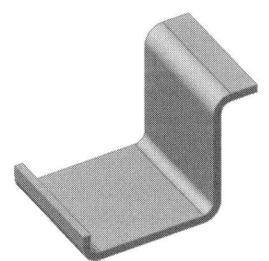

Fig. 8-4. Auto-fillet corners has been turned on.

Auto-fillet corners can be an excellent choice if all sharp corners need to have the appearance of bends. In this case, the advantages are that all bends will have the same radius and the model is not a sheet metal part. Otherwise, you probably will not want to use *Auto-fillet corners*. If you require bends of varying sizes, if the part will be manufactured as a sheet metal part, or if the part needs to be flattened out, use the Base Flange command (discussed in material to follow).

Revolving an Open Profile

It is possible to revolve an open profile, just as it is possible to extrude one. This process functions slightly differently than the extrusion process when it comes to open profiles. The material that follows presents an example that explains the process.

The same sketch that was used in the previous example will be used for creating a revolved thin feature. A centerline will be added for the center of rotation. When the Revolve command is initiated, a window appears warning you that a non thin feature requires a closed profile. This warning is shown in figure 8-5.

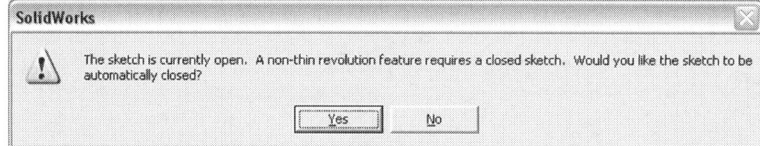

Fig. 8-5. Warning seen while trying to revolve an open profile.

You have two choices at this point. You could click on Yes and let SolidWorks try to close the profile, or you could click on No and revolve the open profile as a thin feature. Usually this message is an annoyance, because if you wanted the sketch to be closed you could have drawn it that way.

When you let SolidWorks attempt to close a sketch automatically, it will add a sketch line between the two open endpoints of the sketch. In some cases this is fine. In other cases, it is not. Take, for example, the sketch shown in figure 8-6. If a line is added between the two open endpoints, a self-intersecting sketch is created.

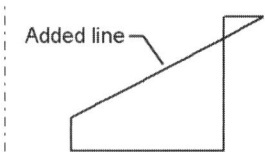

Fig. 8-6. A self-intersecting sketch.

As you should know at this point, a self-intersecting sketch is an illegal operation in SolidWorks. Therefore, the software decides to leave the profile open, displays a message to this effect, and then takes you into the Revolve panel. None of this happens, however, if you click on No when asked if SolidWorks should close the sketch for you.

Fig. 8-7. Thin Feature parameters section.

Getting back to the Revolve panel, note the Thin Feature parameters section, shown in figure 8-7, just like the Extrude panel had. The options you see in the Thin Feature section when revolving an open sketch work exactly the same as they did in the Extrude panel. You simply specify the direction type and the wall thickness, and use Reverse if necessary.

If the same sketch used in the previous examples is revolved 270 degrees, and a wall thickness of .125 inch specified, you would wind up with the part shown in figure 8-8.

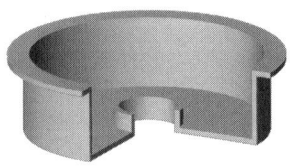

Fig. 8-8. Creating a revolved thin feature.

Closed-profile Thin Features

Yes, it is possible to create thin features from closed profiles. When extruding or revolving open profiles, you do not have a choice. The feature must be a thin feature because there is no alternative. When starting with a closed-profile sketch, the choice is yours. If a thin feature is required from a closed profile, simply check the Thin Feature option and supply the desired parameters for the wall thickness.

If a thin feature is created and you change your mind, you will find the Extrude or Revolve definitions no longer make the Thin Feature option available. The Thin Feature option will be there, but it will be grayed out. This is one of the very few options SolidWorks does not let you go back and edit. The workaround is simple enough, though. If you delete the feature, you will find that the sketch remains in FeatureManager. You can then reuse the sketch as you see fit.

Cap Ends Option

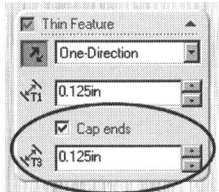

Fig. 8-9. Using the Cap ends option.

When creating a thin feature from a closed profile, there is an option to cap the ends of the thin-feature part. This creates a hollow, airtight part. If you do use the *Cap ends* option, you must also specify the wall thickness for the caps, as shown in figure 8-9. This can be different from the overall wall thickness of the model. Also note that only extruded thin features offer the *Cap ends* option and revolved thin features do not.

✓ **TIP:** *Any feature type can be created as a thin feature, including a swept or lofted feature.*

Defining a Sheet Metal Part

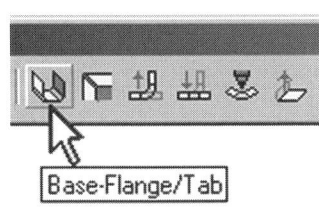

Fig. 8-10. Sheet Metal toolbar.

If you are going to be creating sheet metal parts, make things easy on yourself and first turn on the Sheet Metal toolbar, shown in figure 8-10 (use View > Toolbars). Next, extrude the sketch, assuming that one has already been created, using the Base Flange command, whose icon is also shown in figure 8-10. SolidWorks will then create three features in FeatureManager that directly relate to the sheet metal model.

Typically, you want to start with an open-profile sketch prior to using the Base Flange command. As an example, let's use the same sketch shown in figure 8-1. The Base Flange command will be applied to this sketch, and we will see what happens. How-To 8-1 takes you through the process of creating a sheet metal part using the Base Flange command.

How-To 8-1: Creating a Base Flange

To create a sheet metal part in SolidWorks, perform the following steps. Note that either an open or closed sketch profile can be used, but if closed you will in essence be working with a flat piece of metal. This How-To will use an open profile, which is more common.

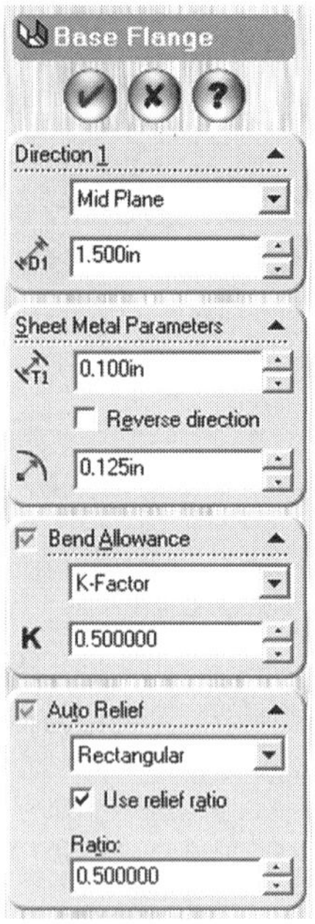

1. Select Insert > Sheet Metal > Base Flange, or click on the Base-Flange/Tab icon. This is a dual-function icon (thus the slash in its name).

2. In the Base Flange panel, shown in figure 8-11, specify the end condition and depth, as for any simple extrusion.

3. In the Sheet Metal Parameters section, specify a thickness for the material.

4. Use the *Reverse direction* option if necessary to change the direction the wall thickness is applied.

5. Specify a value for the default Bend Radius.

6. Specify a value for the default Bend Allowance.

7. Specify the default relief cuts, if any, which should be made.

8. Click on OK to create the base flange. The part will be defined as a sheet metal part.

Figure 8-11. Base Flange panel.

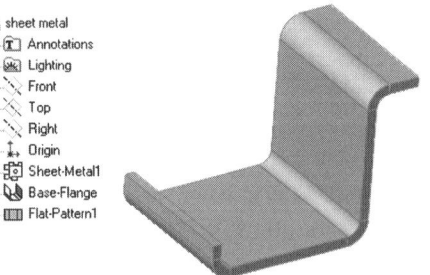

Fig. 8-12. The results of creating a base flange.

Options such as Auto Relief and Bend Allowance are discussed in material to follow. After creating a base flange, a number of features are added to FeatureManager. These are shown in figure 8-12. The sections that follow explain these features in a little more detail.

Sheet-Metal1

Sheet-Metal1 is the name of the feature that defines the basic characteristics of the sheet metal part. The Sheet-Metal1 panel is shown in figure 8-13. Its definition contains information such as the default bend radius, bend allowance, and information related to whether or not automatic relief cuts are made to the sheet metal part when the metal would otherwise tear. There are some aspects of

Defining a Sheet Metal Part

this panel that are not accessible at this point in time, such as the thickness setting. In this case, the thickness is part of the Base-Flange definition and can be changed there. Bend Allowance and Auto Relief are discussed in the following sections.

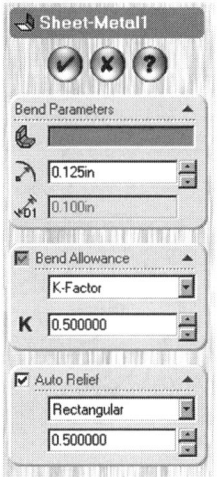

Fig. 8-13. Sheet-Metal1 panel.

Bend Allowance

SolidWorks takes into consideration how much the sheet metal material is going to stretch, which is generally known as bend allowance. There are four options for specifying the method by which this stretch will be carried out. Using the pull-down menu in the Bend Allowance section, select Bend Table, K-Factor, Bend Allowance, or Bend Deduction. Keep in mind that the Bend Allowance option will determine the default bend allowances used for all bends in the sheet metal part. If necessary, bend allowances can be defined on a bend-by-bend basis, which is described later in this chapter.

If Bend Allowance is selected, you will be allowed to enter a value for the bend allowance. If K-Factor is chosen, enter a value for the K-factor. If you are not familiar with using a K-factor for determining the bend allowance of a sheet metal bend, it is not recommended that you use this option. However, because K-factor is the default selection used by Solid-Works for specifying the bend allowance, it will be touched on here.

K-Factor

If you were to take a piece of metal and bend it, there would be some amount of compression on the inside of the bend area, and a bit of expansion on the outside. Somewhere in the middle of that piece of metal would be the neutral plane, where the metal is neither expanding nor compressing. The distance from the surface on the inside of the bend of the sheet metal part to the neutral plane, divided by the total thickness of the metal, is the K-factor.

The K-factor can be set anywhere from 0 to 1, assuming some value has been given for the bend radius. A large K-factor means that the neutral plane is closer to the outside of the bend. As a result, there is more compression of the metal in the bend region, and the flat pattern will need to be larger to accommodate this. If the K-factor is less, there is less compression and the flat pattern does not have to be quite as large.

The good news is that you do not have to figure any of this stuff out, as SolidWorks will do it for you. However, you do need to plug in a valid bend allowance some way or another. (Determining bend allowance is outside the scope of this book.)

Bend Tables

Bend tables are tables that contain the bend allowances for a certain type of material. They can be text files or Microsoft Excel spreadsheets. If using a text file as a bend table, it becomes very cumbersome to plug in all of the various values. In this case it is almost not even worth going to the bother of creating a file. It is highly recommended that if you wish to use a bend table that you use Microsoft Excel rather than a text file.

Fig. 8-14. Sample Excel bend table.

Unit:	Inches								
Type:	Bend Allowance								
Material:	**Soft Copper and Soft Brass**								
Comment:	Values specified are for 90-degree bends								
Radius	**Thickness**								
	1/64	1/32	3/64	1/16	5/64	3/32	1/8	5/32	3/16
1/32	0.058	0.066	0.075	0.083	0.092	0.101	0.118	0.135	0.152
3/64	0.083	0.091	0.1	0.108	0.117	0.126	0.143	0.16	0.177
1/16	0.107	0.115	0.124	0.132	0.141	0.15	0.167	0.184	0.201
3/32	0.156	0.164	0.173	0.181	0.19	0.199	0.216	0.233	0.25
1/8	0.205	0.213	0.222	0.23	0.239	0.248	0.265	0.282	0.299

An Excel spreadsheet is going to be much easier to manipulate than a text file (this is a drastic understatement). SolidWorks includes a number of preformatted tables for plugging in bend allowances. Some even have the bend allowances already plugged in, but they might not match the material you work with. Figure 8-14 shows a portion of an Excel bend table included with the SolidWorks software.

In the case of an Excel bend table, the bend angle of 90 degrees is the constant, and the thickness of the material changes. The bend radius is read down the first column. If you decide to use a bend table rather than entering a K-factor or bend allowance value, How-To 8-2 will show you how.

How-To 8-2: Specifying a Bend Table

The following steps are the process for using a bend table to dictate the bend allowance values for a sheet metal part. Because using Microsoft Excel is outside the scope of this book, that aspect of modifying the bend table will not be spelled out.

1. Using Microsoft Excel, open one of the sample bend allowance tables found in your SolidWorks installation directory. Once you have navigated to the SolidWorks installation folder, look in the lang\English\Sheetmetal Bend Tables folder.

Defining a Sheet Metal Part

☛ **NOTE:** *By default, the SolidWorks installation folder is* C:\Program Files\SolidWorks. *In addition, the English folder will be named after a different language if using a version of SolidWorks in another language.*

2. Edit the bend table to fulfill your requirements. This will require entering bend allowance values, and very likely different thicknesses and bend radius values.

3. Save the file with a meaningful name and make note of the name.

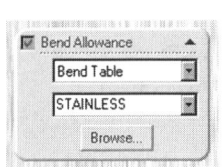

4. In the sheet metal part file, right-click on the *Sheet-Metal1* feature and select Edit Definition.

5. In the Bend Allowance section, specify Bend Table from the Bend Allowance Type drop-down list (see figure 8-15).

Fig. 8-15. Specifying a bend allowance table.

6. Click on the Browse button and select the folder containing the bend table.

7. Using the Bend Table drop-down list, select the bend table file to be used. This should match the actual file name given to the table in step 3.

8. Click on OK.

If you are handy with Excel, it is pretty easy to whip together a few bend tables. If not, you are better off looking up the bend allowance in your handy pocket guide or machinist's handbook. If you happen to have the bend table ready to go when defining the base flange feature, it can be specified at that point in time rather than editing the definition of the *Sheet-Metal1* feature.

Auto Relief

When a bend must be added to a part that would otherwise rip the metal, you need to use Auto Relief. If you do not have Auto Relief turned on and attempt to create such a bend, the bend will fail, giving you a warning message telling you of this fact. You then have the option of manually creating a cut in the area containing the bend or turning on Auto Relief.

To add relief cuts to the sheet metal part automatically, make sure that Auto Relief is checked in the definition of the *Sheet-Metal1* feature (see figure 8-13). The Relief Ratio is the value that determines how deep the relief cut should be. If the material were .100 inch thick, and the Relief Ratio set to .5, the cut would be one-half (.5) the thickness of the material, or .050 inch deep. Additionally, this depth measurement is taken from the point where the bend ends.

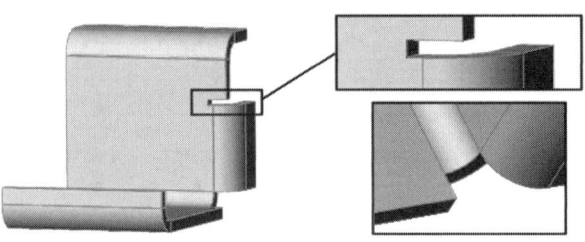

Fig. 8-16. This flange would not accept a bend without tearing the metal.

Figure 8-16 shows a sample part with an edge flange. It would be impossible to insert the bend where the flange meets the rest of the part without creating relief cuts of some sort. The relief cut near the top of the flange is a direct result of turning on Auto Relief. The cut in the bend region is a result of another option (*Trim side bends*), which is specified in the feature definition of the edge flange (which you will learn about in material to follow).

Relief Types

There are three types of relief: rectangular, obround, or tear. Use whichever type suits your purpose. Obround is a term combining the words oblong and round and is used to describe a slot with rounded ends. In the context of sheet metal parts, obround pertains to a cut with a rounded end. Tear relief is similar to creating a cut with a pair of tin snips, rather than a saw blade, plasma cutter, or some other device. All three relief types are shown in figure 8-17, with rectangular on the left, obround in the middle, and tear on the right.

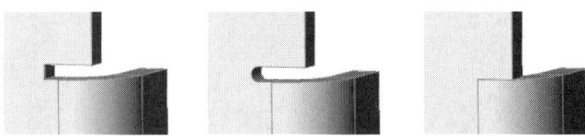

Fig. 8-17. Examples of relief cuts.

Base-Flange

The Base-Flange definition contains a part's flange direction and end condition parameters, along with the default thickness and bend radius parameters (shown in figure 8-10). Although the Bend Allowance and Auto Relief parameters are missing from this feature's definition, these parameters are present in the *Sheet-Metal1* definition (as previously mentioned).

Flat-Pattern 1

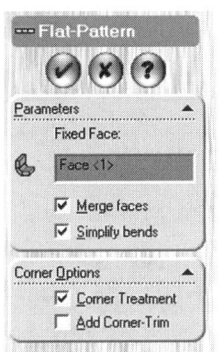

Fig. 8-18. Flat-Pattern panel.

A third feature type has been added to FeatureManager. The *Flat-Pattern1* feature is the key to flattening the part. The Flat-Pattern panel, shown in figure 8-18, will appear when editing the definition of this feature. The options contained in this panel are discussed in material to follow. First, however, let's find out how to flatten the part.

Flattening a Sheet Metal Part

Did you notice that the *Flat-Pattern1* feature was suppressed in FeatureManager (see figure 8-12)? You may not have, in that we have not explored the suppression state of features. That can wait until Chapter 10. It is true that unsuppressing the *Flat-Pattern1* feature will show the model in its flattened, or unfolded, state. But all you really need to know at the moment is that clicking on the Flattened icon, located on the Sheet Metal toolbar, will flatten out the part. The Flattened icon is shown in figure 8-19. Clicking on the Flattened icon a second time will form (or fold) the model back up. Could it be any easier?

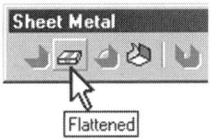

Fig. 8-19. Flattened icon.

When the model is flattened, bend lines will be shown, such as those shown in figure 8-20. These lines will automatically appear in a detailed drawing when the flat pattern view is inserted into the drawing. Detailed design drawings are discussed in Chapter 11.

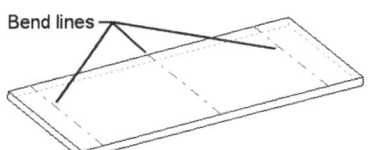

Fig. 8-20. Bend lines in the flattened state.

The model can be flattened at any stage in the design process. Any tabs, hems, jogs, or flanges added to the model will flatten out as well. Use the Flattened icon whenever you need to see the model in its flattened state.

Merge Faces

Checked by default, the Merge faces option treats all coplanar faces on the flattened model as one. Certain situations may require this option to be turned off, but typically it should be left alone. If turned off, the bend regions are shown in the flattened part, as indicated in figure 8-21.

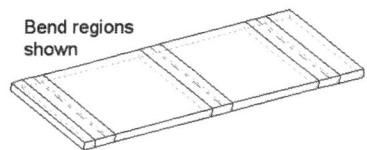

Fig. 8-21. Turning off Merge faces shows the bend regions.

Simplify Bends

The Simplify bends option has to do with how Solid-Works handles the edges of bend regions. The default method is to simplify bend edges, which can most easily be seen through example. Figure 8-22 shows a section of corner molding cut to fit around a 90-degree corner.

Fig. 8-22. Section of corner molding.

If this sheet metal molding is flattened out, we can see how the edges of the bend region appear straight. One area in particular is pointed out in figure 8-23. The example on the left has the *Simplified bends* option checked. The example on the right has the *Simplified bends* option turned off. The edge in the example on the right is noticeably different. Whether or not you decide to turn this option off depends on what you want to send to the patternmakers who will be building this sheet metal part.

Fig. 8-23. Molding in its flattened state.

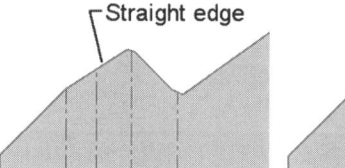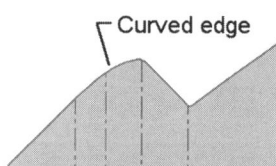

Corner Treatment

In the case where a feature such as a miter flange has been created, the flattened part can display the corners in one of two ways. Figure 8-24 shows a miter flange feature wrapping around the corner of a formed sheet metal part (left). The same part is also shown in its flattened state with corner treatment turned off (center), then again with corner treatment turned on (right). Some designers choose to let the patternmaker create his own corner relief cuts, in which case the designer would check Corner Treatment so that the relief cuts made by SolidWorks would not show up in the flat pattern.

Fig. 8-24. Corner treatment option in action.

Add Corner Trim

Corner trim refers to relief cuts for corners. The Add Corner-Trim option functions somewhat differently than other options, in that turning it on causes another feature to be created and displayed in FeatureManager. Corner trim comes in a number of varieties. To demonstrate, a sheet metal container will be used, which is shown in its unfolded state in figure 8-25.

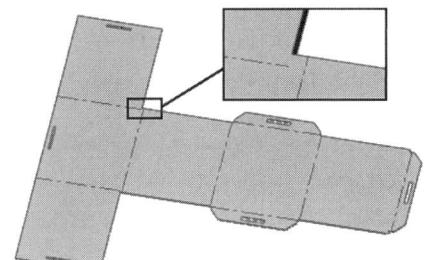

Fig. 8-25. Prior to turning on Add Corner-Trim.

If the Add Corner-Trim option is checked, the panel expands to give the user additional options related to corner trim. An example of the panel you might see (depending on the options selected) is shown in figure 8-26.

Fig. 8-26. Corner Options panel and related options.

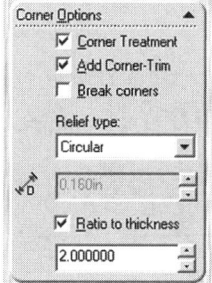

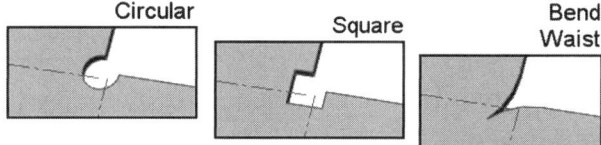

Fig. 8-27. Three versions of corner trim.

So what does this corner trim look like? There are three versions: circular, square, and bend waist. All three versions are shown in figure 8-27. The Corner-Trim feature itself will appear after the *Flat-Pattern1* feature in FeatureManager. Corner trim only appears in the flat pattern. If you wish to see relief cuts on corners in the formed state, use the *Trim side bends* option (described in the next section regarding edge flanges) or create relief cuts manually.

If you change your mind about applying corner trim to the flat pattern, simply delete the feature. To edit corner trim, edit the definition of the Corner-Trim feature. Specify the size of corner trim either by supplying a distance or providing a ratio-to-thickness value. The *Break corners* option, available in the same panel, refers to outside corners. This should not be confused with the Break Corner command discussed in material found later in this chapter.

Edge Flange

An edge flange is the first feature type we will be examining that is peculiar to sheet metal parts. In other words, the Edge Flange command can be used only on a model created as a sheet metal part first, which you discovered could be accomplished by using the Base Flange command.

Figure 8-16 showed an example of an edge flange feature. The convenient aspect of the Edge Flange command is that it does not require the user to create a sketch first. The command will create a sketch for you, and then the sketch can be modified as required. How-To 8-3 takes you through the process of creating an edge flange.

How-To 8-3: Creating an Edge Flange

To create an edge flange feature, perform the following steps. You should not be editing a sketch at this point, and the model should be a sheet metal part.

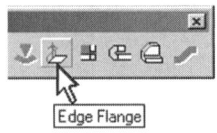

Fig. 8-28. Edge Flange icon.

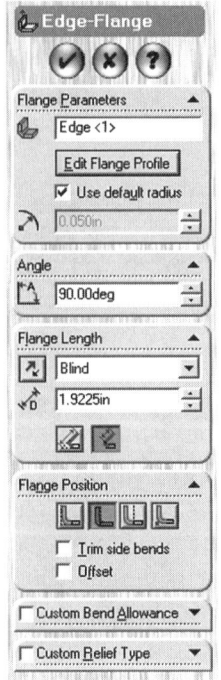

Fig. 8-29. Edge-Flange panel.

1. Click on the Edge Flange icon, shown in figure 8-28, or select Edge Flange from the Insert > Sheet Metal menu.

2. Select a linear edge and drag it in the direction you want the flange to protrude. The angle will default to 90 degrees, but this can be changed later.

3. To use a radius other than the default (optional), in the Edge-Flange panel (shown in figure 8-29), uncheck Use default radius and enter a value.

4. Specify an Angle for the flange.

5. Specify the Flange Length. Accomplish this task by entering a blind length value, modifying the end condition if desired, and choosing where to measure the flange from (Inner or Outer Virtual Sharp buttons).

6. Specify a Flange Position (elaborated upon in material to follow).

7. Optionally specify a custom bend allowance or relief type if desired.

8. Click on OK to create the flange.

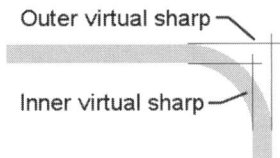

Fig. 8-30. Inner versus outer virtual sharps.

The flange length can be determined in a number of ways, the most basic of which is to enter a value using the Blind end condition. There is also the matter of where the flange length is measured from with regard to the bend. This can be specified using the Inner Virtual Sharp and Outer Virtual Sharp buttons mentioned in step 5. Figure 8-30 shows what is meant by "virtual sharp." The length of the flange will reference one of these virtual sharps, and this will definitely play a role in the final length of the flange, so make the choice accordingly.

Flange Position

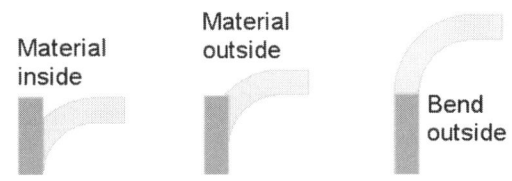

Fig. 8-31. Various flange position settings.

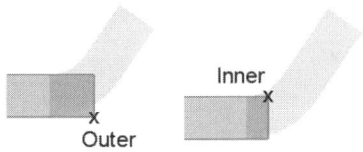

Fig. 8-32. Two possible outcomes using the Bend from Virtual Sharp option.

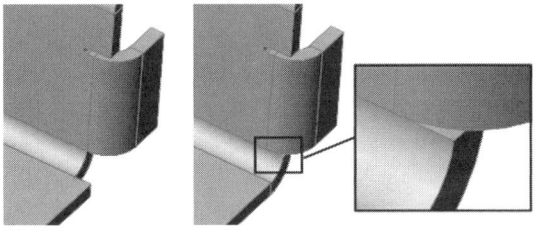

Fig. 8-33. Trim side bends option.

There is more to the Flange Position settings than meets the eye. First, it is very important to choose where the flange should be positioned with regard to the edge from which the flange originated. In its most basic form, it would be necessary to choose whether the flange should be positioned on the inside of the edge, the outside, or if the entire bend should be positioned on the outside of the original edge. This can most easily be seen in figure 8-31, which has been annotated to show the related flange position settings.

There is a fourth setting, titled Bend from Virtual Sharp. This fourth flange position option relies on the flange length virtual sharp setting discussed in How-To 8-3. Figure 8-32 shows two edge flange previews using the Bend from Virtual Sharp flange position option. The outcome of the flange is dependent on whether the Inner Virtual Sharp or Outer Virtual Sharp button is selected in the Flange Length panel.

An example of the *Trim side bends* option was actually shown earlier, in figure 8-16. A side-by-side comparison example is shown in figure 8-33. On the left, the *Trim side bends* option was used. On the right (with inset), *Trim side bends* was disabled. Where the option was disabled, it appears as though the area between the bend and flange was cut with a pair of tin snips, such as when using tear relief. This is the case even when rectangular or obround relief is used.

Flange Position Offset Option

The Offset option allows for performing some remarkable tasks. When checked, more options are presented, which allow for using various end conditions to define the offset distance. End conditions such as Up to Surface make it possible to specify the flange position using some other surface in the part, or even some other externally referenced face of another component in an assembly. You will learn more about defining features in the context of an assembly in Chapter 12.

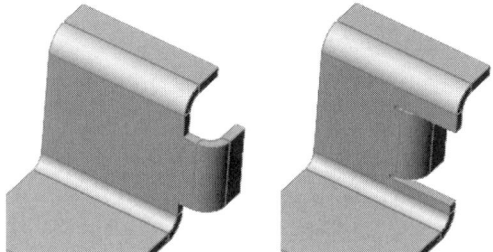

Fig. 8-34. Using the Offset option.

To use a simple example, let's understand what would happen if the Offset option were checked and the Blind end condition were used. In this case, a value could be entered for the offset distance. This serves to push the flange outward from the originating edge in one direction or the other. The direction in which the flange moves from the edge is dependent on the Reverse Direction button, located to the left of the end condition setting. Two examples of adding an offset value to offset the flange position are shown in figure 8-34. The only difference between the two examples is that the Reverse Direction button was used for the example on the right. Relief cuts are added automatically when required. For example, relief cuts were added because the offset pushed the flange into the part.

Custom Relief Type and Custom Bend Allowance

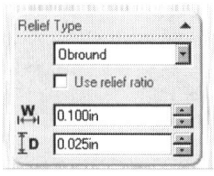

Fig. 8-35. User-defined relief cut settings.

Just because a default relief type, such as Rectangular, is specified in the creation of a sheet metal part does not mean that this setting is required for every bend. Such is the case when creating an edge flange. By checking the Custom Relief Type option in the Edge Flange panel, the user is left with the ability to plug in any values she desires. Figure 8-35 shows an example of this.

Once the Custom Relief Type option is checked, it is possible to set the relief cut's width and depth for that single edge flange feature. Other sheet metal feature types that sometimes require relief cuts have this same functionality. It is always possible to individually control the parameters of a relief cut associated with an individual bend. The option will always be in the definition of the feature in question. The same holds true for customizing bend allowance on a bend-by-bend basis. Simply check the Custom Bend Allowance option and specify the bend allowance as required.

Editing the Flange Profile

One last topic must be discussed before wrapping up this section on edge flanges, and that is the topic of the flange sketch profile. If the default sketch created by SolidWorks specifically for the edge flange feature were always accepted at face value, it would always be the same width as the edge from which it emanated, and you would have control over its length only. That is hardly acceptable, and is certainly not the case.

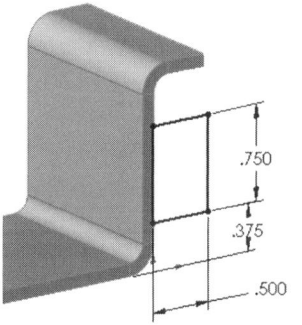

Fig. 8-36. Editing a flange profile.

When creating an edge flange, it is not necessary to define the flange length via the Flange Length panel (see step 5 in How-To 8-3). Instead, try clicking on the Edit Flange Profile button found on the same panel. A message window will appear, presumably informing you that the sketch is valid. Move the window out of the way. You will now have complete control over the sketch, an example of which is shown in figure 8-36.

A flange profile sketch will follow the same principle as any sketch geometry when it is converted from an existing model edge. When first editing the flange profile sketch, the ends of the rectangular sketch must be "loosened" prior to dimensioning them. If this is not done, you risk overdefining the sketch. This behavior is intended and is identical to the workings of the Convert Entities command, described in Chapter 5.

Adding a dimension to control the length of the flange, such as the .500-inch dimension in figure 8-36, will cause the Edge Flange panel to remove the Flange Length dimension parameter from the panel. This makes sense, in that the length is now being controlled by the dimension added to the sketch by the user.

While the flange profile sketch is being edited, the small message window will remain on screen. This window will contain the buttons Back and Finish, as well as the usual Cancel and Help buttons. If the Back button is selected, you will be placed back in the Edge Flange PropertyManager. If Finish is selected, the flange will be created. It does not really matter which button is selected, as you can always edit the definition of the flange anyway. Quite simply, if you have not finished specifying the parameters of the flange, click on Back.

Miter Flange

Unlike an edge flange, a miter flange requires a sketch to be completed by the user. The sketch does not need to consist of much. In its simplest form, the miter flange sketch might be nothing more than a line. A miter flange is similar to an edge flange, but instead of emanating from a single edge, a miter flange can wrap around corners. Those corners can either be sharp corners or rounded corners. If rounded, the miter flange can be made to automatically propagate around the tangent faces.

Fig. 8-37. A miter flange.

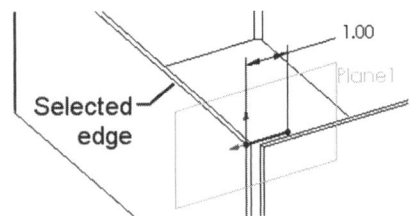

Fig. 8-38. A miter flange sketch.

Miter flanges are essentially edge flanges that fit together nicely where they meet at corners. Some functionality has been added that allows for controlling parameters such as the gap distance between flanges at each corner. Figure 8-37 shows an example of a miter flange at the top of a sheet metal part, with one of the corners shown in greater detail.

To create a miter flange, start with a sketch on a plane that is perpendicular to the edge the flange will run along. The easiest way to start is to pick an edge near the endpoint of where the sketch plane should be positioned. Click on the Sketch icon once the edge has been selected. This neat little shortcut was mentioned in Chapter 4. Figure 8-38 shows the sketch plane used for the miter flange introduced in figure 8-31. It also shows the sketch used to create the flange.

As the figure shows, all that was needed was a simple line. SolidWorks does the rest. How-To 8-4 takes you through the process of creating a miter flange feature on a sheet metal part. It will be assumed that the sketch has already been created.

How-To 8-4: Creating a Miter Flange

To create a miter flange, perform the following steps.

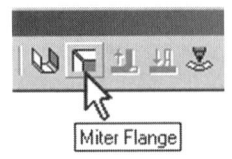

1. Click on the Miter Flange icon (shown in figure 8-39), or select Miter Flange from the Insert > Sheet Metal menu.

Fig. 8-39. Miter Flange icon.

Offset Miter Flange

Fig. 8-40. Propagate icon.

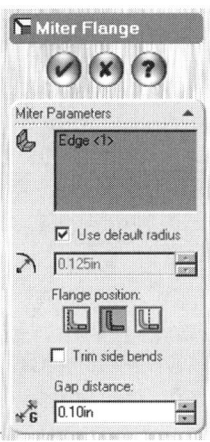

Fig. 8-41. Miter Flange panel.

2. The flange will flow along the first edge by default. To continue the flange around sharp corners, select additional edges. To force the flange to propagate around tangent faces, click on the Propagate icon, shown enlarged in figure 8-40.

3. Specify where the flange position should be relative to the original edge. Like the Edge Flange command, this setting can be Material Inside, Material Outside, or Bend Outside.

4. Specify a Gap Distance, if desired, or just use the default value.

5. Click on OK to create the miter flange.

A portion of the Miter Flange panel is shown in figure 8-41. Occasionally, depending on the nature of the miter flange feature, the default gap distance may not be enough. SolidWorks will then supply an error message that states what the minimum gap distance should be. Using the minimum gap distance as a gauge, you can then specify a new gap distance.

With regard to adjusting the bend angle of a miter flange, there will be no angle option, as was present in the Edge Flange panel. This is because the angle of a miter flange feature is built into the sketch. To edit the angle of the miter flange, edit the sketch of the miter flange.

✓ **TIP:** *Lines and arcs can be used in conjunction when creating a sketch for a miter flange feature.*

Offset Miter Flange

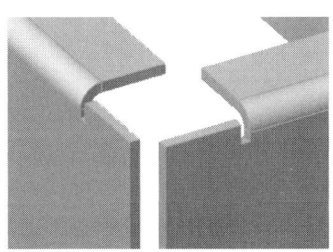

Fig. 8-42. Offsetting a miter flange.

The Start/End Offset option (just out of view in figure 8-41) is available when creating a miter flange. The offset refers to the distance the miter flange is from either the beginning or end of the edges around which the flange is extending. To put it another way, the flange does not have to start at the end of the edge where the sketch plane was positioned.

Figure 8-42 shows an example of using a start and end offset value of 1 inch. The offset values can be different. It just happens that in this case they are the same. The software is smart enough to realize that because an offset distance is being specified the flanges do not require mitering.

When using an offset value for the miter flange, it will more than likely be necessary to add relief cuts. Because this is the case, SolidWorks will present you with these options once either a start or end offset value is specified. Enter the desired auto-relief parameters per your design requirements. (See the section "Auto Relief" earlier in this chapter.)

Tabs

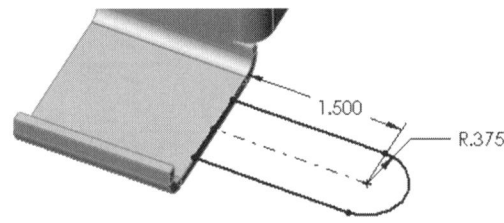

Fig. 8-43. Preparing to create a tab.

Tabs are one of the easiest of all sheet metal features to create. The only manual labor involved is creating the sketch that will define the profile of the tab. A simple example is shown in figure 8-43. The feature in the example was sketched on the top face of the part, but the bottom face could have been selected; it would not have made a difference. When the Tab command is implemented, SolidWorks will automatically figure out the proper extrusion direction. How-To 8-5 takes you through the process of creating a tab.

How-To 8-5: Creating a Tab

Fig. 8-44. The Base-Flange/Tab icon has dual functions.

To create a tab on a sheet metal part, perform the following steps. Note that the Tab icon, shown in figure 8-44, is actually called the Base-Flange/Tab icon, because it has dual functions.

1. Select a face on the model where a tab is to be located, and enter sketch mode.
2. Create a closed profile that describes the shape of the tab.
3. Click on the Base-Flange/Tab icon.

As you can see, there is very little to the creation of a tab. Ninety-eight percent of the work is creating the sketch. An example of a completed tab is shown in figure 8-45.

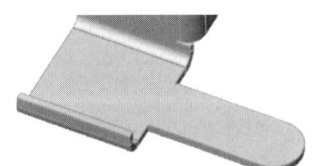

Fig. 8-45. Example of a tab.

Sketched Bends

If an additional bend were required on a sheet metal part, the bend could be defined by first placing a sketch line on a face where the bend is to occur. This bend type is known as a sketched bend. The principle behind sketched bend features is straightforward: place a bend wherever a line has been sketched.

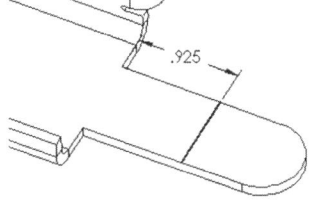

Fig. 8-46. Preparing to add a sketched bend.

Neatness is a desirable trait when creating a sketch prior to defining a sketched bend feature. A good example of a well-drawn sketch line is shown in figure 8-46. Note that the ends of the sketch line terminate on the edges of the tab. The line's endpoints are in fact coincident with the edges of the tab. The Sketched Bend command would still work regardless of whether the sketch line ran short or if it were extending beyond the sides of the tab. However, if underlying geometry were to change, the sketched bend feature could fail. Be neat and future problems will be less likely to arise.

More than one line can be present in a sketch, but bear in mind that each bend will be driven by the same set of parameters. For the greatest amount of flexibility, limit the number of lines in each sketch used to create a sketched bend feature. Do not criss-cross lines or create unreasonable sketch geometry, or SolidWorks will refuse to create the bends. Placing a sketch line too close to a model edge may result in the bend not being created. A single line can span multiple tabs, but should not cross complex regions containing hems or bend regions. How-To 8-6 takes you through the process of creating a sketched bend.

How-To 8-6: Creating a Sketched Bend

To create a sketched bend on a sheet metal part, perform the following steps.

Fig. 8-47. Sketched Bend icon.

1. Create a sketch line on a face of the model where you would like to place a bend.

2. Click on the Sketched Bend icon, shown in figure 8-47, or select Sketched Bend from the Insert > Sheet Metal menu.

3. Select a face to be held stationary when the bend is created. The face will be listed in the Sketched Bend panel, shown in figure 8-48.

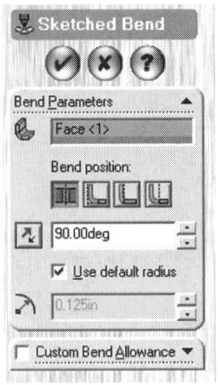

Fig. 8-48. Sketched Bend panel.

4. Specify where the new bend should be positioned with reference to the sketch line. Choices are Bend Centerline, Material Inside, Material Outside, and Bend Outside.

5. Specify a value for the Bend Angle.

6. Click on Reverse Direction if it is necessary to flip the bend direction. The direction will be indicated by a small preview arrow.

7. Click on OK to create the bend.

Figure 8-49 shows an example of a completed sketched bend feature. Regarding the position of the bend with reference to the sketch line, as determined in step 4, there is a new choice that has not been discussed yet. That choice is Bend Centerline. The other three choices were discussed in the section "Flange Position" and were shown in figure 8-31. The Bend Centerline option will ensure that the center of the bend region is at precisely the same position as the sketch line used to define the bend.

Fig. 8-49. A completed sketched bend feature.

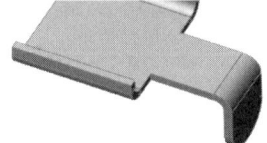

Jogs

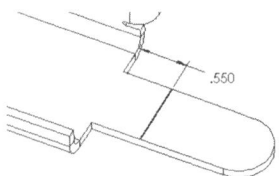

Fig. 8-50. Preparing to create a Jog.

A jog feature is one step beyond simple bends, but starts out the same way a sketched bend feature does. That is, a jog begins with a simple sketch line. For this reason, we will use an example very similar to the one used to demonstrate a sketched bend. The only difference is that the sketch line will be moved back some distance to make room for the jog feature. Figure 8-50 shows the sketch line used to define the jog feature created in How-To 8-7.

How-To 8-7: Creating a Jog

To create a jog feature on a sheet metal part, perform the following steps.

1. Create a sketch line on a face of the model where you would like to place a jog.

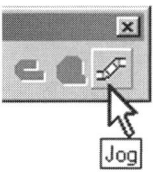

Fig. 8-51. Jog icon.

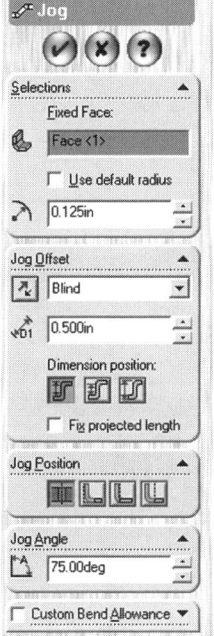

Fig. 8-52. Jog panel.

2. Click on the Jog icon, shown in figure 8-51, or select Jog from the Insert > Sheet Metal menu.

3. Select a face to be held stationary when the jog is created. The face will be listed in the Jog panel, shown in figure 8-52.

4. Specify a value or end condition, which will indicate the Jog Offset distance. You should also specify the Dimension Position for the jog offset. This will be either Outside Offset, Inside Offset, or Overall Dimensions.

5. Click on Reverse Direction if it is necessary to flip the jog direction. The direction will be indicated by a small arrow, and the preview itself should change accordingly.

6. Specify the Jog Position with reference to the sketch line.

7. Specify a Jog Angle. This value can be anything between 0 and 180.

8. Click on OK to create the jog.

Figure 8-53 shows an example of a jog feature. The jog is a result of one simple sketch line, so it is not difficult to see that this feature type could save a lot of time.

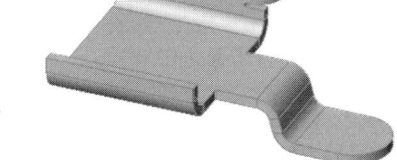

Fig. 8-53. An example of a jog.

Step 6 mentions that the Jog Position should be specified with reference to the original sketch line. This procedure carries with it the same options present when creating a sketched bend. However, a jog contains not just one bend but two. So how is the position of the jog actually being defined? The answer is that the position is defined by the first bend in the jog. The second bend can safely be ignored with regard to the jog position.

When specifying the Jog Offset, there is an option titled *Fix projected length*. This option makes a difference in the final length of the material receiving the jog, so be careful. When *Fix projected length* is checked, SolidWorks extends the length of the material so that the end of the material extends to the same position prior to the jog. With the option off, the jog is added in a more natural fashion, and the length of the material receiving the jog reacts accordingly. Figure 8-54 shows the same jog with the *Fix projected length* option turned off (left) and turned on, or checked (right).

Fig. 8-54. Fix projected length option activated (right) and deactivated (left).

Hems

Sheet metal hems, of which there are four types, are ridiculously easy to create. Hems can be either Closed, Open, Teardrop, or Rolled, examples of which are shown in figure 8-55. No sketch is required prior to creating hems. It is simply a matter of selecting the appropriate edge or edges, and then plugging in the hem parameters. How-To 8-8 takes you through the process of creating hems.

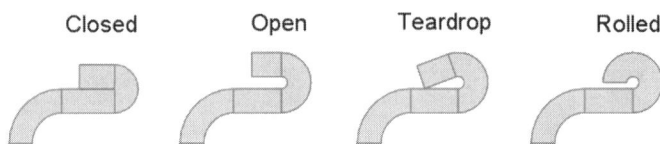

Fig. 8-55. Hems are of four types.

How-To 8-8: Creating a Hem

To create a hem on a sheet metal part, perform the following steps.

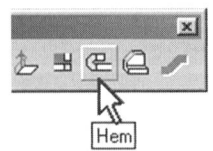

Fig. 8-56. Hem icon.

1. Click on the Hem icon, shown in figure 8-56, or select Hem from the Insert > Sheet Metal menu.

2. Select the edge or edges that are to contain hems. Any selected edges will appear in the Edges list box in the Hem panel, shown in figure 8-57.

3. Using the icons on the Hem panel, specify where the hem should be positioned relative to the selected edge. This will be either Material Inside or Bend Outside.

4. Use the Reverse Direction option, if necessary, to flip the hem from one side to the other.

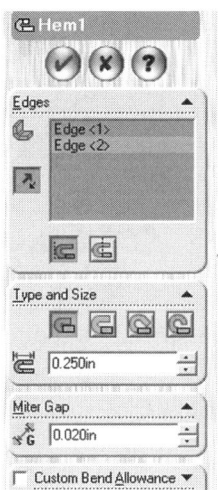

Fig. 8-57. Hem panel.

5. Specify the hem Type.

6. Specify the hem Size. This may be a combination of settings, depending on the hem Type, and may include Length, Gap Distance, Angle, and Radius.

7. Click on OK to create the hem.

The Miter Gap parameter, shown near the bottom of figure 8-57, will not be available unless multiple edges are selected. Even then, the Miter Gap parameter may not be relevant. It depends on whether or not two edges are adjacent. Such an example is shown in figure 8-58. If multiple edges are selected for hemming, and they do not share a common corner, ignore the Miter Gap setting.

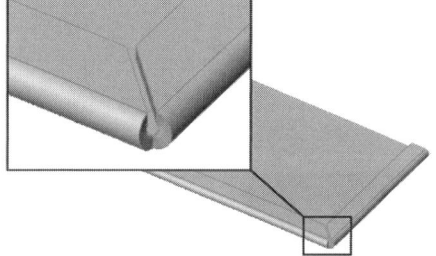

Fig. 8-58. Example of mitered hems.

When an edge is selected during the hem definition process, the hem will attempt to bend toward the selected edge. You can use this to your advantage, and therefore clicking on the Reverse Direction option will not be necessary. It is not possible to form hems in multiple directions at once (in other words, on different sides of the sheet metal part) when selecting multiple edges. The hems will need to be created as separate features in such a case.

Breaking Corners

A special sheet metal command known as Break Corner can be thought of an enhanced fillet or chamfer command designed with sheet metal parts in mind. The primary benefit of the Break Corner command is that it makes selecting the corners to be broken extremely easy. By the way, breaking corners is sheet metal lingo for removing the sharp corners from a sheet metal part, usually for the sake of safety.

As you discovered in Chapter 6, adding a fillet or chamfer as a feature requires selecting the edge on which the fillet or chamfer is to be applied. With regard to sheet metal parts, that edge can be very small. For example, 18-gauge steel sheet metal is only .048 inch thick. That certainly does not present much of an edge to select, so it may require the user to zoom in close in order to pick it, or to use some other selection technique.

Break Corner uses a filtering technique to make selecting very small edges easy. Simply picking near the edge will select it without incorrectly selecting the longer edges adjacent to it. The Break Corner command works on sheet metal parts only, and therefore cannot be used on other thin-feature parts that have not been defined as sheet metal. How-To 8-9 takes you through the process of breaking corners.

How-To 8-9: Breaking Corners

To add chamfers or rounds to (in other words, to break the corners of) a sheet metal part, perform the following steps.

1. Click on the Break Corner icon, shown in figure 8-59, or select Break Corner from the Insert > Sheet Metal menu.

2. Select the edges to be broken.

3. Specify the Break Type (chamfered or rounded).

4. Specify the break Distance.

5. Click on OK to finish breaking the corners.

Fig. 8-59. Break Corner icon.

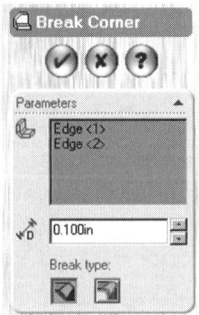

The Break Corner panel is shown in figure 8-60. It should be noted that when *Break type* is set to Chamfer, only one distance parameter is present. The chamfer that is created is limited to a simple 45-degree chamfer (assuming a right-angle corner, though this is not required). If something a little more specific is required, you will need to either use the Chamfer command or create a cut.

Fig. 8-60. Break Corner panel.

Closing Corners

Not to be confused with the Break Corner command is the Closed Corner command. The two commands perform very different functions. Where

Closing Corners

Break Corner operates on the small corner edges of a sheet metal part, Closed Corner operates on gaps, such as those a miter flange feature might leave behind.

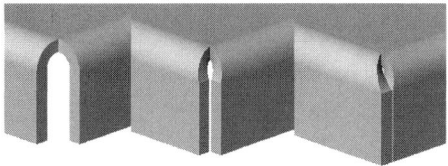

Fig. 8-61. A corner in need of closing, before and after.

When Closed Corner is used, the corner is not truly closed. The Closed Corner command does not create an airtight "weld" between two adjacent flanges. Rather, it creates a butt or overlap region. Figure 8-61 shows the corner of a miter flange prior to using the Closed Corner command (left). Also pictured is the same corner closed using the Butt option (center) and again using the Overlap option (right). How-To 8-10 takes you through the process of closing corners.

How-To 8-10: Closing Corners

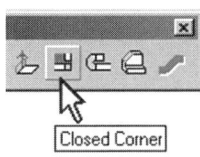

Fig. 8-62. Closed Corner icon.

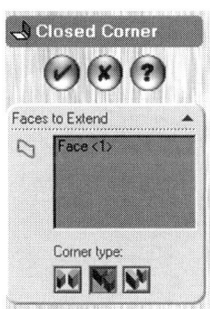

Fig. 8-63. Closed Corner panel.

To close a corner of a sheet metal part, perform the following steps. Note that only planar faces can be selected for closing, and that the corner faces must be perpendicular. Examples of corners that can be closed are those typically resulting from miter flanges or corners left open after executing the Rip command (discussed later in this chapter).

1. Click on the Closed Corner icon, shown in figure 8-62, or select Closed Corner from the Insert > Sheet Metal menu.

2. Select one face on the corner to be closed. You will not need to select the opposing face.

3. Specify how to close the corner. Choices are Butt, Overlap, and Underlap. Use the icons in the Closed Corner panel, shown in figure 8-63.

4. Click on OK to complete the command.

Note that more than one corner can be closed at a time. It should also be noted that Overlap and Underlap are exactly the same thing. The only difference is what face is selected for the corner to be closed. In other words, selecting one face and specifying Overlap is identical to selecting the opposing face and selecting Underlap.

Lofted Bends

The newest addition to SolidWorks sheet metal functionality is the lofted bend feature. It allows for creating a free-form shape that can be flattened. The Lofted Bend command functions as if it were a slimmed-down version of the Loft command (discussed in Chapter 6). There are some rules that must be followed if you wish to create a lofted bend.

Of primary importance is that each profile in the loft be open. Second, there can be two profiles only. Third, all segments of each profile must be tangent. Adding sketch fillets is an acceptable method of overcoming this current limitation. If all of these conditions are met, you can create a lofted bend. How-To 8-11 takes you through the process of creating a lofted bend.

HOW-TO 8-11: Creating a Lofted Bend

To create a lofted bend, perform the following steps. It is assumed that each sketch profile has already been created, and that they meet the requirements dictated in the previous section. Sketch profiles do not have to reside on parallel planes, nor do they have to have the same number of sketch segments.

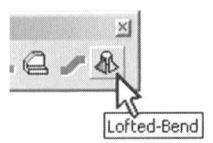

Fig. 8-64. Lofted-Bend icon.

1. Click on the Lofted-Bend icon, shown in figure 8-64, or select Lofted-Bend from the Insert > Sheet Metal menu.

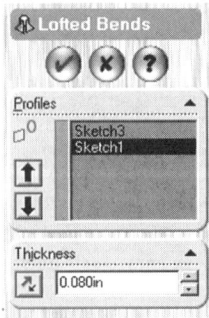

2. Select each sketch from the work area by clicking near common endpoints, which should match up when performing the loft.

3. In the Lofted Bends PropertyManager, shown in figure 8-65, specify a thickness for the new lofted bend feature. You can also reverse the wall direction if necessary.

4. Click on OK to create the lofted bend.

Fig. 8-65. Lofted Bends PropertyManager.

An example of a lofted bend feature is shown in figure 8-66. An open rectangular shape and an arc were used in the example shown. Once a lofted bend feature is created, a *Sheet-Metal1* and a *Flat-Pattern1* feature will appear in FeatureManager. This is similar to when creating a base-flange feature. However, that is where much of the similarity ends.

You will probably find that little can be accomplished once the lofted bend feature has been created. This functionality is still in its infancy. It will evolve in future releases, but for now there are many limitations. Creating an edge flange or miter flange, for example, will prove impossible. Nonetheless, this is a sign of some good things to come in SolidWorks.

Fig. 8-66. Lofted bend feature.

Cutting Sheet Metal Parts

To this point you have learned about all the basic sheet metal functionality present in SolidWorks. If you stopped reading at this point, you would probably be able to hold your own if asked to design a new sheet metal model. However, if you are to gain a more complete understanding of SolidWorks sheet metal commands and options, you should continue with this chapter.

Most sheet metal parts will not have bosses in the normal sense of the word. They will usually have at least a few cuts, though, and often more than just a few. There are some details regarding cut features that are specific to sheet metal parts. These are addressed in the sections that follow.

Link to Thickness

When defining a sheet metal part, the thickness of the material is remembered by SolidWorks. The *Link to thickness* option is a direct result of creating a sheet metal part. When a cut (or a boss, for that matter) is added, the thickness of the cut can be linked to the original material thickness of the part. This option is available for sheet metal parts only.

Figure 8-67 shows the *Link to thickness* option, which appears in the Cut-Extrude panel. Note that the Depth setting normally present when the end condition is set to Blind is not available. This is a direct result of checking the *Link to thickness* option. The Depth setting simply is not needed anymore.

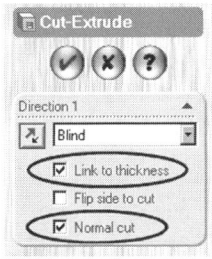

Fig. 8-67. Link to thickness and Normal cut options.

The benefit of using *Link to thickness* is apparent if a design change is required at a later time. You may have spent hours developing a complex sheet metal part, and now your customer decides that the material should be of a heavier gauge. By double clicking on any feature utilizing the *Link to thickness* option, you can access the dimension for the thickness of the material and make the change. This change will be propagated, throughout the part, to any other feature utilizing the *Link to thickness* option.

Normal Cut

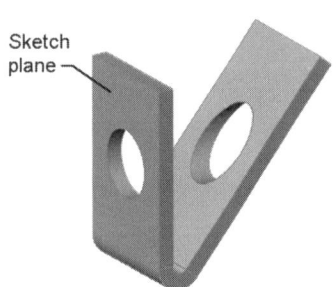

Fig. 8-68. Typical cut through a part.

Fig. 8-69. Normal cut has been turned on.

When the Normal cut option is selected (also shown in figure 8-67), the resultant cut will be perpendicular to the geometry it is cutting through. In contrast, cuts are usually perpendicular to the sketch plane. This can most easily be illustrated with a simple example.

Figure 8-68 shows what you would typically see when creating a cut through a part. A simple circle was sketched on the planar face shown in figure 8-68. The circle was then cut through the model. The *Normal cut* option was not turned on. This allowed the cut to be normal to (perpendicular to) the original sketch plane.

In figure 8-69, *Normal cut* has been turned on. Note the shape of the hole on the angled geometry. All conditions in this second example are identical to the first example (figure 8-68) except for the *Normal cut* option. The sketch plane is the same, and the Through All end condition was used in both cases. With *Normal cut* turned on, the cut will be normal to the geometry, rather than normal to the sketch plane.

It should be noted that the extrusion direction still projects itself in a perpendicular trajectory away from the sketch plane. In this fashion, you can still tell where the cut is going to be on the portions of the model being cut. However, where the cut-extrude feature intersects the model, the cuts will be perpendicular to the sheet metal, as if it were being punched out in its flattened state. The *Normal cut* option is available for sheet metal parts only.

Working in the Flattened State

Before delving into this topic, let's understand exactly what is meant by "working in the flattened state." With sheet metal parts, you basically have two options. The sheet metal part can be designed in the formed state, beginning with a base flange feature, adding tabs, hems, jogs, and so on. You have been learning about these feature types throughout.

A second alternative is to start out with flat stock. This is not necessarily stock, though, in the true sense of the word. If you were to design a sheet metal part in SolidWorks in its flattened state, you would begin with the pattern as if it had been cut from stock, and then form it into shape. This is certainly possible, but not recommended. You would only want to

Cutting Sheet Metal Parts

work from a flat pattern if it were the pattern dimensions that were critical and it was not overly important what the formed part's dimensions were. This is usually not the case.

Usually when designing a sheet metal part, the final dimensions are known. Because SolidWorks figures in the bend allowances for you (which you specify), it is quite easy to design a sheet metal part with precise dimensions in the formed state. It would then be an easy matter of showing the flattened pattern in a design drawing and adding dimensions to it. (Detailed design drawings are explored in Chapter 11.)

The real point of this section is how to create cuts in a sheet metal part while in the flattened state. Because it would not be desirable to create the entire part from a flat pattern, this section will show you how to unfold specific bend regions, add cuts, and fold those regions back up. In this way, cuts can be made at the proper locations on the model, or across bend regions, as if they had been created in the flat pattern from the start.

Proper Technique of Working in the Flattened State

One aspect of sheet metal parts, stated previously, is that feature types are added to FeatureManager when creating a base flange. These features are *Sheet-Metal1*, *Base-Flange*, and *Flat-Pattern1*. You also learned that flattening a part out can be done by clicking on the Flattened icon. When the Flattened icon is selected, the *Flat-Pattern1* feature is unsuppressed, thereby showing the sheet metal part in its flattened state.

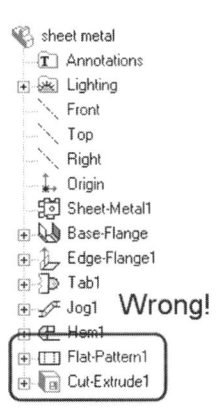

Fig. 8-70. Do not add cuts with Flat-Pattern1 unsuppressed.

It should be strongly stressed that creating cuts with the *Flat-Pattern1* feature unsuppressed is not recommended! The results will almost certainly not be to your liking. What happens is that SolidWorks will add the new feature to the bottom of FeatureManager, as shown in figure 8-70. Take a look at the feature *Cut-Extrude1*. Note in particular that it is below *Flat-Pattern1*. This may seem insignificant, but when the part is formed back up, both the *Flat-Pattern1* and the *Cut-Extrude1* feature will be suppressed. In other words, you will not be able to see the cut except in the flattened state, and there will be no way around it!

The moral of this lesson is simple: do not add features while the model is flattened. This begs another question, however, which is "How can I add cuts to the model in the flat pattern?" This is a common need, and there are special tools for accomplishing this task. These tools are discussed in the following section.

Unfold and Fold Features

Yes, there actually are commands named, appropriately enough, Unfold and Fold. Unfold and fold features appear in FeatureManager. Unfold

314 Chapter 8: Sheet Metal

allows for adding features to the sheet metal part in its flattened state, whereas Fold will form the part back up. Both commands are extremely similar.

The process for adding features in the flattened state is quite simple, as long as the proper technique is followed. First, use the Unfold feature to flatten out any bend regions that will be affected by the cut. Second, add cuts as needed. Third, use the Fold feature to form the part back up. Nearly any feature type can be added to a sheet metal part, but cuts are by far the most common, so that is what is shown in the following example. If adding features other than cuts, make sure the features do not intersect the bend regions. Otherwise, the part may fail to fold back up. How-To 8-12 takes you through the process of using the Unfold and Fold commands.

HOW-TO 8-12: Fold and Unfold Commands

To utilize either the Fold or Unfold command, perform the following steps.

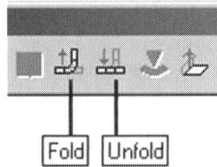

Fig. 8-71. Fold and Unfold icons.

1. Click on the Fold or Unfold icon, shown in figure 8-71, or select Fold or Unfold from the Insert > Sheet Metal menu.

2. Select a face to be held stationary.

3. Select the bend regions to be either folded or unfolded.

4. Click on OK to complete the operation.

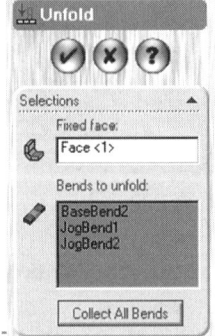

Fig. 8-72. Unfold panel.

The Unfold panel is shown in figure 8-72. The Fold panel looks nearly identical. In step 2 you are required to select a face to be held stationary. This must happen so that SolidWorks will know what to fold or unfold relative to the rest of the part. Additionally, note that there is a Collect All Bends button available. Clicking on this button will select every bend in the entire model.

Use the Collect All Bends functionality when there are a lot of cuts that must be added to the sheet metal part. It is also good practice to get all cuts out of the way prior to folding the model back up. That way, it will not be necessary to unfold portions of the part a second time. If you only have to perform an Unfold and Fold once, it makes for a more streamlined FeatureManager.

Forming Tools

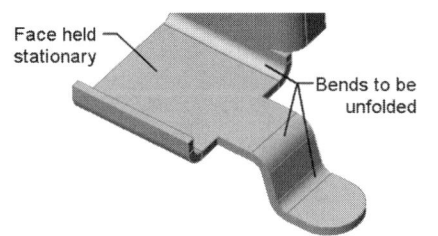

Fig. 8-73. Prior to unfolding.

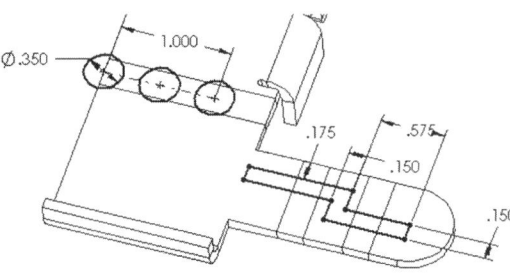

Fig. 8-74. After unfolding and creating a sketch.

Let's see how the Unfold and Fold features are put into practice. In the example that follows, a simple part will be used, but some cuts will be added that would be nearly impossible to add without unfolding the model first. Figure 8-73 shows the part before performing an Unfold on it.

In figure 8-74, the part has been unfolded and some sketch geometry has been added. It would be common practice to break the sketch geometry up instead of trying to put everything into one sketch, and thereby breaking up the features a bit more. But in this case, all sketch geometry has been placed in one sketch for the sake of illustration purposes.

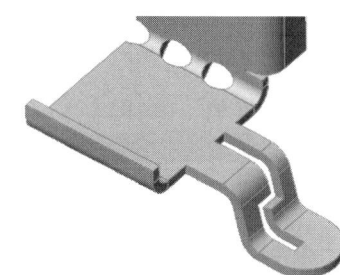

Fig. 8-75. After folding the model back up.

Next, the features are added. Because you know what a cut looks like, that is not shown. However, it is worth pointing out that all cuts have been made through bend regions. Finally, figure 8-75 shows the same model after using the Fold command to form the part back up.

Forming Tools

A big part of working with sheet metal is the ability to perform the punching and deformation operations that typically accompany the manufacturing process. Punching refers not just to cuts, but to creating features such as lances and louvers. Deformations include operations such as adding dimples or embosses. Forming tools will perform these operations, and more, on SolidWorks sheet metal parts. This section shows you how to use forming tools, and how to create your own.

Forming tools are part of a larger aspect of SolidWorks functionality known as the Feature palette, which consist of four categories that carry with them slightly different traits. The Feature palette, in general, allows for reusing geometry. It is a library of features, parts, and assemblies that can be used in other parts or assemblies. The process of reusing these features and parts operates with a simple drag-and-drop interface.

The categories the Feature palette are Forming Tools, Library Features, Palette Parts, and Palette Assemblies. Because this is the first time the Feature palette has been mentioned, it would be prudent to discuss some of the common aspects of its functionality. The intricacies of palette parts and palette assemblies are discussed in another chapter, because they do not relate directly to sheet metal parts. If you would like to explore palette parts and palette assemblies in more depth at this time, turn to Chapter 12.

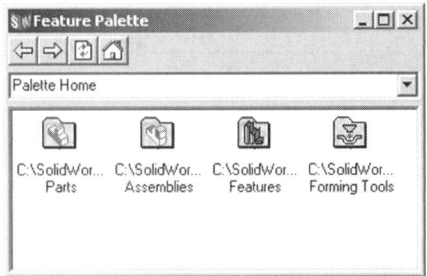

Table 8-1 provides an overview of Feature palette functionality. The table also points out the file types associated with each of the four main categories found in the Feature palette. Figure 8-76 shows the Feature Palette window, along with the four folders that represent the Feature palette categories.

Fig. 8-76. Feature palette.

Table 8-1: Feature Palette Overview

Category	File Type Used	Description
Palette Features, also known as Library Features.	Library feature part files (.sldlfp)	Sketches, features, or groups of features that can be added to part files.
Palette Parts	Part files (.sldprt)	Parts that can be added to assemblies.
Palette Assemblies	Assembly files (.sldasm)	Assemblies that can be added as subassemblies to top-level assemblies.
Palette Forming Tools	Part files (.sldprt)	Features or groups of features that can be added to sheet metal parts.

Using the Feature Palette

Whether inserting palette features, forming tools, parts, or assemblies, the process is always the same. Hold the left mouse button down with the cursor over the item in the Feature palette you would like to use, drag the item into the part or assembly, and let go of the left mouse button. In other words, a simple drag-and-drop operation is all that is required.

When inserting a forming tool or a palette feature, the next step in the process is basically the same. That step is to position the object. What

Forming Tools

amounts to an "orientation sketch" will be displayed. This sketch can be moved or rotated, and positioned, using dimensions or geometric relations.

Once the feature has been positioned, the process has been completed if inserting a forming tool. If inserting a palette feature, you will be given the option of changing dimension values of the feature. This step is not present when inserting a forming tool because the "tool" would be a set size in reality. Even so, the dimensions associated with a forming tool can still be altered after it has been inserted into a model by editing the associated sketch geometry or feature definitions.

Navigating through the folders of the Feature palette is done by double clicking on the folder you wish to access. Double clicking on folders takes you one level deeper into the directory structure. It is also possible to use the arrows located on the Feature Palette window toolbar (refer to figure 8-68) to move backward and forward through previous clicks, similar in functionality to a web browser. How-To 8-13 takes you through the process of using the Feature palette to insert palette features or forming tools.

How-To 8-13: Inserting Features via the Feature Palette

To insert forming tools or palette features via the Feature palette, perform the following steps. If you are looking for instructions on how to insert palette parts or assemblies, see Chapter 12.

1. Select Feature Palette from the Tools menu to open the Feature Palette window.

2. Double click on folders in the Feature Palette window toolbar to navigate to the folder containing the item you would like to use.

3. Using the left mouse button, drag the item onto a planar face of the model. If the item is a forming tool, use the Tab key to punch the feature through the model from one side or the other prior to releasing the mouse button.

4. Position the orientation sketch using dimensions or geometric relations.

5. Click on Finish if inserting a forming tool, or on Next if inserting a palette feature.

At this point the process would be completed if inserting a forming tool. When inserting a palette feature, however, you are given the opportunity to modify dimensions associated with the library feature (or features, as there may be more than one). Figure 8-77 shows a portion of what might be seen when inserting a palette feature. Continue with the following steps.

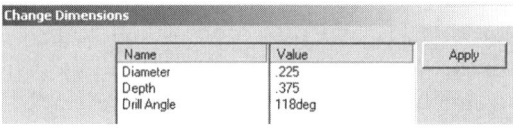

Fig. 8-77. Modifying library feature dimensions.

6. Double click on any dimensions shown in the Change Dimensions window and enter new values.

7. Click on the Apply button to see a preview (if desired), then repeat step 6 if necessary.

8. Click on Finish to complete the process.

A common question when inserting palette features or forming tools is how to rotate the orientation sketch. This task would be accomplished via the Modify Sketch command. Because the Modify Sketch command is fairly extensive, a section has been devoted to it later in this chapter. First, however, let's finish exploring the Feature palette.

Dissolving Palette Features

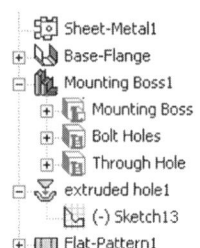

Fig. 8-78. Feature palette items in FeatureManager.

Once a library feature (palette feature) has been inserted from the Feature palette, it appears in FeatureManager as a single feature. In the case of a palette feature, the single feature may be made up of many smaller features. Such is the case with *Mounting Boss1*, shown in figure 8-78. The primary library feature actually consists of three features, named *Mounting Boss*, *Bolt Holes*, and *Through Hole*.

Sometimes it is necessary to modify a library feature to a greater extent than changing a few dimension values. For instance, it may be necessary to delete one or more subfeatures. If this is the case, it is possible to dissolve the main library feature, thereby breaking it down into its constituent components. This can be accomplished by simply right-clicking on the main library feature (such as *Mounting Boss1*, shown in figure 8-78) and selecting Dissolve Library Feature.

Forming tools are not quite as flexible. They are made to mimic tools used in real life. Even if there were numerous features used in the creation of a forming tool, the only items you would see in FeatureManager once the tool were used on a sheet metal part would be the tool itself and the orientation sketch. See figure 8-78 for an example of this, and note the forming tool named *Extruded Hole1*.

Forming Tools

What happens if the forming tool is wrong? Sometimes you can double click on the forming tool feature in FeatureManager and modify its associated dimensions. If this is not satisfactory, you will need to delete the tool and try inserting a different one. To modify an existing forming tool or palette feature, or even to create your own, read on.

Creating Library Features

A library feature part file is how a palette feature begins life. Technically, the Feature palette represents a method of implementing library features. A library feature is nothing more than some item that can be reused over and over, and the Feature palette is just a convenient place to access those features.

A library feature part file contains the geometry constituting the library feature. A library feature can be very simple. In fact, it might be nothing more than a commonly used sketch profile. A library feature might also be a single feature, or half a dozen features. Library features do have their limits. For example, it would not be possible to create threads as a library feature because they are too complex, with too many variables.

To begin creating a library feature, start by creating a part using the knowledge you have gained to this point. If the library feature is to be a cut, start with a block of material the cut will be made through. The block does not have to be part of the library. In other words, you can pick and choose beforehand to determine which features get reused when inserting the library feature at a later time. How-To 8-14 takes you through the process of creating a library feature.

How-To 8-14: Creating a Library Feature

To create a library feature, perform the following steps.

1. Begin a new part and create the features that make up the geometry of the library feature.

2. Select Save As from the File menu.

3. In the Save As window, under the *Save as type* drop-down list, specify Library Feature Part File as the type of file to save as.

4. Enter a name for the file.

5. Click on the Save button.

6. You may be presented with the message shown in figure 8-79. Click on No (this message is explained in material to follow).

Fig. 8-79. A warning about simplifying the library feature part file.

7. You have successfully saved the model as a library feature part file, but must now add features to the library. Right-click on the desired features in FeatureManager and select Add To Library.

8. Save the file a final time.

In step 7, you told SolidWorks to add certain features to the library. Features can be removed from the library using the same technique. In this fashion, you can pick and choose which features make up the complete library feature. Note that any feature added to the library will have a capital L attached to its icon.

When saving a library feature part file, it is possible to preselect features that should be in the library. This keeps you from having to add the features to the library later. The warning message that was displayed in step 6 arises from SolidWorks ability to simplify the file automatically by deleting any unnecessary features. In other words, SolidWorks can delete any features that have not been selected prior to saving the part as a library feature part file.

✗ **WARNING:** *Be careful when clicking on the Yes button when the warning message is displayed! Doing so will force SolidWorks to delete any unselected features. Only the selected features and any parent feature geometry will be left. If nothing has been selected, all features will be deleted. If the original part file was never saved, you will lose everything.*

Palette features are typically stored in one of the folders in the path *data\Palette Features*, found in your SolidWorks installation folder. The location of all palette items is covered in the section "Feature Palette Structure" to follow.

Creating Forming Tools

Forming tools differ from palette features in that they are not library feature part files. Rather, forming tools are regular part files. There is no need to add or remove features from the file's library, and there is no need to save the file as a special "type." There is, however, a particular process involved in creating a forming tool. If you do not follow this process correctly, the forming tool will probably not work when used on a sheet metal part.

Forming Tools

This is a four-step process that is actually quite simple. The process begins by creating a slab of material the forming tool will be modeled on. This "slab," really just a base feature, should be about the same thickness as the average stock from which your sheet metal parts are typically created.

The second stage in the process is to model the tool. The tool should be the same shape you would see on the inside of a piece of sheet metal if the tool were punched into it. This is very true to life, so creating the tool is easy to visualize. Step three would be to cut away the original base feature. Deleting the original base feature is not allowed, because every feature that comes after it would also be deleted. It must be cut away.

Last, an orientation sketch must be created. This orientation sketch is what will be seen when the forming tool is used. This is a necessary part of the process, even if you do not think you will need an orientation sketch. How-To 8-15 takes you through the process of creating a forming tool.

How-To 8-15: Creating a Forming Tool

To create a forming tool, perform the following steps. To help explain the process, an example is used. The example shown in the illustrations is of a knockout, but your forming tool may be different. Figure 8-80 shows the knockout forming tool in its completed state.

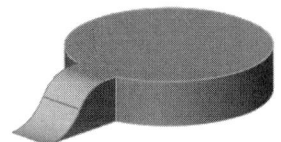

Fig. 8-80. The completed knockout.

1. Create a base feature large enough to accommodate the forming tool. This feature should be similar in thickness to the sheet metal parts the tool will be used on. Figure 8-81 shows an example of a base feature, with a sketch already created for the next feature, which will make up the main body of the forming tool.

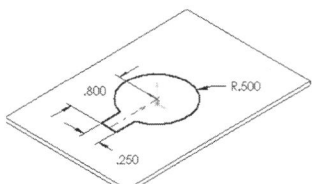

Fig. 8-81. A base feature "slab" and new sketch for the forming tool.

2. Create the shape of the forming tool. Use extrusions, revolved features, fillets, and other techniques you have learned to this point. Figure 8-82 shows the sketch from figure 8-81 extruded, with a few other features added (such as fillets).

Fig. 8-82. The forming tool's geometry.

Fig. 8-83. After cutting away the base feature.

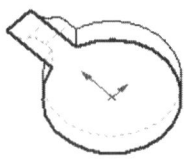

Fig. 8-84. Creating the orientation sketch.

3. Cut away the geometry that makes up the main base feature. The easiest method for accomplishing this is to sketch on one side of the "slab," use the Convert Entities command, and then cut using a Through All end condition. Figure 8-83 shows the result of this operation using the knockout example.

4. Create an orientation sketch by sketching on the bottom of the forming tool and once again using the Convert Entities icon. Figure 8-84 shows the sketch on the underside of the knockout.

5. Once the orientation sketch has been created, exit the sketch.

6. Hide the orientation sketch by right-clicking on the sketch in FeatureManager and selecting Hide Sketch. It will still appear when the forming tool is used.

7. Change to a reasonable view and display the model shaded. You will want to do this because what you see on the screen will represent the preview in the Feature palette.

8. Save the file.

Forming tools are typically stored in one of the folders in the path *data\Palette Forming Tools*, found in your SolidWorks installation folder. As previously mentioned, the location of all palette items is covered in the section "Feature Palette Structure" to follow.

It is a good idea to position a forming tool's main feature sketch on the sketch origin when it is being created. This will centrally locate the tool and make it easier to drag and drop onto a sheet metal model. Otherwise, the tool will be some distance from the cursor as it is being dragged onto the model. In addition, SolidWorks will add centerlines at the origin, for dimensional purposes, when you are using the tool.

One optional step has been left out in How-To 8-15. That step has to do with how faces get removed from the model when the forming tool is used. To better explain this, let's see what would happen if the knockout were used as is. Figure 8-85 shows what the knockout forming tool would look like if it were used on a sheet metal part. The resultant feature is shown from two angles for clarification, but it should be pretty obvious that something is not quite right.

Forming Tools 323

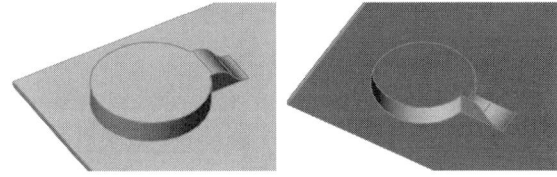

Fig. 8-85. After using the knockout forming tool.

If you noticed that there were no open areas near the sides of the knockout, you are on the right track. It should be possible to slide a screwdriver or something similar under the knockout, but in our case that would not be possible. There are walls surrounding the knockout, and no openings whatsoever.

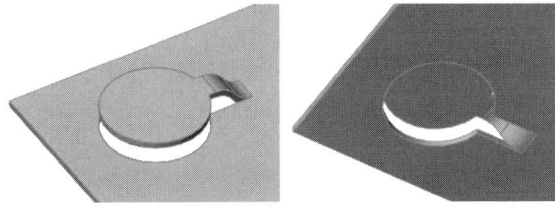

Fig. 8-86. What the knockout operation should look like.

In many cases, forming tools do not break or cut the sheet metal part whatsoever. Many forming tools do just that; they form, but nothing else. Other tools will literally stamp out a piece of material or leave open areas, such as with lancing operations. Figure 8-86 shows what the knockout forming tool should be accomplishing if used on a sheet metal part. Note the open area that results from the operation.

So how do we modify the forming tool to create open areas in a model? The secret is the color red. Specifically, any faces that have been given the color red will be removed when the forming tool is applied to the model. Red is the only color that has any modeling significance in SolidWorks, and this unique functionality only applies to forming tools. Only the true color red can be used, with an RGB (red, green, blue) value of 255,0,0 (the three comma-separated "numbers" representing the three color values, respectively).

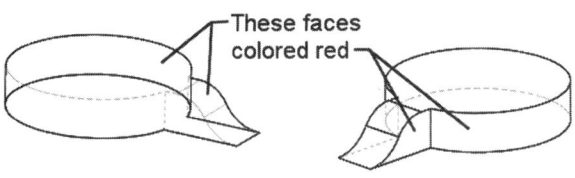

Fig. 8-87. Faces that should be colored red.

Figure 8-87 shows the faces that should be given the color red for the knockout to work properly. There are a total of three (3) faces that should be red: the cylindrical face and either side of the protrusion. The bottom face should not be selected. Other features, or even the part itself, can be assigned other colors to no ill effect on the forming tool. Just be careful what gets assigned the color red with regard to forming tools. Use the Edit Color command to assign colors to a model. The section "Modifying Part Color" in material to follow explains the Edit Color command in detail.

Feature Palette Structure

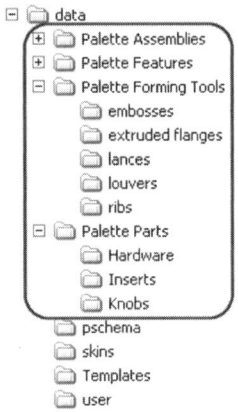

Fig. 8-88. Directory structure of the Feature palette.

SolidWorks stocks the Feature palette with some standard parts, shapes, and features, but you can customize the palette to your heart's content. This is an extremely easy thing to do. If you know how to use Windows Explorer, you know how to customize the Feature palette. Figure 8-88 shows an expanded excerpt from Windows Explorer, which shows the directory structure of the Feature palette. Two of the branches have been expanded to show their underlying structure.

The folders in the Feature Palette window are directly linked to the actual names of the folders stored on the hard drive. These subdirectories contain the various library features and model files, and can be named anything you wish. To put it quite simply, the folders shown in the Feature palette are a direct result of whatever directory structure is on the hard drive. All SolidWorks installations will have a similar directory structure to start with, but this can be customized.

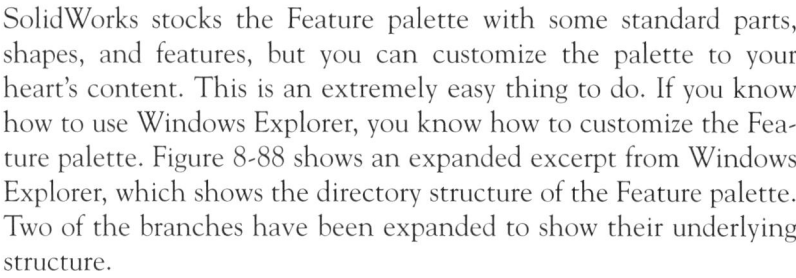

Fig. 8-89. The Feature palette relates to directory names.

Figure 8-89 shows how one of the Feature Palette folders, in particular the *Forming Tools* folders, relate to the directory structure as displayed in Windows Explorer. Additionally, two icons have been circled in the image. These icons represent Reload and Home. The Home icon will display the main top-level folders (*Palette Features, Forming Tools,* and so on). Reload will refresh the display of the Feature Palette window. Use Reload if any additions have been made to the Feature palette or if any new folders have been created via Windows Explorer. Otherwise, the new additions will not be visible.

By default, SolidWorks will look in a particular location for each of the main Feature Palette folders. Table 8-2 outlines the default paths where SolidWorks will look for these folders. Note that <drive> refers to the hard drive letter SolidWorks was installed on, and <installation directory> refers to the SolidWorks installation folder. By default, the installation drive and directory is *C:\Program Files* unless otherwise specified.

Forming Tools

Table 8-2: Feature Palette File Paths

Feature Palette Folder	Path
Palette Features	<drive>:\<installation directory>\SolidWorks\data\Palette Features
Palette Parts	<drive>:\<installation directory>\SolidWorks\data\Palette Parts
Palette Assemblies	<drive>:\<installation directory>\SolidWorks\data\Palette Assemblies
Palette Forming Tools	<drive>:\<installation directory>\SolidWorks\data\Palette Forming Tools

It is possible to change the location where SolidWorks looks for Feature palette items, or even to add folder locations to the path list. In actuality, folder location paths can be established for many objects in SolidWorks, including templates, symbol libraries, blocks, sheet formats, and macros, to name a few. How-To 8-16 takes you through the process of modifying, or adding search paths to, various SolidWorks objects.

HOW-TO 8-16: Modifying Folder Location Paths

To modify, or to add paths to, the paths SolidWorks uses to find Feature palette items or many other SolidWorks objects (such as those listed previously), perform the following steps. Note that file directories should be set up prior to pointing SolidWorks to a particular location. Creating file directories is a function of the Windows operating system and is outside the scope of this book.

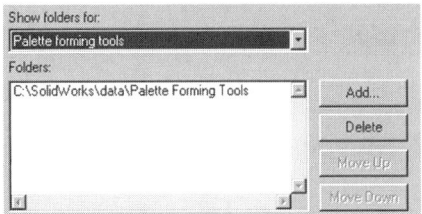

Fig. 8-90. Setting a path for Palette forming tools.

1. Click on Options in the Tools menu.

2. Select the File Locations category in the System Options tab.

3. Using the *Show folders for* drop-down list, select the item to set the path for. In figure 8-90, *Palette forming tools* has been selected from the drop-down list.

4. Select a file location path in the Folders listing and click on the Delete button if it is necessary to remove an existing path.

5. Click on the Add button to add a new or additional file location path.

6. Specify a new file location path and click on OK.

7. Click on OK when finished to close the System Options window.

If additional paths are specified that are above and beyond the default paths created by SolidWorks, they will show up as additional top-level folders in the Feature palette. For example, you may decide to create a folder on a network server with a directory named *Purchased Parts* or *Standard Forming Tools*. These are just examples. Obviously, you can use names appropriate to your company or situation.

Editing Palette Items

Any item found in the Feature palette can be edited by right-clicking on the item and selecting Edit Palette Item. This action will open the item in its own window, where modifications can be made. Generally, anything in the Feature palette can be edited in the same way any other model would be. Add features or alter dimensions as needed to modify the desired palette feature or forming tool. Other than that, there are only a few other things you will need to be aware of, depending on the item being edited.

Restricting Dimension Access

Items found in the Feature palette can have their dimension access controlled to a certain extent. When dimension access is controlled, certain dimensions associated with the palette item can be made off limits to anyone inserting the palette item. In the case of palette features, the dimensions will not appear in the Change Dimensions window when inserting the feature. How-To 8-17 takes you through the process of controlling dimension access.

How-To 8-17: Controlling Dimension Access

To control the access of dimensions when inserting palette items from the Feature palette, perform the following steps. Note that palette assemblies cannot have dimension access restricted.

1. In the Feature palette, right-click on the item to be edited and select Edit Palette Item.

2. Right-click on the name of the item in FeatureManager and select Edit Dimension Access.

3. In the Dimension Access window, a portion of which is shown in figure 8-91, select dimensions, and then use the arrow buttons to move dimensions from one side to the other. Dimen-

Forming Tools

sions appearing on the right are considered internal dimensions and will not be accessible once the feature is inserted into a part.

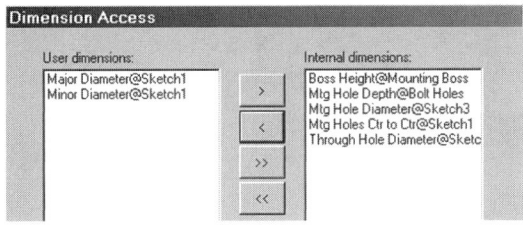

4. Click on Finish when done.

5. Save the file.

Fig. 8-91. Modifying dimensions access.

With regard to forming tools and palette features, once a dimension has been designated an internal dimension, that dimension is completely off limits to anyone inserting that forming tool into a part. The original file can be modified by anyone who has access to that file, but once inserted, internal dimensions are completely inaccessible.

With regard to palette parts, discussed in greater detail in Chapter 12, modifying the dimension access does nothing. All dimensions are still accessible, whether they are internal or not.

Renaming Dimensions

Renaming dimensions is a good idea in the case of palette features because any dimension not designated as internal will appear in the Change Dimensions window when inserting the feature into a model. Renaming dimensions is typically something that would be done while creating a library feature part file that is to become a palette feature. Additionally, renaming features or even sketch names may be something you will also want to do. There are other reasons for renaming dimensions, such as when they are used in equations or design tables, which are discussed later in this book.

Dimensions have names that correspond to the sketch or feature with which they are associated. For instance, a dimension with the name D1 that is associated with Sketch 4 would have the full name of *D1@Sketch4*. The "at" (@) symbol is nothing more than a separator character Solid-Works understands. Dimension names such as *D1@Sketch4* do not mean much by themselves. It would make much more sense to see a descriptive name such as *Diameter@Lower-holes* in the Change Dimensions window.

Indirectly, the section on dimension properties in Chapter 5 discussed the method used to rename dimensions. Access a dimension's properties by right-clicking on the dimension and you will find that the dimension can be renamed quite easily. The area where the dimension name would be changed is shown in figure 8-92. Do not forget to save the file when finished, or renaming the dimensions will be for naught.

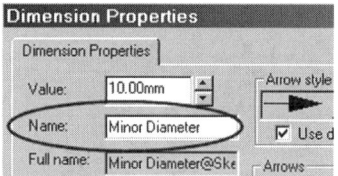

Fig. 8-92. Changing a dimension's name.

Modifying Part Color

You will need to know how to change the color of faces on a model in order to create certain types of forming tools (i.e., those that create openings or cuts in a sheet metal model). For this reason, color editing options are explored here.

Fig. 8-93. Edit Color icon.

Although there are different ways in which to alter the color of a model, one particular method stands out as the easiest. This method involves the Edit Color command, whose icon is shown in figure 8-93. The Edit Color icon allows for changing the color of an entire part, individual features, or even specific faces. Be forewarned that there is no easy way to remove coloring from many faces simultaneously if you change your mind later.

There is a distinct hierarchy followed regarding colors of objects. Specifically, colors of faces override colors of features, colors of features override colors of bodies, and colors of bodies override colors of parts. Just remember the order (faces, features, bodies, parts) and the rest comes easy. For example, if the color of a feature has been changed, it will retain its color if the color of the overall part is altered. How-To 8-18 takes you through the process of changing the color of faces, features, bodies, or parts.

How-To 8-18: Editing Colors

To change the color of faces, features, bodies, or parts, perform the following steps.

1. Click on the Edit Color icon.

2. From the work area or FeatureManager, select the object whose color is to be changed.

3. Select the desired color from the Edit Color window, shown in figure 8-94. Use the Define Custom Colors button for even more choices.

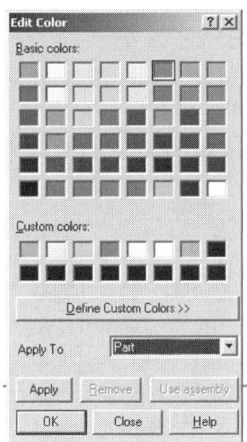

Fig. 8-94. Edit Color window.

4. Select Face, Feature, Body, or Part from the drop-down list.

5. Click on Apply.

6. Repeat steps 2 through 5 as needed and click on OK when finished.

To remove the color from an object, repeat the process described in How-To 8-18, but instead of clicking on Apply per step 5, simply click on Remove. The color of the selected object will revert to whatever the original color of the part happened to be (or to the color of the next higher object in the hierarchy).

The Use Assembly button, which is grayed out in figure 8-94, allows for switching to whatever color is specified for an assembly. With regard to assemblies, there are ways to apply color to a component so that it affects only the component in the assembly and not the original part file. (These topics are covered in Chapter 12.)

You have learned quite a bit about the Feature palette and related items, which can be useful when working with sheet metal parts, but can be applied to other facets of the SolidWorks program. Such is the case with the next section, which deals with modifying sketch geometry. In particular, being able to rotate a sketch is very important when trying to position palette features or forming tools correctly.

Modify Sketch Command

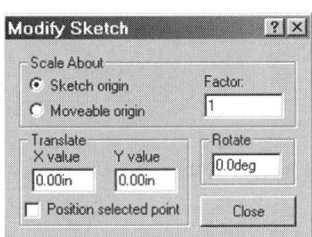

Fig. 8-95. Modify Sketch window.

Sketch geometry can be mirrored, rotated, scaled, and translated using the Modify Sketch command. The Modify Sketch command consists of a small window, shown in figure 8-95.

Any sketch geometry can be modified, unless it has references to external model geometry. For instance, a sketch could not be moved (translated) if it were dimensioned to existing edges of a model. Scaling sketch geometry will work if there are dimensions associated with the sketch, but as with translating, it must not be fixed at one location.

Rotating a sketch is one of the more common uses of the Modify Sketch command, which will work even if the geometry has lines that are horizontally or vertically constrained. This is because the entire sketch, including the origin point, is rotated. Because of this, any entities constrained horizontal or vertical remain so, albeit relative to the movable origin.

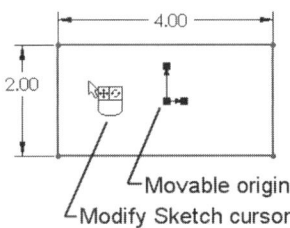

Fig. 8-96. Movable origin and Modify Sketch cursor.

When the Modify Sketch command is used, a symbol that looks like a black origin point with small squares (or "handles") attached to it will be visible. This is known as the movable origin. The handles can be used to mirror or flip the sketch geometry. Additionally, the cursor changes and allows for dynamically translating or rotating the sketch with the left or right mouse button, respectively. The movable origin and Modify Sketch cursor are shown in figure 8-96.

The section that follows covers the capabilities of the Modify Sketch command in a general way. The Modify Sketch command is very simple to use once you understand the basics. All of the functions contained within the Modify Sketch command are implemented in a very similar fashion. How-To 8-19 takes you through the process of using Modify Sketch to perform its basic functions. Following How-To 8-19, some additional functionality is pointed out.

HOW-TO 8-19: Using the Modify Sketch Command

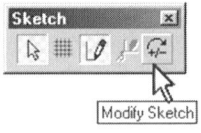

Fig. 8-97. Modify Sketch icon.

To use the Modify Sketch command to translate, rotate, or scale sketch geometry, perform the following steps. Note that you should be editing a sketch in order to use the Modify Sketch command. If not, the sketch to be modified must be selected in FeatureManager. Modify Sketch can also be used when a Feature palette object must be positioned.

1. Click on the Modify Sketch icon, shown in figure 8-97, or select Modify from the Tools > Sketch Tools menu.

2. Click in the applicable area of the Modify Sketch window, depending on what you want to accomplish. For example, click in the Rotate area.

3. Enter a value. For example, enter 90 (degrees) for the Rotate value.

4. Press Enter.

5. Repeat steps 2 through 4 for other modifications as required and click on Close when finished.

As you can see, the Modify Sketch command in its basic form is extremely easy to use. However, there is some additional functionality, along with a few nice tips and tricks, you will want to be aware of. The following sections point out various points of interest and additional functionality of the Modify Sketch command.

✓ **TIP:** *The movable origin can be positioned or related to other geometry by dragging it with the left mouse button. This can play an important role when scaling and rotating.*

Rotating Using the Mouse

In addition to typing in values, you can use the mouse to rotate geometry dynamically. If you hold down the right mouse button, you will notice a line attached to the center of the movable origin. Move the mouse, and a small readout will be displayed on the cursor, informing you how much rotation you are placing on the geometry.

If using the mouse to dynamically rotate, note that the distance of the mouse from the movable origin affects the coarseness of the rotational increments. Closer to the origin will rotate in large increments; farther away will offer smaller incremental values and more precise control. To quickly rotate a sketch (say, 90 degrees), keep the cursor close to the movable origin.

Translating

Both *x* and *y* values can be changed at the same time if entering values. Pressing Enter will apply both values at once. To use the mouse to translate dynamically, pick anyplace on the screen and drag the sketch with the left mouse button. The short arrow (of both the movable origin and sketch origin) always relates to the positive *x* direction, and the longer arrow the *y* direction.

The Position selected point option allows for picking a specific sketch entity point after turning the option on, and then viewing the *x,y* coordinates of that point. Entering a value at that time and pressing Enter will move the selected vertex point to the *x,y* coordinates specified relative to the sketch origin. Note that the sketch will move relative to the sketch origin, not the movable origin.

Using Position selected point is a little confusing because when the sketch is translated the sketch origin moves as well. This is always the case, whether Position selected point is selected or not. This means that if in using this function you were to select the very same point, it would have the same *x,y* coordinates it had prior to being translated.

Scaling

It probably goes without saying that a scale factor greater than 1 increases the size of the sketch, and anything less than 1 will decrease the size of the sketch. If a sketch contains dimensions, the values of those dimensions will change accordingly.

Make sure to specify what you would like to scale about: the sketch origin or movable origin. If the sketch origin is used and the sketch is some distance from the origin, the sketch will move away from or closer to the origin, depending on the scale factor. Positioning the movable origin near the center of the sketch and scaling about the movable origin provides good results.

Flipping

Flipping refers to mirroring geometry to the other side, but not retaining the original. Use the movable origin to flip the sketch left to right, top to bottom, or both. A simple right-click on the applicable black square will accomplish the flip. For example, clicking on the black square at the end of the *x* axis of the movable origin will flip the sketch across the *y* axis.

Converting to Sheet Metal

You may be surprised to discover that SolidWorks can take model geometry from other CAD programs and turn it into a sheet metal part. This includes unrolling or unbending areas of the model to show it in its flattened state. It also includes the ability to rip out corners in order to flatten a part. This added capability gives the user greater flexibility in designing sheet metal parts because the model does not have to start out as sheet metal.

Geometry, whether imported from another CAD system or created in SolidWorks using non sheet metal design techniques, can be "turned into" a sheet metal part by employing the Bends command. However, not every model will lend itself to the Bends command. Use common sense and be practical with what you attempt to convert to a sheet metal part. For example, shelled parts can often be converted to sheet metal parts once the corners are ripped out. Lofted or swept parts usually cannot (swept parts are discussed in the next chapter).

Converting to Sheet Metal

There are two important requirements that must be met by a model that is about to have bends added to it. First, it should be of a uniform thickness. It is not possible to turn a model into sheet metal if it is .050 inch thick at one end and .25 inch thick at the other. Second, the model must physically be capable of flattening, assuming it becomes a sheet metal part. Picture a box, with four sides and open at the top. If the sides are connected, they cannot be unfolded. If there are openings between the sides, essentially making each side a "flap," the box can be unfolded.

When bends are inserted into an existing model, any sharp corners present on the model will have bends applied to them, using a default radius value specified by the user. If relief cuts are required, SolidWorks will add them. Additionally, bend allowance will be figured into the model to account for stretch during the bend (forming) process.

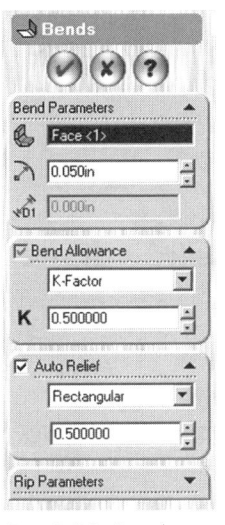

Fig. 8-98. Bends panel.

Inserting Bends

The act of converting a standard SolidWorks part file into a sheet metal part is known as inserting bends. When inserting bends, it is important to remember that a face must be selected that will be held stationary. It is this face that will remain stationary when the part is flattened or folded back up. This is just one of the steps in adding bends to a part. How-To 8-20 takes you through the process of inserting bends. Figure 8-98 shows the Bends panel displayed in PropertyManager when using the Bends command.

How-To 8-20: Inserting Sheet Metal Bends

To insert sheet metal bends, perform the following steps.

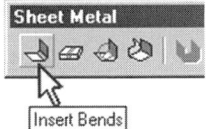

Fig. 8-99. Bends icon.

1. Select Bends from the Insert > Sheet Metal menu, or click on the Insert Bends icon (shown in figure 8-99), located on the Sheet Metal toolbar.

2. Select a face to remain stationary when the part is unfolded.

3. Specify a Bend Radius that will be applied to the inside of any sharp corners.

4. Specify a default bend allowance using the method of your choice.

5. Specify whether or not to incorporate Auto Relief, and if so, what type of relief cuts should be made.

6. Click on OK to accept the settings.

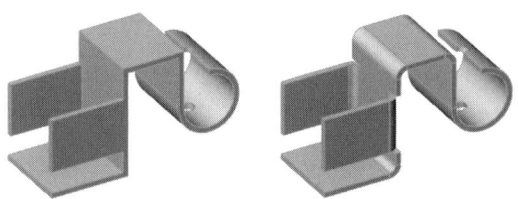

Bends will be added to any sharp corner that existed in your model. The bends' radius will be determined by whatever you specified as the default Bend Radius in the Bends panel. Figure 8-100 shows an example of a thin-feature part both before and after adding bends with a default radius of .125 inch.

Fig. 8-100. After inserting the bends feature.

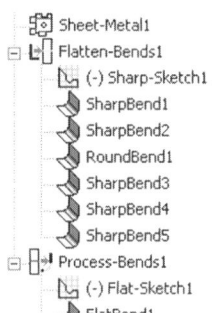

Fig. 8-101. FeatureManager after inserting a bends feature.

The features added to FeatureManager when inserting bends are slightly different than when extruding a base flange. The sections that follow describe what each of these features is, and what it represents from an editing standpoint. Figure 8-101 shows an example of what FeatureManager might look like after adding a bends feature to a model. Note that the *Sheet-Metal1* feature shown in the figure was described previously. Sheet-Metal1 has the same purpose when inserting bends as it does when extruding a base flange, so its description is not repeated here.

Flatten-Bends1

You can think of the *Flatten-Bends1* feature as the place where the bends and the appropriate bend allowances get defined. The part is still in a flattened state (hence the name *Flatten-Bends*). Its definition allows you to change the default bend radius and bend allowance.

Sharp-Sketch1

The *Sharp-Sketch1* sketch is created automatically and is used to control where the bends are placed. SolidWorks adds centerlines on the sheet metal part in its flattened state to show where the bends are. There will be a centerline for every bend listed in the *Flatten-Bends* feature. You cannot edit this sketch because it is generated and controlled by SolidWorks, but it can be shown in a drawing to indicate bend locations and for dimension purposes. The sketch will not be fully defined, but this should not be a concern due to SolidWorks controlling the sketch internally.

SharpBend1

You will see a SharpBend feature listed for every sharp corner in the part. If you edit the definition of this feature, you can specify a bend radius other than the default value. This allows you to individually control the radius of the bends. You can also modify the bend allowance and relief cut options here on a bend-by-bend basis.

RoundBend1

A *RoundBend* listing represents a feature identical to a *SharpBend* listing, except for the fact that round bends are of the rolled variety. The radius of a round bend cannot be changed in its definition because the radius has already been defined in a previous sketch. Round bend features do not start out as sharp corners. Rather, they start out as what would amount to formed or rolled areas (of the model) created prior to inserting the bends feature.

Process-Bends1

Think of the *Process-Bends1* feature as where the bends actually get processed. In other words, this is where the forming of the part takes place. Likewise, by suppressing this feature (or by simply clicking on the Flattened icon), you can see the sheet metal part in its flattened state. Editing the definition of this feature lets you modify the default bend radius and allowance settings, but only for bends added in the *Flat-Sketch1* sketch (see the following section).

Flat-Sketch1

The *Flat-Sketch1* sketch is editable, and there is a good reason you might want to do so. By editing this sketch and adding a line to the flattened sheet metal part, you can add individual bends almost anywhere on the part. Wherever a line is placed, a bend will appear. These bends take names of the form *FlatBend1*, *FlatBend2*, and so on.

FlatBend1

Every time you add a line to *Flat-Sketch1*, one of these *FlatBend* feature listings will appear in FeatureManager. By editing the definition of a *FlatBend* feature, you not only have the ability to change the bend radius but the ability to change the direction and angle of the bend. You can also modify the bend allowance and relief cut parameters.

Figure 8-102 shows an example of some lines added to *Flat-Sketch1*. These lines define where new user-defined bends will be positioned when the model is rebuilt. Note of the point located in the sketch. This is used to indicate what face is held stationary during the forming process (explained in material to follow). Figure 8-103 shows the result of adding the sketch lines to the model.

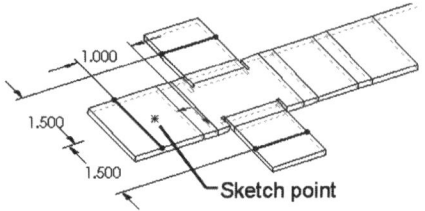

Fig. 8-102. Adding lines to Flat-Sketch1.

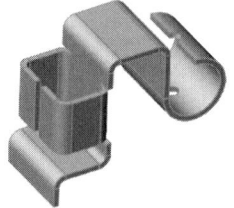

Fig. 8-103. Model with new bends added.

The locating point that defines the stationary face is added by SolidWorks automatically. You cannot dimension or constrain the locating point; SolidWorks will not let you. However, you can drag it to a location of your choosing. The point gets added automatically after (1) a bends feature is added to a part with no sharp corners and (2) you edit the *Flat-Sketch* of the *Process-Bends* feature and sketch at least one line to define the first bend location.

If you edit the sketch of the *Process-Bends* feature for the first time with the intention of adding bend lines, you can beat SolidWorks to the punch and put in your own locating point. SolidWorks will realize this and use the point for defining the stationary face.

Reordering Bends

Often the order in which bends have been created makes a difference with regard to manufacturing the actual part. If this is the case with a sheet metal part you are designing, feel free to reorder the bends using a utility specifically designed for this task.

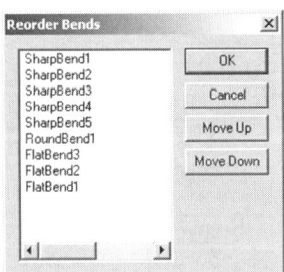

Fig. 8-104. Reorder Bends window.

By right-clicking on the *Process-Bends* feature, access can be gained to the Reorder Bends window, shown in figure 8-104. Selecting a bend from the list and then clicking on the Move Up or Move Down button will reposition the bend in the list. This is more for documentation purposes than anything else, and really has no significant effect on the model.

Rolled Bends

You have already discovered that a face must be selected when inserting bends. This face will be held stationary when the part is flattened. Some of you more observant readers may have noticed that the Bends panel actually asks for a Fixed Face or Edge, as shown in figure 8-105. When would it ever be necessary to select an edge as opposed to a face?

Not all sheet metal parts contain bends. Some are rolled, and some contain combinations of bends and rolled features. Parts that are rolled may not have a flat face that can be designated the stationary face. In such a case, an edge must be selected, as opposed to a face. The part shown in figure 8-106 could go either way. It would be possible to select an edge or planar face.

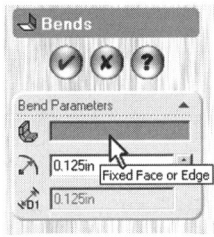
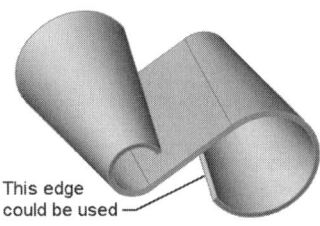
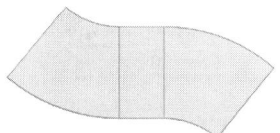

Fig. 8-105. Asking for a face or an edge.

Fig. 8-106. Selecting an edge to be held stationary.

Fig. 8-107. Conical features in a flattened state.

SolidWorks can unroll simple rolled parts and even conical parts. The part shown in figure 8-106 contains an extrusion and a revolved conical feature on either side. If the Bends command is carried out on this model, it can be flattened out in the usual manner. Figure 8-107 shows the part in its flattened state.

Legacy Sheet Metal Parts

The ability to create sheet metal parts in SolidWorks has been around for a while, but this has not always been the case. Older versions of the software used the Insert Bends command to define a model as being sheet metal. A model would first be created using traditional means, such as boss extrusions, and then the model defined as sheet metal after the fact.

The best way to create a sheet metal part is to use the Base Flange command right from the start. This makes for a more flexible model, which will give you many more design options. Occasionally it may be necessary to modify parts that were created using Insert Bends. When this is the case, it is best to use one of the more recent commands on the model in order to "update" the part and give you the ability to benefit from the advanced sheet metal functionality now present in SolidWorks.

If you find yourself working on a legacy sheet metal part and it becomes necessary to add features, try to add a feature that will modernize the model first. This could be accomplished by adding an edge flange, miter flange, jog, hem, sketched bend, tab, or even just breaking a corner. Once that has been accomplished, work on the sheet metal part as if you had created the model as sheet metal from scratch, using the Base Flange command. In this case, ignore the *Flatten-Bends* and *Process-Bends* features in FeatureManager, but do not delete them.

Rip Features

Occasionally it is easier to design a sheet metal part as a shelled part to begin with. This is really more of a convenience than anything born out of necessity. When a part is shelled, what remains is a part that is connected at its edges. Without the ability to separate the walls from each other, there would be no way to flatten the part. This is what the Rip command is for, and it is what allows you to turn a shelled part into a sheet metal part.

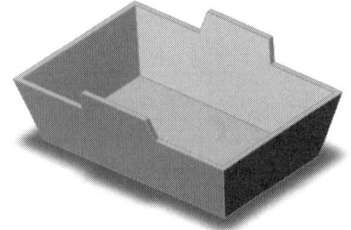

Fig. 8-108. A shelled part.

A shelled part is shown in figure 8-108. There would be no way to flatten this part, not with the walls connected the way they are. However, if you were to perform the Rip command on this model first, it would then be possible to flatten it. You will find the Rip command at two different locations, but do not let this fool you; the commands perform the same function. Rip can be found as a separate command, or it can be found as part of the Bends panel. The latter option can be used to add rips and bends at the same time.

When a rip is performed, a small portion of material is automatically removed from where two walls meet. There is a good degree of control that can be exercised over the Rip command. It is possible to specify which side of the corner (which wall) will be modified during the rip. You can also choose to remove the entire corner. Figure 8-109 shows three possible outcomes of performing the Rip command on a model. The results boil down to taking material away from one side of the corner or the other, or removing the entire corner. The choice is yours.

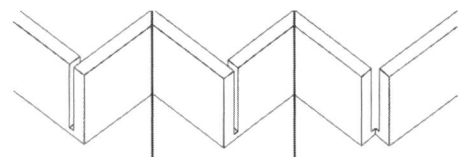

Fig. 8-109. Three possible outcomes of the Rip command.

Rip Features

Adding rips is a pretty simple process. Ripping out corners works the same whether it is done while inserting bends or as a separate command. You may decide it is better to use the separate Rip command if only for the fact that a separate feature is created in FeatureManager. This helps to document what has transpired during the design process. How-To 8-21 takes you through the process of adding a rip as a separate feature.

How-To 8-21: Adding a Rip Feature

To add a rip feature, perform the following steps.

> **NOTE:** *When selecting edges to be ripped, you must select the interior edges.*

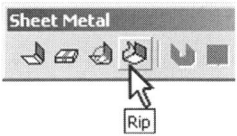

Fig. 8-110. Rip icon.

1. Select Rip from the Insert > Sheet Metal menu, or click on the Rip icon, shown in figure 8-110.

2. Select the edge or edges to be ripped.

3. Click on the Change Direction button as required. Preview arrows will be shown for each edge selected. The arrows will point in the direction of the rip. Two preview arrows indicate that the entire corner will be removed.

4. Uncheck Use default gap if you would like to specify a value larger than the default value.

5. Click on OK when finished.

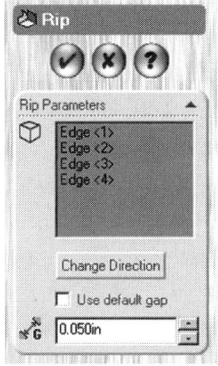

Fig. 8-111. Rip panel.

Figure 8-111 shows the Rip panel. As you can see, there is not much to it. The most significant options in the panel are the Change Direction button, which controls which walls will be ripped for each selected edge, and the Rip Gap setting. All rips do not have to be the same with regard to which side of the edge is ripped out. It is easiest if the Change Direction button is used after selecting each edge, rather than selecting all edges and then going back and using Change Direction afterward.

Figure 8-112 shows the sample part after adding a rip feature. Additionally, bends were inserted, along with two sketched bends and some drill holes. What had once been a shelled part has been defined as a sheet metal part, complete with the ability to be flattened.

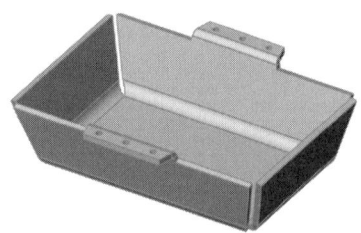

With regard to the gap distance parameter, SolidWorks uses a very small distance (a few thousandths of an inch) when creating the rips. If you are not happy with the default setting, you can specify your own. Simply uncheck the Use default gap option and type in your own value for the gap distance.

Fig. 8-112. After adding a rip feature.

Summary

At the beginning of this chapter, you learned the difference between a thin-feature part and a sheet metal part. Thin-feature parts are normally created from open-profile sketch geometry. Sheet metal parts are defined by creating a base flange feature.

Open profiles can be extruded or revolved. Closed profiles can be used to create thin features as well by specifying the Thin Feature option when extruding or revolving the closed sketch. If it is an open profile you are extruding, you will also have the Auto Fillet option available. If creating a sheet metal part, make it a point to use the Base Flange command in the Insert > Sheet Metal menu.

Once a base flange feature has been created, other sheet-metal-type features can be added. When creating an edge flange feature, SolidWorks automatically generates a sketch. This sketch can be modified by the user via the Edit Profile button found in the Edge Flange panel. Miter flanges require a sketch, though the sketch can be pretty basic. A simple line will suffice.

Other sheet metal features include hems, sketched bends, tabs, and jogs, to name a few. There are also fold and unfold features that should be used if there are cuts that need to be made to areas involving bend regions.

Auto relief can be of three types: rectangular, obround, or tear. Bends can have their individual relief settings modified. This includes changing the type of relief, and the width and depth of the relief cuts. Bend allowance can also be altered on a bend-by-bend basis.

Forming tools are a part of the Feature palette. The Feature palette allows for the reuse of geometry. Palette features, forming tools, palette parts, and palette assemblies make up the Feature palette. Palette features are library-feature part files, whereas forming tools are regular part files. Forming tools can be used on sheet metal parts only, but palette features can be used on any part.

The Modify Sketch tool is an important command when inserting forming tools or library features. It allows for rotating, scaling, translating, and flipping sketch geometry. The fact that the Modify Sketch tool can rotate sketch geometry when inserting forming tools or library features is of utmost importance and is a definite necessity.

Rolled parts can be flattened. In the case of adding a bends feature to a rolled part, it is possible to select an edge as opposed to a face. This will determine the portion of the model that will be held stationary as the part is unfolded and formed.

The ability to rip edges of a part means that you can model a part differently than you otherwise might. A part does not have to start out as a thin-feature part. Rather, it can be a shelled part. This gives you a greater degree of flexibility when modeling the part.

Questions and Topics for Discussion

1. Describe what is meant by the term thin feature.

2. Is it possible to revolve a closed profile in order to create a thin feature?

3. What is the preferred method of defining a sheet metal part?

4. List at least three ways in which bend allowance can be specified.

5. What does Auto Relief accomplish and what is meant by the relief ratio?

6. Link to Thickness is an option that is not always available. Describe what Link to Thickness accomplishes.

7. Would you have to create a sketch for an edge flange feature? Explain your answer.

8. What do the Start and End Offset distances refer to with regard to miter flange features?

9. Can the Break Corner command be used on non sheet metal parts?

10. Describe what happens if features are added to a flattened sheet metal part when the *Flat-Pattern* feature is unsuppressed.

11. Is it possible to insert bends if there are no sharp corners in the model?

12. What is the name of the icon you would click on to change the color of a part?

13. What is significant about the color red in SolidWorks?

14. Describe what the preview arrows represent when adding a rip feature to a part.

Optional Problem

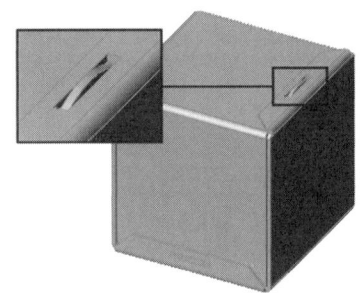

Fig. 8-113. A sheet metal cube.

Create a cube out of sheet metal. Optionally, add flaps that could be used to lock the formed cube together. Perhaps drill holes could be added for rivet placement. There is more than one way to create the sheet metal cube. Be inventive. The size of the cube or thickness of material is not important, but it should be square on each side. Figure 8-113 shows an example of a sheet metal cube.

In the example shown, flaps were added using the Edge Flange command, then small rectangular cuts were made in these flaps. An arc lance forming tool included with the SolidWorks software was added and positioned so that the arc lance falls within each of the rectangular cuts. In this way, the cube could be "snapped" together.

Chapter 9

Springs, Threads, and Curves

SPRINGS ARE A COMMON COMPONENT in many assemblies. They come in all sizes and shapes. Springs are quite easy to model in SolidWorks. Just about every spring begins with a helical curve. Threads also use a helix as a sweep path. Threads might be external threads on a bolt or internal threads on a nut. Helical curves, however, are not the only thing springs and threads have in common.

Both springs and threads must be created as swept features. Swept features can be used for many other purposes above and beyond creating springs and threads. You can also get into some very complex geometry with swept features. This chapter shows you some of the more common things you can accomplish with swept features.

Curves can play a large role in creating swept features. Curves can be used as the path for a swept feature, but curves can also be used for a number of other feature types. You will undoubtedly find many different uses for curves. Thus, even though springs and threads are two of the main topics in this chapter, you will find many applications for what is explored in the material that follows.

Swept Features

To this point, everything you have created in the way of sketched features has required one sketch per feature. For instance, an extruded boss or cut requires only one sketch. A revolved feature also requires just one sketch.

With the introduction of swept features, you will find that one sketch is no longer going to be sufficient most of the time.

Why would you need two sketches to create a swept feature? It is because you must have a sketch you will be sweeping, as well as a sketch you can sweep along. These two sketches are called the profile and the path, shown in figure 9-1. Sometimes the sketch geometry is referred to as the sweep section and trajectory.

You can do quite a bit with swept features. There are a lot of options and physical parameters that can be used to obtain some very interesting geometry. Most of these options are covered in detail this chapter.

A swept feature, taken in its simplest form, is easy to understand. The profile is swept along a path, thereby creating a solid. A swept feature is actually very similar to an extruded boss, except that with an extrusion the path is always perpendicular to the sketch plane. Figure 9-2 shows the result of sweeping a rectangle along the U-shaped path shown in figure 9-1.

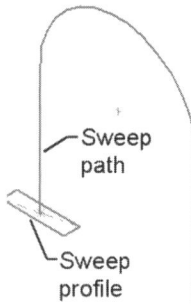

Fig. 9-1. A sweep path and profile.

Fig. 9-2. Previous illustration as a swept feature.

Valid Profiles

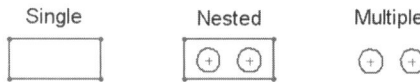

Fig. 9-3. Valid sweep profiles.

A sweep sketch can contain a single closed profile or multiple closed profiles. If the base feature sketch contains multiple profiles, understand that multiple bodies will be created. Sweep profiles can be separate, or nested two deep, as shown in figure 9-3. *Nested* is a term that signifies how objects (in this case, profiles) can exist within other profiles. Creating a thin feature while sweeping is also an option.

With sweep profiles, the general rules of sketching still apply, which were covered in Chapter 3. Mainly, the sketch geometry must not be self-intersecting. In the case of a sweep profile sketch, the profile must be closed. If there is self-intersecting sketch geometry, you must use the Contour Select tool, discussed in Chapter 3.

Valid Paths

When creating a sweep path, you have quite a few options. First, however, there are two rules you absolutely must follow. The first rule is that when creating a sweep path the path must begin on the same plane the sweep profile was sketched on. This does not have to be an actual plane; it can be a planar face. If the path does not start on the sketch plane of the profile, the sweep will not work.

There are two fairly easy methods of making certain a sweep path begins on the plane of the profile. First, if the path was drawn before the profile, create a plane that passes through the start point of the profile if one does not already exist. See How-To 4-13 in Chapter 4.

Second, if the profile was created first, add a coincident relation between the profile sketch plane (or planar face) and the start point of the path. This is a very simple matter of using the Add Relations command, discussed in Chapter 3. It should be noted that the start of the path does not have to touch the profile, though it usually does. If there is some distance between the sweep path and profile, that distance will be maintained during the sweep.

The second steadfast rule you must adhere to is that the path must never self-intersect. This means that if the path so much as touches itself at any point, the sweep will not work. This does not mean to say that a sweep path must be open. For example, a circle can actually be used as a sweep path. Sweeping a circle about a circle is one method of creating a toroid (donut shape).

Another illegal situation that is very similar to a self-intersecting path is a self-intersecting sweep. Not only can the path not intersect itself, but the actual sweep must not create self-intersecting geometry. This can occur in a number of ways. Figure 9-4 shows a simple example of this.

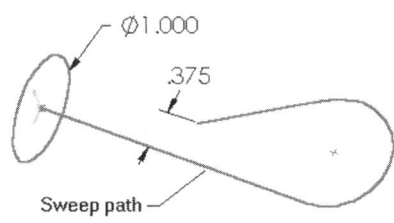

Fig. 9-4. This would result in a self-intersecting sweep.

It should be blatantly obvious that by sweeping the circle along the path shown in figure 9-4, the geometry would self-intersect. The .375-inch dimension could actually be a lot larger, say .750 inch, and the sweep would still self-intersect. This is because the distance between the two closest points of the sweep path should be greater than twice the radius of the circle, or 1 inch. Otherwise, the geometry created by the circle will touch. This is what is meant by self-intersecting geometry.

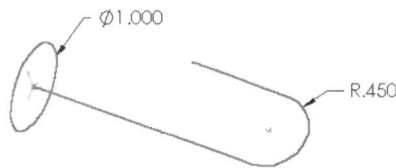

Fig. 9-5. More self-intersecting sweep sketch geometry.

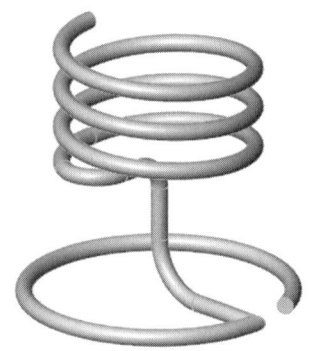

Fig. 9-6. A candle holder.

A situation that is not as obvious regarding self-intersecting geometry can also arise when sweeping along a path with arcs in it. If you are very observant, you may have thought of this when looking at figure 9-4. In figure 9-5, the example has been simplified so that you may more easily visualize what can befall your sweep.

In a situation such as this, the circle's radius is greater than the radius of the arc it is sweeping along. As the profile circle moves along the arc in the path, it overlaps itself. If the arc on the path had a radial dimension of .501 inch instead of .450 inch, the sweep would work.

Just what can you get away with as valid path geometry? Quite a bit, really. The path does not have to be perpendicular where the sweep starts. It does not have to be tangent along its entire length. As a matter of fact, the path can contain angled lines at less than 90 degrees. The path can even be a spline existing in 3D space. Figure 9-6 shows an example of what can be accomplished with a swept feature. The image is of a candle holder. It is a single swept feature, and you will learn how to create this model before this chapter is through.

Now that you are aware of what constitutes valid sketch geometry for a sweep path and profile, we can get down to the business of creating an actual swept feature. How-to 9-1 shows you the actual steps involved in creating a swept feature. Subsequent material explores some of the options available during the sweep process. It should be pointed out that the sweep can be a boss or cut. The process is identical except for the menu picks.

➤ **NOTE:** *Swept features require that the final sketch is exited prior to the Sweep command becoming available. Exit out of any active sketch by clicking on the Sketch or Rebuild icon.*

How-To 9-1: Creating a Sweep

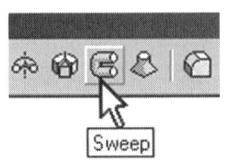

To create a swept feature, perform the following steps. Remember that you must exit out of any sketch you may have created prior to accessing the Sweep command.

Fig. 9-7. Sweep icon.

Swept Features

Fig. 9-8. Sweep panel.

1. Select Insert > Boss > Sweep, or click on the Sweep icon, shown in figure 9-7. If the feature is the first base feature, the menu picks will be Insert > Base > Sweep. If the swept feature is to be a cut, the menu picks will be Insert > Cut > Sweep. This accesses the Sweep panel, shown in figure 9-8.

2. Select the sweep Profile.

3. Select the sweep Path.

4. Click on OK to create the swept feature.

It is possible to select profile and path sketch geometry from either FeatureManager or the work area. This makes absolutely no difference in the outcome of the feature. It is okay to select either the profile or path first. Do whatever is easiest for you, but be aware that the list box that is active (salmon colored) will display whatever you select. It is best to activate a particular list box by clicking inside the box prior to selecting geometry.

Using Existing Edges as a Sweep Path

It is true that both a sweep profile and path are needed to complete a sweep. However, the sweep path does not have to be a sketch. The path can be an existing model edge, which simplifies the process of creating the swept feature because there is one less sketch that must be manually created.

Fig. 9-9. The casserole dish.

Figure 9-9 shows a casserole dish to which a swept cut will be applied. In this example, a small circular cut will be added around the inside of the dish to make room for a seal on the cover. Because the base feature of the dish is centered on the origin point, there are two planes that pass through the center of the model. This will prove very convenient, because it gives us planes on which to sketch. The sweep profile sketch will already be oriented correctly with respect to the path.

The sweep profile sketch is a simple circle in this example. The sketch used is shown in figure 9-10. A dimension controls the circle's size, but what is controlling its location? It is a geometric constraint that holds the circle in position. In particular, a pierce relation has been added between the circle's centerpoint and the edge of the dish. The pierce relation was examined in Chapter 6, so it will not be addressed again here. However,

Chapter 9: Springs, Threads, and Curves

pierce relationships can prove very useful when sweeping, so if you need to refresh your memory on how they work, now would be a good time to do so.

When the Sweep command is initiated, there will be an additional option, which will appear in the Sweep panel when an edge is selected as the sweep path. This option is known as the *Tangent propagation* option, shown in figure 9-11.

When *Tangent propagation* is enabled, the sweep will continue along the selected edge as long as there are tangent edges that come after it. As long as SolidWorks keeps finding tangent edges, the sweep will keep on going. When it comes to the end of the last tangent edge, the sweep will terminate at that point. Without the *Tangent propagation* option turned on, the sweep will continue until it gets to the end of the selected edge only.

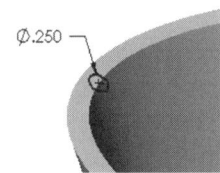

Fig. 9-10. Sweep profile sketch.

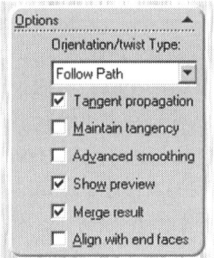

Fig. 9-11. Tangent propagation option.

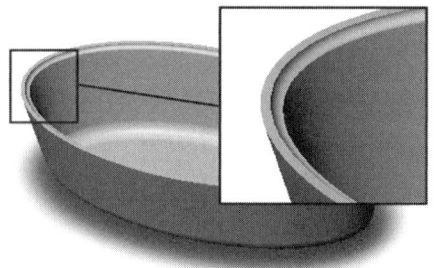

Fig. 9-12. Completed swept cut.

Because more than one edge cannot be selected for a sweep path, the *Tangent propagation* option often plays an important roll. Figure 9-12 shows the swept cut created in the casserole dish. In this particular example, the *Tangent propagation* option was not required because the edge being used as the path was one continuous edge anyway.

The edge used as the path is noteworthy in that it is a closed profile. As was mentioned earlier, sweep paths do not need to be open, as is evident by the casserole dish example.

Align with End Faces Option

The *Align with end faces* option is useful for situations in which the sweep path does not completely allow the profile to cut the desired material away during a swept cut. The following example demonstrates this situation. Figure 9-13 shows a typical ogee fillet cut, which can be made with a router. Note the leftover material at the end of the swept cut.

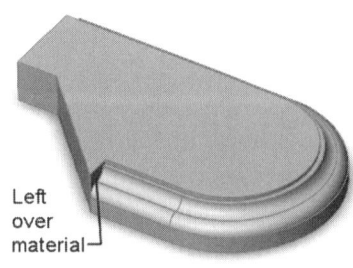

Fig. 9-13. Remaining material after a swept cut.

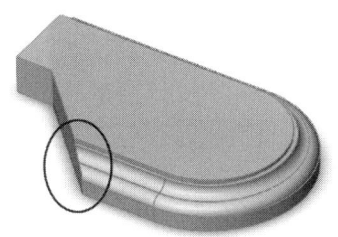

Fig. 9-14. After turning on Align with end faces.

When the *Align with end faces* option is turned on, the sweep continues beyond the end of the sweep path. The end face is typically the next face encountered on the part. Note that turning on the *Align with end faces* option cleans up the remaining scrap of material on the part, as shown in figure 9-14. Without the *Align with end faces* option, it would have been necessary to manually create another feature to perform this task for us.

Additional Sweep Options

Most of the other basic sweep options are easy to explain, and do not require any additional examples or illustrations. The options being referred to are those that can be seen in figure 9-11. One exception would be the Orientation/twist Type option, explained in detail in material to follow. First, however, let's explore the functionality of remaining options.

Maintain Tangency

This is a commonly misunderstood option. When *Maintain tangency* is checked, tangent segments in the sweep profile will result in the corresponding faces of the model being tangent. In other words, if a line and arc in a sweep profile are tangent before being swept along some path, the corresponding faces that are created will also be tangent. This would not necessarily be so if *Maintain tangency* were not checked.

Advanced Smoothing

When *Advanced smoothing* is turned on, any faces created from circular and elliptical arcs in the sweep profile will be created using a different mathematical process. This process is one of approximation. This will result in smoother faces in the swept feature, but is not as mathematically accurate. Sketch arcs may be converted into splines, which may make the geometry more difficult to work with later on. The general rule is to leave *Advanced smoothing* off unless it is necessary for cosmetic reasons.

Show Preview

Just as it sounds, the *Show preview* option will force SolidWorks to show a preview of the swept feature. During complex sweeps, it may take some time to complete the sweep, and the preview will also take a proportionately long time to generate. In such cases, once you become more familiar with the software you will probably want to leave the *Show preview* option off. For now, however, leave it on so that you can see what the feature will look like.

Merge Result

Merge result is only something to worry about if working with multiple bodies. By default, this option is left on, and that is the way you should leave it unless working with a design process that requires otherwise. If it becomes necessary to create a swept feature as a separate body from the rest of the model, uncheck *Merge result*.

Orientation and Twist Control

The Orientation/twist Type option allows for controlling how the sweep profile is oriented as it is swept along the path. It also controls the twist exhibited by the swept feature if guide curves are used. Guide curves were discussed in Chapter 6, in reference to lofted features, and are further explored in this chapter. Guide curves can play a very significant role in the outcome of a swept feature.

The first two options in the Orientation/twist Type drop-down list will be options employed most often when sweeping, with Follow Path being the most common. The last two options are used much more rarely, but are described for your reference. All four of the options are described in the following material. A portion of the Sweep panel is shown in figure 9-15.

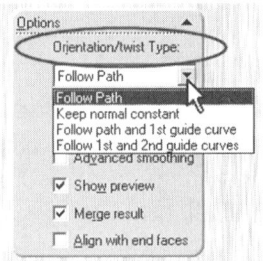

Fig. 9-15. The Orientation/twist Type option.

Follow Path

The Follow Path option is the default when performing a sweep. It is the safest option to use, meaning that it is less likely to create an error in the geometry while sweeping. It is also by far the most commonly used Orientation/twist Type setting. When Follow Path is used, the sweep profile remains at a constant angle to the path along the entire length of the path. An example of this option in use is shown in figure 9-16. The original sketch geometry of the path and profile is shown in the same image.

Fig. 9-16. Using the Follow Path option.

The Follow Path option is used when creating springs or threads. It is also used when creating piping or wiring, or in any situation in which a constant cross section is required.

Keep Normal Constant

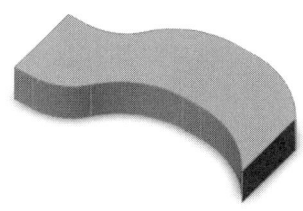

Fig. 9-17. Using the Keep normal constant option.

When the *Keep normal constant* option is used, the sweep profile remains parallel to its original sketch plane as it is swept along the path. This option is not quite as common as the Follow Path option, but it does have its uses. An example of a swept feature created with the *Keep normal constant* option is shown in figure 9-17. The same sketch geometry used in the previous example (figure 9-16) is being used here.

As is evident in figure 9-17, the orientation of the profile plane never changes. The profile is always parallel to its original sketch plane along the entire path. This option sometimes makes it impossible to complete a sweep. For example, if a bend in the path were greater than 90 degrees, the sweep would intersect itself. This is not allowed in SolidWorks. Use a little common sense with the *Keep normal constant* option.

Follow Path and 1st Guide Curve

Follow path and 1st guide curve requires the use of at least one guide curve. The sweep profile remains at a constant angle to the path as it moves along the length of the path, just as it does with the Follow Path option. The twist of the sweep profile is based on a vector between the path and the first guide curve.

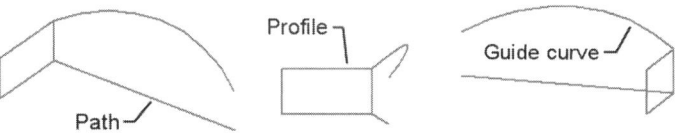

Fig. 9-18. Sketch geometry for Follow path and 1st guide curve example.

Until an example is shown, it is difficult to imagine what is happening when this option is employed. To help with the visualization process, figure 9-18 shows what the sketch geometry looks like that will be used. A couple different angles are used to help give an idea of how the three sketches are positioned in space in relation to one another.

To best understand how the *Follow path and 1st guide curve* option affects the outcome, let's first look at the geometry if the default Follow Path option were used. Figure 9-19 shows the resultant swept feature geometry from a few different angles. Note in particular how the guide curve changes the shape of the swept feature.

Fig. 9-19. Results using the Follow path *option.*

The geometric relations on the original sketch profile geometry will play an important role in the final shape. For example, there was a horizontal relation at the top of the rectangle, but there was no vertical relation on the right side. When the rectangle was swept, the guide curve forced the rectangle to change shape within the constraints applied to it. Another significant contributing factor is a pierce relation that was placed between the top right-hand corner of the rectangle and the guide curve, which forced that point on the rectangle to follow the guide curve.

Now let's examine what happens when the *Follow path and 1st guide curve* option is used. Figure 9-20 shows the same part as in figure 9-19, with the only change being the *Follow path and 1st guide curve* option used instead of Follow Path. Once again, we will examine the model from a few different angles. The changes are easily noticeable.

Fig. 9-20. Results using the Follow *path and 1st guide curve* option.

The rectangle's left side in the example in figure 9-20 is twisting upward. For the most part, figure 9-20 looks very much like the geometry in figure 9-19, except for the twisting taking place.

Follow 1st and 2nd Guide Curves

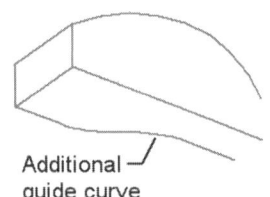

Fig. 9-21. Adding a second guide curve.

The *Follow 1st and 2nd guide curves* option requires the use of two guide curves. Once again, the sweep profile remains at a constant angle to the path as it moves along the length of the path. The twist of the sweep profile is based on a vector between the first and second guide curves. Once again, let's use illustrations to help understand what is happening during the sweep process.

Swept Features

Because a second guide curve will help demonstrate this option, some new sketch geometry is in order. Figure 9-21 shows the additional guide curve. Although it is difficult to tell from the illustration, the second guide curve is actually coplanar with the path.

Fig. 9-22. Sweeping with two guide curves using the Follow path option.

Fig. 9-23. Sweeping with two guide curves using the Follow 1st and 2nd guide curve option.

The three views in figure 9-22 show the outcome of the sweep using the Follow Path option. Figure 9-22 points out the difference between using the Follow Path and the *Follow 1st and 2nd guide curve* options. It also gives you a chance to see what effect multiple guide curves can have on a sweep. Keep in mind that the original sweep profile was nothing more than a simple rectangle, and you begin to appreciate what guide curves can accomplish.

Finally, let's look at one last example, in which the *Follow 1st and 2nd guide curve* option has been used. Figure 9-23 shows the same swept feature, now using the *Follow 1st and 2nd guide curve* option instead of Follow Path. In this example, the twist is determined via a vector between the first and second guide curves, as opposed to a vector between the path and the first guide curve (shown in figure 9-20).

Sweeping with Guide Curves

This section will not go into a lot of detail, in that guide curves were discussed in Chapter 6. However, guide curves are important enough when sweeping that at least one more demonstration is in order. It should be noted once again that although guide curves are commonly used when sweeping, the *Follow path and 1st guide curve* and *Follow 1st and 2nd guide curve* options are not.

Guide curves have the ability to change the size and shape of a sketch profile as it is being swept along a path. The path itself can be a simple line, and as long as guide curves are used, the outcome can be dramatic. Guide curves work best if the pierce relation is used in conjunction with the guide curves. Often, the pierce relation is a requirement.

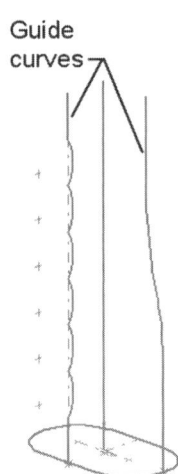

Fig. 9-24. Sketch geometry for the shampoo bottle.

Probably the most important aspect of using guide curves that you should try to commit to memory is that the sketch profile should be pierced by the guide curves, not the other way around. This means that you should create the guide curves prior to the sketch profile. That way, the curves are already defined and in place. When the sketch profile is created, points on the profile can then be pierced by the guide curves.

To understand how much of an impact guide curves can have on a swept feature, we will create a hypothetical shampoo bottle. The design intent of this bottle will be that the bottle should retain symmetry front to back and side to side. The bottle will be less wide at the top than at the bottom, as seen from the front. Seen from the side, the bottle will not taper. Rather, there will be a set of large scallops centered on the bottle vertically. These scallops will be positioned on the front and back of the bottle only. This can all be done in one sweep, and the sketch geometry that will be used is shown in figure 9-24.

The sweep profile is a simple obround shape, and the path is nothing more than a straight line, as can be discerned from figure 9-24. The guide curves define the precise shape the bottle will take. If the outer shape for the bottle is known and can be defined through dimensioned sketch geometry, the shape of the sweep can be precisely controlled and the exact desired shape can be obtained. Obviously, this can be quite significant to a designer. Figure 9-25 shows the outcome of the sweep operation.

One more point worth mentioning is that guide curves can be created in a variety of ways. Guide curves can be simple 2D sketch geometry, or curves created using the 3D sketcher, 2D or 3D splines, created via user-defined coordinates, or by some other means. You will learn how to create curves in general using a variety of techniques in this chapter, some of which will be useful for establishing guide curves.

Fig. 9-25. Base sweep feature of the shampoo bottle.

Helical Curves

A helical curve is a prerequisite to creating a spring or a thread, both of which require a sweep operation. Before you can create a helix in SolidWorks, you must first make a circle. After that, it is pretty much just a matter of plugging in the parameters that define the helix.

The circle defines the diameter of the helix and helps to determine the start point of the helical curve. Like everything else in SolidWorks, a helix is parametric. You can change the number of revolutions and the

Helical Curves

pitch and height of a helix by editing its definition or accessing its dimensional values by double clicking on the helix feature.

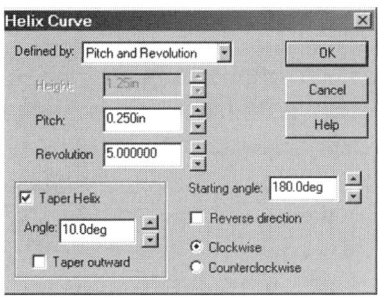

The Helix Curve window, shown in figure 9-26, is used to define a helix. Usually the first thing you would do after bringing the Helix Curve window up is to determine how you want to define the helix. You have the following three options.

- Pitch and Revolution
- Height and Revolution
- Height and Pitch

Fig. 9-26. Helix Curve window.

Creating Helixes

Once you have determined how you want to define the helix, it becomes a matter of plugging in the appropriate information. The pitch is the distance between revolutions. Height is the overall height of the helix curve, and revolutions are the number of total turns the helix makes.

The Reverse Direction option specifies in what direction the helix is going away from the circle, such as up or down. The Clockwise and Counterclockwise options are self-explanatory. If creating threads from a helix, you might equate the clockwise or counterclockwise options with regular or reverse (left-hand) threads, respectively.

SolidWorks also provides the ability to create a tapered helix, as shown in figure 9-27. This is very convenient for creating pipe threads. If you do create a tapered helix, make sure you specify an angle for the taper, and whether the taper is inward or outward.

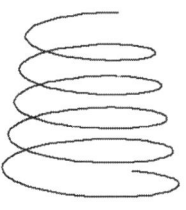

Fig. 9-27. Creating a tapered helix.

The starting point of the helix is more important than it might first seem. For example, if you were creating threads, you would probably want the threads to begin at a particular point. Using the *Starting angle* setting, you could rotate the helix so that it would coincide with a precise location. By clicking on the arrows to the right of the *Starting angle* value, you can see the helix rotate clockwise or counterclockwise about its axis. It is also possible to simply type in a value from 0 to 360.

If creating a simple spring, adjust the start point of the helix so that it coincides with one of the default planes. This assumes that the circle defining the helical curve is centered on the origin. The reason for this has to do with one of the steadfast rules of sweeping: the sweep path must begin on the plane of the sketch profile. If you will be using the helix as a

356   Chapter 9: Springs, Threads, and Curves

sweep path, which is the typical reason for creating a helix, this becomes very important.

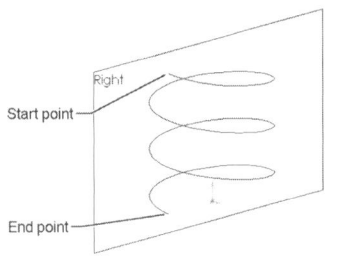

Fig. 9-28. Start and end points of a helix.

Continuing with a simple spring as an example, it is possible to set the start angle of the helix to 0, 90, 180, or 270 degrees. The start point of the helix will then fall on one of the standard planes (such as Front or Right). In figure 9-28, the start angle was set to 0 degrees. Because of this, the start point of the helix coincides with the Right plane. It would be possible to sketch on the Right plane in this case and use the helix as a sweep path. Incidentally, the circle used to define the helix in figure 9-28 was sketched on the Top plane.

What happens when it is not possible to create the helix so that it starts or ends on one of the default planes? Actually, this situation is common when creating threads, and has an easy fix. Create a plane at the end of the helical curve where you want to start the sweep. Use the Normal to Curve method of creating a plane, shown in figure 9-29. This command option is one of the Reference Plane creation options discussed in Chapter 4.

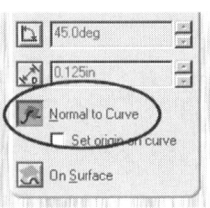

Fig. 9-29. Using Normal to Curve to create a reference plane.

If you recall from Chapter 4, there is a shortcut that makes creating a sketch plane on a curve easy. It is only necessary to select the helix near (but not on) the endpoint where you want the plane to exist, and then click on the Sketch icon, to establish a plane at that location. Using this technique, it is not even necessary to start the Reference Plane command.

It should be noted that if the cross section of the swept feature being created is critical, or is part of the design intent, a plane normal to the helical curve would be required. You will step through the process of creating a spring later in this chapter. How-To 9-2 takes you through the process of creating a helical curve.

How-To 9-2: Creating a Helical Curve

Fig. 9-30. Helix icon.

To create a helical curve, perform the following steps. Note that the Helix icon, shown in figure 9-30, can be found on the Curve toolbar.

1. Create a sketch that contains a circle. The circle will define the diameter of the helix.

2. Select Curve > Helix/Spiral from the Insert menu, or click on the Helix icon.

3. In the *Defined by* section, specify how you would like to define the helix.

4. Specify the Height, Pitch, or Revolution parameters as required.

5. Specify a *Starting angle* of 0 to 360 degrees. If 0 degrees is required, the value must be typed in.

6. Select CW or CCW.

7. Check *Reverse direction* if necessary.

8. Click on OK to create the helix.

If it is necessary to create a tapered helix, make it a point to place a check in the Taper Helix option. It will then be possible to enter an angle for the taper value and to specify whether or not the taper is outward or inward.

Spirals

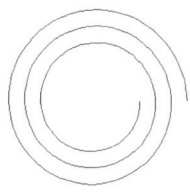

Fig. 9-31. A spiral curve.

While exploring the Helix Curve window, you may have noticed that there is an option for creating a spiral curve, shown in figure 9-31. This option is hidden away in the *Defined by* list. Spirals are similar to the coils of an electric stove burner or the mosquito coils some people burn to keep those pesky critters at bay. Spiral curves can also be used to create coil springs, such as those found in a carpenter's tape measure or a mechanical watch.

Spirals are defined by pitch and revolution only, and have no height. The pitch, in the case of a spiral, is the distance between coils. The *Reverse direction* option refers to whether the spiral winds its way outside the circle or inside the circle. If inside, you must use some common sense as to the number of revolutions and the length of the pitch. If the spiral winds its way in too far, it will have a negative radius. This is not allowed. This problem does not exist in the case of an outward-flowing spiral.

Springs

Once you know how to create a swept feature and define a helix, creating a spring is a piece of cake. Follow your fundamental rules regarding sweeping and you should do fine. You actually have already learned everything you need to know to create a simple spring. For the sake of convenience, the entire process is broken down into segments in the sections that follow, for easy reference so that you can try working through the process. If necessary, refer to previous sections of this chapter to gain a more complete understanding of each phase of the process. It is suggested that you follow along in the sections that follow.

Phase 1: Starting with a Circle

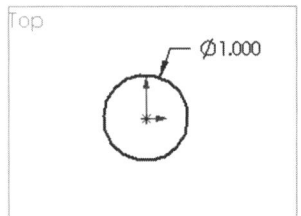

Start a new part and set the units to inches. Start a new sketch on the Top plane, and then sketch a circle with a diameter of 1 inch, as shown in figure 9-32.

Fig. 9-32. The circle that will define the helix.

Phase 2: Defining the Helix

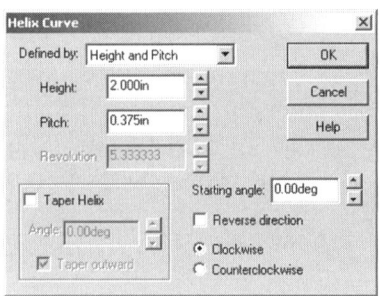

Fig. 9-33. Settings used to define the helix.

Enter the Helix command using either the Helix icon or the menus, and define the helix. Usually, you would be editing the sketch where you just created the circle prior to initiating the Helix command. If you have exited the sketch, just make it a point to select the sketch first from FeatureManager, which will make the Helix command accessible.

In the Helix Curve window, enter the parameters to define the helix. Let's use height and pitch settings, with a height of 2 inches and a pitch of .375 inch. Type in a starting angle of 0 degrees, and accept the default setting of Clockwise. Leave all other settings at their default values. The settings used in this example are shown in figure 9-33.

Phase 3: Creating a Sketch Plane

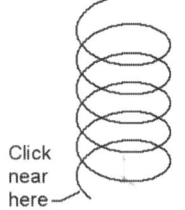

Fig. 9-34. Preparing to establish a sketch plane.

At this point, you should have a helix on your screen. It is necessary to establish a sketch plane next. If you have never actually tried the Normal to Curve plane creation shortcut we are about to use, you will enjoy this next sequence of steps. To create a new sketch plane for the sweep profile, select near the start point of the helical curve, as shown in figure 9-34.

Phase 4: Sketching a Sweep Profile

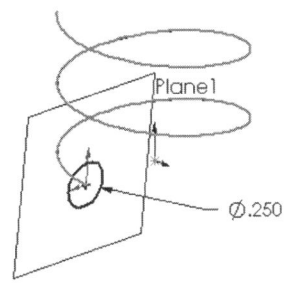

Fig. 9-35. Creating the sweep profile.

Click on the Sketch icon immediately after performing phase 3. This will do two things: create a plane at the start point of the helix and place you in sketch mode. At this point, the sweep profile for the spring can be created. Figure 9-35 shows the new sketch plane and the completed profile. For our example, a circle with a diameter of .25 inch is used. Make sure to add a pierce constraint between the circle's centerpoint and the helix so that the circle will be properly positioned and fully defined.

Phase 5: Exiting the Sketch and Creating the Sweep

Fig. 9-36. The completed spring.

You are almost done at this point. Exit the sketch, as that is a prerequisite to creating a swept feature, and then enter the Sweep command. Pick the circle as the sweep Profile, and the helix as the sweep Path. Click on OK to complete the sweep and you should wind up with a spring on your screen, such as that shown in figure 9-36.

The helix is parametric, just like anything else in SolidWorks. By double clicking on the helix feature, access to the underlying dimensions is gained. This includes the helical curve's height, pitch, and number of revolutions. Even though only two of those parameters were used when defining the helix, any of the parameters can be modified by double clicking on the helix feature in FeatureManager. Because the helix has been absorbed by the newly created swept feature, expanding the swept feature will allow for gaining access to its constituent components. Use the small plus sign (+) to the left of the Base-Sweep feature.

Composite Curves

There are many occasions, especially when working with swept features, when the ability to join a number of sketches would prove beneficial. Being able to join existing feature edges to form a single sweep path would also prove useful. This capability exists within SolidWorks and is the domain of the Composite Curve command.

The Composite Curve command, found in the Curve menu, allows for joining feature edges, sketch geometry, or various curve types to form a single curve. This single curve can then be used for other SolidWorks

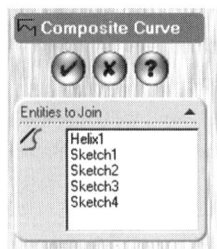

functions. For example, a composite curve is commonly used as a sweep path, but could be used as a guide curve as well.

Implementing the Composite Curve command is ridiculously simple. The Composite Curve panel, shown in figure 9-37, is very basic. When the items to be joined as a composite curve are selected, they appear in the list box area. To create a composite curve, follow the steps outlined in How-To 9-3.

Fig. 9-37. Composite Curve panel.

How-To 9-3: Creating a Composite Curve

Fig. 9-38. Composite Curve icon.

To create a composite curve, perform the following steps. The Composite Curve icon, which is found on the Curves toolbar, is shown in figure 9-38.

1. Select Curve > Composite from the Insert menu, or click on the Composite Curve icon.
2. Select the items to be joined into a composite curve.
3. Click on OK to create the curve.

There is only one major requirement when creating a composite curve, and that requirement is that all selected objects must form one continuous string. In other words, the objects must be connected end to end. There can be no openings between selected objects. The order in which items are selected is not a consideration and makes absolutely no difference in the outcome.

In Exercise 9-1 you will create the candle holder shown in figure 9-6. This will require creating a helical curve and using the Composite Curve command to add sketch geometry to the helix. The resultant geometry will be used as a sweep path. This exercise will help strengthen your knowledge in the area of sweep paths and swept features.

Exercise 9-1: Creating the Candle Holder

Start a new part and save it with the name *Candle Holder*. Set the units to inches. You will begin by creating the individual sketches that make up the main base of the candle holder. Each sketch will be a simple 2D sketch, although you will learn how to use the 3D Sketch command later in this chapter. When joined using the Composite Curve command, the

Composite Curves

2D sketch geometry and a helix will form a complex 3D path for the swept feature. Use the illustrations as guidance for completing this exercise.

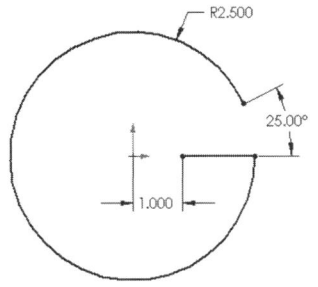

Fig. 9-39. Creating the first sketch for the candle holder.

1. Create the sketch shown in figure 9-39. The sketch consists of a line and an arc. Make sure to fully define the geometry.

2. Exit the first sketch.

3. Create a new sketch on the Front plane. Sketch the geometry shown in figure 9-40. Add tangent relationships to the arc to fully define it. The arc must be coincident to the end of the first sketch, which is also visible in figure 9-40.

4. Exit the second sketch.

5. Start a new sketch on the Right plane and create the simple sketch shown in figure 9-41, which shows the completed sketch as seen from a Left view. Once again, tangent relationships between the arc and adjacent geometry are critical to the design intent of the model. Make sure the start-point arc is coincident to the endpoint of the line in the second sketch.

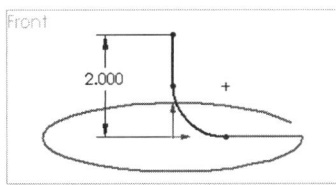

Fig. 9-40. Creating the second sketch on the Front plane.

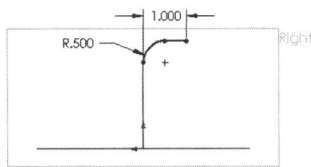

Fig. 9-41. Developing the third sketch.

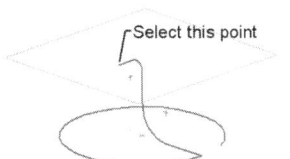

Fig. 9-42. Defining a new reference plane.

6. Exit the third sketch.

You should be noticing a common theme by this point. The process is fairly repetitive, and the sketch geometry is simple, which makes the entire process very straightforward and easy to carry out. The procedure begins to get a little trickier at this point. A plane must be created for the next sketch, and the sketch geometry itself will be slightly more complex. A construction circle will be used in this next sketch, which will serve two functions. It will serve to define the location of an arc, and it will be used to define a circle for the helix. Continue with the following steps to see how this will pan out.

7. Define a plane using the Parallel Plane at Point option. (Hint: if you forgot how to access the Reference Plane command, click on Insert > Reference Geometry.) Use the Top plane and point shown in figure 9-42 to define the new plane.

8. Rename the new plane *helix plane*.

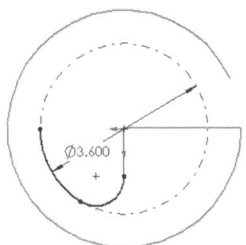

Fig. 9-43. Developing the fourth sketch.

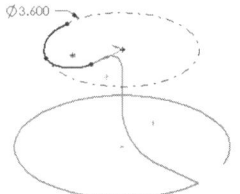

Fig. 9-44. The fourth sketch from another angle.

9. Start a sketch on the helix plane.

10. Create the sketch geometry shown in figure 9-43. The existing geometry from the first three sketches is also shown, so do not confuse that with what is new. The new sketch should contain two (2) arcs and a construction circle. (Hint: to draw a construction circle, sketch a circle, and then use the Construction Geometry icon.) Figure 9-44 shows the same sketch from another angle, to help you understand how the geometry is positioned.

11. Exit the sketch.

12. Start a new sketch on the helix plane once again.

13. Select the construction circle from the fourth sketch and convert it to a sketch circle using the Convert Entities sketch tool. The construction circle will be converted into a regular circle at this point.

14. Enter the Helix command.

15. Use the following parameters to define the helix:
    — Pitch = .75 inch
    — Revolution = 3
    — Starting angle = 270 degrees
    — Clockwise is selected
    — Reverse Direction and Taper Helix are both unchecked

16. Click on OK to create the helix once all parameters have been specified.

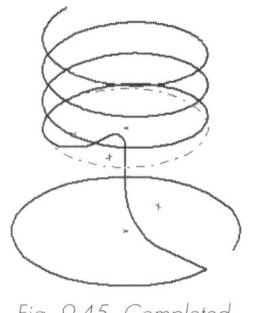

Fig. 9-45. Completed helix and sketch geometry.

You should now have something similar to that shown in figure 9-45. FeatureManager should list four separate sketches, along with a helix. The helix plane and other usual items will be listed also, but they do not concern us. What needs to happen next is to join the four sketches and the helix to form a single curve.

17. Enter the Composite Curve command.

18. Select the four sketches and the helix. This can be done via the work area or FeatureManager. (Hint: you can use the flyout FeatureManager by clicking on the name of the command, Composite Curve, at the top of PropertyManager.)

Composite Curves

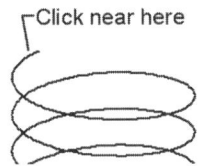

Fig. 9-46. Establishing a sketch plane for the sweep profile.

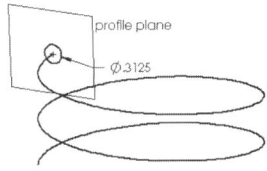

Fig. 9-47. Creating the sweep profile.

Fig. 9-48. Final swept feature.

19. Click on OK to complete the composite curve. The result will look similar to that shown in figure 9-45, but will be a single curve.

20. Using the by-now-familiar shortcut to establish a sketch plane, click near the end of the composite curve shown in figure 9-46, and then click on the Sketch icon.

21. Create a circle with a diameter of .3125 inch. The circle will be used as the sweep profile, and is shown in figure 9-47. The sketch plane has been renamed profile plane in the illustration.

22. Add a pierce constraint between the center of the circle and the helix. It is important to select the helix near the end of the helix but not the actual endpoint. The area where the helix is selected is also significant because theoretically there are different locations on the helix that could pierce the circle's centerpoint.

23. Exit the sketch.

24. Enter the Sweep command.

25. Select the circle as the profile, and the composite curve as the path.

26. Leave all default settings and click on OK to complete the sweep.

27. The final feature should look similar to that shown in figure 9-48. Feel free to add any other finishing details, or make some modifications to the model if you like. For example, you may decide to add a sketch fillet to the first sketch to remove the sharp corner from the sweep path. Just make sure you use a radius larger than the object being swept (the .3125-inch-diameter circle). You could also add a slight outside taper to the helix, or perhaps round off the upper end of the swept feature with a revolved feature. This would help keep wax from being scraped off as a candle is rotated into the holder.

Threads

The only real difference between springs and threads is the complexity of the sketch geometry for the profile. It does not take much to create a spring profile, especially when it is a round-wire spring. Threads, on the other hand, usually require a little more finesse.

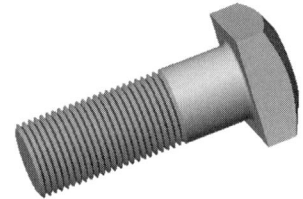

Fig. 9-49. Threads: real or fake?

The first question is: Do you really need to show actual threads? Most often the answer is no. There are alternatives to creating actual helical threads that require much less processing power and that render features that look like threads for all intents and purposes. Figure 9-49 shows the result of one such alternative.

The discriminating person can tell that the threads in this illustration are indeed simulated. There is no helical sweep being used. Real helical threads are only needed for very accurate visual renditions of a product, at least for parts such as fasteners. They are not usually needed for rapid prototype models, because threads are often machined into the prototype anyway.

Creating Threads

The threads in figure 9-49 were created by sketching the profile of one thread near the end of the bolt. The profile was then revolved 360 degrees as a cut-revolve feature, then patterned linearly up the bolt. In this case, the width of the thread and the distance between instances in the pattern were the same. This makes for a sharp thread, with no flats at the peak of each thread.

There are times, however, when nothing short of real threads will do the trick. Creating the threads on the neck of a soda bottle might be a good example. These are usually threads of a particular design, and cannot be called out with a simple note. Rather, they need to be modeled and shown in a design drawing. The procedure for creating this type of thread is similar to creating a spring. The mechanics are the same.

Assume you have a new soda bottle or something similar you are designing. You can use a simple cylinder for practice if you want to follow along with the book. Your dimensions and profile might be different, but the procedure is one you can use anytime you need to create threads.

You already know the beginning of the routine. You must create a circle, and then define a helix (the sweep path). After that, you can create the thread profile (sweep section), and then create the threads. If the threads are to start at a specific position from the top of the bottle, that is where you must sketch the circle to define the helix. Therefore, create an offset plane (per Chapter 4) where you want the threads to begin.

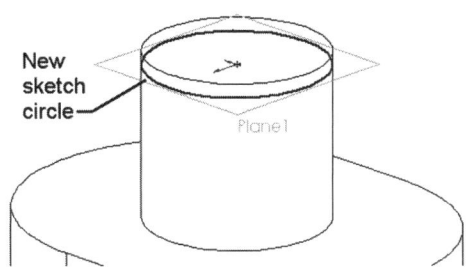

Fig. 9-50. Creating a circle at the proper height.

Figure 9-50 shows the offset sketch plane and the sketch circle. The plane was offset .075 inch below the top of the neck. The circle that will define the helix was created using the Convert Entities icon. This is by far the easiest method to use for creating the circle, and the threads will always be the right diameter, even if the diameter of the neck changes. The selected top edge of the neck was projected to the sketch plane, which is the nature of the Convert Entities command.

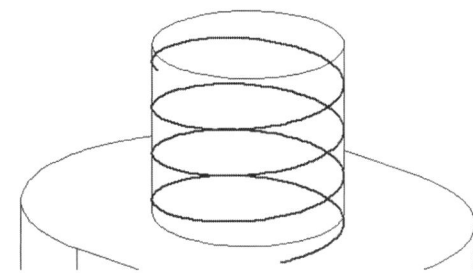

Fig. 9-51. A helix on the neck of a bottle.

Now define the helix. You have been through this a couple of times if you have read through the previous sections in this chapter. Plug in the values necessary to define the helical curve that meets your requirements. Figure 9-51 shows a possible example. In the example shown, the helix was given enough revolutions so that any threads created will completely embed themselves into the rest of the model. This is certainly not a requirement, and the length of the threads you create will of course be dependent on your design requirements.

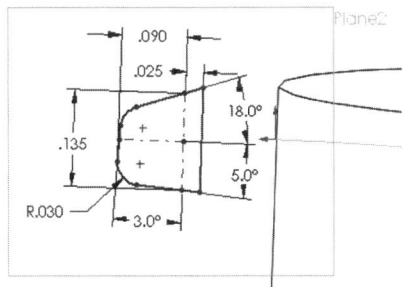

Fig. 9-52. Sketch profile for the threads.

Next it is simply a matter of defining a plane at the end of the helix and creating a profile for the threads. The profile should match the thread cross section, obviously. Figure 9-52 shows an example of what a thread cross-section profile might look like. What still needs to be accomplished in the example shown is to locate the profile in some way. A pierce relation is not required, but it works well for situations such as this.

A point that should be made has to do with the way in which the sketch profile was created in figure 9-52. Note in particular how construction geometry (centerlines) are being used. One centerline has been dimensioned 25 thousandths of an inch from the right side of the profile. The top endpoint of this centerline is where the helix will pierce the centerline, and the centerline has also been made parallel to the silhouette edge of the bottle's neck.

All of this effort to create the profile may seem unnecessary, but in fact it is very important. As the profile is swept along the helix and around the neck of the bottle, it is of the utmost importance that no air gaps occur under the threads. Any miniscule twisting of the sweep profile that might occur during the sweep would cause air gaps to appear. Because of the 25-thousandth-inch overlap built into the sketch, air gaps are not a concern.

What is the overlap area actually accomplishing? It is a built-in safety net. Software is not perfect, and there is a certain tolerance built into every CAD program. By establishing an overlap area, we allow for any built-in error factor within the software to be of no consequence. The thread feature will easily blend into the neck of the bottle, and the possibility of any air gaps will be nonexistent. Figure 9-53 shows the sketch after adding a pierce constraint and exiting the sketch preparatory to performing the sweep command.

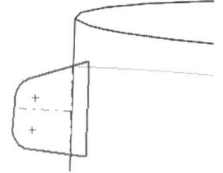
Fig. 9-53. After properly positioning the thread profile.

At this point the threads are all but done. Completing a swept feature should be familiar territory to you by this time. Complete the sweep, leaving all of the settings in their default state. The only setting that deserves some mention is the *Align with end faces* option. Turning this option on may cause the sweep to have undesirable results, usually because there are no end faces to align with. It depends on the situation, but a general rule of thumb is to leave *Align with end faces* off whenever creating threads or springs. Figure 9-54 shows what the completed thread feature would look like. Read on to see how to add some finishing touches to complete the threads.

Fig. 9-54. Completed thread feature.

Rounding Threads

Typically, you would not want to leave the ends of the thread feature with flat faces. It would be best to round the faces off somehow, or have the ends of the threads gently taper into the neck. There is a very easy technique that can be used to accomplish this.

By creating a revolved feature at the end of the threads, you can gradually taper the threads instead of having it cut off abruptly. This is accomplished by sketching on the end of the thread and converting the feature edges into sketch geometry. A centerline can then be added, which the converted geometry can revolve about. A general outline of how to accomplish this procedure runs as follows.

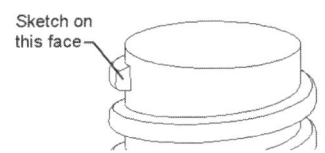

Fig. 9-55. Selecting a sketch plane.

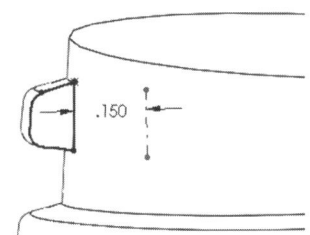

Fig. 9-56. Preparing to create the tapered end of the thread.

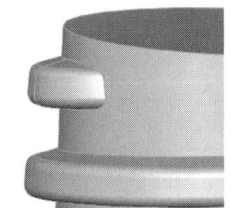

Fig. 9-57. Completed tapered thread.

Begin by selecting the face on the end of the threads where you would like to create a taper. The face at the beginning of the threads previously created, shown in figure 9-55, is a perfect example.

Next, convert all of the edges on the same face. Use the Convert Entities command. Because the face was already selected when you established a sketch plane, it would be easy to click on the Convert Entities icon after clicking the on Sketch icon. Afterward, a centerline should be created for a revolved feature. The entire flow of this process is very smooth and requires minimal effort once you have been through the routine a few times. Figure 9-56 shows the converted geometry, and the added centerline required for the revolved feature.

A dimension was created between the added centerline and the top right-hand corner of the converted geometry. A larger value for the dimension will make the thread taper off more gradually. You can experiment with this setting to obtain a look you prefer. An example of the revolved feature is shown in figure 9-57.

It may be necessary to check the Reverse option if the feature revolves in the wrong direction. As far as the total angle of revolution is concerned, a value of 90 degrees or so is fine. The main concern is that you want to make sure the taper feature makes contact with the rest of the part.

Springs and Threads: Final Comments

When creating springs, ask yourself if you really need to show them in the assembly. They usually make for a large part file. The simple spring previously shown had a file size of over 2 megabytes. This is large, especially considering there was only one feature in the part file! An alternative is to use a cylinder to represent a spring, or even just a helix itself, without creating the swept feature.

Threads will also have a tendency to make a part file quite large, and springs and threads are difficult to compute and to display because of their high polygon count. The technical reasons for this are beyond the scope of this book. The point of the matter is that you may notice rebuild times getting longer and model rotation speeds begin to slow after threads have been created, depending on the speed of your machine.

Springs and threads are fun to create, but consider whether you need true helical threads. Often you can get by with a patterned cut or even simple cosmetic threads. The decision should be made with regard to the level of performance you desire from your computer, and with regard to whether or not these threaded parts or springs will be placed in higher-level assemblies.

3D Sketcher

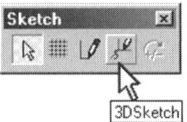

Fig. 9-58. 3D Sketch icon.

The 3D Sketch icon, shown in figure 9-58, is another way to create a curve that exists in 3D space. It should not be confused with the Sketch icon used for 2D sketch geometry. The 3D Sketch tool is different from 2D sketching in implementation, operation, and functionality.

The first major difference between the 3D sketcher and the 2D sketcher is that with the 3D sketcher a plane or planar face does not have to be selected. Click on the 3D Sketch icon and you are ready to roll. When this happens, the available sketch tools on the Sketch Tools toolbar will become active. The key word here is available. Out of all sketch entities you have learned, the only entities that can be created while using the 3D sketcher are points, lines, centerlines, and splines. A few other tools, such as Convert Entities and Sketch Fillet, are also available.

The 3D sketcher is great for creating paths for wiring or piping. As a matter of fact, the 3D sketcher is the main tool used for routing in conjunction with the Piping module. The Piping module is an add-on program for pipe layouts. Whereas the Piping module would be an extra investment, the 3D sketcher is still included with the SolidWorks software.

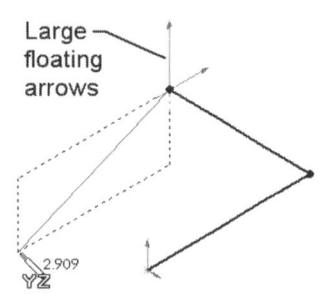

Fig. 9-59. What might be seen while using the 3D sketcher.

When using the 3D sketcher, planes can still be selected to use as reference planes when sketching, but they are selected differently than with the 2D sketcher. When the actual sketch process begins, two large red arrows appear to help show what virtual plane is being sketched on. Figure 9-59 shows an example of these arrows, which can be thought of as an oversized floating sketch origin that changes position as the sketch is being created. (Solid-Works on-line help refers to these arrows as a "space handle.") The virtual plane being sketched on can be changed by pressing the Tab key while sketching a line, for example.

Another way of altering the sketch plane being referenced is to select other planes while in the 3D sketcher. This involves clicking on the plane to be referenced in FeatureManager. This can be done after beginning a sketch, and it does not matter if a

3D Sketcher

sketch tool (such as the Line command) is active or not. The important thing to remember is that the plane must be selected in FeatureManager and not in the work area, due to the fact that you can be in the middle of a sketch command at the time you are switching reference planes.

All of this probably sounds quite confusing. In all truthfulness, the 3D sketcher is a bit cumbersome to use the first few times. It takes practice, and until you get the hang of it you may find it a bit frustrating. Keep in mind that the 3D sketcher does not function at all like the 2D sketcher. Treat the 3D sketcher as a completely new command, because it might as well be for all intents and purposes. How-To 9-4 takes you through the process of creating a 3D sketch.

How-To 9-4: Using the 3D Sketch Command

Perform the following steps to gain a better understanding of how to use the 3D Sketch command. It is highly recommended that this command be experimented with to a large degree because there just is no other way to get a good feel for working with the 3D sketcher. You may find it necessary to run through the steps a few times, because the techniques for referencing different planes are not ruled by logic, and basically require memorization.

1. Select 3D Sketch from the Insert menu, or click on the 3D Sketch icon, shown in figure 9-58.

2. Select a plane to be referenced when sketching. The virtual sketch plane will align itself orthogonally to the reference plane.

3. Select the sketch tool to use, such as the Line icon.

4. Begin sketching using traditional techniques. For example, pick and drag the endpoints of a line, always paying attention to system feedback.

5. While sketching, press the Tab key on the keyboard to change virtual planes. For example, press the Tab key while dragging to position the endpoint of a line.

6. To change the referenced plane, select a plane from Feature-Manager. This can be done while a command (e.g., Line) is active and you are between sketching line segments.

7. Add any dimensions or relations necessary. As always, it is good policy to fully define the sketch.

8. Exit the sketch when finished.

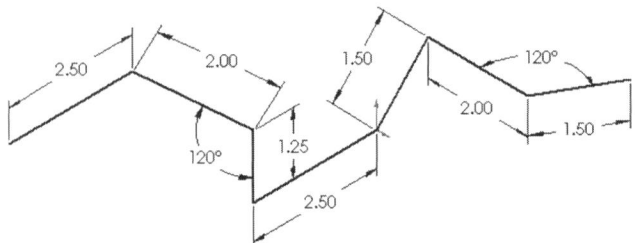

Fig. 9-60. A 3D sketch.

Figure 9-60 shows an example of what a 3D sketch might look like with all dimensions applied to it. Although it is difficult to appreciate from a 2D image in a book, the 3D sketch shown in figure 9-60 does indeed weave its way back and forth through 3D space across the computer screen. The dimensions shown are added in the same way they are added to a 2D sketch, so the topic of dimensioning does not need to be expanded on here.

Quite often when creating simple extruded or revolved features it is not necessary to exit the sketch. Simply complete the sketch, and then create the feature. A 3D sketch cannot be extruded or revolved. Quite the contrary, a 3D sketch is usually created with a higher purpose, such as for a swept feature. Thus, there is no reason to remain in the sketch after it is completed. It is a requirement, however, that you exit the sketch before it can be used toward creating a more substantial feature.

Sketch fillets can be added in a 3D sketch in the exact manner they are added in a 2D sketch. Sketch chamfers can be added as well. The Sketch Trim and Sketch Extend sketch tools will still function in the 3D sketcher, as does the Split Curve command. In that these commands have been discussed in previous chapters, they are not discussed here. Geometric relations, however, host some new options when creating a 3D sketch. These are explored in the section that follows.

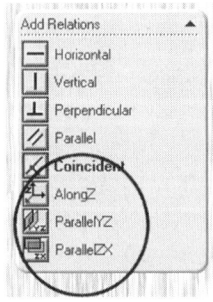

Fig. 9-61. Additional relations available in a 3D sketch.

3D Sketch Relations

If the Add Relations icon is depressed while creating a 3D sketch, three new relations will be made available, though you may not see them unless the correct combination of objects is selected. The three relations, shown in figure 9-61, are only available when working in the 3D sketcher.

The three additional relations can be used along with any of the existing and more well-known relations. 3D sketch geometry can be overdefined, just like any other sketch, so use the same care when adding relations that would be employed when working in 2D. Table 9-1 outlines the three new relations available only when working in the 3D sketcher.

Table 9-1: 3D Sketch Relations

Relation	Description
AlongZ	Two or more points can be aligned along the world Z axis. If lines are selected, the lines will align with the Z axis and be parallel to it.
ParallelZX	A line and plane must be selected. The selected plane will represent the XY plane of a coordinate system whose ZX plane the line will be parallel to.
ParallelYZ	A line and plane must be selected. The selected plane will represent the XY plane of a coordinate system whose YZ plane the line will be parallel to.

The AlongZ relation option is the most commonly used of the three relation options particular to a 3D sketch, and is fairly easy to understand. If you are one of those people that want to fully understand what the ParallelZX and ParallelYZ relation options accomplish, create some angled planes, and then in a 3D sketch draw a line that has an endpoint locked onto the sketch origin. Add one of the parallel relations between the line and an angled plane, and then try moving the line or its endpoint. This should help give you an idea of how these relations can be used.

Drawing Splines in 3D

Drawing a 3D spline using the 3D sketcher is definitely an interesting experience. It is fun, but in order to get any type of real control out of the spline, construction geometry should certainly be used. A skeleton of construction lines (centerlines) can be created, and the control points of the 3D spline can then be dimensioned or related to the skeleton in some way.

Sometimes a fully controlled spline may not be what is needed. For instance, maybe a wire going from point A to point B needs to be shown, and a free-form 3D spline would make a perfect sweep path. The path does not need to be fully defined, because the wire may just be routed through an assembly in some fairly loose fashion anyway.

If this is the case, use the 3D sketcher to roughly shape a route for the wire in your assembly and use it for a sweep path to show the wire. You may need to insert a new component into the assembly to do this, which is an advanced topic discussed in Chapter 13.

The 3D sketcher can be used for many things other than just creating wires. Exercise 9-2 takes you through the process of creating a 3D sketch used for a sweep path. You will explore some of the more common functions accessed when using the 3D sketcher.

Chapter 9: Springs, Threads, and Curves

EXERCISE 9-2: Creating a 3D Sketch

This exercise takes you through a simple example of using the 3D sketcher to create a bent tube used in an imaginary assembly. All details of specific steps for the exercise are not listed, but a general outline of the process is provided as guidance. You should have the particulars of individual steps "under your belt" from previous material. Begin by starting a new part. Make sure the Units for the part are in inches.

1. Change to an Isometric view.

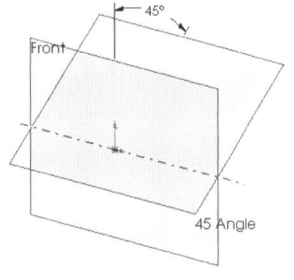

2. Using the traditional Sketch icon (for 2D sketching), sketch on the Front plane and draw a horizontal centerline out from the origin point. This will be used to create an angled plane.

3. Create a plane at an angle of 45 degrees from the Front plane and passing through the centerline, such as that shown in the isometric view displayed in figure 9-62.

4. Click on the 3D Sketch icon.

Fig. 9-62. Creating an angled plane.

5. Select the Front plane from FeatureManager.

6. Click on the Line icon.

7. Draw a line upward from the origin along the positive Y axis, approximately 2 inches. (Hint: watch the system feedback and this will be an easy exercise.)

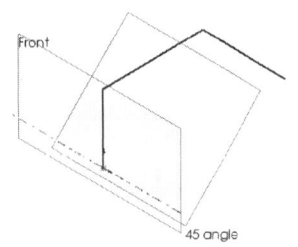

8. From the end of the last line, draw another line approximately 2 inches along the Z axis away from the screen. (Hint: you will have to hit the Tab key once.)

Fig. 9-63. The partially completed 3D sketch.

9. From the last line, draw a third line toward the right along the X axis. At this point your screen should look similar to that shown in figure 9-63.

10. Select the angled plane from FeatureManager.

11. Draw a fourth line downward along the Y axis, approximately 4 inches.

12. Draw a fifth line to the left along the X axis, approximately 5 inches.

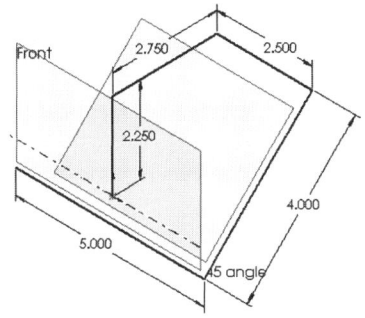

Fig. 9-64. Dimensioning the 3D sketch.

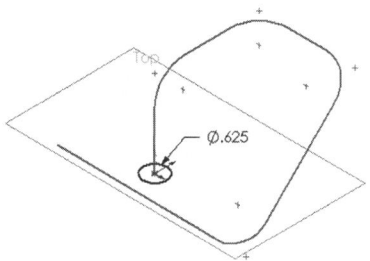

Fig. 9-65. Adding a .625-inch circle.

13. Add dimensions to the sketch, as shown in figure 9-64. The planes are shown for reference.

14. Add .750-inch fillets to all four corners of the sketch. Make sure the Keep constrained corners option is checked. There is a good chance the fillets' dimensions will look odd. That is the nature of filleting in the 3D sketcher.

15. Exit the sketch.

16. Hide the sketch containing the centerline, and any planes being shown, by right-clicking on these items in FeatureManager.

17. Using the traditional Sketch icon, sketch on the Top plane.

18. Create a circle at the origin, with a diameter of .625 inch, as shown in figure 9-65.

19. Exit the sketch.

20. Access the Sweep command.

21. Make sure Show preview is checked, and check the Thin Feature option.

22. Set the wall thickness to .080 inch inward. Use the Reverse Direction icon so that you can see which direction the wall thickness is being applied.

23. Create the swept feature.

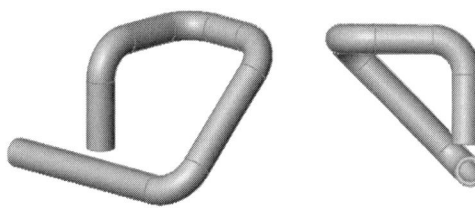

Fig. 9-66. Two views of the swept feature.

The swept feature is shown in figure 9-66. Both a dimetric and a right-side view are shown. Note in particular the right-side view, and how any object connecting with the downward facing end of the tube will almost certainly interfere with the first pipe segment. Let's remedy this problem, and add another tube branch, prior to closing the file.

24. Double click on the angled plane and change the angle from 45 degrees to 62.

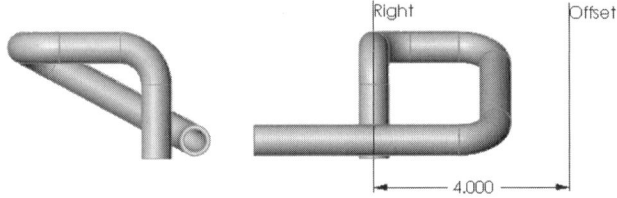

Fig. 9-67. Adding an offset plane.

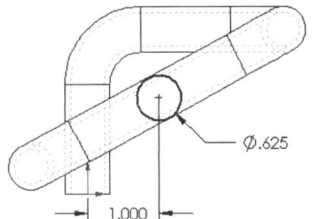

Fig. 9-68. Sketch for the additional tube branch.

25. Rebuild the model. The resultant part should look similar to that shown on the left in figure 9-67.

26. Add a plane offset 4 inches to the right of the Right plane. Refer to the image on the right in figure 9-67.

27. Start a sketch on the offset plane and create the sketch shown in figure 9-68. Try to fully define the circle. (Hint: add a relation to the temporary axis of the existing tube.)

28. Extrude the sketch up to the main body of the tube. Use the Up To Next end condition.

29. Rename this new feature *T branch*.

30. Sketch on the end of the T branch.

31. Using Offset Entities, offset a circle .080 inch from the circular edge at the end of the T branch.

32. Cut the circle into the model, again using Up To Next as the end condition.

33. Save your work.

Fig. 9-69. Completed tube model.

Figure 9-69 shows what the finished part should look like, as seen from a trimetric view. This is a pretty simple part, but you can have quite a bit of fun creating much more elaborate 3D sketches. One thing to be mindful of, especially when creating 3D sketches, is that a feature must not intersect itself during a sweep. If intersecting geometry is a requirement for a part, create the part in stages. There is no rule regarding independent features intersecting each other.

Projected Curves

Sometimes it would be desirable to create a sketch on a nonplanar face. SolidWorks does not allow this directly. What it does allow, however, is to first create a sketch on a planar face and then project the sketch onto the nonplanar face. The projected curve can then be used as a sweep path, or for a number of other functions. The command for performing this function is named, aptly enough, Projected Curve.

Another function of the Projected Curve command is its ability to take two existing 2D sketch profiles and project them together. The resultant curve is a combination of the original two curves. Projecting a sketch onto a face is a much more common action in SolidWorks than projecting a sketch onto another sketch. However, the section that follows presents an example of why it might be necessary to project a sketch onto a sketch, and how the functionality might be beneficial.

Projecting a Sketch onto a Sketch

This topic refers to two 2D sketches being joined mathematically. It is required that two sketches are created that are typically open profiles. The resultant curve will take on the shape of each original sketch as seen perpendicular to their original sketch planes. This is somewhat difficult to picture, even for those used to working in 3D. Following are some illustrated examples that should help to explain.

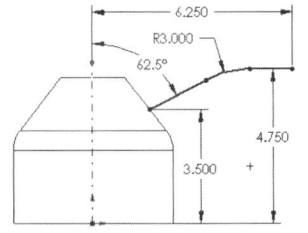

Fig. 9-70. Supply line path from the front.

Let's use a hypothetical situation in which you have been given the task of designing a new vacuum fitting for your company's next big project. The size of the main feature has been determined, and the design intent for a supply line has been given. You have been given the dimensional requirements for the path the supply line must take with respect to the front and top views, but that is all you know at this point. Figure 9-70 shows the design requirements for the path as seen from the front.

What complicates matters is that there are other objects the supply line must route around. Those other objects are not shown here, as this is a hypothetical situation, but imagine that the supply line must take an odd path as seen from the top to avoid objects already in place. Figure 9-71 shows, from the top this time, the path the supply line must take.

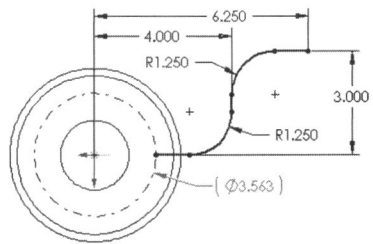

Fig. 9-71. Supply line path from the top.

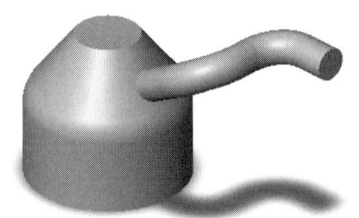

Fig. 9-72. A projected curve used in a swept feature.

Knowing what the path must be from both the front and top allows us to create the third curve. The resultant curve will have the same profile as the sketch shown in figure 9-70, as seen from the front, and will have the same profile as the sketch shown in figure 9-71, as seen from the top. If you were to then use the resultant curve as a sweep path, the result might be a feature similar to that shown in figure 9-72.

Although the name of the command is Projected Curve, what we have discovered from this section is that sketches are actually what are being projected. A sketch is projected onto another sketch to create a projected curve, but do not get bogged down in semantics.

As was mentioned earlier, creating curves by this method is probably not as common as other curve creation methods in SolidWorks, but the command does have its place. From time to time, under certain conditions, you may find that a projected curve proves to be a very valuable asset. This is especially true of projected curves created from projecting a sketch onto a face, discussed in the following section. You will also learn the steps required to carry out the Projected Curve command.

Projecting a Sketch onto a Face

To project a sketch onto a face, a sketch must first be created on a plane or planar face. The sketch can contain various geometry, such as lines, arcs, or whatever else is required. However, the sketch must contain one contour only. Once the sketch has been created, it can then be projected onto either a planar or nonplanar face.

A sketch to be projected onto a face will usually be a closed profile, though this is not a requirement. If an open profile is used, the sketch being projected must completely cross the geometry it is being projected onto. If an open profile sketch falls short of completely crossing a face, it will fail to project onto that face.

Let's take a closed profile sketch and project it onto the front face of a perfume bottle for use as a label outline or cosmetic enhancement. Figure 9-73 shows the sketch that will be used. Now is also an ideal time to walk through the steps of the Projected Curve command. How-To 9-5 takes you through the process of creating a projected curve.

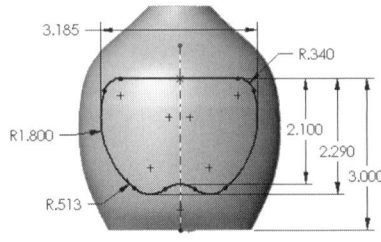

Fig. 9-73. Preparing to project a sketch onto a face.

How-To 9-5: Creating a Projected Curve

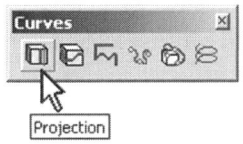

Fig. 9-74. Projection icon.

To project a sketch onto another sketch or face, perform the following steps. It is assumed that at least one sketch has already been created. In the case of projecting a sketch onto a sketch, two sketches are required.

1. Exit out of the completed sketch if you have not already done so.

2. Select Curve > Projected from the Insert menu, or click on the Projection icon, shown in figure 9-74, located on the Curves toolbar.

3. Select either the Sketch onto Sketch or Sketch onto Face(s) option, as required.

4. Select the sketch to be projected. Select both sketches if Sketch onto Sketch was selected.

5. If the Sketch onto Face option was used, select the face or faces the sketch is to be projected on. The Projected Curve panel will appear (as shown in figure 9-75) if the Sketch onto Face(s) option was selected.

6. Click on the Reverse Direction option if required (pertains to the Sketch onto Face(s) option only).

7. Click on OK to complete the command.

Fig. 9-75. Projecting a sketch onto a face.

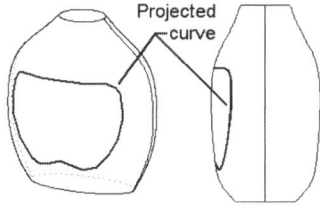

Fig. 9-76. Completed projected curve.

Continuing with the perfume bottle example, in figure 9-86 we can see that the sketch has been projected onto the front face of the bottle. Note that the sketch conforms to the face, as you would expect it to. The curve could now be used as a sweep path to create, for example, a raised outline, which could serve as a border for the label.

Curves Through Points

The last couple of curve-related commands we will explore in this chapter sound very similar. These commands are named Curve Through Reference Points and Curve Through Free Points. The first allows for creating a curve by simply picking points on the screen, and the latter allows for typing in *xyz* coordinates. Let's start with the easier of the two commands to use, described in the following material.

Curve Through Reference Points

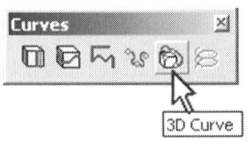

Fig. 9-77. 3D Curve icon.

SolidWorks commands do not get much easier than the Curve Through Reference Points command, which can be found in the Insert > Curve menu. The icon for this command is shown in figure 9-77. The icon is named 3D Curve, which differs from its menu counterpart. Do not let this fool you, as they are the same command.

The Curve Through Reference Points panel is shown in figure 9-78. To use this command, simply pick points off the screen that will define a new curve. The order in which the points are selected is important, because the new curve will pass through the points in the order you pick them. The points can be anything, such as sketch points, endpoints of sketch geometry, or vertex points.

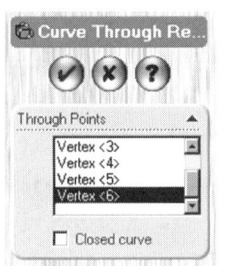

Fig. 9-78. Curve Through Reference Points panel.

Where would you use this command? Use it anyplace you need to define a curve when you already have points established. Usually, points are not already conveniently positioned right where you need them. A more likely scenario is that points would have to be created first, and by the time you have finished positioning points to define a curve, you could have used some other command to create the desired curve. Nonetheless, this command is here if you need it.

Curve Through Free Points

If you need to create a curve based on precise *xyz* coordinates, Curve Through Free Points is the command for you. It allows for typing in *xyz* coordinates that dictate where the curve will pass through. The coordinates that are entered can be values you pick on the fly, or coordinates generated through some other program. How-To 9-6 takes you through the process of creating a curve using the Curve Through Free Points command. Subsequently, the command is examined in a little more detail.

How-To 9-6: Creating a Curve Using Curve Through Free Points

Fig. 9-79. Curve Through Free Points icon.

To create a curve through a set of *xyz* coordinates using the Curve Through Free Points command, perform the following steps.

1. Select Curve Through Free Points from the Insert > Curve menu, or click on the Curve Through Free Points icon, shown in figure 9-79.

2. In the Curve File window that appears (shown with some coordinates already specified in figure 9-80), double click on the first available button below the word *Point* (using the left mouse button) to add a set of *xyz* coordinates.

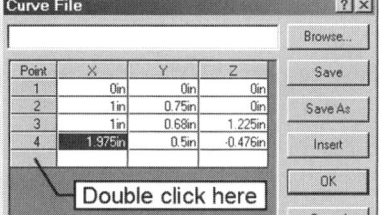

Fig. 9-80. Curve File window.

3. Type in a value for the X coordinate.
4. Press the Tab key.
5. Type in a value for the Y coordinate.
6. Press the Tab key.
7. Type in a value for the Z coordinate.
8. Double click on the button in the next available row.
9. Repeat steps 3 through 8 as required.
10. When you are finished entering coordinates, click on OK to create the curve.

Step 2 states that you should double click on the first available button below the word *Point*, but in reality, double clicking anywhere in a blank row will suffice. Additionally, double clicking in any occupied cell will allow for altering that cell's value.

Clicking on a row number will highlight the entire row. If the Insert button is selected at this time, a new row will be inserted above the highlighted row. In this way, additional control points can be added anywhere along the curve.

Curve Files

There is a special file type directly related to the Curve Through Free Points command and the Curve File window. This file type is known as a curve file, indicated by the file extension *.sldcrv*. Curve files can be saved directly from the Curve File window via the Save or Save As button. You may very well want to save a curve file if you have taken the time to manually enter a large number of coordinates and are likely to need the same set of coordinates at some point in the future.

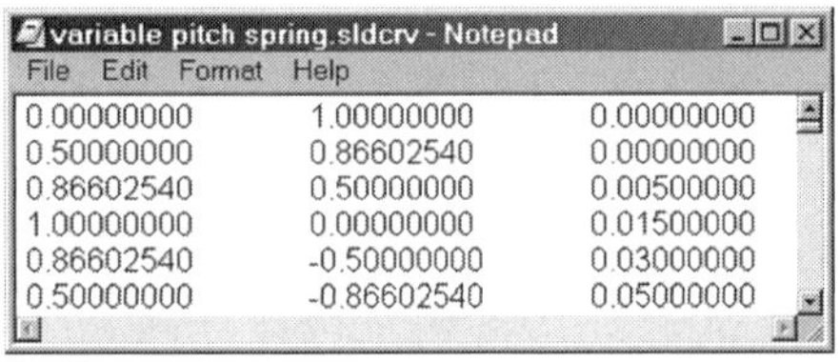

A curve file is nothing more than a simple text file. The file should be a space-delimited file, and each set of *xyz* coordinates should be on a separate line. If more than one space separates each coordinate, that is fine, as it will not pose a problem. Figure 9-81 shows an example of a curve file opened in the basic Notepad text editor included with the Windows operating system. The file being displayed has the standard *.sldcrv* SolidWorks curve file extension, but a simple *.txt* text file extension is also compatible with SolidWorks.

Fig. 9-81. Curve file.

Along with saving curve files, you can also import them, as you would expect. Use the Browse button in the Curve File window to import either text or curve files with the *.txt* or *.sldcrv* file extensions, respectively.

How to generate the *xyz* coordinates is a problem that can be solved in a number of ways. One option is to use a spreadsheet to generate the coordinates. As an example, it is possible to use a Microsoft Excel spreadsheet to generate the *xyz* coordinates of a variable-rate pitch spring. Those coordinates can then be saved as a space-delimited text file, or copied and pasted into a text file, for instance. The math and formulas behind such an endeavor can be very intimidating, especially if your math or Excel skills are a little rusty. The point, however, is that it can be done.

Fig. 9-82. A variable-rate pitch spring created from a curve file.

Once the *xyz* coordinates have been generated and the space-delimited file has been created, the point data can be brought into the Curve File window, using the Browse button previously mentioned. This process is exactly what was done to create the variable-rate pitch spring shown in figure 9-82. The curve was used as a sweep path to create the spring, which serves to illustrate what the end result of this process can be.

Section Properties

It is possible to derive information from a solid model in the form of section properties and mass properties. The Section Properties and Mass Properties commands are found in the Tools pull-down menu. Although

Section Properties

these commands do not have any direct link to curves, threads, or springs, they are explored here toward understanding material to follow.

How Section Properties Work

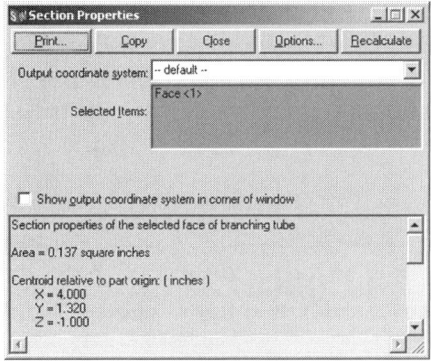

Fig. 9-83. Section properties.

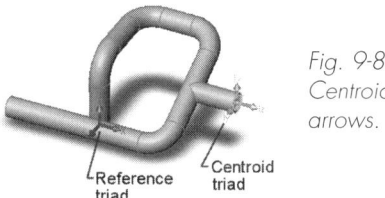

Fig. 9-84. Centroid arrows.

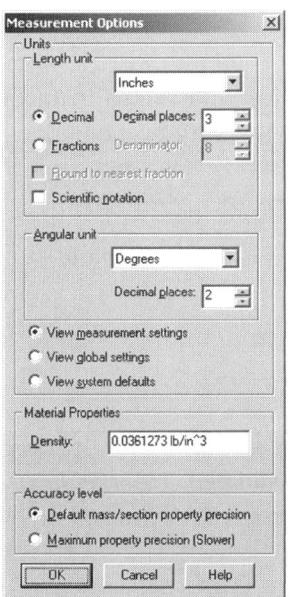

Fig. 9-85. Measurement Options window.

For a selected face (or faces), SolidWorks returns the information shown in figure 9-83, plus other data, such as the moments of inertia. Selected faces must be planar. Only a portion of the Section Properties window is shown in figure 9-83. In SolidWorks, you would have the capability of scrolling down to view the rest of the data, or of resizing the Section Properties window. Explaining statistics such as moments of inertia is outside the scope of this book.

A centroid for the selected face will be shown, as can be seen in figure 9-84. In the example being shown, the section properties of the face at the end of the tube were required. The purple arrows being displayed (also known as the centroid triad) represent the face's center of mass. Multicolored arrows represent the coordinate system, which represents the point (by default, the model's origin point) referenced by the software when calculating the section properties. This does not have to be the model's origin point, but can be any user-defined coordinate system (see "Coordinate Systems" in Chapter 7). Think of the multicolored arrows as the reference triad.

You have the ability to copy the section properties' data to the Windows clipboard and to print the data out to hardcopy. When you are finished, simply click on the Close button. In addition to the Print, Copy, and Close buttons, there are also the Options and Recalculate buttons. If changing options or selecting a different face, you may need to click on the Recalculate button. Clicking on the Options button allows for changing the units used by the Section Properties window. The Measurement Options window, shown in figure 9-85, is what would be seen after clicking on the Options button.

Measurement Options Window Settings

Only the units for the Section Properties display window will be changed if you change the units from the Measurement Options window. The units for the part itself will remain unchanged. Most of the optional unit settings in the Measurement Options window should be self-explanatory. If not, review the section in Chapter 3 titled "Units of Measurement." The three radio button settings near the bottom of the Measurement Options window may require additional explanation. The following material describes these settings.

View Measurement Settings

View measurement settings is the default setting. Any change made to the units when this option is selected will only affect the display of the units in the Section Properties window. Changes made will not affect the units used by the part.

View Global Settings

When *View global settings* is selected, the current units being used by the model are shown. The units displayed will match those shown in the Units section of the Document Properties window (Tools > Options). Note that the units of the model cannot be changed from the Measurement Options window, and that this setting is only for informational purposes or for determining the units used by the Section Properties window.

View System Defaults

View system defaults shows what the system default settings for the units happen to be. This setting is determined during the SolidWorks installation process. The setting is also used for defining the very first templates created during the software installation process, which may not be the same as those specified in the template used to create the current model.

✎ **NOTE:** *Changing the Density setting will affect the model's document properties and will be saved with the file.*

Mass Properties

Prior to obtaining the mass properties of a part, you should define the density of the part material. The part's density can be defined either by clicking on the Options button while in the Mass Properties window or by performing the steps outlined in How-To 9-7, which takes you through the process of defining a part's density via the Document Properties settings.

How-To 9-7: Defining the Density of a Part

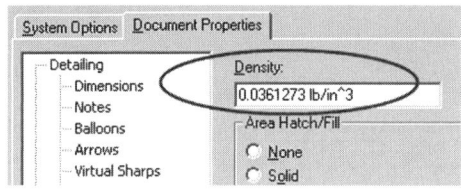

Fig. 9-86. Changing a part's density.

To define the density of a part, perform the following steps. Figure 9-86 shows the option involved in this process, found in the Material Properties section of the Document Properties window.

1. Select Options from the Tools menu.
2. Select the Document Properties tab.
3. Select the Material Properties section.
4. Click in the area marked Density and specify the density of the part.
5. Click on OK to accept the changes.

✓ **TIP:** *If working in one set of units, but the density is only known in another set of units, specifying the density in the known units will make SolidWorks convert the density to the current operating units. For instance, if working in inches and the density of .0024 gram/mm^3 is entered, the value will be converted to .0867 lb/in^3.*

SolidWorks does not currently give you a choice of common density settings. The reason may be because there are just so many different alloys used in the world today. You must look up this information on your own. When you are ready to view the mass properties of the part, click on Mass Properties in the Tools pull-down menu. SolidWorks will return the same information found in the Section Properties window, plus the density value, mass, volume, and surface area.

The Mass Properties window is nearly identical to the Section Properties window, except for the information it provides. You will still have the options for copying the data to the clipboard or printing the data out to hardcopy. The Options button brings up the same window you would see if pressing the button of the same name found in the Section Properties window. The center of mass of the model is displayed with the same purple centroid triad.

An option in the Mass Properties window that does not appear in Section Properties is the option Include Hidden Bodies/Components. This option, shown in figure 9-87, is only relevant if working with parts containing multiple bodies or if working with assemblies. If a body or component (in an assembly) is hidden, the body or component can still be included in the mass properties calculation. Be forewarned that if a body or component is suppressed, it will not be included in the calculations under any circumstances. Suppressed objects are removed from memory, whereas hidden objects are simply removed from view.

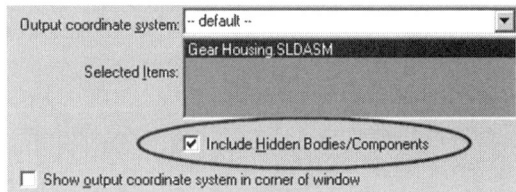

Fig. 9-87. Include Hidden Bodies/Components option.

Changing the Coordinate System

When checking the section or mass properties of a face or model, the referenced coordinate system is the coordinate system of the model by default, located at the model's world origin point. Sometimes it is necessary to use a different coordinate system for obtaining moments of inertia data from a particular point, for instance.

To analyze section or mass properties from a different coordinate system, a new coordinate system must first be defined. If you need to learn how to accomplish this task, see "Coordinate Systems" in Chapter 7. Once the coordinate system has been defined, it can be selected from the Output coordinate system drop-down menu in the Section Properties or Mass Properties window.

Mass Properties in Assemblies

The area marked Selected Items in the Mass Properties window shows the items selected. In an assembly, this can be particularly important. The mass properties of individual components can be selected, in any combination, rather than being limited to viewing mass property data on the entire assembly.

In the case of any model, some of the mass property data is only going to be accurate if you assign the proper density to the model first. This holds true for assemblies as well, of course, but in an assembly the possibility of error is increased due to multiple components being present in the assembly. In that each component can have its own density setting, you will need to make it a point to set the density value for each component.

✓ **TIP:** *It is good practice to set the density and crosshatch properties of a part when it is being created (Material Properties section of Document Properties) so that you do not have to worry about setting these values later.*

Placing a Point at the Centroid

A common question SolidWorks users ask is "How can I position a point at the centroid?" The answer is pretty simple. Use the Mass Properties command to find out what the *xyz* coordinates are for the centroid and jot the values down on a piece of paper. Use the extent of decimal place precision necessary.

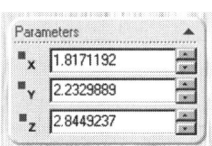

To create the point, click on the 3D Sketch icon and, using the Point command, place a point somewhere in the model. Do not place the point on a face. Rather, place it in space somewhere. With the point selected, use PropertyManager to change the value of the point's *xyz* position. The Parameters panel for points is shown in figure 9-88. Change the values to match the *xyz* coordinates you made note of.

*Fig. 9-88.
Changing a point's xyz coordinates.*

Once you have a point at the same location as the centroid, it can be used to take measurements or to locate a user-defined coordinate system, among other things. The choice is yours.

Summary

This chapter taught you how to create some fairly impressive-looking geometry. You have seen how you can use the Sweep command to create springs and threads. A swept feature typically consists of two sketches. The sweep profile is guided along a sweep path to create the swept feature.

There are two important guidelines you must keep in mind when creating a swept feature. The first is to make sure the sweep path start point is coincident with the sketch plane of the sweep profile. The other requirement is that the geometry created by the sweep does not intersect itself. If the sweep profile is created first, make sure a coincident relationship is added between the start of the path and the profile plane (do this when sketching the path). If the path is created first, a plane can be made to pass through either end of the path, using the Normal to Curve plane creation method.

A very important geometric relationship is the pierce constraint, which in essence locks a specified point onto an existing edge or curve. The point can move along the curve, but will not veer off the curve it is pierced by. The pierce constraint can be very useful for creating swept features because it can be used to help guide the sweep profile along a path or guide curve.

Guide curves can be used when sweeping in much the same manner they are used when lofting. A guide curve will help control the shape of the profile as it is being swept. Guide curves should be created prior to the sweep profile, which then allows for piercing the profile geometry with the guide curves. Pierce relations are required if using guide curves when sweeping.

The Helix command can be used in the creation of springs and threads. You may want to gauge the relative importance of creating true threads with respect to system performance. It is often easier to create "fake" threads that look very much like the real thing but do not require as much computational power. Use revolved cuts and linear patterns to create the appearance of threads without the performance overhead.

Curves are a necessary part of SolidWorks and can be created in a variety of ways. The Composite Curve command allows for joining edges or sketch geometry to form a continuous curve. The Curve Through Free Points command generates a curve through a set of *xyz* coordinates that can either be specified by the user or imported via the Curve File window. Curve files can have either a *.txt* or *.sldcrv* file extension, and are nothing more than basic text files.

Projected curves can be used when it is necessary to establish a curve on a nonplanar face. The Projected Curve command can also be used to project one sketch onto another, thereby creating a new curve that takes on the appearance of the previous two sketches relative to each sketch's original sketch plane.

Use the Tools menu to access Section Properties of a planar face, and Mass Properties to access the physical properties of a part or assembly. Make sure you select a planar face before obtaining the sectional properties. If accessing Mass Properties, make sure you specify the part's density. One way this can be done is through the Material Properties section of the Document Properties window.

Questions and Topics for Discussion

1. What two objects are required by the Sweep command?
2. Is it possible to use an existing edge for a sweep?
3. Is it possible to sweep more than one profile at a time?
4. Can a closed profile be used for a sweep path?
5. Why is it sometimes not feasible to create true helical threads on a part?
6. Describe an alternative method of creating the appearance of threads on a part.

7. Can you sweep an open profile? Explain your answer.
8. What command is used to create a spiral?
9. What is the name of the command that allows for joining multiple sketches end to end?
10. How does one obtain the centroid of a part?
11. How does one determine the surface area of a planar face?
12. What is the difference between the Split Line command, discussed in Chapter 6, and the Projected Curve command when projecting a sketch onto a face?
13. What type of sketch entities can be created when using the 3D Sketch command?
14. What function does the Tab key perform when creating a 3D sketch?

Optional Problem

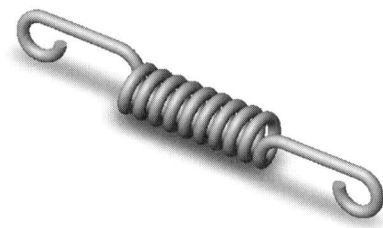

Fig. 9-89. Tension spring.

Simple springs are easy to create, so let's try something a little more challenging. Create the tension spring shown in figure 9-89. The most difficult problem will be figuring out how to create the proper sketch geometry so that it will blend in with the helical curve at the center of the model.

It may be difficult to know where to start with this particular problem. This exercise represents a challenge, and is not a "How-To" guide, so there is no step-by-step list to follow. You are largely on your own with this one, but the following material outlines the general steps you should adhere to in working through this challenge.

You will want to begin with a sketch that represents one end of the spring, excluding the helix portion. The sketch should look similar to that shown in figure 9-89, but feel free to be creative. Use whatever dimension values you wish. Of particular importance is the included angle of the arc on the left. When it is time to create the helix, you will want it to blend nicely with the end of the arc. Figure 9-90 illustrates this by showing the soon-to-be-created helix in gray.

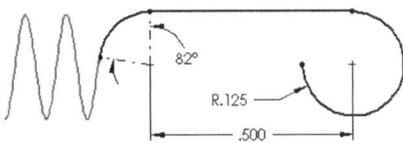

Fig. 9-90. A good sketch to start with.

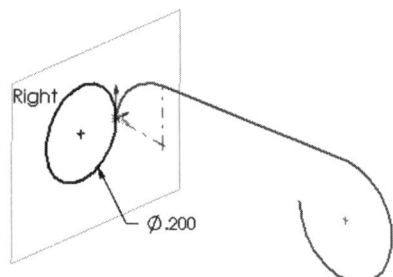

Fig. 9-91. Circle for defining the helix.

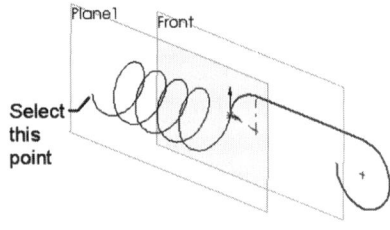

Fig. 9-92. Creating the next sketch plane.

Once the first sketch is done, create a plane that passes through the end of the arc. If you positioned the first sketch properly, you may be able to use one of the default planes. You will need to create a circle that can be used to define the helix, but the circle will need to be positioned in the correct place. A position similar to that shown in figure 9-91 should work. Add the proper relations to fully define the geometry.

You know what to do next: define the helix. Make sure to specify the correct starting angle for the helix so that the helix begins where the arc in the first sketch left off. If the circle defining the helix was positioned correctly relative to the arc in the first sketch, you should not have a problem.

Assuming you defined the helix, now it is a matter of creating another plane, positioned correctly for the final sketch and passing through the endpoint of the helix. The process for accomplishing this may be different, depending on how many revolutions your helix has. One solution might be to create a plane parallel to another plane that passes through the end of the helix. In figure 9-92, the new plane is parallel to the Front plane.

On the new plane, create a sketch that is basically a mirror image of the first sketch. As a matter of fact, one option would be to copy and paste the first sketch into a new sketch, and then use the Modify Sketch tool (discussed in Chapter 8) to flip it into the correct orientation. Add a pierce relation to lock the sketch onto the end of the helix.

To copy and paste a sketch, select the sketch from FeatureManager, press Ctrl-C, click on the plane you want to paste the sketch onto, and then press Ctrl-V. Make sure you are not in a sketch when performing this operation.

From this point forward, you are just about home free. Join the helix and sketch geometry using Composite Curve. Pick near the start of the new composite curve and start a sketch. Create the profile, and then create your swept feature. By the way, there is more than one way to create a spring of this type. If you do not want to use the suggestions here and have some ideas of your own, that is fine.

So how did you do? If you managed to create the tension spring, give yourself a pat on the back. You have just managed to accomplish something that most new SolidWorks users would be envious of.

Chapter 10

Part Configurations

PARTS CONSIST OF FEATURES, AND ASSEMBLIES CONSIST OF PARTS. It is possible to turn individual features within parts on and off. It is also possible to turn components within an assembly on and off, which you will learn more about in Chapter 15. This ability to turn features or components off is known as *suppression*. When a feature is suppressed, it has been turned off. Likewise, unsuppressing a feature turns it back on.

It is possible to save a part with specific features suppressed. This is known as a *configuration*. You can have as many configurations in a part as you want. Each configuration can have different sets of features suppressed or unsuppressed, depending on your requirements. This chapter shows you how to create and manage these configurations within part files.

Another more powerful aspect of configurations has to do with design tables. If you have Microsoft Excel on your computer, you can create family part tables in which one part file can have literally hundreds of configurations, all driven by a single spreadsheet. Design tables are discussed in detail in Chapter 16, but it is mandatory that you have a good understanding of configurations first.

Reasons for Configurations

Whatever area of manufacturing or design work you are in, configurations can be very valuable. You might need to take advantage of configurations for a variety of reasons. If you fall into any one of the following categories discussed in the sections that follow, you should seriously consider using configurations.

Assembly Performance

If you work with large assemblies, it is beneficial to create simplified versions of parts. These simplified versions can be configurations used in the assembly. This has a tendency to increase the performance of the assembly, in that the fine details of individual components do not have to be shown. A perfect example would be threads on fasteners. It may have been necessary to model the threads, but the threads for every fastener do not need to be displayed in the assembly.

Sheet Metal Forming Operations

Using configurations, the entire forming process of a sheet metal part can be shown. Each bend-forming operation can be shown, step by step and in the proper sequence, until the forming process is completed.

Part Families

Different versions of the same parts can be saved as separate configurations. For instance, if your company designs components that have different size requirements, configurations can be used to represent the various sizes. Instead of maintaining and tracking 20 different part files representing the various sizes, only one part file containing 20 configurations is required. The logistics of tracking one file is undoubtedly easier than maintaining 20 files.

Application-specific Requirements

It may be necessary to have different versions of a part for use in different applications. One version may be used for finite element analysis, whereas another version may be used for kinematic studies.

Design-specific Requirements

Those working in the medical industry may have different design requirements than those working in the military, aeronautics, or civilian sectors. Configurations could be used to distinguish between these requirements. The examples could go on and on. You may have other reasons for utilizing configurations. By the end of this chapter you will have a much better idea of what can be accomplished with configurations and why you may find them0 useful.

ConfigurationManager

There are three tabs at the bottom of FeatureManager. The second tab from the left is the PropertyManager tab, discussed in Chapter 2. The third tab will open ConfigurationManager. If you have access to SolidWorks, try to follow along on your own screen. Click on the ConfigurationManager tab, shown in figure 10-1, to open ConfigurationManager.

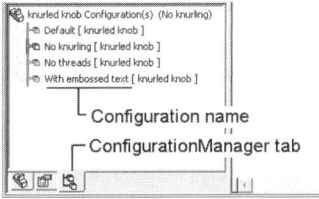

Fig. 10-1. ConfigurationManager tab.

Once in ConfigurationManager, the first thing you will notice is that there does not appear to be much going on. ConfigurationManager, also shown in figure 10-1, will have the name of the part at its top, followed by the word *Configuration(s)*. Below this will be the names of all configurations. When you begin, you will see the default configuration only, which is present in every part and assembly file. How-To 10-1 takes you through the process of adding a configuration to a part or assembly file.

How-To 10-1: Adding a Configuration

To add a configuration to a part or assembly file, perform the following steps.

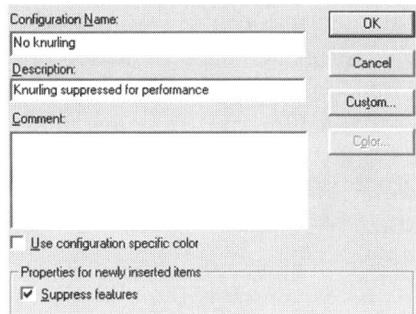

Fig. 10-2. A portion of the Add Configuration window.

1. Click on the ConfigurationManager tab.

2. Right-click on the name of the model at the top of ConfigurationManager and select Add Configuration.

3. Type in a name for the configuration in the Add Configuration window, partially shown in figure 10-2.

4. Click on OK to create the configuration.

Once you click on the OK button, the new configuration should appear in ConfigurationManager. When there are multiple configurations, SolidWorks automatically alphabetizes everything for you. There is no way around this, not that it really matters. To get back to the window that appeared in step 2, right-click on any configuration in ConfigurationManager and select Properties. The configuration does not have to be active in order to access its properties.

Only one configuration can be active, or current, at a time. The current configuration will have a yellow icon, whereas all other configurations will be gray. To change which configuration is currently being displayed, double click on the configuration you want to see. The upper limit of the number of configurations possible in SolidWorks is beyond any reasonable amount you would ever need. What can be altered and stored in configurations is discussed in material to follow.

When the Add Configuration window pops up, as after performing step 2 in How-To 10-1, there is a spot for typing in a comment or description if you are so inclined. Comments and descriptions are not required. Because there is more to descriptions than meets the eye, they will examined in material to follow.

Placing a check in the *Use configuration specific color* box activates the Color button. You can then pick a color for the model and it will change the overall model color, but only in that configuration.

FeatureManager and ConfigurationManager Descriptions

Descriptions can be assigned to both configuration names and features. These descriptions can then be displayed in FeatureManager, in ConfigurationManager, or in both. By default, the description of a feature is the same as its corresponding feature name. Likewise, configuration descriptions are the same as the corresponding configuration name.

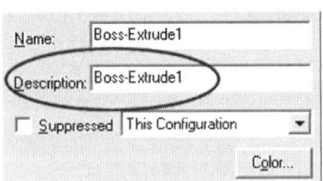

Fig. 10-3. A portion of the Feature Properties window.

To see the description of a configuration or feature, right-click on the configuration or feature and select Properties (or Feature Properties, in the case of a feature). A portion of the Feature Properties window is shown in figure 10-3.

To display descriptions in FeatureManager or ConfigurationManager, right-click on the name of the model at the top of either manager and select Tree Display. You will see a set of options, the content of which depends on whether you are working with a part or assembly. In either case, the options basically amount to whether object names or descriptions are being displayed. The inset in figure 10-4 shows what the Tree Display menu might look like. Also shown is a sample of what ConfigurationManager might look like after turning on configuration descriptions.

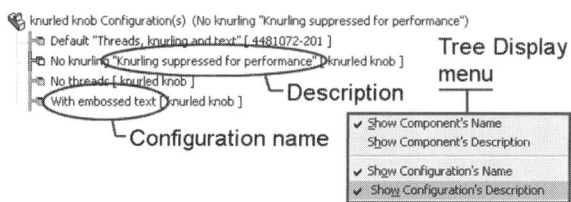

Fig. 10-4. Tree Display menu and sample ConfigurationManager with descriptions turned on.

Because names of features or configurations can be basically whatever you want, and because descriptions can be added, FeatureManager and ConfigurationManager can wind up being quite informative. They can theoretically contain more information than most people require. If you do not feel the need to use descriptions, by all means do not bother with them. After all, configuration or feature names can be fairly "descriptive" in their own right.

Incidentally, the text in the brackets at the end of each configuration name is what appears in an automatically generated bill of material (BOM). BOMs, discussed in detail in Chapter 12, can be inserted into assembly drawings.

Properties for Newly Inserted Items

Suppress features is an important option you will want to pay attention to. This option is in a section titled *Properties for newly inserted items*, visible at the bottom of figure 10-2. The *Suppress features* option is checked (on) by default. If checked, a feature added to another configuration will be suppressed in the configuration where *Suppress features* is checked. In other words, you may want features added in the current configuration to be suppressed in other configurations. This is exactly what the *Suppress features* option does for you.

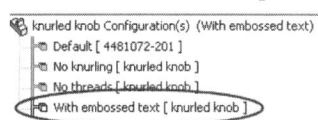

Fig. 10-5. Another configuration has been added.

The following is an example of using the *Suppress features* option. In this example, some embossed text will be added to a knurled knob. A configuration will be added to the part, appropriately named *With embossed text*, as shown in figure 10-5. In the Properties section of each configuration, *Suppress features* will be checked.

Next, the embossed text feature will be added. The technical details of how to accomplish this task are not presented here, but you can read how this is done in Chapter 21. The important point with regard to this discussion is what happens next. Figure 10-6 shows the part with the new text feature added.

Fig. 10-6. Adding a feature to the model.

With embossed text is the configuration that was active when the text feature was added. If you were to return to any of the other configurations, you would see that the new feature is suppressed. It is worth noting that even the *With embossed text* configuration had the *Suppress features* option checked. However, the text feature appeared anyway. This is because the *Suppress features* option only affects other configurations that are not active, not the one you are currently working with. This makes sense, of course. SolidWorks assumes you would want to see the feature you are adding in the current configuration.

The name of the current configuration can be seen after the name of the part at the top of ConfigurationManager (and FeatureManager). Examine figure 10-5 and you will see the configuration name in parentheses. The configuration name will appear in parentheses only if there is more than one configuration present in the part (or assembly).

As a final word of advice regarding the *Suppress features* option, you may want to make sure that all configurations have the option checked or that all have this option unchecked. The reason is that with numerous configurations it can be very difficult to keep track of which features are being added to which configurations, and which features are being suppressed. Temporarily turn *Suppress features* on or off as required, and then set it back so that all configurations are the same before closing the file and going home for the night.

Copying Configurations

Essentially, when a new configuration is added, you are really just creating a copy of the current configuration. This fact can be used to your advantage. If an additional configuration is required, and the new configuration happens to be very similar to an existing configuration, it would make sense to make the existing configuration current before creating the new configuration. A little forethought can save you steps in the long run.

When a new configuration is created, all settings of the current configuration are also transferred to the new configuration. This even holds true for the *Suppress features* option in the Properties for newly inserted items section. It also holds true for any configuration-specific color changes that were made, and of course for any features that were suppressed.

Configurations can be copied using the Windows clipboard. This has the advantage of being able to create many copies very quickly. To use this technique, the pattern to be copied must be active, and must be selected. In other words, first double click on the configuration to activate it, and then click on it once more to select it. Pressing the hot key combination of Ctrl-C will copy the configuration to the clipboard. You can then press Ctrl-V to paste a copy into ConfigurationManager. Pressing Ctrl-V again will paste a second copy in, and so on.

✓ **TIP:** *Configurations can be renamed, like any other feature in SolidWorks, by performing a slow double click on the configuration.*

Suppression States

What does it mean to suppress something in SolidWorks? It means to temporarily turn the display of that object off. To be more technically precise, a suppressed object is removed from memory. SolidWorks acts as though any suppressed items do not exist. You can unsuppress items at any time. Features can be suppressed in part files, and components can be suppressed in assembly files. You can read more about suppressing assembly components in Chapter 15.

We have already discussed a number of reasons configurations can be important. The reasons for suppressing or unsuppressing features go hand in hand with this. The act of suppressing a feature removes it from view and from memory. The feature continues to be included in the part file's database, but SolidWorks ignores it.

✗ **WARNING:** *Suppressing a feature is not the same as deleting a feature. When a feature is deleted, it is removed from all configurations and is no longer present in the model.*

The whole idea of being able to turn a feature off (which will be known as suppressing from this point forward) raises an interesting issue. What happens to an object that is dependent on an object that has been suppressed? A simple example would be suppressing a block that contains a hole. With the block suppressed, what becomes of the hole? This question is answered in the following material.

Parent/Child Relationships

Because of the dependencies involved with a feature-based modeler such as SolidWorks, relationships are established between features. These relationships can be described as parent/child relationships. When a feature is dependent on another feature, it is considered a child of that feature. When a feature has other features dependent on it to exist, it is known as a parent feature. If the block-and-hole example were used, the block would be a parent of the hole, and the hole would be a child of the block.

With this simple block-and-hole example, it is easy to understand the parent/child relationships. However, not all relationships are quite as cut and dry.

Many seemingly innocent actions can create parent/child relationships. For example, adding a constraint to an object causes a relationship to exist to that object. If a circle were sketched on one face of a model, and the center of the circle made coincident with an edge on the other side of the part, the feature the edge belongs to would be considered a parent of the feature created by the circle.

Parent/child relationships are something a SolidWorks user grows accustomed to over time. At first, such relationships seem to be a cumbersome annoyance. In time, you will find relationships to be a very powerful way of imparting your design intent to a model.

With regard to suppressing features, parent/child relationships mean that suppressing a parent will also suppress any dependent, or child, features. There is no way around this, except for somehow removing the relationship back to the parent. Removing dependencies to other features sometimes needs to be done for various reasons, and the removal process often involves the Display/Delete Relations command (Chapter 3) or some other editing process.

It is often necessary that the parent/child relationship information be available in some straightforward manner. The steps you perform to find out what the parent/child relationships are for a particular feature are simple and involve using the right mouse button. How-To 10-2 takes you through the process of accessing parent/child relationship information.

How-To 10-2: Accessing Parent/Child Data

To access parent/child data, perform the following steps.

1. Right-click on the feature in question (usually via FeatureManager).

2. Select Parent/Child.

3. Click on Close when done viewing the data.

There is nothing that can be accomplished from the Parent/Child window, shown in figure 10-7. It is strictly there for the convenience of the SolidWorks user. The left-hand portion of the Parent/Child window will show the parents of the feature in question, and the right-hand portion shows the children. The window can be resized if necessary.

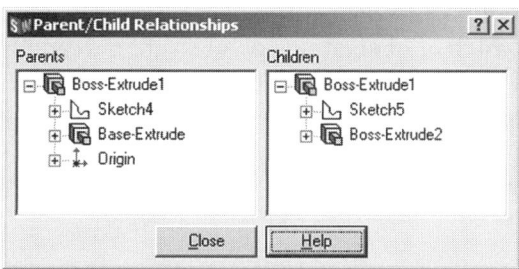

Fig. 10-7. Parent/Child window.

Now that you understand parent/child relationships, you know more of what can be expected when suppressing a feature. This takes us to the next topic: techniques for suppressing or unsuppressing features.

There are a number of simple methods of changing the suppression state of features. One method would be to use the Edit menu, which is the method described in How-To 10-3.

HOW-TO 10-3: Changing Feature Suppression States

To suppress or unsuppress features, perform the following steps. Do not forget to hold down the Ctrl key if it is necessary to select more than one object.

1. Select the feature or features to be suppressed or unsuppressed.
2. From the Edit menu, select Suppress, Unsuppress, or Unsuppress with Dependents, as required.
3. Select This Configuration, All Configurations, or Specified Configurations, as required.

When features are suppressed, their icons appear grayed out in FeatureManager. As mentioned earlier in this chapter, if a feature is suppressed, all of its children will be suppressed. Child features cannot exist without the parent feature, though it is possible to suppress a child without its parent.

As can be seen from step 2, it is possible to unsuppress a parent and its dependents simultaneously. The choice is up to you. Use the Unsuppress option to unsuppress the selected feature, or use Unsuppress with Dependents to bring back the selected feature, along with any dependent features.

Using the Edit menu to change the suppression state of features allows for being particular about exactly which configurations will be changed. Options include Current Configuration, All Configurations, and Specified Configurations. If there is only one configuration, the option This Configuration will be the only one available. Selecting the Specified Configurations menu option brings up the window shown in figure 10-8.

Chapter 10: Part Configurations

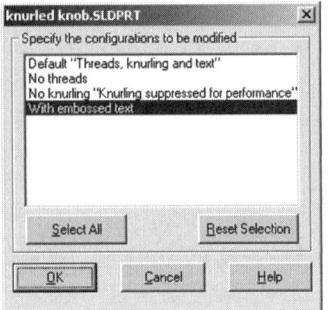

From the window shown in figure 10-8, it is possible to pick and choose exactly which configurations of a feature or features are to be suppressed or unsuppressed. This is extremely convenient, and can be something of a necessity when there are many configurations being used. With numerous configurations, manageability becomes an issue.

Fig. 10-8. Selecting configurations to be modified.

Other Methods of Changing Suppression

Probably the easiest method for changing the suppression state of a feature would be to use the right mouse button menu. Right-clicking on a feature will show the Suppress or the Unsuppress option in the menu, depending on the feature's current state. There is no Unsuppress with Dependents option available with this method.

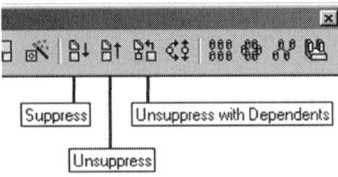

Fig. 10-9. Changing suppression via icons.

Another choice is to use the icons usually found on the Features toolbar, shown in figure 10-9. The icons may not be present on your SolidWorks workstation. If this is the case, they can be added with a little customization. Customizing toolbars is discussed in Chapter 23.

Finally, by right-clicking on a feature and selecting Properties from the menu, the Feature Properties window appears, shown in figure 10-10. There will be a checkbox option for changing the suppression state of the feature in question. There will also be a drop-down list containing the options This Configuration, All Configurations, and Specified Configurations.

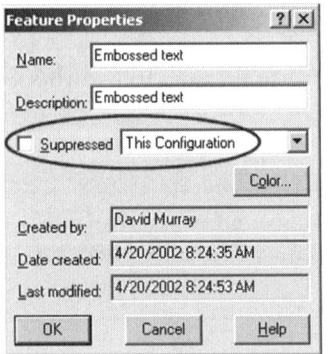

Fig. 10-10. Feature Properties window.

The Feature Properties window has more to it than just the ability to suppress or unsuppress a feature. It also contains the information of who created the feature and when, along with when it was last modified. More accurately, the name in the Created by box is the log-in name (user name) of whoever logged onto the computer before the SolidWorks session was begun.

Feel free to use the Feature Properties window to suppress or unsuppress multiple features at a time. If more than one feature is selected, accessing the properties of any one of those features will bring up the Feature Properties window. Only common properties will be displayed. None of the name, creation, or modified times will be displayed because those bits of data will likely be different for each feature.

Changing the color of a feature can also be accomplished through a feature's properties. See Chapter 8 for discussion of changing the color of parts and features.

Dimensional Configurations

Suppressing and unsuppressing features is just one aspect of configurations. Another capability has to do with being able to control dimension values within a configuration. There must be a minimum of two configurations before this functionality will be present.

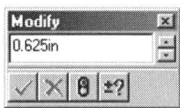 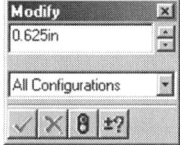

Fig. 10-11. The Modify window changes when there are multiple configurations.

If more than one configuration is present in the part file, and a dimension value is double clicked on with the left mouse button, the Modify window will appear. This is a basic editing technique, discussed in Chapter 4. However, the Modify window will look different when more than one configuration exists. With more than one configuration present, the Modify window will appear as it does on the right in figure 10-11.

The implementation of this functionality is easy enough, and has been explained previously. Simply select This Configuration, All Configurations, or Specify Configurations for the dimensional change. If This Configuration is selected, the dimensional change will only affect the current configuration. When Specify Configurations is selected, the window shown in figure 10-8 will appear.

This is an ideal way to create multiple versions (or sizes) of a part. If there are many dimensions to be changed, or perhaps numerous configurations to be made, you may opt to create a family-of-parts table. This functionality requires what is known as a design table, discussed in Chapter 16.

Other Configurable Objects

Certain commands or functions can be made to affect particular configurations. So far, we have discovered this to be true for suppression states, colors, and dimension values. There are still other items that are configurable. These other items include end conditions, geometric relations, and equations (discussed in Chapter 17), to name a few.

Many items that are configurable use the same process used to specify which configurations are affected when changing feature suppression states. The object being configured may be different, but the process is generally the same. The process usually involves editing some item in some way (such as an end condition in a feature's definition), and then specifying one of the three previously examined options: This Configuration, All Configurations, or Specify Configurations.

It should be noted that keeping track of configurations can be complicated enough without worrying about what geometric relations are suppressed in this configuration, or what end condition was used in that configuration. Make absolutely certain you fully understand configurations before treading into the next section.

Configuring End Conditions

End conditions are applied to extruded or revolved features. They are configurable, but the process must occur after the feature has been created. For example, to specify which configuration an end condition should apply to, the feature must first be created, and then the feature's definition edited.

For this next example, an extruded cut will be modified so that it has different end conditions for three different configurations, which have already been established. The part used in the example, shown in figure 10-12, has a series of fins. The cut, which cuts through the top fin by way of an Up To Next end condition, has already been created.

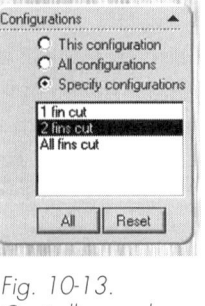

Fig. 10-12. Sample part containing a cut.

If the definition of the cut is edited, a new panel becomes available, which allows for dictating which configurations will be affected when the end condition is altered. The panel is shown in figure 10-13. If you have not seen this panel before, it is probably because you have not had more than one configuration present in a part at one time.

If the *2 fins cut* configuration is specified, the end condition can then be changed to Up To Surface, and the cut can be made to slice through the first two fins, as opposed to just the top fin. The order in which these parameters are changed makes no difference. Changing the end condition before specifying the configurations to be affected is perfectly acceptable. The result is shown in figure 10-14.

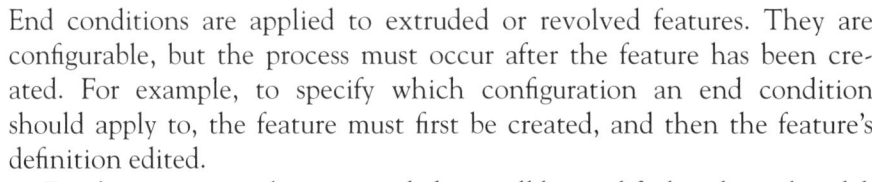

Fig. 10-13. Controlling end conditions for certain configurations.

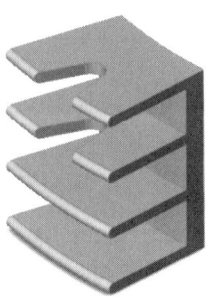

Fig. 10-14. After changing the end condition for the second configuration.

Other Configurable Objects

Continuing with this same scenario, we could then switch to the final configuration, titled *All fins cut,* edit the definition of the cut, change the end condition to Through All, specify what configuration is to be affected, and then click on OK to accept the changes. Again, it is all quite simple, really, at least on a small-scale basis. How-To 10-4 takes you through the process of configuring end conditions.

How-To 10-4: Controlling End Conditions in Configurations

To specify an end condition for a specific configuration, perform the following steps. You should be familiar with editing feature definitions and configurations in general prior to attempting this procedure. More than one configuration must be present in the part file.

1. Right-click on an extruded or revolved feature and select Edit Definition.

2. Modify the end condition as desired.

3. In the Configurations panel, select This Configuration, All Configurations, or Specify Configurations.

4. If Specify Configurations is selected, select the configurations the end condition modifications should apply to.

5. Click on OK to accept the changes.

If the Specify Configurations option is used, the current configuration cannot be deselected. This means that it is not possible to alter other configurations without changing the current one. It is possible to change an end condition in a specific configuration that also happens to have the relevant feature suppressed. You may not know (or perhaps you forgot) that the relevant feature was suppressed, because that other configuration is not the one you are currently working with. So what happens in that situation? The end condition is changed, but it will not be evident until that configuration is made active and the respective feature is unsuppressed.

The default when editing any configurable object is to affect all configurations. In other words, it is the All Configurations option that is used. If you want to alter the current configuration only, or specific configurations, you have to make the conscious decision to do so.

Configuring Geometric Relations

When adding geometric relations to sketch geometry, the Configurations panel will appear in PropertyManager, assuming there is more than one configuration present in the part file. This is the same panel shown in figure 10-13, and it works the same way.

In that it is possible to add a geometric relation simply by selecting a sketch entity, it should be stated that the Configurations panel will only appear if the Add Relation command is selected (see "Geometric Relations" in Chapter 3). Using the Display/Delete Relations command will also display geometric relation configuration information.

Between the use of the Add Relations and Display/Delete Relations commands, it is possible to control the suppression states of individual geometric relations on a configuration-by-configuration basis. You should be warned that this can get seriously confusing very quickly! It is not for the fainthearted. How-To 10-5 takes you through the process of controlling the suppression state of geometric relations and applying them to specific configurations.

How-To 10-5: Controlling Geometric Relations in Configurations

To specify a geometric relation for a specific configuration, perform the following steps. You should be familiar with adding relations, editing sketch geometry, and configurations in general prior to attempting this procedure. More than one configuration must be present in the part file, and you must be editing a sketch.

1. Click on the Add Relations icon.
2. Select the item or items to add a relation to.
3. In the Configurations panel, select This Configuration, All Configurations, or Specify Configurations.
4. If Specify Configurations is selected, select the configurations the geometric relation should apply to.
5. Repeat steps 2 through 4 as required.
6. Click on OK when finished.

It is most important to remember that when adding geometric relations the relation is not applied upon clicking on the green check. Rather,

Other Configurable Objects

the relation is applied as soon as the geometric relation icon of choice is selected.

If you have decided to control geometric relations in configurations, you almost certainly will be required to modify those relations via the Display/Delete Relations command. This goes hand in hand with the first six steps, so continue with the following steps.

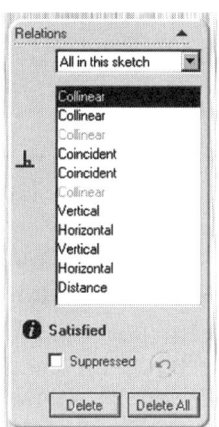

Fig. 10-15. Controlling the suppression of relations.

7. Click on the Display/Delete Relations icon. You should see a panel similar to that shown in figure 10-15.

8. Select a sketch object whose relations require suppression state modification. The panel displayed in figure 10-15 will change to show the selected entity, along with any associated geometric relations.

9. Select a geometric relation from the Relations panel. Keep a close eye on the graphics area to see exactly what objects are being highlighted, so that you know what the relation is associated with. Otherwise, you may modify the wrong relation.

10. Repeat step 3 (and step 4 if step 3 requires this action).

11. Check or uncheck the Suppressed option as required.

12. Click on OK to accept your changes.

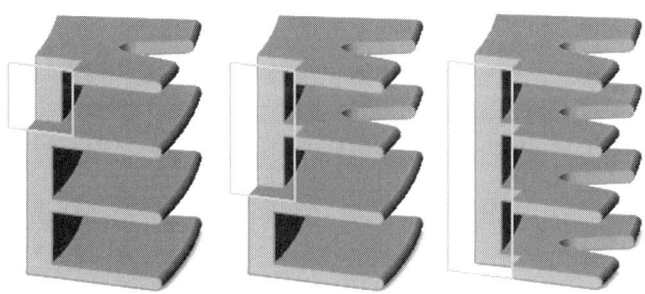

Fig. 10-16. An example of controlling relations through configurations.

You may very well decide that all of the confusion is not worth it, or perhaps you are very good at staying organized and will not have a problem. Configurations are a very powerful tool, but how far they are taken is really a personal choice. It is recommended that you be judicious when controlling geometric relations via configurations. Figure 10-16 shows an example of a part in which geometric relations are being controlled through configurations. The cut on the left side of the part is one feature created from a rectangle. The bottom of the rectangle was made to be collinear with different edges of fins in various configurations. The underlying sketch geometry is shown for clarity.

Nested Configurations

Nested configurations do not require a lot of explanation. To put it quite simply, configurations can exist within other configurations. These are known as nested configurations, and the depth to which configurations are nested is up to the user.

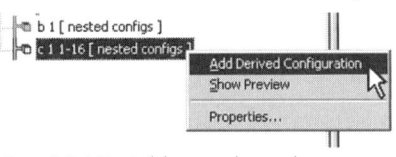

Fig. 10-17. Adding a derived (or "nested") configuration.

To create a nested configuration, right-click on the configuration you want to add a nested configuration to, and select Add Derived Configuration. This process is depicted in figure 10-17. Once that is done, the Add Configuration window will appear, as it does when adding any configuration (see figure 10-2).

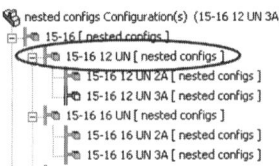

Fig. 10-18. Configurations nested three deep.

You know what to do at this point. Type in a name for the new configuration and click on OK. The short process is almost exactly what you learned in How-To 10-1. As far as the result is concerned, nested configurations establish parent/child relationships. The derived configuration is a child of the parent configuration. What exactly does this mean when you start changing dimensions? Let's use figure 10-18 as an example, which contains configuration names for a bolt. This will serve as a good example because the nested configurations are three deep.

Assume "15-16 12 UN" is the active configuration, and the length of the bolt is changed. In the Modify window, we would be presented with the options in the pull-down menu shown in figure 10-19. Three of the options you should be familiar with, in that they have been mentioned numerous times in this chapter. The fourth option is highlighted in figure 10-19 and reads Link to Parent Configuration.

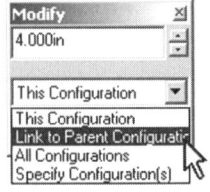

Fig. 10-19. Nested configuration options.

When a parent configuration dimension is altered, it affects the children as well. This means that if a dimension was changed when "15-16 12 UN" was active, "15-16 12 UN 2A" and "15-16 12 UN 3A" would both change. However, "15-16" would not. Parent configurations do not change when a dimension of a nested configuration is changed, at least not if This Configuration is selected from the drop-down menu.

If a dimension of a nested configuration has been changed and This Configuration is selected from the drop-down menu, the link to the parent is broken. This means that the parent can have the same dimension altered and will no longer affect the configuration derived from it.

If a nested configuration has had a dimension changed previously, Link to Parent Configuration will be one of the options in the drop-down menu if the dimension is changed again. As a matter of fact, the dimension value itself does not have to be modified by the user to link it back to the parent. By double clicking on a dimension and selecting Link to Parent Configuration, SolidWorks reassociates the value to the parent.

Using nested configurations for the sole sake of categorizing is a poor idea, because the only legitimate configurations would probably be the lowest-level configurations. All others would be categorical configurations, yet SolidWorks will still see them as separate configurations, which serves to complicate the model unnecessarily. In other words, what was pictured in figure 10-19 serves as an excellent example of what you should not use nested configurations for!

If you have managed to follow along with all of this nested configuration business, you are doing well. Perhaps you will find a good use for nested configurations. If not, do not worry, because nested configurations can add confusion very quickly and their usefulness remains in question. However, that should not scare you away from taking advantage of configurations in general.

Summary

There are many reasons for adding configurations to a part. Most reasons boil down to the same situation, and that is one in which having multiple versions of the same part within the same file would be a convenience. Parts can have different features suppressed in different configurations, and even different dimensional values between configurations.

There can exist as many configurations as necessary within the same file, but only one configuration can be active at any one time. Double click on a configuration name to show that configuration (make it active). Suppress or unsuppress features while in an active configuration and the configuration will remember the suppression state of the features.

Features can be suppressed quite easily via the Edit menu or right mouse button menu. Right-clicking on a feature in FeatureManager or the work area will gain access to either the Suppress or Unsuppress command. Unsuppressing with dependents will unsuppress a parent, along with any child features.

Parent/child relationships exist between features when dependencies are created. Not all features are dependent on each other. When a feature is suppressed, its children will be suppressed as well. There is no way around this, as child features cannot exist without the parent feature.

Items other than features and dimensions are configurable. End conditions and geometric relations are two other item types that can be altered on a configuration-by-configuration basis. You must be careful to keep track of what is being altered in each configuration if taking configurations to this level. To avoid making unwanted changes to individual configurations, avoid the Configuration panel when making changes to feature definitions, for example.

Once configurations have been added, dimensional changes can be made to individual configurations. Changes will take effect according to selection of one of the options This Configuration, All Configurations, or Specified Configurations. The default is All Configurations, but Solid-Works remembers what was selected last on individual dimensions. Therefore, once you have started making dimensional changes to specific configurations, you will have to use caution from that point forward whenever changing dimension values.

Questions and Topics for Discussion

1. How can a new configuration be added in SolidWorks?

2. Is there a maximum number of configurations that can be created?

3. Describe three ways in which a feature can be suppressed.

4. What is the difference between suppressing and deleting a feature?

5. If a feature is suppressed, what happens to the dependent features?

6. Can a child feature be suppressed without its parent? Explain your answer.

7. Other than dimensions and features, specify at least two items that can be saved within a configuration.

8. Describe, in your own words, three uses you think you could find for configurations in SolidWorks. Feel free to use, as examples, projects you might someday want to accomplish using the SolidWorks software.

Chapter 11

Design Drawings

YOU MAY BE USING CUTTING-EDGE SOLID MODELING SOFTWARE, creating complex parts and designing intricate assemblies. Yet in today's world, most people still must have the ability to create design drawings. No matter how advanced the CAD industry becomes, there is still a need for drafters.

Two-dimensional CAD systems have been around for quite some time, and many improvements have been made. Creating design drawings has gotten easier, with much of the process now automated. From a Solid-Works standpoint, adding a typical top, front, and right side view to a standard engineering drawing is literally a drag-and-drop operation. You will see how to do this in material to follow.

There are many view types that can be created in SolidWorks. These views include auxiliary views, section and detail views, broken views, and many others. Annotations of all sorts can be added to a design drawing to help explain your design intent. Notes, surface finish, and weld symbols are only a few examples.

Much of the tedium of adding dimensions has been eliminated from creating drawings. Most dimensions are added automatically for the user. It is basically up to the drafter to decide where to place the dimensions, and on which views. This is simply a matter of dragging the dimensions around to position them.

You will learn how to perform all of these tasks in this chapter. All of the SolidWorks view types, along with how to add dimensions and annotations, are explored. The parts used in the examples throughout this chapter are parts created in earlier chapters, so they should look familiar to you.

File Associativity

Design drawings are a distinct file type within SolidWorks. Part files, which you have been learning about to this point, have an *.sldprt* extension. Drawing files have a different extension type, which is *.slddrw*. Assemblies use yet another extension, which is *.sldasm*. This is only important from the standpoint of knowing what file types to browse for when opening an existing file.

Bidirectional associativity is a complicated-sounding term that describes a very easily understood concept. What it means is that all three basic SolidWorks file types (parts, drawings, and assemblies) are associative with one another.

If a part is altered, whether it be by changing dimensions or adding or deleting features, the design drawing for that part will be updated the next time it is opened. Likewise, a part file can be altered from within a drawing file by modifying the dimensions in the design drawing. Design drawings of assemblies exhibit the same associativity.

Parts can be directly edited within an assembly. This can even be taken one step further. That is, parts can be built up from scratch completely within the context of an assembly, if you need to go to that extent. If a part has been placed into an assembly, the assembly will automatically update if the part file is altered in any way.

As you can see, associativity goes both ways between each pair of file types; that is, between parts and drawings, between drawings and assemblies, and between assemblies and parts. Hence, this functionality is described as bidirectional associativity.

✒ **NOTE:** *Existing CAD operators may take the term* drawing *to mean any CAD document. Regarding SolidWorks, the term* drawing *refers strictly to the 2D design drawing of a part or assembly, not part or assembly documents.*

New Drawings

You begin a new drawing the same way you begin a new part; that is, by selecting New from the File menu. You then specify the drawing template you want to use for the new drawing. Templates were first mentioned in Chapter 2, and can exist for any of the three SolidWorks document types. Do not confuse the term *template* with the term *drawing format*. With regard to this book, drawing format refers to the title block and border data, whereas a template is the file used when opening a new SolidWorks document.

New Drawings

How-To 11-1 takes you through the process of creating a new drawing. Subsequent material shows you how to create a drawing template and drawing formats, and addresses a few other relevant issues.

How-To 11-1: Starting a New Drawing

To create a new drawing, perform the following steps.

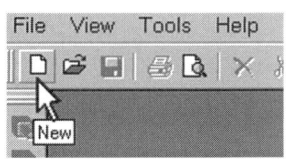

Fig. 11-1. New icon.

1. Select New from the File menu, or click on the New icon, shown in figure 11-1.

2. Select the tab that contains the template you want to use. In a new installation of SolidWorks, this would be the Templates tab.

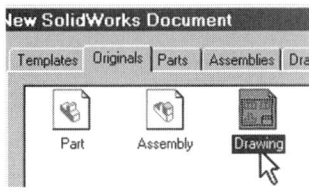

Fig. 11-2. Selecting a drawing template.

3. Select a drawing template to use, such as that shown in figure 11-2.

4. Click on OK.

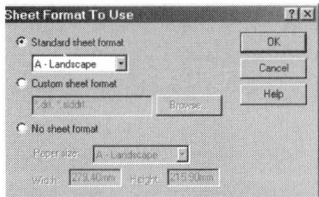

Fig. 11-3. Specifying what, if any, sheet format to use.

If the drawing template you selected does not contain a sheet format, you will be prompted to select one. Otherwise, you have just successfully started a new drawing. If the Sheet Format To Use window appears, shown in figure 11-3, continue with the following steps.

5. Specify whether to use a Standard or Custom sheet format. Optionally, select No Sheet Format and specify a paper size for the new drawing.

6. Click on OK to begin the new drawing.

At this point, SolidWorks will open a new drawing. The interface will differ only slightly from the part files you are used to working with. FeatureManager has been replaced with DrawingManager, and the menu content will be slightly different. These developments are discussed in the following section.

When beginning a new drawing, it may not be necessary to use a format at all, for whatever reason. If this is the case, select the No Sheet Format option mentioned in step 5 of How-To 11-1. All of the standard English and metric paper sizes will be available, or you could choose a custom paper size and enter specific values.

Drawing Interface

Whether you are working with a part, drawing, or assembly document, the pull-down menu headings will always be the same. These menu headings are shown in figure 11-4. Add-in programs such as PhotoWorks or Feature-Works will add their own menus, but these programs are outside the scope of this book.

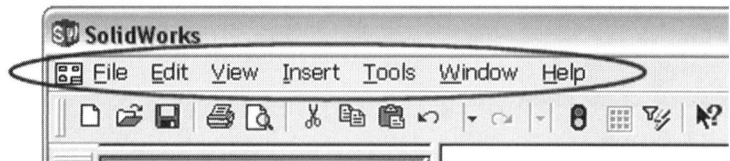

Fig. 11-4. Pull-down menus.

Where you will see the most difference is in the Insert menu itself. Some of the other menus, such as the Tools menu, will undergo a change in their content as well. This occurs as you alternate between the Solid-Works file types, such as part and drawing documents. Most of the menus will remain fairly consistent, even when alternating between file types. This makes learning the program that much easier.

Certain toolbars should appear automatically when starting or opening a drawing file, such as the Drawing and Line Format toolbars. If not, there is a useful way you can "customize" toolbar layouts for each of the three SolidWorks file types. In that we are now dealing with additional Solid-Works document types, now is a good time to discuss this. How-To 11-2 takes you through the process of setting and saving toolbar layouts.

How-To 11-2: Setting and Saving Toolbar Layouts

Actual toolbar locations are saved globally; not on an individual document basis. Whether individual toolbars are turned on or off, however, they are saved by document type. That is, certain toolbars will be turned on when in a part, and different toolbars can be turned on for drawings. This is also true for assembly documents.

Knowing this, there is a simple procedure you can use to set up your toolbars so that they always appear in the same place, depending on what type of SolidWorks file is open. This procedure involves opening one of each type of SolidWorks file.

- **NOTE:** *Although we have not explored assembly files yet, you can still perform this procedure. A typical assembly toolbar layout would be the same as for a part, with the addition of the Assembly toolbar.*

New Drawings

To establish toolbar layouts for all three SolidWorks file types, perform the following steps. If you do not feel comfortable with setting up the toolbar layout for a SolidWorks assembly at this time, skip steps 6 and 7.

1. With no SolidWorks files open, turn off any extraneous toolbars you do not care to see. You could turn them all off, or leave just those common to all three document types.

2. Start a new SolidWorks part using a part template of your choice.

3. Turn toolbars on or off, and position them as desired. (Hint: right-click on a docked toolbar.)

4. Start a new SolidWorks drawing using a drawing template of your choice.

5. Turn toolbars on or off as desired. Any toolbar that will be common to both parts and drawings should not be repositioned.

6. Start a new SolidWorks assembly using an assembly template of your choice.

7. Turn toolbars on or off as desired. Any toolbar common among parts, drawings, and assemblies should not be repositioned.

8. Press Ctrl-Tab to cycle between the open part, drawing, and assembly files. Fine-tune toolbar placement as necessary while switching between document types.

9. Close all open documents. You need not save them.

10. Close SolidWorks. This will ensure that toolbar positions and layout are remembered.

From this point forward, the toolbars associated with the various SolidWorks document types will appear when that document type is opened or activated. There are a few things that will affect toolbar layouts, however.

If you would like to maintain your current toolbar layout, do not shrink the SolidWorks window down to a different size. Leave it maximized. Otherwise, the toolbars will reposition themselves to fit the screen, but they will not snap back to their original positions when SolidWorks is maximized again.

Sketch toolbars can be made to automatically turn on whenever a sketch is being edited. The option is on by default. If you would like to turn the auto-activate option off for the sketch toolbars, How-To 11-3 will show you how. This will make it so that the sketch toolbars are on all the time, whether or not a sketch is being edited, resulting in a more consistent interface.

HOW-TO 11-3: Sketch Toolbar Auto-activation

To turn the sketch toolbar auto-activate option on or off, perform the following steps. Note that this setting will affect the Sketch Relations and Sketch Tools toolbars.

   1. With any file open, click on Customize from the Tools menu.
   2. Make sure the Toolbars tab is selected.
   3. Change the option titled *Auto-activate sketch toolbars* as required.
   4. Click on OK.

It is not recommended that any of the other customization tabs be accessed at this time. If you wish to customize SolidWorks, it is highly suggested that prior to any experimentation you read Chapter 23, which deals with the issue of customization.

Other parts of the SolidWorks drawing interface you need to be aware of are the paper sheet border, the ruler bars, and the sheet tab. The border gives you an indication of where drawing entities are in relation to the paper. The ruler bars help you position objects, such as entities and drawing views, on the sheet. The sheet tabs allow for easy switching between sheets in the drawing. There may also be a format present. These are all shown in figure 11-5.

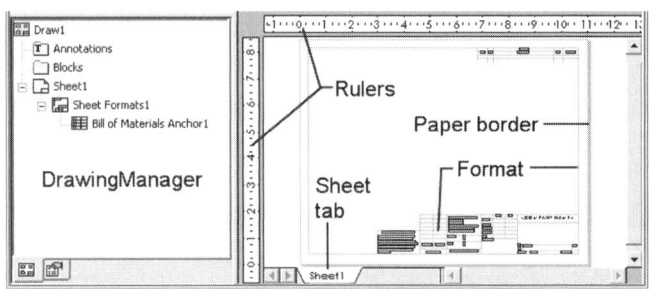

Fig. 11-5. Example of the SolidWorks drawing interface.

✓ **TIP:** *Use the View menu to turn off the ruler bars. The status bar can be turned off the same way.*

Drawing Sheet Formats

First, just as a reminder, sheet formats and templates are two different objects. Formats contain title block and border geometry; templates contain document settings. Drawing templates may or may not contain sheet formats. It all depends on whether or not a format was present when saving the template. See Chapter 2 if you need to refresh your memory on how to create or save template files.

There are two types of sheet formats: standard and custom. Standard formats are those SolidWorks loads on your hard drive when you install the software. Custom formats are any formats you create or import into SolidWorks. The Sheet Format To Use window, shown in figure 11-3, will appear whenever a new drawing template is used that does not already contain a sheet format.

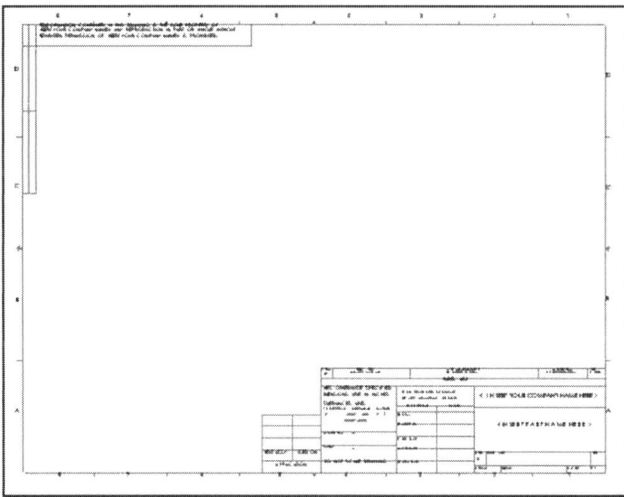

Fig. 11-6. Example of a standard drawing sheet format.

The standard formats included in the software are actually a very good place to start if you want to customize and create your own formats. They provide a basic starting point for the most common sizes of sheet formats, with title block and additional information already added. Figure 11-6 shows a standard drawing sheet format.

Creating a Sheet Format

To edit a sheet format, you must first tell SolidWorks you wish to do so. To create a new sheet format from scratch, you must, in essence, "edit" a blank sheet format. Format geometry and drawing geometry are kept separate. This is a good thing, as it helps reduce the risk of human error. The steps of How-To 11-4 takes you through the process of editing a drawing sheet format.

How-To 11-4: Editing a Drawing Sheet Format

To edit a sheet format, perform the following steps. Use these steps as well to create a new sheet format from scratch.

1. Select Sheet Format from the Edit menu, or right-click anywhere in a blank area of the drawing sheet and select Edit Sheet Format.

2. Edit the format as needed. If creating a new format, use various sketch tools learned in Chapter 3 to create the geometry that will constitute the format.

3. When finished, select Sheet from the Edit menu, or right-click anywhere in a blank area of the drawing sheet and select Edit Sheet.

Step 3 will take you back to editing the sheet, as opposed to the sheet format. You can tell this is the case because the sheet format geometry will turn gray. When editing a format, any drawing views present on the sheet will temporarily disappear. Do not be alarmed, as this is meant to happen. Drawing views will reappear when editing the sheet.

When editing a format, note that the format geometry turns blue or black, instead of gray. This just means that the format geometry is now available for editing. As mentioned in step 2, use any of the tools on the Sketch toolbar. You can draw lines or any other geometry, just as you can when creating a sketch. You can also use the offset, trim, extend, mirror, or other sketch tools. The Linear Step and Repeat command discussed in Chapter 7 works well for creating rows or columns of lines.

Use the Add Relations icon to constrain geometry in the format. Add dimensions if you want to control the exact size and shape of title blocks or anything else in the format. If you do add dimensions, you have a few options for hiding them. One option is to simply delete them. After all, once you create the title block, chances are you will not need the dimensions again.

✓ **TIP:** *If format dimensions are deleted, lock geometry in place using the Fix relation.*

A second option would be to place all dimensions on a layer that can then be hidden. This second option is a good choice, as it is certainly the easiest option to implement. A third option is to hide the dimensions. Layers, dimensions, and hiding dimensions are topics covered later in the chapter.

Almost certainly it will be necessary to add text to the format as well. This can be accomplished by selecting Note from the Insert > Annotations menu. Adding notes is covered in greater detail later in this chapter, but briefly the process works as follows. Access the Note command, click on the sheet to position the note, type in your text, and then click in a blank area of the drawing. To move a note, drag it with the left mouse button.

Adding a Company Logo

When creating a customized format, it is often desirable to add a company logo to the format. This can best be accomplished by using the Windows clipboard copy-and-paste function.

If you already have a company logo, chances are you have it stashed away somewhere on a computer for safekeeping. The logo must be in electronic format, and you must be able to open it for viewing in your favorite computer photo editor. Once you are viewing the logo on your computer, the rest is easy.

All paint and photo editing programs are different, and therefore it is not practical to list the steps for performing this procedure for your specific arrangement. However, all you need to do is copy the logo to the Windows clipboard (typically by pressing Ctrl-C). Then, once you are editing your drawing format in SolidWorks, click somewhere on the screen and paste the image into it by pressing Ctrl-V. Move the object by dragging it with the left mouse button. Scale the object by dragging the handles that appear when the object is selected.

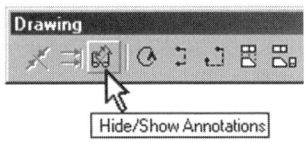

Fig. 11-7. Adding a logo to a format.

Figure 11-7 shows a sample title block with an attached company logo. Anytime the format is used, the logo will appear, because it is embedded in the format. This assumes you have saved the format for use on other drawing sheets or in other drawings. The key is to save the format separately from the drawing or template. Optionally, it would be possible to save the template, in which case the format would be saved within the template file. However, if it is your intention to use the format on other drawing sheets (or in other drawing files), which it almost certainly is, you must save the format as a separate file. How-To 11-5 takes you through the process of saving a sheet format.

How-To 11-5: Saving a Sheet Format

To save a sheet format, perform the following steps.

1. Select Save Sheet Format from the File menu. The Save Sheet Format window, shown in figure 11-8, will appear.

2. Select Custom Sheet Format.

3. Click on the Browse button to save the format in a particular directory of your choosing.

4. Type in a name for the format.

5. Click on Save.

6. Click on OK.

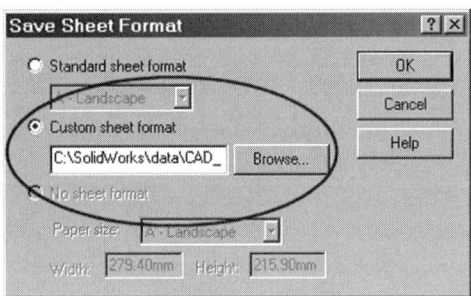

Fig. 11-8. Save Sheet Format window.

The Browse button allows you to place the format in any directory through the use of a typical Windows file save window. If you do not specify a directory, the format will be saved in the default SolidWorks location, where all standard formats are kept. This location is typically the *Program Files\SolidWorks\data* directory. This path may be different on your computer, depending on the SolidWorks installation folder. For your reference, drawing sheet formats have an *.slddrt* file extension.

Although it is possible, you probably will not want to save your customized format on top of an existing SolidWorks standard format. This will overwrite the standard format with your own, and you will no longer have access to the standard format. This is why you should always select the Custom Sheet Format option when saving a format, or at least back up the standard formats before overwriting them.

Drawing Templates

Templates were discussed in Chapter 2, and therefore will not be extensively dealt with here. Because templates have significant bearing on drawings, however, this topic is being discussed here in a somewhat different light.

Keep in mind that templates can exist for parts, assemblies, or design drawings. In the case of parts and assemblies, it would not be uncommon to have two or three templates for each. For example, you might have one part template for working in inches, and one for working in millimeters. Some companies create part templates for materials of varied densities, in which case there may be dozens.

In the case of drawing templates, you might decide to create templates for various sheet sizes, various dimensioning standards (such as ISO or ANSI), metric or English dimensioning units, or any number of other reasons. It may also be necessary to create templates that contain various sheet formats.

Creating a template for a drawing is no different than creating a template for a part or assembly, for the most part. A new SolidWorks drawing is begun, and then the Document Properties are modified to reflect the characteristics the template is to have. Once that is finished, save the drawing as a template file by clicking on Save As in the File menu. The complete steps are listed in Chapter 2 if you need to review this material further.

With regard to drawing templates, an additional step would usually be to add a drawing sheet format to the template. You learned how to do this in How-To 11-1, during the process of starting a new drawing. If a drawing was begun without a format (a blank sheet of paper, in other words), and it is now decided that a format should be added, the next section will show you how this can be done. The "Sheet Setup" section also examines how to change to a different sheet format.

It is very common that a company will already have drawing formats set up, and it would be to their benefit to import these formats into Solid-Works, where they can be reused instead of recreated from scratch. Importing drawing formats is discussed in Chapter 22.

Sheet Setup

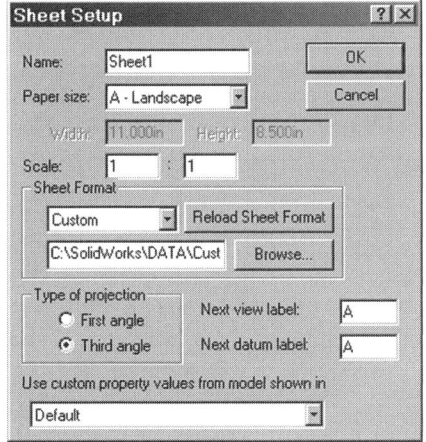

Fig. 11-9. Sheet Setup window.

Almost everything in SolidWorks has properties, whether it is a feature, line, dimension, or drawing sheet. The properties of an object are accessible via the right mouse button and can be used to modify the given object in various ways. In the case of a drawing sheet, properties can be used to modify sheet scale, the format used, paper size, and other options.

Accessing the properties of a drawing will open up the Sheet Setup window, an example of which is shown in figure 11-9. How-To 11-6 takes you through the process of accessing the Sheet Setup window and modifying the sheet's properties.

How-To 11-6: Modifying a Sheet's Properties

To access the Sheet Setup window and modify a sheet's properties, perform the following steps.

1. Select Properties from the Edit menu, or right-click in a blank area of the drawing sheet and select Properties.

2. Modify the sheet's properties as required.

3. To change the paper size, select a new size from the *Paper size* drop-down list.

4. To change the sheet format, specify a new format from the Sheet Format drop-down list.

5. Click on OK to accept the changes.

Most of the various properties within the Sheet Setup window are self-explanatory. If there are any items that may still be unclear, the following material provides some useful notes that should help in regard to the Sheet Setup window.

Paper Size

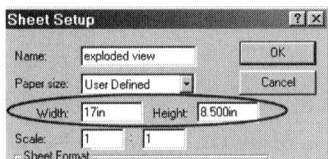

Fig. 11-10. Entering a custom paper size.

Changing the paper size of a drawing in midstream is not a problem. Just specify a new paper size, or define a custom size of your own. To define a custom paper size, select User Defined in the *Paper size* listing and type in values for the paper Width and Height, as shown in figure 11-10. The units suffix can be left off, but make sure to add the suffix if entering values other than the current working units. For example, type *mm* following the value if working in inches and all you know are the metric values.

Sheet Scale

Sheet scale is basically controlled from two settings. One setting is in the Sheet Setup window. It is simple enough to change and to understand. For example, a setting of 1:2 is half scale. All of the views on the sheet will be half their actual size. You need not worry about dimension values, as they will read correctly, regardless of the sheet scale.

The second setting that affects scale is an option titled *Automatic scaling of 3 view drawings*. This option, shown in figure 11-11, will automatically set the scale when the standard three views are inserted. This would correlate to the top, front, and right side views if using third-angle projection. To access this option, Select Options from the Tools menu, and then select the Drawings section (see figure 11-11).

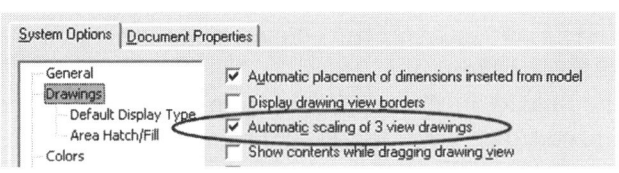

Fig. 11-11. Automatic scaling of 3 view drawings *option.*

Note that the Automatic scaling of 3 view drawings option is not retroactive, and that it only changes the scale upon creating the three standard views, which are covered in material to follow. When checked, the *Automatic scaling of 3 view drawings* option changes the scale found in the Sheet Setup window if Solid-Works deems it necessary. Using the *Automatic scaling of 3 view drawings* option in no way makes the scale value permanent. It is always possible to change the sheet scale at any time.

If the scale of a drawing sheet is set to a particular value, and that drawing is saved as a drawing template, the Scale setting will carry forward to the next new drawing that uses that template. This assumes that *Automatic scaling of 3 view drawings* is turned off. To keep things simple, one recommendation would be to leave *Automatic scaling of 3 view drawings* on and let SolidWorks worry about the scale.

Changing the Sheet Format

If you change the sheet's paper size, it would be a very good idea to change the format as well. You would not want to wind up with an A-size format on a C-size sheet of paper. If you forget to match the paper and format sizes, you will know there is a problem as soon as you exit Sheet Setup. The border of the paper will not match the format size, and will be immediately noticeable.

One of the options in the Sheet Format drop-down list is None, which allows for using a blank sheet of paper. Another option is Custom, which requires clicking on the Browse button and selecting a format.

Sheet formats are treated as OLE objects. OLE is an acronym for Object Linking and Embedding, a programming technology compatible with the Windows operating system that allows the user to insert data from one application to another. More importantly, what this means to you is that sheet formats are embedded objects, not linked to individual files. Once a sheet format has been specified, a copy of the original drawing format file is placed within the drawing. This is being explained because it is directly related to the Reload Sheet Format button. Examine figure 11-10 and you will see the button in question.

The Reload Sheet Format button can be important for two reasons. If changes are made to a drawing format, and then you decide not to keep those changes, reloading the sheet format will remove any changes by reloading a fresh copy of whatever format was originally specified. This requires that the original format exists on the computer or network where the drawing is being edited.

A second reason the Reload Sheet Format button may be important is due to formats being embedded, and not linked. If a change is made to the original drawing format file, it will not automatically propagate to any drawing containing that format. This is the nature of embedded documents. However, if the Reload Sheet Format button is pressed, the changes will transfer from the modified format to the drawing.

Type of Projection

The *Type of projection* section of Sheet Setup allows you to change between the *First angle* and *Third angle* projection options. *Third angle* is the default, and is what most industries in the United States use. This will give the user a top, front, and right side view of the part in a standard engineering drawing.

First-angle projection is used in many European countries. If someone familiar with third-angle projection were to view a drawing using first-angle projection, views would appear to be in the wrong position, and a left-side view would replace what is typically a right-side view. With first-angle projection, the projected view is directly opposite the model with respect to the viewer's line of sight.

Additional Sheet Properties

The next time a section or detail view is created, the letter shown in *Next view label* will be used on that view. The next time a new datum is added, the letter shown in *Next datum label* will be used. The option at the very bottom of the Sheet Setup window relates to custom properties, discussed in detail in Chapter 16. Linking to custom properties is discussed later in this chapter, in the section titled "Notes" (see subtopic "Text Format Panel").

Adding Drawing Sheets

On a single drawing sheet you can mix and match parts and assemblies in any fashion. It is also possible to add as many sheets to a single drawing file as necessary. One scenario might be a model drawing that contains one sheet with the three standard views, a second sheet with section and detail views, and a third sheet with additional auxiliary or projected views.

It is probably not a good idea to place too many sheets in a single drawing, for the same reason it is not good to place all of your eggs in one basket. Additionally, drawings containing too many sheets can become unwieldy due to SolidWorks file associations. Keep in mind that typically SolidWorks will search for a model when opening a drawing because the drawing is referencing the model.

✓ **TIP:** *It is perfectly acceptable to name drawings the same as the part or assembly they reference, because the file extensions will be different. This makes finding drawings that much easier later on.*

How you organize your drawings is a matter of personal or company policy. A drawing of a complex part might require half a dozen or more sheets, which is acceptable. A 20-part assembly drawing with multiple sheets for each component would be a bad idea. Use some common sense when it comes to how many sheets you are adding to the drawing document. How-To 11-7 takes you through the process of adding drawing sheets.

HOW-TO 11-7: Adding Sheets to a Drawing

To add sheets to a drawing, perform the following steps.

1. Select Sheet from the Insert menu, or right-click on the drawing sheet and select Add Sheet. The Sheet Setup window will appear.

2. Specify a name for the sheet, or leave the default name (e.g., Sheet2).

3. Specify paper size, template to use, scale, and other relevant information.

4. Click on OK to add the sheet.

A new sheet tab will appear at the bottom of the drawing, and the new sheet will become the current (top) sheet. An example of sheet tabs is shown in figure 11-12. You will have a tab for as many sheets as exist in your drawing. To switch between sheets, simply click on a tab.

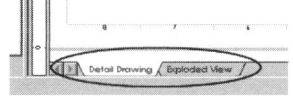

Fig. 11-12. Drawing sheet tabs.

DrawingManager will reflect any sheets added to the drawing or any templates being used for each sheet. Figure 11-13 shows an example of DrawingManager. This particular drawing contains two sheets: Detailed and Tabulated. You can also discern that a format is being used for both drawing sheets. The Detailed sheet currently contains seven views, which are listed as well, although these views do not follow a consecutive numbering scheme, which is common.

Any active sheet's icon will appear in yellow, whereas all other sheets in DrawingManager will appear ghosted (in lighter shades). Also note in figure 11-13 that there is an item under the second sheet format named *Bill of Materials Anchor2*. This is used for setting the anchor point of a BOM that might be used with this particular sheet. Bill of material creation is discussed in Chapter 12, which relates to the topic of assemblies.

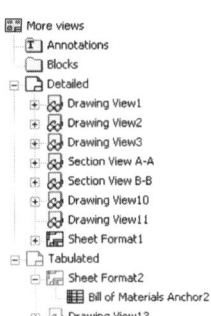

Fig. 11-13. DrawingManager.

Inserting Views

Nearly any view type you could possibly want can be inserted into a SolidWorks drawing. This includes standard three-view layouts, section and detail views, and projected and auxiliary views, to name a few. This section explores bringing these views into your design drawing.

The entire process of inserting views has been automated to a very large degree. You no longer need to spend hours developing a section view. The same section view now takes a matter of seconds because SolidWorks generates the view from the solid model.

Most views require some sort of user input or sequence of events you must perform before the view can be inserted. All of the steps for every view are spelled out for you in the material that follows. By the end of this chapter you will be creating professional-looking drawings.

Bringing the standard three views into a design drawing is normally the first step of any basic design drawing. Whether you use third-angle or first-angle projection is probably dependent on what part of the world you are from. For the sake of this book, third angle will be used. This means that the standard three views will be top, front, and right-side views of the model.

There are four ways to insert the standard three views into a drawing. The four methods are outlined in How-To 11-8. Two methods require you to have the part (or assembly) file open first, and another requires the use of Windows Explorer. You should use whichever method is easiest for you, or most convenient at the time.

How-To 11-8: Inserting the Three Standard Views

To insert the three standard views of a part or assembly into a drawing, perform the steps outlined in any one of the four methods described in the following material.

Method 1: Using the Insert Menu

This method works well if the part or assembly document is not already open. This method also works for named views and relative-to-model views, discussed in material to follow.

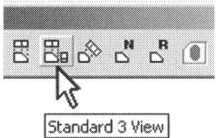

Fig. 11-14. Standard 3 View icon.

1. Select Insert > Drawing View > Standard 3 View, or click on the Standard 3 View icon (shown in figure 11-14), located on the Drawing toolbar.

2. Right-click over the drawing sheet and select Insert From File.

3. Select the file and click on OK.

Method 2: Using Windows Explorer

This method is easy if you are comfortable with the Windows Explorer.

1. Open Windows Explorer.

2. Navigate to the directory that contains the file you want to create the three standard views of.

3. Drag the file onto the drawing sheet using the left mouse button.

Method 3: Drag and Drop from File

If the part or assembly is already open, this method is convenient because it uses standard drag-and-drop functionality.

1. Open the file you want to create the three standard views of (typically the file would already be open for some other reason).

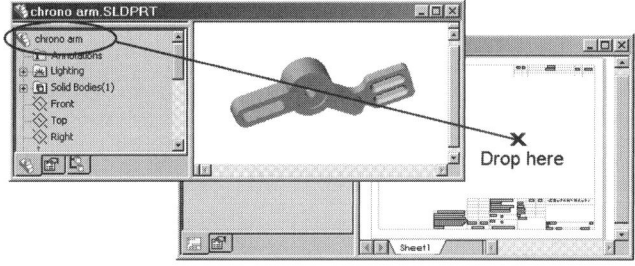

2. Select Tile Vertically or Tile Horizontally from the Windows menu.

3. Drag the model name from the top of its FeatureManager listing into the drawing, as shown in figure 11-15.

Fig. 11-15. Creating the three standard views.

Method 4: Referring to Existing Views

This last method requires either that the file whose views are to be inserted into the drawing be open, or that the drawing already contain a view of the model. The latter is usually the case, because if the model is already open, method 3 is definitely more convenient. This method is very common when creating named views and relative-to-model views. The following steps assume there is already one view of the model (i.e., a named view) present in the drawing.

1. Select Insert > Drawing View > Standard 3 View, or click on the Standard 3 View icon.
2. Select the view that contains the model you are inserting the three standard views of.

In place of step 2, it would be possible to activate the model window and click anywhere inside the work area of the model. At that point, the three standard views would be inserted. However, as mentioned earlier, it is much easier to simply drag and drop the model into the drawing per method 3 than to manipulate the menu structure.

Moving Views

The next thing you will more than likely need to do after inserting views is move the views. It will be easier to explain how to position the views if you can see the view borders. The borders can be toggled on and off through a checkbox in System Options. How-To 11-9 takes you through the process of hiding or showing drawing view borders.

How-To 11-9: Hiding/Showing View Borders

To hide or show drawing view borders, perform the following steps.

1. Select Tools from the Options menu.
2. Select the System Options tab.
3. Select the Drawings section.

Inserting Views

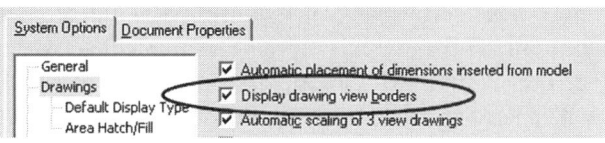

Fig. 11-16. Toggling the display of drawing view borders.

4. Check or uncheck *Display drawing view borders* as required, shown in figure 11-16.

5. Click on OK to accept the changes.

Even if the drawing view borders are displayed, they will never print or plot. They are strictly there for your convenience. To move a view, place the cursor directly over the view border, hold the left mouse button down, and then drag the view to a new position. If a parent view is moved, its dependent views will move as well. In the case of a Standard 3 View, moving the front view moves the top and right side with it. It should be noted that moving a view is still possible even if its border is not visible. Moving the cursor over a view will highlight its border, in which case you could position the cursor over the highlighted border and drag the view to its new position.

If you move the right-side view, it will move by itself. The same is true of moving the top view. If you move the top view, it will be vertically aligned with the front view. If you move the right-side view, it will be horizontally aligned with the front view. This makes it very easy to move the views around while maintaining their alignment, which also maintains a neat drawing.

Modifying View Alignment

It is sometimes desirable to place a view at some other location on the drawing sheet, rather than leave it in its default aligned condition. This is sometimes the case with section views, which you will learn about later in this section. This operation requires breaking the alignment of a view, which is just a right mouse click away. How-To 11-10 takes you through the process of breaking the alignment of a drawing view.

How-To 11-10: Breaking View Alignment

To break the alignment of a view, perform the following steps.

1. Right-click inside the border of the view whose alignment you want to break.

2. Select Break Alignment from the Alignment menu.

3. Move the view to the desired location.

Once a view's alignment has been broken, it is possible to change your mind and return its alignment to the default state. The procedure is the same, but you must select the Default Alignment option instead of Break Alignment. The view will then snap back into position to maintain its default alignment.

Another option you have is to align a particular view horizontally or vertically with another view. Most of the time, the software takes care of view alignments just fine. Normally you will not need to manually align to other views. However, it is nice to have the options available for those times you need them.

You probably already noticed the additional alignment options for aligning a view horizontally or vertically to another view, also found in the Alignment menu. The mechanics behind the procedure are the same as for breaking a view's alignment, with one extra step. How-To 11-11 takes you through the process of horizontally or vertically aligning one view to another.

How-To 11-11: Horizontally or Vertically Aligning a View

To horizontally or vertically align a view, perform the following steps.

1. Right-click inside the border of the view whose alignment you want to change.

2. Select Align Horizontal or Align Vertical from the Align menu. Alignment can be by center or by origin, whatever the task calls for.

3. Click on the view you want to align with.

When you attempt to move the newly aligned view, you will find it will move only horizontally or vertically with the view it is aligned with, depending on your menu selection. When aligning views, there exists the ability to align by center or origin. In most cases, which method is used will make no difference.

Aligning by origin refers to the origin of the part. Aligning by center refers to the center of a bounding rectangle; that is, a rectangle just large enough to enclose the model. In most cases, the views being aligned are of the same model, so either option provides the same result.

Right-clicking Precautions

It should be noted that some care should be taken when right-clicking while working on a drawing. When you are right-clicking, remember that you are accessing a context-sensitive menu. This means that the right mouse button is sensitive to what you click on.

One of three menus will appear when right-clicking in a drawing. Right-clicking on the drawing sheet will result in a menu appearing with various options pertinent to the drawing sheet. Right-clicking on a drawing view will result in a menu with options pertinent to that view. Last, right-clicking on geometry within a view will result in a menu with options pertinent to the geometry over which the cursor was positioned when the right mouse button was clicked. Obviously, it is important to be aware of the cursor's location whenever accessing the right mouse button menu.

✏ **NOTE:** *Always be careful where the cursor is positioned when accessing the context-sensitive menu. This is always important with any SolidWorks document, but even more so in the case of drawings.*

Projected Views

Projected views are very similar in nature to the standard three-view arrangement. When you get right down to it, the three standard views are nothing more than two views projected from a front view: one from the top of the front view and one from the right side.

A projected view can be projected from any view, not just orthographic views. This means that you can project from isometric or auxiliary views as well, though this is not typically done. The mechanics behind the process are very easy. How-To 11-12 takes you through the process of projecting views.

How-To 11-12: Creating a Projected View

To create a projected view, perform the following steps.

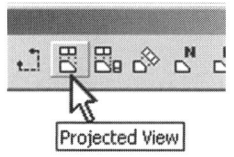

Fig. 11-17. Projected View icon.

1. Select Insert > Drawing View > Projected, or click on the Projected View icon, shown in figure 11-17.
2. Select the view to project from.
3. Specify the side to project to. This must be left, right, above, or below. You will see a rectangular preview, which will help you position the new view.

NOTE: *If a view is selected prior to starting the Projected View command, SolidWorks will not prompt you to select a view first.*

Auxiliary Views

Auxiliary views are similar to projected views in that they are also projected. The difference is in what they are projected from. Projected views must be projected orthogonally (to the left, right, above, or below). Auxiliary views can be projected orthogonally, but usually they are not. Rather, auxiliary views are typically projected from an angled edge. How-To 11-13 takes you through the process of creating an auxiliary view.

How-To 11-13: Creating an Auxiliary View

To create an auxiliary view, perform the following steps.

Fig. 11-18. Auxiliary View icon.

1. Select Insert > Drawing View > Auxiliary, or click on the Auxiliary View icon, shown in figure 11-18.
2. Select an edge to project the auxiliary view from.
3. Position the auxiliary view using the preview as a guide, as shown in figure 11-19.

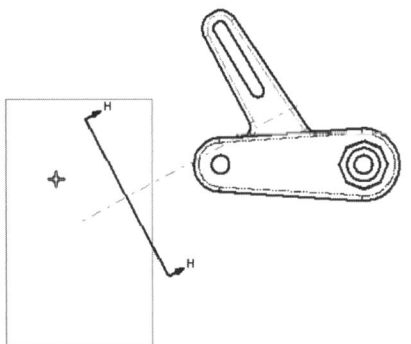

Fig. 11-19. Positioning the auxiliary view.

Auxiliary views are typically used to show a face or feature straight on; that is, as a plan view. This is because the face does not appear flat in any of the other standard views. When the face or feature is shown in the auxiliary view, it can then be dimensioned, and the manufacturer can view the face or feature in its true perspective. This helps distinguish any details that might otherwise be indistinguishable on the face or feature in question.

Named Views

In drafting terminology, a named view is not really a particular type of view. This is a term coined by SolidWorks to describe the view type. A named view can be any view listed in the Orientation window, including views you add. See How-To 3-11 in Chapter 3 for an explanation of how to add user-defined views to a model.

Named View is the function you would use if you wanted to place an isometric view in a drawing. The view does not have to be an isometric view, however. As mentioned previously, it can be any view listed in the Orientation window. How-To 11-14 takes you through the process of creating a named view using one of two methods.

How-To 11-14: Creating a Named View

To create a named view of a part or assembly, perform the following steps, using either method 1 or method 2. Use whichever method is most convenient at the time, depending on your situation.

Method 1: Referring to Existing Views

Use this particular method if there are already views in the drawing of the model you would like to insert a named view of. Because views typically already exist in the drawing, this method is the one you will probably use the most.

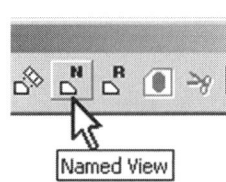

Fig. 11-20. Named View icon.

1. Select Insert > Drawing View > Named View, or click on the Named View icon, shown in figure 11-20.

2. Select any view in the drawing that contains the model you would like to insert a named view of.

3. In PropertyManager, select the named view to be inserted.

4. Click somewhere on the drawing sheet to position the view.

At this point, PropertyManager will still be open. To close PropertyManager, either click on the green check or click in a blank portion of the screen. It makes no difference.

Step 2 states that a view containing the model should be selected. It is also possible to click inside the graphics area of the part or assembly file. You might need to do this if, for example, the drawing did not yet contain any views of the model. You can also select a part or assembly from the FeatureManager of another open document.

If you need to gain access to an open SolidWorks document, use the Window pull-down menu. All open documents will be listed. Simply select the name of the desired file you wish to bring to the top of the screen so that it can be accessed.

✓ **TIP:** *The Ctrl-Tab hot key combination will also allow you to alternate between open documents in SolidWorks.*

Method 2: Inserting from a File

If inserting a named view, the three standard views, or a relative-to-model view (discussed in the next section) and the model file is not open and there are no existing views in the drawing on which to click, there is another option. This option is known as Insert From File.

When inserting any of the aforementioned view types, Insert From File can be selected via the right mouse button menu. This option becomes available in place of step 2 in the previously listed steps for creating a named view. A window will appear, allowing you to select the file from a list. This window looks almost exactly like the standard File > Open window. The rest of the steps are identical.

Relative to Model View Method

Relative to Model is a view insertion option that lets you control how the part is oriented when you insert it into the drawing. The process involves selecting faces on the model and telling the software how you want each face oriented, such as toward the front or the top. How-To 11-15 takes you through the process of creating a view using Relative to Model. The model document should be open when implementing this function, so that faces on the model can be selected.

How-To 11-15: Creating a View Using Relative to Model

To create a view using Relative to Model, perform the following steps. It is suggested that these steps be read through completely before attempting this procedure, so that you have a better understanding of what this command accomplishes.

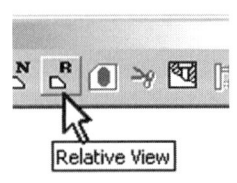

Fig. 11-21. Relative View icon.

    1. Select Insert > Drawing View > Relative to Model, or click on the Relative View icon, shown in figure 11-21.

Inserting Views

2. Select a face on the part or assembly whose view you want to place in the drawing. If a drawing view is selected, that model will be opened in its own window, whereupon a model face can be selected.

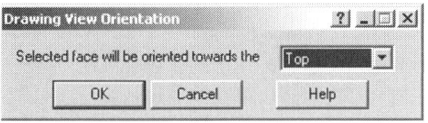

Fig. 11-22. Specifying the orientation for the selected face.

3. In the Drawing View Orientation window that appears, shown in figure 11-22, specify how you want the selected face to be oriented.

4. Click on OK.

5. Select a second face on the part or assembly.

6. Once again, specify how you want the selected face to be oriented.

7. Click on OK.

8. Select a position on the drawing sheet where the view is to be positioned.

The Relative to View method is probably not one of the more commonly used view creation methods. However, it is convenient from time to time.

Section Views

SolidWorks handles section views very nicely. You will find that the Section View command will save you countless hours of manually drafting such views. Section views require that a line be added to the view to be sectioned. This line, or set of lines, is what becomes the section line.

Because geometry is being added to a view, there must be some way to tell the software that the geometry is associated with the view. Otherwise, when the view is moved or repositioned, the geometry will not know enough to move with it. There is a way to associate geometry with a view, which is covered in the next section. It is a very important section and one you should pay close attention to.

Activating Views

Whenever geometry is added to a view, you must first activate that view. This process, described in material to follow, is different from simply selecting a view. The term *geometry* is taken to mean any lines, arcs, or other sketch-type entities. Dimensions will be associated with a view whether or not the view is activated.

When a view is activated, any geometry added to the drawing will be associated with that view. If the view is moved, the geometry will move with it.

✏ **NOTE:** *It is important to know which view is activated when adding geometry! Otherwise, the geometry will be associated with the wrong view. Additionally, geometry is only accessible when the appropriate view is activated.*

To activate a view, double click it. Another option is to right-click within the view's border and select Activate View from the menu. Either method is fine, so use whatever method you find easiest.

When a view is activated, it will have a gray-shadowed border. An example of an active view is shown in figure 11-23. Note the gray-shadowed border. Once again, note that an activated view and a selected view are not the same. Just because a view is selected does not necessarily mean it is activated, and vice versa.

A drawing sheet can also be activated. Making the sheet active allows you to associate geometry with the sheet, instead of any one particular view. You would want to activate the sheet whenever adding geometry that should remain in the same position, whether or not any of the views are moved.

Fig. 11-23. Activated view.

Dynamic Drawing View Activation

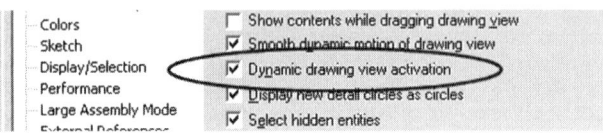

Fig. 11-24. Enabling dynamic drawing view activation.

If you prefer, SolidWorks can dynamically activate views automatically. This can simplify things and makes it unnecessary to always keep track of which view is active. To enable automatic view activation, make sure the option Dynamic drawing view activation is checked. You will find this option in the Drawings section of the System Options tab (Tools > Options menu). The option is shown in figure 11-24.

Some may prefer to leave dynamic view activation off, as it tends to cause a lot of flashing on the screen. This is a result of the views activating automatically as the cursor passes over them. This is strictly a personal preference. If you decide to enable this function, be aware that you can temporarily override the automatic view activation by locking the focus on a particular view or the sheet.

To lock the focus on a particular view, right-click on the view and select *Lock view focus*. By right-clicking on the drawing sheet, it is possible to use

Inserting Views **433**

Lock sheet focus. Finally, if the focus has been locked onto the sheet, it is also possible to use the *Unlock sheet focus* option to unlock it. Locking the focus on a drawing view and activating it mean the same thing.

Now that you understand how to activate views, we can get on with creating section views. You can begin by drawing a line that will become the section line. Section lines can consist of lines and arcs in a wide variety of combinations. They can also be construction entities. It makes no difference. How-To 11-16 takes you through the process of creating a simple section view.

How-To 11-16: Creating a Section View

To create a section view, perform the following steps.

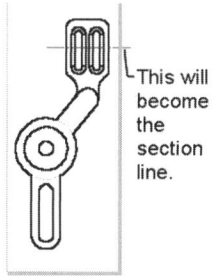

Fig. 11-25. Creating the section line.

1. Activate the view you want to create a section of.

2. Create geometry that will represent the section line, such as shown in figure 11-25. Use constraints if desired.

3. Select the section line. If the section line contains more than one segment, select the segment you want to project the section view from.

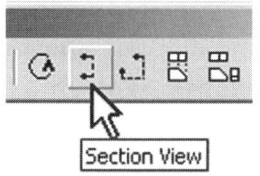

Fig. 11-26. Section View icon.

4. Select Insert > Drawing View > Section, or click on the Section View icon, shown in figure 11-26.

5. Position the section view using the preview as a guide. The section line arrows will flip, depending on the position of the new section view.

The first thing you will notice is that the geometry you added to the view to be sectioned now looks like a section line, as shown in figure 11-27. Figure 11-28 shows what the section view itself would look like.

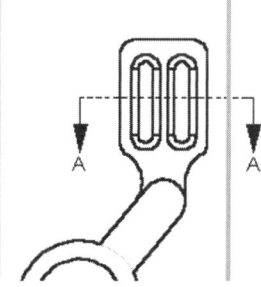

Fig. 11-27. The line has been transformed.

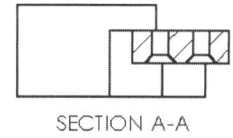

Fig. 11-28. Section view.

✓ **TIP:** *Instead of creating geometry first, just click on the Section View icon. You will automatically be placed in the line command, and can then sketch the section line.*

Flipping Section Line Arrows

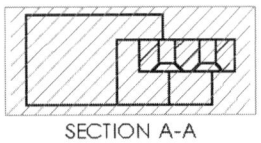

Fig. 11-29. A view in need of updating.

One item you may decide to alter after creating a section view is the direction of the arrows on the section line. To flip these arrows in the opposite direction, double click on the section line. This will result in the section view looking like it has crosshatch lines running completely through the view, as shown in figure 11-29.

Whenever these lines are crossing a view, it means the view requires updating. Updating is necessary anytime you make a modification that alters the way a view will appear. You have two choices. If you want to rebuild the entire drawing, click on the Rebuild icon and be done with it. This can sometimes take a little while, depending on the complexity of the drawing and other variables.

Your other choice is to update only the one view. You can do this by right-clicking over the view and selecting Update View. This will rebuild only the selected view, rather than the entire drawing, and is the preferred rebuild method for a single view.

Section View Properties

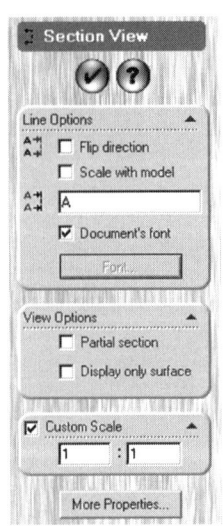

Fig. 11-30. Section View panel.

Selecting either a section line or its associated view will display the Section View panel in PropertyManager, shown in figure 11-30. This assumes you have the Auto-show PropertyManager option enabled, discussed in Chapter 2. The arrow direction can be altered here (*Flip direction* option), along with the section line label and the font type and size used for the label.

The *Scale with model* option will scale the section line with the model. This is opposed to having the section line remain in one particular location if the model changes size or shape. Assuming there are no geometric constraints on the underlying section line geometry, you should be able to drag it to new positions using the left mouse button. If you wish to alter the geometry, perhaps to add or remove relations, right-clicking on the section line will provide access to the Edit Sketch command.

You would want to use the *Partial section* option when creating a section line that does not cut completely across the part. An example of this is shown in figure 11-31. Under these circumstances, if *Partial section* were not checked, the view would not appear correctly. The view would appear

Inserting Views

brown, indicating that there is a problem. In such a case, make sure *Partial section* is checked.

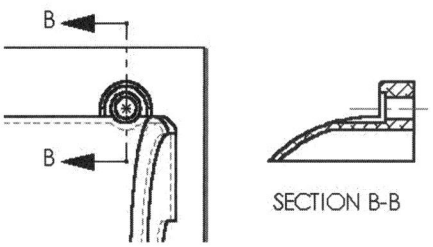

Fig. 11-31. Showing a partial section.

Fig. 11-32. Using the Display only surface cut option.

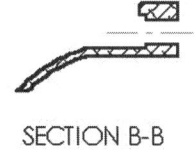

If creating a section line that does not cut completely across the part, SolidWorks will inquire if you would like to create a partial section view. You should answer yes by clicking on the Yes button on the prompt window. However, occasionally the software asks when it does not need to. This may happen if the section line cuts completely through geometry, but does not necessarily extend from one side of the model to the other. When prompted to create a partial section under these circumstances, it would be acceptable to select No.

When *Display only surface cut* is checked, only the sectioned surface is shown in the view. An example of this is shown in figure 11-32. Compare this to the section view shown in figure 11-31 and you will notice the difference.

Finally, there is the option to change the view's scale. All views can have their individual scales set independently of the sheet. Just check the *Custom scale* option and enter whatever value is desired.

Section Line Positioning

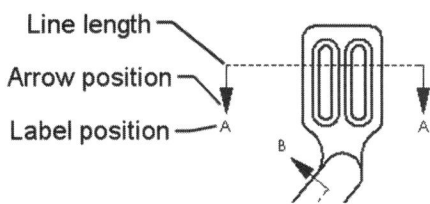

Fig. 11-33. A section line's drag points.

When the section line is selected, small green handles appear that allow for dragging (moving) the arrows. Other drag points would be at either end of the section line itself, which allows for changing the length of the section line, as long as the underlying line's endpoints were not fully defined. No green handle will appear, but the section line length can be changed nonetheless. Figure 11-33 shows where the various drag points are located, which also includes the labels.

Aligned Sections

Another type of section view is an aligned section. In this case, the section line must consist of two segments. The section line is "unfolded" and the resultant view is displayed as if it were projected perpendicularly from each segment in the section line.

The illustrations that follow help to elucidate how an aligned section works. First, as with any section view, geometry that will define the section lines must be added to an active view. Again, this must be a two-segment section line if using the Aligned Section View command. Figure 11-34 shows the first of two related illustrations. Note that the resultant view will be a partial section due to the upper line not extending all the way through the geometry.

Next, the line to project from is selected. Make sure to select one line segment only, not both. This is important not just for aligned views but for any section view with a section line that has more than one line segment. In figure 11-34, the vertical line was selected.

Now it would just be a matter of carrying out the Aligned Section View command. Figure 11-35 shows the result of creating an aligned section view using the section line geometry shown in figure 11-34. In contrast, a standard section view is shown on the right of figure 11-35. Aligned section views are unfolded, whereas the standard section view is not.

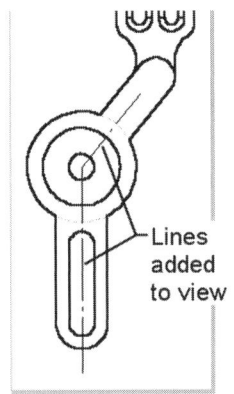

Fig. 11-34. Creating the section line geometry.

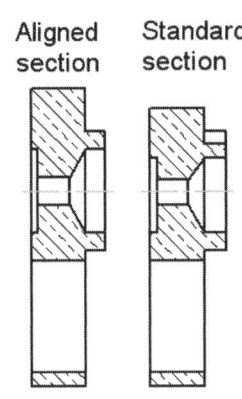

Fig. 11-35. An aligned section view (left).

How-To 11-17 takes you through the process of creating an aligned section view.

How-To 11-17: Creating an Aligned Section View

Fig. 11-36. Aligned Section View icon.

To create and aligned section view, perform the following steps.

1. Activate the view you want to create a section of.
2. Add two line segments that will become the section line.
3. Select one line to project the new section view from.
4. Select Insert > Drawing View > Aligned Section, or click on the Aligned Section View icon, shown in figure 11-36.
5. Click to position the aligned section view on the sheet.

More elaborate section views can be created from more complicated section line geometry, a sample of which is shown in figure 11-37. You would need to use the Section View command in this example, rather than Aligned Section View, due to the number of segments in the section line.

Inserting Views

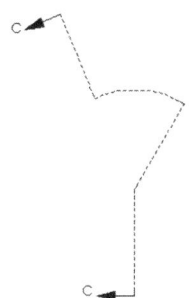

Fig. 11-37. Example of section line geometry.

If it is ever necessary to create a section line without generating a section view, use the Make Section Line command, found in the Insert menu. Because no view is generated from the resultant section line, you will probably not often use this command.

Modifying Crosshatch

There are two places it is possible to modify the hatch style being used. You should first ask yourself if you want to modify the hatch pattern associated with the part file, or if you want to modify the hatch strictly for a particular view. The following material explains why this question is so important.

If the hatch pattern is modified in the drawing, the section view will look correct. However, will the part being sectioned be used in assemblies? If so, it would be more prudent to change the hatch associated with the part itself. This will make the crosshatching automatically appear correctly in other drawings. Furthermore, if the part is used in an assembly and the assembly later appears in a drawing section view, the crosshatch for the part will already be set. The moral of this story is to set the crosshatch properties up in the part once and be done with it. How-To 11-18 takes you through the process of modifying crosshatch associated with a part.

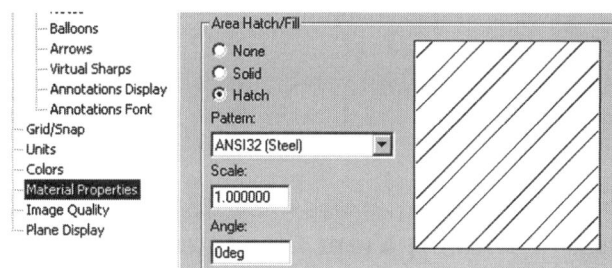

Fig. 11-38. Changing a model's crosshatch.

How-To 11-18: Changing Part Crosshatch

To change the crosshatch of a part so that any section view of that part will always display the correct hatch pattern, perform the following steps. Bear in mind that this must be done in a part file, not a drawing, with the result that the change will propagate to the drawing.

1. Select Options from the Tools menu.

2. In the Document Properties tab, select Material Properties.

3. Select the appropriate crosshatch pattern from the drop-down list.

438   **Chapter 11: Design Drawings**

4. Optionally, specify the desired scale and rotation.

5. Click on OK.

✓ **TIP 1:** *While modifying the crosshatch, it is a good idea to specify the density of the part if this information is known. There is no Material Properties section for drawings (or assemblies), so if you attempted the steps listed previously and could not locate this section, it is probably because a part file was not open.*

✓ **TIP 2:** *If you consistently use parts of a specific material, it is a good idea to incorporate that information into a part template.*

Any changes made to the crosshatch in the Material Properties section of the Document Properties window will immediately affect the part. Anytime a section view of the part is shown in a drawing, the crosshatch will take on the characteristics specified in the Material Properties section. Knowing this, why would it ever be necessary to modify crosshatch properties from the drawing view? One common reason is associated with the case of assembly section views. It is often necessary to rotate the hatch patterns to make it easier to discern individual components. Another reason is simply flexibility.

If you are interested in changing just the crosshatch properties of a specific view, the procedure is different than that for a part file. How-To 11-19 takes you through the process of changing the crosshatch properties of a view.

How-To 11-19: Changing View Crosshatch

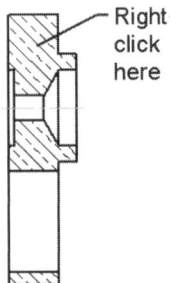

Fig. 11-39. Modifying view crosshatch.

To change the crosshatch properties of a view, perform the following steps.

1. Right-click on any crosshatch area in the drawing view, as indicated in figure 11-39.

2. Select Properties.

3. In the Area Hatch/Fill window, shown in figure 11-40, make any necessary changes to the crosshatch pattern, scale, and angle.

Inserting Views

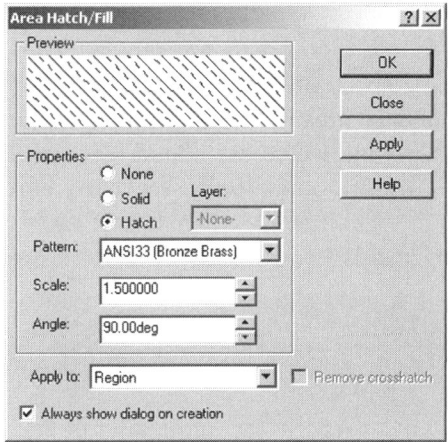

4. In the Apply To drop-down list, specify either View or Region. If you are in an assembly section view, applying changes to the part will also be an option.

5. Click on OK to accept the changes.

Fig. 11-40. Area Hatch/Fill window.

The Area Hatch/Fill window is almost identical to the Material Properties section in the Document Properties window. The biggest difference is the *Apply to* option. If View is selected, the crosshatch will be changed for any hatching in the entire view. If Region is selected, only the boundary area selected will have its pattern characteristics altered.

The Part option is only available in the *Apply to* listing if you are working with a section view of an assembly. Specifying Part will alter any crosshatch associated with the entire part whose crosshatch you are changing in the specified view. All regions (hatch boundary areas) of the part will be affected.

Crosshatch can be solid, if desired. It can also be placed on a specific layer. In this way, crosshatch can be a specific color, taking on whatever color is assigned to the layer. For those not familiar with layers and their uses, there is a section devoted to layers later in this chapter.

The *Remove crosshatch* option removes any changes made to the crosshatch properties. It does not actually remove the crosshatch itself. If you employ this option, whatever settings have been set in the Material Properties parameters for the part will take effect. Earlier modifications made to the crosshatch in the drawing view will be discarded. The *Always show dialog on creation* option refers to area hatch, described in the following section.

✓ **TIP:** *Modifying crosshatch in the part's document properties is the preferred method of changing crosshatch, because the setting is saved with the model.*

Area Hatch

When section views are created, crosshatch is automatically applied where necessary. Area hatch is simply crosshatch applied manually. Area hatch can be applied to any surface or boundary area you would like hatch to appear. In many cases, area hatch can be applied to nonplanar faces, with the limitation that the face does not wrap around itself, displaying both inside and outside faces (for example, a cylinder). Area hatch can be applied in drawings only. How-To 11-20 takes you through the process of setting up and adding area hatch.

How-To 11-20: Adding Area Hatch

Prior to adding area hatch, you will want to set the appearance (or characteristics) of the hatch pattern. To set the default characteristics of area hatch, and then to apply area hatch, perform the following steps.

1. Select Options from the Tools menu.
2. In the System Options tab, select the Area Hatch/Fill category found under Drawings.
3. Specify the Pattern, Scale, and Angle for the area hatch. You can also choose to use a Solid fill pattern, if desired.
4. Click on OK when finished. Now you are ready to add area hatch.
5. Select a face or faces on which to apply hatching.
6. Select Area Hatch from the Insert menu.

During step 6, you may see the Area Hatch/Fill window, shown in figure 11-40. If the *Always show dialog on creation* option in the Area Hatch/Fill window is checked, the window is displayed every time area hatch is added. Otherwise, the default settings are used for the area hatch. Editing area hatch is performed the same way crosshatch is edited, but there is no *Apply to* option in the Area Hatch/Fill window because the crosshatch is added manually. Figure 11-41 shows a drawing view in which area hatch has been applied to one face.

✓ **TIP:** *Sometimes area hatch can be used to simulate qualities such as knurling.*

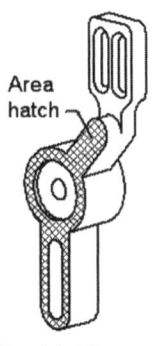

Fig 11-41.
Drawing view
with area hatch.

Broken-out Section Views

Similar to section views, broken-out section views display a cross section of a model. In the case of a broken-out section, the cross section is taken by removing a portion (or "breaking out") a bit of material on the model, to see what is underneath. A depth must be specified in some way to indicate the depth of the broken-out section. Closed profiles are used rather than the open-profile sketch geometry that would be used with typical section views. To create a broken-out section view, perform the steps outlined in How-To 11-21.

HOW-TO 11-21: Creating a Broken-out Section View

To create a broken-out section view, perform the following steps.

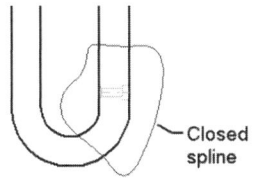

Fig. 11-42. Closed spline sketch geometry.

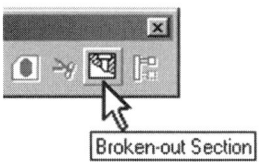

Fig. 11-43. Broken-out Section icon.

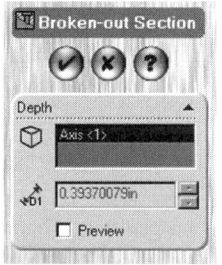

Fig. 11-44. Broken-out Section panel.

1. Activate the view on which to create the broken-out section.
2. Add sketch geometry in the shape of a closed profile, which will dictate the shape of the broken-out section. Figure 11-42 shows a closed spline being used as an example.
3. Select the profile geometry.
4. Select Insert > Drawing View > Broken-out Section, or click on the Broken-out Section icon, shown in figure 11-43.
5. In the Broken-out Section panel, shown in figure 11-44, either enter a value for the section depth or select an object that will dictate the depth. Examples of objects that can be selected are model edges and axes.
6. Click on OK to create the broken-out section view.

An example of a broken-out section view is shown in figure 11-45. Crosshatch is added automatically by SolidWorks. Sometimes broken-out sections can be used in the place of section views, which would have the advantage of not requiring a parent view. Using a spline to indicate the break-out profile is not required. Any geometry can be used, including a rectangle that surrounds the entire drawing view.

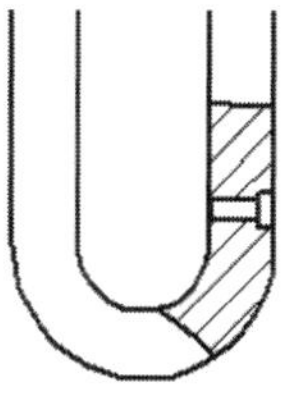

Fig. 11-45. Broken-out section view.

Detail Views

Like section views, detail views require that you add geometry to the view you want to create a detail of. The geometry required can take any shape. The detail view itself can be a circle or some other geometric shape. As with section views, the view being used to create the detail view must be activated before adding the geometry. The geometry created, which will represent the detail area, should be a closed profile.

Creating a Detail View

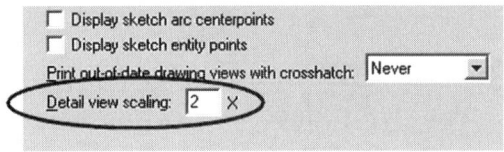

Fig. 11-46. The detail view default scale setting.

Like any view, the scale of a detail view can be changed independently of the sheet. However, you can establish the default scale for detail views in the Drawings section of System Options (Tools > Options). Figure 11-46 shows the *Detail view scaling* option.

The value set in the *Default view scaling* option is what the sheet scale will be multiplied by. For example, if the sheet scale is set to 1:2 and *Default view scaling* to 2X, the detail view will have a scale of 1:1. How-To 11-22 takes you through the process of creating a detail view.

HOW-TO 11-22: Creating a Detail View

To create a detail view, perform the following steps.

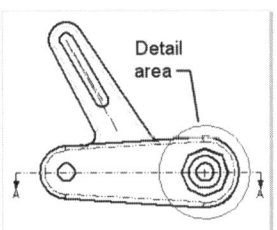

Fig. 11-47. Creating a detail view profile.

1. Activate the view you will be creating a detail of (see section "Activating Views").

2. Draw a circle (or shape) around the area to be detailed, as indicated in figure 11-47.

3. Select the circle. If other geometry was used, such as a rectangle, select one line segment.

4. Select Insert > Drawing View > Detail, or click on the Detail View icon, shown in figure 11-48.

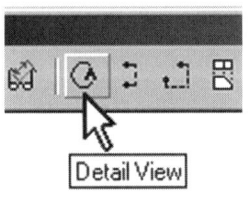

5. Click on the drawing sheet (left mouse button) to position the detail view. A detail view is shown in figure 11-49.

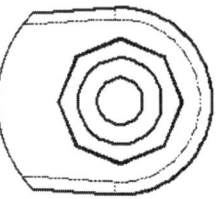

Fig 11-48. Detail View icon.

Fig. 11-49. Detail view.

Detail View Properties

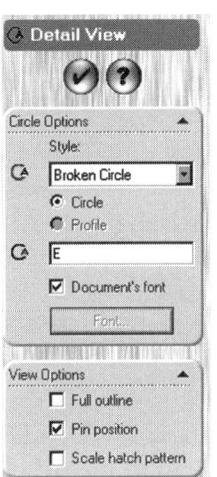

Fig. 11-50. Portion of the Detail View panel.

Similar to section lines and section views, selecting a detail view or its associated geometry displays the detail view's properties. A portion of the Detail View panel, displayed in PropertyManager, is shown in figure 11-50.

The Detail View panel provides access to a number of options, such as the detail view Style, Label, and Font. With regard to the Style setting, the Per Standard option will display the detail view circle per whatever drafting standard the drawing is set to (e.g., ANSI).

You also have the ability to show the actual profile geometry used to distinguish the detail area, as opposed to showing a circle only. If Profile is not checked, SolidWorks takes whatever geometry was used to define the detail area and turns it into a circular area. Selecting the Profile option, however, allows for displaying any geometric shape for the detail area. Note that only the With Leader, No Leader, and Connected styles allow for using the Profile option. Figure 11-51 shows an example of using the Connected detail style with the Profile option selected.

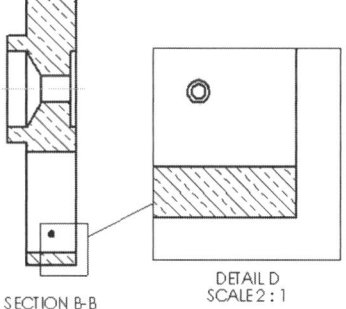

Fig. 11-51. A connected detail view using a square detail area.

✓ **TIP:** *A detail view's label is bidirectionally associative with the label associated with the detail circle. Changing one automatically updates the other.*

It is possible to drag the detail area in order to resize it, if the underlying geometry is not fully defined. If you drag the centerpoint of the detail circle, you can also alter its location. When moving or resizing detail geometry, the detail view is updated automatically and does not require a rebuild.

Any constraints placed on the original detail geometry will limit your ability to alter that geometry. For example, if a detail circle has been constrained to a particular position, you are not going to be able to move it. This behavior is identical to that exhibited by section line geometry. Right-clicking on detail geometry will provide you with access to the Edit

Sketch option. Full editing options can be carried out on the geometry. Once finished, rebuild the drawing, as usual.

There are three other options that require explanation. These options are *Full outline*, *Pin position*, and *Scale hatch pattern*, described in the sections that follow.

Full outline Option

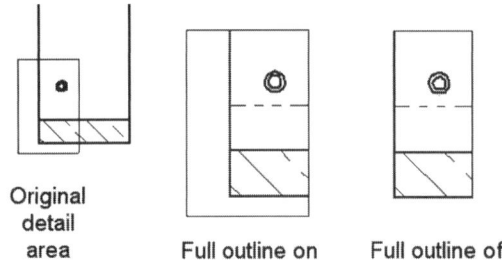

Fig. 11-52. Effects of the Full outline option.

If the detail area does not completely fall on the part, *Full outline* will draw a border around the detail view. This will work with any profile geometry. An example is shown in figure 11-52.

Pin position Option

With the *Pin position* option on, the detail view will stay locked in its current position, even if the parent view changes size. The only time this might be a problem is when the parent view changes to a large extent and runs into the detail view.

Scale hatch pattern Option

When you have detail view of an area on a section view, checking *Scale hatch pattern* will increase the scale of the hatch using the same scale value as the detail view.

Broken Views

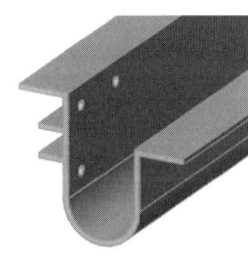

Fig. 11-53. A long, extruded part.

When a part is too long to fit on a drawing sheet, you would typically create what is known as a broken view. Broken views can be created in SolidWorks with minimal effort. The break lines can be adjusted or modified in appearance, and the break gap can be altered. You will learn how to perform all of these functions.

As an example, the extruded part shown in figure 11-53 will be used. This part is 32 inches long, and could fit quite nicely on an E-size sheet of paper. However, an E-size sheet would be overkill for a part of this nature. There are no detailed features on this part, except for holes drilled in the center of the part and three holes on each end.

Figure 11-54 shows the same part on a piece of paper it is obviously too big for. Once the part is broken, it will fit just fine on the A-size sheet of paper, even at 1:2 scale. How-To 11-23 takes you through the process of breaking a view.

Inserting Views

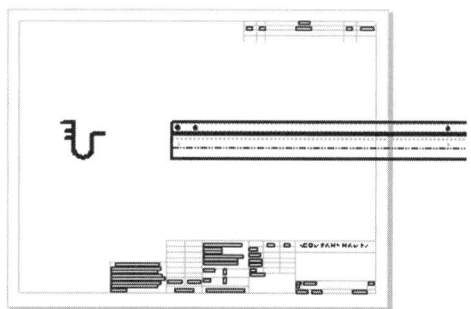

Fig. 11-54. Prior to breaking the view.

HOW-TO 11-23: Creating a Broken View

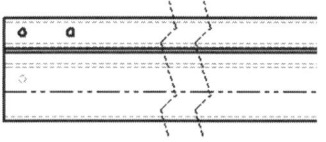

Fig. 11-55. After inserting break lines.

To create a broken view, perform the following steps.

1. Select the view to be broken.

2. Select Horizontal Break or Vertical Break from the Insert menu, depending on the situation. Break lines will appear on the view, as shown in figure 11-55.

   There are a few details worth mentioning at this point. First, break lines are not considered sketch geometry. Therefore, it is not necessary to activate a view prior to inserting break lines, as you may have deduced from the first two steps. Second, break lines can be added to a view to create more than one break. Simply repeat the first two steps as necessary. In our example, two sets of break lines will be added.

3. Drag the break lines where the break (or breaks) are to occur. Use the left mouse button for this.

4. Right-click on the view to be broken (but not on part geometry) and select Break View.

5. Reposition the view or break lines as needed. A completed broken view is shown in figure 11-56.

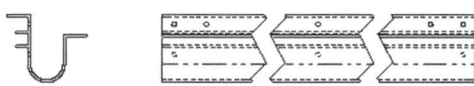

Fig. 11-56. Completed broken view.

As mentioned in step 5, break lines can be repositioned by dragging them to a new location. The appearance of the break lines can be changed as well. Right-click on one of the individual break lines and you will see four options at your disposal. These options include straight, curved, zigzag, and small zigzag break lines, which are shown in this order in figure 11-57.

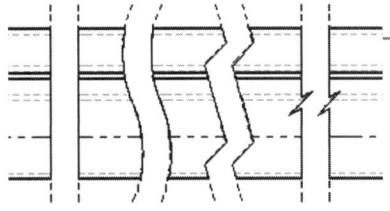

Fig. 11-57. Break line choices.

You also have control over the break line gap and extension. The extension is how much the break lines extend out from the model. How-To 11-24 takes you through the process of establishing the break gap and break line extension.

How-To 11-24: Establishing Break Gap and Break Line Extension

To establish the break gap and break line extension, perform the following steps.

1. Select Options from the Tools menu.

2. Select the Detailing section from the Document Properties tab.

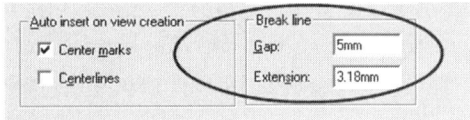

Fig. 11-58. Setting the break line gap and extension.

3. Modify the break line Gap setting, shown in figure 11-58, as required.

4. Modify the break line Extension setting. This value can be zero if desired.

5. Click on OK.

Dimensions, of course, will be inserted with the proper value, regardless of whether they were added before or after breaking the view. Adding dimensions is covered in material to follow.

Cropped Views

When a view is cropped, portions of the view that are not critical to the drawing are hidden. A profile is created around an area of the view that is to remain. When the view is cropped, everything outside the profile is removed. How-To 11-25 takes you through the process of cropping a view.

How-To 11-25: Cropping a View

To crop a view, perform the following steps.

1. Activate the view, in that you will be adding sketch geometry to it.

2. Sketch a closed profile around the area of the view that is to remain visible. An example of this process is shown in figure 11-59.

3. Select any single segment in the crop profile sketch.

4. Select Tools > Crop View > Crop.

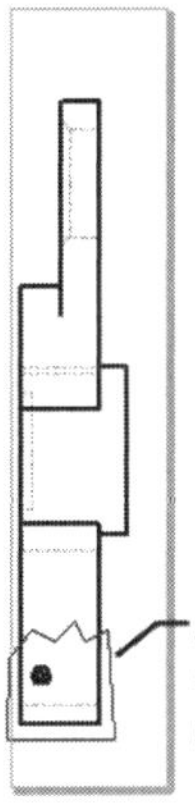

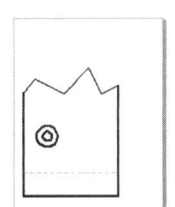

Fig. 11-59. Preparing to crop a view.

A completed cropped view is shown in figure 11-60. If it becomes necessary to edit a cropped view, possibly to alter the underlying sketch geometry or to change the cropped area, select the view, and then use the Edit Crop option found in the Crop View menu (accessed in step 4). The same menu also has an option for removing a cropped view. Actually, the underlying view will remain, but it will no longer be cropped. If you edit a cropped view, make sure to rebuild the drawing afterward.

Fig. 11-60. Cropped view.

Empty Views and Creating Tables

At first you may wonder why you would ever need to create an empty view. The view would not so much be an empty view as it would be a placeholder for geometry you create. An empty view allows you to add geometry to the view while it is active, and then move the geometry as you typically would move a view. This is a way of "grouping" geometry.

You may find it necessary to add geometry for a number of reasons. Possibly you would like to create a simple detail of a part, and it may actually be easier to just sketch it in by hand rather than create a new view. Creating an empty view gives you that option, and groups the geometry because all of the geometry is associated with the view.

To create an empty view, select Empty from the Insert > Drawing View menu, and then use the preview to help position the view. You may want to make sure the drawing view borders are being displayed; otherwise, the empty view will not be visible unless the cursor is positioned over it, or until it is activated or some geometry is sketched in it. Make certain the empty view is activated before adding geometry to it, or you will defeat the purpose of using the Empty View command.

Empty views can come in handy when it is necessary to create a table in a drawing. Additionally, the linear step-and-repeat sketch tool works well for creating tables. The table shown in figure 11-61 was created in about 60 seconds. It can easily be moved about by dragging the view's border to another location. The linear step-and-repeat function was examined in Chapter 7.

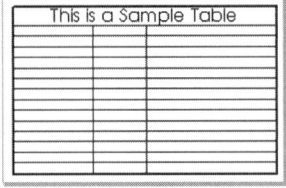

Fig. 11-61 Creating a table in an empty view.

That wraps it up for the various view types that can be created in SolidWorks. However, it will undoubtedly be necessary to tweak the appearance of individual views to meet certain requirements, or for the sake of clarity. These topics are discussed in the following section.

View Appearance

There are a number of options that can be used to change the appearance of views. These are options such as hiding tangent edges, controlling what lines are being shown, setting default display characteristics, and so on. Some views may need to have their hidden lines removed, others may require that hidden lines be shown, and yet others need to be shaded.

How is the appearance of individual views controlled? The answer is the View toolbar, of course. What could be easier? Select a view, and then click on the appropriate icon to show the view in the desired state. The View toolbar was discussed in Chapter 3, and is shown in figure 11-62 for your reference. Icons that control view appearance are circled.

Fig. 11-62. Controlling view appearance from the View toolbar.

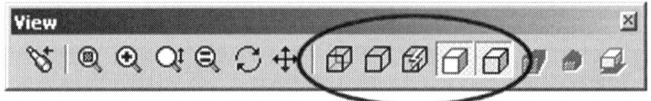

Hiding Individual Edges

Individual edges can be hidden when necessary. Simply right-click over the edge to be hidden. In the menu that appears, you will see the option Hide Edge. If you change your mind, position the cursor over where the edge used to be. The hidden edge will highlight, and you will be able to right-click over it. The menu that appears will contain a Show Edge option.

View Appearance

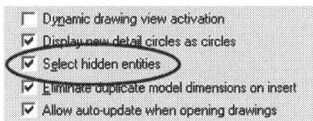

Fig. 11-63. Activating Select hidden entities.

To be able to show edges that are hidden, you must turn on the *Select hidden entities* option, shown in figure 11-63. This option is found in the Drawings section of the System Options window (Tools > Options).

Tangent Edges

One display option you may commonly need to change is the appearance of the view's tangent edges. Tangent edges may need to be hidden, or perhaps the tangent edges should be displayed with a lighter, noncontinuous line type, for instance. This really depends on the part, though.

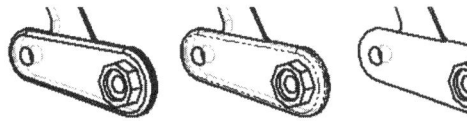

Fig. 11-64. Tangent edge display options.

You have three choices when it comes to displaying tangent edges: display them, turn them off, or display the tangent edges with a specific style of line (also known as a line font). These options are found in the right mouse button menu. To access them, select the Tangent Edge menu after right-clicking on a view. Figure 11-64 shows the pivot arm part with Tangent Edges Visible activated (left), Tangent Edges with Font activated (middle), and Tangent Edges Removed activated (right).

You can control the default behavior of how tangent edges are displayed for new views added to the drawing. This setting is in the Default Display Type section of System Options. There is also a setting for what display mode is used for new views. How-To 11-26 takes you through the process of changing the default display characteristics of new views (not existing views).

How-To 11-26: Setting Default Display Characteristics

To change the default behavior of tangent edges in new views, and to set how new views are displayed, perform the following steps.

1. Select Options from the Tools menu.

2. Select the Default Display Type section.

3. In the area labeled Default display mode for new drawing views, shown in figure 11-65, specify how new views should initially appear.

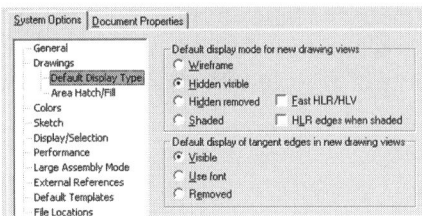

Fig. 11-65. Setting default view display characteristics.

4. In the section labeled *Default display of tangent edges in new drawing views*, specify how tangent edges are displayed in new views.

5. Click on OK when finished.

Default settings should be settings that are most frequently used. View appearance can always be changed later, as you have recently discovered. In the *Default display mode for new drawing views* section, there is a setting titled *Hidden visible*. When used, the hidden lines can be solid gray or dashed, depending on line font options in Document Properties (discussed in material to follow). The line font used for tangent edges can also be specified by the end user.

Hidden edges in drawings are, by default, displayed with a thin dashed line type. For tangent edges, the default line type when the Tangent Edges with Font option is used is a thin phantom line. These settings work well, and it is not advised you make modifications to the line fonts used by SolidWorks unless you feel confident you understand what you are doing. How-To 11-27 takes you through the process of changing what line fonts are used for various objects, such as tangent edges and hidden lines.

HOW-TO 11-27: Modifying an Object's Line Style

Actual line styles cannot be modified and new line styles cannot be created by the end user, at least not easily. However, there is a list of line types and thickness settings that can be altered to meet particular requirements.

Line styles are known as line fonts in SolidWorks. Any changes made to line fonts are saved with the document, and do not affect any previously created files or new documents. It is possible to incorporate line font changes into a drawing template if desired. To modify the line font used for specific SolidWorks objects or operations, perform the following steps.

✗ **WARNING:** *Changing line font settings indiscriminately can be detrimental to your current drawing. Use caution when performing these steps! It is recommended you create a copy of a drawing and use that as a test bed.*

1. Select Options from the Tools menu.

View Appearance

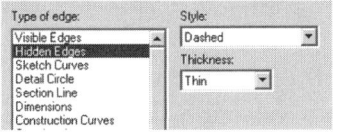

Fig. 11-66. Changing an object's line style.

2. Select the Line Font section in the Document Properties tab. A portion of the Line Font section is shown in figure 11-66.

3. In the *Type of edge* list, select the object you wish to change the line font for. A preview of the line type for that object will be displayed.

4. Change the Style and Thickness (line weight) if desired.

5. Click on OK to accept the changes.

Note that much more than hidden lines and tangent edge display can be altered (see figure 11-66). It would not be a bad idea to print some test plots both before and after making any line font changes to see what the results of your modifications will look like. Be careful changing too much. On the positive side, anything changed with regard to line font settings will only affect the current document, as long as it is not saved as a template. In other words, use a nonproduction drawing as a test bed. That way, nothing gets ruined if you do not like the changes.

Changing a View's Scale

You learned how to change the scale of a drawing sheet earlier in this chapter (see "Sheet Setup"). However, any view's scale can be altered independently of the sheet's scale. To do this, you would access the properties of the view. How-To 11-28 takes you through the process of modifying a view's scale independently of the sheet's scale.

How-To 11-28: Modifying View Scale

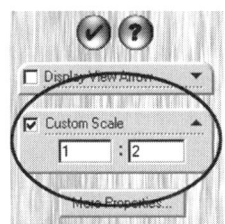

Fig. 11-67. Changing a view's scale.

To modify the scale of any view independently of the sheet scale, perform the following steps. Note that the Auto-show PropertyManager option must be enabled for this process to work.

1. Select the view whose scale is to be changed.

2. In PropertyManager, select Custom Scale, shown in figure 11-67.

3. Enter the desired scale.

4. Click on OK.

If the view whose scale you want to change is either a section or detail view, it is possible to change the scale of the view by modifying the associated note (otherwise known as the view label). This technique demonstrates the bidirectional associativity between a view and its label. Double clicking on the view's label gains access to the text, which can then be altered. Clicking outside the label edit box will exit text edit mode. Adding and modifying view labels and notes in general are covered later in this chapter.

If a section or detail view is using the same scale as the sheet, no note will be displayed calling out the scale in the view's label. This is because the sheet scale is typically called out in the title block area. It is possible, though, to add scale value (e.g., 2:1), which is different than the sheet's scale. SolidWorks will correctly interpret the scale entered and use it for the view. For this trick to work, it is not necessary that the value be prefixed by the word *scale*.

Dimensioning

You first got a good look at dimensions in Chapter 3, and dimension properties in Chapter 5. The properties of the dimensions, as you discovered, could be changed. This altered the appearance or behavior of dimensions.

When dimensioning a drawing, you do not have to manually insert all dimensions needed to build the part. SolidWorks can do it for you. More specifically, the dimensions you add while creating sketches for features can be reused. There is no need to recreate them.

All dimensions (and any other annotations) can be reused and inserted into a drawing. It does not matter whether they were created by the user while sketching or by SolidWorks during an extrusion, while adding draft, or any other process that involves creating a feature. Dimensions from the part file can easily be added to the drawing. The user's biggest task is to position those dimensions.

When inserting dimensions, ask yourself which view should be the main view in which most of the dimensions will be placed. If a view is not selected prior to inserting dimensions, SolidWorks will attempt to place as many dimensions as it can on any detail or section views first. It will then place the rest of the dimensions on any other available views. Duplicate dimensions will typically not be added automatically. In addition, dimensions are only added to a view if they are parallel with the view. These facts should be considered when inserting dimensions.

The example used here to illustrate how to insert dimensions is a drawing for the *Pivot Arm* part created in Chapter 5. You will take a look

Dimensioning

at how to bring dimensions into your design drawing. Note that more than just dimensions can be inserted. Therefore, SolidWorks uses the term *model items* to describe these elements. How-To 11-29 takes you through the process of inserting model items (such as dimensions) into a drawing.

How-To 11-29: Inserting Dimensions

To insert into a drawing any dimensions or other annotations that were placed on the model in a part or assembly document, perform the following steps.

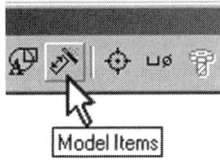

Fig. 11-68. Model Items icon.

1. Select the sheet or view you want to insert the dimensions on.

2. Select Model Items from the Insert menu, or click on the Model Items icon (shown in figure 11-68), located on the Annotations toolbar.

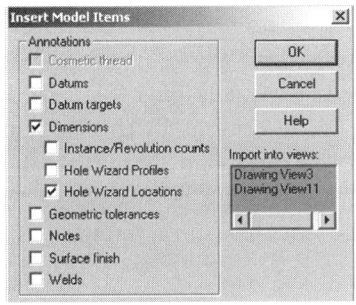

3. From the Insert Model Items window, a portion of which is shown in figure 11-69, check the items you want to bring into the drawing.

4. Click on OK to insert the model items.

Fig. 11-69. Inserting model items.

The items that were checked in step 3 will be brought into the drawing. That is, assuming they existed in the part in the first place. Dimensions are checked by default. Typically, other annotation types are added in the drawing and do not exist in the part. Model items will be imported into the entire drawing if no views were selected. In the case of figure 11-69, two views were selected prior to inserting the model items.

☞ **NOTE:** *Change the size of the font used by dimensions or notes by clicking on the Font button found in either the Dimensions or Notes sections of Document Properties (Tools > Options menu).*

If you decide to add other types of annotations to the part, for whatever the reason, there would be no sense in recreating them in the drawing. For this reason, SolidWorks gives you the opportunity to bring those annotations into the drawing. This is the purpose of the other options in the Insert Model Items window.

Most of the listed annotations are self-explanatory. For instance, *Surface finish* relates to surface finish symbols added to the model. Some annotations may require elaboration. *Hole Wizard profiles* and *Hole Wizard locations* refer to the dimensions associated with holes created with the Hole wizard (discussed in Chapter 6).

The *Instance/Revolution counts* option has to do with the number of instances in a pattern or revolutions in a helix. Normally you would not want that information called out in a drawing, or you might go about it in a different way. For example, if you patterned holes on a metal plate, you would not list the number of holes. Instead, you would typically insert the dimension for one of the holes and prefix the dimension value with 6X, or something similar (assuming there were six holes).

Cosmetic threads are treated differently than any other annotation (see the section "Cosmetic Threads" later in this chapter for a full description). When added, cosmetic threads attach themselves to the associated feature in FeatureManager. Cosmetic threads will appear automatically in part drawings. The *Cosmetic thread* option in the Insert Model Items window is grayed out when inserting model items into views of parts, but available when inserting model items into views of assemblies.

The following section explores the remaining options in the bottom half of the Insert Model Items window. For the most part, leaving these options alone will not hurt the drawing, and you will not be any worse off for ignoring them. However, they are discussed here for your reference.

Reference Geometry

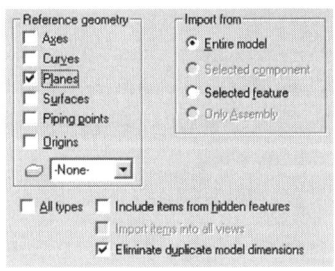

Fig. 11-70. Bottom half of the Insert Model Items window.

The bottom half of the Insert Model Items window is shown in figure 11-70. The Reference Geometry section allows for bringing in other geometry that would not typically be shown in a drawing. There are exceptions, which is why the option exists. Perhaps it becomes necessary to show a plane to denote some sort of datum, or maybe you need to show axes for dimensioning purposes. The reasons vary.

As a special side note to inserting reference geometry into a drawing, there is a more practical way to display reference geometry in particular views, which involves DrawingManager. Material to follow expands on this topic, but suffice it to say for the moment that individual items, such as a surface, can be shown by right-clicking on that item in DrawingManager.

There is a drop-down list (the only one in the Insert Model Items window) that is used for placing items on a particular layer when inserting them. This option is only available if there are layers in the drawing. Layers are discussed in more detail later in this chapter (see the section "Layers"). If None is selected, imported items are not placed on any layer.

All types Option

The *All types* option is used when every annotation needs to be brought into the drawing from the part or assembly. Checking *All types* will automatically select every annotation type and reference geometry option. Use this with caution.

Include items from hidden features Option

Unchecking the *Include items from hidden features* option will prevent dimensions from being inserted if they are associated with features hidden by other geometry.

Import items into all views Option

Import items into all views is only checked if no views were selected. Feel free to uncheck this option and select individual drawing views, even while the Insert Model Items window is open, though you may need to move the window out of the way. What this option amounts to is a way of changing your mind after opening the Insert Model Items window.

Eliminate duplicate model dimensions Option

When checked, *Eliminate duplicate model dimensions* attempts to keep duplicate dimensions from appearing on the drawing. It may not appear to function with 100% accuracy, but that is due to the nature of feature-based modeling. What appear to be duplicate dimensions to you may be individual dimensions for two separate features as far as the software is concerned. At any rate, leaving this option checked will allow SolidWorks to remove a few duplicate dimensions, and can sometimes cause dimensions to be spread out a little more evenly between views.

Import from Option

The *Import from* section comes into play when you want to import just the dimensions of selected features in a part or components within an assembly drawing view. The steps for performing this operation require that specific features, components, or views be selected. DrawingManager can play a role in this process. The following section explores DrawingManager, including an examination of *Import from* in more depth.

DrawingManager

You first heard about DrawingManager earlier in this chapter. You learned how sheets and formats were listed in this manager. Now we will take it one step further and see how views and other items are listed within DrawingManager.

DrawingManager items are expanded just as directory folders are in Windows or features in FeatureManager. Use the small plus (+) or minus (–) symbols to the left of the item's icon to expand or collapse, respectively, DrawingManager objects.

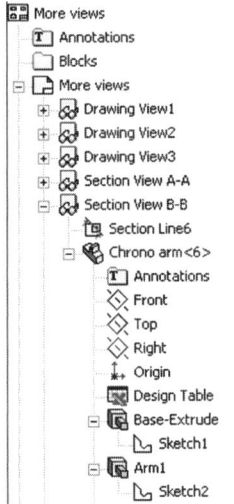

Fig. 11-71. DrawingManager expanded to show the feature tree.

As you can see from figure 11-71, DrawingManager is very similar to FeatureManager, especially once views have been added. Below each view is the name of the part or assembly contained in that view. If you expand that further, you will also see the features that make up the part. Expanding features will also show the sketches used to create the features. In short, the entire history tree is present for every model in every view.

If a view contains an assembly, all parts and subassemblies contained in the assembly will be accessible from DrawingManager. This is important when it comes to inserting dimensions, because it is possible to select specific features or assembly components you want to insert the dimensions for.

Selecting a drawing view's border from the work area will highlight its associated listing in DrawingManager. Specifically, the view's icon will turn blue. This is a big help when trying to discern which view is which when you have numerous views in the design drawing.

Knowing what you now know about DrawingManager, the *Import from* option in the Insert Model Items window begins to make sense. Dimensions for a particular feature or part can be preselected prior to inserting the model items. Then, when the Insert Model Items window is called up, the *Import from* option will allow for inserting dimensions from just those selected features. This will be indicated by the *Import from* setting, shown in figure 11-72.

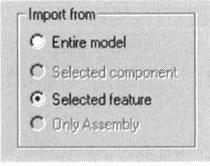

Fig. 11-72. Import from *option*.

Alternatively, a feature or part can be selected on the view itself. Sometimes this is much simpler, and may be the preferred method. However, use care when selecting, and zoom in if you need to. Depending on the complexity of the drawing view, you could easily select the incorrect item for dimensioning. When selecting from DrawingManager, features are spelled out for you and selection is sometimes easier.

Moving Dimensions

SolidWorks' method of adding dimensions to drawing views works well, but it is not perfect. This is the nature of programs such as SolidWorks. When dimensions are added to the model, extension lines terminate at places that may not be suitable when those same dimensions are shown in a drawing.

Sometimes dimensions are brought in that would be better suited to another view. Others just do not belong at all. A good example would be in the case of a revolved feature. If you insert the dimensions on a view containing a revolved feature, you will see the angular dimension indicating the total angle of revolution (e.g., 360 degrees). That is one dimension you could do without.

To delete a dimension, simply select the dimension value and press the Delete key. This is no different than deleting anything else in SolidWorks. You can delete more than one dimension at a time by control-selecting multiple dimensions. Window-selecting dimensions is also an option.

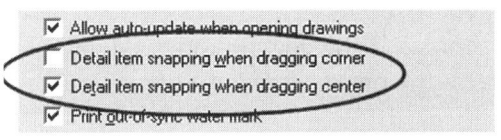

Fig. 11-73. Detail item snapping.

To move a dimension, place the cursor over the dimension value, hold down the left mouse button, and drag the dimension to a new location. You may notice what is known as inferencing lines trying to snap the dimension to a particular location (see figure 11-73). This inferencing action is controlled by settings in the System Options window.

The inferencing action is better known as detail item snapping. The options are found in the Drawings section of the System Options window (Tools > Options). The two options are Infer when dragging corner and Infer when dragging center. Which you use, if either, is totally up to you. By the way, "corner" and "center" refer to the corner or center of a dimension value or a note. When a dimension value or note is dragged (for repositioning), either the corner or center of the object being dragged will align with other stationary dimensions or notes.

You may decide you do not like item snapping, or you may decide you like the way it helps you line up dimensions and notes. Experiment with the settings to see what is best for you.

✓ **TIP:** *You can temporarily override item snapping by holding down the Alt key. This gives you the best of both worlds.*

Moving Dimensions Between Views

SolidWorks gives you an easy way of moving or copying dimensions between views. Normally, a move would be the better choice, because you would not want to duplicate dimensions. However, both methods are examined.

The difference between moving and copying dimensions is actually quite minor. Both processes involve dragging the dimension with the left mouse button. To move a dimension, hold down the Shift key while dragging the dimension. To copy a dimension, use the Ctrl key. When dropping the dimension, make it a point to drop it directly on top of the view you are moving it to. Otherwise, the drag-and-drop operation will not work.

The order in which the mouse button and keyboard are depressed is not important. In addition, it should be mentioned that you cannot move any dimension to any view arbitrarily. Use some common sense. If the dimension cannot be accurately displayed in its new view, SolidWorks will not allow it to be moved there.

Sometimes it also helps to have the drawing view borders turned on when performing this function. This is because the dimension value must be dropped within the new view's border. If you do not drop the dimension within the view you are moving it to, the process will not work.

Hiding Dimensions

After importing dimensions on various views, and then moving them around, you will probably find that some of the dimensions just are not needed. Deleting a dimension is one option, but that removes it completely. Hiding dimensions may be a better choice, because if you change your mind it is easy to bring the dimension back.

Once dimensions are hidden, you will not be able to see them and they will not plot. You will, however, be able to access them if they are needed, such as if a dimensional change must be made from the drawing. Hidden dimensions can be shown using the same process used to hide them. How-To 11-30 takes you through the process of hiding and showing dimensions.

How-To 11-30: Hiding and Showing Dimensions

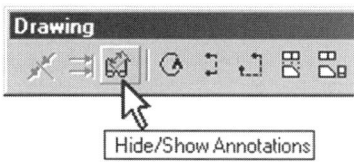

Fig. 11-74. Hide/Show Annotations icon.

To hide and show dimensions in a drawing, perform the following steps. Note that cosmetic threads, discussed later in this chapter, can also be hidden or shown.

1. Select Hide/Show Annotations from the View menu, or click on the Hide/Show Annotations icon (shown in figure 11-74), located on the Drawing toolbar.
2. Select the dimensions to be hidden. They will turn gray.
3. Select any previously hidden dimensions to show them.
4. Select Hide/Show Annotations a second time to exit the command, or simply press the Esc key.

If it becomes necessary to hide all dimensions for a particular reason, the best choice would be to place all dimensions (and possibly other annotations) on a separate layer. That way, the layer can easily be turned off. Layers are discussed in material to follow.

Dimension Favorites

Adding a dimension favorite is a way of transferring a set of dimensional attributes to other dimensions in the same drawing or even other drawings. It is a way of transferring properties between dimensions so that they have the same characteristics.

Prior to establishing a dimension favorite, you should modify the properties of at least one dimension so that it has the appearance you wish. This may consist of changing the decimal place precision, adding text or a tolerance value, or any other attributes appropriate to the situation.

Dimension favorites fall into three categories: part favorites, assembly favorites, and drawing favorites. Part favorites, for example, would relate to those dimensions associated with a part. This is true whether or not those dimensions are shown in a drawing. In other words, if the dimensions associated with a part are brought into a drawing, they will still be considered part dimensions. Likewise, if a dimension favorite is added to a part dimension in a drawing, that "dimension favorite" is considered a part favorite. Because the dimension favorite was added to a part dimension in a drawing, the dimension favorite travels with the dimension and will be available in the part file.

Because dimension favorites can travel between documents, it is necessary to distinguish between them. Believe it or not, this actually makes keeping track of your dimension favorites a lot easier. Just do not get lost in the lingo, and you should be okay. How-To 11-31 takes you through the process of adding dimension favorites.

How-To 11-31: Adding Dimension Favorites

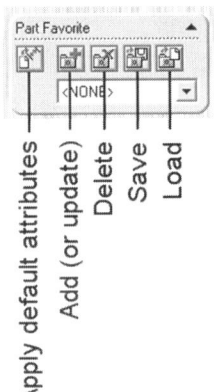

In general, utilizing dimension favorites works the same way for dimension favorites created in a part, assembly, or drawing. The example used in this How-To relates to part model dimensions brought into a drawing, and therefore some of the illustrations reflect the use of part favorites. This is an incidental detail. To create a dimension favorite, perform the following steps.

1. Select the dimension that contains the attributes you wish to retain as a dimension favorite.

2. Click on the Add or Update a Favorite icon, shown in figure 11-75. (Note: figure 11-75 does not show true labels for each icon, for reasons of brevity.)

Fig. 11-75. Part (dimension) Favorite panel.

3. In the Add or Update a Favorite window, shown in figure 11-76, enter a name for the favorite. The name should be short but descriptive.

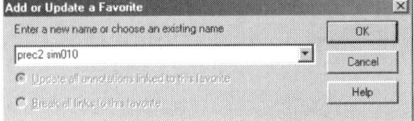

4. Click on OK to create the favorite.

Fig. 11-76. Add or Update a Favorite window.

As shown in the Add or Update a Favorite window in figure 11-76, the name used here was *prec2 sim010*. This refers to two-decimal-place precision for the primary dimension value, along with the fact that a symmetric tolerance is being applied with a tolerance value of .010 inch. Use whatever naming convention makes sense to you, but try to be succinct, as there is not a lot of room in the dimension favorite display panel drop-down list.

Once you have some favorites established, applying those favorites to other dimensions is a snap. Simply select the dimensions you wish to apply the favorite to, and then select the appropriate dimension favorite from the drop-down list, shown in figure 11-77.

Dimensioning

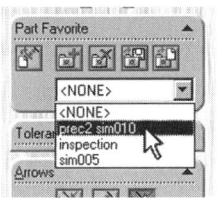

Fig. 11-77. Dimension favorite listing.

To update a favorite so that it contains modified attributes, select the dimension that contains the desired attributes, and then click on the Add or Update a Favorite icon. In the drop-down list in the Add or Update a Favorite window, select the favorite you wish to update and click on OK. If you wish to remove attributes attached to a dimension via a dimension favorite, select the dimension and click on the Apply Default Attributes icon (shown in figure 11-75).

Saving or loading favorites is quite simple and requires little explanation. If the Save a Favorite icon is selected, whatever favorite is listed in the Favorites panel drop-down list will be exported as an SLDFVT file. Likewise, clicking on the Load a Favorite icon opens a window in which a an SLDFVT file can be selected for importing into the drawing (or other SolidWorks document). If you are going to be exporting (saving) or importing (loading) dimension favorite files, it is suggested you create a folder specifically in which to store these files.

Reference Dimensions

Reference dimensions are nothing more than driven dimensions. They will still update if geometry is altered (just like dimensions inserted via the Model Items function, discussed earlier), but you cannot use them to modify geometry. Reference dimensions do not drive model geometry. Rather, they are driven by the model geometry.

You can add reference dimensions to the drawing the same way you add dimensions to a sketch. The mechanics involved are exactly the same. Reference dimensions appear differently than regular dimensions. By default, they appear gray and are enclosed in parentheses. How-To 11-32 takes you through the process of globally turning off the use of parentheses for reference dimensions.

How-To 11-32: Turning Off Reference Dimension Parentheses

To turn off parentheses used with reference dimensions, perform the following steps.

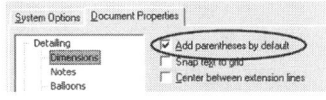

Fig. 11-78. Toggling the use of parentheses.

1. Select Options from the Tools menu.

2. Select the Dimensions section in the Document Properties tab.

3. Check or uncheck the *Add parentheses by default* option (see figure 11-78) as required.

4. Click on OK to accept the changes.

Modifying the appearance of reference dimensions in this way will affect only the current document. You may want to incorporate this change into your drawing templates. Also be aware that changing whether parentheses are shown on reference dimensions is not a retroactive setting. Any existing reference dimensions will not be affected.

It is possible to add or remove the parentheses from individual dimensions by use of the right mouse button. Right-click on a reference dimension (or any dimension) and you will see a Display Options menu. This menu contains the toggle option Show Parentheses. If it is checked, the parentheses will be shown.

Extension Lines

Another important aspect of cleaning up design drawings is modifying where the dimensions' extension lines are terminating. Most dimensions are added when sketching and creating the part. In the sketch, a dimension may look fine. The trouble is that in the drawing it may not. This is because after a feature is created the extension lines are still terminating at the same place. Thankfully, modifying the extension lines is an easy task, albeit tedious.

Extension Line Placement

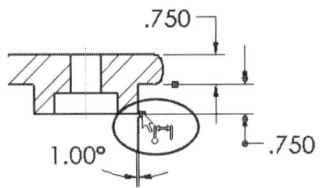

Fig. 11-79. Modifying extension lines.

To alter the terminating location of a dimension's extension lines, you must first select the dimension. You will then see small green "handles" attached to the extension lines. Place the cursor over a green handle and drag it to the desired location. This is illustrated in figure 11-79. The circled area shows the cursor as it is dragging the extension line handle to a new location.

The green handles on the tips of the arrows are for flipping the arrows to either side of the extension lines. This is very convenient. Additionally, there is a handle on dimensions that contain a bent leader, such as the two .750-inch dimensions shown in figure 11-76. The lower of these two dimensions has been selected, and therefore the small handle at the leader's "elbow" can be seen. Clicking on this handle will flip the dimension value and bent leader line from one side to the other.

Breaking Extension Lines

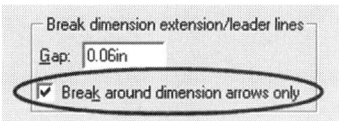

Fig. 11-80. Uncheck Break around dimension arrows only.

Fig. 11-81. Break Dimension Lines panel.

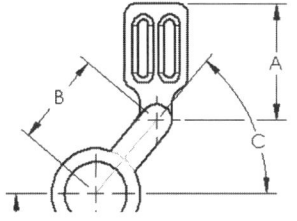

Fig. 11-82. Broken extension line.

Current ANSI standards specify that dimension extension lines should break only around arrows. Default settings reflect this specification, so if you choose to break extension lines around dimension line leaders, for example, the default settings must be altered first. To accomplish this task, access Document Properties (Tools > Options) and select the Dimensions section. Locate the option *Break around dimension arrows only*, shown in figure 11-80, and uncheck it.

Also shown in figure 11-80 is the gap setting used for the break. This value defaults to .060 inch, but can be increased to create a more noticeable gap. To complete the process, it is necessary to specify exactly which dimensions should have their extension lines broken. Incidentally, dimension leaders can be broken as well. Selecting a dimension will display its properties in PropertyManager. Note the Break Dimension Lines panel near the bottom of PropertyManager, shown in figure 11-81.

By turning on the Break Dimension Lines option for a particular dimension, that dimension will break around whatever other dimension it intersects. Note that this option will not work by itself, and that the *Break around dimension arrows only* option mentioned previously must be turned off first. An example of a dimension with a broken extension line is shown in figure 11-82. (This image was taken from a tabulated drawing, discussed at the end of the chapter.) Note the lower extension line for dimension A.

Center Marks

Adding center marks is one of the easiest tasks when annotating a drawing. There is not much involved when it comes to adding center marks. However, for your reference, the following section describes the process.

Creating Center Marks

How-To 11-33 takes you through the process of adding center marks.

How-To 11-33: Adding Center Marks

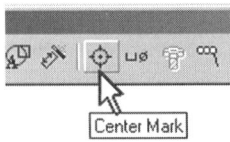

Fig. 11-83. Center Mark icon.

To add center marks to a drawing, perform the following steps.

1. Select Insert > Annotations > Center Mark, or click on the Center Mark icon (shown in figure 11-83), located on the Annotation toolbar.
2. Select any arc or circle where you want a center mark to appear.
3. Press the Esc key to exit the command when finished.

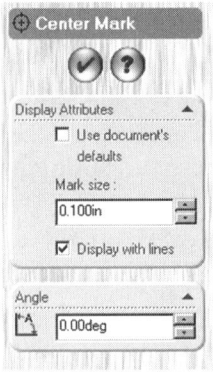

Fig. 11-84. A center mark's properties.

Selecting a center mark displays its properties in PropertyManager. Figure 11-84 shows the Center Mark properties panel. Unchecking the *Use document's defaults* option provides access to the other settings that allow for adjusting the size and appearance of the individual center mark. Unchecking the *Display with lines* option will cause the center mark to display as a small plus sign. The Angle setting will rotate the center mark. Negative or positive values can be used.

Automatic Center Marks and Centerlines

Center marks and centerlines can automatically be added during the view creation process in drawings. You are already aware of what center marks are. Centerlines are centerline entities that can be added to views of cylindrical parts. As an example, the part shown in figure 11-85 will be used in the creation of a drawing.

To enable the automatic addition of center marks and centerlines, access the Detailing section of Document Properties (Tools > Options). There, you will see options for turning on the automatic creation of center marks and centerlines, as shown in figure 11-86. Because these options are part of Document Properties, they can be incorporated into a template if desired.

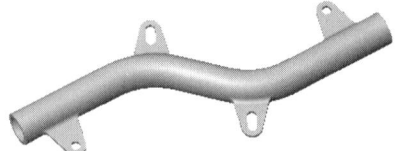

Fig. 11-85. Preparing to create a drawing of this model.

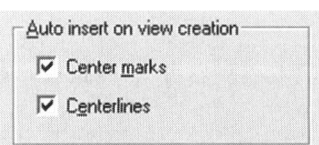

Fig. 11-86. Automatic insertion options for center marks and centerlines.

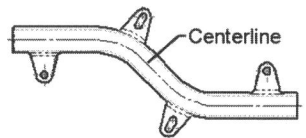

Fig. 11-87. Center marks and centerlines were added automatically.

Once the options have been enabled, create your drawing views as usual. Figure 11-87 shows the sample part in a drawing after inserting a front view. Note the centerline. There are also center marks at each of the holes, including one center mark at either end of the rounded slots. This was all done automatically.

Layers

Nearly anyone who has ever used a 2D CAD program understands what layers are. For those not in the know, layers are a way of keeping track of objects in a drawing. The biggest benefit of using layers is seen in the 2D drawing world, in which objects can be separated into various layers. An example is the case of an architectural drawing. The foundation may be on one layer, the plumbing on another, and so on. There may be layers for walls, windows, electrical, landscaping, notes, dimensions, and whatever else is necessary.

Layers are not as important in SolidWorks as they are in non-solid 2D programs, but they can still be useful. Layers are only available in drawings, not parts or assemblies. It may be convenient to place all dimensions on a particular layer, notes on another layer, and perhaps weld symbols and annotations on yet another. This makes turning certain groups of annotations on and off very easy. Your reasons for creating layers may vary.

Each layer has its own individual settings for color, line type, line weight, and whether or not the layer is visible. Objects can be moved from one layer to another. Layers can have different names assigned by the user, and descriptions. Layer properties can be changed at any time, and layers can be deleted if they are no longer necessary. They can also be useful when exporting drawing files to other CAD systems.

Using layers is certainly not a requirement. If you find no advantage to using layers, do not feel obligated to make use of this functionality. If you decide to use layers, you should first know how to create some new layers and set up their properties. How-To 11-34 takes you through the process of creating a new layer.

How-To 11-34: Creating Layers

To create a new layer, perform the following steps.

466 Chapter 11: Design Drawings

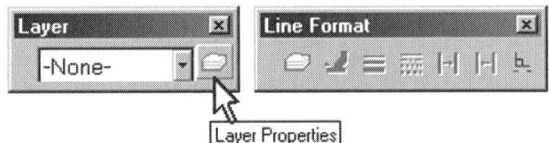

Fig. 11-88. Layer Properties icon.

Fig. 11-89. Layers window.

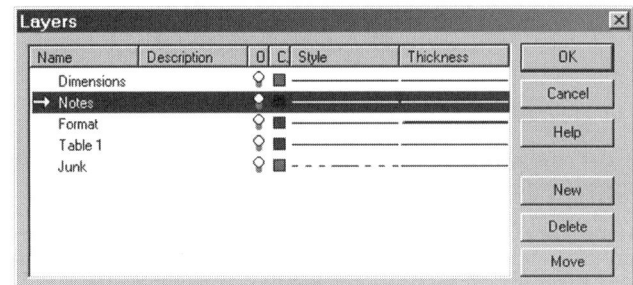

1. Click on the Layer Properties icon, shown in figure 11-88. The Layer Properties icon is found on both the Layer toolbar and the Line Format toolbar.

2. In the Layers window that appears (shown in figure 11-89), click on the New button.

3. Type in a name for the new layer and press the Enter key.

4. Perform a slow double click in the area under the Description heading and type in a description if desired.

5. Click on the color square under the Color heading and select a color if desired. Black is the default.

6. Click on the line under the Style heading and select the desired line style.

7. Click on the line under the Thickness heading and select the desired line thickness.

8. Repeat step 3 for as many layers as you would like to create. Steps 4 through 7 are optional, as you can always use the default values for color, line style, and thickness.

9. Click on OK when finished.

The Layers window (shown in figure 11-89) in the example has a few layers that have been created. The *Notes* layer happens to be the current layer. When dimensions or any other annotation is added, or when geometry is added to the drawing, it will automatically be placed on the current layer. It is definitely important to know which layer is current at any point in time. The following material explains how to set a layer to be the current (or active) layer, as well as what the various buttons in the Layers window accomplish.

Making a Layer Current

With the Layers window open, simply click to the left of the name of any layer. When a small yellow arrow appears to the left of the layer name, it is current. Whatever is added to the drawing in the way of annotations or sketch geometry will be placed on the current layer.

The Layer toolbar contains a drop-down list (shown in figure 11-90) that can be used to set the current layer. This is a quick and easy way to make a layer current without opening the Layers window, and is the preferred method. Note in this example the layer named *-None-*. This is the selection that would be used when adding geometry or annotations that should not be on any layer.

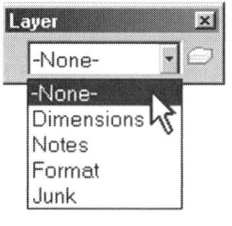

Fig. 11-90. Making a layer current.

Deleting a Layer

Obviously, the Delete button in the Layers window deletes a layer. Select a layer name from the Name column and press the Delete button. You will be asked if you are sure you want to delete the layer, in which case you can then make the appropriate reply, Yes or No.

When a layer is deleted, the objects on that layer are not deleted. Any annotations or geometry on the layer will be transferred to layer *-None-*, which really is not a layer at all. Rather, it is a designation for any objects that do not belong to a particular layer.

Turning Layers On and Off

With the Layers window open, note the light bulbs associated with every layer. Clicking directly on a light bulb will toggle the associated layer on or off. If the light bulb is yellow, the layer is on. If the light bulb is dimmed, the layer is off. Turning a layer off takes effect immediately. It is not necessary to click on the OK button to have these settings take effect.

SolidWorks will let you turn off the active (current) layer. It will not issue a warning message, so make sure you know which layer is current. Otherwise, the objects added to your drawing will not be visible.

Moving Objects Between Layers

Placing objects on the wrong layer is something that has happened to every CAD operator on the planet; that is, assuming the CAD software uses layers. Sometimes objects need to be moved from one layer to another, for various reasons. Whatever the reason, you will invariably find the need to move objects from one layer to another. There are a few ways to accomplish this. The method used in How-To 11-35 makes use of the Move button in the Layers window.

How-To 11-35: Moving Objects Between Layers

There is a button in the Layer window named Move. This button allows for moving objects onto a specific layer (if those objects are not currently on a layer), or from one layer to another. The implementation of this function is very simple, and is outlined in the following steps.

1. Click on the Layer Properties icon.
2. Make sure the current layer is the layer objects are to be moved to.
3. Select items from the drawing, such as annotations or sketch geometry, to be moved to the current layer. You may have to reposition the Layers window to accomplish this.
4. Click on the Move button.
5. Click on OK when finished.

It is possible to select the objects to be moved either before or after opening the Layers window. If selecting objects with the Layers window open, it is not necessary to hold down the Ctrl key to select multiple objects. This is the case with every SolidWorks command.

✓ *TIP: With assembly drawings, individual components can be selected from the work area and placed on layers.*

Just about every object in SolidWorks has what are known as "properties." As you discovered in Chapter 5, a dimension's properties can be changed to alter the appearance of the dimension. Other annotations have properties as well. The layer an object is on is considered one of its properties. Right-clicking on any type of annotation in SolidWorks gains access to its properties.

Properties of Multiple Objects

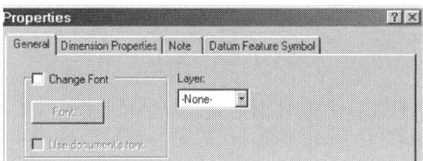

Fig. 11-91. Common properties of multiple annotations.

Accessing the common properties of many objects at a time is common practice when creating design drawings. If more than one dimension (for instance) is selected at one time, and their properties are accessed, the resultant window will show only the properties common to all selected dimensions. If multiple annotations (e.g., dimensions, datum feature symbols, and notes) are selected simultaneously, a window will appear containing tabs for the various annotations selected, and there will be a tab marked General that will contain the common properties of all selected annotations. An example of this is shown in figure 11-91.

If annotations and geometry are selected simultaneously and their properties accessed, only the properties of the annotations will be shown. This is the case even if it is the geometry that is right-clicked on in order to access the Properties option. For this reason, make it a point to select geometry separately from any annotations when the geometry's properties must be accessed. This is true under any circumstances, not just when moving the geometry to a different layer. By the way, "geometry" in this case is referring to sketch geometry (such as lines and arcs) added to the drawing.

Line Formatting

The Line Format toolbar has options for changing line color, thickness, and style. These functions are fairly straightforward. Select the objects that are to have their formatting changed, click on one of the Line Format icons, and specify the desired traits, such as color, thickness, or style.

Fig. 11-92. Line Format toolbar.

Line formatting is reserved strictly for lines, as the name implies. Sketch geometry of any type can be formatted, including edges of feature geometry and even crosshatch. You can even change the color of annotations with the Line Format toolbar, although the other formatting options are off limits. The Line Format toolbar is shown in figure 11-92.

The icons on the Line Format toolbar pictured are, from left to right, Layer Properties, Line Color, Line Thickness, Line Style, Hide Edge, Show

Edge, and Color Display Mode. How-To 11-36 takes you through the process of changing the appearance (format) of a sketch entity or model edge.

How-To 11-36: Using the Line Format Toolbar

To change the appearance of sketch entities, model edges, and the like, perform the following steps.

1. Select the object to be formatted.
2. Click on the Line Color, Line Thickness, or Line Style icon.
3. Select the desired formatting.

That about sums it up. There is not a lot to it, as you can see. It should be noted that line formatting is better reserved for just a few entities that should have a unique formatting. Otherwise, it would be best to create a new layer, with the desired formatting, and place the desired objects on that layer. Line formatting will override any layer characteristics associated with the object.

The Hide Edge and Show Edge icons perform the same function as hiding or showing individual edges via the right mouse button (see the section "Hiding Individual Edges"). The Color Display Mode icon is a toggle switch. When the icon is depressed, sketch geometry in the drawing will display, with the sketch color codes you are familiar with when creating sketch geometry in a part. This refers to blue, black, red, and the other less frequently displayed sketch color codes, covered in Chapter 2. When the Color Display Mode icon is not toggled on, all sketch geometry will appear black, unless assigned specific colors via the Line Format toolbar.

Annotations

There are many items you might need to add to a design drawing besides dimensions and center marks. Notes, geometric tolerances, weld and surface finish symbols, and other annotations all go into the making of a complete detailed drawing. Most of these annotation types are discussed in this section. Any remaining annotations are covered in the next chapter, the main topic of which is assemblies. This is because certain annotations are better suited to assemblies, such as balloons.

Annotations

It is not the intention in this book to teach what goes into making a good surface finish symbol or how to create geometric tolerancing (and other) symbols. That is up to you to learn if you or your employer deems it necessary. This section deals with employing the various annotation commands in SolidWorks. The general steps for creating annotations are presented, but it will be up to you to type in the correct parameters. More detailed descriptions are provided for certain annotations or annotation parameters if necessary. Some samples are provided as guidance.

Most of the annotations that can be added in SolidWorks are pretty much self-explanatory. As previously mentioned, you should be familiar with how to use the individual types of annotation symbols; otherwise, they are meaningless. For example, if you have never seen a surface finish symbol, entering the proper parameters to create a surface finish symbol that makes sense will be difficult. The technical procedure for actually creating one on a SolidWorks drawing, however, is simple. The sections that follow explore the basic procedures for adding various annotation types in SolidWorks.

Notes

Adding text to a drawing is almost always a necessity. Perhaps you want to add text to the title block. Maybe you are adding notes to a template. It makes no difference as far as the application is concerned. You will have total flexibility as to where the note or leader is located, even after the note has been inserted. You can move the note by dragging it, and you can position the leader arrow wherever you want, if a leader is used.

If an entity such as an edge or face is clicked on prior to adding a note, a leader will automatically be added. The leader's arrow will point to the location clicked, and attach itself to that same point. If the underlying object is modified in any way, the leader will move with the object, remaining attached to it. This keeps leader arrows from becoming disassociated with the model, if, for example, the model undergoes size alterations. How-To 11-37 takes you through the process of adding a note.

How-To 11-37: Adding Notes

To add a note to a drawing, perform the following steps.

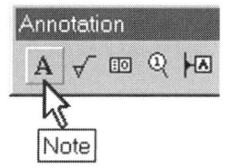

1. Select Insert > Annotations > Note, or click on the Note icon, shown in figure 11-93.

Fig. 11-93. Note icon.

2. Click on the drawing sheet where the note is to be positioned.

3. Type in the desired text. Pressing Enter will drop down one line.

4. Click in a blank area of the drawing when finished.

5. Repeat steps 2 through 4 for additional notes if needed.

6. Click on OK or press the Esc key when finished.

If clicking on an object, such as an edge in a drawing view, a leader will automatically be added. A second pick will be required in this case to position the note. Leader arrows are intelligent and will display arrows if attached to an edge, and small dots if attached to a face of a model in a view. To edit a note, simply double click on it.

Aligning Notes

If a drawing contains a series of notes that should be aligned in some way, SolidWorks provides you with plenty of options. First, you must select the notes to be aligned. This can be accomplished by Ctrl-selecting, or by dragging a selection window around the notes. Once that has been done, right-clicking on any one of the notes will display an Align menu. Within the Align menu are alignment options, including Leftmost (aligns notes vertically to the leftmost note), Uppermost (aligns notes horizontally to the uppermost note), Compact Horizontal, and others. These options should be self-explanatory.

There are a number of other options available to you in PropertyManager when adding notes. The various options can be changed prior to adding a note, but also appear if a previously created note is selected. These options are explained in the following section.

Arrows/Leaders Panel

The Arrows/Leaders panel, shown in figure 11-94, contains a lot of icons, but they make sense once you understand their layout. The first row of icons determines if there will be a leader attached to the note or not. Using one of the first two icons overrides the intelligent behavior normally inherent in the Note command. The small yellow star in the third icon signifies that "smart" mode is enabled and that a leader will be added automatically if the first mouse click is on an object when placing the note in the drawing.

Fig. 11-94. Arrows/Leaders panel.

✓ **TIP:** *Whenever a small yellow star is shown on an icon, it means a "smart" (or "automatic") mode is available. Using automatic modes is preferable to other modes unless a particular setting must be overridden for some reason.*

The second row of icons determines if a straight or bent leader will be used for the note. Obviously, if there is no leader, this setting will not matter. The third row of icons determines what side of the leader the text is on. Leave this set to Leader Nearest (automatic mode) unless it becomes necessary to specify one side or the other for some reason.

The type of arrow present on the leader can be specified. Here, as is typically the case, you should leave this setting in automatic mode. In other words, select from the drop-down list the arrow with the star attached in order to let SolidWorks automatically set the arrow according to what the leader is pointing to.

Arrow "smart" mode is only as smart as what the document defaults tell it to use. In other words, the arrows used in automatic mode are determined in the document's properties. If these default settings are changed, it is possible to change smart mode to, well, stupid mode! If you find it necessary to alter the default settings for arrows, access the Arrows section of Document Properties (Tools > Options).

The *Apply to all* option pertains only to notes with multiple leaders. Because different leaders could have different arrows, *Apply to all* would allow for making them all the same. How does one obtain multiple leaders on a note? One way would be to select a note containing a leader, and then Ctrl-drag the tip of the leader to an additional attachment point.

Text Format Panel

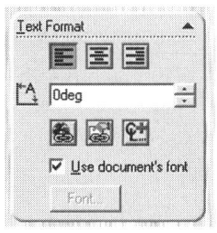

Fig. 11-95. The Arrows/Leaders panel.

The Text Format panel, shown in figure 11-95, contains icons for left, center, and right text justification. Changing the angle of the text is an option, along with modifying the type of font used. These options are self-explanatory.

The row of icons in the middle of the panel requires a more elaborate explanation. They are, in order from left to right: Insert Hyperlink, Link to Property, and Add Symbol. If you have ever surfed the Web, you should be familiar with what a hyperlink is. Clicking on the Insert Hyperlink button allows for adding a hyperlink to a note. The link can be to a file on your computer or company network, or it can be a web address, also known as a uniform resource locator (URL).

Hypertext links are commonly shown as underlined blue text. Solid-Works will make the text blue, but you will have to underline it yourself. Use the Font button to display the text underlined, or in bold or italics.

When the cursor is placed over a hyperlink, it will change into a small hand. If you hold the cursor still for a moment, a yellow box will appear that contains the address to the link or to the file and file location. Keep in mind that the person clicking on the link must have the capability to open the link. In other words, they must have the appropriate software to open the file. If the link is to an Excel spreadsheet, the person must have Excel loaded on her computer.

Moving a hyperlink is tricky, because if you click on it you will open the link. Try placing the cursor at one corner of the hyperlink, just out of range so that you cannot see the hand. You can then move the link by dragging it. You can also right-click on the link and access its properties for editing purposes.

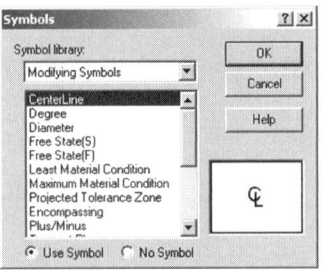

Fig. 11-96. Symbols window.

Clicking on the Add Symbol button will open the Symbols window, shown in figure 11-96. Many different symbols are available, so you should never find yourself lacking for a particular symbol. The No Symbol option (shown in figure 11-90) is for removing a symbol previously added to a note.

The Link to Property button opens the door to another topic. All SolidWorks documents contain file properties, as do all other files on your computer. SolidWorks files can contain information a user can specify. This information can be literally anything, such as mass properties, size characteristics, cost, vendor, or any other information deemed important.

Once a file's properties have been established, those properties can be linked to a note. If the properties change, the note updates accordingly. Creating custom properties within a file is explored in detail in Chapter 16. For now, this section will show you how to link to a file's properties.

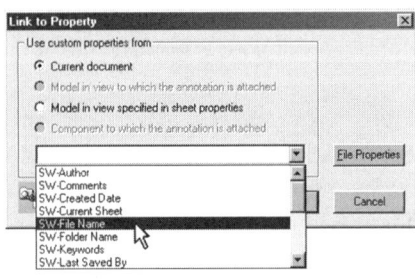

Fig. 11-97. Link To Property window.

Clicking on the Link to Property button opens up the Link to Property window, shown in figure 11-97. As can be seen in the figure, SolidWorks includes a number of custom properties with every document. The property highlighted in the illustration happens to be SW-File Name. The *SW* signifies that this is a custom property created by SolidWorks for the convenience of the user. *File Name* obviously refers to the name of the file. In other words, if this link is used, the name of the file will appear in the note. If the file name is changed, the note automatically updates, which is the beauty of custom properties.

When linking to a property, there are four options that control what file the property is linked from. If *Current document* is selected, the proper-

Annotations

ties of the current document will be used. This option is available for all SolidWorks document types, meaning parts, assemblies, and drawings.

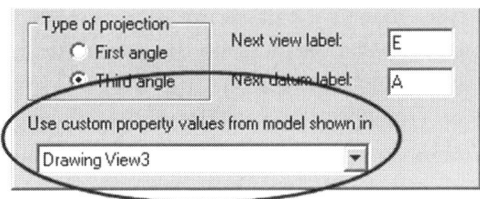

Fig. 11-98. Specifying a view from which properties will be extracted.

The *Model in view to which the annotation is attached* option will link to the properties of the part or assembly in the drawing view with which the note is associated. If *Model in view specified in sheet properties* is selected, the link looks to the model in whatever view is specified in the sheet's properties. This is important if there are views of different parts or assemblies in the same drawing. Accessing the sheet's properties will gain access to the option by which the view can be selected (see figure 11-98). See the previous section "Sheet Setup" for instructions on accessing or altering a sheet's properties.

Finally, the fourth option, *Component to which the annotation is attached*, applies to assemblies or assembly drawings only. It allows for attaching a note to a component in an assembly and having the note link to (access) the properties of the component rather than the assembly. This option, in essence, allows SolidWorks to traverse documents, reach into an associated document file, extricate the file's properties, and then display that information in the original file to which the note has been added.

Border Panel

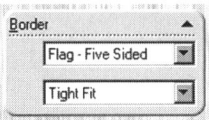

Fig. 11-99. Border panel.

The Border panel, shown in figure 11-99, allows for placing a border around a note. These are often referred to as balloons. The border style can take many shapes, including circles, diamonds, rectangles, five-sided flags, and others. The border size setting should typically be left at Tight Fit, so that it will automatically adjust to whatever note is present inside the border. Otherwise, the border size will be relegated to accepting a certain number of characters, and will stay at that size even if the number of characters in the note changes.

Hole Callouts

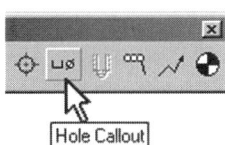

Fig. 11-100. Hole Callout icon.

When a hole is created, the information regarding the hole is remembered by the software. When a hole is created via the Hole wizard, information such as depth, diameter, counterbore dimensions, and any other data associated with the hole is stored in the part file. The data associated with a simple circular cut can be inserted as a hole callout in a drawing. If the Hole wizard is used to create a hole, the information can be quite extensive. Figure 11-100 shows the Hole Callout icon.

To create a callout for a hole, select the Hole Callout icon, and then click on the outermost edge of the hole. A leader will attach itself to the hole, and then the callout can be positioned. It is important to remember to select the outermost edge of the hole, or the callout will not be complete.

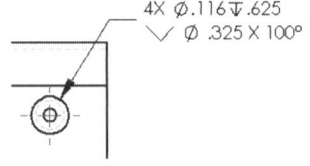

An example of a hole callout is shown in figure 11-101. Move the annotation to a new position by dragging the callout, just as you would move a dimension or note. Because the hole in figure 11-101 was created with the Hole wizard, there is quite a bit of data associated with the feature. The callout is associative, but it does not drive the geometry.

Fig. 11-101. An example of a hole callout.

Hole callouts are added via the Hole Callout icon, but they are treated as reference dimensions. Text can be added to the hole callout, along with tolerance values and symbols. The PropertyManager displayed when a hole callout is selected is shown in figure 11-102.

Of particular significance to PropertyManager for a hole callout is the Variables button, which can be seen near the bottom of figure 11-102. This button is what allows the user to insert a particular dimension associated with a hole created with the Hole wizard. Clicking on the Variables button opens a window (not pictured) that allows for selecting specific dimensions associated with the hole you wish to appear in the callout. In this fashion, the callout can be customized quite extensively.

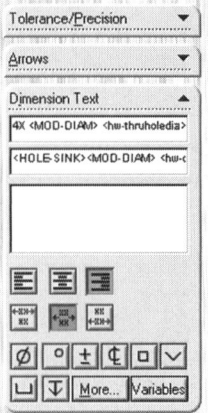

Fig. 11-102. PropertyManager when selecting a hole callout.

Note that the Hole Callout command can be used to display information on any hole. However, only holes created with the Hole wizard will contain the Variables button in PropertyManager, allowing for the creation of an associative customized callout. Non–Hole-wizard hole callouts can be customized to some extent, but the associativity may be broken. If this is the case, SolidWorks warns you of the fact first and allows you to change your mind before customizing the callout.

Cosmetic Threads

When it is necessary to show threads in a drawing, it can be very cumbersome to show actual detailed threads (true helical threads). Showing true threads is almost never a necessity when it comes to design drawings. Even showing the illusion of threads with a linearly patterned V groove makes for too much ink on a plotted drawing. It is likely the threads will simply appear as black areas (see figure 11-103). For this reason, it is often desirable to create what are known as cosmetic threads.

Annotations

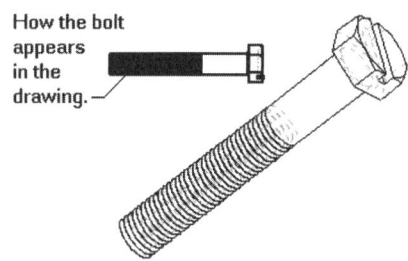

Fig. 11-103. Threads in a drawing are too dense.

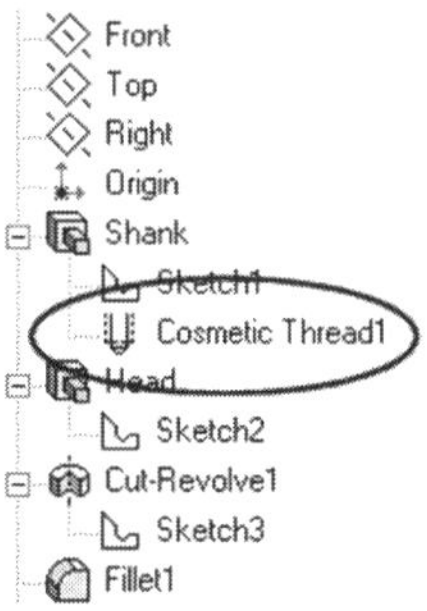

Fig. 11-104. A cosmetic thread in FeatureManager.

It is possible to place cosmetic threads in parts, assemblies (assembly features only), or drawings. Many designers opt to place the cosmetic threads in the part file. This relieves the drafter of the burden of having to call out the thread specifications. The specs are already in the part, and can be inserted along with the rest of the dimensions when that time comes.

Cosmetic threads work a little differently than most other annotations. They are more substantial, in a manner of speaking, than most other annotations that would typically exist in a strictly 2D drawing. If a cosmetic thread is added to a drawing, it will traverse to the part and will actually appear in the part's FeatureManager. This is illustrated in figure 11-104.

Cosmetic threads associate themselves with the feature they are attached to. This is evident in figure 11-104. Unlike all other annotations, cosmetic threads are inserted into a drawing automatically if they are already present on the part model. There is no need to use the Insert Model Items window to accomplish this task.

To add cosmetic threads, a circular edge must be selected prior to beginning the command. There are only two annotations that require preselection (the other are weld symbols). Whether inserting cosmetic threads in a part, assembly, or drawing, the process is the same. How-To 11-38 takes you through the process of adding cosmetic threads.

How-To 11-38: Inserting Cosmetic Threads

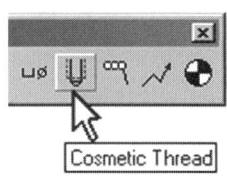

Fig. 11-105. Cosmetic Thread icon.

To add cosmetic threads to a SolidWorks part, assembly, or drawing, perform the following steps.

1. Select a circular edge that defines where the cosmetic threads should start. In the case of a drawing, a circular edge may appear as a line, such as when viewing a cylinder from the side.

2. Select Insert > Annotations > Cosmetic Thread, or click on the Cosmetic Thread icon, shown in figure 11-105.

3. In the Apply Thread section of the Cosmetic Thread window, a portion of which is shown in figure 11-106, specify the length of the thread or end condition.

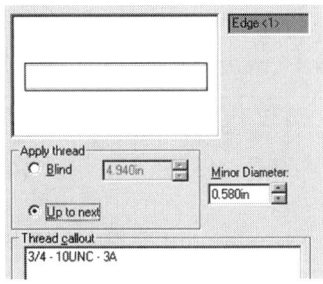

4. Specify the Minor Diameter or Major Diameter parameter, depending on whether the feature is a cylinder or hole, respectively.

5. Specify a callout, if desired.

6. Click on OK to create the cosmetic threads.

Fig. 11-106. Cosmetic Thread window.

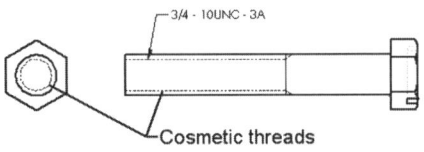

Fig. 11-107. Cosmetic threads.

An example of cosmetic threads is shown in figure 11-107. If a callout has been added, the callout will not appear in a part, but will appear in the drawing of the part. The cosmetic threads themselves will look the same way in both part and drawing.

After adding a cosmetic thread to an edge of a model in a drawing view, cosmetic threads will appear on the other views of that model in the drawing. If you do not wish to see the cosmetic threads in a particular view, it will be necessary to find the feature associated with the cosmetic thread in the history tree for the model in that particular drawing view. Selecting the view first will highlight its associated icon (blue) in DrawingManager, which will help you find the feature. Right-clicking on the cosmetic thread will gain access to a Hide or Show toggle option.

✓ **TIP:** *Hiding all cosmetic threads, or many other annotation types, can be accomplished by right-clicking on the* Annotations *folder at the top of Drawing-Manager and selecting Details. Use the display filter to check only what should be shown.*

Editing the text of a cosmetic thread callout can be accomplished by double clicking on the callout. This action will also change the text in the Cosmetic Thread window, which you would see if editing the cosmetic thread's definition. This happens due to the associative nature of cosmetic threads. Be aware that the callout will lose this associativity if it is moved to another view, or if the leader is detached from the cosmetic thread. Editing the definition of a cosmetic thread is only available from within the part, not the drawing. This is true even if the cosmetic thread was added to the drawing initially.

Deleting a cosmetic thread must be done from within the part, although a cosmetic thread callout can be deleted from the drawing.

Deleting the callout is not recommended because it is not possible to get the callout back if you change your mind. It would be much better to use Hide/Show Annotations, discussed earlier, which is much more forgiving.

Weld Symbols

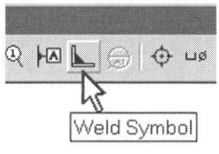

Fig. 11-108. Weld Symbol icon.

Weld symbols require that a position for the symbol be determined prior to gaining access to the Weld Symbol icon, shown in figure 11-108. This is similar to adding cosmetic threads in that an object must be selected prior to accessing the command. In the case of weld symbols, an edge or surface must be selected; otherwise, the Weld Symbol icon will remain grayed out (inactive).

How-To steps will not be listed for adding a weld symbol. Just remember to first select where to position the symbol, and then click on the Weld Symbol icon. The rest is just a matter of entering the proper parameters in the Weld Symbol Properties window, a portion of which is shown in figure 11-109. Obviously, you should be familiar with how a weld symbol should appear before attempting this process.

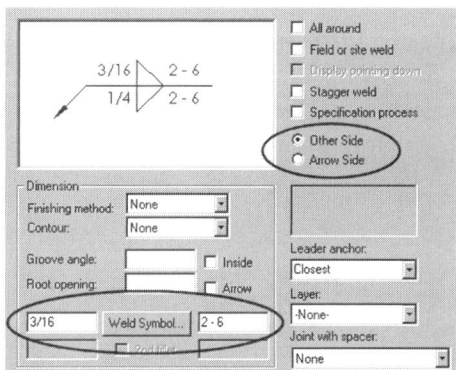

Fig. 11-109. Weld Symbol Properties window (preview area has been enlarged).

There are a few areas of the Weld Symbol window to note. These areas are circled in figure 11-109. Let's use the example shown in the preview area (enlarged for the sake of clarity). We can see that the weld on the far side will be a fillet weld with a radius of 3/16 inch. The length of the weld will be 2 inches, and each segment will be spaced 6 inches apart, center to center.

To input the values for the far side, first make sure the Other Side option is selected. This allows for entering the values for the far side by clicking on the Weld Symbol button to specify the weld symbol and entering the appropriate parameters in the white boxes on either side of the button.

To fill in the values for the near side, select the Arrow Side option and enter the relevant parameters, once again by clicking on the Weld Symbol button to pick the appropriate weld symbol and then specifying the parameters in the white boxes. All other options in the Weld Symbol window should be self-explanatory to anyone familiar with weld symbols. For example, to create a stagger weld, check the *Stagger weld* option and the symbol will be updated in the preview automatically. To specify a particular process, check the *Specification process* option and enter the desired data. Again, the preview will update accordingly.

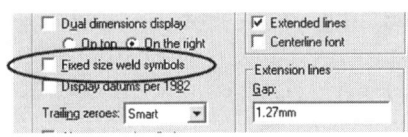

Fig. 11-110. Fixed size weld symbols option.

An option that controls the size of weld symbols is shown in figure 11-110. When *Fixed size weld symbols* is left unchecked, which is the default setting, weld symbols change size according to the size of the dimension font. Otherwise, the weld symbol stays the same size even if the dimension font size is altered.

The *Fixed size weld symbols* option is found in the Detailing section of Document Properties (Tools > Options menu). To change the dimension's font, access the Dimensions section of Document Properties and click on the Font button.

Geometric Tolerancing

The art of using geometric tolerancing could probably be a book in itself. As a matter of fact, it is, and is better known as ASME Y14.5M-1994, a standard established by the American Society of Mechanical Engineers. It is not the purpose of *Inside SolidWorks* to teach you how to use geometric tolerancing, but to show you how to implement geometric tolerancing through the SolidWorks interface.

Fig. 11-111. Geometric Tolerance icon.

Adding geometric tolerances works much the same way as adding most other annotations. Click on the Geometric Tolerance icon (shown in figure 11-111), type in the desired parameters, and then pick where to position the tolerance on the drawing. Picking on an object first, such as a model edge, will automatically add a leader, whereupon a second pick is required to position the geometric tolerance. Picking in a blank area simply places the geometric tolerance on the drawing, with no leader.

A preview of the geometric tolerance will be shown inside the Geometric Tolerance Properties window. The preview is exactly what will appear on the screen when you are done, minus the optional leader. Because some of the abbreviations in the Geometric Tolerance window may not be readily understood, they are briefly explained in to the sections that follow.

GCS

GCS is an acronym for geometric characteristic symbol. Clicking on this option will open the Symbols window. This would typically be a symbol for cylindricity, flatness, or some other characteristic.

MC

MC stands for material condition. Like the GCS button, this too opens the Symbols window, but usually will default to the various material condition symbols, depending on what geometry is selected.

Options

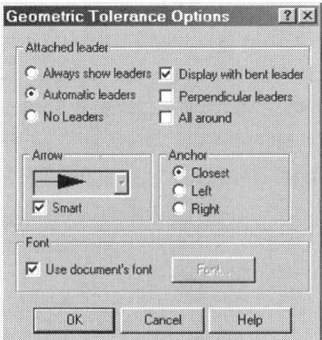

Fig. 11-112. Geometric Tolerance Options window.

The Options button opens the Geometric Tolerance Options window, which allows for controlling the leader and font characteristics of the geometric tolerance. The window is shown in figure 11-112. This option is typically not utilized, in that leaders and arrows are automatic anyway, depending on how the tolerance is added.

A geometric tolerance symbol can be attached to a dimension. This is done by dragging the geometric tolerance to a dimension and dropping it on the dimension value. If the geometric tolerance has a leader, the leader will automatically be turned off. When the dimension now containing the geometric tolerance is moved, the geometric tolerance symbol will move with it. Figure 11-113 shows the result of performing this action.

Fig. 11-113. A geometric tolerance attached to a dimension.

Surface Finish Symbols

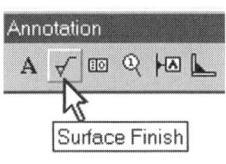

Fig. 11-114. Surface Finish icon.

It is not necessary to list the steps involved in creating a surface finish symbol. Like most other annotations, you simply click on the Surface Finish icon (shown in figure 11-114), type in the appropriate parameters, and then pick some location in the drawing where the symbol should be placed. As is standard operating procedure, multiple symbols can be added without ever leaving the Surface Finish Symbol Properties window.

If you are familiar with surface finish symbols, the parameters contained in the Surface Finish Symbol Properties window should be meaningful to you. Other than that, there is the usual section for overriding automatic leaders and arrow style, but you are usually better off leaving these settings alone. The Rotated option will rotate the symbol 90 degrees counterclockwise. The Layer drop-down list will allow for placing the surface finish symbol on a particular layer.

Datum Features

Datum feature symbols are easily added, consisting of nothing more than clicking on the Datum Feature Symbol icon, shown in figure 11-115, entering the appropriate parameters, and clicking on the geometry where the symbol should be placed. Once the datum feature symbol has been added, it can be moved around, just like any other annotation.

Some companies still use the outdated style of datum feature symbols, shown in figure 11-116. The symbol on the left is the current standard. The symbol on the right shows the 1982 standard for datum feature symbols.

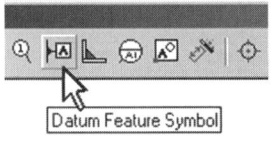

Fig. 11-115. Datum Feature Symbol icon.

Which version of the datum feature symbol you use will depend on the standards dictated by your company. Make sure you are aware of these standards before using datum feature symbols on your drawings. The option for changing which standard is used is titled *Display datums per 1982* and is found in the Detailing section of Document Properties (Tools > Options). This option is shown in figure 11-117.

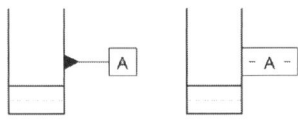

Fig. 11-116. Two types of datum feature symbols.

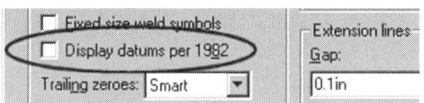

Fig. 11-117. Display datums per 1982 option.

Datum Targets

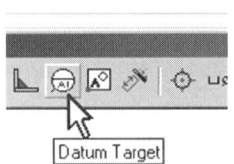

Fig. 11-118. Datum Target icon.

When adding datum targets, SolidWorks exhibits a slightly different behavior than when adding most other annotations. After clicking on the Datum Target icon, shown in figure 11-118, the actual Datum Target Properties window will not appear until a position for the target is selected. Datum targets can be placed on a point, edge, or surface. Once one of these objects is selected, it is business as usual. Enter the appropriate parameters that will be used to establish the datum and click on OK when you are done.

Multi-jog Leaders

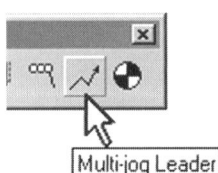

Fig. 11-119. Multi-jog Leader icon.

Multi-jog leaders are leaders with multiple segments, typically with a note at one end. The icon for creating a multi-jog leader is shown in figure 11-119. Multi-jog leaders can be attached to notes, surface finish symbols, geometric tolerances, and balloons (covered in Chapter 12). The key is to have the note or other annotation created prior to creating the multi-jog leader. How-To 11-39 takes you through the process of creating a multi-jog leader and attaching it to a note (or one of the other annotations mentioned previously).

How-To 11-39: Adding a Multi-Jog Leader to an Annotation

To add a multi-jog leader to a note, surface finish symbol, geometric tolerance, or balloon, perform the following steps. Be aware that the annotation must already exist in the drawing.

1. Select Insert > Annotations > Multi-jog Leader, or click on the Multi-Jog Leader icon.

2. Pick where the leader will be pointing to. This will be the arrow side of the leader.

3. Pick points to establish the various segments in the leader.

4. To end the leader at the appropriate annotation, place the cursor over the attachment point of the annotation and left-click. This attachment point will appear as small red squares, and is vaguely discernible in figure 11-120.

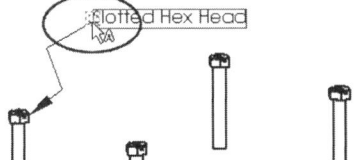

Fig. 11-120. Attaching a leader to a note.

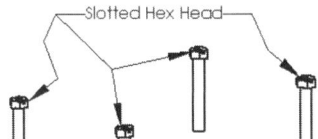

Fig. 11-121. Horizontal bend and multiple attachment points.

Once a multi-jog leader has been attached to a note, the note can be moved to a new position, and the leader will follow. Right-clicking on the leader segment that touches the note will gain access to a menu option titled Add Horizontal Bend. Selecting this option adds a small horizontal leader line prior to connecting to the note. This is shown in figure 11-121.

Figure 11-121 also shows the branches that have been added. To add a branch, right-click on a jog point and select Insert New Branch. As can also be seen in the figure, a multi-jog leader has been added to the right side of the note as well.

Jog points can be added and deleted as needed. To delete a jog point, right-click on the point in the leader and select Delete Jog Point. To add a jog point, right-click somewhere on one of the leader segments and select Add Jog Point.

It should be noted that additional branches can be added to any annotation that already contains leaders. The leaders do not have to be multi-jog leaders. Selecting an annotation that contains a leader will result in small green handles appearing on the annotation, on the end of the leader, and on the point that attaches to the model. Hold down the Ctrl key, and then drag the green handle at the "arrow" end of the leader. This results in a new leader that can be attached to some other object.

Dowel Pin Symbols

Fig. 11-122. Dowel Pin Symbol icon.

Dowel pin symbols win the award for being the easiest annotation type to create. Click on the Dowel Pin Symbol icon, shown in figure 11-122, and then select an arc or circle where the symbol should appear. Right-clicking on the symbol gains access to only one option, which is to flip the symbol.

Blocks

Blocks are essentially collections of entities grouped as one object. They can be inserted into drawings and positioned as needed as many times as necessary. A block can be a separate file, which takes the file extension *.sldsym*. Think of a block as a symbol file.

If there are certain symbols or tables used in your drawings on a regular basis, blocks will probably work in your favor. Blocks can be linked to external files and can be made to update if the original block is redefined.

The following section shows how to create, insert, and edit a block. It also shows how the default location for a symbol library folder can be defined in the SolidWorks Options window. First, however, let's take a look at how a block is created.

Creating and Editing Blocks

Fig. 11-123. Recycle symbol.

The most difficult part of creating a block is sketching the geometry that will be included in the block. You should now be fairly familiar with using the various sketch tools, so sketching the block geometry should not prove that difficult. In the example that follows, a recycle symbol will be used. The recycle symbol is shown figure 11-123.

This shape can be inserted into drawings, rather than being modeled on the solid model as a feature. As a block, the recycle symbol can be rotated and scaled. This gives us some added flexibility for positioning the symbol. Any shape or collection of geometry can be used, and any text added to the block can easily be edited later. How-To 11-40 takes you through the process of defining a block.

How-To 11-40: Defining a Block

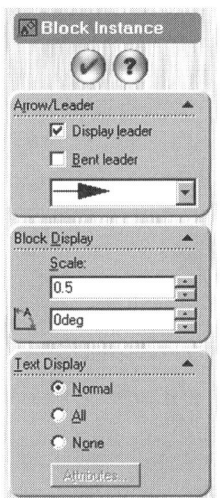

To define a block, perform the following steps. You should be working in a drawing file at this time.

1. Create the geometry that will make up the block symbol. Add text as required.
2. Select the geometry.
3. Select Tools > Block > Make. The Block Instance panel, shown in figure 11-124, will appear.
4. Modify the leader parameters if necessary.
5. Click on OK to close the Block Instance panel.

Fig. 11-124. Block Instance panel.

Once a block has been created on a drawing, copying and pasting allows for easily adding multiple copies of the block throughout the drawing. If any instance is edited, all instances of that same block will update. Leaders, however, can be independently altered between block instances.

The Block Display panel allows for rotating and scaling the block, and obviously the Arrow/Leader panel allows for changing the arrow and leader. These functions do not require further explanation, as they are self-explanatory. If a leader is used, where the leader attaches to the block can be modified. By default, the attachment point is in the lower left-hand corner of the block. To change this attachment point, place the cursor over the current attachment point so that a small green handle is displayed. Hold down the Ctrl key, and drag the attachment point to another location with the left mouse button.

The Text Display panel can be used to turn off the text display within the block. In actuality, Text Display controls whether the invisibility attribute usually associated with imported AutoCAD blocks determines the visibility of the text. Using the All setting or the None setting overrides the invisibility attribute, but also benefits SolidWorks users as a simple text display toggle for text within the block.

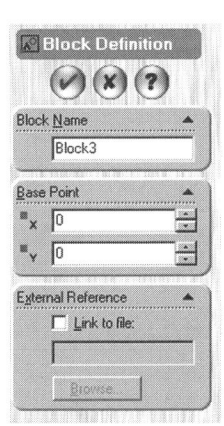

Fig. 11-125. Block Definition panel.

To edit a block, right-click on the block and select Edit Definition. The Block Definition panel, shown in figure 11-125, will appear. When a block is being edited, it opens up in its own window in an exploded state. When exploded, a block is broken down into its constituent components.

Otherwise, all entities in the block are treated as a group. Block geometry can be modified during the editing process, and the Block Definition panel allows for changing other aspects of the block, such as its name.

An alternative to defining a block via the method outlined in How-To 11-40 is to select New from the Tools > Block menu. This will open up a window in which a new block can be created from scratch. The Block Definition panel will be displayed at the same time.

The insertion point (base point) of the block defaults to the bottom left-hand corner of the block geometry, but this can be changed easily enough. It is typical to pick some spot on the drawing where the block should go prior to inserting the block. It is the insertion point that is positioned on that user-selected location when inserting or pasting a block into the drawing. Therefore, set the insertion point to a convenient reference point on the block and you will not have to move it after insertion.

Another aspect of the Block Definition panel is the option for linking a block to a file. You do not want to link a newly created block to a file. This would destroy your block by overwriting it with whatever is in the file. Linking a block to a file and saving blocks pretty much go hand in hand, and are examined in material to follow.

Exploding Blocks

One way to edit a block is to explode it by right-clicking on the block and selecting Explode. As mentioned earlier, this breaks a block up into separate entities, such as sketch lines, arcs, and text (for example). If a block is exploded, it loses its association to the other instances of the same block (assuming there are some) in the drawing. If the block is linked to a block file, the external link is broken.

The only time you would want to explode a block is if it does not need to update any longer if another instance is modified, or if an externally referenced block file is edited. If you wish to use the block geometry for something else, or to use the geometry to define a new block, go ahead and explode it.

Inserting Blocks

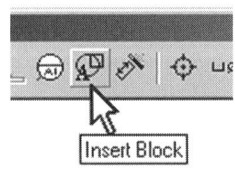

Fig. 11-126. Insert Block icon.

Clicking on the Insert Block icon, shown in figure 11-126, brings up the Insert Block panel (shown in figure 11-127). You could also select Block from the Insert menu. If a block has already been defined in the drawing, it can be inserted again at an alternate location once it is selected from the drop-down list, also visible in figure 11-127. All blocks present in the drawing will be listed. Additionally, the Browse button can be used to browse for a block file to import into the drawing.

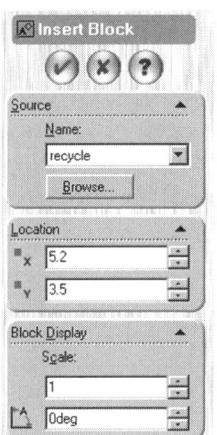

Fig. 11-127. Insert Block panel.

The x-y location of the block can be precisely entered in the Insert Block panel, as can the rotation or scale of the block. You must pick a location somewhere on the sheet to set the block down prior to positioning it using x-y coordinates. Modify the scale and rotation, if necessary, and then click on OK, or continue to add instances of the block. The entire process is quite user friendly.

Saving and Linking Block Files

Blocks can be saved as separate files with the file extension *.sldblk* or *.sldsym*. This can be done most easily by right-clicking on a block and selecting Save to File. The alternative would be to select the same command from the Tools > Block menu. You will then be able to give the block file a name and SolidWorks will add the *.sldblk* file extension to the file name. Obviously, a block must be defined prior to saving the block as a file. To edit a block file whether or not it is in the current drawing, select Edit File from the Tools > Block menu.

Linking to a block file really needs to be done after a block has been defined, saved, and inserted into the current drawing, though not in any particular order. Just because a block has been inserted into a drawing does not necessarily mean it is associated to a separately saved block file. After all, you could define a block in the drawing without ever saving it as a file.

If you have defined a block in your drawing, and have taken the time to save it as a separate block file, linking the inserted block to the block file is certainly an option. Be forewarned that it is highly recommended a specific directory be set up for the sole purpose of storing your block files. You can also tell SolidWorks to look in a particular location or locations for block files. See How-To 8-16 in Chapter 8 for how this is accomplished.

Why even bother linking to an external file? The reason is because the external file could be altered, thereby causing the blocks in the drawing to update. Multiple drawings could all be affected in this way. This could be dangerous, but it could also be beneficial. Do not add a link unless you want the drawing to update if the referenced block file changes.

Imported Block Attributes

Those readers who have worked with AutoCAD blocks are more than likely aware of what attributes are. In that this is a SolidWorks book, we will only touch upon this topic as it bears on SolidWorks. Basically, attributes represent a means of attaching information to a block. For example, an attribute tag named Manufacturer may contain the value ACME Castings, Inc.

When a block is imported from AutoCAD, and that block contains attributes, those attributes can be accessed and modified in SolidWorks. The typical scenario would play out as follows. An AutoCAD file is opened via the File > Open command (see Chapter 22 for more on importing files). Once the file is open, selecting a block causes the Block Instance panel (shown in figure 11-124) to appear. If the block contains attributes, the Attributes button will be available, shown in figure 11-128.

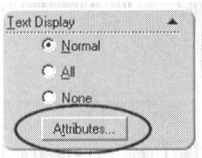

Fig. 11-128. Attributes button in the Block Instance panel.

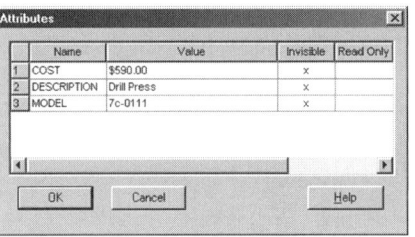

Fig. 11-129. Attributes window.

Clicking on the Attributes button opens up the Attributes window, shown in figure 11-129. The attribute name cannot be altered, but the values can. Click on a value to alter it as required. The Invisible and Read Only attribute options cannot be modified.

RapidDraft Files

Typically when creating a SolidWorks drawing, the parts or assemblies shown in the various drawing views are required in order to open the drawing. This has to do with the associative aspect of the SolidWorks software. It allows changes made to a part or assembly to automatically propagate to the drawing, eliminating the need to manually keep track of these changes or edit the drawing.

There are some drawbacks to file associativity, however. For example, sending a drawing to another individual over the Internet would mean that any parts or assemblies referenced by the drawing would also need to be sent. If an assembly were referenced by a drawing, it would mean that now all part files referenced by the assembly must also be sent. This can wind up making for some very large e-mail attachments, or would even require burning a compact disk to send via snail mail.

Enter RapidDraft, which allows sending drawing files without the need to send the referenced parts or assemblies. Furthermore, the drawing can be synchronized to the referenced documents, thereby updating geometry once the referenced files are available. In essence, the user can decide exactly if and when the referenced documents are referenced!

Another big advantage of RapidDraft drawings is shorter load time. Large assembly drawings no longer have to wait for the referenced assembly files to load first. This should prove a big time saver for those with large drawings with many views and sheets, or for drawings of assemblies with many components.

Creating RapidDraft Drawings

It is interesting the way in which a RapidDraft drawing is created. If one were to guess, it would be fair to assume a RapidDraft drawing could be saved using the Save As option under the File menu. This is not true, however. RapidDraft drawings can be created when opening an existing drawing.

✥ **NOTE:** *Once converted to RapidDraft format, a drawing cannot be converted to its original format.*

Once a drawing has been converted to a RapidDraft format, there is no going back. Is this bad? Not necessarily. Are there disadvantages to the RapidDraft format? Not really, but the file association back to the part or assembly is not quite as "tight" as standard SolidWorks drawing files. Those accustomed to drawings automatically being updated might get caught off guard if they are not careful.

There is no harm in converting drawings to RapidDraft format. The only word of warning would be to know when you are working in a RapidDraft file. From that fact you would know that the associated solid model may need to be loaded prior to the drawing views showing the correct and updated geometry. You may only want to convert to RapidDraft format when necessary, and not as a general procedure. With that said, How-To 11-41 takes you through the process of converting to RapidDraft.

How-To 11-41: Converting to RapidDraft

To convert an existing drawing to RapidDraft format, perform the following steps.

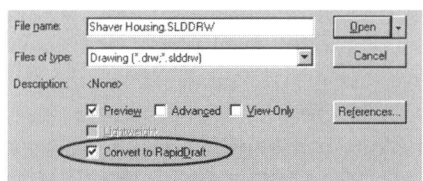

Fig. 11-130. Converting to RapidDraft.

1. Select Open from the File menu.
2. Select the file you wish to open.
3. Make sure the Convert to RapidDraft option is checked, as shown in figure 11-130.
4. Click on the Open button.

5. Once the file has been opened, save the file to complete the conversion process.

When a file is first converted to a RapidDraft drawing, the load time may not be as fast as expected. However, the next time the file is opened, you will notice a significant reduction in the time it takes to open the file. The reduction in time may not be noticeable with simple part file drawings, but with drawings of complex parts and large assemblies you should notice a substantial difference.

In some cases, the file size of a RapidDraft file may increase over the original. This could be due to a number of factors. However, file size is not the reason for creating RapidDraft files. You create RapidDraft files to shorten file access time and to transport drawings across networks or the Internet.

Synchronizing Model Geometry

There are certain operations that require a model be loaded first once a drawing has been converted into RapidDraft format. The following is a partial list of these operations.

- Adding new views
- Inserting design tables
- Inserting model items
- Changing dimension values

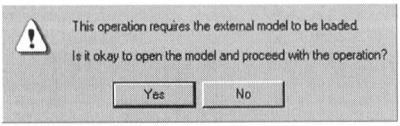

Fig. 11-131. Should the model be loaded?

Basically, anything that requires access to the original solid model will require that the part or assembly be loaded first. This is all very logical, and you can usually guess what will require that the model be loaded first. SolidWorks will prompt you and ask if the model should be loaded if it senses something is being done that requires it. In this case, the warning message shown in figure 11-131 will appear, asking if you would like the model loaded.

In this case, you would more than likely just click on the Yes button and let SolidWorks load the model. In addition, it is possible to manually load the model geometry as well. To manually load the model, simply right-click on the drawing view that contains the model to be loaded and select *Load model*. That is all there is to it.

Tabulated Drawings

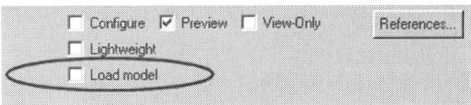

Fig. 11-132. Load model *option*.

How do you know when you are opening a RapidDraft drawing? If using the Open command in the File menu to open a drawing, there will be a *Load model* option available (shown in figure 11-132) if the drawing is a RapidDraft drawing. If not a RapidDraft drawing, there will be a Convert to RapidDraft option, described previously. If opening the file some other way, such as by double clicking on the file in Windows Explorer, there is no way to discern that you are opening a RapidDraft file. However, once the file is open, there are other indicators that you are in a RapidDraft file.

Once a drawing file has been opened, one way to discern that it is a RapidDraft file is to right-click on any of the drawing views. If there is a *Load model* option, it must be a RapidDraft file. Another indicator is the color of the view highlighting. Running the cursor over a RapidDraft drawing view will highlight the view's border in blue. (Once the model is loaded, the borders highlight in yellow.)

> **NOTE:** *If creating drawings of large assemblies with hundreds of components, be aware that there is an option for automatic conversion to RapidDraft drawings when in large assembly mode. See the section "Large Assembly Mode" in Chapter 18 if this applies to you.*

Tabulated Drawings

	Description	A	B	C	D	E
13112-0139	3/4 fillister 50	2.75	2.25	50	3.00	1.50
13112-0140	3/4 hex head 50	2.90	2.10	50	3.10	1.53
13112-0141	3/4 socket 50	3.10	1.95	50	3.25	1.67
13113-0142	3/4 fillister 30	2.75	2.25	30	3.00	1.50
13113-0143	3/4 hex head 30	2.90	2.10	30	3.10	1.53
13113-0144	3/4 socket 30	3.10	1.95	30	3.25	1.67

Fig. 11-133. Tabulated drawing table.

A tabulated drawing is a drawing in which dimension values have been replaced with labels (often simply letters of the alphabet). A table is then created, in which the dimension labels are called out at the top of each column (see figure 11-133), and dimension values can be referenced via the table. Drawings of this type are created when it becomes necessary to list dimensions for a family of parts or when there are numerous features or holes whose *x-y* coordinates must be given.

The table itself can be created a number of ways. The geometry for the table can be created manually through the use of basic sketch tools. This method was discussed in an earlier section (see "Empty Views and Creating Tables"). Another method would be to insert an Excel spreadsheet as an object. This would be accomplished by selecting Object from the Insert menu. Finally, a family-of-parts table, better known as a SolidWorks design

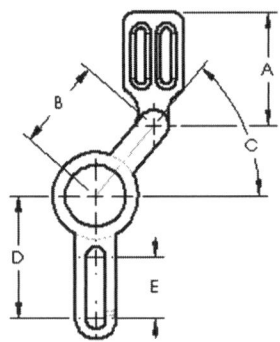

Fig. 11-134. Dimensions driven by the design table.

table, could be inserted into the part file to drive configurations. A snapshot of the design table could then be placed in the drawing.

A design table is what was used to create the table shown in figure 11-133. Design tables are covered in detail in Chapter 16. They are used to drive part or assembly configurations. The table shown in figure 11-133 performs two functions. It both serves to display certain dimension values for different variations of the part model and actually drives those same dimensions in the model, which is shown in figure 11-134.

The majority of the work that was done to create the tabulated drawing used in our example involved the creation of the design table itself. As mentioned earlier, you will learn how to perform this task in Chapter 16. Once the design table has been created in the part file, inserting the snapshot of the table can be done by selecting a view, and then selecting Design Table from the Insert menu. The table can then be scaled or moved to a particular position on the sheet by simple left mouse button drag techniques.

Another aspect of creating a tabulated drawing would be to change the dimension values to appear as labels. This process is a very easy one, and consists of selecting Properties from the menu after right-clicking on a dimension value. Once in the Dimension Properties window, change the dimension name to the label of your choice. Figure 11-135 shows where the dimension name can be altered. In this case, the letter A was used, but any text can be entered for a dimension name.

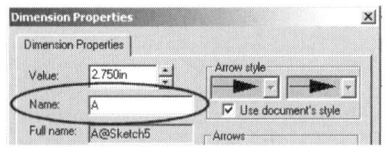

Fig. 11-135. Changing the name of a dimension.

Summary

In this chapter you have learned that there is a lot you can do with design drawings. Many views can be created, with a good portion of the view creation process being automated by SolidWorks. A perfect example is the drag-and-drop functionality when creating the three standard views. You can drag a part or assembly file from Windows Explorer into a drawing, or drag the file from the top of FeatureManager if the file is already open.

When beginning a new drawing, specify the template you want to use to start your drawing. Templates contain various settings, such as dimensioning standards and work units (i.e., English or metric). Templates may also contain a drawing sheet format. If you do not want to use a format, you must at least specify a sheet size for the paper. SolidWorks will ask what format you would like to use only if the selected template does not already contain one.

If creating a new sheet format, it is best to save the format as a separate file. This can be done by using the Save Sheet Format command found under the File menu. Once a format has been saved as a separate file, it can be used for other drawing sheets that may be added to the same drawing.

You can alternate between editing the sheet and editing the format by right-clicking in an empty place on the drawing or by using the Edit menu. Edit the properties of the sheet (again, by right-clicking) to change the sheet size, scale, or format on the fly. If you alter the scale while editing the format and then save the format, the scale setting will be saved with the format.

To move a view, drag its border. A parent view will move with its dependent views. However, if the dependent view is moved, it will move by itself. Views will remain aligned with their parent views unless you break the alignment. If you change your mind, you can return to the default alignment condition. Once again, this can all be done with the right mouse button.

Creating certain views sometimes requires specific geometry or selections to be made. For instance, detail or section views require circles or lines to be sketched on the parent view before the detail or section views can be created. Remember that a view must be active before geometry can be added to it. An active view has a gray shadow border. Geometry added when a view is active will be associated with that particular view. This holds true for notes as well.

Auxiliary views require that you select an edge to project from when creating the view. Projected views require that you select a view to project from. When creating a named view, select somewhere in the part or assembly's work area to indicate that it is the part or assembly you want to insert a named view of. The act of clicking in the part or assembly's work area indicates to SolidWorks that it is the part whose view you are creating. If a view of that particular model is already in the drawing, selecting the view will suffice.

Dimensions used to create a part can be brought into a drawing and reused. This is known as inserting model items. Any type of annotations can be inserted from a part or assembly into a drawing in this manner. Most annotations, however (other than dimensions), are typically created in the design drawing and not in the part.

Moving dimensions is a simple drag operation. You can also drag the ends of extension lines by selecting the dimension, and then dragging the small green handles that appear at the ends of the extension lines. The green handles attached to the dimension arrows can be used to flip the arrows inside or outside the extension lines. To move a dimension to another view, hold down the Shift key.

Edit a dimension's properties (right mouse click) to change its appearance. Many characteristics (such as arrow display, diameter, radial options, symbols, and text) can be altered through a dimension's properties. Some of these options can also be modified in PropertyManager.

Other annotation types (such as hole callouts, center marks, and notes) can be added via the Annotations toolbar. Most annotations, such as notes, allow for placing multiple annotations without issuing the command repeatedly. You can add a leader to the note, or reposition the note or leader once it has been created. Attaching a leader to an edge or surface of the model associates the leader with the model. If the model changes size or shape, the leader will remain pointing to the desired location. Add multiple leaders by holding down the Ctrl key when positioning the leader.

Use blocks when there are symbols or groups of sketched objects that need to be reused. Blocks can be defined in a drawing, and then added multiple times at various locations. If one block is edited, all related blocks will update as well. Blocks can also be externally linked to a separate file. Editing the block file results in drawings updating to reflect the change. All blocks can be scaled and rotated as required.

RapidDraft drawing formats can be created from existing drawings when the drawing is opened. A RapidDraft drawing does not require the referenced model geometry in order to open and view the drawing. However, most functions, such as modifying dimensions or adding views, will require that the referenced model geometry be loaded. This can be done by right-clicking on a drawing view and selecting *Load model*.

Questions and Topics for Discussion

1. Explain what is meant by the term *bidirectional associativity*.

2. What are file extensions? What file extensions are used for the three main types of SolidWorks documents?

3. What temporarily happens to the views in a drawing when you edit the drawing sheet format?

Questions and Topics for Discussion

4. Describe the difference between a drawing sheet format and a drawing template.

5. Describe the process of inserting a company logo into a drawing.

6. When editing a drawing sheet format, can you add constraints to template geometry? Can you use the Trim and Extend icons on format geometry?

7. Describe two ways of inserting the standard top, front, and right-side views into a drawing.

8. When is it necessary to activate a view? Describe two methods of activating a view.

9. Name two view types that require adding geometry to the parent view.

10. What must be selected in order to create an auxiliary view?

11. What must be selected in order to create a projected view?

12. When creating the geometry that will define a section line, what sketch entities can legally be used?

13. How can multiple leaders be added to an annotation? Describe the process in your own words.

14. How can the tangent edges be hidden or shown in a drawing view?

15. How can the borders of drawing views be shown?

16. If drawing view borders are turned on, will they appear when printed or plotted?

17. Is it possible to change a single view's scale? Explain your answer.

18. What is a partial section view, and how do you obtain one?

19. Describe the process of inserting dimensions into a view.

20. How are the dimensions inserted into a view as model items different from reference dimensions?

21. In your own words, describe what a block is.

22. What happens when a block is exploded?

23. Describe the advantages of RapidDraft drawings.

Optional Problem

Use the shaver housing, begun in Chapter 6 and completed in Chapter 7, to create a design drawing. To get you started, use a B-size standard sheet format (with a B-size sheet of paper, of course). Modify an existing drawing template if necessary. Insert the three standard views, a section view, a detail view, and an isometric (named) view.

Add dimensions, both inserted as model items and reference dimensions, as needed. You will also need to add a hole callout. Modify the title block to suit your needs. An example of a completed drawing is shown in figure 11-136. Your drawing may look different, but that is okay. The purpose of this exercise is to give you a little practice inserting views and annotations. Feel free to take a little artistic license.

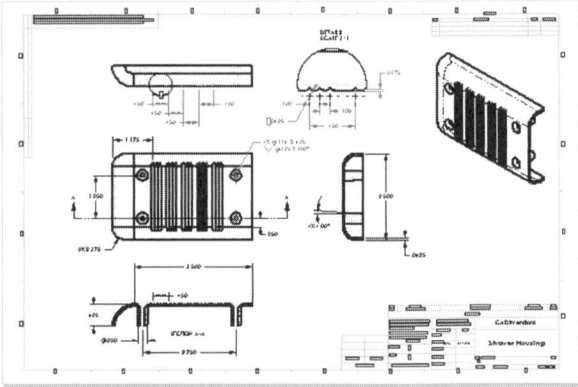

Fig. 11-136. Completed design drawing.

CHAPTER 12

Assemblies

PART FILES, AS YOU HAVE LEARNED, typically contain one contiguous solid model. Assembly files can contain more than one part. Assembly files give you the capability of assembling the parts you have created, putting the parts together as if you were actually building the assembly in real life. Figure 12-1 shows an example of an assembly. It is the example you will use to learn about assemblies in this chapter.

Assemblies vary in complexity over a very wide range. They can be as simple as a two-part assembly, or as complex as an assembly containing thousands of components. The only limiting factor on the size of an assembly is the type and amount of computer hardware you have. The faster the processor, the better off you will be, and a lot of computer memory is a very good thing. If you will be creating large assemblies with thousands of parts, you would be wise to purchase extra memory. On the order of 512 MB of RAM is common, and over 1024 MB would not be considered overkill.

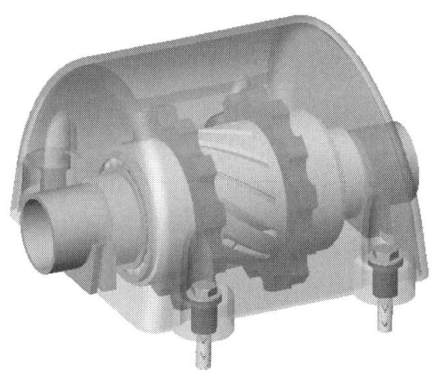

Fig. 12-1. A SolidWorks assembly.

Assemblies fall into two categories. The first and most common assembly is known as a *bottom-up assembly*. A bottom-up assembly is what you will be reading about in this chapter. Think of a bottom-up assembly as a table with parts lying on it. You pick the parts up and place them together, building the assembly. The parts themselves have already been created.

498   *Chapter 12: Assemblies*

The other type of assembly is known as a *top-down assembly*. Imagine that you have a partial assembly. A part that is needed for the assembly must be built by referencing other parts in the partially completed assembly. The new part is, in essence, created from the inside out. They are designed from within the context of the assembly. (This topic is discussed in detail in Chapter 18.)

This chapter first deals with how to begin an assembly and then insert the components into the assembly. There are various methods you can use for component insertion. You will learn all of them, and then can decide for yourself which method is most convenient for you.

Starting a New Assembly

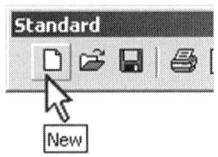

Fig. 12-2. New icon.

Start a new assembly just as you would start a new part. Select New from the File menu, or click on the New icon, shown in figure 12-2. The only difference is that you should specify an assembly template instead of a part template, as shown in figure 12-3. You can then click on the OK button and be on your way to building an assembly.

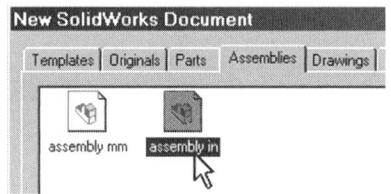

Fig. 12-3. Specifying an assembly template.

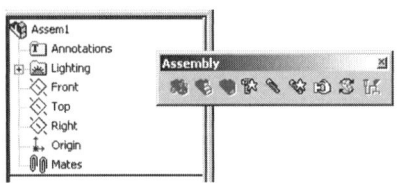

Fig. 12-4. Assembly FeatureManager and toolbar.

You will not notice many differences between the part and assembly interfaces. There are only a few. First, you may notice that there is something new in FeatureManager, shown in figure 12-4. An item named *Mates* is present in FeatureManager, and the icon to the left of the assembly file name is slightly different. The *Mates* item is used as a storage area for the geometric relations between components in an assembly, known as mating relationships, or simply mates.

Another new item that will make its appearance is the Assembly toolbar, also shown in figure 12-4. Its icons will be introduced to you gradually throughout this chapter. This toolbar is your key for moving and rotating individual assembly components.

The most significant change in the interface comes with the Insert pull-down menu. Where the menu items Boss and Cut used to be are now the words *Component* and *Mate*. These are the most commonly used menu picks, so they are placed at the top of the menu. There are other changes to this particular menu as well. The most significant of these new menu items are covered in this chapter.

Inserting Components

Now that you have a new assembly begun, it is time to bring some components into the assembly. There are three basic ways of accomplishing this. One method is by way of Windows Explorer. This is literally a drag-and-drop procedure. For instance, by dragging and dropping a SolidWorks part from Windows Explorer into the work area of an open assembly, the part will thereby be inserted into the assembly.

The benefit of using the drag-and-drop technique is that it allows you to insert more than one component at a time. Additionally, if mate references are used, components can be mated into their proper position with nothing more than a simple drag-and-drop maneuver. Mate references are explored in material to follow.

Another alternative requires that the part file already be open. Again, this second method is a drag-and-drop procedure. Place the cursor over the name of the part in the part's FeatureManager, and then drag the part to the assembly work area. This procedure is depicted in figure 12-5. If you happen to have the part open, this method is most convenient. Note the system feedback from the cursor. This indicates that the part will be dropped directly on the assembly's origin point.

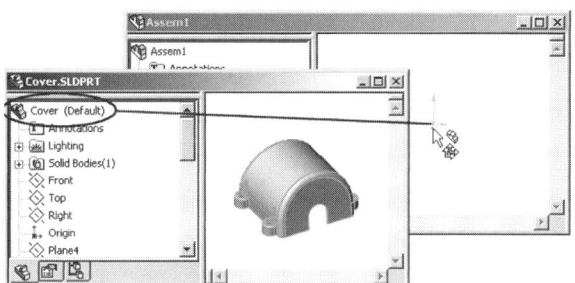

Fig. 12-5. Dragging and dropping from an open part file.

This is good technique, as it makes sense that the first component is centrally located within the assembly's coordinate system.

A third option is very common and will probably wind up being the method you use the most. This method makes use of the pull-down menus. It is advisable to gain familiarity with this method of inserting a component before trying out some of the "shortcut" methods, as there are some additional points you should be aware of that go along with drag-and-drop functionality. These points you will learn, but all in good time. How-To 12-1 takes you through the process of inserting a component via the Insert menu.

How-To 12-1: Inserting Components Using the Insert Menu

To insert components using the pull-down menus, perform the following steps.

1. Select Insert > Component > From File.
2. Select the file you want to insert into the assembly.
3. Click on Open.
4. Pick with the cursor to indicate where you want to position the component.

Fig. 12-6. Centrally positioning the first component.

This particular method of inserting a component has some benefits, one of which is the ability to preview the component. Another benefit has to do with positioning the component. When you are ready to position the part, note what happens when you place the cursor over the origin point. You should see two small origin symbols displayed next to the cursor. This system feedback indicates that the new component's origin and the assembly's origin will be coincident if placed at that point. The symbol is shown enlarged in figure 12-6. Although this symbol presents itself when dragging and dropping from a part file, it does not when using Windows Explorer.

✓ **TIP:** *It is good technique to position the first component directly on the assembly's origin point. This means the component will be centrally located with respect to the assembly's coordinate system.*

It can often prove beneficial if the first component is centrally located. An example might be if you are creating some sort of symmetrical assembly. It may make editing the assembly easier at a later time. Having planes that pass through the center of the assembly, and simply locating the assembly near the world origin point, can be useful later on.

Fixed or Floating?

It would benefit you to take some time to consider what should be the first component inserted into the assembly. When building an assembly, the first component should be the component you will be attaching (mating) other components to. It should be the main component in the assembly,

Inserting Components

generally speaking. Sometimes this is not a cut-and-dry decision, in which case the first component inserted into an assembly should be one that will remain stationary at all times.

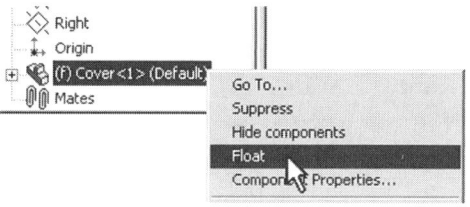

Fig. 12-7. Fixed components can be floated.

By default, the first component inserted into an assembly is fixed in space. That is, its location is fixed, or locked. It should be noted that this is not necessarily a permanent condition. Any component can be made to be fixed in space or floating. This is accomplished with the right mouse button. By right-clicking on the part's name in FeatureManager, you will gain access to the toggle that controls the fixed or floating condition. This is shown in figure 12-7.

You will always want to have at least one component fixed in space. Think of it as a real part you could hold in your hand. During the assembly process, you would want to hold the part steady so that it would not slide around as other components were attached to it. This is what the Fix option accomplishes.

FeatureManager Symbology

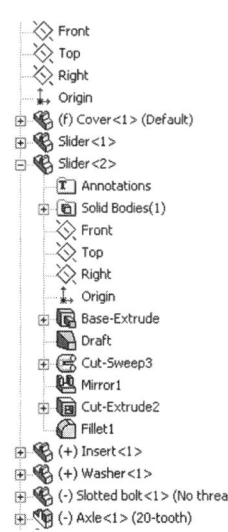

Fig. 12-8. FeatureManager symbols.

You can tell a few things about an assembly just by looking at its FeatureManager. For instance, if a component is fixed, it will have (f) prefixing the component's name. This can be seen in figure 12-7.

Before a component is mated to other components in the assembly, you will see a minus sign (–) before its name. If a component is fully mated to other components, you will not see any symbol before its name.

A plus sign (+) is bad news, and means you have overdefined a component. This happens if you add too many mating relationships between components, or if mates are added to a component that is fixed in position. If you encounter this situation, you should find a solution to the problem before going any further. Some of FeatureManager's symbols are shown in figure 12-8. Some components were intentionally overdefined for the sake of displaying the plus symbol in FeatureManager for this book.

The same component can be inserted into the same assembly as many times as required. SolidWorks uses a numbering scheme to keep track of the components. Note in figure 12-8 that the slider component has been inserted into the assembly twice, once as *Slider<1>* and again as *Slider<2>*. These numbers are known as instance identification numbers.

Instance ID numbers are necessary, because although the two sliders are different components, they are the same part. The same part file is referenced for each of the two components. End users have no control over instance ID numbers, nor does it matter that they should.

It should be noted that there is no real significance between the number of components in the assembly and the instance ID number. For example, if another slider were brought into this assembly, it would be named *Slider<3>*. It is a good possibility there would be three sliders in the assembly, but this is not necessarily true. If *Slider<1>* were deleted from the assembly, *Slider<3>* would still be named *Slider<3>*, even though there are only two sliders in the assembly.

All of a part's features are accessible from FeatureManager (see figure 12-8). You can see that *Slider<2>* has been expanded so that the features are viewable. You could go one step further and expand the features to gain access to the underlying sketch geometry. Collapse or expand FeatureManager as necessary, using the small plus or minus signs located just to the left of FeatureManager icons.

- **NOTE:** *Do not confuse the small plus or minus signs to the left of an object's icon with the plus or minus symbol in parentheses used to signify an overdefined or underdefined part.*

Moving and Rotating Components

The Assembly toolbar contains two icons specifically designed for moving and rotating components. This proves invaluable when trying to build an assembly. In the case of adding mates, it is best to position the components close together and in an orientation that is similar to how they should wind up after adding the mates. This is so that you can zoom in to the faces being mated and more easily select them, but also to help Solid-Works pick the correct alignment condition, discussed in material to follow.

Moving or rotating components is extremely helpful when trying to determine which mates have or have not already been added to the assembly. This is similar to dragging sketch geometry to see how it will behave. By moving components, you can discern how the assembly will behave and whether or not more mating relations should be added.

Assembling with motion in mind is the first step in creating a realistic model that moves as it would in the real world. What mates you add between components makes a difference in how components will behave when moved. The first and most common icon used for manipulating a

Moving and Rotating Components

component is the Move Component icon, shown in figure 12-9. Any component can be moved, as long as it is not fixed or fully mated. How-To 12-2 takes you through the process of moving a component.

How-To 12-2: Moving a Component

To move an individual component, perform the following steps.

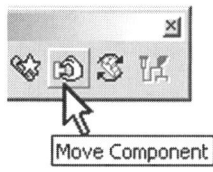

Fig. 12-9. Move Component icon.

1. Select the component to be moved.
2. Select Tools > Component > Move, or click on the Move Component icon (figure 12-9).
3. Hold down the left mouse button and drag the component to its new location.

When selecting a component to be moved, you may select it from either FeatureManager or by selecting a face on the part in the work area. By adding mates, you begin to limit the degrees of freedom of movement a part has. It would benefit you to experiment with moving a part after adding mates. This will allow you to see how the mates affect its movement.

✓ **TIP:** *A component can also be moved by clicking on the Move Component icon, and then positioning the cursor over the component. The component can then be dragged to a new location without being selected first.*

The Rotate Component icon is shown in figure 12-10. The steps used to rotate a part are almost identical to moving a component. How-To 12-3 takes you through the process of rotating a component.

How-To 12-3: Rotating a Component

To rotate an individual component, perform the following steps.

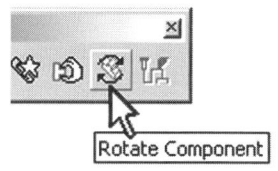

Fig. 12-10. Rotate Component icon.

1. Select the component to be rotated.
2. Select Tools > Component > Rotate, or click on the Rotate Component icon (figure 12-10).
3. Hold down the left mouse button and rotate the component.

As noted in the previous Tip (and similar to moving a component), it is possible to click on the Rotate Component icon first, and then position the cursor over a component and rotate the component by dragging with the left mouse button. This eliminates having to select the component first. Sometimes adding a mate will position a component within other components, where it is out of sight. In this case, you would likely want to move the component by selecting it from FeatureManager and following the steps listed in How-To 12-2.

Moving Components in Specific Directions

Fig. 12-11. Move and Rotate panels.

If you want a little more control over how a component is moved or rotated, use the drop-down menus found in the move or rotate panels, shown in figure 12-11. It should be noted that both the Move and Rotate panels cannot in actuality be opened at the same time. Either one or the other can be shown. Clicking on the arrow at the top right of a panel expands or contracts that panel. This is a nice way of switching between the Move Component and Rotate Component commands quite easily. For example, expanding the Rotate panel while in the Move Component command will switch to the Rotate Component command.

The options in the Move and Rotate panel pull-down menus vary, depending on which command you are in. Both panels contain the option Free drag, which is the default setting and by far the most commonly used. It is rare that any of the other available options are used, though they have their place. For example, the Rotate panel contains the option *About entity*. By preselecting a model edge along with a component, that component can be rotated about the selected edge. The Move panel's pull-down menu contains an option for moving a component a specific amount in the x, y, or z direction.

These various options are nice, but none of them will incorporate design intent into the assembly. It is much more important to add the proper mates between components. That way, components will move with respect to the mates that have been applied to them, and none of the specific move or rotate options will usually be necessary.

Occasionally, special move or rotate options (such as *About entity*) are necessary, but only to gain a little more control over the item you are moving. The item, ideally speaking, should already be mated in position. If movement is desired, the mates should be added in such a way to give the component the proper degrees of freedom of movement it requires.

Mating Relationships

Once you have begun bringing components into the assembly, you must then mate them to each other. It is not enough to just position the components. You must establish a set of conditions that describes how the components relate to one another. This should sound familiar to you. After all, it is the same thing as adding geometric relations when creating sketch geometry. However, in this case it is applied to 3D geometry instead of 2D sketch geometry.

How do you know what types of mating relationships to add between components? Ask yourself how you would put the components together if it were a real-life assembly, and you will have your answer. Generally speaking, create an assembly with motion in mind. Unless, of course, the assembly has no moving components, in which case the assembly process is simplified even more.

There are six degrees of freedom of movement with regard to the individual parts in an assembly. Any component, assuming it is not fixed, can translate in the x, y, or z direction. Each component can also rotate along the x, y, or z axis. By restricting the direction a part can translate or rotate, you can control how the assembly behaves.

Different mates affect assembly components in different ways. A concentric mate, for instance, will allow a component to translate in one direction and rotate about the axis of translation. That is the nature of the concentric mate.

Another example might be coincident mates. Say, for example, you want a part to be able to slide along a slot on another part. It would make sense in this case to add two coincident mates for positioning the first part in the slot of the second part, but leave the third direction free so that the part could move along the slot.

Adding mates between components in an assembly is similar to adding geometric relations in a sketch, but with two main differences. When creating a sketch, it is good technique to fully define the sketch. When dealing with an assembly, you do not need to worry about fully defining the components. On the contrary, you may want to leave components intentionally underdefined, thereby enabling the component to move or rotate in the desired direction, whatever that may be. On the other hand, fully mating components that should not move is perfectly acceptable.

A second difference is that you are now dealing with 3D geometry. When working with a sketch, you needed to select points or sketched entities to define relations. Because you are now dealing with 3D parts, it is usually best to select faces. This is not required, but is often good technique. Mate relationships can be added between many things, including, but not limited to the following.

- Faces
- Planes
- Axes
- Model edges
- Sketch geometry
- Origin or vertex points

Fig. 12-12. Coincident mate between vertices.

Fig. 12-13. Coincident mate between edges.

Let's take a simple example to help illustrate what might happen when adding a coincident mate. Assume there are two cube-shaped building blocks in an assembly. What happens if a coincident mate is added between two corners (vertex points) of the blocks? Assuming also that one of the blocks is fixed (anchored in position), the remaining block will be free to rotate in any direction, but will pivot at the mating point. This is illustrated in figure 12-12.

You could think of this mate as a ball-and-socket type of relationship. This is perfectly legal, but care must be taken to select the proper objects to mate between. If a plane were selected on the anchored block as opposed to a vertex point, only one degree of freedom would have been removed, which would have been the upper block's ability to translate perpendicular to the selected planar face.

Now consider what would happen if a coincident mate were added between two edges. This is shown in figure 12-13, and creates a hinge-type mate. This leaves only two degrees of freedom of movement: the ability to translate along the mating edges and to rotate about the mating edges.

Finally, what happens when a coincident mate is applied between two faces? Figure 12-14 shows the outcome of this mate. The upper block is free to rotate about one axis but can slide along the other two axes. (Note that although the illustrations show single-ended arrows, the translation and rotations directions are bidirectional.)

Mating Relationships

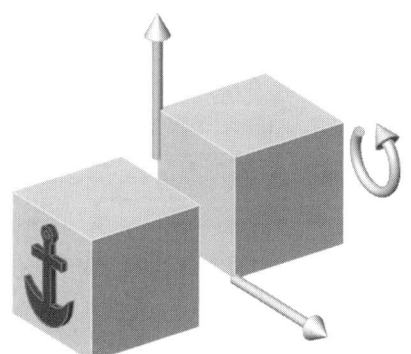

Hopefully, what you gather from these illustrations is that it is okay to mate between various objects, but care must be taken. A good general rule is to add mates between faces or planes when possible. Next, you will take a look at the mechanics behind adding mating relationships between components. How-To 12-4 takes you through the process of adding mate relationships.

Fig. 12-14. Coincident mate between faces.

HOW-TO 12-4: Adding Mate Relationships

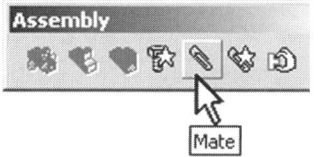

Fig. 12-15. Mate icon.

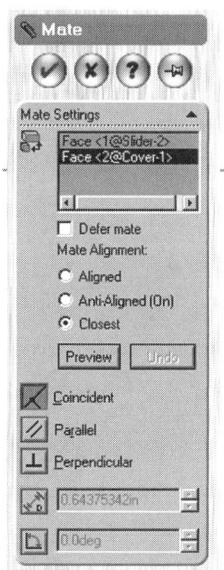

Fig. 12-16. Mate panel.

To add mate relationships, perform the following steps.

1. Select Insert > Mate, or click on the Mate icon, shown in figure 12-15.

2. Select the objects between which to add a mate relationship. For instance, select a face on each of the two components being mated.

3. From the Mate panel, shown in figure 12-16, select the mate type to be added.

4. Click on Apply to add the mate.

Components being mated will only move as far as they have to in order to satisfy the mating condition. A common question is "Which component will move, and is the order in which the faces are selected important?" Selection order is not important. Additionally, the component that will move depends on existing mates. If one component is fixed or already fully mated in position, it will not be able to move.

Mate Options

There are other functions in the Mate panel you should be aware of. There are also other mating relationships that can be performed with the Mate panel beyond the coincident mate mentioned in the earlier examples. These topics are explored in the material that follows.

Similar to adding geometric relations in 2D geometry, only the mate relations that can physically be added to the objects selected will be available. Other mate relations will not be available. A spherical face and a planar face, for instance, could not be made parallel. SolidWorks knows this, and does not show the Parallel mate option because it would be geometrically impossible.

There are many mate conditions that can be applied between components, and most of them are self-explanatory, such as perpendicularity and parallelism. Not everyone may be familiar with all of the terms used to describe mates. If you fall into this category, the material that follows will help. The various mate types are examined in the sections that follow.

Concentric Mate

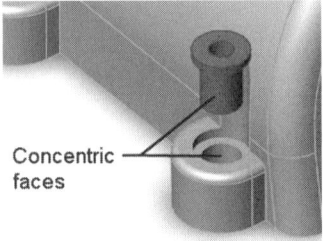

Fig. 12-17. A concentric mate could be used here.

Concentric mates are useful when inserting cylindrical objects into holes, and similar situations. An example of faces that could be made concentric is shown in figure 12-17. Cylindrical, conical, and even spherical faces can be selected when adding this mate relation. Some examples of objects that can be mated concentric follow.

- Cylindrical faces
- Conical faces
- Spherical faces
- Axis and a conical face

Tangent Mate

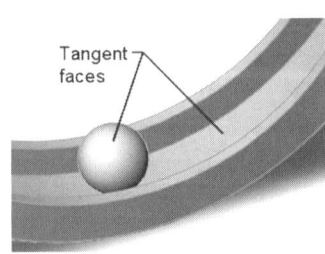

Fig. 12-18. A good candidate for a tangent mate.

Tangent mates often occur between objects such as a cylindrical and a planar surface, or a sphere and some other object. Figure 12-18 shows a mate between a cylindrical face and a sphere. Incidentally, the parts shown in figure 12-18 are on their way to becoming a bearing assembly.

Because tangent mates can usually have two conditions that satisfy the mate relation, make sure you preview the mate first (via the Preview button). The following are examples of the types of objects between which a tangent mate could exist.

- Cylindrical face and plane
- Two cylindrical faces
- Conical and planar faces
- Two spherical faces

Mating Relationships

- A spherical face and a nonplanar face

Tangent mates can be used to create a special type of mate condition known as a cam mate. This is illustrated in figure 12-19. To add a cam mate, add the tangent mate in the usual method, but make it a point to select all of the perimeter faces of the cam component. All of the faces on the cam should be tangent, and the end of the follower component should be spherical or cylindrical in nature for this mate type to work best.

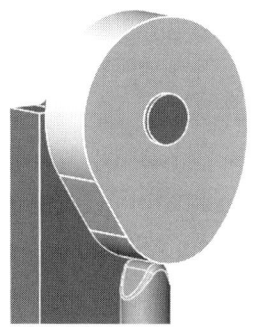

Fig. 12-19. A Cam mate was used here.

Coincident Mate

Coincident mates will position one planar face next to another. Technically, the correct term for this would be *coplanar*. However, coincident mates can be used to mate two edges or two vertex points, as well as a variety of other combinations of objects. Therefore, the term *coincident* is used as a catch-all term, instead of differentiating between coincident, collinear, and coplanar. The following are examples of objects that can be mated coincident.

- Two planes or planar faces
- Edge and plane
- A point and a face

Distance Mate

If you want to leave some clearance between two faces, use the distance mate instead of coincident. Generally, anything that can be mated coincident can also be mated using a distance mate. The only difference is that there will be some user-specified distance between the two items being mated. Typically, distance mates are applied between two planar faces.

To enable the distance value setting of the distance mate, you must first click on the Distance mate icon, shown in figure 12-20. Once the distance mate has been activated, the value can be adjusted. By default, the value will be whatever the actual distance is between the two objects being mated.

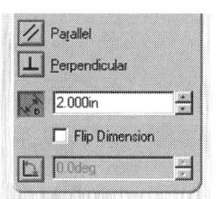

Fig. 12-20. Adding a distance mate.

The option for flipping the distance dimension, also visible in figure 12-20, allows for flipping the side the dimension is on. In figure 12-21, a gear is being mated to the flat face at the end of a cylindrical component via a distance mate. What can be seen is the position of the gear relative to the cylindrical part both before and after the dimension was flipped to the other side.

Fig. 12-21. Before and after flipping the distance mate dimension.

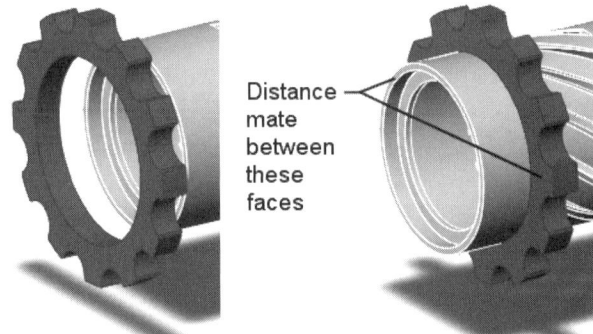

The dimension of a distance mate can be accessed in the same way dimensions for any feature are accessed. That is, double clicking on a distance mate in the *Mates* folder will display the associated dimension on screen and allow for easy editing.

Angle Mate

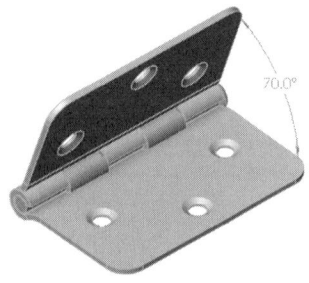

Fig. 12-22. Adding an angle mate.

The angle mate, as its name implies, will place two planar faces at a specific angle. Angle mates, which can be used to create hinged assemblies, are implemented in exactly the same manner as distance mates. In figure 12-22, the two faces of the hinge assembly had an angle mate added between them. Each hinge is also mated concentric about the hinge pin. Additional coincident mates are added to position the hinges correctly so that they mesh, and to seat the pin completely into the female hinge. The pin is fixed, and the female hinge has been completely mated in position.

Without the angle mate, the male hinge would be able to freely move about the hinge pin. Adding the angle mate, however, gives the user complete control over the precise opening angle of the hinge assembly.

Parallel and Perpendicular Mates

You should be familiar with what parallel and perpendicular mates accomplish. These mate types can come in handy for certain things, but should not be overused. It is possible to add plenty of parallel and perpendicular mates to a single component, and never actually fully define it. If your intention is to fully define a component so that it has no movement, accomplish the task in as few mates as possible. This reduces the overall number of mates in the *Mates* folder and simplifies the assembly.

When finishing up an assembly that contains many nonmoving parts, it sometimes helps to fully define components such as hardware that would not otherwise necessarily need to be fully defined. It is easy to gaze down through FeatureManager and make sure everything is fully defined by looking for the absence of the little minus signs that precede an undefined component. As an example, parallel or perpendicular mates can be convenient for stopping a bolt from spinning in a hole.

Symmetric Mate

Symmetry mates are often misunderstood. Symmetry itself is easy enough to understand, but what to select when adding a symmetry mate is what causes some problems. By adding a symmetric mate, you can make two components behave in a mirror-like fashion about a plane or planar face. Symmetric mates often do not react well when used by themselves, but when used with perhaps one additional mate, they work very nicely.

When adding a symmetric mate, three things need to be selecting in total. These items are a plane or planar face, as previously mentioned, and then similar entities from either of the two components that are to be made symmetric. The key element is to use similar entities, and note that points work best. When faces or edges are used, it is sometimes difficult to predict the behavior of the mated components.

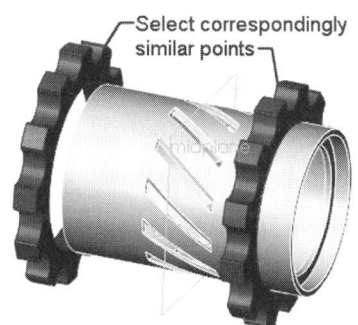

Fig. 12-23. Adding a symmetric mate.

Consider an example. Figure 12-23 shows two identical components, gears in this case, which have both been mated concentric to a cylindrical sleeve. The gear on the right additionally has a distance mate to the end of the sleeve. Because the degrees of freedom have already been reduced for the second gear, all that is needed is to add symmetry between one point on each gear. Using anything more than points would result in overkill. Unfortunately, points are exactly what many SolidWorks users do not use, which results in conflicts of some sort.

Once the symmetric mate has been added, the gear on the left slides and rotates into the proper position, as shown in figure 12-24. If either gear rotates, its counterpart will rotate as well. If the distance mate positioning the gear on the right changes, the gear on the left will update accordingly. It is obvious that symmetric mates have advantages.

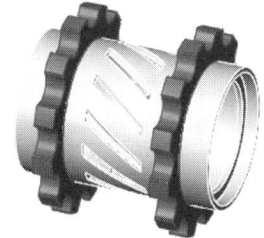

Fig. 12-24. The gears will now move symmetrically.

Previewing Mates

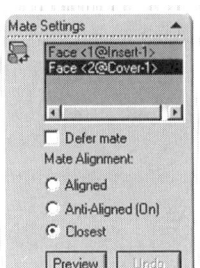

Fig. 12-25. Preview button.

When adding mates, it sometimes helps to see what the results will look like before actually accepting the mate. The Preview button allows you to do this. Figure 12-25 shows the portion of the Mate panel, which contains the Preview button.

After clicking on Preview, and you are happy with the outcome, click on Apply to accept the new mate. If not, click on the Undo button and try a different mate, or exit out of the Mate panel.

Alignment Conditions

When two components are mated, such as with a coincident mating relationship, the components could either be aligned or anti-aligned. This alignment refers to the side of the selected faces the part geometry is on. If the geometry is on the same side, the components are aligned. If the geometry is not on the same side, the components are anti-aligned. This is depicted in figure 12-26.

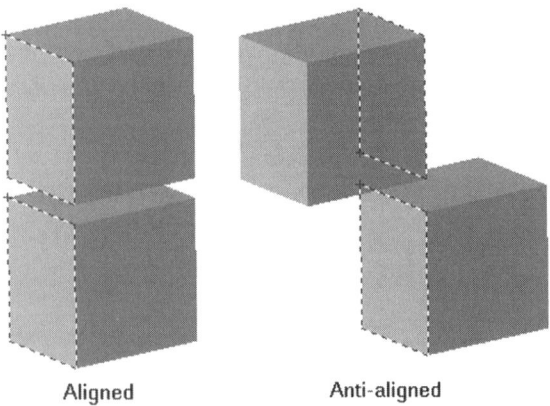

Fig. 12-26. Aligned versus anti-aligned components.

With regard to the coincident mate, alignment conditions are very easy to understand. Regarding other mates, such as tangent, alignment conditions are not as straightforward. The best advice regarding alignment conditions is to not worry about them. The reasoning behind this is due to how SolidWorks functions and an option named Closest.

The default setting for the alignment condition is Closest. SolidWorks basically guesses what condition would be best for the parts being mated. The alignment will wind up being either aligned or anti-aligned. If you are not sure what the alignment should be, let SolidWorks decide. After all, you have a 50-percent chance of it being correct. If the alignment winds up being incorrect, edit the mate's definition (via the right mouse button) and change the alignment to the opposite setting.

The software is essentially lazy with regard to adding mates. It will only move a component as far as it has to in order to satisfy the mate condition. If you make it a point to position a component in the general position of where it should wind up anyway, the alignment condition will almost always be correct.

Finding Mates

Yes, we are still discussing SolidWorks, and have not changed to the topic of a new dating service. One of the strengths of the SolidWorks program is its consistency. To edit a feature's definition, right-click on the feature. To access a feature's dimension, double click on the feature. This functionality carries over to mating relationships as well. Right-clicking on a mate will gain access to the Edit Definition command, which will bring you back to the same Mate panel used to create the mate. Finding a particular mate, however, can be tricky.

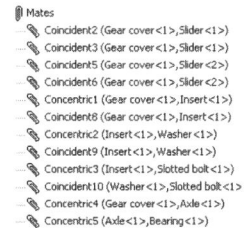

Mentioned at the beginning of this chapter was an object in the assembly FeatureManager called *Mates*. This item is commonly referred to as the *Mates* folder. The *Mates* folder is where the mating relationships you have added to the assembly are stored. Figure 12-27 shows an expanded view of the *Mates* folder.

Fig. 12-27. The Mates folder houses mates.

As you can see from the illustration, each of the mates is given a name, which is followed by the names of the two components being mated, in parentheses. If you were to select a mate within the *Mates* folder, the objects used for the mate would be highlighted. This makes it much easier to find a specific mate, which is something you often need to do when editing an assembly.

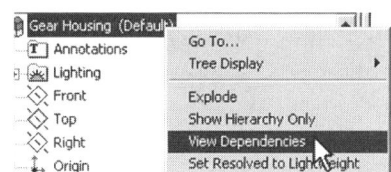

Fig. 12-28. Viewing dependencies.

When there are literally hundreds of mates in the *Mates* folder, clicking through all mates to find the one you are searching for becomes next to impossible. A simple procedure for finding a specific mate is to view FeatureManager by dependencies rather than by features. This can be done with a right mouse button click on the assembly name, as shown in figure 12-28. Mates are dependent on components to exist, and hence the name View Dependencies.

When you view dependencies, any mate attached to a certain component is shown under that component. Mates will be displayed more than once because each mate is associated with two components. Viewing dependencies makes it very easy to find a mate attached to specific components. This aids in troubleshooting assemblies. To return to the usual way of viewing FeatureManager, right-click once more on the assembly's name and select View Features.

Editing the definition of a mate sometimes becomes necessary if, for instance, a coincident mate needs to be changed to a distance mate. You might also edit the definition of a mate to flip the dimension of a distance or angle mate, or possibly to alter the alignment condition. If it is deter-

514   *Chapter 12: Assemblies*

mined that a mate should be deleted, simply select it and press the Delete key.

Viewing Mates with PropertyManager

Another alternative to viewing what mates are associated to which components, or between specific components, is to take advantage of PropertyManager. This is extremely convenient, and a simple process that should be remembered.

Right-clicking on a component in FeatureManager and selecting View Mates will display the mates for the selected object in PropertyManager. Multiple components can be selected by holding down the Ctrl key, and then accessing the View Mates option. Furthermore, if two components are selected, PropertyManager will list, in bold, any mates between the two selected components. This plays a very important role when troubleshooting assembly mate conditions.

Smart Mates

It is best that mates in general be understood before addressing this next topic. Smart mates provide a means of automating the assembly process. When adding a mate relationship, certain mate types can be added very quickly and easily through the use of smart mates. There are a few ways smart mates can be implemented, some of which will be shown here.

Smart mates cannot be used for all mate relationships. For instance, a cylinder cannot be made tangent to a plane via smart mates. A cylinder could be mated concentric to a hole using smart mates, however. Let's take a look at one method of adding a mate relationship using smart mates functionality. How-To 12-5 takes you through this process.

How-To 12-5: Using Smart Mates

To add a mate using smart mates, perform the following steps.

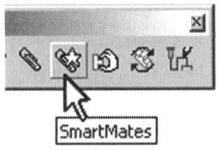

Fig. 12-29. SmartMates icon.

1. Click on the SmartMates icon, shown in figure 12-29. The cursor will change to a move cursor (four arrows).

2. Double click on the surface to be used in the mating relationship. The component whose surface was double clicked on should turn a translucent color.

3. Click on the surface to be mated to.

4. Click on the SmartMates icon when finished to turn off the SmartMates command.

There are a number of combinations of objects that can be used when employing smart mates. Different combinations of objects produce different mate relationships. They are all very logical if thought about for a moment. Table 12-1 provides a partial listing of objects that can be mated, and the mates that are added as a result of using smart mates.

Table 12-1: Smart Mate Combinations

Entities Selected	Mates Added
Two points	Coincident mate
Two planar faces	Two cylindrical faces
Two cylindrical faces	Concentric mate
Two edges	Coincident mate
Two circular edges	Coincident and concentric mates

As you can see from the table, mating a circular edge to another circular edge will result in two mates being added simultaneously. This can indeed be a time saver.

An alternative to step 3 in How-To 12-5 is to drag the component to the object you want to mate to. As the component is dragged around the screen and the cursor is paused over various compatible mating surfaces, you will see the component momentarily snap into position. When this occurs, the cursor will also indicate the type of mate being added, through its system feedback.

As is always the case, system feedback is very important. When this snapping behavior takes place, and the system feedback is informing you of the mate about to be added, it is safe to let go of the left mouse button. Rest assured that the mate will be added. If the alignment condition is incorrect, pressing the Tab key will flip the alignment. In this way you can make sure the alignment is correct prior to releasing the mouse button.

Inserting Components with SmartMates

Earlier in this chapter we discussed some methods of inserting components into an assembly. These methods involved using the menu structure in order to select the component being inserted, or dragging and dropping a component via either the open part file or Windows Explorer.

If a part or assembly file has already been opened, and it needs to be inserted into the assembly, SmartMates technology can be implemented. The process is so ridiculously straightforward it is almost too easy. No geometry needs to be selected first and no commands need to be initiated. The process of inserting a component and simultaneously mating it in position involves nothing more than a simple drag-and-drop procedure.

The key to correctly inserting a component and taking advantage of the SmartMates functionality involves how you use the mouse. When the part is dragged into the assembly, it is important to drag it from the edge or face that will be referenced for the mate. When the part is dropped into the assembly, it is important to drop it on the edge or face the part will be mated to. If you can keep those two facts in mind, you are well on your way to building assemblies using smart mates.

Dragging a part in by a cylindrical face and dropping it on a cylindrical face of a component in the assembly will add a concentric mate between the two faces. Dragging a part in by a circular edge and dropping it on a circular edge of a component in the assembly will add both a coincident and concentric mate. These are only two examples, but all of the various smart mates can be employed via this method.

Is there any disadvantage to using SmartMates? Absolutely not. SmartMates is just a fancy name for a method of adding mates between components. The talented SolidWorks programmers have made the process very easy for us SolidWorks users. But how can a person take advantage of SmartMates technology when dragging in components from, for example, Windows Explorer? It would not seem possible, in that there is no geometry visible that we would drag the part in by. There is only a file name in a folder on the hard drive someplace. Read on to discover a method of enabling SmartMates functionality using Windows Explorer.

Mate References

When a file is dragged into the assembly window from Windows Explorer, the ability to reference specific geometry is not available because the file is not open. A way to overcome this apparent shortcoming is to add what is known as a mate reference. A mate reference tells SolidWorks what object should be referenced for mating purposes when the part is inserted into the assembly.

To add a mate reference to a part, select Mate Reference from the Insert menu. The Mate Reference panel, shown in figure 12-30, will appear. At this point, it is just a matter of selecting the object that should

Palette Parts and Palette Assemblies

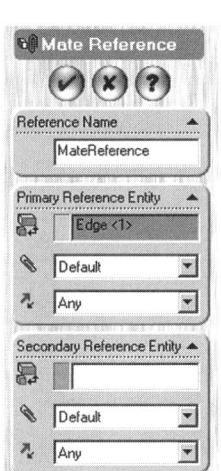

Fig. 12-30. Mate Reference panel.

be referenced for mating purposes, and then clicking on OK. Note that adding a mate reference must be done in a part. The command is not present in an assembly document menu structure.

Once a mate reference has been added, it will appear near the top of FeatureManager. All of the usual rules that apply to SmartMates technology will apply when a part with a mate reference is inserted into an assembly. If the part is dragged into the assembly from Windows Explorer, it will appear in a translucent color and will "snap" into position (complete with system feedback) when positioned over another object, which the mate reference can associate itself to.

It is not necessary to add secondary or tertiary references, but the option is available if you wish to use it. If SolidWorks finds that it cannot make use of the primary reference, it will use the secondary. If that does not work either, it will attempt to use the tertiary reference. What it will not do is add mates using all three references. That assumption is a common mistake, so do not let yourself be fooled.

Mate references are typically added to parts that frequently get used numerous times in multiple assemblies. For example, it would not necessarily be worthwhile to add a mate reference to a component that is only going to get used once in one assembly. Mate references are also an ideal companion to palette parts, discussed in the following section.

Palette Parts and Palette Assemblies

Palette parts are nothing more than part files that can be conveniently inserted into an assembly. Palette assemblies are assemblies that would typically become subassemblies in a top-level assembly. They are generally items that will be used over and over again in various assemblies. There is no special need to save a file with a special extension, no need to rename dimensions, and no special routine used to insert palette parts or assemblies into an existing assembly. Simply drag the item from one of the palette folders into the assembly. (The Feature palette was first discussed in Chapter 8.)

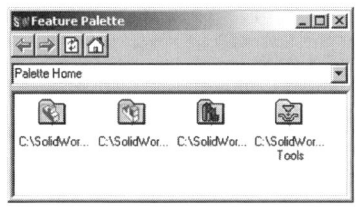

Fig. 12-31. Feature palette.

Figure 12-31 shows the Feature palette. The text below each folder is the path to that folder. These folder paths can be changed to point to different or multiple locations, depending on where you decide to store your palette items. Use the method discussed in Chapter 8 (see "Modifying Folder Location Paths") to modify the paths to palette parts or palette assemblies. Although the path names are truncated, the palette folders are, from left to right, *Palette Parts*, *Assemblies*, *Features*, and *Forming Tools*.

It is very common to add mate references to palette parts or palette assemblies. When the item is dragged into the assembly, drop the component on what it should be mated to. As is standard operating procedure, a translucent image of the part or assembly will be shown, along with system feedback, which will help you in positioning the component prior to releasing the mouse button.

An aspect of using palette parts is whether or not the original palette part should be the part being referenced by the assembly. It all depends on what you want updating in your assembly. The alternative would be to save a copy of the part and reference the copy instead. Just to mention quickly what the options are, let's consider a couple of scenarios.

If a palette part is used in an assembly, and the original palette part is altered in some way, it will affect every assembly in which that part is used. This is sometimes a desirable situation, such as in the case where a revision is made to a part and the revised part should then be used in every assembly where it is located. The down side to this is that there may be some assemblies for which the revised part should not be used. Additionally, if something were to happen to the original palette part, every assembly using it would be affected.

The alternative is to make a copy of the palette part and have the assembly reference the copy. This can make for easier organization with regard to all assembly part files. If all parts being used in an assembly are located in one area, housekeeping becomes a much easier chore.

If you wish to save a copy of a part that was just inserted from one of the palette part's folders, open the part in its own window and perform a Save As. As you should be aware, this command is in the File menu. When you click on Save As, SolidWorks will issue the cryptic warning shown in figure 12-32.

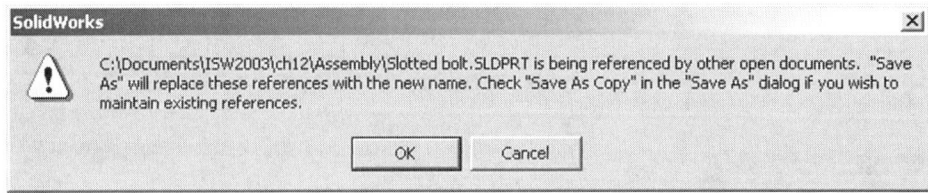

Fig. 12-32. A warning message about saving copies.

It is very important that you understand exactly what this message is actually trying to say. First, go ahead and click on OK when you see the message. When the Save As window appears, decide where you want to save the file and give the file a name. You can change the name or leave it as is, whatever you choose. Now here is where you need to pay close attention.

If you do not check the Save As Copy option, shown in figure 12-33, the assembly will reference the copy, not the original file. In other words, if you were copying a file named *Bolt*, and the copy is being named *Bolt Revision2*, the assembly that was using the file named *Bolt* will use the copy, *Bolt Revision2*, instead. If you do check the Save As Copy option, the assembly will continue to use the original file.

Fig. 12-33. Saving a copy.

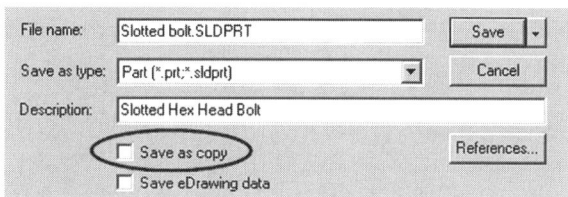

You will only see the warning message if the assembly referencing the file you are copying is open. If the assembly is not open, you will not see the message when performing a Save As because the message does not apply. The assembly would not use the copy even if the Save As Copy option were not checked.

In our particular case, we want the assembly to begin referencing the copy, so make it a point not to check the Save As Copy option. Once the file has been saved to the hard drive, you can feel free to close the part window and return to the assembly. You will find that FeatureManager now references the copy, and not the original part that was dragged out of the *Palette Parts* folder and placed in your assembly. See Chapter 18 to understand more about how to copy files being used in an assembly, and how to control file references.

Component Patterns

Some components that will be brought into the gear housing assembly will be a rubber insert, long bolt, and washer. Each of these items is used four times in this assembly. It is not necessary, however, to insert all of the items individually. Instead, they can be patterned using the Component Pattern command.

Creating Component Patterns

To begin, one instance of the rubber insert, bolt, and washer are brought into the assembly. Next, the components are mated into position. The mated components are shown in figure 12-34.

520   **Chapter 12: Assemblies**

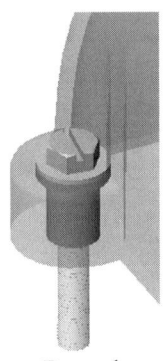

Fig. 12-34. The bolt, washer, and insert components.

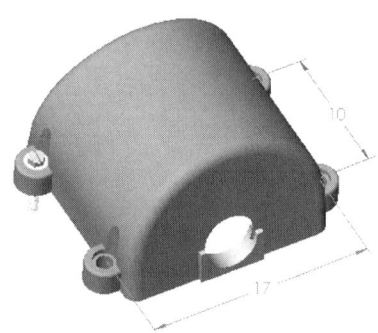

Fig. 12-35. Hole spacing is known.

But what of the other three holes in the gear cover? It is known that the distance between the holes in the cover is 17 inches by 10 inches (shown in figure 12-35), because the person who created the cover designed it that way. This data can be used to create a component pattern. How-To 12-6 takes you through this process.

How-To 12-6: Creating a Component Pattern

To create a component pattern, perform the following steps. In this particular example, you will create a linear pattern, but circular component patterns can also be created.

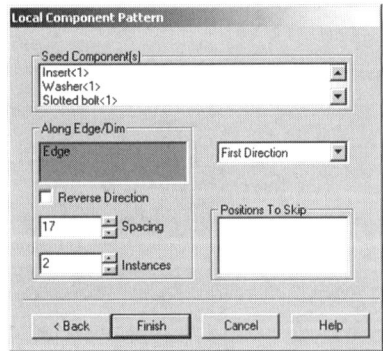

Fig. 12-36. Local Component Pattern window.

1. Select Insert > Component Pattern.
2. Select *Define your own pattern (local)*.
3. Specify either a Linear or Circular pattern.
4. Click on Next. The Local Component Pattern window, shown in figure 12-36, will appear.
5. Select the components to be patterned.
6. Click in the Along Edge/Dim list box and select an edge or dimension that will be used to dictate the pattern direction. A preview arrow will appear.
7. Check Reverse Direction if necessary. Doing so will flip the preview arrow.
8. Specify the Spacing and Instances. The Instances are the total number of items for pattern direction 1, including the original, so a value of 2 will be used for our example.

9. Because you will need to pattern in a second direction as well, select Second Direction from the drop-down list.

10. Repeat steps 6 through 8 for the second direction.

11. Click on Finish.

Where once there were three empty holes in the gear cover, there are now holes filled with three components each. If you were expecting to see a preview when creating the component pattern, you may have been disappointed. There is no preview available. The completed pattern is shown in figure 12-37.

In our particular case, a linear pattern was created. It would have been just as easy to create a circular pattern. There really is not much difference, except for the fact that an axis or angular dimension would need to be selected to dictate the pattern direction, as opposed to a linear edge or linear dimension.

Fig. 12-37. Finished component pattern.

Let's consider what has just been accomplished. A locally defined component pattern has been created. This means that it has been defined "locally" within the assembly. The pattern is not based on anything else other than a few edges of the cover component, which dictated the pattern directions. Consider what would happen if the positioning of the holes were to change.

It is not too difficult to guess what would happen if the hole position, or perhaps the size of the cover, were changed. The patterned components would continue to reside at their current locations without knowing any better. This does not make for a flexible assembly, nor does it incorporate our design intent.

It would be best if we could somehow base the position of the patterned components on the position of the holes themselves. If you had the foresight to have created the holes using a feature pattern, the component pattern can be based on the feature pattern. This is ideal, because if the underlying feature pattern changes in any way the component pattern will automatically update. This includes changing the number of instances in the original feature pattern.

The type of component pattern we are discussing is known as a *derived pattern*. It is the type of pattern you should always use if circumstance permits it. How-To 12-7 takes you through the process of creating a derived pattern. The process is much less involved than defining a local pattern.

HOW-TO 12-7: Creating a Derived Pattern

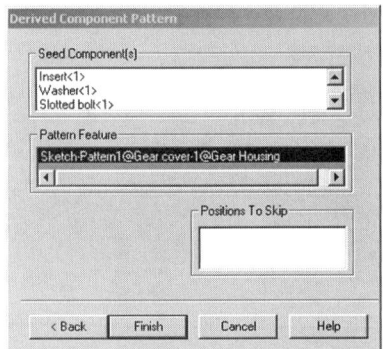

Fig. 12-38. Derived Component Pattern window.

Because derived patterns are based on existing feature patterns, an existing feature pattern belonging to a component in the assembly is a requirement. The Derived Component Pattern window is shown in figure 12-38. The steps that follow outline the process of creating a derived component pattern.

1. Select Insert > Component Pattern.
2. Select Use an existing feature pattern (derived).
3. Click on Next.
4. Select the Seed Component(s), which is the component you will be patterning. Specify more than one component if necessary.
5. Click in the Pattern Feature area and select a feature pattern to base the component pattern on. Perform this task using FeatureManager.
6. Click on Finish.

There are benefits to using a derived pattern as opposed to a local pattern. The first benefit is that there is less information to plug into the Derived Component Pattern window. This is because the relevant data has already been supplied when the feature pattern was created.

Another important benefit of using a derived component pattern is associativity. The derived pattern is based on an existing feature pattern. Therefore, if the feature pattern changes, the derived pattern components will update accordingly. In other words, if the number of holes increased or decreased in the original feature pattern, the patterned components would increase or decrease.

An important aspect of derived component patterns is the flexibility present in such patterns. The feature pattern used to define the derived component pattern can be any of the following.

- Linear or circular feature pattern
- Sketch-driven feature pattern
- Table-driven feature pattern

- Curve-driven pattern feature

With all of these options available, there is little left out. If you are looking to position components in 3D space using a set of *x-y-z* coordinates in table format, a workaround will be needed. That functionality is not currently available, but will probably be included in a future release of the software.

Deleting Patterned Components

Deleting a component in an assembly component pattern is accomplished in the same manner features are deleted from feature patterns. If a face is selected on a patterned component and the Delete key is pressed, Solid-Works will ask you to confirm whether the component should really be deleted.

Once a component or series of components has been deleted from a component pattern, it will appear in the pattern's definition. Specifically, the deleted components will be listed in the Positions to Skip list box. By removing a listing from this box, the deleted pattern component can easily be recovered. To edit a pattern's definition, right-click on the pattern in FeatureManager and select Edit Definition.

If a series of components has been patterned, it is not possible to delete just one of the components from one instance in the component pattern. Applying this to our gear housing example, it would not be possible to delete just the washer from one of the patterned instances. The bolt and insert would wind up being deleted as well. You could, however, hide or suppress one component from one instance, such as the washer.

Hiding Components

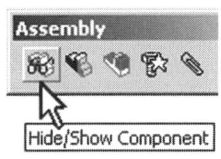

Fig. 12-39. Hide/ Show Component icon.

Sometimes it is very difficult to add mating relationships because other parts get in the way. It would be very convenient indeed if it were possible to hide components from time to time. This possibility does exist, and it is just a right mouse click or icon click away. The icon used for hiding or showing a component is shown in figure 12-39.

Right-clicking on a component either in FeatureManager or in the work area will gain access to either a *Hide components* or *Show components* menu selection. Because this is a toggle switch, the choice presented will either be one or the other. When a component is hidden, its icon will appear ghosted, as shown in figure 12-40. If you prefer to use the icon instead of the right mouse button menu, simply select the desired component and click on the Hide/Show Component icon. If you want to hide or show more than one component at a time, make it a point to hold down the Ctrl key in order to select more than one item.

Alternatively, you can use the pull-down menus instead of the icon. Under the Edit menu, you will find a selection for either Hide Component or Show Component. There is also a selection titled Show with Dependents, which will show a hidden component, along with any dependent components hidden with it.

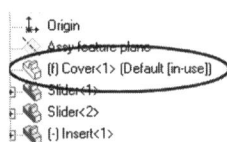

Fig. 12-40. Hidden component icons appear ghosted.

When using the Edit menu to hide or show components, there will be a submenu containing the options *This Configuration*, *All Configurations*, and *Specified Configurations*. These are made available because a component's visibility is a configurable trait. In other words, a component can be hidden in one configuration but not in another. (See Chapter 10 for additional information on configurations.)

Component Suppression States

You can suppress components in assemblies using the same techniques learned in Chapter 10 to suppress features. These options included using the Suppress/Unsuppress icons, using the Edit pull-down menu, and accessing a component's properties. There is also a multifunction icon on the Assembly toolbar that can be used to suppress components (discussed in material to follow).

When a component is suppressed, it cannot be seen. You may be wondering, then, what would be the difference between suppressing a component and hiding one. When a component is hidden, it cannot be seen by the user. It can, however, still be "seen" by SolidWorks. When a component is suppressed, it is invisible to both user and SolidWorks.

Suppressing versus hiding can have its advantages and disadvantages. For example, suppressing components removes them from memory, which will reduce rebuild times and increase overall system performance. It will also suppress and dependent mates. This may cause the assembly to move apart unexpectedly if too many key components are suppressed. On the other hand, hiding a component does not affect its associated mates.

Lightweight Components

When discussing features of a part, the terms *suppressed* and *unsuppressed* suffice to describe a particular condition. This is because there are only two states a feature can be in. This is not the case with assembly components, which actually have three states.

Component Suppression States

Fig. 12-41. Lightweight components appear with blue feathers.

When an assembly is opened in SolidWorks, its components can be loaded in a state known as *lightweight*. A component loaded lightweight is not fully loaded into memory. Only the visual information needed to display the component is loaded. This situation results in two very important facts. The first is that assemblies opened lightweight can open very quickly. This has huge benefits to those who work with assemblies containing many components. When a component is loaded lightweight, it is displayed with a blue feather in FeatureManager, as indicated in figure 12-41.

Another important aspect of lightweight components is that they must be loaded first before they can be edited. Simply clicking on a component in the work area is enough to load it. This also brings up the third state of an assembly component, which is resolved. When a component is resolved, it has been fully loaded. It is neither suppressed, nor is it lightweight.

Now that you are beginning to understand that components can exist in three states, let's first take a look at how to set the suppression state of a component. How-To 12-8 takes you through the process of using the Assembly toolbar to control the suppression state of components.

How-To 12-8: Changing the Suppression State of Components

To change the suppression state of a component using the Assembly toolbar, perform the following steps.

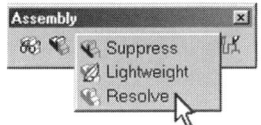

Fig. 12-42. Changing the suppression state of a component.

1. From FeatureManager, select the component whose state you wish to change.

2. Click on the Change Suppression State icon on the Assembly toolbar. This is the second icon from the left, as seen in figure 12-42.

3. Specify the suppression state the component should be changed to.

When suppressed, items in FeatureManager appear gray and cannot be expanded to show the features or components they contain. Because the items being suppressed are physically removed from memory, system performance will typically improve, though this depends on other factors as well.

As mentioned previously, it is also possible to use the Edit pull-down menu to suppress or resolve components. Using the Edit menu has the added advantage of being able to dictate if the state change is applied according to one of the options *This Configuration*, *All Configurations*, or *Specified Configurations*. The suppression state of an assembly component is a configurable item, but we will wait until Chapter 15 to expand on this particular topic.

The Suppress or Unsuppress icons, typically found on the Features toolbar, can also be used. These icons are usually reserved for feature suppression, but will work for components as well. And if that were not enough options, a simple right-click on a component in FeatureManager will allow for changing its state. If a component is resolved, you will have the options Set to Lightweight and Suppress, for instance.

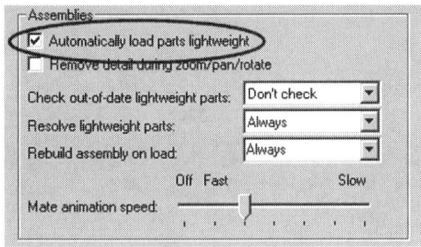

Enabling an assembly to be loaded lightweight can be done by enabling lightweight components in System Options. This option is shown in figure 12-43. How-To 12-9 takes you through the process of enabling lightweight components.

Fig. 12-43. Enabling lightweight components.

How-To 12-9: Enabling Lightweight Components

To enable lightweight components, perform the following steps.

1. Select Options from the Tools menu.

2. Click on the Performance section in the System Options tab.

3. Place a check in the option titled *Automatically load parts lightweight*.

4. Click on OK.

➥ **NOTE:** *An assembly must be saved in shaded display mode for it to be loaded lightweight.*

The next time an assembly is opened, you will notice that it opens much more quickly. The time difference will be more dramatic with assemblies containing many components. As mentioned earlier, when an assembly is loaded lightweight, there will be small blue feathers on the assembly's components listed in FeatureManager. If there are red stripes on the blue feather of a component's icon, it indicates that the component is out of date.

When a component is out of date, it has changed in some way since the last time the assembly was opened. This might raise the question of whether or not you are looking at current geometry. In this case, it would be a good idea to make sure the component is completely resolved, so that any changes can be seen. An option for whether or not SolidWorks checks to see if lightweight components are out of date can be seen in figure 12-43.

✓ **TIP:** *You can choose whether an assembly is loaded lightweight or resolved when opening an assembly from the File > Open command.*

Any component that requires editing must first be resolved. This is not an option. What is optional, however, is which components get resolved. The user has full control over this. The section that follows further explores resolving components.

Resolving Components

You have a few options in regard to resolving components. In the same section in which lightweight components were enabled, there is a setting that helps to tell SolidWorks when lightweight components should be resolved. This setting is titled *Resolve lightweight parts*, which can be set to Prompt or Always. It is recommended that you make this setting Always. This allows a part to be resolved simply by clicking on it. This is tremendously convenient. If a lightweight part needs editing, just click on it. The part will be loaded into memory and you can then perform any required actions on the part.

Just about anything short of rotating the assembly will require that one or more components be resolved. You might also decide it would be easier if the entire assembly were resolved, a reasonable option. By right-clicking on the assembly in FeatureManager, the option for Set Lightweight to Resolved will appear. Selecting this option will completely resolve every component in the assembly. Be forewarned that this may take a while if working with a large assembly.

The other option available is Set out-of-date Lightweight to Resolved. This only resolves the out-of-date components, as the name implies. As mentioned earlier, out-of-date components are indicated with red stripes on the feather. After resolving them, you can then rest assured that the current part geometry is being displayed on screen.

In summary, remember that almost anything done to an assembly requires that components be resolved. This includes mating components, adding assembly features, creating exploded views, taking measurements, and modifying dimensions. It does not include rotating, zooming, or panning the assembly. If you have a fast computer, or do not typically work with large assemblies, do not bother using lightweight assemblies.

Working with Subassemblies

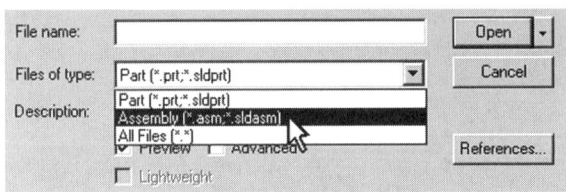

Fig. 12-44. Specifying the proper file type.

It is not always parts that get added to an assembly. Sometimes other assemblies must be added. Assemblies that get inserted into other assemblies are known as subassemblies. Inserting a subassembly works just like inserting a part, except that you must browse for the correct file type. Follow the same procedure you would for inserting any component. When the Insert Component window opens, specify assemblies in the *Files of type* drop-down list box, shown in figure 12-44.

Component Subassemblies

After an assembly has been inserted as a component into another assembly, it is then considered a subassembly. Subassemblies are no different than actual assembly files we have been discussing throughout this chapter. It is just a matter of hierarchy and terminology.

An important concept you should try to keep in mind is that a subassembly is considered a single component. Subassemblies will (by default) move and react as a single component. It typically takes just three mates to fully mate a subassembly into position, even if it has hundreds of components.

If you expand a subassembly, you can see the components that belong to it. This is shown in figure 12-45. Note that the *Bearing* subassembly contains three components, and has its own *Mates* folder. The bearing is in turn a component in the *Axle* assembly. FeatureManager can get quite complicated in an assembly. Parts can exist within subassemblies, which in turn exist within subassemblies, which in turn exist within higher-level assemblies, and so on. Incidentally, this hierarchical tree is known as an assembly's "depth."

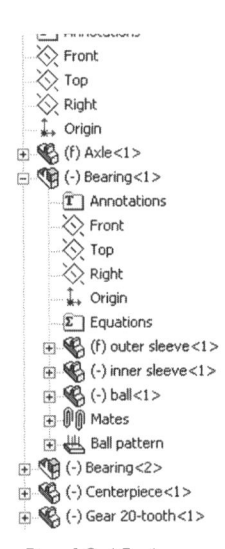

Fig. 12-45. A subassembly in FeatureManager.

All of this relates to component properties. All components have properties, but subassemblies have more properties a user must be concerned with; that is, in contrast to individual part components. Therefore, the section that follows explores component properties.

Component Properties

There are a few functional parameters, as well as general information, that can be obtained from a component's properties. Accessing a component's properties is done via the right mouse button menu. By this time, you should be getting used to the right mouse button menu process. Right-click on a component and select Component Properties.

Whether you right-click on an individual part component or a subassembly component will make a slight difference as to what is shown in the Component Properties window. In figure 12-46, the window represents the properties for the *Axle* subassembly.

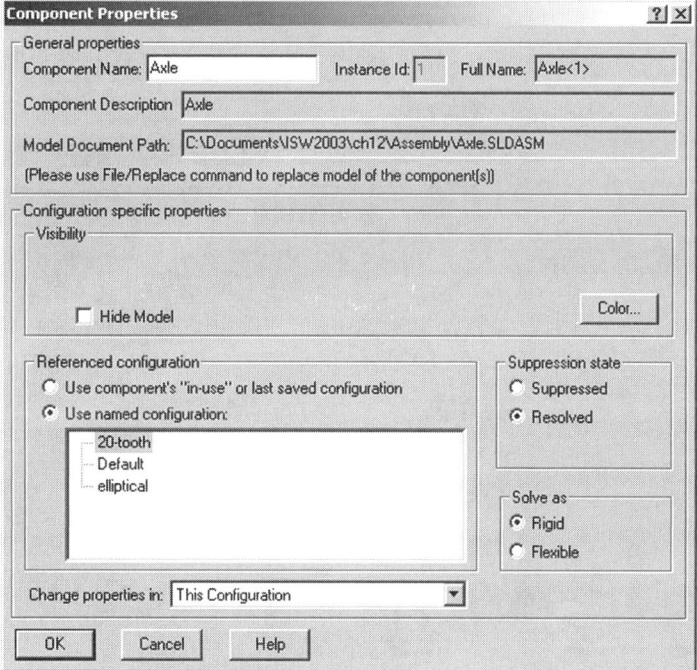

Fig. 12-46. Component Properties window.

The following material describes the options found in the Component Properties window. Note that the term *component* refers to individual part components and subassemblies as well. Where a specific option or function is only available for a particular type of component, that fact is clearly noted.

Component Name

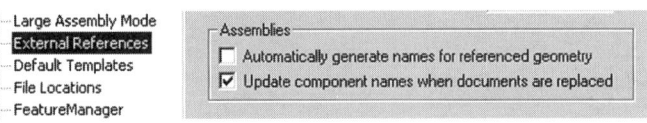

Fig. 12-47. Update component names when documents are replaced option.

This is the name of the component, which is typically the same as the actual name of the part or assembly file being referenced. However, it does not have to be. In the External References section of System Options (Tools > Options), there is an option titled *Update component names when documents are replaced*, as shown in figure 12-47. When this option is unchecked, the component name can be anything typed in by the user.

You need to be very careful with this option! It is recommended that *Update component names when documents are replaced* be left on (checked). Otherwise, names of components in FeatureManager do not have to reflect the names of the actual files being referenced. This can cause a lot of confusion and make file tracking particularly nasty. Use this option only if you have a proper document tracking system in place, such as PDM software.

Instance ID

Sometimes a part exists numerous times in an assembly. In other words, the same part was inserted multiple times as a component. They may all be the same part, but they are different components. SolidWorks needs a way of keeping track of different components, so it assigns an instance ID number to each. These same numbers appear in FeatureManager to the right of each component, and cannot be changed by the user. (See the earlier section "FeatureManager Symbology" for more on instance ID numbers.)

Full Name

A component's full name is simply the option for component name followed by its instance ID (see previous sections).

Component Description

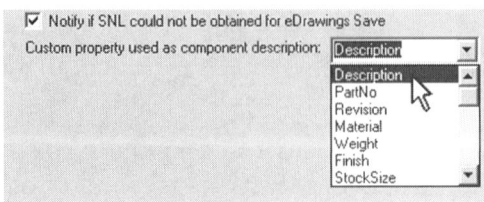

Fig. 12-48. Specifying which property to generate descriptions from.

An optional description of a component can be added via Custom Properties. If this is the case, which property is used to derive the description can be chosen in the General section of System Options (Tools > Options), such as shown in figure 12-48.

Descriptions appear in a number of places, including the Open window, shown in figure 12-49. If a category other than Description is chosen in System Options, that category title is what is displayed in the Open window. In figure 12-49, the category chosen was Material. Component descriptions can be made to appear in FeatureManager as well, similar to feature or configuration descriptions, discussed in Chapter 10.

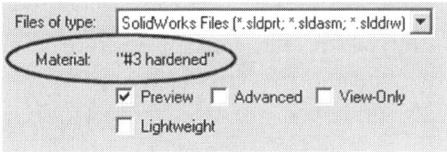

In the case of a component, a description must be added as a custom property in the file's properties. The description name must then match the category chosen in System Options (see figure 12-48). See Chapter 16 for a complete explanation of how custom properties are added to a model.

Fig. 12-49. Component description category changed to Material.

Model Document Path

Model Document Path points to the location of the component. In other words, the path displays the location of where the model is stored on the hard drive, or perhaps over the network.

Hide Model

The Hide Model checkbox is directly related to the Show/Hide function discussed earlier in this chapter. It is just another way of showing or hiding a component.

Visibility Options

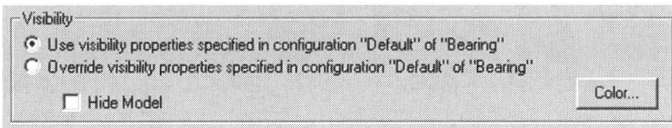

Fig. 12-50. Visibility options.

If accessing the properties of a component within a subassembly, the Visibility section will include the options shown in figure 12-50. When "Use visibility properties" is selected, the component whose properties were accessed will inherit the visibility properties of its parent assembly.

When "Override visibility properties" is selected, the component whose properties were accessed can have visibility properties different from those of the parent subassembly. What this allows for is being able to hide an individual component within a subassembly even if the rest of the subassembly is being shown. The reverse would also be true.

Color Button

The Color button allows for changing the color characteristics of an individual component. The color change, however, will not affect the original part file. This is useful when you want to change the color of a component but not have other assemblies affected that might also contain the same part file.

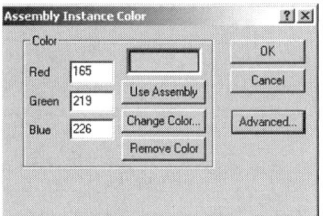

Fig. 12-51. Assembly Instance Color window.

That may sound somewhat confusing, so let's put it another way. If a bolt has been inserted into an assembly 10 times, changing the color of one bolt via the Color button in its component properties will not affect the other bolts. In contrast, if you wanted all bolts to change color, open the bolt in its own window and change the color of the part itself. Be aware that component color always overrides part color. Clicking on the Color button opens the Assembly Instance Color window, shown in figure 12-51.

The Use Assembly button forces the component to use the same color settings set for the assembly in the Color section of Document Properties (Tools > Options). The Remove Color button forces the component to revert to its color settings as designated within the original part file. Remove Color also disables any adjustments made in the Advanced settings (discussed in the upcoming section on component transparency). Use Change Color to change the component's color.

Suppression State

Just like it sounds, this is just another way of controlling the suppression state of the component. Like hiding and showing components, this topic was already discussed, so it will not be covered again here.

Solve As Rigid or Flexible

This option is only available for subassembly components. Subassemblies, by default, will act as a single component. That is, movement between individual components in a subassembly is typically not possible. This behavior would be known as a rigid subassembly. By setting the *Solve as* option to Flexible, individual components within a subassembly can move. This option does increase the complexity level of an assembly, so use this only when necessary.

Referenced Configuration

Parts can have different configurations, as you learned in Chapter 10. In the *Referenced configuration* section, it becomes possible to control what configuration is being used by the assembly. Every configuration within a part or assembly will be listed in the list box next to the *Use named configuration* option. Simply pick the configuration you wish to use within the assembly.

Change Properties In

The component whose properties are being accessed may have multiple configurations, but the top-level assembly may have multiple configurations as well. If any configurable component properties are changed (otherwise known as configuration-specific properties), those changes can take place in any top-level assembly configuration you choose.

Let's use a simple bolt as an example once again. You access the properties of the bolt and decide to begin using a different configuration of the bolt within the assembly. For example, a longer version of the bolt is deemed necessary. The *Change properties in* drop-down menu contains the options *This Configuration, All Configurations, and Specified Configurations*. By selecting *All configurations*, you would be telling SolidWorks to begin using the longer version of the bolt in every configuration of the top-level assembly. This can be a huge time saver when working with assemblies that contain multiple configurations.

Component Replacement

It is typical to replace assembly components from time to time with a revised component, usually a component with a better design. Maybe the new component is lighter or stronger, or cheaper to build, or maybe it fulfills the stringent requirements set out by the customer. Whatever the reason, the component replacement function allows you to do this.

You might think that replacing a component with another in an assembly would wreak havoc with your mating relationships. Sometimes, when the new component originates from a completely different part file, mate relationships to the original component do require editing. However, if the replacement component is a modified version of the original, SolidWorks will reuse the existing mates.

Fig. 12-52. Prior to replacing the 12-tooth gears.

The Replace command performs a variety of functions. It can be used to replace just one occurrence of a component in an assembly, or all occurrences. It can also be used to replace, for example, an assembly with a part.

In the following example, both of the 12-tooth gears used in the axle assembly will be replaced with 20-tooth gears. The Replace command will be used to accomplish this. How-To 12-10 takes you through the process of performing a component replacement. Figure 12-52 shows a "before" screen shot of the assembly used in the example.

How-To 12-10: Replacing Multiple Components

To replace a component (or every instance of a component) in an assembly, perform the following steps. The steps listed are the basic steps you will need to perform. Other options are discussed in material to follow.

534 Chapter 12: Assemblies

Fig. 12-53. Replace panel.

Fig. 12-54. Mated Entities panel.

1. Right-click on the component to be replaced and select Replace.

2. In the Replace panel, shown in figure 12-53, click on the Browse button and select a replacement component.

3. Place a check in the All instances option if you wish to replace every instance of the component in the assembly.

4. Click on OK. The Mated Entities panel, shown in figure 12-54, will appear.

5. At this point, you could choose to try to repair any problematic mates. Mates prefixed with an X are those with problems. Green checks indicate mates that transferred to the replacement component without problems.

6. Click on OK to complete the replacement process.

In the example used here, both instances of the 12-tooth gear were replaced with the 20-tooth gear. Figure 12-55 attests to this. The 20-tooth gear was a reworked version of the original 12-tooth gear, so the mates were reapplied by SolidWorks successfully. Component replacement does not always work as cleanly.

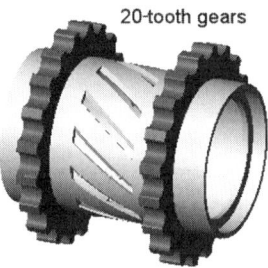

Fig. 12-55. After component replacement.

If mates do not reapply themselves to the new component, there is only so much that can be done in the Mated Entities panel. Selecting a mate from the list allows you to select a replacement (for the referenced face) that is not present in the new component. If it becomes necessary to change the mate type (e.g., coincident rather than distance), edit the definition of the mate after completing the replacement.

Following step 3, you could select *Match name* if SolidWorks should try to use a configuration of the same name in the replacement component as the component being replaced. This is the default setting, and should be left alone if neither the original or replacement components have any additional configurations. If *Manually select* is chosen, you will be presented with a list from which a configuration should be selected for the replacement component.

Reloading Components

Reloading is quite different from replacing components, and can most easily be explained by using a hypothetical situation between two workers, who we will call Ed and Bob. Ed arrives at work first and opens a part he has been working on over the past week. The part Ed is working on is a component in an assembly Bob is working on. When Bob arrives at work, he opens the assembly that contains the part and is faced with a message that reads something to the effect of "Document-filename is being used by Ed. Do you want to make a copy?"

Bob would have a choice at this point. He could either have Solid-Works create a copy of the part or open the part read-only. If a copy is created, Bob's assembly now contains a copy of the part Ed is working on. This is usually not a desirable situation because the assembly would no longer contain the original part.

If Bob were to specify "No" when asked if he would like to make a copy, the assembly would be opened with that particular part in read-only mode. He could make changes to the assembly, continue to add components, or whatever the case may be. He just would not be able to edit the component, which is read-only. Because Ed has the job of working on that part anyway, this is probably exactly what you would want to have happen.

Some time later, after Ed has made numerous modifications to the part, Bob may want to see what types of changes have been made to it. This is where the Reload function is useful. By reloading the read-only component, Bob can see the changes made to the part.

Accessing the Reload function is done by right-clicking on a component and selecting Reload. The Reload window, shown in figure 12-56, will appear. Pay special attention to the *Open as read-only* option when performing this function. Under normal circumstances, such as the scenario described between Ed and Bob, you would want this option to be checked.

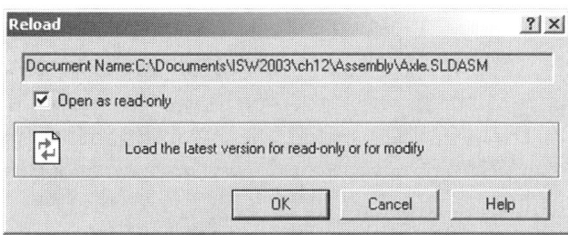

Fig. 12-56. Reload window.

To uncheck the *Open as read-only* option, you must have write access to the file being reloaded. This means that nobody else can have the file open at that time. You cannot be given write access to a read-only file, for the main reason that a single document cannot be opened by multiple users at the same time.

Mate Troubleshooting and Repair

When replacing components, it is common to have mating problems arise. In a worst-case scenario, the old mates to the original component are no longer valid with the new replacement component. This has to do with the way SolidWorks recognizes the faces or other objects being mated to. For example, faces have internal identifiers, and a new replacement component will have faces with different internal identifiers than the original component. This is why mates fail with replacement components and errors appear in FeatureManager.

Usually, the types of problems that arise from component replacement are easy to fix, once you know the drill. This is because it is rare to replace a component with one that is drastically different than the original. For instance, a hex-head bolt might get replaced with a square-head bolt, but a hex-head bolt would not get replaced with an external retaining ring. Well, at least not usually!

Chapter 2 first introduced you to error symbology. Red circles in FeatureManager indicate problems that should be cleared up at the first possible opportunity. Chapter 4 discussed methods of editing the definition of a sketch or feature. Mates have definitions just like any feature, and those definitions can be edited just like any feature. That is, right-clicking on a mate provides access to the Edit Definition command.

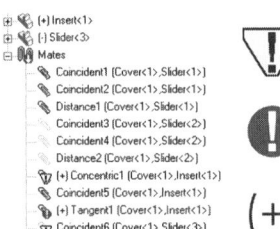

When mate errors arise, the symbology is more advanced than the symbology used for basic sketch or feature problems. The three main symbols that are commonly shown consist of red circles, yellow triangles, and plus signs. All of these symbols can be seen in figure 12-57. Red circles indicate a primary problem, and the yellow triangles indicate secondary conflicts.

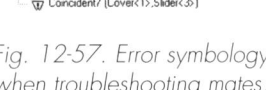

Fig. 12-57. Error symbology when troubleshooting mates.

There is more than one way to mess up mates in SolidWorks. Some of these problems are easier to fix than others. The following section categorizes and explores some of the problems that can occur with mates.

Missing References

Fig. 12-58. Fixing a mate with a missing reference.

When replacing a component with another, a missing reference is exactly the type of problem that generally crops up. A few mates in the assembly are looking for faces on a component that is no longer present. Look for error symbols in the *Mates* folder and edit the definition of the mates with problems. When a reference is missing, the Mates panel will look something like that shown in figure 12-58.

There are three things that will help you in this situation. First, look for the invalid object in the Mates panel. You cannot miss it, because it will be labeled *Invalid*. Delete that object from the list because it is not in the model anymore. Second, note the type of mate you are editing. This is written in big letters at the top of the Mates panel. In figure 12-58, note the text *Coincident?*, followed by some number. This will help in picking something to replace the invalid object you just deleted.

The third helpful aid would be to look in the work area for the currently highlighted object. This will be one of the objects the mate is between, or in this case coincident with. Now all that remains is to pick something, such as a face, on the replacement component. Continue this process until you have repaired all of the missing references.

Overdefined Mates

When too many mates are present, they overdefine components. This will be immediately noticeable due to the red circles and plus signs in FeatureManager and the *Mates* folder. Look for the components in FeatureManager with plus signs preceding the component names and you will have found the overdefined components.

As long as overdefining mates are caught early on and you do not let them go, they are not difficult to fix. If they are ignored and more accumulate, it gets more difficult to track down the problems. The moral: do not put off fixing errors or they will just get worse.

Typically, an overdefining mate is found between a couple of components that have a few too many mates. Select the mates in the *Mates* folder and this will highlight in the work area the objects used for the mate. For example, selecting a concentric mate may highlight two cylindrical faces. Repeat this process for any mates containing error symbols. Once you get a handle on which mates are conflicting, delete the mate or mates you have decided you do not need.

Improper Alignment

When one mate has its alignment condition improperly set, the mate may show up as an overdefined mate. However, this often results in only one mate showing up in FeatureManager as being overdefined. How can one mate be overdefined? Doesn't it take two mates to cause a conflict and show up as overdefining mates? The answer is not necessarily, when it is a mate's alignment condition that is backward.

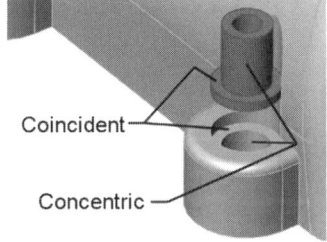

Fig. 12-59. Preparing to add a coincident mate. The concentric mate is already present.

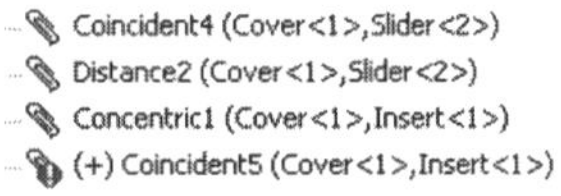

Fig. 12-60. Alignment conditions can cause mate errors.

If we use the insert and cover components shown in figure 12-59 as an example, it will be easy to understand how alignment conditions can cause mate errors. Assume that there is already a concentric mate between the cylindrical surface of the hole in the cover and the outer cylindrical surface of the insert. This reduces the movement of the insert to only two degrees of freedom.

Next, a coincident mate will be added between the surface of the insert and the top planar face surrounding the hole. This is also illustrated in figure 12-59. The insert is currently flipped in the wrong direction, so the alignment condition is set to anti-aligned. Because the insert cannot flip around to be in the correct direction to satisfy the coincident mate, an error occurs, such as that shown in figure 12-60. This is because the concentric mate has already reduced its degrees of freedom of movement. What to do?

To reiterate: the coincident mate needs to be anti-aligned. This causes the insert to want to flip, but it cannot because of the concentric mate. The solution: because the concentric mate is not technically causing errors, it contains no error symbol as the coincident mate does. Nonetheless, changing the alignment condition of the concentric mate causes the insert to flip around so that the coincident mate can be satisfied.

Mate Diagnostics

With all of the problems that can occur with mate relations, it is good to know there is a little extra help available when needed. This extra help is known as *mate diagnostics*, and will aid in troubleshooting assemblies with mate problems. A simple right-click on the problem mate will gain access to the Mate Diagnostics command.

Figure 12-61 shows a problem with overdefining mates. With an assembly containing a couple dozen parts, this would be easy to track

Component Editing

down and repair. Crank up the complexity to a few hundred components and things get a bit more difficult. In the example shown, a tangent mate has been added, which is causing a conflict. The faces used for the mate are highlighted in the image.

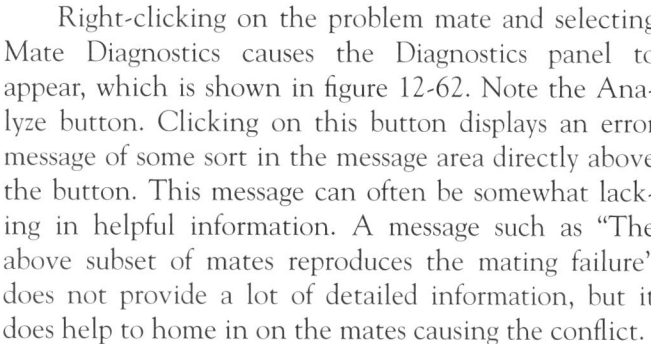

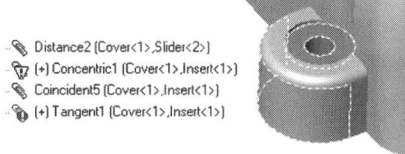

Right-clicking on the problem mate and selecting Mate Diagnostics causes the Diagnostics panel to appear, which is shown in figure 12-62. Note the Analyze button. Clicking on this button displays an error message of some sort in the message area directly above the button. This message can often be somewhat lacking in helpful information. A message such as "The above subset of mates reproduces the mating failure" does not provide a lot of detailed information, but it does help to home in on the mates causing the conflict.

Fig. 12-61. Mate diagnostics could help with this problem.

In the lower area of the Diagnostics panel is a section titled Not Satisfied Mates. The primary mate causing the conflict is displayed in bold. If selected, this mate will display an additional message. In our case, the message is "Cylinder and planar face are not tangent. Separation distance is 1.5in." This helps quite a bit. Now it is just a matter of deleting one of the mates causing the conflict, or editing the definition of the tangent mate. The choice would be yours.

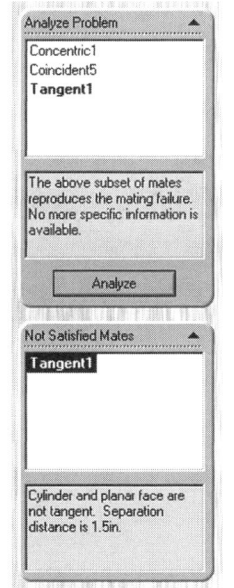

Fig. 12-62. Mate Diagnostics panel.

Component Editing

If it is necessary to make a few dimensional changes to an assembly component, it is quite an easy process to access the dimensions. The process involves nothing you have not already learned. To access a dimension, double click on the part's feature whose dimensions you want to access, and then double click on the dimension to edit it.

> **NOTE:** *Altering a dimension on a component in an assembly will affect the original part.*

Components can be expanded in FeatureManager as needed to gain access to their respective features. Clicking on a component in the work area will highlight the component's name in FeatureManager. Turning on an option named *Scroll selected item into view* will cause FeatureManager to scroll up or down as necessary to show the component when it is selected from the work area. This makes hunting down a component in the feature tree much easier. Look in the FeatureManager section of System Options (Tools > Options) for this setting, shown in figure 12-63.

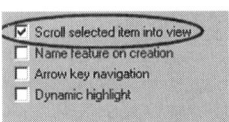

Fig. 12-63. Scroll selected item into view *option*.

Often it is necessary to go one step further than simply modifying dimensions on assembly components. There are occasions when feature geometry must be added to components. This requires editing the component and adding features to it, either as a separate part in its own window or in the context of an assembly. In either case, you would use the right mouse button to access these functions.

Opening Component Files

If you need to open a part being used by an assembly, it is much easier to use the right mouse button as a shortcut rather than using the Open icon or File menu. To open a part in an assembly, right-click on the part to be opened. This can be done in the work area or FeatureManager. You will see the option *Open filename*, where *filename* is the name of the part you right-clicked on.

The part (or subassembly) will open in its own window. You may then work on the part without all of the clutter of the assembly around it. This accomplishes two important things. It allows for easily working on the part without the additional overhead of an assembly, and it keeps external references to other components from occurring. You first read about external references in Chapter 7.

Because external references can be a complicated topic, we will only touch on the matter here in relation to assemblies. Chapter 18 will deal with the topic on a much grander scale. For now, accept the fact that external relations to other geometry is sometimes necessary, but can complicate the assembly and reduce restructuring flexibility. External references should be used when needed, but not indiscriminately.

There are times when it is necessary to edit a part while still in the assembly, so that geometry can be extracted from other assembly components in order to help build new features on the part being edited. Be aware that this process will establish external references. The process is examined in the following section.

Editing Components in Context

The act of editing an assembly component within an assembly is known as *editing in context*. In context means in association with. The meaning in SolidWorks is identical to that associated with speech. For instance, words can be taken out of context, which means their original meaning is lost.

Component Editing

When the same words are read in context within the original text, their true meaning is discovered. The same holds true with features that are out of context or in context with respect to the original assembly in which they were created.

When you wish to edit a part in the context of an assembly, it is usually because some other component must be referenced. For example, a hole in one component must be aligned with the position of a hole on another component in the assembly. In such a case, it would be necessary to use the Edit Part option. The Edit Part option is located in the right mouse button menu, just as the Open option is. How-To 12-11 outlines the method by which a part can be edited within the context of an assembly.

How-To 12-11: In-context Component Editing

To edit a component in the context of an assembly, perform the following steps.

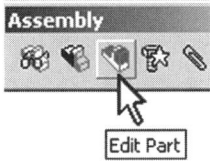

Fig. 12-64. Edit Part icon.

1. Right-click on the component to be edited. This can be a part or subassembly.

2. Select Edit Part (or Edit Sub-assembly), or click on the Edit Part icon, shown in figure 12-64.

3. To revert to editing the main assembly, right-click on the top-level assembly and select Edit Assembly, or click on the Edit Part icon again.

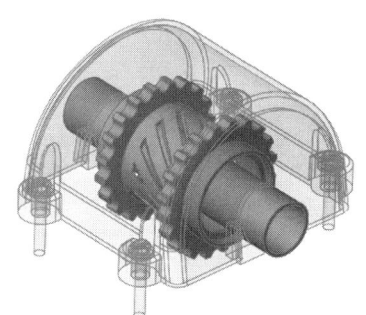

Fig. 12-65. Editing the Axle subassembly.

When you edit a part in the context of an assembly, two things happen right away. First, if the components are shaded, you will see all of the geometry in the assembly turn either gray or become semitransparent, except for the part you are editing. The part being edited will turn a salmon color. Figure 12-65 shows the gear housing assembly with the axle being edited.

Second, note that the title bar of the SolidWorks window states the part being edited, followed by the assembly name, whereas before it stated the assembly name only. This is an ideal way of discerning what you are editing at any point in time.

✏ **NOTE:** *Always be aware of what you are editing. The title bar will inform you of what is being edited while working with an assembly.*

It cannot be stressed enough how important it is to be aware of what you are editing while working in an assembly. Are you editing a sketch? Are you editing a component? A component within a subassembly? It is critical you understand this. You could easily waste a lot of time if you neglect to pay attention to what is being edited.

What you are editing is important for a number of reasons. Parent/child relationships have a lot to do with it. For example, if you began a sketch with the intention of creating a feature on a part, but realized too late that you were actually editing the assembly and not the part, the sketch would belong to the assembly, not the part. The sketch could not be used to create the feature you had intended. It is easy to tell what is being edited because of the color changes that take place and the title bar text changing to inform you what is being edited. Just make it a point to learn to recognize these signs and what they mean.

When editing a part in the context of the assembly, the Insert menu will change to reflect the same menu you would see when working on a part in its own window. Nothing is different. Use the skills you have learned to this point regarding designing and editing SolidWorks parts. There are no special tips or tricks that need to be employed. You now have the benefit of accessing geometry from other components within the assembly. Feel free to use your Convert Entity or Offset Entity commands, or any of the other sketch tools.

Controlling Transparency

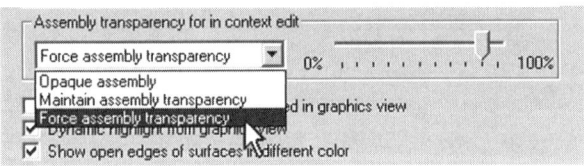

Fig. 12-66. Controlling assembly transparency.

When editing components in context, the assembly usually turns a transparent gray color. You have control over this, and can choose whether transparency is used, and to what degree. The options for controlling the transparency effect are in the Display/Selection section of System Options (Tools > Options). Figure 12-66 shows these options.

There are three options in the pull-down menu. The first, *Opaque assembly*, will turn the components in the assembly gray whenever a component is being edited, but the assembly components will remain opaque. The second, *Maintain assembly transparency*, will continue to use whatever transparency settings have already been established for the assembly components. How to control transparency for individual components is explored in material to follow.

Component Editing

The last option, *Force assembly transparency*, will force every component to be shown with transparency, except for the component being edited. When this option is used, the slider bar, which can be seen in figure 12-66, can be used to control the transparency. A higher setting results in more transparency. A transparency setting of 100% will result in an invisible assembly.

Transparency Selection Techniques

When working with transparency enabled, sooner or later you will find it necessary to either select something that is transparent or something behind a transparent object. There are special techniques that can be used to accomplish this.

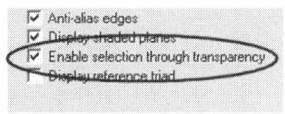

Fig. 12-67. Controlling selection through transparency.

The option that will have the greatest impact on how objects are selected when transparency is in use is shown in figure 12-67. The option, *Enable selection through transparency*, is found in the Display/Selection section of System Options (Tools > Options).

When *Enable selection through transparency* is checked, faces on the component being edited can be selected through transparent objects. If turned off, the transparent faces are selected. Holding down the Shift key will temporarily reverse the *Enable selection through transparency* option, whatever it is setting.

It is recommended that the *Enable selection through transparency* option be left on. This will force SolidWorks to exhibit the behavior that might be described as "if you can see it, you can select it." This type of behavior is logical for most users, and works out well. Remember, you can always hold down the Shift key if it becomes necessary to select something on a transparent component.

Component Transparency

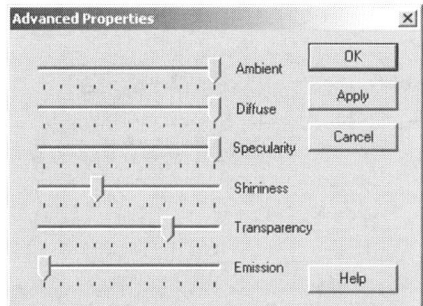

Fig. 12-68. Color Advanced Properties window.

A command mentioned earlier when discussing component properties was the Color button. It was learned that the Color button can be used to change the color of a component. Only the component was affected, and the actual part file would remain unchanged.

Clicking on the Color button opened the Assembly Instance Color window, which contained another button, named Advanced. Clicking on the Advanced button would open the window shown in figure 12-68. There are settings for much more than just transparency in the Advanced Properties window. These settings are outlined in Table 12-2.

Table 12-2: Color Advanced Property Descriptions

Advanced Property	Characteristic of Light Property
Ambient	Light is reflected and scattered by other objects
Diffuse	Light is scattered omnidirectionally from component surfaces
Specularity	Component's ability to reflect light
Shininess	Glossiness of component; another reflective characteristic
Transparency	Component's ability to pass light
Emission	Component's ability to project (emit) light

Interference Checking

At certain times during the assembly process, or when the assembly is completed, you may want to see if there are components that interfere with each other. There is a tool that will allow you to perform an interference check, and it is aptly named Interference Detection. It is found in the Tools menu. Suspected components or the entire assembly can be tested for interference. How-To 12-12 takes you through the process of testing an assembly for interference.

How-To 12-12: Testing an Assembly for Interference

To test an assembly for interference, perform the following steps.

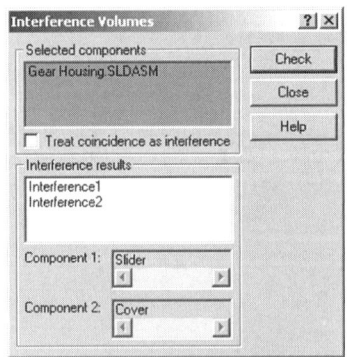

Fig. 12-69. Interference Volumes window.

1. Select Interference Detection from the Tools menu. The Interference Volumes window, shown in figure 12-69, will appear.

2. Select the components (parts or subassemblies) to be checked for interference. By default, the entire assembly will be selected, but you can pick and choose.

3. Click on Check. Occurrences of interference will be listed in the *Interference results* section.

4. Click on Close when finished.

Collision Detection

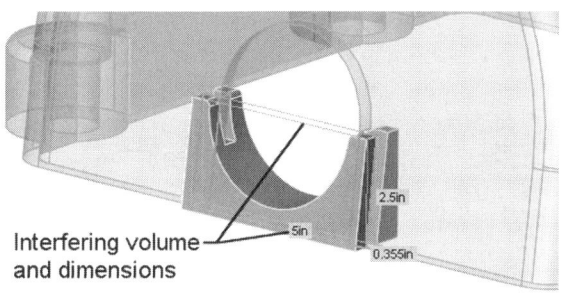

Fig. 12-70. Interference bounding box and dimensions.

If interference is found, clicking on the interference listed in the *Interference results* section will display which components are interfering in the lower portion of the window. This can be seen in figure 12-69. The interfering areas (or volumes) are indicated by a bounding box, such as that shown in figure 12-70. Dimensions will be given describing the bounding box volume in which the interference resides.

It is up to you to fix the interference problems. There is no magic button to push. SolidWorks is intelligent, but it cannot guess at your design intent, nor would you want it to make such assumptions. The next logical step in this series of events would be to edit at least one of the components in order to remedy the interference problem. If we use the illustrations as an example, it is clear the slider components must be modified so that they will fit in the openings at either end of the cover. Once that has been completed, it would be prudent to perform another interference check to make sure the interference problem has been remedied.

There is one other option in regard to interference checks, and that is the *Treat coincidence as interference* option, found in the Interference Volumes window. If this option is checked, any faces on assembly components will be defined as interfering if they so much as touch each other. If you do not want certain assembly components to touch (in other words, be coincident), check *Treat coincidence as interference* and run the interference check.

Collision Detection

In the solid modeling industry, there is something known as collision detection. It sounds like something you might want for a 16-year-old newly licensed driver, but it is not. It has to do with whether or not components can sense the surface of another component and keep from colliding with it. SolidWorks does employ this technology, but it must be enabled.

Another name for collision detection, at least in the context of Solid-Works, is dynamic interference detection. This is something that can be checked in an assembly that has moving parts. As an example, imagine that a crank is being turned on an assembly, and other components are moving while the crank is turned. At some point in time, components may start grinding together. How would you know if this were happening without testing for interference while the components were moving?

This is precisely what collision detection allows the user to see. Without having to check for interference with the components at every conceivable position, collision detection allows for moving the components dynamically and observing whether or not they interfere, or collide. How-To 12-13 takes you through the process of performing collision detection.

HOW-TO 12-13: Performing Collision Detection

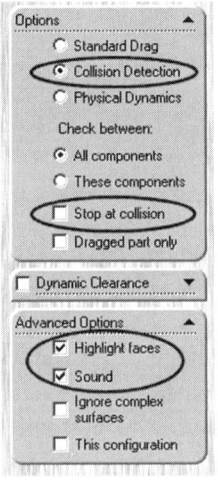

To perform collision detection, which can also be considered dynamic interference detection, perform the following steps.

1. Click on the Move Component icon on the Assembly toolbar. (Note that the Rotate Component command can also be employed for collision detection.)

2. In PropertyManager, select Collision Detection, shown in figure 12-71. Options that pertain to collision detection are circled.

3. Optionally, check the option *Stop at collision*.

4. Move the component you would like to check collision for.

Fig. 12-71. Enabling collision detection.

Once the Collision Detection option is enabled, moving the component as usual will result in any colliding faces turning green. There will also be audio confirmation of collision. *Highlight faces* and Sound are the two options that control this behavior (refer once again to figure 12-71).

> **NOTE:** *The sound used by SolidWorks to indicate collision is the same sound used for the default beep in your Windows operating system. Consult Windows help if you wish to change this sound.*

The *Stop at collision* option does exactly what its name implies, which is to stop the component being moved if it collides with another. This takes a lot of processor power because of the great deal of computations involved. There is no way around this, but the computations can be limited in scope by telling SolidWorks to only check for collision between specific components.

Limiting the scope of what is checked when enabling collision detection is especially useful when using the *Stop at collision* option, because it narrows down the number of computations that need to be performed by the computer. To limit the scope of the components being checked for collision, select *These components* rather than *All components*. You can then select directly from the work area the components between which you wish to check for collision, and then click on the Resume Drag button. Check for collision as usual.

With a slower computer system, trying to check for collision, especially with *Stop at collision* checked, may result in the computer appearing to be frozen. This is due to the computer needing to process all of the data before it can be rendered to video. This is just the nature of the beast, but will only get better as computer hardware catches up with what we are trying to put it through.

Dynamic Clearance

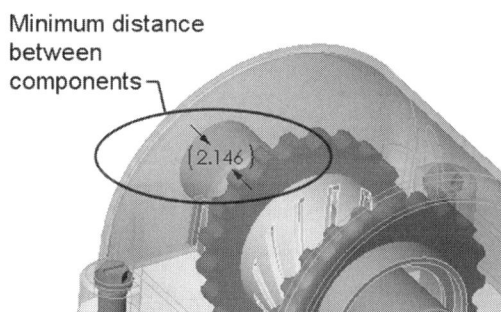

Fig. 12-72. Checking dynamic clearance.

Another function that goes hand in hand with these last few topics is the ability to view clearance as a dynamically changing value as parts are moved. An example of this is shown in figure 12-72. The 20-tooth gears have been replaced with gears with a slight eccentricity. (Incidentally, the gears have a rotational offset 90 degrees from each other.) As one of the gears is rotated, the minimum distance between the gear and the cover can be detected. At the particular time the image in figure 12-72 was captured, the gear was at a distance of 2.146 inches from the cover.

Because dynamic clearance is dependent on two components for determining the minimum clearance distance, it is mandatory that the components be selected prior to being able to calculate this information. The selection technique used is identical to the steps used when narrowing the scope for collision detection. The steps for checking dynamic clearance, in their entirety, are presented in How-To 12-14.

How-To 12-14: Checking Dynamic Clearance

To check dynamic clearance between two components, perform the following steps.

548   **Chapter 12: Assemblies**

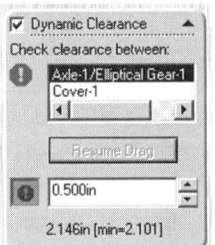

Fig. 12-73.
Enabling the
Dynamic
Clearance option.

1. Click on the Move Component icon on the Assembly toolbar. (Note that the Rotate Component command can also be employed for checking dynamic clearance.)

2. In PropertyManager, select the Dynamic Clearance option, shown in figure 12-73.

3. Click in the list box area directly below the Dynamic Clearance option.

4. Select two components between which you want to check for clearance.

5. Click on the Resume Drag button.

6. Move the desired component as usual.

The minimum distance will be displayed in the work area, and in PropertyManager at the bottom of the dynamic clearance list box. Note that there is another number in brackets after the first number. This is the minimum distance found while moving the component. What this means is that after moving a component through its entire range of motion the minimum amount of clearance for the entire range of motion can be determined.

An additional setting near the bottom of the Dynamic Clearance panel allows for specifying a minimum clearance distance. If you examine figure 12-73, you will see that the value for this parameter has been set to .500 inch. What this means is that while moving the components through their range of motion they will not be allowed to get any closer than half an inch from each other.

In summary, SolidWorks does not prohibit a person from incorrectly creating an assembly. If you are not careful, you can place components inside other components, create interference, or make other mistakes. However, if you use the tools SolidWorks provides, building a complete assembly with no interference, and moving components, can be easier than ever thought possible with previous CAD software.

Physical Dynamics

One of the most innovative functions to come along in a CAD program in years is the ability to see how components will react to each other as they

Simulation

would in the real world. SolidWorks does not offer in its base software what a full kinematics package might offer, but dynamic assembly motion is still a very nice feature.

Dynamic assembly motion, also known as physical dynamics, will use a combination of two processes to exhibit what is very similar to real-world movement. Collision detection is one process, and analyzing each individual component's center of gravity is the other. Physical dynamics is very resource intensive and may not function ideally with large assemblies, slower computers, or with complex geometry, such as worm gears.

Physical dynamics, the settings for which are shown in figure 12-74, is implemented in a manner similar to collision detection. It is possible to narrow the scope of which components will be interacting with each other by selecting These components. Narrowing the scope is not a bad idea because dynamic assembly motion is very resource intensive.

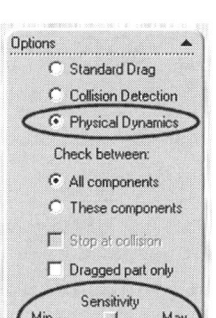

Fig. 12-74. Physical dynamics settings.

The Sensitivity setting sets the time interval sampling rate between components. Increasing sensitivity will result in more accurate physical dynamics, but will require greater computations. It is recommended that you experiment with a sensitivity setting near the minimum side of the scale to start with, and then gradually turn up the sensitivity if all goes well.

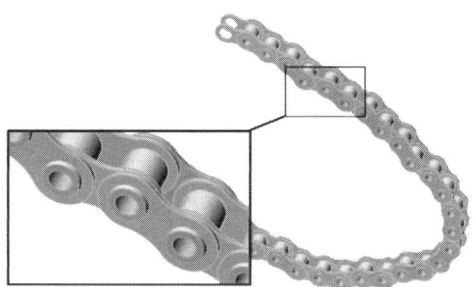

Fig. 12-75. Physical dynamics exhibit real-world movement.

When employing physical dynamics, move components the same way you would when using collision detection. It is unfortunate that physical dynamics cannot be displayed in this book. However, an example can be given. Physical dynamics would allow for moving the individual linkages in the chain shown in figure 12-75, and watching how the individual links react over the entire length of the chain. The effect is very dramatic, but only if you have a computer that can handle the computational demands. But then, this is the nature of cutting-edge software; it drives the PC hardware market.

Simulation

The simulation capabilities of SolidWorks takes physical dynamics to the next level. Four physical simulation effects can be added to an assembly model. These four effects are Linear Motor, Rotary Motor, Spring, and Gravity. Adding any of these simulation effects can most easily be accomplished via the Simulation toolbar, shown in figure 12-76.

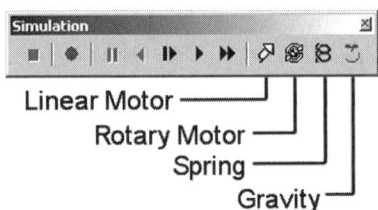

Fig. 12-76. Simulation toolbar.

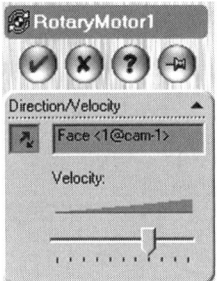

Fig. 12-77. Rotary Motor panel.

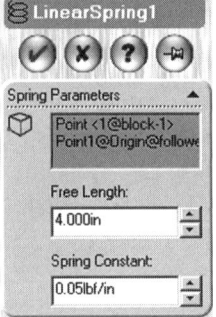

Fig. 12-78. Linear Spring panel.

Although the simulation tools are fairly rudimentary, they can be used to add animation-type movement to a SolidWorks assembly. Gears can be made to turn, a box can roll down a conveyor, or a spring-loaded mechanism can be shown in operation. Most of the simulation tools have the same basic interface. The Rotary Motor panel is shown in figure 12-77.

All that is really required to add a simulation effect is to select an object (such as a face or edge) to indicate the force direction, and to specify the strength of the force via a slider bar. What needs to be selected is all very logical. For example, a cylindrical face will suffice for adding a rotary motor. The Spring effect is the only odd man out, in that objects must be selected from two different components. Edges or vertex points work nicely. Also regarding springs, a Free Length and Spring Constant should be given as well. These settings are found in the Linear Spring panel, shown in figure 12-78.

Simulation effects can be added in multiple combinations, but your assembly must be put together correctly before any of the simulation tools can reasonably be expected to function. It will not be possible, for example, to spin a fan blade if it has been fully mated into position. Components must have some degree of freedom of movement left.

In the very simple example shown in figure 12-79, you can see the various forces applied to the assembly. A rotary motor is turning the cam, a spring is holding the follower in position, and the green arrow is representing gravity. To create the simulation, click on the Record icon on the Simulation toolbar. The components will begin reacting according to the effects that have been placed upon them. To stop recording, click on the Stop icon.

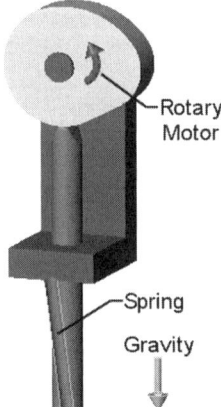

Fig. 12-79. Forces present on the Cam assembly.

The rest of the icons on the Simulation toolbar (other than Record and Stop) are for playing the simulation back. From left to right, the options are Pause Replay, Reverse Replay, Slow Replay, Replay Simulation, and Fast Replay. The icons are easy enough to figure out. It is making sure your assembly has been put together correctly that can cause problems. Test your assembly using Physical Dynamics before applying forces and attempting to record a simulation. If you cannot obtain dynamic assembly motion, it is a sure bet simulation will not work for you either. Double check your mate relationships and check for interference, and then try the dynamic motion again.

Assembly Features

Anytime you create a feature that affects the assembly but not the original part files, it is considered an assembly feature. To put this another way, assembly features only affect components within the assembly. This is why they are termed assembly features.

Assembly features are limited in their complexity. For example, sketched assembly features can only remove material from an assembly, never add material. This means that sketched assembly features can be cuts only, not bosses. Other assembly features, such as weld beads, are not assembly features from a technical standpoint. The Weld Bead function is in the Assembly Feature menu. The Weld Bead function, however, actually creates an individual part file, not an assembly feature. (Weld beads are discussed in Chapter 14.)

An assembly feature must be created by first sketching a profile, as with any other sketched feature. It is also possible to create holes using the Hole Wizard command. All of this aside, the important thing to remember is to let SolidWorks know what components you wish the assembly feature to affect. This is done via the Feature Scope function.

Feature Scope

It is the job of the Feature Scope window to keep track of which components in the assembly should be affected by any given assembly feature. Each assembly feature has its own feature scope, and each feature scope can be edited at any time. It would be best to walk through an example to see how this is accomplished. The gear housing assembly will be used.

To begin creating a typical assembly feature, a plane will be created at some distance above the gear cover. A sketch will be started on the plane, and this sketch will be used to create a cutaway view of the gear housing.

552 Chapter 12: Assemblies

A quarter chunk of the assembly will be cut away to show its internal mechanism, but not all components will be affected by the cut. Figure 12-80 shows the results so far.

Next, the assembly feature can be created, but first SolidWorks must be told what components in the assembly should be affected by the assembly feature. To do this, edit the feature scope. How-To 12-15 takes you through the process of editing the feature scope.

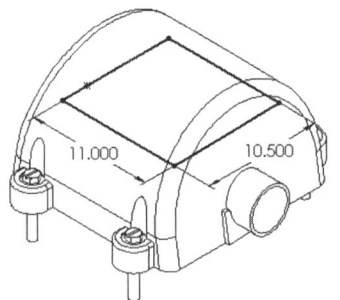

Fig. 12-80. Preparing to create an assembly feature.

How-To 12-15: Editing the Feature Scope

To edit the feature scope in order to determine which components will be affected by an assembly feature, perform the following steps.

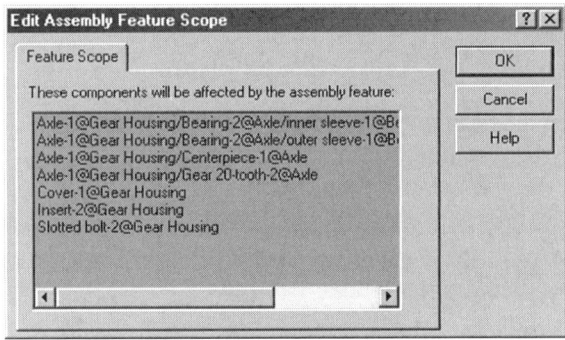

1. Select Feature Scope from the Edit menu. This can be done while in a sketch prior to creating the feature.

2. Select the components to be affected by the assembly feature. The components will be listed in the Feature Scope window, shown in figure 12-81.

3. Click on OK.

Fig. 12-81. Feature Scope window.

During step 2, when it is necessary to select the components, be aware that the components can either be selected in FeatureManager or directly from the work area. If a component is selected by accident, clicking on it a second time will deselect it. Also note that component instances that have been generated by component patterning (covered earlier in this chapter) can also be selected.

Now that you have edited the feature scope and told the software which components to include in the assembly feature, it is time to create the assembly feature itself. Although your options are limited as to the types of features that can be created, the outcome of assembly features can

Assembly Features

be dramatic. How-To 12-16 takes you through the process of creating an assembly feature.

How-To 12-16: Creating an Assembly Feature

To create an assembly feature, perform the following steps. These steps assume you have already created a sketch.

1. Select Insert > Assembly Feature > Cut > Extrude.

2. Specify the desired parameters in the Cut Extrude panel. These parameters are identical to those discussed in Chapter 4, or when creating any basic extrusion.

3. Click on OK to create the assembly feature.

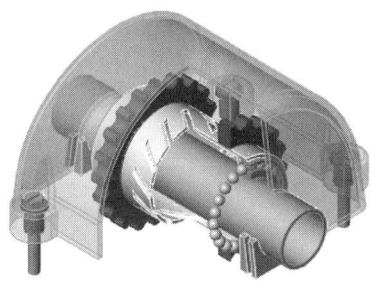

Fig. 12-82. An example of an assembly feature.

An example of a completed assembly feature is shown in figure 12-82. Some, but not all, of the components in the assembly have been affected by the cut. The individual balls from the bearing subassembly, the axle, and the slider component the axle rests on were not included in the feature scope. Therefore, the cut simply ignores these components.

It is important to edit the feature scope prior to creating a feature, because it is necessary to specify which components will be affected and to limit which components are considered for the cut. If a feature scope is not specified, SolidWorks automatically considers every component in the assembly. This is not very efficient, especially when there are many components.

If an assembly feature has already been created, using the Feature Scope command in the Edit menu will not suffice. The feature scope accessible from the Edit menu is only for future assembly features. Existing assembly features have their own feature scopes, and every assembly's feature scope can be different.

To edit the feature scope of an existing assembly feature, right-click on the assembly feature in FeatureManager and select Feature Scope. You will be presented with the Feature Scope window and can add or remove components from the listing as desired.

Assembly Feature Patterns

Any of the assembly features mentioned in the previous section can be patterned. The process of creating an assembly feature pattern is extremely similar to creating regular feature patterns. The only real difference is in menu picks. To access assembly feature pattern commands, select Assembly Feature from the Insert menu. The pattern types that can be used to pattern assembly features are as follows.

- Linear and circular pattern
- Table-driven pattern
- Sketch-driven pattern

If your memory needs refreshing on how to create feature patterns, turn back to Chapter 7 and review feature patterns. There is one difference when it comes to patterning assembly features, and that is the feature scope. Whatever the Feature Scope option was set to when the original assembly feature was created will have bearing on what components are affected when the pattern is created. If the feature scope is edited for the assembly feature, the pattern of that assembly feature will update accordingly to affect only those components listed in the modified feature scope.

✎ **NOTE:** *Every assembly feature can have its own feature scope, but assembly feature patterns cannot have a scope different from their parent feature.*

Assembly Drawings

There are a few differences between creating a design drawing of a part versus creating a drawing of an assembly. Some of these differences are covered in this section, along with some other significant aspects of assembly drawings. This includes creating a bill of materials (BOM), which is an automated process requiring Microsoft Excel. Some of the various annotation options not covered in Chapter 11 are also covered here.

The mechanics behind creating views of an assembly are exactly the same as for creating views of parts. You can create standard three-view layouts, detail views, named views, or any other view type. When it comes to creating a section view of an assembly, you can specify which components in the assembly are sectioned. This process is covered in the material that follows.

Assembly Section Views

There is a special consideration when working with section views in relation to assemblies. This has to do with whether or not you want every component in the assembly sectioned, or if you want to choose which components are sectioned. Components that are not sectioned appear whole, with no crosshatch. Here you will learn how to control which components are sectioned in the section view.

To begin creating an assembly section view, use the same process you would when creating a section view of a part. The procedure is the same, except for one additional step, which requires you to add components to the section scope. The section scope is very similar to the feature scope. With the section scope, you select the components you want to exclude from being sectioned.

When selecting the components to exclude from being sectioned in the section view, you can either select the components from the work area or from DrawingManager. Often, some components are obscured by others in the assembly view being sectioned. For this reason, it may be easier to select components from DrawingManager. Use whatever method is easiest for you, as selection technique will not affect the outcome whatsoever. How-To 12-17 takes you through the process of creating an assembly section view.

How-To 12-17: Creating an Assembly Section View

To create an assembly section view, perform the following steps. All required steps are listed here for reference, but details regarding basic section views will not be repeated, as that topic was covered in Chapter 11.

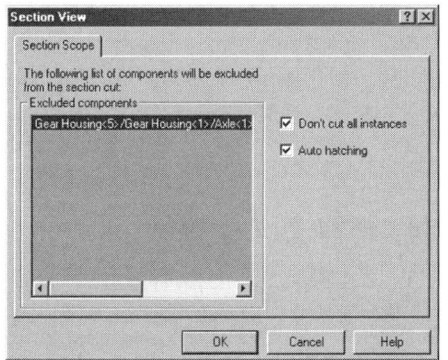

Fig. 12-83. Section Scope window.

1. Activate the view you wish to generate a section view from.

2. Sketch geometry representing the section line.

3. If the section line geometry consists of more than one segment, select a line segment to project the section view from.

4. Select Insert > Drawing View > Section, or click on the Section View icon. The Section Scope window, shown in figure 12-83, will appear.

5. Select the components to exclude from being sectioned.

6. Click on OK to create the assembly section view.

7. Pick somewhere on the drawing sheet to position the new view.

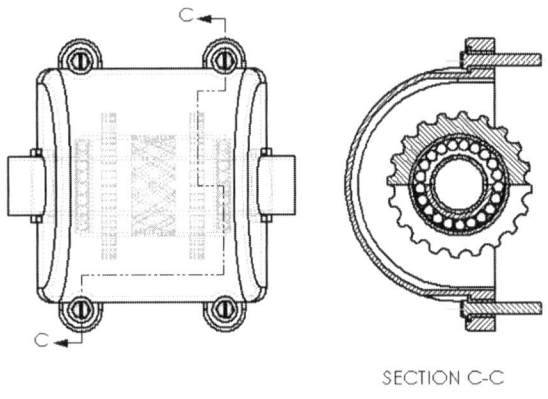

Fig. 12-84. An assembly section view.

An example of an assembly section view is shown in figure 12-84. It is interesting to note that a multisegmented section line was used. This has the effect of showing the internal workings of the assembly at different "depths." Incidentally, the bearing subassembly was selected so that it would not be sectioned.

None of the ball bearings from the bearing assembly have been sectioned. This is the case, even though only one of the ball bearings was selected. Within the Section Scope window is an option titled *Don't cut all instances*. What this refers to is other instances of a selected component. For example, if there are 18 ball bearings in the assembly drawing view, and only one of the ball bearings has been selected, checking the *Don't cut all instances* option will exclude all 18 of the ball bearings from being cut in the resulting section view. This keeps the user from having to select, in this case, all 18 instances.

There is an additional option titled *Auto hatching*. If the *Auto hatching* option is selected, sectioned components will have alternating hatch patterns, making it easier to distinguish them from one another. To modify the section scope after the section view has been created, right-click on the section view and access its properties.

Hiding Components in Drawing Views

Sometimes it becomes necessary to hide a particular component for the sake of clarity, or for other reasons. The process is quite simple and can be accomplished via the right mouse button. Right-click on the view containing the components you wish to hide and select Properties. Drawing views have properties, just like most objects in SolidWorks.

Assembly Drawings

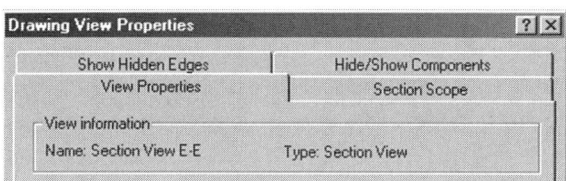

Fig. 12-85. An example of a Drawing View Properties window.

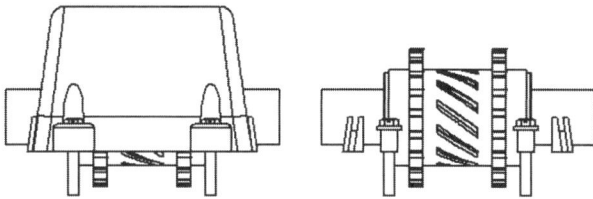

Fig. 12-86. Before and after hiding the cover.

Fig. 12-87. Automatic hiding of components on view creation option.

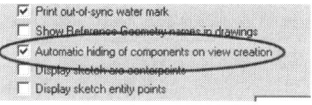

Once the view's Drawing View Properties window has been opened, note the tabs present. If the view is a section view, there is a tab named Section Scope, and the top of the Drawing View Properties window would look as it does in figure 12-85. This is where the section scope for the assembly section view, discussed previously, can be modified.

In the same Drawing View Properties window is a tab named Hide/Show Components. Selecting this tab and then selecting components, either from DrawingManager or from the drawing view, will result in those components being hidden. Figure 12-86 shows an example of an assembly drawing view. On the left is the assembly prior to hiding the cover. On the right is the exact same assembly with the cover hidden.

An option that goes hand in hand with hiding components in an assembly drawing view is an option titled *Automatic hiding of components on view creation*. This option, shown in figure 12-87, can be used to access the Drawings section of System Options (Tools > Options). When enabled, all components that are behind other components will automatically be hidden. If you intend on showing the drawing views with hidden lines removed anyway, this option is an outstanding choice.

Using *Automatic hiding of components on view creation* can decrease the amount of time it takes to generate assembly views containing many components. If you then choose to show the view with hidden lines dashed, it is possible to remove from the Hide/Show Components tab listing only those components you wish to show the hidden lines for.

Show Hidden Edges

If you would rather not show any hidden edges, but still find it necessary to show the hidden edges for one or perhaps just a few components, the Show Hidden Edges tab is for you. This is the final tab in a drawing view's properties. Similar to using the Hide/Show Components functionality, showing the hidden edges of particular components in a view is just a matter of selecting the components. However, selecting the components may be a bit more difficult, in that they cannot be seen in the first place!

There is no secret process for selecting a hidden component in order to have its hidden lines shown. You will just have to select the component from DrawingManager instead of the drawing view. No big deal. If the view is selected first, its associated icon will turn blue in DrawingManager. You can then drill down through DrawingManager to find the desired component or components.

Bill of Materials

A bill of materials (BOM) is a list of the components being used in the assembly. One of the basic BOM templates included with the SolidWorks software has four columns, one each for Item Number, Quantity, Part Number, and Description. Once a BOM is inserted into an assembly layout, it will keep track of components automatically if you add or remove components from the assembly.

> **NOTE:** *You must have Microsoft Excel installed on your computer in order to take advantage of BOM functionality.*

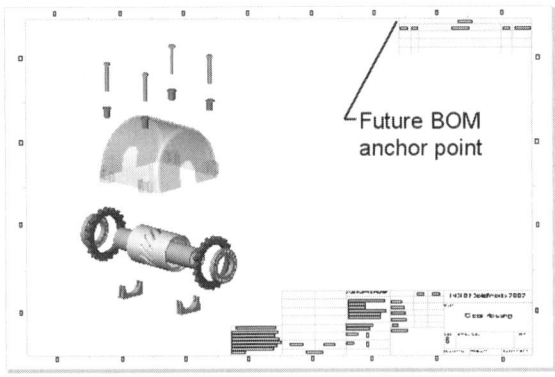

Fig. 12-88. A BOM will be added to this drawing.

An exploded view of the gear housing assembly will be used as an example when inserting a BOM. You will learn how to create an exploded view in material to follow. For now, let's focus on bill of materials functionality. The drawing on which a BOM will be added is shown in figure 12-88. For future reference, the BOM attachment (anchor) point is being shown as well. Establishing an anchor point is another aspect of BOMs you will learn in material to follow.

Inserting a BOM is easy, but there are a fair number of steps in the process. How-To 12-18 takes you through the process of inserting a BOM. The section that follows How-To 12-18 explores the various options present during the insertion process.

Bill of Materials

How-To 12-18: Inserting a BOM

To insert a BOM, perform the following steps.

1. Select a view. (This is required because theoretically there may be different assemblies or parts within the same sheet and the software needs to know what it should generate the BOM from.)

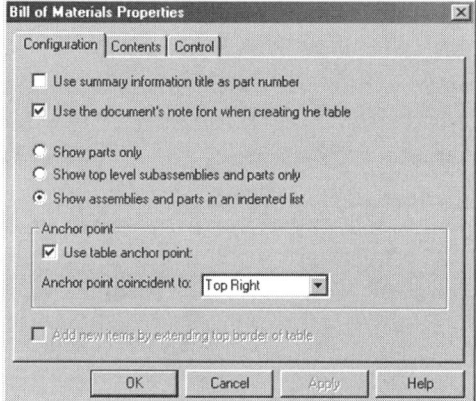

2. Select Insert > Bill of Materials.

3. Select the desired template to base the BOM on. Templates included with SolidWorks are found in the *SolidWorks\lang\<language>* directory, where *<language>* is the installation language. If not familiar with the various templates, it is suggested the basic *bomtemp.xls* file be used.

4. Click on Open. The Bill of Materials Properties window will appear, as shown in figure 12-89.

Fig. 12-89. Bill of Materials Properties window.

5. Specify the desired parameters in the Bill of Materials Properties window. These parameters are explained in detail in material to follow.

6. Click on OK to create the BOM.

There will be some flashing on screen while SolidWorks and Excel work together to create the BOM. When things settle down, it is time to go back to work. A portion of the drawing containing the BOM is shown in figure 12-90. The BOM is illegible in the image, but we will take a closer look at it in material to follow.

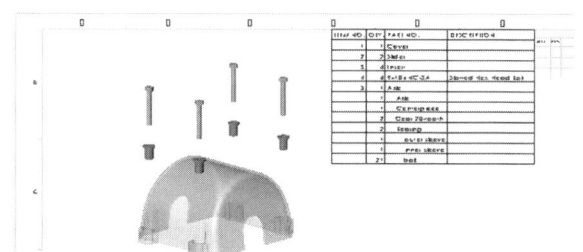

Fig. 12-90. After inserting the BOM.

Note where the BOM is attached to the drawing. This is because an anchor point was established.

Bill of Materials Properties Window Options

Before discussing how to create an anchor point in more detail, let's look at the options found in the Bill of Materials Properties window. These options, described in the sections that follow, control the appearance and placement of the BOM, so you will want to read these sections carefully.

Use summary information title as part number Option

The summary information title has to do with the individual part file's properties. To view the Summary Information window, open any part and select Properties from the File menu. If data is typed into the area labeled Title in the Summary tab of the Summary Information window, that data will be used in place of the file's name for the Part Number column in the BOM if *Use summary information title as part number* is checked.

The down side to this option is what the results will be if everyone in the CAD department does not fill out the Summary Information window's Title option for every part they create. In such a scenario, the BOM may not be consistent. On the other hand, if the Title area is not filled in, SolidWorks will use the part file name as the part number, rather than leave the part number column blank. Using the part file name as the part number is the default when *Use summary information title as part number* is not checked.

Use the document's note font when creating the table Option

By checking the *Use the document's note font when creating the table* option, through Document Properties (Tools > Options) you can specify what font will be used by the BOM. In the Document Properties tab, select the Note section, and then click on the Font button and specify a font, font size, and any other characteristics required for the note font. The benefits of this are that the BOM font can match the font in the rest of the document. If this option is not checked, the default font saved with the BOM template will be used.

Show parts only Option

When the *Show parts only* option is selected, all parts will be listed as separate components in the BOM, whether those parts are top-level components or whether they belong to a subassembly. Subassemblies themselves are not listed.

Show top level subassemblies and parts only Option

Show top level subassemblies and parts only means that top-level subassemblies will not have their individual components listed in the BOM. Top-level components, meaning both top-level parts and subassemblies, will be listed. Use this option if subassemblies will be represented on other drawings with their own BOMs.

Show assemblies and parts in an indented list Option

Show assemblies and parts in an indented list will break down subassemblies into their separate components, much like the previous option, but will show those separate components in an indented list. Only top-level components (parts and subassemblies) will actually be assigned item numbers.

Use table anchor point Option

- The table anchor point is the point where the table (the BOM, in other words) gets anchored. If the *Use table anchor point* option is checked, the "corner" of the BOM table attached to the anchor can be selected from the drop-down list. The drop-down list contains options such as *Top right*, in which case the top right corner of the BOM will be attached to the anchor.

  How-To 12-19 takes you through the process of setting an anchor point. Keep in mind that a sheet format must be used in order to establish an anchor point. Blank sheets do not allow for anchor points.

How-To 12-19: Setting an Anchor Point

To set an anchor point for a bill of materials, perform the following steps.

1. In DrawingManager, right-click on the Bill of Materials Anchor object located under the name of the sheet format, as shown in figure 12-91, and select Set Anchor.

2. Pick a point entity, endpoint of a line, or something similar to indicate where the anchor point should be.

3. Select Sheet from the Edit menu.

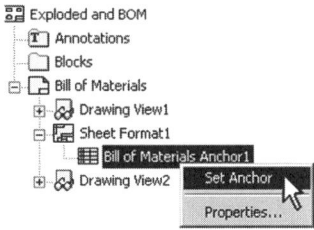

Fig. 12-91. Setting the BOM anchor point.

During step 2, you will find yourself editing the sheet format. You are placed in this mode automatically by SolidWorks. That is why it is necessary to return to editing the sheet in step 3. If the Bill of Materials Anchor object is selected in DrawingManager, a small asterisk will appear on the sheet indicating where the anchor point is.

Add new items by extending top border of table Option

The *Add new items by extending top border of table* option is only available if an anchor point is not being used. It is one of the more commonly misunderstood BOM options. What it means is that if new components are added to the assembly the top of the BOM will be extended. What it does not mean is that newly inserted items will be placed at the top of the BOM.

This option works well if the BOM is positioned directly above the title block. Newly inserted items will always be inserted into the BOM automatically by SolidWorks at the bottom of the BOM list, but the BOM will "grow" upward, adding new rows to the top as needed.

Table 12-3 shows a recreation of a SolidWorks BOM. (A table was created for the sake of this book to retain visual integrity of the data, rather than using an image. However, every effort has been made to maintain the originality of the data for explanation purposes.)

Table 12-3: Recreation of a SolidWorks BOM

ITEM NO.	QTY.	PART NO.	DESCRIPTION
1	1	Cover	
2	2	Slider	
3	4	Insert	
4	4	3/4-10UNC-3A	Slotted Hex Head Bolt
5	1	Axle	
	1	Axle	
	1	Centerpiece	
	2	Gear 20-tooth	
	2	Bearing	
	1	outer sleeve	
	1	inner sleeve	
	21	ball	

Note in Table 12-2 that five items have been assigned item numbers. These five items are all top-level components, with Axle being a top-level subassembly. Axle has its separate components listed below it in an indented list. Bearing is in turn a top-level component of Axle, and is in itself a subassembly. The Bearing component contains three components (outer sleeve, inner sleeve, and ball), which are indented further to illustrate that they are components of Bearing.

It is not necessary to show components in this fashion. It just happens that the *Show assemblies and parts in an indented list* option was used. If, for example, the *Show parts only* option were used, The *Axle* and *Bearing* subassemblies would not be listed, and all of the remaining components would be assigned item numbers.

Editing a BOM

Common reasons for editing a BOM are to add a description or to add an item such as paint to the bottom of the BOM. Colors and text formatting can be applied as well. Do not change things in the BOM such as part number or quantity, as those items are directly controlled by SolidWorks. These items will update when changes to the assembly are made.

To edit a BOM, either double click on it or select Edit Bill of Materials after right-clicking on the BOM. This action will open Excel within SolidWorks and allow for making manual text changes to the BOM. There is a problem associated with adding text to the BOM in this manner. If the BOM is regenerated for any reason, any text added manually to the BOM will be lost. There are a number of processes that cause BOM regeneration (covered in material to follow), but this is not the point. The point is that there is a proper way of adding data to a BOM that will associate itself with a component and remain even through a regeneration.

As an example, let's discuss how a description can be added to a BOM. This method, involving the file's properties, can be used to add any type of custom data to a model whereby the data can then be made to populate a BOM. This includes data such as cost, vendor, mass, size parameters, or anything else imaginable.

Because getting custom data to appear in a BOM is basically a two-part process, we will look at both processes separately. First, a custom property will be added to a part. As mentioned previously, a description will be added as an example. How-To 12-20 takes you through the process of adding a description that will appear in the BOM. Following this section, you will learn how BOMs can be customized.

How-To 12-20: Adding Custom Properties

To add a custom property to a part (in this case, a description) so that it will automatically appear in a BOM, perform the following steps. Note that this process will also work with assembly and drawing files. It could also be noted at this point that custom properties are added for other reasons above and beyond BOM generation. See Chapter 16 for a more in-depth look at custom properties.

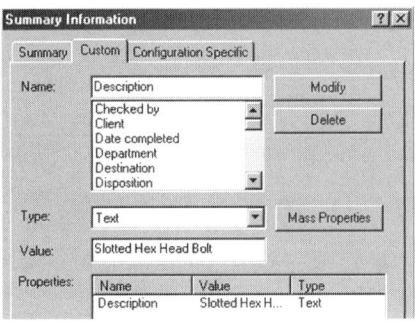

Fig. 12-92. Adding a custom property.

1. Select Properties from the File menu. The Summary Information window will appear.

2. Click on the Custom tab. You should see something similar to that shown in figure 12-92.

3. In the Name field, type in the word *Description*.

4. In the Value field, type in the description you want to use.

5. Click on the Add button. The new *Description* field and the value for that description will appear in the Properties list box.

6. Click on OK.

7. Save the file.

Back in the design drawing, a description will appear in the BOM for the part you added the custom property to.

If everyone in the CAD department became accustomed to adding descriptions to part files, drawings would consistently have informative BOMs. In addition, any data can easily be automatically added to a BOM as long as the BOM template has been modified to accept the data. This task is described in the following material.

Customizing a BOM Template

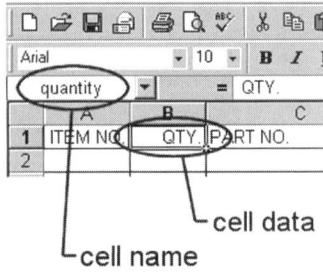

Fig. 12-93. Cell names versus cell data.

The key to customizing a BOM template is to make sure to name the cell at the top of a column in the BOM template exactly the same as the name entered in step 3 of How-To 12-20. Quite simply, the name at the top of a column in a BOM must match the name of the custom property.

It is very important to realize the difference between naming a cell and adding data to a cell. The cells must be named, or the customized BOM template will not work! This book is not meant to teach Excel, but this topic deserves a bit of special attention. Figure 12-93 shows a small portion of an Excel window in which a BOM template is being edited. The first three column headings of the template can be made out.

Focus on cell B1 in figure 12-93. Note the data within the cell. The data is QTY. This data is not the same as the name of the cell. The name, as we can see, is the word quantity. It is the name of the cell that is most important. It is the name that must match the name of the custom property created in a file's properties. Spelling is critical for the custom property to transfer to the BOM.

Now you know the technique for creating your own BOM templates. Be aware that formulas can also be added to increase a BOM's functionality even further. When designing your own template, it is best to begin with one of the default templates provided by SolidWorks, and go from there. Once the template has been modified, save it with a different name so as not to overwrite the original.

BOM Properties

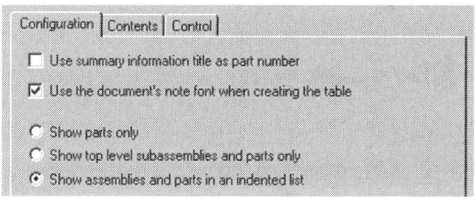

Fig. 12-94. Accessing a BOM's properties.

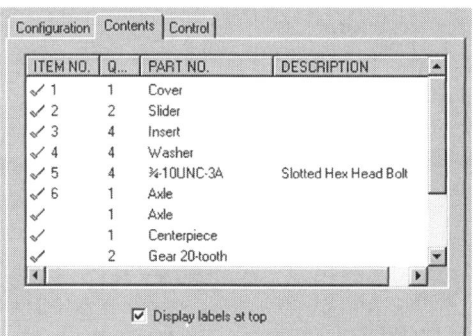

Fig. 12-95. Contents tab of a BOM's properties.

Double clicking on a BOM will allow for making formatting changes or manually adding text, but accessing the properties of a BOM provides access to other, more powerful, options. Right-clicking over the BOM and selecting Properties will bring up the same Bill of Materials Properties window displayed when first creating the BOM. Figure 12-94 shows a portion of this window. Note in particular the three tabs at the top.

In that the options in the Configuration tab were already explained in previous material, we will focus on the other two tabs, Contents and Control. The Contents tab, shown in figure 12-95, is used for removing components from being listed in the BOM without having to remove them from the assembly.

Through the use of the Contents tab, the BOM can be sorted in various ways. For instance, clicking on the Part No. heading sorts the contents alphabetically. A second click on the same heading reverses the order. Unchecking the Display labels at top option places the BOM labels on the bottom.

The Control tab, shown in figure 12-96, gives additional functionality to customize the appearance of the BOM even further. If *Row numbers follow assembly ordering* is checked, the order in which items are displayed in the BOM will be the same order the components are listed in the assembly's FeatureManager. This can be advantageous. By dragging components to different positions in the assembly FeatureManager, the order of items in the BOM can be altered.

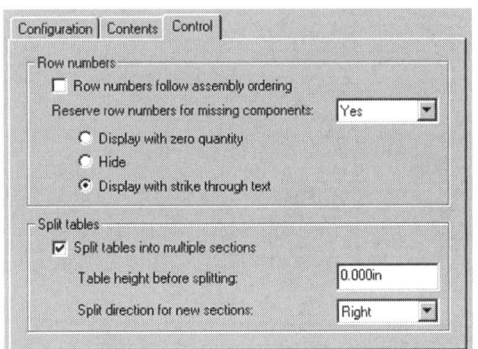

When *Row numbers follow assembly ordering* is turned off, it becomes possible to reserve row numbers for missing components. The missing components can either have their rows removed entirely or be displayed with a zero quantity or strike-through text.

If a BOM is long enough to flow off the bottom of the sheet, use the *Split tables into multiple sections* option. You could then specify how often the split should occur, such as every 10 inches.

Fig. 12-96. Control tab of a BOM's properties.

Overriding BOM Part Number

One other way to further customize a BOM is to specify a particular configuration name, or any name at all, as the part number. Not everyone wants to accept the SolidWorks default of using the part's file name in the part number column. Through the use of ConfigurationManager, you can control what appears in the part number column. This is true whether or not you actually have more than one configuration.

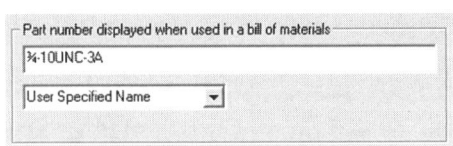

Fig. 12-97. Deciding what to use as the part number in a BOM.

You have already learned a good deal about configurations in Chapter 10. This includes how to access a configuration's properties (right-click on a configuration's name and select Properties). When the Configuration Properties window appears, you will find an area at the bottom of the window shown in figure 12-97. This area is where it becomes possible to override the file name being used as the part number in the BOM.

The drop-down list pictured in figure 12-97 contains three settings. The default setting is Document name, which is what tells SolidWorks to use the file name as the part number. The second setting is *Configuration name*. As you would guess, this forces the name of the configuration to be shown in the BOM. Finally, there is *User specified name*, which allows for typing in anything you wish. This data will then be transferred directly to a BOM's part number column and will be displayed in place of the file name.

Exploded Views

An exploded view is a view of an assembly where the components have been separated, typically in the reverse order in which they were assembled. These view types are often created for illustrating how assemblies are built, or even just to show all components in a clear manner without using numerous cutaway views. Often assemblies are exploded for BOM drawings as well. Once an assembly has been exploded, it can also be shown in a design drawing in this exploded state, which is discussed near the end of this section.

There are two ways to create an exploded view in an assembly. You can take the easy way out and let SolidWorks explode the assembly, or you can use the more entertaining (and useful) method and explode the assembly yourself. Whichever method you choose, the command to begin creating an exploded assembly is the same. How-To 12-21 takes you through the process of exploding an assembly.

How-To 12-21: Exploding an Assembly

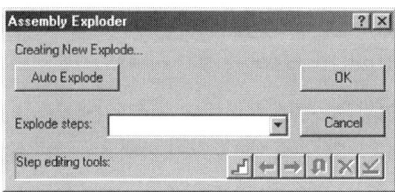

Fig. 12-98. Assembly Exploder window.

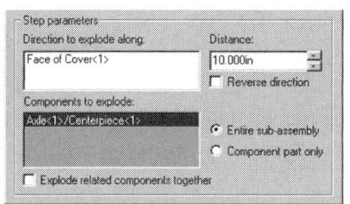

Fig. 12-99. Bottom half of the Assembly Exploder window.

To explode an assembly, perform the following steps. This must be done in an assembly file, not a drawing.

1. Activate the configuration you wish to explode.

2. Select Exploded View from the Insert menu. This will open the Assembly Exploder window (or at least the top half of it), shown in figure 12-98.

   NOTE: *If the Exploded View menu item is grayed out, try rebuilding the assembly first.*

3. To let SolidWorks automatically explode the assembly, click on the Auto Explode button. Otherwise, click on the New (explode step) icon. This is the only active icon and resembles steps or stairs. The Assembly Exploder window will expand to reveal its bottom half, shown in figure 12-99.

4. Click in the *Direction to explode along* area and specify an edge, plane, or cylindrical face that will indicate the explode direction. A preview arrow will appear previewing the explode direction.

568   *Chapter 12: Assemblies*

5. Click on the *Reverse direction* option if the preview arrow is not pointing in the correct direction.

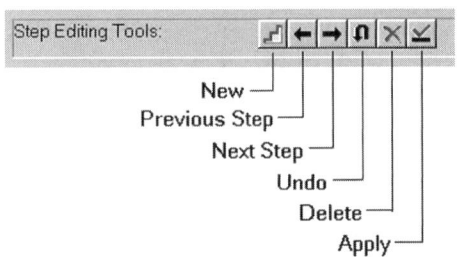

6. Click in the *Components to explode* list box and select the components to be included in the first explode step. Multiple components can be selected.

7. Specify a distance for the explode step.

8. Click on the Apply icon, shown in figure 12-100. Apply is one of the step-editing tools.

Fig. 12-100. The explode view step-editing tools.

9. Click on the New icon and repeat steps 4 through 8 until you have exploded the assembly to your liking.

10. Click on OK when finished.

SolidWorks occasionally does a pretty good job of exploding an assembly automatically, and an exploded view can be edited. All the same, it is usually more fun just to go ahead and do the job yourself. The exploded view usually looks much better when manually created.

✓ **TIP:** *If Auto Explode was used and you do not like the outcome, click on the Undo icon on the Standard toolbar (not the step-editing tools) and try exploding the assembly manually instead.*

During step 4, if an edge is selected to indicate the explode direction, the explode step will be in the direction of the edge. If a plane is used, the explode direction will be perpendicular to the plane. If a cylindrical face is used, the explode direction will be in the direction of the cylinder's axis. Use whatever geometry is most convenient at the time.

After clicking on the Apply icon, you will see your explode step take effect. The components will explode according to the explode direction and distance specified. If you are not pleased with the results, you can change the distance or direction by modifying those parameters as you see fit. You can also change the *Direction to explode along* option by deleting the selection from the area of the same name (select the item and press the Delete key), and then selecting a new direction vector.

If you are unhappy with the results of the last explode step, you can click on the Undo icon and try again. Typically you will have more than one step, in which case the Previous Step or Next Step icon will cycle you through the list of explode steps. The Delete icon will allow you to delete the current explode step. You do not want to click on the OK button until you are completely done exploding the assembly. If you do, you will exit from the Assembly Exploder window.

If at any time during the process of exploding the assembly you select a subassembly component as one of the objects to explode, you will have a choice of exploding either the selected component or the entire subassembly. These options are shown in figure 12-99. Note the options for *Entire sub-assembly* and *Component part only*. Choose whatever best suits your needs at the time.

The option *Explode related components together* (also visible in figure 12-99) will keep components together if they have been mated coincident. If you check this option, you will see all of the related components listed in the *Components to explode* area after the explode step has been applied.

Editing an Exploded View

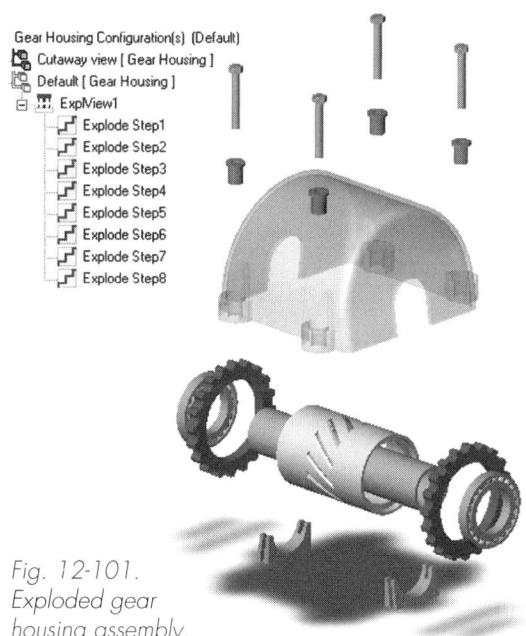

Fig. 12-101. Exploded gear housing assembly.

Exploded views are stored in ConfigurationManager. Figure 12-101 shows the gear housing assembly after creating eight explode steps. Inset is what the exploded view appears like in ConfigurationManager.

The name of the exploded view is *ExplView1*. Each of the exploded steps is then listed below the name of the view. It is possible to edit the definition of the *ExplView1* object by right-clicking on that object in ConfigurationManager. For that matter, right-clicking on any of the explode steps will also allow you to edit the exploded view's definition. Editing the definition of the exploded view allows for modifying any aspect of the exploded view. If it is strictly explode distances that need to be tweaked, there is a much more direct way of making changes.

If any of the explode steps listed in ConfigurationManager are selected, a green dashed line and a small green triangle attached to the dashed line will appear. The green triangle acts as a handle that can be dragged. By placing the cursor over the handle and holding down the left mouse button, you can move the mouse and dynamically drag the part or parts exploded in that step.

Expanding and Collapsing an Exploded View

Collapsing and expanding an exploded view is a simple process and can be accomplished with the right mouse button. Right-click on the *ExplView* object in ConfigurationManager and select Explode to explode (expand) the assembly. Once the assembly has been exploded, right-clicking on the same *ExplView* object will provide access to the Collapse option. These menu commands act as a simple toggle switch between the exploded and collapsed state.

When collapsing an exploded view, it is not actually necessary to right-click on the *ExplView* object. Right-clicking anyplace in the work area will provide access to the Collapse option. This is because when collapsing an exploded view you do not have to be specific. However, when you explode an exploded view it is necessary to specify exactly which view should be exploded, since there can be more than one exploded view in an assembly.

Copying Exploded Views

There can be only one exploded view per configuration, but because it is possible to have as many configurations as you want, the number of exploded views is unlimited. Well, at least theoretically. The upper limit of the number of configurations that can exist in a part or assembly is so high that it would never be reached.

If another configuration exists and another exploded view is required, there is a way to copy one exploded view to another configuration. The duplicate exploded view can then be altered if necessary. The routine for copying an exploded view to another configuration is very particular; it must be carried out just right. How-To 12-22 takes you through the process.

How-To 12-22: Copying an Exploded View

To copy an exploded view to another configuration, perform the following steps.

1. Activate the configuration that contains the exploded view to be copied.

2. Expand the configuration so that the *ExplView* object can be seen.

3. Hold down the Ctrl key, and with the left mouse button, drag the *ExplView* object to the configuration the exploded view is to be copied to.

Exploded Views 571

4. Activate the configuration containing the copied exploded view. Only then will the *ExplView* object be shown in the new configuration.

5. Right-click on the new *ExplView* object and select Edit Definition. Make any required changes.

☛ **NOTE:** *It may be necessary to edit the definition of the newly copied* ExplView *object before it can be exploded.*

If the configuration the exploded view was copied into is quite a bit different than the original exploded view configuration, it will be necessary to either add or delete some explode steps. This is due to the configuration possibly having either more or less components available for inclusion in the exploded view.

As a final word, an assembly cannot be edited while in an exploded state. SolidWorks will allow you to make modifications to the assembly, but as soon as you attempt to make any alterations to the assembly it will be forced to collapse. Because of this, it is best to collapse the assembly first manually, just to be on the safe side.

Exploded Views in Drawings

Exploded views are typically shown with the assembly in some sort of isometric viewing angle. This works best for showing the assembly components in their exploded state. Additionally, a named view could be created that shows the exploded assembly in its best light. If you need to refresh your memory on how to save a named view, see Chapter 3.

You do, of course, have to create the exploded view in the assembly before you can show it in the design drawing. The following How-To assumes you have done this, and assumes you have inserted a named view into the drawing. How-To 12-23 takes you through the process of showing an exploded view in a drawing.

How-To 12-23: Showing an Exploded View in a Drawing

To show an exploded view in a drawing, perform the following steps.

1. Right-click on the drawing view you wish to show in an exploded state.

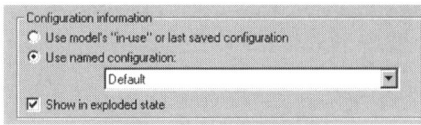

Fig. 12-102. Configuration information section of the Drawing View Properties window.

2. Select Properties. The Drawing View Properties window will appear.

3. In the Configuration information section, shown in figure 12-102, specify the configuration that contains the exploded view. This is only necessary if there is more than one configuration present.

4. Place a check in the option titled *Show in exploded state*.

5. Click on OK.

SolidWorks will show the view using the exploded state for that particular view. You may need to modify the scale so that the view will fit on the sheet, which you learned how to do in Chapter 11.

It might be worth noting that once an assembly view has been exploded its view border gets very large. Try shrinking it as much as possible by selecting the view and then dragging the border's green handles inward toward the model. The border can be shrunk only so much. The reason for wanting to shrink the border has to do with being able to access the sheet's context-sensitive menu. It is a common problem to right-click in a seemingly blank area of the sheet to access the sheet's menu, but instead the menu for the exploded view appears. This is not an issue as long as you make sure the cursor is outside the view's border before attempting to access the context-sensitive menu for the sheet.

Balloons

A BOM is quite often accompanied by balloons. The balloons usually incorporate leaders that point to the various components or subassemblies in the assembly layout. This correlates the items in the BOM with those items shown in the drawing.

Working with Balloons

Balloons can be customized to meet most requirements. This is done through the Document Properties tab of the Options window (Tools > Options). Figure 12-103 shows the Balloons section of Document Properties.

Balloons

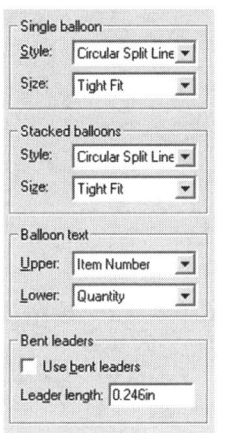

Fig. 12-103. Balloons section of Document Properties.

The Balloons section of the Document Properties tab has many ways of customizing how you want the balloons to appear. You can specify the style and size of balloons for both single and stacked balloons independently. The section marked *Balloon text* allows for customizing what appears inside the BOM balloons. For example, if a circular split line style is used, both the upper and lower areas of the balloons can be specified. Figure 12-104 shows an example of circular split line balloons. In this case, the bolt is item number 5, and there are four bolts in the assembly.

Balloons can be customized to fit your company's standards. For the example that will be used in this book, the top half of the balloon will contain the item number and the bottom will contain the quantity. The size of the text used by the balloons is controlled by the size of the note font. This you already know how to change. The size of the balloons can also be controlled. Tight fit is the setting used for the examples shown in this book.

When adding balloons, the order in which you place the balloons is not important. SolidWorks will number the balloons correctly, corresponding with the item's item number as designated in the BOM. For that matter, balloons can be added prior to inserting a BOM, and SolidWorks will still get it right. How-To 12-24 takes you through the process of adding balloons.

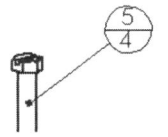

Fig. 12-104. A circular split line balloon.

How-To 12-24: Adding Balloons

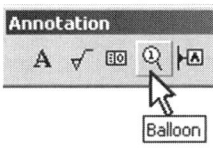

Fig. 12-105. Balloon icon.

To add balloons, perform the following steps.

1. Select Insert > Annotations > Balloon, or click on the Balloon icon (shown in figure 12-105), located on the Annotation toolbar.

2. Select an item in the drawing on which to place the balloon. Add as many balloons as necessary.

3. Exit the Balloon command when finished.

4. Position the balloons as necessary by placing the cursor over a balloon and dragging it to a new location.

A sample exploded assembly view with balloons is shown in figure 12-106. It should be noted that if balloons are added to an assembly drawing that does not contain a BOM, the balloons will still number according to a certain scheme. This scheme is dependent on the order in which parts were added to the original assembly file. The first part inserted into the assembly will be balloon 1, the second part balloon 2, and so on.

✓ **TIP:** *Reattaching a balloons leader to another component will make the balloon's contents automatically update.*

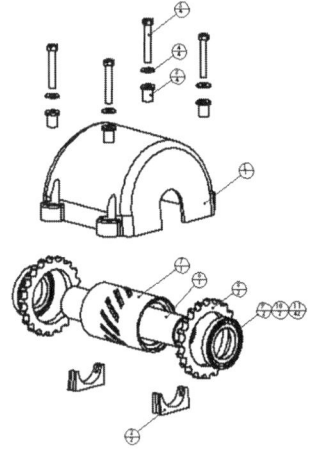

Fig. 12-106. Exploded view with balloons.

Balloons and Component Names

A common request among SolidWorks users is the ability to place the name of a component inside a balloon on a drawing. You have actually learned what is necessary to accomplish this task, which involves linking to the properties of a component.

Assuming you are working in a drawing containing at least one view of an assembly, begin by adding a note. It is necessary to add a note and not an actual balloon mainly due to the ease in which the component name can be linked to a note. Format the border of the note if you wish. Next, add the note as usual (you learned all of this in Chapter 11).

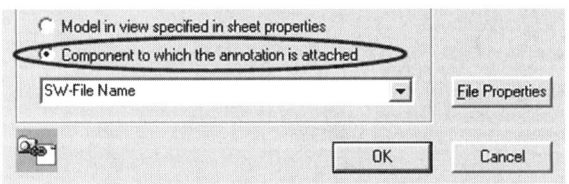

Fig. 12-107. Portion of the Link to Property window.

For the text of the note, use the Link to Property icon found in the Text Format panel of the Note PropertyManager. The Link to Property window, a portion of which is shown in figure 12-107, will appear. Select the *Component to which the annotation is attached* option and then pick the property named *SW-File Name* from the drop-down list.

Once completed, you should have a balloon style note in the drawing that points to a particular component. The text will indicate the file name for whatever component the leader is attached to. For the example used in figure 12-108, a five-sided flag was used as the border for the note.

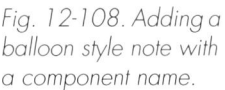

Fig. 12-108. Adding a balloon style note with a component name.

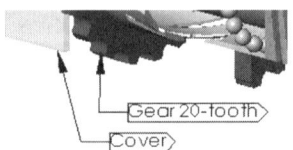

✓ **TIP:** *Component balloon notes can be copied and pasted. If a leader from a copied note is reattached to another component, the text will update with the correct component name automatically.*

Summary

Looking back on this chapter, there is quite a bit of information that goes along with assemblies. The first aspect of dealing with assemblies is getting the components into the assembly. The most common method of accomplishing this is by using the Insert > Component > From File menu selection.

The first component brought into an assembly is fixed in location by default. Generally, this first component should be the main component other parts will be attached to. Any component in an assembly can be fixed in space or allowed to float freely in space. Use the right mouse button and click on the component to fix or float a component.

Components can be moved and rotated independently of other components in the assembly. Use the Move Component or Rotate Component icon to move or rotate individual components. Any part can be moved or rotated, as long as the part is not fully defined.

Add mates to place components together and build the assembly. Always try to assemble with motion in mind if your assembly will have moving parts. Remember that every component has six degrees of freedom of movement prior to being mated, as long as it is not fixed. These degrees of freedom are in reference to translation and rotation about the x, y, and z axes. Never add a mate to a part that is fixed, and vice versa, as this will overdefine the assembly.

Hide components by right-clicking on the component. You can also suppress components using the Edit menu or the Assembly toolbar. When a component is suppressed, it is removed from memory, and any mates associated with the component will be suppressed also.

Use interference checking to establish whether or not there is interference between selected components. Treating coincidence as interference means that the components will be considered interfering even if they are touching. If you have assembled components that can describe motion, use dynamic interference detection, otherwise known as collision detection, to check for interference. Faces of components that collide and interfere with each other will turn green when collision takes place.

Assembly features are features that affect the assembly only. Assembly features are limited to revolved or extruded cuts. Use the feature scope to specify which components will actually be affected by an assembly feature.

Patterns of assembly features can be created as well. Creating assembly feature patterns is very similar to creating part feature patterns, so the process should be familiar to you.

Components can be patterned in an assembly. These patterns can be locally defined patterns, which can be circular or linear, or derived patterns. A derived pattern is based on an existing feature pattern. If the feature pattern changes, so will the derived component pattern.

When creating assembly drawings, use the same techniques you learned for creating the various views of parts. Assembly section views are different only in the regard that you must specify what components you want to have excluded from being cut during the section, if any. This is known as the section scope.

To add a bill of material, make sure you select a view first. This is because drawing sheets can contain views from different parts or assemblies. When adding balloons, the item number in the bill of materials will automatically match the item the balloon is attached to on the view geometry.

Exploded views can be added to an assembly through the use of the Insert menu. Create exploded steps manually or let SolidWorks explode the assembly. Letting SolidWorks explode the assembly should be reserved for smaller assemblies. Only one exploded view can exist per configuration. An easy way to edit an exploded view is to select the explode step from ConfigurationManager and drag the green handle displayed in the work area. Show exploded assemblies in an assembly design drawing by accessing the drawing view's properties.

Questions and Topics for Discussion

1. What is different about the assembly FeatureManager versus the part FeatureManager?

2. What is significant about the first component inserted into an assembly?

3. If more than one of the same part is inserted into an assembly, how does SolidWorks keep track between the two different components?

4. Name five objects that can be used for adding mate relationships.

Questions and Topics for Discussion

5. Describe in your own words what is meant by alignment conditions.

6. Is it possible to use the Move Component icon once a component has been mated? Explain your answer.

7. What is an excellent technique used to discern which types of mating relationships have been, or may need to be, added to a component?

8. What is the function of the *Mates* folder?

9. When creating an assembly feature, how do you go about specifying which components should be affected by the assembly feature?

10. Describe two ways of suppressing a component in an assembly.

11. What is the difference between hiding and suppressing a component?

12. How is a derived component pattern different from a locally defined pattern?

13. Describe the sequence of events you would perform in order to carry out a single-component replacement.

14. Is it possible to conduct an interference check between more than one component at a time?

15. Describe two indicators that signify you are editing a part within the context of an assembly.

16. Describe a shortcut for opening a component when in an assembly containing that component.

17. Once in the Exploded View command, what would be the next series of actions for manually creating an exploded view step?

18. How would you collapse an exploded assembly? How would you expand one?

19. How would an exploded view be shown in a drawing? Describe the process in your own words.

20. Is it possible to make the name of a component appear in a balloon? Explain your answer.

CHAPTER 13

# Cavities, Cores, and Mold Making

BECAUSE IT IS POSSIBLE TO CREATE CAST PARTS IN SOLIDWORKS, it would also be convenient if molds could be created for those cast parts. Many companies do nothing but make precision molds for the use of creating cast parts. A shape is hollowed out of two or more pieces of material, and then typically ejector pins, gates, and a variety of other mold elements are added.

SolidWorks gives you the ability to create a mold using various methods. One mold making method works by subtracting material from another block of material, the block being what becomes the mold. This process requires a minimum of two components and must be carried out in an assembly. The function being described is known as the Cavity command. In this chapter you will learn the entire process of starting with a cast component and then design a mold from which to cast that component.

Another technique must be used when working with parts that contain more complex parting lines. Not all parts have a simple parting line, or even a parting line that is planar. In such a case, a more advanced technique requiring surface manipulation must be used. This is covered in its entirety as well.

The Cavity Command

To carry out the Cavity command, you must first fulfill a few requirements. First, you must have at least one component that will be representative of the shape of the mold's cavity. This is the component or components you will be casting. You are not limited to one component.

Second, an assembly must be started. In this new assembly you will be required to add the block or shape that will become the mold, and you will add the design component used to create the mold cavity. You will want to insert the mold block (called the die from this point on) into the assembly first. As you should recall from the last chapter, the first component inserted into an assembly is fixed. You will want the die fixed so that it will not move around and ruin the resultant mold.

Third, it is important to mate the design component inside the die. If the two components were not firmly mated, in theory it would be possible to accidentally move the design component to an undesirable location. If the design component were not properly positioned, the results could be disastrous. You might find the design component is actually cutting a hole in the side of the die and that molten material is leaking out of the mold.

Only when these three requirements are met should you attempt to run the Cavity routine. In the next few sections, you will see firsthand how these three requirements are met. You have learned how to start a new assembly, insert components, and mate components, so this should be nothing new to you. The pivot arm will be used as the design component, and a die will be created that will eventually become the mold.

Creating the Die

The approximate size of the design component is usually known before the die is created. In this case, the overall size of the pivot arm, shown in figure 13-1, is nearly 11.5 inches wide and 10.5 inches tall. The die's dimensions will be set to 14 inches by 12 inches. It is also known that the pivot arm is only 1.5 inches deep, so the die will be made a total of 4 inches thick.

Now that the dimensions of the die have been determined, a sketch can be created and the die can be built. Figure 13-2 shows the completed die. Because you already know how to create extruded parts of this basic nature, the details of how the die was modeled will not be elaborated upon.

The main two components needed for the Cavity routine are now completed. The next objective is to bring both the die base and the pivot arm into a new assembly file. Use a combination of coincident and distance mates to position the design component within the die base. (See Chapter 12 for assembly creation.) Figure 13-3 shows the completed assembly with the pivot arm correctly positioned within the die base. The die base is

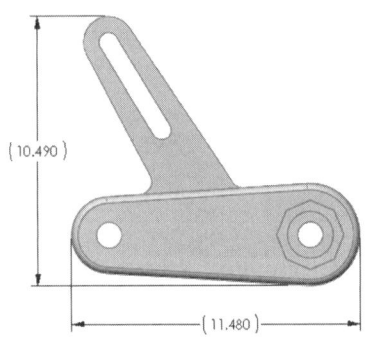

Fig. 13-1. Overall dimensions of the pivot arm.

Fig. 13-2. Die component.

The Cavity Command

shown transparent so that the design component can be seen.

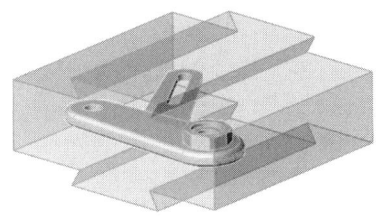

Fig. 13-3. Completed assembly.

The assembly has been created and we are just about ready to perform the Cavity command, but how can we be certain the finished cast parts will actually come out of the mold? To ensure the design component has the proper amount of draft, use the Draft Analysis tool, discussed in Chapter 6. Draft Analysis will tell if there is positive or negative draft on the model faces, and can also classify straddle faces. You could then rework the part as necessary prior to creating the mold.

Creating the Cavity

Take a moment to understand exactly what is happening when you perform the Cavity routine. To put it simply, the Cavity routine allows you to subtract the solid geometry of one or more components from another. In the case of creating the mold in the example this chapter is using, the pivot arm is essentially being subtracted from the die base.

If the die base is the component that will be undergoing modifications, you must first be editing that component. You learned how to edit an assembly component in Chapter 12. Here is a quick reminder as to how this editing is done. Right-click on the component you want to edit (in this case, the die base), and then select Edit Part. You would then be placed into edit mode for that particular component.

How-To 13-1 takes you through the process of creating a cavity using the Cavity command. Remember that you must be editing the component you want to create the cavity in. Only then will the Cavity command be available to you.

How-To 13-1: Using the Cavity Command

To create a cavity, perform the following steps. Be aware that the Cavity command can be used to subtract the volume of one or more components from another. It does not necessarily have to be used strictly for the sake of creating a cavity.

1. Right-click on the component that will contain the cavity, and select Edit Part.

2. Select Insert > Features > Cavity. The Cavity window will appear, as shown in figure 13-4.

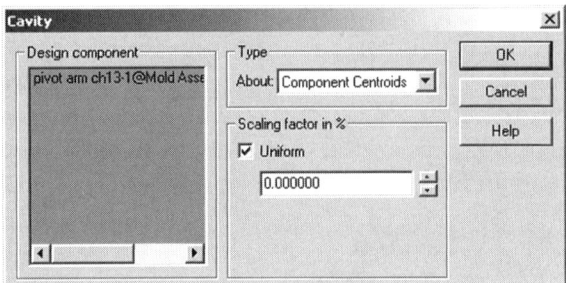

Fig. 13-4. Cavity window.

3. Specify the Design component with which to create the cavity. More than one component can be selected.
4. Specify the Scaling Type, if desired.
5. Specify the Scaling Factor as a percentage (%). For no scaling, specify zero (0) as the scaling factor.
6. Click on OK to create the cavity.
7. To return to editing the assembly, right-click on the name of the assembly in FeatureManager and select Edit Assembly, or click on the Edit Part icon.

The Scaling Type option you noticed when running through the Cavity routine can be specified a number of ways. The Scaling About field can be set to Component Centroids, Component Origins, Mold Base Origin, or Coordinate System. Different parts and materials cool differently, so use your best judgment. If the Coordinate System option is used, a user-defined coordinate system must be selected. Coordinate system creation was discussed in Chapter 7.

The scale factor is determined by specifying a percentage. If you know that the cast component created with the mold will be shrinking by 2%, you will want to increase the scale factor of the cavity by 2% to account for the shrinkage.

You have a total leeway of plus or minus 50%. These are the maximum and minimum values you can specify as a scale factor. It is also possible to perform a nonlinear (nonuniform) scale. If the Uniform option is unchecked, the scale factor for all three axes can be determined.

Now that you have formed the cavity, it is time to create the mold. You would want to take the die base component, at this point, and cut it into two pieces, thereby creating the top and bottom halves of the mold. This process can be accomplished via one of two methods, both of which are covered in the following section.

Inserting Parts

There are several ways to create the separate sections of the mold. One is to create copies of the die base that can be cut in half, and another would be to use the Split command. First let's look at the method that involves copying the die base part.

Entire part files can be inserted as bodies into other part files. They can then be used as features (often, base features) in the new part file to aid in the development of a new part. A body created in this fashion is dependent on the original part for its dimensions. For this reason, the new body will have no dimensions of its own. To modify the newly inserted body, the original part file must be modified.

Inserting a part will be convenient in the application of creating the mold due to the inherent file associativity common to inserted parts. If the die base is used as the basis for both the top and bottom halves of the mold, any change made to the original die base will affect both halves of the mold.

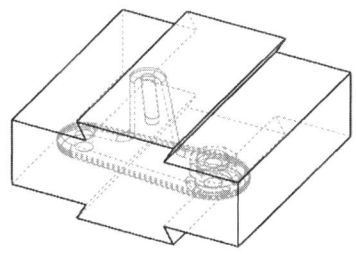

When inserting a part, you have the opportunity to position the part at a particular location. Rotating the inserted part is also possible. The inserted part that will be used in this next example is the die base component modified with the Cavity routine earlier in this chapter. The die base part is shown in figure 13-5.

It should be noted that the image in figure 13-5 is not the mold assembly. It is the die base part. The outline of the pivot arm you see within the die base is the cavity. This will become apparent when the mold halves are created. Before that can be done, a new part will be started and the die base will be inserted into the new part. How-To 13-2 takes you through the process of inserting a part into another part file.

Fig. 13-5. Die base part file.

How-To 13-2: Inserting a Part

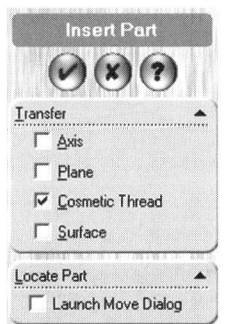

To insert a part into an existing part file, perform the following steps.

1. Begin a new part.
2. Select Insert > Part.
3. In the Open window, select the part to be inserted.
4. Click on Open. The Insert Part panel will appear, shown in figure 13-6.
5. Specify any additional geometry that should be included with the part file you are inserting.

Fig. 13-6. Insert Part panel.

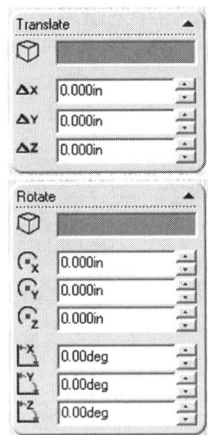

Fig. 13-7. Locate Part panel.

6. Check the Launch Move Dialog option if you wish to translate or rotate the part you are inserting.

7. If the Launch Move Dialog option was checked, the Locate Part panel will appear, shown in figure 13-7. Locate the part as required. Use the preview available in the work area as a guide.

8. Click on OK to insert the part.

With regard to the previous steps, step 7 will not be necessary if the Launch Move Dialog option was not checked. It is common not to check this option if inserting a part as a base feature. If other features already exist in the part file, you will almost certainly find it necessary to position the inserted part at a specific location.

Both the Translate and Rotate panels will not be open simultaneously, as in figure 13-7, and you will only be able to access one or the other at a given time. However, the inserted part can continue to be repositioned using the Move/Copy command found in the Insert > Features menu, or in the Insert > Surface menu. In either case, the panel displayed is identical, and very similar to the panel displayed when inserting a part.

When translating, geometry can be selected to help position the inserted part. For example, selecting two vertices will position the inserted component along the vector established by the two vertices. When rotating, the first three settings dictate the x-y-z point (with reference to the origin point) the part will rotate about. The second set of three settings determines the rotation amount about each axis (see figure 13-7).

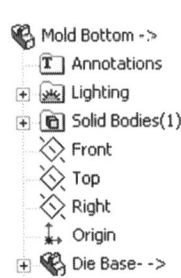

Fig. 13-8. FeatureManager after inserting a part.

Once the part (the die base, in this case) is inserted, an exact duplicate of the original part will appear on screen. The new part, which we will call mold bottom, contains an external reference back to the original die base. It is very easy to tell whether or not a component or feature has external references. If you look at FeatureManager, you will see a small arrow symbol (->) following any object that contains external references. This is depicted in figure 13-8, which shows FeatureManager for the mold bottom part. Note that the icon of the new feature is identical to that of a part, and that the new feature takes on the name of the part it references.

The newly inserted part is now a solid body in the mold bottom part. If additional parts were inserted, they would be separate solid bodies. You would not want to build an "assembly" in this way, as it would not be a true assembly and would contain none of the functionality inherent in SolidWorks assemblies.

There are no dimensions associated with the new body, which for all intents and purposes can be considered a base feature. Double clicking on the base feature causes no dimensions to appear. The base feature does not even contain a sketch that can be accessed. This is because the base feature is externally referencing the die base part.

Modifying the die base will update the mold bottom part. In our case, that is what we wish to have happen. However, if you neglected to remember this fact, you could wreak havoc on any components referencing the die base. Document management software can help remedy this situation, and there are plenty of choices on the market today that integrate with SolidWorks.

Referencing Configurations

Chapter 7 discussed external references in a fair amount of detail. In Chapter 7, you discovered that external references were created as a result of the Mirror Part command. Mirrored parts and inserted parts are extremely similar with regard to their external references. It is possible to reference a particular configuration when using either of these commands. In that you are now familiar with configurations and what they are, the topic of referencing configurations can be discussed.

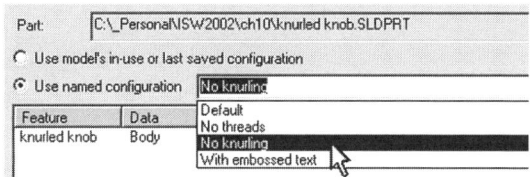

Fig. 13-9. Specifying which configuration to reference.

When a file being referenced has more than one configuration, the software allows for specifying which configuration to use. This is found in the External References window, shown in figure 13-9. As a reminder, this window can be opened by accessing the List External Refs menu item after right-clicking on a feature with an external reference.

Implementing this function is easy enough. Just select from the dropdown list the configuration you want to use. If the option *Use model's in-use or last saved configuration* is specified, the part with the external reference may change unexpectedly. This would almost always be an undesirable situation. Typically it is best to lock in a particular configuration. Whatever configuration is being referenced will always appear in parentheses following the external reference symbol visible in FeatureManager.

Creating the Mold

Before going off on the tangent of external references, we were getting ready to create the bottom half of the mold. You learned how to insert a part and now must finish what was started by cutting away half of the die. To do this, you will sketch on the front face of the die.

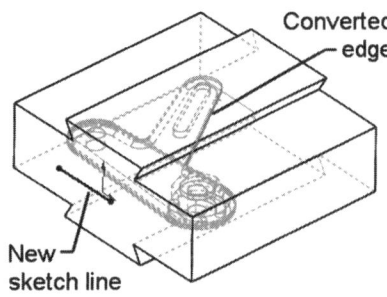

Fig. 13-10. Preparing to cut the die.

The sketch you create next will determine where the die gets cut. For this reason, it is very important that you use the Convert Entities icon and select an existing edge from the cavity area that will be representative of the parting line. This way, if the original pivot arm is modified and the parting line changes position, the line used to cut the die will move accordingly. Figure 13-10 helps illustrate which model edge was used in this case, and the resultant line to be used to cut the die to obtain the mold halves.

Before the new sketch line can be used to cut through the part, make sure it stretches completely across the part. Drag the edges of the line so that it stretches up to or past the edges of the die, as shown in figure 13-11. Ideally, if you terminate the endpoints of the sketch line at the edges of the die, you will not have to worry about the sketch line being too short if the die is increased in size. This is because the sketch line would be coincident with the edges of the die and would increase or decrease in length to correspond with the die. Now you are ready to create the cut.

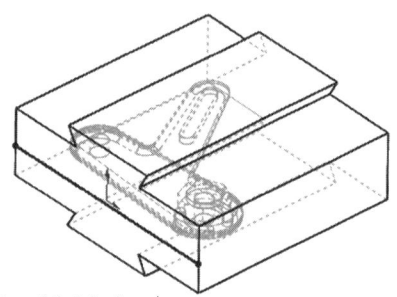

Fig. 13-11. Final modification to the sketch line.

Open-profile Cutting

When cutting with an open profile, your options are limited to using the Through All end condition only. A good analogy is using a handsaw to cut through a board. If you cut through the board only half way, the board remains as one piece. SolidWorks would not understand what you are trying to accomplish with such a cut. Therefore, you must cut completely through the entire board, or in the case of SolidWorks, through the entire part.

The mechanics of creating a cut with an open profile versus using a closed profile are exactly the same. In that basic extruded features (whether they are bosses or cuts) were covered in earlier chapters, the details will not be reiterated here. We can, however, observe some of the particulars of open-profile cutting. Note the arrows in figure 13-12, which appear when creating a cut. (The image has been edited for clarity and for reasons of size.)

Derived Parts 587

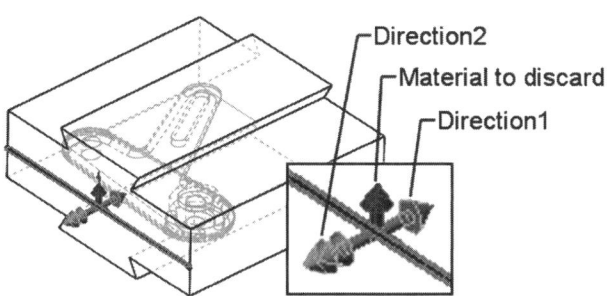

Fig. 13-12. Extruded cut preview arrows.

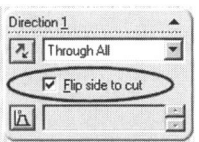

Fig. 13-13. Flip side to cut *option.*

Fig. 13-14. Completed mold bottom part.

Any basic extrusion will show the standard single and double arrows representative of the *Direction1* and *Direction2* end conditions. When cutting with an open profile, there is a third arrow. This third arrow points toward what is being thrown away. Clicking on the *Flip side to cut* option, or clicking on the third arrow itself, will change what side gets thrown away during the cut operation. The *Flip side to cut* option is shown in figure 13-13.

The completed mold bottom part is shown in figure 13-14. Obviously, it would be necessary to repeat the process of starting a new part and inserting the die base part a second time to create the other side (top half) of the mold. For the mold top part, as we will call it, a slightly different process will be used, but one with a result very similar to that of the Insert > Part routine discussed in this section.

Derived Parts

Using the Derive Component Part command is essentially another means of accomplishing the same thing as inserting a base part. The result is almost the same, but with one difference. This difference has to do with assembly features propagating to the new parts, discussed in material to follow.

The process of creating a new part via the Derive Component Part command takes place from within an assembly. In contrast, inserting a base part has to be performed while editing a part. The Derive Component Part command is easier to use than the base part method because there are fewer steps involved. How-To 13-3 takes you through the process of creating a new part using the Derive Component Part command. Then we can explore the result a little more closely.

HOW-TO 13-3: Derived Component Parts

To create a new part derived from an existing component within an assembly, perform the following steps. Be aware that you must be editing an assembly in order to complete this command.

1. Select the component to be used to create a new part.
2. Select Derive Component Part from the File menu.
3. You may be prompted to select a template. If so, select the part template you wish to use and click on OK.

SolidWorks does the rest for you automatically. In many cases, using the Derive Component Part command is the same as inserting a part into another part file. The new part will have external references to the component it was derived from, and the base feature will inherit the name of the part it is referencing. Any changes made to the original file will propagate to the new part with the external reference.

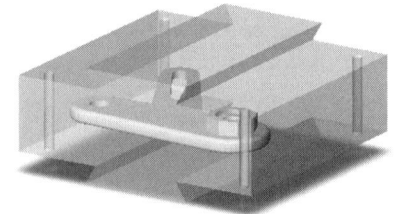

Fig. 13-15. An assembly feature has been added (four holes).

So what exactly is different between inserting a part and using the derived component method? As was hinted at earlier, the difference has to do with assembly features. If we imagine for the sake of argument that four holes have been drilled through the die base component in the assembly, the results might look something like that shown in figure 13-15.

Keep in mind that the holes in the die base are part of an assembly feature, so the die base part itself has not been affected (if you need to review, see "Assembly Features" in Chapter 12). When inserting a part, it is the original part file that is being externally referenced by the new part. Therefore, the new part will not contain any features added at the assembly level.

When using the Derive Component Part command, the key word to keep in mind is component. The new part will reference the component in the assembly and not the original part file. Therefore, the new part will inherit any features created at the assembly level. In other words, the new part is derived from (and externally references) the component, not the part.

If we use the die base component from which to derive a new part (namely, the mold top part), and then cut the mold top part using the same technique used on mold bottom, the results shown in figure 13-16 would be obtained. Note in particular the holes at the four corners of the new part. Additionally, there is an inset that displays what the base feature will look like when Derive Component Part is used.

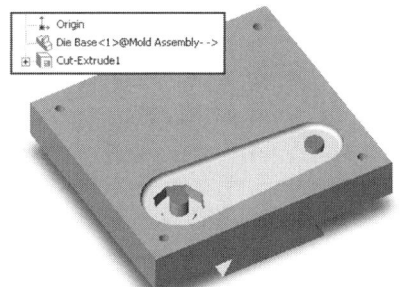

If no assembly features exist or will ever be created in an assembly, there is no reason not to take the easy way out and make use of the Derive Component Part command. If there are assembly features, you will have to make the decision as to whether or not those features should propagate to the new parts that will be created.

Fig. 13-16. A derived component part.

Splitting Parts

Another tool SolidWorks users have at their disposal is a command aptly named Split. It allows for taking a part and breaking it into separate pieces. The Split command can often make fast work of building a mold, but it can be used for other situations as well.

The Split command goes beyond simply breaking a part into smaller chunks. It can also then take those chunks and automatically build an assembly out of them, placing every piece into its proper position. You could then continue to develop the assembly or the individual components from that point forward, adding ejector pins and various other mold elements.

Using the die base component as an example, we will split the die base into two separate pieces and then have SolidWorks build an assembly out of those pieces. How-To 13-4 takes you through the process of carrying out the Split command.

How-To 13-4: Using the Split Command

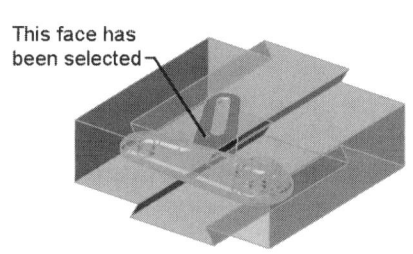

Fig. 13-17. Face selected as the trim tool.

To split a part into multiple pieces, perform the following steps. Note that the process of splitting a part could be performed within an assembly, as long as the component being split was being edited. For this example, the part will be opened in its own window as a standalone part.

1. Select Insert > Features > Split.

2. Select the face or faces to be used as the trim tool. In other words, the faces that will determine where the part is split. Figure 13-17 shows the face used in the example.

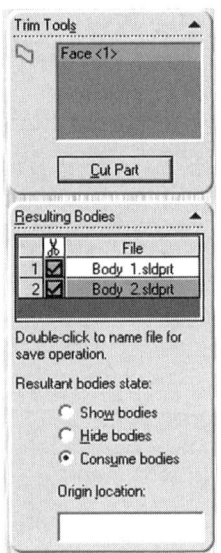

Fig. 13-18. Split command panel.

3. Click on the Cut Part button on the Split panel, shown in figure 13-18.

4. Once SolidWorks has cut (or "split") the part, you must give the individual pieces names. Double click on the newly created parts in the Resultant Bodies panel to name the files (see figure 13-18), or click on the callouts displayed in the work area.

5. Specify what should happen with the resultant pieces. They can be shown, hidden, or consumed (removed from the original file).

6. Click on OK when finished naming all of the newly created parts.

When finished, a new feature (named *Split1*) will appear in FeatureManager. If you chose the Consume option in step 5, the model itself has disappeared from view. This is what is meant to happen. Whatever pieces are split from the model become separate part files, leaving behind whatever is left. If there is nothing left, such as in our case, there is nothing to display.

If the Hide or Show options are used, there will be multiple bodies present in the part file. Each portion split from the original will be a separate body. You should have a separate body listed in the *Solid Bodies* folder in FeatureManager for each chunk split from the original part, as well as one solid body if there was anything remaining. Individual bodies can be hidden or shown by right-clicking on the body in the *Solid Bodies* folder and selecting Hide or Show Solid Bodies.

Although it would be perfectly acceptable to take the resultant bodies created from the Split command and modify them independently, it would probably be best to bring the new parts into an assembly and finish things up there. SolidWorks will create the assembly for you. Because the assembly creation process is really just an extension of the Split command, let's continue from the previous steps to complete the overall process.

7. Right-click on the Split feature in FeatureManager and select *Create an assembly*.

8. In the Split Assembly panel, shown in figure 13-19, click on the Browse button.

9. Type in a name for the assembly.

10. Click on Save.

Creating a Core and Cavity

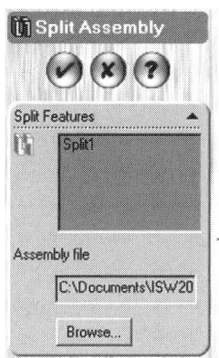

Fig. 13-19. Split Assembly panel.

11. Click on OK to complete the process.

At this point, a new assembly will be built with the components in the correct positions. The assembly will actually look like the original part before it was split. From here, it would be possible to add assembly features, create an exploded view, generate drawings, or whatever else needs to be done.

As an added note, when SolidWorks builds the assembly, it really is not quite smart enough to mate all components. What happens instead is that each component is fixed in position. You can change this if you deem it necessary, by floating all components except one, and then adding the appropriate mates.

✏ **NOTE:** *When using the Split command as a follow-up to the Cavity command, you may find that the part that was split disappears from the original assembly, or at the very least causes mate errors to arise. To prevent the mate errors, mate to the planes of the part being split, rather than to faces.*

Creating a Core and Cavity

SolidWorks has some very powerful tools that are designed to make a mold maker's job much easier. Creating a core and cavity, for instance, can be accomplished by utilizing surfacing commands well suited for the task. The entire process of creating a mold with a core can be a little intimidating if you have not been through it a few times. This section examines the entire process, with step-by-step instructions that will get you up to speed in no time.

There are a number of steps that must be performed to create a core and cavity. The cavity being created in this section will be created differently than the one created earlier with the Cavity command. Here, the actual Cavity command will not be used. Instead, an alternative process will be employed. Summarized, the steps for the core-and-cavity process are as follows.

1. Scale the part that will be used to create the mold cavity.

2. Create any shutoff surfaces necessary, so that no holes are present in the part.

3. Radiate a surface outward from the parting line.

4. Knit all surfaces, including the radiated surface, to create a single surface.

5. Extrude up to the knit surface to create the cavity or core component.

The last two steps must be performed in an assembly because there will be more than one part involved. In addition, the knit surface must belong to the part that contains the sketch that will be extruded up to the knit surface. This is another reason for using an assembly to accomplish this feat. If all of this does not make sense at this point in time, do not worry, because it will as you read further. Read on to understand the entire process, which begins with scaling the part.

Scaling

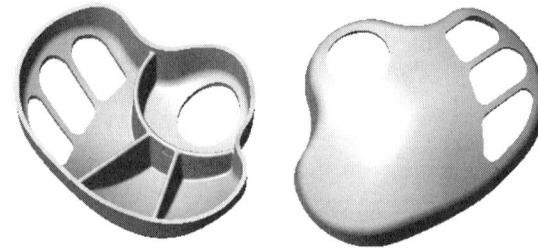

Fig. 13-20. A core and cavity will be created for this model.

There is no built-in scaling command in the process you will be using to create the core and cavity, as there was in the process of using the Cavity command, which you learned earlier in this chapter. For that reason, the part must be scaled first. Figure 13-20 shows the part that will be used throughout this section. It is a free-form shape and is the top cover for a trackball three-button mouse.

In addition to the challenge the free-form shape of the mouse cover will present, there are also ribs on the underside of the part. Pay attention to the open areas for the mouse buttons and trackball. These areas will pose an interesting problem when creating the core and cavity.

Let's move on to the business of scaling the part. The command is named Scale, appropriately enough. Scaling does not have to be uniform, but can be differing values in each of the three axes. How-To 13-5 takes you through the process of scaling a part.

How-To 13-5: Scaling a Part

To scale a part, perform the following steps. The Scale icon, typically located on the Mold Tools toolbar, is shown in figure 13-21.

1. Select Insert > Features > Scale, or click on the Scale icon. The Scale panel, shown in figure 13-22, will appear.

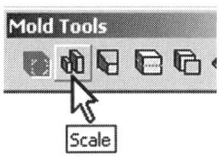

Fig. 13-21. The Scale icon on the Mold Tools toolbar.

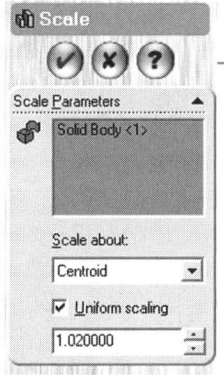

Fig. 13-22. Scale command panel.

2. If there are multiple bodies in the model, select which bodies to scale. Surfaces can be selected as well.

3. Specify a scale factor. If a nonuniform scale, uncheck the *Uniform scaling* option and specify the scale factor values for the *x*, *y*, and *z* axes.

4. Specify the scaling origin. The part's centroid is used in our example.

5. Click on OK to scale the part.

If there are not multiple bodies or other surfaces in the part file, there will be no need to select anything when performing the Scale command. As a matter of fact, the list box area shown in figure 13-22 will not even be present. Keep in mind that the scale value is a factor, not a percent value, such as that present in the Cavity command. Therefore, if the final part will have a shrink factor of 2%, the value to type in for the scale factor would be 1.02. Now that the easy part is out of the way, we can move on to the next stage in the cavity/core process.

Creating Shutoff Surfaces

Depending on the part, shutoff surfaces may need to be created in a variety of ways. Shutoffs are only needed when the model has holes in it. Why would shutoff surfaces even be necessary? The reason is that if they were not used, the mold would be one piece as opposed to two. The mold would not be able to break in half because each side of the mold would contact the other through the holes in the model. There needs to be solid geometry, or at least surface geometry, that will physically divide the core from the cavity.

One common method of creating shutoff surfaces is to use the *Fill surface* command. Lofted or swept surfaces can also be used. In extremely simple cases, extruded or planar surfaces can be used, but rarely is anything that simple.

There are a number of surface commands, and their command panels are nearly identical to those used to create their counterpart solid features. For instance, selecting Insert > Surface > Extrude opens a panel with parameters found in the panel displayed when selecting Insert > Boss > Extrude. The only real difference is the result. Instead of winding up with solid geometry, you wind up with a surface. Surfaces are different from solids because they have no thickness.

594 Chapter 13: Cavities, Cores, and Mold Making

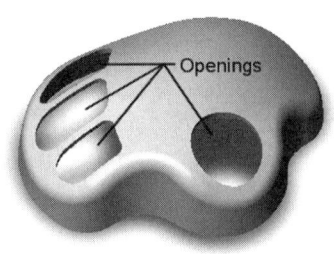

Fig. 13-23. Openings that need to be closed.

If you look at any of the areas requiring a shutoff on our sample part, you can see the obstacles. Figure 13-23 shows the openings that require shutoff surfaces. There are multiple edges that make up each hole, and none of the holes can be closed with anything as simple as a planar face. What would be ideal is if somehow a surface could be used to "patch" the hole while remaining tangent to all surrounding faces. This is exactly what the Fill command will accomplish.

When employing the Fill command, all of the edges surrounding the opening should be selected. This can be accomplished quite easily by using the Select Tangency or Select Loop command, discussed in Chapter 6. You are then given the ability to specify whether the new surface to be created should be tangent to these edges, or simply contact them. Mixing things up is allowed, so some edges can be contact edges and some can be tangent. In our case, we would want all tangent conditions. Let's put the Fill command to the test in How-To 13-6.

HOW-TO 13-6: Using the Surface Fill Command

Fig. 13.24. Flat Surface icon.

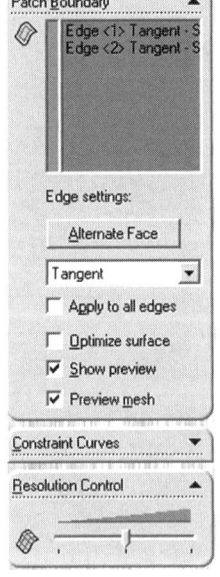

Fig. 13-25. Surface Fill panel.

To fill in gaps in surface or solid geometry, use the Fill command as follows. In worst-case scenarios, when the Fill command does not work, you may have to use swept or lofted surfaces to fill gaps in geometry. Look to Chapter 20 for additional surface commands.

1. Select Insert > Surface > Fill, or select the Filled Surface icon (shown in figure 13-24), located on the Surface toolbar.

2. From the drop-down menu in the Surface Fill panel, shown in figure 13-25, select either Tangent or Contact as the default setting for the edges to be selected in the next step. Edge settings can be changed on an edge-by-edge basis later, if required.

3. Select the edges that surround the area to be filled in.

4. If a Tangent edge setting is used, the new surface will be tangent to a face on one side of a selected edge or the other. If the wrong face is highlighted for a particular edge, it needs to be changed. To most easily accomplish this task, as needed, click

on the various edges in the Patch Boundary listing, and then click on the Alternate Face button.

5. Using the same technique as in step 4, additionally change the edge settings to either Contact or Tangent for specific edges as required.

6. Click on OK to complete the process.

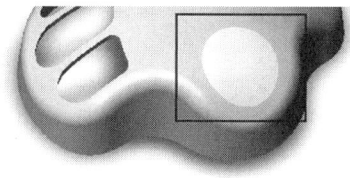

Figure 13-26 shows the trackball cover after one of the shutoffs has been created. The newly created surface is tangent to all of the surrounding faces. The surface is smooth, did not require one bit of sketch geometry, and was a breeze to create. But for the Fill command, the shutoff would have not been nearly as easy to create.

Fig. 13-26. After adding a shutoff.

When using the Fill command, there is an optional setting named Resolution Control. This option is in the form of a slider bar whose default value is in the center. Moving the slider to the right will often result in a more refined surface. The surface will typically be smoother, but this comes with a price. The surface will be more complex, which manifests itself in larger file size and longer rebuild times. Alternatively, the slider can be moved to the left. If irregularities in the surface occur, you may then decide to try a higher setting.

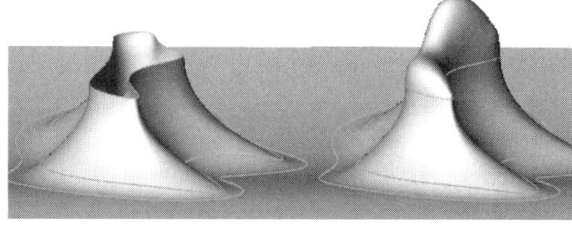

If you happen to spot the Reverse Surface button (not shown) sometimes present in the Surface Fill panel, it is because conditions are just right. For example, when using tangent edge settings to fill the top of the lofted surface shown in figure 13-27, the Reverse Surface button makes its appearance. Clicking on Reverse Surface has the effect seen in the illustration. It is apparent that the fill surface is flipping its tangency condition with the surrounding surfaces.

Fig. 13-27. The effects of the Reverse Surface button.

The *Show preview* and *Preview mesh* options are there to help you see what the resultant surface will look like. The *Optimize surface* option creates a simplified surface that should result in faster rebuild times. It would be beneficial to check this option whenever possible. If the resultant surface does not meet your demands, uncheck *Optimize surface* and use the

596   Chapter 13: Cavities, Cores, and Mold Making

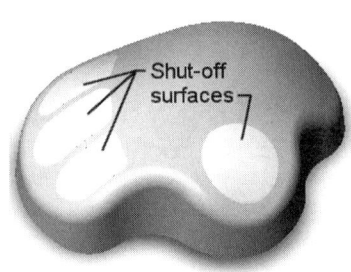

Fig. 13-28. All shutoffs have been completed.

Resolution Control slider bar instead. The trackball cover with all of the openings filled is shown in figure 13-28.

Radiated Surfaces

Eventually there will come a time when a sketch will need to be extruded to create both the core and cavity. There will actually be two parts, of course, and two separate extrusions will need to be performed. When this occurs, the idea is to have a surface to extrude up to. Imagine an object sitting on a table top. Take a large piece of putty and push it down over the object. The putty takes on the shape of the object. That is essentially what will be happening.

Where the problem arises has to do with where the putty stops. It does not know enough to stop at the parting line, so it just keeps going. Something is needed to keep the putty (the extrusion) from continuing past the parting line, and this is where the Radiate Surface command comes in.

If a surface can be made by extending the parting line outward radially, the extrusion would then know where to stop. There is actually a little more to the process than creating a radiated surface, but that is covered in material to follow. The area where the extrusion is going to terminate should be larger than the sketch for the extrusion. You could think of this in terms of keeping the extrusion from flowing over the edges of the radiated surface. How-To 13-7 takes you through the process of using the Radiate Surface command.

How-To 13-7: Creating a Radiated Surface

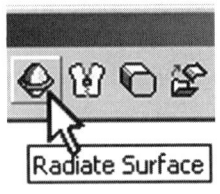

Fig. 13-29. Radiate Surface icon.

The Radiate Surface icon, shown in figure 13-29, is found on both the Mold Tools and Surfaces toolbars. To create a radiated surface, perform the following steps.

1. Select Insert > Surface > Radiate, or click on the Radiate Surface icon.

2. In the Radiate Surface panel, shown in figure 13-30, select a plane that will determine the direction of radiation. The radiate direction will be parallel to the selected plane.

Creating a Core and Cavity

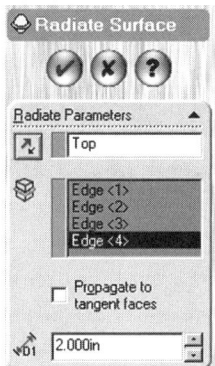

Fig. 13-30.
Radiate Surface
panel.

3. Click in the *Edges to radiate* list box and select the edges to be radiated.

4. Click on the Flip Radiate Direction icon if necessary. This will reverse the radiate direction indicated by the small preview arrows.

5. Optionally, select the *Propagate to tangent faces* option. This reduces the number of edges the user has to pick. SolidWorks will select all tangent edges automatically.

6. Specify the radiate distance.

7. Click on OK to create the radiated surface.

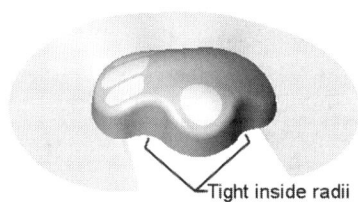

Fig. 13-31. Radiated surface.

The option Flip Radiate Direction is typically not needed because by default the selected edges will radiate outward from the part. The large preview arrow shows the normal of the face selected, which helps to indicate the radiate direction. This direction will be parallel to the selected face and perpendicular to the large preview arrow. An example of a radiated surface is shown in figure 13-31.

If you are looking at figure 13-31 and thinking that the model looks incomplete, you would be correct. There is a large piece missing from the radiated surface. With this particular part, a surface cannot radiate completely around the entire part using the desired radiate distance because of the inside curves present on the model. The radiated surface would have to be capable of folding over on itself, and possibly even intersecting itself. This would not be allowed in SolidWorks.

This particular model poses some interesting challenges, but then again, that is why it was used. There are holes in the model, it has a non-planar parting line, and the parting line arcs inward at two places, making geometry creation difficult. These are the types of real-world problems mold makers face every day when using SolidWorks. So the question remains, how can the rest of the surface geometry be added to this model so that the mold making process can continue? We find our answer with lofted surfaces, discussed in the next section.

Lofted Surfaces

In Chapter 6 you learned about lofted features. Here we will apply that knowledge to creating a lofted surface. We will also build upon what you learned about guide curve functionality introduced in that same chapter. Additionally, the Composite Curve command discussed in Chapter 9 will be used, along with the 3D Sketch command.

When creating lofted features, a minimum of two sketches is required. When creating lofted surfaces, creating a sketch is not a requirement, though sketches can be used if necessary. Two surfaces can be lofted together to create a "bridge" between those two surfaces. This can be done by clicking on the edges to be lofted between, and then carrying out the surface loft command, which is exactly what you will do in material to follow.

If a surface loft is attempted strictly between the two edges where the gap exists on our sample part, the loft will not attach itself to the parting line of the model. To overcome this situation, a guide curve will be used. A perfect tool for creating the guide curve in this case is the Composite Curve command (see "Composite Curves" in Chapter 9).

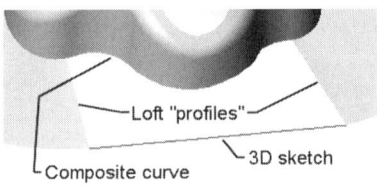

Fig. 13-32. The components that will make up the lofted surface.

If a composite curve is employed as the only guide curve, the resultant lofted surface will attach itself to the parting line, but will also result in excessive waviness on the perimeter of the lofted surface. This would not result in an ideal situation for the finished mold. For this reason, a second guide curve will be created. Using the 3D Sketch tool (see "3D Sketcher" in Chapter 9), a simple line can be created that connects the two outer points of the edges to be lofted between. This 3D sketch line can be seen in figure 13-32, along with the other components that will make up the lofted surface.

The loft "profiles" the loft will bridge are not profiles in the typical sense, but rather two surface edges serving as profiles. The composite curve will be made up of the edges of the model that run from where the loft will start to where it will end. The 3D sketch line will make up the second guide curve, which can be created by adding a line while in a 3D sketch, and then adding coincident relations between the ends of the line and the ends of the loft profile edges. Once that has been accomplished, it is time to create the lofted surface. How-To 13-8 takes you through the process of creating a lofted surface.

How-To 13-8: Creating a Lofted Surface

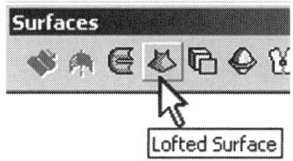

Fig. 13-33. Lofted Surface icon.

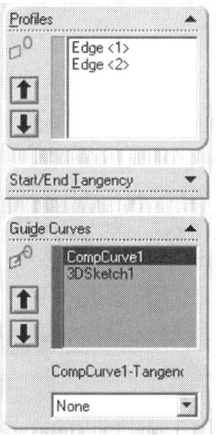

Fig. 13-34. Part of the Surface Loft panel.

To create a lofted surface, perform the following steps. Bear in mind that the process of creating a lofted surface is nearly identical to creating lofted features. Consequently, how the profiles are selected is critical, and guide curves are optional, though they will be used in the example shown here.

1. Select Insert > Surface > Loft, or click on the Lofted Surface icon (shown in figure 13-33), located on the Surfaces toolbar.

2. Select sketches or surface edges to loft between. You are not limited in the number of profiles that can be lofted between, even when using surface edges.

3. Modify the Start/End Tangency conditions as necessary.

4. Optionally, click in the Guide Curves list box and select model edges, curves, or sketch geometry to use as guide curves. A portion of the Surface Loft panel is shown in figure 13-34.

5. If using guide curves, modify the side tangency conditions as required by selecting a guide curve from the list box and specifying the appropriate tangency condition (explanation in material to follow).

6. Click on OK to complete the lofted surface.

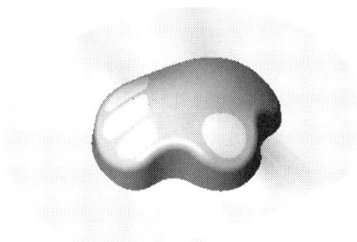

Fig. 13-35. All required surfaces have been created.

When using guide curves, an additional option for controlling the side tangency of the loft is available, as mentioned in step 5. Controlling the side tangency conditions is actually similar to controlling the start and end tangencies, as described in Chapter 6. For example, one option would be to select the composite curve object from the Guide Curves list box, and then specify All Faces for the tangency condition. This would force the lofted surface to be tangent to the faces the guide curve runs adjacent to. In our example, that would not be desirable. After completing the lofted surface, the model now looks as it does in figure 13-35.

Creating the shutoffs and the radiated and lofted surfaces are just steps in the process whose final outcome is to obtain a core and cavity. Sometimes the surface types required change, depending on the geometric nature of the model.

The surfaces previously created "belong" to the trackball cover part. That is, the surface was created in the part file and its geometric data is contained within that part file. This is a given, but the thinking behind this is related to another important aspect of performing the next step in obtaining the final objective.

When the act of extruding a sketch to create the core or cavity is performed, the Up To Surface end condition will be used. At that time, a surface will need to be selected. The important point is that one surface only can be selected. For this reason, the surfaces must all be knitted in order for there to be one selectable surface. This is the basis for performing the next command.

The knitted surface must belong to (i.e., be owned by) the new part where the extrusion will be taking place. Because of this, it is necessary to create an assembly, insert the component the mold is being created for, insert a new component that will become either the core or cavity, and then edit that new component and knit the surfaces. This seems like a lot, but take it one step at a time and everything will work out fine.

Creating the Mold Assembly

Before knitting surfaces, and before the cavity or core can be created, it is necessary to begin a new assembly. The following steps are included for the sake of thoroughness. Because the procedure for carrying out these steps has already been covered in Chapter 12, it will not be necessary to go into detail.

1. Create a new assembly.
2. Insert the design component into the assembly.

The design component in our case is the trackball cover. It is suggested the component be positioned at the assembly's origin point. There is no special reason for this. It is just that it makes sense to position the main part at a central location with respect to the assembly file's coordinate system.

For the example used here, Core and Cavity will be used as the assembly's name. Next, a new component must be added to the assembly. This will be the cavity component.

Inserting a New Part into an Assembly

There are times when it is not known what a part will look like until the assembly it will go into is partially completed. Sometimes it is necessary to build most of the assembly, and then create the part inside the assembly. This is known as *in-context editing*, also referred to as top-down design or top-down assembly modeling. This is in contrast to bottom-up assembly creation, in which the parts used to build the assembly have already been created.

Chapter 12 taught you how to edit a component in the context of an assembly. Here you will learn how to insert a brand new component and create it from scratch in the context of an assembly.

When it is necessary to create a new component in the context of an assembly, SolidWorks first requires that a new part be given a name and then saved to the hard drive. Once this has been accomplished, the new part can be edited within the context of the assembly. Features can be added to the part in the usual manner. The added benefit is that geometry from the assembly can be used to help create the new part. The action of basing a new part on other components in an assembly creates external references to the assembly.

InPlace Mates

When a new part is added to the assembly, a special mate relationship is added by SolidWorks. This is an automatic process the user has no control over. Part of the process of inserting a new component consists of picking a plane to sketch on. When the plane is selected, the software adds an *InPlace* mate between the Front plane of the new part and the selected plane or planar face.

The reason for the *InPlace* mate has to do with external references. Because it is almost a sure thing that external references will be developed, the last thing you would want is to have the new part moving around. This would affect the new part in drastic ways. For this reason, the *InPlace* mate is added, which essentially locks the new part in position.

What happens if movement is a desired trait of the new part? If that is the case, it is best to design the part outside the context of the assembly, and then insert the part as an existing part file. The part can then be mated with the desired motion in mind.

If the part has already been created in the context of the assembly, and you change your mind about how the part should have been created, the *InPlace* mate can be deleted. In this case, the external references should also be deleted, which could primarily be accomplished via the Display/ Delete Relations command. Finally, the component can be mated to existing components in the assembly. Deleting an *InPlace* mate, breaking external references, and remating the component should only be done as a last resort. It is additional work that can be avoided by a little planning on the part of the designer.

In-context parts are typically created as single-use parts. They are not used outside the assembly they were created in. How-To 13-9 begins the process of inserting and creating a new part in the context of an assembly.

How-To 13-9: Creating In-context Parts

To create a new part in the context of an assembly, perform the following steps. Note that this is not the same as inserting an existing component.

1. Select Insert > Component > New Part.

2. Type in a name for the new part (*Cavity* will be used for the example used in this book).

3. Click on Save.

4. Select a plane or planar face to sketch on for the first feature of the new part.

It is usually best to have a plane or planar face in mind before inserting a new part. For the *Cavity* component being created in this book, the Top plane of the assembly was used.

Once a plane or face is selected on which to sketch, the new part will appear in FeatureManager. Note also that the new part is actively being edited. This is indicated by the title bar of the SolidWorks window, and from the salmon-colored listing of the part in FeatureManager. Typically, the part would be salmon-colored in the work area as well, but there are no features present in the part yet. For all intents and purposes, *Cavity* is an empty part file. That will change shortly.

If you were to take a peak inside the *Mates* folder in FeatureManager, you would notice the *InPlace* mate that has been added. This is shown in figure 13-36. Leave the *InPlace* mate alone, as it is not something you should mess with anyway. *InPlace* mates can be edited, but this is rarely necessary.

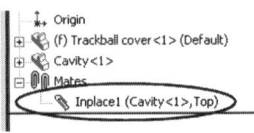

Fig. 13-36. An InPlace mate relationship.

Knitting Surfaces

Now that the new component has been inserted into the assembly, the first order of business is to create the knit surface. The act of knitting surfaces is a very easy process in its simplest form. However, there are some added features to the Knit Surface function that are not readily apparent. One of these features is the ability to select a "seed face" when a radiated surface is selected as one of the surfaces to be knitted.

The basic operation of the Knit Surface panel, shown in figure 13-37, is to select the faces or surfaces to be knitted. This will cause the selected faces to be knitted, resulting in a new surface that is a compilation of the selected faces. The knitted faces will appear as a new surface in FeatureManager.

When one of the selected faces is a radiated surface, the Knit Surface panel expands to include the *Seed face* area for selecting a seed face. This area can be seen at the bottom of figure 13-37. Once this occurs, a seed face can be selected, allowing SolidWorks to select most of the faces automatically. This is crucial. If not for this capability, the user would have to individually select every face on one side of the model. This could literally be hundreds of faces, many of which are very small. The prospect of having to perform this task is daunting. Creating a knit surface for a mold is often the point in time when a mold maker will cross his fingers and hope for the best.

When a seed face is used, any faces adjacent to the seed face will be selected. All adjacent model faces are used, up to the radiated surface. This keeps the user from having to select every face on the model. If shutoff surfaces are present, they need to be selected as well, and should appear in the same listing in which the radiated surface appears. How-To 13-10 takes you through the process of creating a knitted surface.

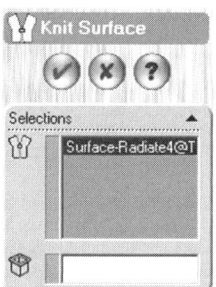

Fig. 13-37. Knit Surface panel.

◆ **NOTE:** *It is not possible to create a knitted surface while in a sketch. If you have just inserted a new component, click on the Sketch icon to exit the sketch prior to proceeding with the next section.*

HOW-TO 13-10: Creating a Knitted Surface

To create a knitted surface, perform the following steps. The Knit Surface icon, shown in figure 13-38, is found on both the Mold Tools and the Surfaces toolbar.

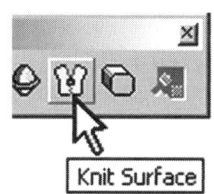

Fig. 13-38. Knit Surface icon.

1. Select Insert > Surface > Knit, or click on the Knit Surface icon.

2. Select the surfaces to be knitted (which should include any shutoff surfaces).

3. If one of the surfaces is a radiated surface, optionally click in the *Seed face* list box and select a seed face on the model.

4. Click on OK.

Fig. 13-39. The newly created knitted surface.

Once the knitted surface has been created, the model will look very bizarre, as shown in figure 13-39. This has to do with the geometry of the part and the knitted surface occupying the exact same position in space. The computer's graphics card is attempting to show both objects at the same time, and it does not know how to figure out which object to show. It would be a good idea to hide the design component at this time, leaving only the knitted surface visible.

Common Errors with Knitted Surfaces

A very common error is to receive a message that states "The radiate surfaces cannot subdivide the body." This is a very frustrating error, because it usually means you will be stuck having to pick every face, no matter how miniscule, in order to create the knitted surface. Incidentally, this very error occurred while working with the trackball cover.

Subdivision errors will most certainly occur if shutoff surfaces are necessary but have not yet been created. This goes back to the situation in which it is necessary to create both the core and cavity without them touching each other. In other words, it must be possible to create one complete surface that divides what will become the core with what will become the cavity.

The subdivision error can also occur if the shutoff surfaces do not completely connect with every edge of the holes they are filling. This can happen in a number of ways, such as failing to use guide curves if lofting or failing to include every edge of the hole when creating the shutoff surface. Try going back to the shutoff surface definitions and checking your work, or possibly recreating them using a different process.

If it happens that the *Seed face* capability is not going to work out for you, knuckle down and select all faces to be knitted. You will want to select all faces on one side of the part. Select the shutoff surfaces and the radiated surface from FeatureManager, as those surfaces may consist of numerous surface patches and it will make the selection process easier.

Checking the Knitted Surface

It is especially important to check the knitted surface visually if it was your misfortune to be forced to select all faces one by one. This is because any missed faces will cause holes to appear in the geometry, making the next step in the process (the extrusion) fail. Rotate the knitted surface on screen, checking it for holes. If there are any holes, make note of the hole's location, edit the definition of the knitted surface, and additionally select any surfaces that were missed the first time around.

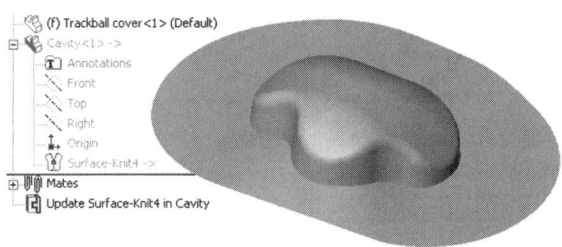

Fig. 13-40. The knitted surface and FeatureManager.

Figure 13-40 shows the *Cavity* part and FeatureManager at the current stage of developing a core and cavity. There are a few things worth mentioning. First, the icon for the trackball cover is ghosted because that part is currently hidden. This is to make it easier to check the knitted surface. Second, the *Cavity* component is currently being edited. For that reason, it appears as a salmon color in FeatureManager, though this detail is lost in the black-and-white image.

If you have made it this far, it is almost time for the final stage. Because a sketch plane will be needed, it is necessary to create one before creating the sketch geometry. The sketch plane will be offset some distance from the knitted surface. This is because eventually you will want to extrude up to the knitted surface, as was already pointed out.

Use skills learned previously to create a sketch plane. A good recommendation would be to offset a plane from one of the planes present in the *Cavity* component. Bear in mind that the offset distance will determine the thickness of the mold.

The Cavity Base Feature

This is the final stage of the cavity creation process. The process will need to be repeated for the core, starting from the point of inserting a new part into the assembly. From this point on, here is a synopsis of what will take place. Once a sketch is started on the new offset plane, the profile sketch for the cavity portion of the mold will be established. An extrusion will be created, using the knitted surface to terminate the extrusion. Afterward, the knitted surface can be hidden from view, as it will no longer be needed. How-To 13-11 takes you through this final process.

How-To 13-11: Creating the Cavity Component

To complete the process of creating the cavity, perform the following steps. The steps will not be explained in detail, as they involve procedures discussed in previous chapters. It is assumed a sketch plane has already been created, as suggested in the preceding material.

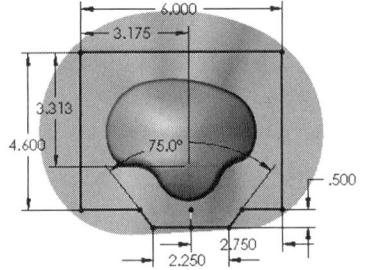

Fig. 13-41. Sketch for the cavity mold base.

1. Start a sketch on the new offset plane just created.
2. Create sketch geometry to describe the shape of the cavity mold base. An example of this is shown in figure 13-41.
3. Create the extrusion (Insert > Base > Extrude).
4. For the extrusion's end condition, select Up To Surface.
5. Select the knitted surface from FeatureManager.
6. Click on *Reverse direction* if necessary.
7. Click on OK to complete the extrusion process.
8. Return to editing the assembly.
9. In FeatureManager, right-click on the knitted surface to hide it.
10. Show the original design component (in this case, the trackball cover).
11. You may wish to hide additional surfaces that belong to the design component if the mold has been completed.

Congratulations, you have just finished creating the cavity. Do not forget to save your work. The finished *Cavity* component is shown in figure 13-42, as seen from the underside. Of course, the actual shape of the cavity mold base can be any shape, depending on your requirements.

Fig. 13-42. The completed cavity mold base.

Completing the Core

To create the component that will be the core, the same procedure should be used as that to create the cavity. The only difference is that you will be carrying out the procedure on the opposite side of the design component. The shutoff surfaces, if required, have already been created, but another knitted surface will be needed.

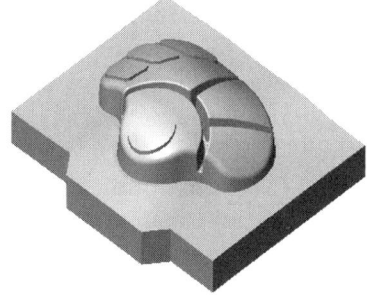

In summary, begin by creating a new component in the context of the assembly. You will be placed in edit mode for the new component once it has been given a name and once a plane or planar face has been selected to sketch on. You can immediately exit the sketch, because creating the knitted surface requires that you not be in a sketch. Next, create the knitted surface. Following that, create a plane to sketch on, create a sketch on that plane that will describe the shape of the mold, and then extrude the sketch up to the knitted surface. Figure 13-43 shows an example of a completed core, once again using the trackball cover as the design component.

Fig. 13-43. The completed core.

A good technique for creating the sketch geometry that defines the shape of the other side of the mold when one half has already been completed is to convert the edges from the completed half of the mold. This results in a sketch that will update automatically if the sketch for the first half of the mold is modified.

Final Touches

The finished *Core and Cavity* assembly is shown in figure 13-44. An exploded view was created and the cavity component was made transparent to more clearly see the internal detail.

The assembly really is not finished yet, but the toughest part is over. From this point it is mostly downhill. Of course, more elaborate molds might require some fairly innovative modeling techniques, but at least now you should be able to work through creating a core and cavity.

Other features, such as gates and ejector pins, would be fairly easy to add. Use the knowledge you have gained in previous chapters to create these additional features yourself, and complete your own core and cavity.

Fig. 13-44. The completed Core and Cavity *assembly in an exploded view.*

Summary

The creation of a mold begins with an assembly. If using the Cavity command, first insert the component that will become the mold. This is good practice due to the fact that the first part brought into an assembly is always fixed.

Any components you want to use as the design components should be mated within the assembly so as to fully define those components. A design component is the component that will be used to create the cavity. You are not limited to using only one component as a design component.

When creating a cavity, you must decide whether or not there will be shrinkage or expansion on the component that will eventually be cast from the mold. Make sure you specify a scale factor in the form of a percentage specifying the degree of shrinkage or expansion.

Inserting parts into other part files can be used in many situations for which two or more parts are based on the same basic shape. Any change made to the original part will propagate to the parts referencing it. This association is one-way only. Making changes to the part that references the original will never affect the original part in any way. If an assembly is open, another way of beginning a new part with a component of the assembly is to use the Derived Component Part command. This is different from inserting an existing part using the Insert > Part method in that any assembly features are inherited by the new part.

A small arrow symbol to the right of a part or subassembly in FeatureManager indicates an external reference. To check an item's external reference information, right-click on the item with the external reference and select List External Refs. External references are created whenever the Cavity command is used, or when using base parts or derived component parts.

It is possible to create cuts using an open profile. When doing so, make sure the profile stretches completely across the part to be cut. In addition, when using an open profile for a cut, the only end condition type available is Through All. Use the preview arrow to determine which side of the cut will be discarded. Use the Convert Entities command and convert the parting line when preparing to cut the part that will become the mold.

When creating a core and cavity, it is best to scale the design component first. Unlike the actual Cavity command, which uses a percentage value for scale, the Scale command uses a scale factor. This means that a scale factor of 1.04 (for example) would scale the part up 4%.

Once a part is scaled, any shutoff surfaces needed to close up any holes in the part must be created. In addition, a radiated surface is necessary so that there will be enough surface area to extrude up to when creating either the core or cavity component. A knitted surface is necessary because during the extrude process only one surface can be selected.

When creating a knitted surface, selecting a radiated surface allows a special function of the Knit Surface panel to become available. This special function is the ability to select a seed face, which greatly simplifies the face selection process. The seed face and all adjacent faces up to the radiated surface will be selected. If there are any shutoff faces, they must be selected individually.

Questions and Topics for Discussion

1. Why is it important to fully define an assembly if it is going to be used to create a cavity using the Cavity command?

2. If the part you will be casting is going to have a shrinkage value of 1.5%, what number should you plug into the Scale factor box when using the Cavity command?

3. What is the name of the resultant feature when a part is inserted into another?

4. Is it possible to insert more than one part into the same file?

5. How do you know if a feature is externally referencing something else?

6. How do you determine what a particular external reference is pointing to?

7. What end condition type is available when you use an open profile to create a cut?

8. When you have completed the pieces of a mold, are they associated with the original part used to create the mold cavity? Explain your answer.

9. What action causes an *InPlace* mate to be created?

10. When creating a knitted surface, selecting what object causes the *Seed face* list box area to appear?

11. Describe the steps involved in inserting a new component into an assembly.

12. What is one problem that can cause the error message "The radiate surfaces cannot subdivide the body" to appear when trying to create a knitted surface?

CHAPTER 14

Welded Assemblies

THE TOPIC OF WELDED COMPONENTS comes at this stage of the book because, similar to creating a mold or core and cavity, welding components requires you to work within an assembly document. The act of welding components does not literally form one component of two (at least as far as SolidWorks is concerned). There is another command, called the Join command, which will perform the function of turning multiple components into a single component.

The command that will be used to create welded components is known as the Weld Bead command. The Weld Bead command is found in the Assembly Features submenu under the Insert pull-down menu. Technically, though, creating a weld bead does not create an assembly feature. The Weld Bead command will create a separate part file, with its own file name and location on the hard drive.

Creating weld beads is a very specific function, meaning that if you do not do any welding, chances are you will not care about how SolidWorks handles weld beads. The Join command, on the other hand, can be used for many situations other than joining welded parts. The point is that even if you are not a welder you may still want to check out the section regarding the Join command.

Adding Weld Beads

To reiterate, adding weld beads must be done in an assembly. This would be logical, because you normally would not need to weld a single part. Prior to adding weld beads, you should fully define the components in your assembly. There is no need to assemble with motion in mind when adding welds. It would be a very unique assembly, or a very poor job of welding, if

Fig. 14-1. An assembly in need of weld beads.

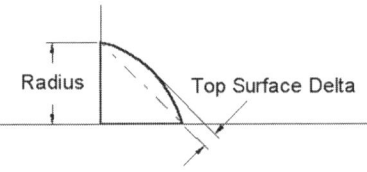

Fig. 14-2. Fillet weld bead dimensions.

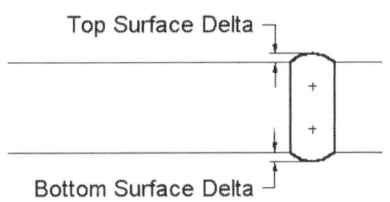

Fig. 14-3. A backing run weld bead.

you could move components after welding them. For this reason, components to be welded should be fully mated.

The mechanics behind adding weld beads is a very streamlined process. There are a series of four windows that step you through the entire process. Each window requires that you plug in some bit of information for one reason or another. Different weld bead types require different information. There are quite a few types of weld beads Solid-Works lets you create. This chapter covers a few of them so that you can get a good feel for how weld beads are created. For demonstration purposes, an assembly has already been built and the components fully mated in position. The assembly, shown in figure 14-1, is called *Welding*.

Before diving into the steps used to create a weld bead, you should first understand the way in which SolidWorks dimensions weld beads. Three different measurements are used. Depending on the type of weld bead you are creating, different combinations of these measurements will be required. The measurements are top surface delta, bottom surface delta, and radius.

A radius would be used for weld beads such as fillet welds. You would also need to specify the top surface delta for a fillet weld bead. Figure 14-2 shows a fillet weld bead as seen from the side.

Figure 14-2 shows how the radius and top surface delta dimensions are used to describe the size of the fillet weld bead. These measurements are used on other weld bead types in the same fashion. On occasion, as with a backing run weld bead, a bottom surface delta must be specified. An example of a backing run weld bead is shown in figure 14-3.

These three measurements (radius, top surface delta, and bottom surface delta) are the only three needed to describe all of the weld bead types. With this knowledge, you are on your way to creating weld beads. In the following material, a fillet weld will be added. How-To 14-1 takes you through this process.

How-To 14-1: Adding a Weld Bead

To add a weld bead, perform the following steps.

Adding Weld Beads

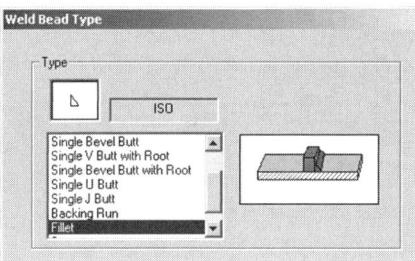

Fig. 14-4. Weld Bead Type window.

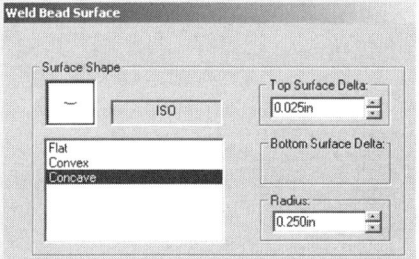

Fig. 14-5. Weld Bead Surface window.

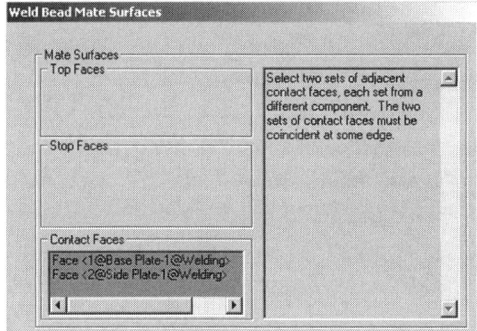

Fig. 14-6. Weld Bead Mate Surfaces window.

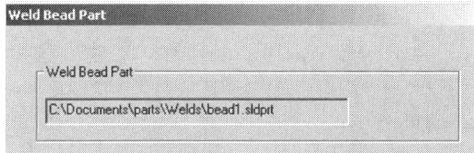

Fig. 14-8. Weld Bead Part window.

1. Select Insert > Assembly Feature > Weld Bead. The Weld Bead Type window will appear, a portion of which is shown in figure 14-4.

2. From the list, select the weld bead you wish to create.

3. Click on Next. The Weld Bead Surface window, shown in figure 14-5, will appear.

4. Select the Surface Shape, such as convex or concave.

5. Specify the Top Surface Delta, Bottom Surface Delta, and Radius, as required.

6. Click on Next. The Weld Bead Mate Surfaces window, shown in figure 14-6, will appear.

7. Specify the appropriate faces that will describe the location of the weld bead. This will include Contact Faces, and possibly Top Faces and/or Stop Faces, depending on the weld bead type. Figure 14-7 shows which faces were selected for the fillet weld bead being created in the current example.

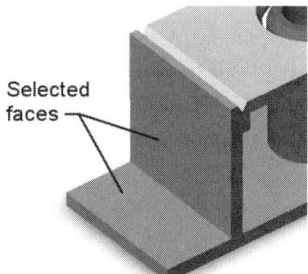

Fig. 14-7. Faces selected for the fillet weld bead.

8. Click on Next. The Weld Bead Part window will appear, a portion of which is shown in figure 14-8.

9. The Weld Bead Part window will display the default file name and location given to the weld bead. This can be changed by clicking on the Browse button, but there is usually no reason to do so. Click on OK to accept the SolidWorks default file name and location.

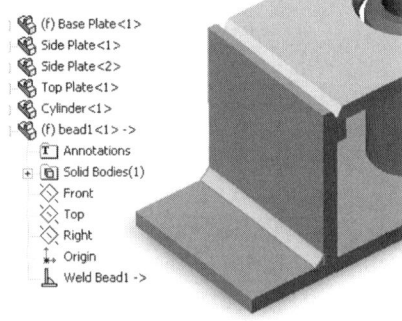

Fig. 14-9. The new weld bead and its appearance in FeatureManager.

After completing the last step, a weld bead will be created and added to the assembly. As stated previously, the weld bead will be a part file. As a matter of fact, you will be able to see it added to FeatureManager, just like any other assembly component. Figure 14-9 shows what the new fillet weld bead looks like, along with an insert that shows the weld bead in FeatureManager.

•◦ **NOTE:** *When objects such as faces are selected in an assembly and you are working in one of the wireframe display modes, you will see the part outlined by a box to show that a face on that component is selected. If working in shaded mode, you will not see a box surrounding the part because the face turns green instead.*

The name of the component in FeatureManager is *bead1*. As more weld beads are added, you will see the file names for those weld beads automatically increment to *bead2*, *bead3*, and so on. You can name the weld beads anything you want, and place the beads at any location on the hard drive. To do this, click on the Browse button found in the Weld Bead Part window. However, there are few reasons you would need to change the default name or file location for the bead. It is suggested that you accept the SolidWorks default settings for the weld bead file.

When a weld bead is created, it is fixed in position automatically. Common sense dictates that you would never want to move a weld bead. You certainly would not be able to move a weld bead if this were a real part. Furthermore, if you try to float a weld bead by way of the right mouse button, you will find that SolidWorks will not let you. Weld beads are fixed, and are made to stay that way.

To edit a weld bead, edit the definition of the weld bead feature via the right mouse button. Weld bead files contain one feature only, so the feature is very easy to find. By right-clicking on the weld bead feature and selecting Edit Definition, you will be placed in edit mode for that particular weld bead. You can then cycle through the same four windows displayed originally, changing any parameters necessary.

Adding Weld Beads

When the Weld Bead Part window appears while editing a weld bead's definition (the final of the four windows), use the default name specified by SolidWorks. This file name will be the same as the original file name and will overwrite the existing file. There is no reason you would want to give the file a different name when you are editing an existing weld bead.

If you do decide you need to edit a weld bead's definition, remember that you are in an assembly. When you are finished editing the definition of the weld bead, you will still be in edit mode for that particular part. Make it a point to return to editing the assembly, by right-clicking on the name of the assembly in FeatureManager and selecting Edit Assembly, or by clicking on the Edit Part icon on the Assembly toolbar.

Weld Bead Mate Surfaces

The trickiest part of mastering weld beads is determining what faces to select for the mate surfaces. Sometimes you must select what SolidWorks refers to as "stop faces," and sometimes "top faces" are required. These faces, known as mating surfaces, are used to determine where certain types of weld beads will be positioned. SolidWorks knows that selecting these faces can be confusing, and therefore provides descriptions of the specific faces required. If another weld bead type is added, you can see for yourself how SolidWorks handles the act of selecting the required faces. A Single V Butt weld will be added in the example that follows.

Taking your cues from the steps outlined earlier in this chapter, you know that you must select Insert > Assembly Feature > Weld Bead to start things off. The following data is supplied for this particular weld bead.

```
Convex
Top Surface Delta = .025 inches
```

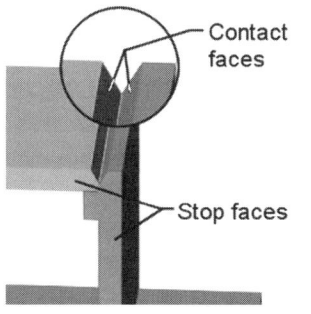

Fig. 14-10. Faces selected for a Single V Butt weld.

Eventually, you will find yourself at the Weld Bead Mate Surfaces window. For the Single V Butt weld type you will be required to select both contact faces and stop faces. Figure 14-10 shows the faces that were selected to satisfy these requirements.

The stop faces are the faces where the weld bead will terminate. Because the weld bead has two ends, you must select two sets of stop faces. The first set of stop faces is shown in figure 14-10. The second set of stop faces will be the same set of corresponding faces on the opposite side of the part. Make sure you select the second set of stop faces as well, or the

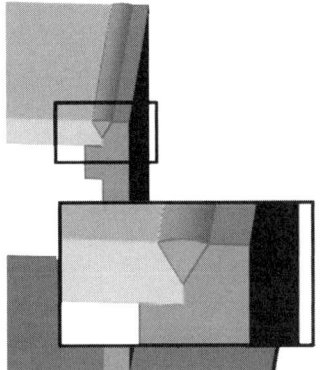

weld bead function will not work. The completed Single V Butt weld is shown in figure 14-11.

Because it is sometimes difficult to understand what faces should be selected for the various weld beads, SolidWorks gives you a description of what those faces are. Take your time and read through the description, because they can be helpful.

Fig. 14-11. Completed Single V Butt weld.

Weld Bead Special Situations

Fig. 14-12. The cylinder and top plate would prove troublesome for adding a weld bead.

Because weld beads are basically defined by sets of faces, you will find certain locations between components that lend themselves poorly to defining weld beads. Take the case of where the cylindrical part protrudes up into the top plate, as shown in figure 14-12. This would appear to be an area in which you would have little trouble adding a weld bead. It is, however, impossible to add a weld bead between the cylinder and top plate without making a small modification first.

The geometry between the cylinder and top plate appear to be likely candidates for a Single V Butt with Root weld bead. This type of weld bead requires two sets of contact faces: two faces for the V portion of the weld and two for the root. The two contact faces for the root are what pose the problem. The cylindrical face below the chamfer would have to be selected as a contact face, but this face does not create the desired results. The weld bead winds up extending all the way to the bottom of the cylinder.

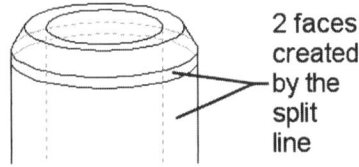

Fig. 14-13. Using Split Line on the cylinder.

For the weld bead to see the contact face as ending .125 inch below the chamfer, a split line is added to the cylinder (see "Split Lines" in Chapter 6). As a quick reminder, a split line will split a face into two faces by projecting a sketch onto a face. Figure 14-13 shows the cylindrical face separated into two faces via the split line command.

Adding Weld Beads

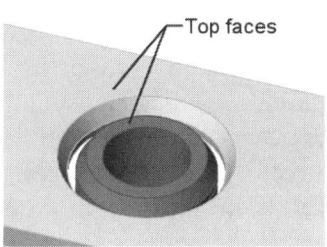

Fig. 14-14. Selecting the top faces.

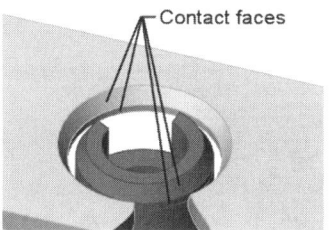

Fig. 14-15. Selecting the contact faces.

The odds of successfully creating a weld bead between the cylinder and the top plate have just drastically increased. Let's examine the process of creating one last type of weld bead. This time the Single V Butt with Root weld bead will be created. You already know how to begin the command. For the weld bead type, select Single V Butt with Root. For the weld bead surface, select Flat. You will not have to specify a surface delta if the surface is flat. Next, supply the mate surfaces. The top faces are the two shown in figure 14-14.

For the contact faces, you will need two sets of faces, as previously mentioned. The contact faces are shown in figure 14-15, using a cutaway view to make it easier to see what faces are being selected.

For a Single V Butt with Root weld bead, it is usually necessary to also select stop faces. Stop faces would normally be the faces that tell the weld bead where to stop. In this case, there are no stop faces because the weld bead is circular. Instead of issuing a warning message telling you to select the stop faces, SolidWorks decides it can create the weld bead without any stop faces and finishes the task without complaining. Figure 14-16 shows how the Single V Butt with Root weld bead turned out. A cutaway view is used to better show the cross section of the new weld bead.

Fig. 14-16. The finished Single V Butt with Root weld bead.

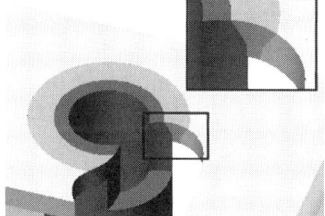

Weld Bead Associativity

Like almost everything else in SolidWorks, weld beads are associative. This means that if the underlying geometry on which the weld bead was applied is modified, the weld bead will update as well. Take the Single V Butt with Root weld bead just recently added as an example. It is partially between two chamfers: one on the top plate and one on the cylinder. What happens to the weld bead if the chamfers are modified? The weld bead updates. What happens if a component is repositioned? Again, the weld bead will update.

If drastic changes are made, a weld bead will not update. For example, if a component is mated to a different face on another component, any weld beads associated with that component will not update. For that matter, the weld bead would fail because the faces being used as mating surfaces for the weld bead definition are no longer valid. It would be necessary to redefine the weld bead at that time, or simply delete it from the assembly and create a new weld bead.

Update Holders in FeatureManager

There are certain functions you can perform in an assembly that will create update holders in FeatureManager. Adding weld beads is one of those functions. An example of these update holders is shown in figure 14-17. It is SolidWorks' way of keeping track of things, and a way of letting the user know that external references are being developed in the model. If there were a large number of these icons, you might want to think twice about how geometry was being created.

Fig. 14-17. Update holders.

The reason these icons are appearing in the *Welding* assembly has to do with the faces being selected for establishing the weld beads. In this situation, there is no way around keeping the external references from occurring. The references are what will allow the weld beads to update if model geometry is altered.

External references can be a good thing at times, but these same references can hinder you at other times. For example, creating a component in the context of an assembly may be desirable because the component should update automatically if the assembly changes. However, this same reference will get in the way when trying to reorganize components at a later time (such as regrouping subassemblies). Assembly restructuring is something you will learn more about in Chapter 18.

Aside from understanding that the update symbols are warning you of the external references being developed in the assembly, there is nothing of value the SolidWorks user can accomplish with update holders. Therefore, they are best left alone. SolidWorks will not let you delete them, although they can be hidden. If you right-click on the name of the assembly at the top of FeatureManager, you will see the option Hide Update Holders. Likewise, if they are hidden, you will see the option Show Update Holders.

Weld Bead Files

If you were so inclined, you could open one of the weld bead files in its own window. After all, weld beads are individual part files in their own right. There are not many good reasons you would want to open a weld bead part file. You really would not want to modify it because it is dependent on the rest of the assembly in which it was created. Figure 14-18 shows the first weld bead added to the welding assembly.

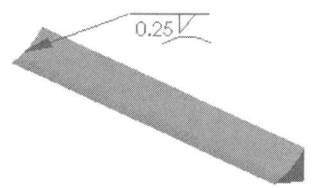

One point of interest that arises from seeing the weld bead in its own window is that the weld bead has a weld symbol attached to it. SolidWorks adds these symbols automatically. You can leave these symbols as is, or you can edit them. To edit the weld symbol, simply double click on it.

Typically you would add a weld symbol in a detailed design drawing, but weld symbols, like almost any annotation, can be added in a part or assembly file as well.

Fig. 14-18. A weld bead part file.

The process for adding weld symbols is the same whether in a part or drawing. The mechanics of adding weld symbols were covered in Chapter 11. It is not the purpose of this book to explain welding terminology or weld symbols, so further elaboration is not justified.

The main point to remember when adding weld symbols is that you must select an edge or surface first. Otherwise, the Weld Symbol option will not be available to you. The reason for this is that there will always be a leader on the weld symbol that must point to something. The edge or face you select is what the leader will point to.

The Join Command

To summarize the Join command, you might say that joining components in an assembly adds those components together. Components that had been touching in the assembly can be made into one contiguous solid part. This new part is literally a part file representing the sum of all original assembly components, or perhaps specific components only. There are a number of reasons a SolidWorks user might want to employ the Join command, a few of which are examined in this section.

One reason you may want to join components has a lot to do with welded components. If you were to weld components, you would normally want to show that welded assembly in a drawing layout as a single component. Many of the edges and hidden lines that would have been present in the assembly would not be shown if that same assembly had been welded. Joining the welded assembly components removes the edges not normally seen on a welded assembly and allows you to create a drawing layout that is more representative of the welded assembly. You will see an example of a joined component's hidden lines later in the chapter.

Many analysis programs require that you use a single part as opposed to an assembly. Sometimes this is a requirement, and other times it is a convenience. It is usually much easier to analyze a single component rather than an entire assembly. Some situations, however, do not always permit this. Without going into too much detail regarding analysis software, suffice it to say that it is sometimes necessary to join components before analyzing them. Consult your vender or reseller if you have questions regarding any analysis software you may be using.

Another important reason to join components has to do with communicating and performance issues. For example, assume an assembly has been joined and the newly created part is named *Phantom Assembly*. Now also assume that the external references back to the original assembly are broken. This leaves us with a standalone part with no external references that is an exact duplicate of the original assembly.

One use of the *Phantom Assembly* would be to insert it into a higher-level assembly. If the *Phantom Assembly* were a purchased item, or if it were a completed unit that would not change, inserting it into a higher-level assembly would increase performance. The *Phantom Assembly* would appear as a component. It could be mated to like any other component, but the top-level assembly would have a much reduced overhead on computer system resources.

A second use of the *Phantom Assembly* scenario is that it could be sent to a customer, client, or perhaps an international sister company overseas (whatever the case may be) without the need to send every individual component. Keep in mind that if it became necessary to send an assembly via e-mail, all components contained in the assembly would also have to be sent. With *Phantom Assembly*, this is not the case.

Similar to using the Cavity command, you must be in an assembly to gain access to the Join command. In addition, a new component must be added to the assembly before attempting the Join command. This new component will become the joined component. You must also be editing this newly added component. This all may sound somewhat confusing, but in practice the procedure is very straightforward.

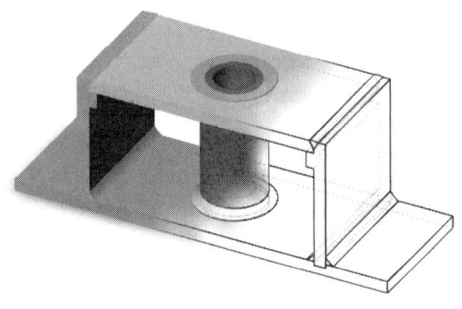

How-To 14-2 takes you through the process of joining components. As an example, the assembly containing the weld beads shown in figure 14-19 will be joined. Every component in the entire assembly will be joined in this procedure. The figure is shown in part with hidden edges because the hidden edges will change once the assembly is joined.

Fig. 14-19. This assembly will be joined.

HOW-TO 14-2: Joining Components

To join components, perform the following steps. Even though inserting a new part into an assembly was covered in the last chapter, the steps are included here because they are integral to creating a joined component.

1. Select Insert > Component > New.

2. Give a name to the new component being inserted. For this example, the joined component will be named *Weldment*.

3. Click on Save.

4. Select a face to sketch on for the new component.

Even though selecting a face to sketch on is a requirement whenever you create a new component in the context of an assembly, you should not be in a sketch when joining components. Correspondingly, the face being selected is not critical because you will not actually be sketching on it. In that an *InPlace* mate will be added, you may choose to select something that will not change in any way. For example, the *Weldment* component may be machined at some future point. Therefore, it is safest to select a plane that will never change, such as the Front plane of the top-level assembly.

5. Click on the Sketch icon to exit the sketch.

6. Select Insert > Features > Join. The Join panel, shown in figure 14-20, will appear.

7. Select the components to be joined.

8. Click on OK to join the components.

9. Click on the Edit Part icon on the Assembly toolbar to return to editing the assembly.

✓ **TIP:** *When selecting components to be joined, all components can easily be selected by selecting the first component in FeatureManager, and then Shift-selecting the last component in FeatureManager. Use the fly-out FeatureManager to accomplish this by clicking on the name of the command (Join) at the top of PropertyManager.*

After joining the components, assuming the default options were used, you will be viewing the newly joined component. As is evident from step

Fig. 14-20. Join panel.

9, you will still be in edit mode for the joined component. Because it is now one component, it will appear a salmon color until you go back to editing the assembly. In addition to using the Assembly toolbar, you can also right-click on the assembly name in FeatureManager and select Edit Assembly.

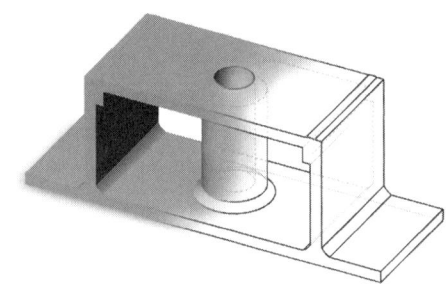

Figure 14-21 shows the *Weldment* component. It is all one color because it is all one component. The hidden lines shown are fewer than in the image shown of the original assembly (see image 14-19). Try comparing the illustrations and you will see the differences.

Fig. 14-21. The joined Weldment *component.*

The Join window is very basic, containing just two options. The first option is Hide Parts, which refers to the design components being joined. The general rule of thumb regarding this option is that if you have more work to do on the assembly, do not hide the input components. If you are finished with the assembly, as is usually the case when joining components, you will want to leave the Hide Parts option checked, which is the default setting.

If you decide to hide or show the input components later, you can always change your mind. You can hide or show assembly components with the Hide/Show icon, which you learned about in Chapter 12, or by right-clicking on components.

The other option, Force Surface Contact, will add a small bit of material to components that just touch each other. For example, if a cylinder is tangent to a planar face, a miniscule amount of material will be added so that the components can physically be joined. Otherwise, a zero thickness error would occur. Without delving too deeply into technicalities, zero thickness is what a surface would have. Solids must have some thickness, even if very small.

A Common Mistake When Joining

One common mistake made when attempting the Join command is to additionally select the part that will become the joined part as one of the design components. This is not something you want to do, and you will receive an error message if you try. When you insert the new part into the assembly (for the Join command), that part is acting as a placeholder for the geometry that will become the joined component.

Disassociating Joined Components

Joined components are dependent on the assembly in which they were created. On occasion, you may want to break this associativity. You can do this very easily by breaking the external references (see "External References" in Chapter 7), or by exporting the joined component as a file other than a SolidWorks part file, such as a Parasolids or STEP file. These are translation options discussed in more detail in Chapter 22. For now, bear in mind that exporting a file removes all intelligence from the file (such as parametric capabilities) but leaves the solid geometry intact.

Exporting a joined component can be a better solution than breaking external references. Once broken, external references cannot be reestablished. Leaving the references intact means that if the original assembly is modified the joined component will automatically update. Likewise, a new file encompassing these changes could be exported. This provides the best of both worlds yet still allows for the creation of a disassociated phantom assembly.

Multiple Bodies in Joined Parts

Because multiple bodies can coexist in the same SolidWorks part file, joining components in an assembly that do not physically touch is not a problem. This particular fact makes it much easier to create phantom assemblies from assemblies that contain large numbers of parts.

When joining components and implementing the Force Surface Contact option discussed earlier, components that touch will be formed into a solid body. If there happens to be other components that do not touch any other components, they wind up becoming solid bodies as well. This is not an issue, but you should be aware of how SolidWorks is handling the situation.

Saving Assemblies as Part Files

An alternative to joining assembly components is to save the assembly as a part file. There are benefits to this method if your sole objective is to create a phantom assembly. For example, only exterior components can be saved as a part, thereby limiting what might be numerous components housed within the assembly.

624   *Chapter 14: Welded Assemblies*

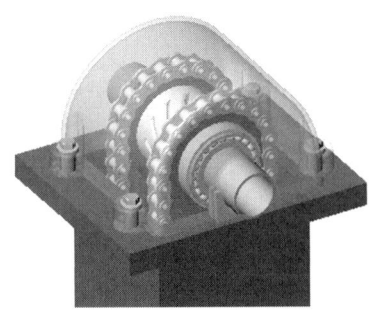

Fig. 14-22. Gear housing assembly.

A variation of the gear housing assembly used as an example in Chapter 12 will be used here. The gear housing has been placed on a stand, and drive chains have been added to the pulley gears, along with some other hardware. The assembly is shown in figure 14-22, with the cover semi-transparent so that interior components are visible.

All told, there are 237 components in the gear housing assembly: 206 parts and 31 subassemblies. Imagine an individual designer creating a factory layout with pieces of machinery at various locations on the factory floor. It would be very unwieldy to be forced to work with assembly files of this nature when all that is really needed are phantom assemblies. Working with single part files that represent the completed assemblies would be much more efficient.

An assembly file does not have to be joined into a single part to obtain a single part file. Rather, the assembly can be saved directly as a part file. The Save As command would be used to accomplish this task. How-To 14-3 takes you through the process of saving an assembly as a part file.

How-To 14-3: Saving an Assembly as a Part

To save any assembly file as a single part, perform the following steps. The possible results of performing this operation are discussed in material to follow.

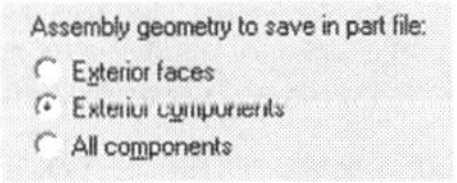

Fig. 14-23. Selecting the type of geometry to be saved in the part file.

1. With the assembly open, select Save As from the File menu.
2. In the *Save as type* drop-down list, select Part.
3. Enter a name for the new part file. By default, the part file name will be the same as the original assembly, but this can be changed.
4. Specify the type of geometry to be saved in the part file. Use one of the three selections shown in figure 14-23.
5. Click on Save to complete the process.

Depending on the type of geometry saved and the size of the original assembly file, it may take a moment to save the part file. When Solid-Works is finished saving the file, you will find yourself back in the original assembly.

Saving Assemblies as Part Files

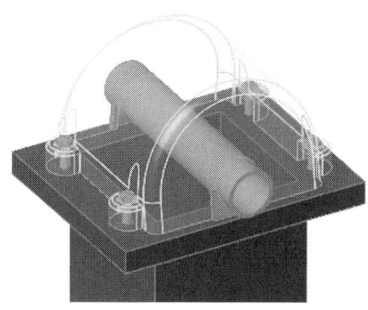

Fig. 14-24. The results of saving exterior faces.

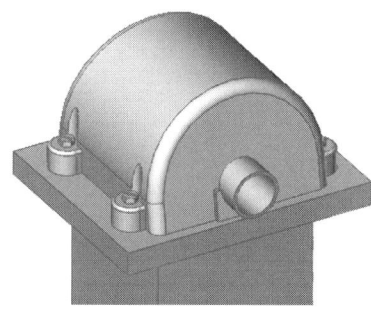

Fig. 14-25. A single body part file.

The important factor when saving an assembly as a part is what type of geometry is to be saved in the part. There are three options for saving: *Exterior faces, Exterior components,* and *All components*. If saving just the exterior faces, the resultant part contains surfaces that basically represent all exterior components in the assembly. Figure 14-24 shows the result of using the *Exterior faces* option.

Note in the figure that the interior components have disappeared from within the gear housing cover. The resultant part file is much smaller in size than the original assembly. The down side to using the *Exterior faces* option is that the part contains surfaces, not solid geometry. This could be advantageous to some software programs that function better with surfaces, but not necessarily to SolidWorks.

The second option is often the better of the three choices. As the name implies, saving *Exterior components* will save a part that displays only those components that could originally be seen in the assembly. The resultant part file contains solid bodies, not surfaces. No image is shown because it would look exactly the same as that shown in figure 14-24.

When the exterior components of an assembly result in the components becoming solid bodies, the part file becomes much more manageable. As an example, additional features could be added quite easily. This would not be so if the solid bodies were surfaces. Another option would be to combine the solid bodies into one, thereby simplifying the phantom assembly (now a part) even more. Such is the case with the single body part shown in figure 14-25. The colors have been reduced to a single color to better represent the model, which is no longer an assembly of separate components.

Originally, the assembly contained 206 parts. The part file that resulted from saving the exterior components contained 21 solid bodies. This relates to the 21 components that could be seen from the outside of the assembly. One final option not yet discussed is to save the assembly as a part while employing the *All components* option. This results in a part file containing the same number of solid bodies as there were parts in the assembly. The part file turns out to be quite large, and there are few reasons for making use of this option.

Table 14-1 outlines some interesting statistics in regard to saving an assembly as a part. Note in particular the file sizes. Examine the data and you will be able to make an intelligent decision which option would best to suit your purposes. Keep in mind that these statistics are for one particular assembly. Other assemblies may react differently.

Table 14-1: Assembly Saved as Part Statistical Comparison

Description of Document	Number of Components, Solid Bodies, or Surfaces	Number of Files Required	Total File Size
Original Assembly	237 components (206 parts, 31 subassemblies)	16 parts, 6 assemblies	7.82 MB
Saved as part (exterior faces)	33 faces	1 part	760 KB
Saved as part (exterior components)	21 solid bodies	1 part	1.25 MB
Saved as part (all components)	206 solid bodies	1 part	17.1 MB

Summary

SolidWorks gives you the ability to add weld beads to assemblies. Weld beads come in a variety of types. When creating a weld bead, you would typically define the weld bead by a specific set of dimensions and faces that determines where the weld bead will be positioned. Each weld bead type requires different sets of faces that define its boundaries.

Weld beads are associative. If you modify the dimensions of the components that contain the weld bead, the weld bead will update. The Weld Bead command, although found in the Assembly Features menu, is actually a separate part file SolidWorks creates. When a weld bead is added to an assembly, it is fixed in space automatically. You cannot float a weld bead and move it independently of the rest of the assembly.

Weld symbols are automatically added to weld beads. You can edit weld bead symbols by double clicking on them. You can also add your own weld bead symbols. You do not have to insert a weld bead to do this. The only prerequisite to adding a weld bead symbol is to select an edge or face first. Only then will the Weld Bead symbol command become available.

Ask yourself if it is really necessary to model weld beads. In that each weld bead is a separate component, the file size of a welded assembly can grow fairly quickly. If weld bead annotations will suffice, it is much more efficient to simply call out the welds on a design drawing than to model the welds.

The Join command allows for creating a new part that is a conglomerate of components in an assembly. The new part may consist of specific components, or even every component in the assembly. Use joined components as phantom assemblies, for showing welded components, or for various other reasons.

Assemblies can be saved as part files, resulting in a file that is much more simplistic in nature. This is a better solution than the Join command if the objective is to create a phantom assembly. Of the three options available when saving an assembly as a part, the *External components* option is usually the optimal choice. Only the visible external components of the assembly will be saved as solid bodies in a new part file.

Questions and Topics for Discussion

1. How does SolidWorks name weld beads if you accept the default settings for the weld bead names?

2. Name the three types of mate surfaces (faces) you may need to select for defining weld beads.

3. What do update holders in FeatureManager mean to the user, and are they anything you should be concerned with?

4. What must you do before the Weld Symbol annotation command is available?

5. When carrying out the Join command, are you editing a part or an assembly?

6. Are joined components associative? If yes, what are they associative with?

7. Describe a process for turning an assembly file into a part file that does not employ the Join command.

8. What dictates the number of solid bodies that will be present in a part file that was saved from an assembly if the *All components* option is used?

9. If an assembly containing subassemblies is saved as a part, what becomes of the subassemblies?

CHAPTER 15

Assembly Configurations

IN CHAPTER 10 YOU LEARNED HOW TO SUPPRESS FEATURES WITHIN A PART FILE. SolidWorks also gives you the capability to apply this same functionality to assemblies. Instead of features, though, it is components that are usually suppressed. It is possible to add configurations to an assembly and control which components are suppressed for each configuration, allowing for different "versions" of the assembly within the same file.

If a part file has multiple configurations, which version of the part is being used in the assembly can be specified and saved in assembly configurations. You will also explore how certain dimensions that belong to the assembly can be controlled via configurations.

Adding Assembly Configurations

The process of adding a configuration in an assembly is exactly the same as adding a configuration in a part. As a quick refresher, the following is a synopsis of how it works.

1. Click on the ConfigurationManager tab at the bottom of FeatureManager.

2. At the top of ConfigurationManager, right-click on the name of the model.

3. Select Add Configuration, type in a name, and click on OK.

630   Chapter 15: Assembly Configurations

Fig. 15-1. Sliding brace assembly.

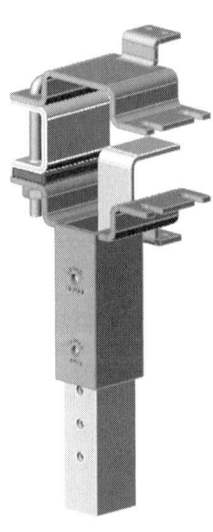

Assume for a moment you had an assembly on your hands such as the sliding brace shown in figure 15-1. The inner component is a post, which can be positioned at various heights with the aid of a pin. Now imagine that you wanted a separate "version" of the assembly with certain components removed so that the post could be shown more clearly.

Two options come to mind immediately. You could hide certain components, or you could suppress them. Hiding components would remove them from view. Suppressing components would remove them from memory. Suppressing components also has the by-product of removing the components from view as well. Because of this, suppressing and hiding seem very similar on the surface, but actually they are not. This you learned in Chapter 12.

The main point to keep in mind is that suppressing a component will also suppress any dependent items, such as mates. If you do not care that the mates are suppressed, and the assembly will not be adversely affected, suppression is a better alternative. This is due to the suppressed components being removed from memory and increasing overall system performance.

Every assembly is different and assemblies should be considered on a case-by-case basis. Often, if the assembly contains movable components, you may want to hide certain components instead of suppressing them. This is so that the assembly will not fly apart unexpectedly if other components are put through their motions. You may find that a certain combination of hidden and suppressed components works best for a particular assembly. Figure 15-2 shows the sliding brace assembly with a few components suppressed.

In the case of our example, it would be a simple task to suppress the components and then reverse the process (unsuppress them) once again when you wanted to see them. However, it would be much more convenient if configurations were used. Because configurations remember which components have been hidden or suppressed, it would be very easy to revert to the original assembly, where nothing is hidden or suppressed, simply by switching which configuration were being shown.

Fig. 15-2. Assembly with a few components suppressed.

Assembly Configuration Properties

If you were to add a new configuration to an assembly, you would find that the window for an assembly configuration's properties is more complex than that for a part. There are more options, as can be seen in figure 15-3, which shows the bottom half of an assembly's Configuration Properties

Assembly Configuration Properties

window. To access the Configuration Properties window, right-click on the configuration's name and select Properties.

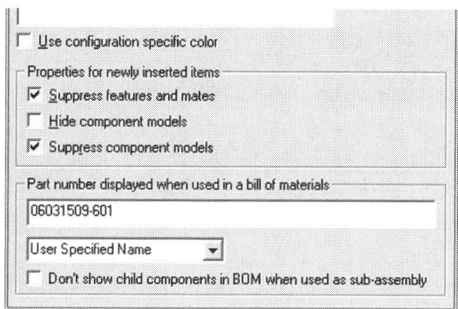

The various options available in the Configuration Properties window for an assembly are discussed in material to follow. Some properties common to both part and assembly configurations will not be discussed here, as they have already been explained in Chapter 10.

Fig. 15-3. Properties for an assembly configuration.

Use configuration specific color Option

This has the same meaning as it does when working with a part file's configuration properties. If checked, a button for modifying the color of this specific configuration becomes active. Clicking on the Color button allows for specifying a color for the configuration. This makes it easier to differentiate between configurations that have minimal changes between them.

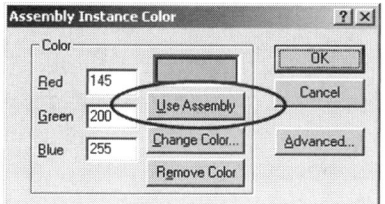

Fig. 15-4. Assembly Instance Color window.

Applying a specific color to an assembly configuration requires some extra effort from the user and is not as easy to implement as with solitary part configurations. If *Use configuration specific color* is implemented for an assembly and a new color is chosen, no change will be seen unless another action is taken. Specifically, access Component Properties for a particular component and click on the Color button (see "Component Properties" in Chapter 12). The Assembly Instance Color window, shown in figure 15-4, will appear.

By default, the color of an assembly component will be whatever the color of the actual part is. The Color button can override this. If the Use Assembly option is selected, the component will use the colors set forth for the assembly via the Colors section of the assembly's Document Properties (Tools > Options). Additionally, the Use Assembly option also allows the *Use configuration specific color* option in the Configuration Properties to work.

You may very well find that the *Use configuration specific color* option is not worth the bother when working with assembly configurations. It is just too much work to edit the properties of every component whose color needs to be configuration dependent, and is not necessary in most cases. Configuration-dependent colors are occasionally useful if certain key components are being modified in various assembly configurations. Having the key component change colors in the various configurations is sufficient to draw attention to it. This helps remind the user which configuration he is viewing.

Suppress features and mates Option

This option will suppress assembly features and mates that are newly inserted in other active configurations. If an assembly feature or mate were being added to another configuration, but that feature or mate should be suppressed in the configuration in question, *Suppress features and mates* should be checked.

Hide component models Option

Check the *Hide component models* option if components newly inserted into other active configurations are to be hidden in the configuration in question. Remember that hiding components merely removes them from view, and does not suppress them.

Suppress component models Option

Check the *Suppress component models* option if components newly inserted into other active configurations are to be suppressed in the configuration in question. If this option is checked, components added to the assembly in another active configuration will not appear in this configuration.

BOM-related Options

There are a few options specific to assembly configuration properties that deal strictly with bill of material (BOM) manipulation. These are the options that appear at the bottom of an assembly's Configuration Properties. Mainly, the options dictate what appears in a BOM's part number (Part No.) column.

There are three options that can be chosen from the drop-down list. The default is Document Name, which displays the file name in the Part No. column of the BOM. A second option is Configuration Name. As the name implies, this option will display the configuration name under the Part No. heading of the BOM. If neither of these options fit your requirements, there is the option User Specified Name. With this last option selected, anything at all can be entered in the area directly above the drop-down list and the text entered by the user is what will appear in the BOM's Part No. column.

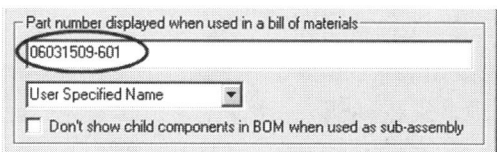

Fig. 15-5. Designating a user-specified name for a BOM part number.

Figure 15-5 shows an example of how the User Specified Name option can be implemented. In the example shown, the part number has been entered by the user and is the value 06031509-601. The text or numbering convention will obviously be different depending on the company's naming practices.

A final option related to BOM manipulation is the *Don't show child components in BOM when used as sub-assembly*. In other words, if the assembly is part of a higher-level assembly, it would be designated a subassembly. If that is the case, and assuming the *Don't show child components in BOM when used as subassembly* option is checked, none of the components in the subassembly will appear in the BOM. This option would be useful if, for instance, the subassembly had its own drawing and its own BOM.

Controlling Component Configurations

A very important aspect of assembly configurations is the ability to control what configuration of a part is being used in an assembly. This capability allows you to use one part file containing different configurations in different assemblies. It even allows for using different configurations of the same part in the same assembly. These conditions would also apply to subassemblies.

As a simple example, imagine a scenario in which different configurations of U-bolts and C-links are used in the sliding brace assembly. The C-link is a component made up of a single part, whereas the U-bolts are subassemblies consisting of the U-bolt itself, two washers, and two nuts. Figure 15-6 shows three versions of the sliding brace. They are all the same assembly, but are different configurations.

Fig. 15-6. Three different configurations of the sliding brace assembly.

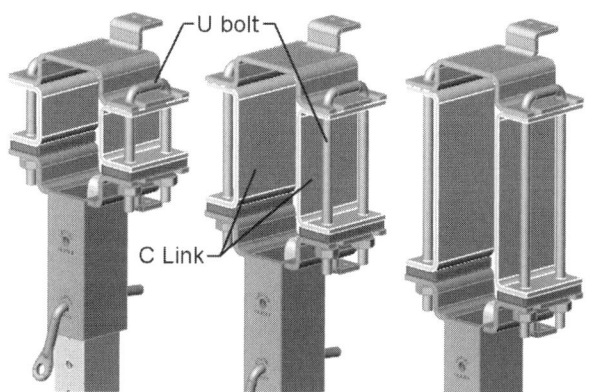

It is important to understand that both part configurations and assembly configurations are being used. For example, the *C-link* part contains multiple configurations that control the height of the C-link. The configurations used by the C-link can be called out from the assembly in which the *C-link* component resides. This is particularly significant because individual component feature dimensions cannot be controlled or saved within top-level assembly configurations.

The entire concept of configurations, when related to assemblies, takes on a new complexity. However, once it is understood how component configurations can be controlled from within an assembly, the situation becomes easier to comprehend. Although the idea of controlling component configurations was touched on in Chapter 12 (see "Component Properties"), How-To 15-1 will take you through the entire process of specifying a configuration for a particular component within an assembly.

How-To 15-1: Controlling a Component's Configuration

To control a component's configuration within an assembly, perform the following steps.

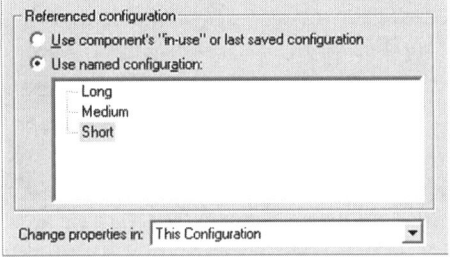

Fig. 15-7. Using a named configuration.

1. In FeatureManager, right-click on the component whose configuration you want to specify, and select Component Properties.

2. In the area titled *Referenced configuration*, shown in figure 15-7, select Use named configuration.

3. From the configuration listing, select the named configuration to be used.

4. Using the *Change properties in* drop-down list, select whether the configuration change should apply to This Configuration, All Configurations, or if you would rather Specify Configuration(s).

5. Click on OK.

If there are a hundred configurations present in a particular component, a scrollable list of those configuration names will be presented. Step 4 refers to what assembly configurations should be affected by the component configuration change.

Changing which configuration a particular component is using is trivial, and it is easy enough to specify which assembly configurations will be affected by the change. One component can have its configuration changed in 50 different assembly configurations at the same time. But what if a component has been inserted into an assembly 50 times? How could 50 instances of that component in the assembly be switched to use a different configuration? Would you have to access Component Properties for each of those 50 components? No, you would not.

There is a bit of a "cheat" involved with replacing numerous configurations of identical components simultaneously. You could use design tables, covered in Chapter 16, but there is another way. Use the Replace command, discussed in Chapter 12. Replace one of the components whose configuration you wish to change with the same component. Make sure you place a check in the *All instances* option, and select the *Manually select* option. It is this last option that allows you to pick a different configuration for the components being replaced. End result: you have replaced multiple instances of a component with a different configuration of the very same component.

Using Assembly Configurations Successfully

If you started at the beginning of this chapter and have been reading through all of the material, you have already read about the options in the Component Properties window. You have read the descriptions of options such as *Suppress features and mates* and *Suppress component models*, and you may even have a good understanding of these options, but how can they best be put to use? This section hopes to show you how.

A common use for configurations in an assembly is to show components mated in different arrangements. In actuality, we have already covered the material necessary for you to show components in alternative arrangements. However, How-To 15-2 takes you through the process of carrying out this task. This should strengthen your understanding of the various options available when working with assembly configurations.

How-To 15-2: Developing Assembly Configurations

This How-To examines a real-world scenario you may very well encounter on your own. It is not a "How-To" in the usual sense because instead of showing steps for a particular command or function, an entire process is examined. To set this scenario up, let's first examine what model we will be working with and what it is we need to accomplish.

636   **Chapter 15: Assembly Configurations**

Fig. 15-8. Sliding Brace assembly.

Figure 15-8 shows the *Sliding Brace* assembly. The *Lynchpin* component is also shown, but for sake of argument, let's assume it has not been inserted into the assembly just yet. The *Bracket with Tube* component (one part welded from two) contains holes, and bears stamped text. This text, shown in figure 15-8, reads *Locked Open* and *Locked Closed*.

The assembly contains a component named *Post*, which can slide up and down within the *Bracket with Tube* component. The post contains holes along its length. With the aid of the lynchpin, the post can be made to lock at certain heights, or in the case of this example in an "open" or "closed" position.

There is currently one configuration in our sample assembly. The configuration is named *Small Free* because the assembly utilizes the small versions of the *C-link* and *U-bolt* components and because the post is free to slide up and down.

Now that it is understood what we are starting with, let's examine what is needed. What is desired in this assembly are two additional configurations. Each configuration will include the lynchpin, but the lynchpin is not to appear in the original configuration. One configuration will be named *Small Closed*, where the lynchpin will be mated in the *Locked Closed* hole (with the post in a raised position). The second configuration will be named *Small Open*. In the second configuration, the lynchpin will be mated into the *Locked Open* hole, with the post in its lowest position.

It is known what we are starting out with, and it is known what we need to accomplish. Now it is just a matter of understanding how to get to our final objective of having multiple configurations that show the assembly in its various states. Follow the steps outlined here, paying attention to the illustrations, and you should walk away from this How-To with a very good understanding of how to use assembly configurations to your advantage.

1. In the assembly's ConfigurationManager, access the Properties of the original *Small Free* configuration (via the right mouse button).

2. Modify the Properties for newly inserted items as required. In our case, the *Suppress component models* and *Suppress features and mates* options should be checked, as in figure 15-9.

3. Click on OK when finished.

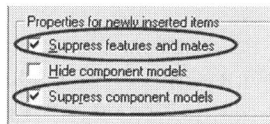

Fig. 15-9. Properties for newly inserted items.

When the new configurations have been created and the lynchpin is inserted into the assembly, we do not want the lynchpin to appear in the *Small Free* configuration. The *Suppress component models* option accomplishes this for us. In other words, the lynchpin will be suppressed automatically in the *Small Free* configuration (when the *Small Free* configuration is inserted into the assembly) if one of the other configurations is active.

The *Suppress features and mates* option is incidental, because any mates dependent on the lynchpin will be suppressed anyway. *Suppress features and mates* refers to any new assembly features or mate relationships. Therefore, any new mates added to the current configuration will be suppressed in any other configuration. The *Suppress component models* option has already ensured that the lynchpin will be suppressed in the *Small Free* configuration, which means any dependent items (mates) will be suppressed as well. However, we will leave the *Suppress features and mates* option checked anyway.

4. Select the *Small Free* configuration and copy it to the clipboard (i.e., Ctrl-C hot key).

5. Paste a copy of the configuration into ConfigurationManager (i.e., Ctrl-V).

6. Rename the new configuration *Small Closed* (hint: slow double click).

7. Paste in a second copy, in that two new configurations are needed.

8. Rename the new configurations *Small Open*.

Copying and pasting is a very easy way of creating new configurations. The copy contains the same settings as the original. For example, the *Small Open* configuration will have the *Suppress component models* option checked because the *Small Free* configuration did.

9. Switch to one of the new configurations, such as *Small Closed* (hint: double click).

10. Right-click on the *Small Open* configuration's Properties and uncheck *Suppress component models*.

11. Click on OK to close the Properties window.

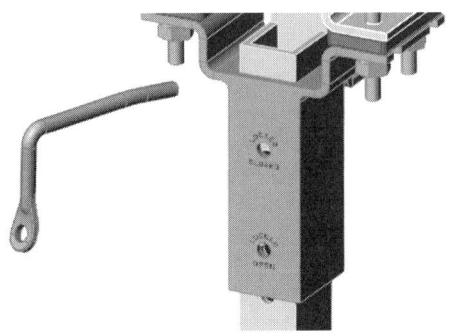

Fig. 15-10. The newly inserted lynchpin.

Although step 10 appears to be setting back an option you made sure was set correctly earlier, keep in mind that it is a different configuration. We will soon be adding a new component, *Lynchpin*, to the *Small Closed* configuration. We will want the new component to be unsuppressed in the *Small Open* configuration as well, which is the reason for performing step 10.

12. Insert the new component. The lynchpin is shown in figure 15-10.

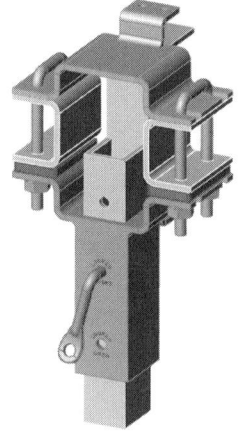

13. Mate the component into position. Three mates were used for the lynchpin: a concentric mate to the *Locked Closed* hole, a concentric mate to a hole in the post, and a distance mate to keep the pin from moving in and out unexpectedly. The pin can still rotate in the hole, but that does not matter. Figure 15-11 shows the newly mated lynchpin.

14. Switch to the *Small Open* configuration.

Upon switching to the *Small Open* configuration, you should see that the lynchpin is still displayed. This is due to turning off the *Suppress component models* option in step 10. The *Suppress features and mates* option, on the other hand, was left on. Therefore, the three mates just added to the *Small Closed* configuration are suppressed. This is good, because the pin needs to be mated to different holes for the *Small Open* configuration.

Fig. 15-11. After mating the lynchpin.

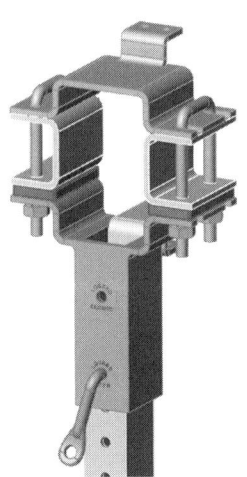

15. Move the lynchpin so that it is closer to the lower hole, and then add two (2) concentric mates that will position the pin in the lower *Locked Closed* hole, with the post in its lowermost position.

16. Unsuppress the distance mate added in step 13. It can be reused for the *Small Open* configuration, so there is no sense in creating a new mate to keep the pin from sliding in and out of the hole. Figure 15-12 shows the lynchpin mated in the lower hole.

Fig. 15-12. After mating the lynchpin into the lower hole.

17. Access the Properties of the *Small Open* configuration and turn *Suppress component models* back on.

18. The assembly configurations would be complete at this point. Save your work.

It may seem unimportant that the *Suppress component models* option is turned back on. Why even bother with it? In a word, consistency. When working with multiple configurations, it is of the utmost importance to be consistent. If the properties of all configurations are the same, it becomes easy to predict what will happen in each of the configurations. When different configurations contain a variety of settings, configurations become unmanageable.

Depending on your personal situation, the settings you use in each of your configuration's properties may be different than what was used in this How-To. However, if the settings between your various configurations are the same, keeping track of what will happen if new components or mates are added will be much easier. Occasionally, it will be necessary to change certain property options for one or more configurations. This is as it should be, but unless you have a photographic memory, reset those options so that all configurations are consistent before saving the assembly and going home for the night.

Showing Alternate Components

Showing certain components in one configuration and different components in another configuration is accomplished using the steps outlined in How-To 15-2. A component is inserted into one configuration (named *Rev1*, for example), but is suppressed in the other configurations. The component would be mated into position in the usual manner.

If you were to switch to another configuration (call this *Rev2*), another new component could be inserted and mated into position. If you were to continue on in this vein, you would wind up with a number of configurations all containing different components. The key to making this scenario work would be to have both the *Suppress component models* and *Suppress features and mates* options checked in every configuration.

Controlling Configurations in Drawings

It is possible to control which configuration is being shown for a part or assembly within a design drawing. The configuration is controlled for each

640 Chapter 15: Assembly Configurations

view, so design drawings containing multiple views can show different configurations for each view. How-To 15-3 takes you through the process of specifying which configuration to show for any particular drawing view. The newly created configurations of the *Sliding Brace* assembly are used in this example (see figures 15-11 and 15-12).

HOW-TO 15-3: Drawing View Configuration Selection

To specify which configuration is to be shown in a particular drawing view, perform the following steps. This assumes, of course, that the part or assembly already contains more than one configuration. The process is exactly the same whether working with a part or assembly drawing view.

1. Right-click on the view you want to specify the configuration for and select Properties.

2. In the Configuration Information section of the Drawing View Properties window, shown in figure 15-13, select *Use named configuration*.

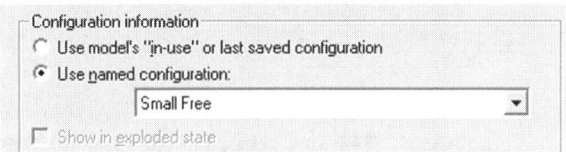

Fig. 15-13. Using a named configuration for a drawing view.

3. From the drop-down list, select the configuration to be shown in that view.

4. Click on OK.

Figure 15-14 shows the results of specifying three different configurations for each of the three drawing views. What may be discernible from the image is that two of the drawing view borders have a dashed appearance and one view uses solid lines for the border. This is to indicate which configuration is currently active in the assembly file. This is true even if the assembly file is not open.

There are many additional benefits to being able to show specific configurations in drawings. Consider some of the uses this functionality could be put to. A configuration may show a particular cut through a part, for instance, which allows for a precise cutaway view of the model. A configuration could show a housing of an assembly with a portion cut from it in

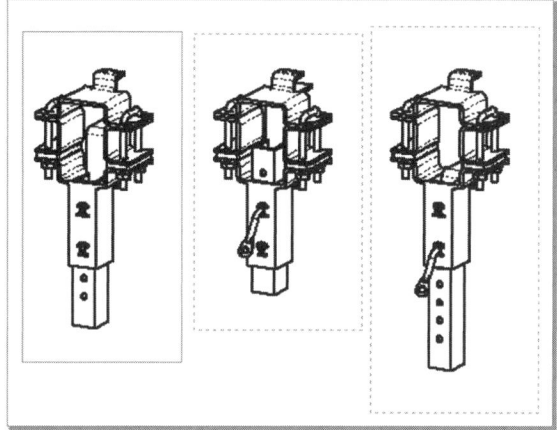

Fig. 15-14. Showing three different configurations in a drawing.

order to show the detailed internal components. Springs could be shown in the compressed and free states. There is no end to the possibilities.

Summary

The configuration a part is using while in an assembly can be controlled from within the assembly. This is accomplished by accessing the Component Properties feature of the part while in the assembly. You can then specify a named configuration to use for the component within the assembly. A similar situation exists in which you are in a drawing and need to see a particular configuration for a part or assembly in a view. By accessing the view's properties, you can specify a named configuration to be shown in that view.

Adding configurations in assemblies works the same as it does for part files. You reviewed this process early in this chapter, and explored what the various options mean when accessing the properties of an assembly configuration. Two very important options are *Suppress component models* and *Suppress features and mates*. By manipulating these options, it becomes possible to easily add new components and mates while developing multiple configurations.

The name that appears for a component in a BOM can be controlled via a configuration's properties. By default, the name of the document will be used in the BOM. Additional options would be to use the name of the configuration itself, or even a user-defined name, which can be anything you decide to type in.

Questions and Topics for Discussion

1. If a part with multiple configurations has been inserted into an assembly, how can a configuration for the part be controlled within the assembly?

2. What is the difference between hiding and suppressing a component?

3. If there are five configurations in an assembly, and a new component needs to be added that will only appear in one of the configurations, which option(s) should be enabled in the properties of the other configurations?

4. Describe one method of overriding the default document name being used in a BOM.

5. Describe three ways in which configurations might prove useful with regard to drawings or drawing views.

6. Describe how multiple instances of a component can have their configurations changed within an assembly simultaneously.

CHAPTER 16

Design Tables

THE ABILITY TO CREATE DIFFERENT CONFIGURATIONS for parts or assemblies is very convenient. The configurations you have learned so far involve controlling feature or component suppression and dimension values on a relatively small scale.

Of course, there is nothing stopping you from creating 100 configurations, but it would be exceedingly time consuming using only the tools you have been given to this point. For greater flexibility and to control dimensions and suppression states on a larger scale, a family-of-parts table is necessary. This functionality requires the use of a design table, which you can create via an Excel spreadsheet.

Be aware that you must have Excel 97 or a later version to take advantage of SolidWorks' design table functionality. No earlier version of Excel is supported, and you cannot use an alternate spreadsheet program and save the spreadsheet as an Excel file. Additionally, it is highly recommended that Service Release 2 be installed if using Microsoft Excel 97. Microsoft Office 2000 and XP are also supported.

Adding a design table to SolidWorks is a simple process, but it is very sensitive to typographical errors. You will learn all of the dos and don'ts regarding design tables in this section. The Windows clipboard can play an important role in the creation of design tables. By copying data to and from the clipboard, you greatly reduce the possibility of human error. In addition, SolidWorks can generate a design table with a minimum of human intervention.

Linking Versus Embedding

In previous versions of SolidWorks, design tables were embedded objects. This is in contrast to linked objects, which retain their connections to the original files. Beginning with SolidWorks 2003, design tables can be linked objects that can be made to update if configured parameters in the model are altered. The terms *embedded* and *linked* are terms that have been created as a by-product of OLE functionality, which stands for Object Linking and Embedding.

If an Excel spreadsheet is embedded in a SolidWorks model, it is saved with the file it is embedded in. There is no connection back to the original spreadsheet. A link to the original file is not stored in the host document.

If an Excel spreadsheet is linked to a SolidWorks model, a link is made back to the original file elsewhere on a hard drive or over a network. The original file (such as a spreadsheet) is not stored within the host document. Rather, it resides at a separate location. These facts are important to note, as they will play an important role later on.

Preparations for Design Table Creation

When creating a design table, the only preparation you really need to be concerned with is renaming your dimensions or features. You already know how to rename features. As far as renaming dimensions is concerned, you need to rename only the dimensions you want to drive through the spreadsheet. Renaming dimensions is not a requirement, it is just good practice. The process of renaming dimensions was covered in Chapter 8, and is reviewed in material to follow. First, let's take a moment to understand how SolidWorks names dimensions.

The full name of a dimension depends on whether you are in an assembly or part. You can almost think of dimension names as being similar to human names. They have first names, second names, and sometimes last names. A dimension's full name might look like either of the following two examples.

```
D1@Sketch2
D2@Boss-Extrude2@Widget<1>
```

D1 or *D2* is the actual name of a dimension. The second part of the name (*Sketch2* or *Boss-Extrude2*) is the name of the sketch or feature the dimension belongs to. In the first example, the dimension's full name consists of just two portions because it belongs to a part file. In the second example, the component name (*Widget<1>*) is added to the dimension's full name because the dimension is part of an assembly. Theoretically, there could be two *D2@Boss-Extrude2* dimension names in an assembly, so the third name is required to differentiate them.

Do not let dimension names scare you. The "at" (@) symbol is nothing more than a character SolidWorks understands as a separator. Other than that, it has no importance from a user perspective. The full assembly dimension names are only required when adding a design table to an assembly, which we will tackle in the second half of this chapter.

Because dimension names must be supplied to the spreadsheet, it helps to have descriptive names for the dimensions instead of the nondescriptive names SolidWorks uses. For instance, names such as Length or Depth have a lot more meaning than do SolidWorks' names of *D1* or *D2*.

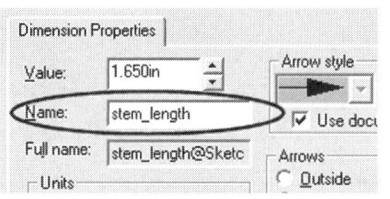

Fig. 16-1. Changing a dimension's name.

To rename a dimension, first access the dimension in the usual way (double clicking on a feature), so that it is displayed on screen. Second, right-click on the dimension value and access its properties. The area in the Properties window containing a dimension's name is shown in figure 16-1.

Note that the Name area of the window has been changed to *stem_length* as opposed to one of SolidWorks' default names, which would be *D1*, *D2*, or something similar. The dimensions that will be driven by the design table have all been renamed in this fashion so that they will make more sense when placed in the design table. Figure 16-2 shows a screen shot of the valve stem with some of its dimensions renamed, along with the dimension names themselves. See the next section to discover how dimension names can be displayed.

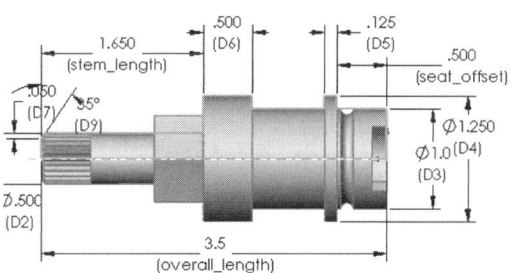

Fig. 16-2. Showing some of the renamed dimensions.

Turning on Dimension Names

There is an option that exists specifically for displaying dimension names. The option is titled *Show dimension names*, located in the General section of System Options (Tools > Options). This option is shown in figure 16-3. Simply check it to turn on dimension names, and then access dimensions as you normally would.

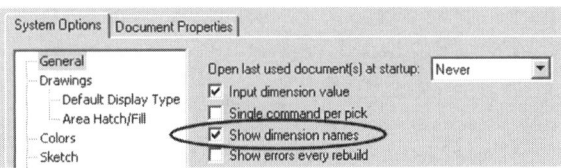

Fig. 16-3. Show dimension names option.

It should be mentioned that turning on dimension names is not a requirement when creating design tables. It is just a convenience that some users may prefer. Sometimes it helps to be able to see the dimension names when creating the design table.

Displaying All Model Dimensions

This is another option that will make creating design tables somewhat easier. Because getting the dimension names into the design table is of prime importance, it would be beneficial to be able to see all dimensions at the same time. This would make the dimensions easily accessible.

In contrast, however, if the model is a complex one and there are many dimensions, it may not be advisable to show all dimensions at once. Too many dimensions on the screen at the same time and you wind up looking at a black blob. This certainly would not be a productivity enhancement!

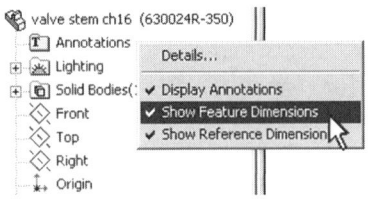

Fig. 16-4. Enabling the Show Feature Dimensions option.

The Annotations object in FeatureManager is a shortcut for performing a few simple functions. One of these functions is the ability to show all dimensions in the model. By right-clicking on Annotations and selecting the option Show Feature Dimensions, shown in figure 16-4, you can turn on all dimensions in the model. To turn the dimensions back off, simply uncheck the option.

Creating a Design Table

There are a couple of ways in which a design table can be created. Much of the process involves getting dimension names into the table. Much of this tedium has been automated. You should find that creating design tables in general is a fairly streamlined process.

Once you have renamed relevant dimensions, you are ready to begin the design table. The aspects of working with Excel are outside the scope of this book, but the process of creating a design table is simple and you should be able to get by if you have even the most basic understanding of Microsoft Excel software.

	A	B	C	D	E	F
1	Design Table for: valve stem ch16					
2		$COMMENT	overall_length@Sketch1	stem_length@Sketch1	seat_offset@Sketch1	$state@O-ring
3	630022R-325		3.250	1.650	0.500	U
4	630022R-350		3.500	1.650	0.500	U
5	630024R-350	The 630024R	3.500	1.750	0.825	S
6	630022R-400		4.000	1.650	0.500	U
7						

Fig. 16-5. The beginnings of a design table.

Configuration names will go down the first column, from cell B1 on down. Dimension names (and other table parameters discussed in material to follow) are typically placed in the second row. To the right of the configuration names and below the dimension names go the dimension values themselves. These are the dimension values that will be used to drive the dimensions in the newly generated configurations. Figure 16-5 shows an example of a design table.

When naming configurations it is best to stick with alphanumeric characters, such as A through Z, lower- or uppercase, and numbers 0 through 9. You can also use the hyphen (-) or underscore (_), but it would be in your best interest to stay away from using other symbol types. Using nonstandard characters in configuration names might have unexpected results. Officially, the only characters that cannot legally be used to name configurations are the forward slash (/) or the "at" symbol (@).

How-To 16-1 takes you through one method of creating a design table. There are a number of options that go along with creating a design table, explained in material to follow.

HOW-TO 16-1: Creating a Design Table

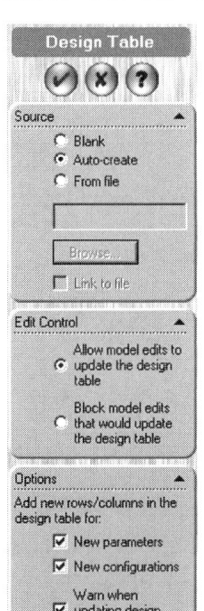

To add a design table to a SolidWorks part or assembly document, perform the following steps.

1. Select Design Table from the Insert menu.
2. Select Blank or Auto-create from the Design Table panel, shown in figure 16-6. The Auto-create option will be used for this How-To example.
3. Click on OK.
4. In the Dimensions window that appears, shown in figure 16-7, select the dimensions you wish to control from the design table.
5. Click on OK.

Fig. 16-6. Design Table panel.

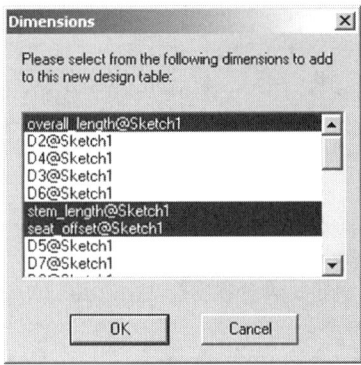

Fig. 16-7. Dimensions window.

6. Add parameters or configurations to the design table as necessary.

7. Click outside the design table to complete the process of adding the design table to the model.

During step 4, you noticed that the Dimensions window appeared. If you renamed the dimensions you wished to drive from the design table, you discovered that selecting those dimensions from what can be a very extensive list becomes an easy task. Clicking on OK places any selected dimensions into the design table for you.

Place down the first column the names of the configurations you want to create. The default configuration will be added automatically because it is already present in the model (unless it was deleted). You do not have to retain the default configuration if you do not want to, though it would not hurt anything leaving it there. Keeping the default configuration can sometimes be a good idea, in case the new configurations do not turn out quite as expected (not an uncommon occurrence with new users).

✓ **TIP:** *Formulas can be used within the design table to control dimension values. For example, one cell can always equal half another cell by typing in the Excel formula for that operation (i.e., =C3/2).*

In the body of the design table, enter the dimension values for the various configurations listed in the first column. Incidentally, any formatting can be used when creating a design table. This includes formatting cells for a particular decimal place precision, adding color to the table, rotating text, and so on. Feel free to use a little artistic license to add a little flair to your design table.

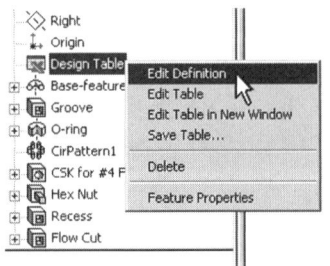

Fig. 16-8. Design Table object in FeatureManager, and related menu.

Once the design table has been generated, a new Design Table object will appear in FeatureManager. This is shown in figure 16-8. Right-clicking on the Design Table object will display a menu, also shown in figure 16-8. Selecting Edit Definition will bring you back to the Design Table panel, shown in figure 16-6. Edit Table will open the design table itself.

The Edit Table in New Window option opens the design table in Microsoft Excel rather than the Excel application running within SolidWorks. This has the advantage of giving you an unfettered view of your SolidWorks model and menu structure. Use the Alt-Tab hot key combination to switch between SolidWorks and Excel.

Adding a Blank Design Table

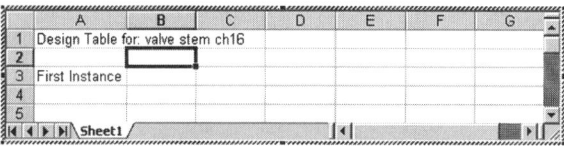

Fig. 16-9. A blank design table.

The Blank option found in the Design Table panel allows for creating a design table in a somewhat more manual manner. If used, a blank table is created such as that shown in figure 16-9. The first row of a design table will contain the text Design table for:, followed by the name of the part.

The name *First Instance* is a dummy name for a new configuration. Think of it as a placeholder. Whether this configuration is renamed or kept the way it is depends strictly on the user's preference. Almost certainly, more configurations will need to be added, which can be done by simply typing them in the first column. Make sure not to skip rows, as SolidWorks stops processing data as soon as it hits a blank row. For that matter, skipping columns is also prohibited.

Obviously, a blank design table will not do you much good unless you complete it by adding some dimensions (or other parameters) to the table. This can be accomplished by double clicking on a dimension name. Double clicking on a dimension places the name of the dimension into the first available column, along with the current dimension value directly below it. Make absolutely certain the second cell down in the first available column is highlighted, or this process will not work!

If the dimensions have been shown, double clicking on the dimension value to be placed in the design table is a simple matter. But what if the dimensions are not being shown? As you are aware, double clicking on a feature will show the dimensions related to that feature. However, when a design table is being created, double clicking on a feature places that feature into the design table, along with the $STATE heading. This is a special heading that allows for controlling the suppression state of a feature. The $STATE heading is discussed in greater detail in the next section.

If you do not wish to use the $STATE heading, temporarily access an alternate cell prior to double clicking on a feature to access its dimensions. For example, the second cell down in the first available column will always be highlighted automatically by SolidWorks. This is necessary, but if the third cell down is highlighted (for example) nothing happens within the design table when double clicking on model dimensions or features. In this way it becomes possible to trick SolidWorks into not adding anything to the table while you access the appropriate dimensions. Just make sure to highlight the second cell down in the next available column prior to double clicking a dimension in order to add it to the table.

✒ **NOTE:** *Double clicking on items to add parameters to a design table also works if the table was created using the Auto-create option.*

Feature Suppression in Design Tables

You have already learned how to suppress or unsuppress features in a part. You can also control the suppression of features from a spreadsheet, and the process for doing so is quite simple. The process amounts to inserting the name of the feature into the design table, along with a special heading.

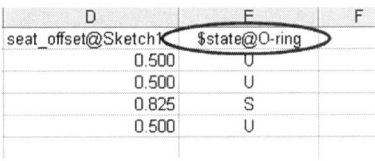

Fig. 16-10. Controlling the suppression of the O-ring feature.

There are a few special headings that can be used at the top of a column in a design table (or sometimes at the beginning of a row) that will add functionality to the design table. One such heading is the $STATE heading. When this heading is used, feature suppression can be controlled. The controlling characters in this case are U to unsuppress a feature and S to suppress it. Figure 16-10 shows the design table for the valve stem using this functionality.

O-ring is the name of a feature whose suppression is being controlled. Looking at the sample design table in figure 16-10, you can see that the O-ring feature will be turned on in all but one configuration. Casing is not important. Additionally, you could spell out the entire word unsuppressed or suppressed as well, but why bother?

The $STATE heading can be used to control the suppression of items other than features. For example, lighting sources and equations (discussed in Chapter 17) can have their suppression state controlled. If you wish to control the state of an equation, use the special syntax of $STATE@ equation_number@EQUATIONS, where equation_number is the number of the equation you wish to control the state of.

There are other special headings in addition to the $STATE heading (though this is the most commonly used). These are outlined in Table 16-1. The table includes descriptions of special heading functionality, with more in-depth explanations of their usage in material immediately following the table. (To reiterate, special headings are not case sensitive. They are just shown that way so that they will stand out from other text.)

Table 16-1: Design Table Headings for Parts

Heading	Function
$STATE@feature_name	Controls the suppression of features via the controlling characters S and U.
$CONFIGURATION@part_name	Controls which configuration of a part is referenced when that part has been inserted into another part.
$PARENT	Allows for creating derived ("nested") configurations.
$PARTNUMBER	Allows for adding text in a column. The text will be used as the part number in a bill of materials.
$COMMENT	Allows for adding text in a column. Additionally, text will appear in the configuration's properties in the Comment area.
$USER_NOTES	Allows for adding text in a column or row.
$COLOR	Allows for controlling the color of the model in each configuration.

As is indicated by Table 16-1, the $CONFIGURATION heading must be followed by the name of a part. Below the heading in the design table, enter configuration names the model will reference. These names must be valid configuration names that exist within the part inserted into the model containing the design table. For more on the topic of inserting parts into an existing part file, see Chapter 13.

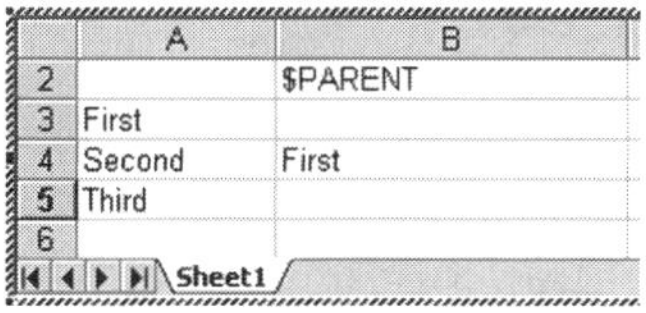

Fig. 16-11. Proper use of the $PARENT heading.

The $PARENT heading is one of the more obscure design table headings. It allows for creating nested configurations, but must be used in a particular way. Figure 16-11 shows an example of how this heading can be put to use. The three configurations being generated are named *First, Second,* and *Third*. The configuration *Second* will be nested under the configuration First once the configurations are generated. This works because SolidWorks creates the configurations from top to bottom.

If an attempt was made to type in the word *Third* in cell B4, SolidWorks would issue an error message. This is because it would be like instructing SolidWorks to make the configuration Second a nested configuration of the configuration Third before the configuration *Third* existed, which is not possible. See Chapter 10 for more on configurations.

Using the *$PARTNUMBER* heading at the top of a column will force whatever data is in the cell below it to appear in a BOM in the Part Number column. This assumes that the part configuration is used in the assembly the BOM is being generated for. By default, an assembly component's file name is used as the part number. There are other parameters that can be used in the column cells below the $PARTNUMBER heading. All parameters for the *$PARTNUMBER* heading are listed in Table 16-2.

Table 16-2: Part Number Heading Parameters

Parameter	What the BOM Will Display
Any text	Text is displayed in the BOM as a custom name.
Cell is left blank	Configuration name.
$DOCUMENT	Document file name.
$CONFIG	Configuration name.
$PARENT	Parent configuration name.

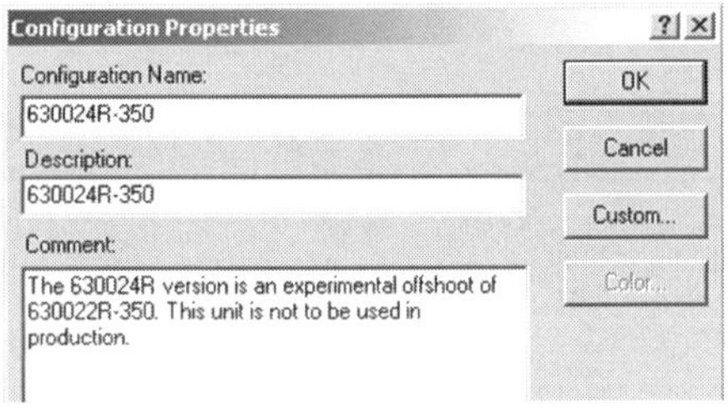

Fig. 16-12. These comments were inserted via a design table.

If the $COMMENT heading is used, any text added in the cells below the heading will automatically appear in the appropriate configuration's properties. An example of this is shown in figure 16-12. The $USER_NOTES heading is a more generic version of the $COMMENTS heading and is used strictly for adding text. It is a way of warning SolidWorks not to try to process any data, as the data is only text for the user's reference.

If the $COLOR heading is used, a 32-bit integer must be specified to dictate the red, green, and blue (RGB) values for the color. Instead of trying to understand how the computer program interprets the RGB values, it is easiest if you stick with understanding the simple math behind the final number that relays the color information to SolidWorks.

RGB values are typically specified as values ranging from 0 to 255. A value of 255-0-0 means that the red component is "turned up" all the way, and green and blue are both set to 0. This would give you the pure color red. Likewise, 0-0-255 would be blue.

To convert the RGB value to a 32-bit integer, we must multiply the green and blue values by a certain amount. The formula is as follows.

```
Red + (Green * 16^2) + (Blue * 16^4)
```

In other words, red plus green times 16 squared plus blue times 16 to the fourth. Let's try an example and see how it works. A nice aquamarine color would probably have a small red component value, and higher green and blue component values. Let's make red = 60, green = 220, and blue = 220. Therefore, the formula can be worked out as follows.

```
60 + (220 * 256) + (220 * 65536)
60 + 56320 + 14417920 = 14474300
```

The number 14474300 is exactly what you would type in the cell below the $COLOR heading to obtain the desired color for that particular configuration. It certainly helps if the RGB value is known ahead of time. Many paint programs will give the RGB values for colors, and there are many freeware programs that can be obtained on the Internet.

Design Table Options

Whether or not a design table will update to reflect dimension changes in the model is something you have control over. This is known as edit control. There are only two options in this particular category, and they can be accessed in the Design Table panel. Edit the definition of the design table to gain access to any of the various options discussed in this section. Refer back to figure 16-6 to see the options discussed herein.

Edit control is very straightforward. Allowing model edits to update the design table means that any dimensional changes made to the model will propagate to the table. Obviously, this assumes that the dimensions being edited are in the table. If they are not, whether or not they update in the table is not an issue.

The alternative to seeing dimensional changes propagate to the table is to lock dimensions so that they cannot be changed if they are already being driven via the design table. After all, you would not want to develop a conflict of interest, which is exactly what would happen if dimensions could be changed from two different locations (the table or the model). If you choose to block model edits that would update the design table, you must edit the design table to change the dimension values.

If you choose to allow model edits to update the table, you can have SolidWorks warn you when a dimensional change or other edits are about to be made that will alter the table. This option is labeled Warn when updating design table. If enabled, the message shown in figure 16-13 will appear if model modifications will affect the table.

Fig. 16-13. Warning that a change to the table will occur.

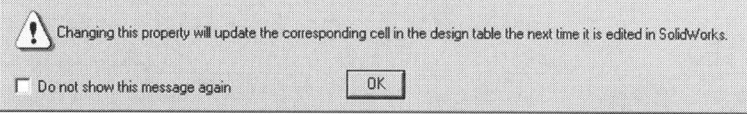

Adding Rows and Columns Automatically

Aside from dimensional changes updating in a design table, you can also have new rows or columns added to the table automatically if the software deems it necessary. For example, if a new configuration is manually added in ConfigurationManager, a new row will be added to the design table listing the new configuration. If a new dimension is altered, or if a new fea-

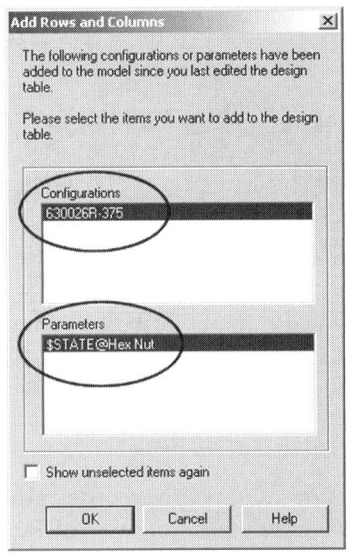

ture has had its suppression state altered, new columns can be added to reflect these additional parameters.

To have newly added configurations automatically added to the design table, check the *New configurations* option. To have modified dimension or suppression state parameters added to the table automatically, check the *New parameters* option. If any manual changes were made, the Add Rows and Columns window, shown in figure 16-14, will appear the next time the design table is edited.

When the Add Rows and Columns window appears, you must select the new configurations or parameters you wish to have automatically added to the table. SolidWorks will automatically add the appropriate rows or columns, but then it is up to you to finish adding the information in the body of the design table. For example, it would be necessary to type in the dimension values for a newly added dimension parameter, or to enter the letter S or U to represent the suppression state of a newly added $STATE heading.

Fig. 16-14. Add Rows and Columns window.

Saving and Linking to Design Table Files

You have so far discovered that generating a new design table can be a somewhat automated process, and that alterations made to a model can be reflected in the design table. The design table is an embedded object at this point. The table can be made to update to reflect model changes, but it is not a separate file.

To save the design table as a separate file, right-click on Design Table in FeatureManager and select Save Table. By default, the name of the table will be the same as the name of the SolidWorks document. The spreadsheet will also be placed in the same folder as the model. It is a good idea to accept these defaults, as it makes finding the table much easier.

Once the design table has been saved as a separate file, you will more than likely want to link the model to that file so that the table will retain its associativity. This is accomplished by checking the *Link to file* option, shown in figure 16-15, and then using the Browse button to select the previously saved file. If a link is not established, the spreadsheet is just a standalone copy with no association back to the model. However, breaking the associativity to a design table saved as a separate file may be desirable for documentation purposes. For example, a table might need to be sent to a different department.

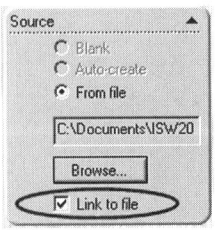

Fig. 16-15. Link to file *option.*

It should be noted that it is possible to insert an existing design table into a model. This can only be accomplished if there currently is no design table in the model. By selecting Insert > Design Table, and then selecting the *From file* option, it becomes possible to click on the Browse button and select a previously created design table that will generate new configurations within the model document.

The method of browsing for a previously created design table is not as commonly used as it once was. The reason for this is that it requires a lot more work to manually enter the dimension names and other headings into a table via the keyboard. Even copying and pasting dimension names into the table is more laborious than letting SolidWorks do it for you. Hence, browsing for a previously created design table is not necessary.

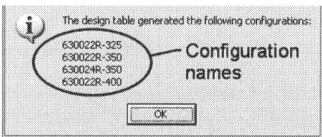

On the off chance you wind up inserting a previously created design table into a model file, SolidWorks will open the table inside a SolidWorks session. You could then continue to modify the table manually. Once you exit the table in the usual fashion, a confirmation will appear that states what configurations were added. A sample of this confirmation window is shown in figure 16-16.

Fig. 16-16. Confirming new configurations were generated.

ConfigurationManager Symbology

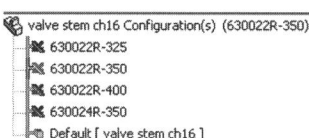

Fig. 16-17. Symbols in ConfigurationManager.

Configurations generated via a design table are displayed differently in ConfigurationManager than those added manually. An example of ConfigurationManager is shown in figure 16-17. Note the appearance of the configuration icons. Compare this to the default configuration, whose icon is also visible, and you can see the difference.

Figure 16-18 shows four new configurations generated from a design table. They have been annotated so that they can be compared to the data shown in the design table sample shown in figure 16-5. Note that the *O-ring* feature is missing from configuration 630024R-350 due to it being suppressed in the design table.

Fig. 16-18. New configurations generated with a design table.

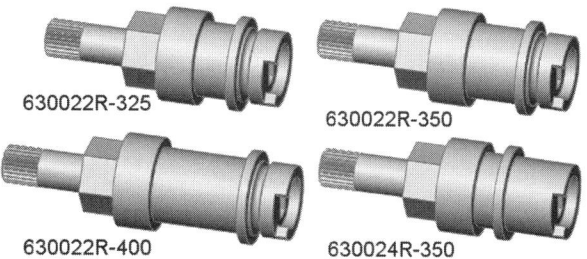

Design Table Editing

When a design table is being edited, the Excel menus and toolbars appear once again. During the process of editing the design table, make sure you do not click outside the table in a blank portion of the work area. If you do, you will be kicked out of edit mode and placed back into the Solid-Works environment. How-To 16-2 takes you through the process of editing a design table.

HOW-TO 16-2: Editing a Design Table

To edit a design table, perform the following steps.

1. Right-click on the design table listed in FeatureManager and select Edit Table.

2. Make the necessary modifications to the design table.

3. Click outside the design table in a blank portion of the work area.

4. If a new configuration was added and the confirmation window appears, click on OK to acknowledge that the new configurations have been created.

When editing a design table, you can rename or add configurations, modify dimensions, or add new dimension names for other dimensions you want to control within the design table. Feel free to add new column headings, change the color formatting of the cells, or basically anything needed, as long as the changes do not break the fundamental rules of design tables.

As mentioned previously, new configurations or parameters can be added automatically, with only a minimum amount of user intervention. Additionally, dimension names can be inserted into the table by double clicking on them. Features can be added, along with the *$STATE* heading, by double clicking on a feature. If you wish to add parameters in this fashion, bear in mind that when the design table is edited a cell will be highlighted. This cell will be the second cell down in the next available column.

> **NOTE:** This is extremely important! If the second cell down in the first available column is not highlighted, automatic insertion of design table parameters will not function.

If a design table was inserted from an existing file, inserting parameters by double clicking does not work. In such a case, you would either need to manually type in new parameters or copy and paste parameters (such as dimension names) using the Windows clipboard.

When opening old SolidWorks documents, you may not know how the design table was created. If this is the case, look to the first row in the table. If the first row contains the text *Design table for:*, you know new parameters can be added via double clicking. If the first row contains parameters, the table was created manually.

> ✓ **TIP:** If working with a legacy design table created manually, you can delete a table and then recreate it using Auto-create (see How-To 16-1). This allows for easier editing of the table.

Deleting a Configuration

If you remove a configuration from a design table, such as by deleting an entire row, SolidWorks will realize this and ask if you want it deleted from the ConfigurationManager list as well. If a design table was not used to create configurations, deleting a configuration can be accomplished by selecting it from ConfigurationManager and pressing the Delete key.

If a design table is used, deleting the configuration from ConfigurationManager will not permanently remove the configuration from the model file. The act of editing the design table forces SolidWorks to reevaluate the data, thereby reinstating the previously deleted configuration. This is true even if nothing was changed in the design table. To permanently delete a configuration, delete the row containing that configuration in the design table.

If a row is deleted from a design table, SolidWorks asks whether or not you want to retain the associated configuration. An example of this is shown in figure 16-19. All that is necessary at this point is to click on the Yes button, and the configuration will be permanently removed from the model file. If No is selected, the configuration will be retained, but it will not be driven via the design table.

Fig. 16-19. Deleting a configuration.

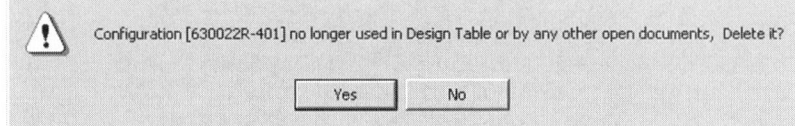

✓ **TIP:** *If you want to temporarily delete a configuration from the design table, replace the configuration name with the heading $USER_NOTES. This will disable the configuration in the design table (and ConfigurationManager if desired) yet retain the data in the table in case it is needed in the future. The configuration can then easily be reinstated by reverting to the original, or another, configuration name.*

Assembly Design Tables

When used in an assembly file, design tables have a different functionality than they do when used in a part file. A very good place to start would be with a quick review of what design tables can accomplish within a part. In contrast, what can be accomplished within an assembly design table will also be listed. Following this comparison is an in-depth look at the various controls used in an assembly design table. This includes examples of an assembly design table that will help to reinforce the topics learned.

Table 16-3 will give you a better understanding of just what can be accomplished in part design tables in comparison to assembly design tables. The differences are significant, so you will want to understand the capabilities of both types of design tables to have a better understanding of what your options are. When a particular parameter does not apply, that parameter is noted as being "not applicable" (N/A) for the part or assembly design table.

Table 16-3: Part and Assembly Design Tables Compared

Part Design Tables	Assembly Design Tables
Can control feature dimensions	Cannot control feature dimensions (associated with parts)
Can control suppression of features	Cannot control suppression of features (associated with parts)
Can contain user-defined notes	Can contain user-defined notes
Can add comments to configurations	Can add comments to configurations
Can control part number used in a BOM	Can control part number used in a BOM
N/A	Can control if assembly components are listed in a BOM
Can define custom properties	Can define custom properties

Part Design Tables	Assembly Design Tables
Can control part color	N/A
N/A	Can control mate dimension values (i.e., distance and angle)
N/A	Can control assembly feature dimension values
N/A	Can control component suppression
N/A	Can control suppression of assembly features and mates
N/A	Can control component visibility
N/A	Can control component configuration usage

Assembly design tables cannot be used to control the individual feature dimensions within a part, nor can they control the suppression state of individual features. However, an assembly design table can control what configuration of a component is being used in an assembly. In combination, part and assembly design tables can offer some extremely powerful tools for creating a family of parts or assemblies. Additionally, assembly design tables can be used to control many aspects of an assembly above and beyond what can be accomplished in a part. The section that follows describes the headings and control characters that can be used within an assembly design table.

Controlling Assembly Feature and Mate Dimensions

The only types of dimensions that can be controlled from an assembly design table are mate dimensions and any dimensions associated with the assembly itself. Examples of assembly dimensions (other than mate dimensions) would be dimensions associated with assembly features or perhaps a layout sketch created in the assembly. There are only two mate types that contain dimensions. These are distance mates and angle mates. The syntax used when adding a dimension to an assembly design table is the full dimension name. Therefore, the dimension name might appear as follows.

```
D1@Distance1
```

Because dimension names can be renamed, and because mate relationships can also be renamed, you need not suffer with the nondescriptive default dimension name. As an example, it would be possible to perform a slow double click on a mate relationship in the *Mates* folder and rename the mate to something such as clearance. Additionally, the name of the dimension associated with this same mate could be altered.

There are a few dimensions that can have the value of zero, and this may aid you when defining the design table. For instance, when creating an offset plane, the offset can be 0. When creating a distance mate, the distance can be 0. This can be very useful, because 0 can be used in a distance mate to create what amounts to a coincident relationship. If necessary, a value can be specified in place of 0 to establish an offset distance.

Design tables are created in an assembly in the same fashion they are created within parts. Assembly design tables are also edited in the exact same fashion as part design tables. The primary difference between part and assembly design tables centers on the parameters that can be used. Table 16-3 outlined these parameters, which are explained in detail in the following material.

Controlling Component Suppression

The controlling characters used to indicate whether a component is being suppressed or not in an assembly are different from those when specifying feature suppression in a part design table. The heading is still the same. The proper terminology to use when indicating suppression of features in a part is suppressed or unsuppressed. The terminology with regard to assembly components is suppressed or resolved. This has to do with lightweight components, discussed in Chapter 12.

When the suppression state of a component needs to be controlled within an assembly design table, the heading used is $STATE. Another important factor is to use the full name of the component being controlled. That is, the component's ID number must be appended to the component's name. An example would be as follows.

`$STATE@Lynchpin<1>`

The *<1>* suffix is the ID number that tells SolidWorks which component is being controlled in the table. If there are multiple components in an assembly, and only certain components are to be controlled, the numbers between the <> symbols can reflect those components using dashes and commas. For instance, *<1-12,14,18>* would mean that components 1 through 12 and components 14 and 18 will be controlled in the table. Controlled how? That depends on your intentions. This formatting works with any of the headings that require that a component name be included in the heading. The following is an example regarding this topic.

`$STATE@Bolt<1-12,14,18>`

Assembly Design Tables

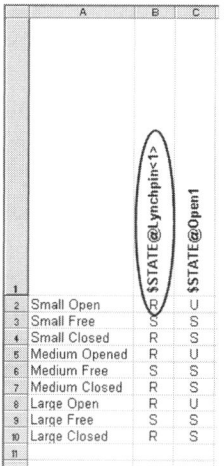

This example would control the suppression state of all components named *Bolt* in the assembly whose ID numbers are 1 through 12, 14, and 18. An asterisk can also be used to control every occurrence of a component in the assembly, such as <*>. An example of the *$STATE* heading in use is shown in figure 16-20. The text in the top cells has been rotated to conserve space.

The controlling character when using the *$STATE* heading for components is either the letter *S* for suppress or *R* for resolve. Do not use the letter *U*, for unsuppress, as this will not work. The letter *U* is reserved for unsuppressing assembly features or mates in assembly design tables. This is discussed in the next section.

Fig. 16-20. Controlling the suppression of the lynchpin.

Controlling Assembly Feature and Mate Suppression

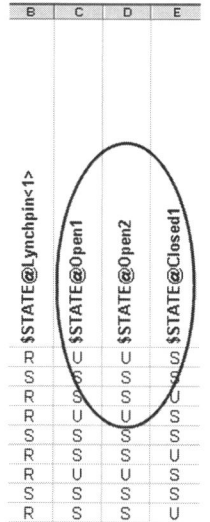

The heading used to control the suppression state of assembly features and mates is exactly the same as that used to control the state of components. In that the *$STATE* heading is the same heading used to control component suppression, care should be taken to use the proper controlling character, which is *S* for suppressed and *U* for unsuppressed. Some examples of the proper syntax would look like the following.

```
$STATE@assembly_feature_name
$STATE@mate_name
```

Remember, the letter *R* is reserved for components only. Figure 16-21 shows an example of mates being controlled in various configurations of an assembly. What is occurring in this example is different combinations of mates are either suppressed or unsuppressed, resulting in certain components being mated in different positions.

Fig. 16-21. Controlling mate suppression.

Controlling Component Visibility

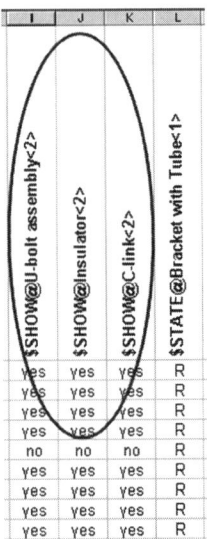

Fig. 16-22. Hiding some components in a configuration.

What is the difference between visibility and suppression? When a part is suppressed, it is removed from memory. When it is hidden, it is only removed from view. If hidden, a component's mates are still present and accounted for. This is not the case when suppressing a component. This can be significant, because you would not want your assembly to fall apart, which could be the case if too many components are suppressed.

The heading used to control a component's visibility is $SHOW. The controlling character is either Y for yes (show the component) or N for no (do not show the component). The words *yes* and *no* can also be spelled out, if desired. An example of the $SHOW heading in use is shown in figure 16-22.

There are three components in figure 16-22 having their visibility controlled. The three components are visible in all configurations (whose names are not being shown) except one. The *U-bolt assembly<2>*, *Insulator<2>*, and *C-link<2>* components will all be hidden in the fifth configuration. Their mate relationships, however, will still be active and holding things in place.

Specifying a Component's Configuration

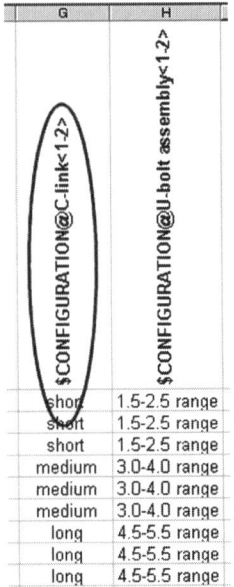

Fig. 16-23. Controlling the C-link configurations.

This is an extremely important aspect of assembly design tables. It is very important to be able to specify what configuration of a part an assembly is using. This is even more important considering the fact that feature dimensions cannot be controlled within an assembly design table, nor can feature suppression states.

The heading used to control which configuration of a component is being used is $CONFIGURATION. This should be followed by the name of the component. An example of the full heading might look like the following.

$CONFIGURATION@C-link<1-2>

This heading will control which configurations are being used for the C-link components whose ID numbers are <1> and <2>. In other words, the components being controlled are named *C-link<1>* and *C-link<2>*. In the cells below the heading, the actual names of the configurations to be used in the assembly must be spelled out. Spelling is critical here. If a configuration name is misspelled, SolidWorks will not recognize it.

It should go without saying that the part configurations must be created before this function will work. In figure 16-23, the configurations of the C-link components are being called out in the design table. There are three different configurations being specified. These are short, medium, and long. Incidentally, configurations for the *U-bolt* assembly are also being controlled.

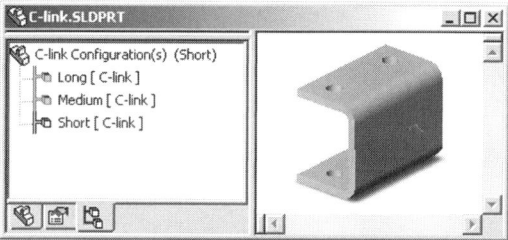

Where do the names short, medium, and long come from? They are the names of the configurations in the *C-link* part. If the *C-link* part is opened, and you look at ConfigurationManager, these are the names of the configurations that would be found. This is shown in figure 16-24.

Fig. 16-24. The C-link's configurations.

Adding Notes and Controlling Part Numbers

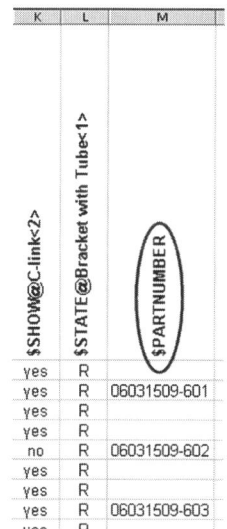

The act of adding notes to a design table is a simple process, and functions exactly the same as it does when creating part design tables. The heading *$USER_NOTES* can be used at the top of a column or at the beginning of a row. The *$COMMENT* heading can be used only at the top of a column. Whatever appears in a cell below the *$COMMENT* heading will automatically appear in the associated configuration's properties in the Comment section.

The *$PARTNUMBER* heading (one word, no spaces) also functions in both assemblies and parts the same way. To reiterate, it is possible to specify what part number will appear in a BOM if the assembly configuration is a component in an upper-level assembly. Typically, the part number is the name of whatever the file's name is. With multiple configurations, though, the file name does not always suffice.

In figure 16-25, only three of the configurations have any data specified for the part number. The part number column of the BOM will revert to whatever the configuration name happens to be for those configurations, with no data under the *$PARTNUMBER* heading. For those configurations where data has been specified, that string of text will appear in the BOM part number column.

Fig. 16-25. Controlling what part number is used in a BOM.

Controlling Subassembly Expansion in a BOM

Another heading that has to do with a BOM is the heading *$NEVER_EXPAND_IN_BOM*. It should be used in a situation in which a subassembly should not be expanded to show its components in a BOM. You would not want to use this heading when it is desirable to show the components contained within the subassembly.

To implement this function, place the $NEVER_EXPAND_IN_BOM heading at the top of a column and use the controlling characters Y for yes or N for no. The full words *yes* and *no* can be used as well. Specifying yes will turn on the option and will never expand the subassembly's components, regardless of any options used when inserting the BOM.

A Pictorial Overview

Assembly design tables can be confusing to the uninitiated. For this reason, it would be beneficial to see some of the design table options in action. The design table shown in figure 16-26 will be used. Use it as a reference to the illustrations and descriptions that follow. This will help give you a more clear understanding of assembly design tables and what they can accomplish.

Fig. 16-26. A sample assembly design table.

	$STATE@Lynchpin<1>	$STATE@Open1	$STATE@Open2	$STATE@Closed1	$STATE@Closed2	$CONFIGURATION@C-link<1-2>	$CONFIGURATION@U-bolt assembly<1-2>	$SHOW@U-bolt assembly<2>	$SHOW@Insulator<2>	$SHOW@C-link<2>	$STATE@Bracket with Tube<1>	$PARTNUMBER
Small Open	R	U	U	S	S	short	1.5-2.5 range	yes	yes	yes	R	
Small Free	S	S	S	S	S	short	1.5-2.5 range	yes	yes	yes	R	06031509-601
Small Closed	R	S	S	U	U	short	1.5-2.5 range	yes	yes	yes	R	
Medium Opened	R	U	U	S	S	medium	3.0-4.0 range	yes	yes	yes	R	
Medium Free	S	S	S	S	S	medium	3.0-4.0 range	no	no	no	R	06031509-602
Medium Closed	R	S	S	U	U	medium	3.0-4.0 range	yes	yes	yes	R	
Large Open	R	U	U	S	S	long	4.5-5.5 range	yes	yes	yes	R	
Large Free	S	S	S	S	S	long	4.5-5.5 range	yes	yes	yes	R	06031509-603
Large Closed	R	S	S	U	U	long	4.5-5.5 range	yes	yes	yes	R	

The assembly used in the following examples is called the *Sliding Brace*. It contains a total of nine configurations. One configuration, *Default*, will be ignored. The other configuration names attempt to reflect either the size of some of the components in the assembly or the positioning of components. Other than that, the configuration names are not important.

In figure 16-27, the configurations *Small Free*, *Medium Free*, and *Large Free* are all shown (left to right). There are two major points you should notice about these configurations. One main difference is that the *C-link* and *U-bolt* assembly components are changing. To be more technically accurate, it is the configurations of the *C-link* and *U-bolt* assembly components that are changing. In the assembly design table (see figure 16-26), the small, medium, or large version of the *C-link* assembly is being specified, and therefore used, within the overall assembly. The same holds true of the *U-bolt* assembly, which uses descriptive configuration names that describe its working range.

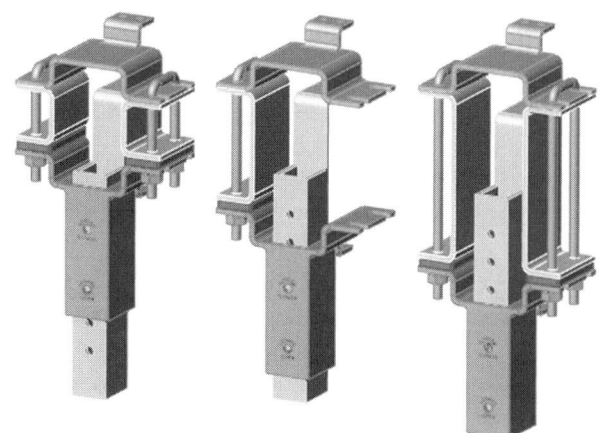

Fig. 16-27. Three different configurations of the Sliding Brace *assembly.*

The second major point worth noting in figure 16-27 is that in the *Medium Free* configuration (middle image) there are some components not being shown. This is due to the three *$SHOW* headings in the design table. Note the word *no* below these headings for the *Medium Free* configuration. This is the reason certain components are not visible.

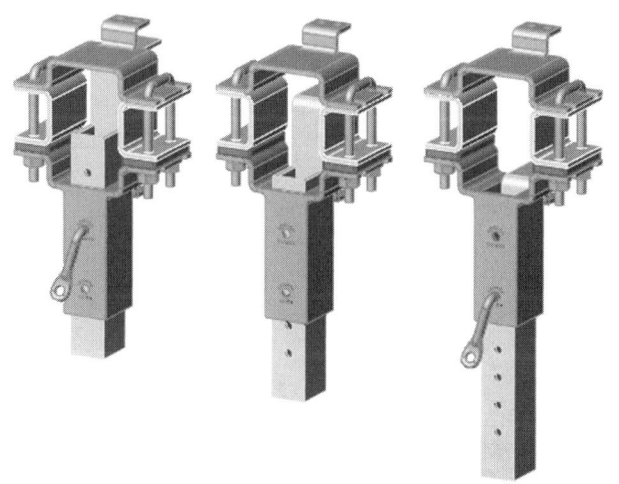

Fig. 16-28. Three configurations in which mates are being controlled.

Next, let's look at three other configurations in which different parameters are being controlled. Looking at figure 16-28, we see three configurations, one of which does not contain the lynchpin. This is a direct result of the *$STATE* heading being used in the design table. Additionally, the lynchpin is in different positions in the first and third configurations. This is a result of different sets of mates being either suppressed or unsuppressed. For instance, one concentric mate is being used to position the pin in the upper hole, and in another configuration a different concentric mate positions the pin in the lower hole.

As is evident from the previous illustrations, there is a great deal that can be accomplished through the use of design tables. There is no realistic limit to the number of configurations that can be created, but there is a limit as to how many parameters are controlled by the table. The current limit is set at 255 parameters. This is due to an Excel limitation. Spreadsheets can contain up to 256 columns. If we reserve one column for configuration names, that leaves 255 columns for parameters.

Most SolidWorks users will not have to be concerned with hitting the 256 column limitation. Only under extreme situations or isolated instances will this problem be faced. If this limitation does present itself, there are ways around it, such as using Visual Basic scripting to achieve design-table-style functionality. This topic is beyond the scope of this book.

Showing Design Tables in Drawings

Both design tables and BOMs use Excel spreadsheets. It is sometimes easy to confuse the two, but they are two very different objects. Try to keep the two functions separate in your mind. Design tables drive parameters in part or assembly files, whereas a BOM is a list of the components in an assembly.

Design tables can easily be shown in a drawing. As a matter of fact, the process is so simple that the steps need not be spelled out as a How-To. The process is a matter of selecting a view in the drawing and selecting Design Table from the Insert menu. SolidWorks places the table in the drawing, whereby it can be scaled or moved as required. To move the table, drag it to a new location. To scale it, drag the small black handles on the perimeter of the table that appear after it has been selected.

There is one problem in particular that is often faced when displaying a design table in a drawing. There is a size limitation to embedded objects, and this limitation can sometimes present itself with design tables because they can often be quite large. There is not much you can do in this case, but there are a few things that might help.

Double clicking on a design table in a drawing will open up the table in the part or assembly it is associated with. It could then be edited, but that is not the issue. You will see small black handles on each side of the table's border and at each corner. Use them to drag the border so that only the relevant information is being shown, and then close the table and save the model. Back in the drawing, click on the Rebuild icon and see if that helps.

If you still cannot see the entire table, you have a few other options. Try compacting the columns as much as possible. Rotate text 90 degrees in the row containing the headings so that the columns can be made smaller.

Use letters (such as S) instead of spelling out *Suppressed*, or Y instead of *Yes*. Finally, if that does not work, reduce the font size of the spreadsheet and re-compact (auto fit) the column widths. If you are creating a lot of configurations, you may need to re-compact the row heights as well. Use the on-line Excel help to aid you in these tasks if necessary.

Design Table Summary and Recommendations

Design tables are arguably a very efficient way of generating configurations, which in themselves are a powerful function of SolidWorks. In light of this, what really is the best way to put a design table together? The following summarize the content to this point in this chapter, with recommendations as to how a design table can most easily be created.

- *Rename dimensions.* By renaming dimensions, you make it much easier to pick those dimensions from a list when generating the design table later. You should also rename any mates (if an assembly), features, or other objects that will be driven in the table. This also helps to document what is happening within the table for future reference or for others working with the model.

- *Use Auto-create.* It is taking the easy way out when generating a design table, but so what? There are no disadvantages to using the Auto-create option, but there are plenty of advantages. It is not necessary to show all dimensions in the model because they can be selected from a list.

- *Know your Excel shortcuts.* Taking advantage of a few simple shortcuts in Excel can make short work of a design table. An example would be selecting a series of cells and dragging that selection downward to completely populate the rest of the table. You could easily populate the body of a table representing 100 configurations in a matter of seconds.

Custom Properties

If you examined Table 16-3, you may have noticed that one of the design table parameters mentioned earlier has not been touched on yet. The parameter is the custom properties parameter. Although custom properties can be created via a design table, the functionality of custom properties extends far beyond design tables. To begin, let's first come to understand exactly what custom properties are.

A SolidWorks document can contain certain information above and beyond model data. For example, a part file can contain relevant information such as who created the file, or even that person's phone number and work extension. This information is known as the file's properties, and they are totally customizable. You can create your own custom properties, which can contain literally any information you want them to.

The really beautiful thing about custom properties is that they can be transferred to other documents. For example, linking to a part file's properties gives you an easy way of updating notes scattered throughout a drawing. When the properties of the original part document are altered, these changes are reflected to the notes linked to these properties.

What are some of the reasons for creating and linking to a file's properties? One reason is to have the name of a part automatically appear in the title block of a drawing. If the file name changes, so does the text in the title block. Another reason might be to have the total number of sheets in the drawing title block automatically update. Yet another reason might be to show a note on a part that lists the mass properties for the part. If and when the mass properties change, the note automatically updates. The reasons vary. You may find many uses for customizing and linking to a file's properties. First, however, you need to know how to create them. How-To 16-3 takes you through the process of creating custom properties.

How-To 16-3: Creating Custom Properties

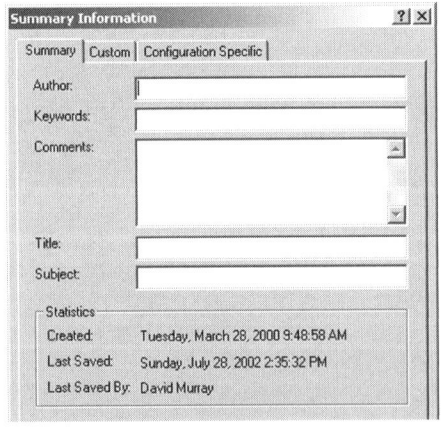

Fig. 16-29. Summary Information window.

To create custom properties, perform the following steps. Note that custom properties can be created for all SolidWorks document types (parts, assemblies, and drawings). It is also possible to create configuration-specific properties, meaning that a property for one configuration can have a value different than the same property for another configuration.

1. With a SolidWorks document open, select Properties from the File menu. This opens the Summary Information window, a portion of which is shown in figure 16-29.

2. Select the Custom tab.

3. Click in the section labeled Name and type in a name for the property you wish to create. In the example shown in figure 16-30, the word *Mass* was used.

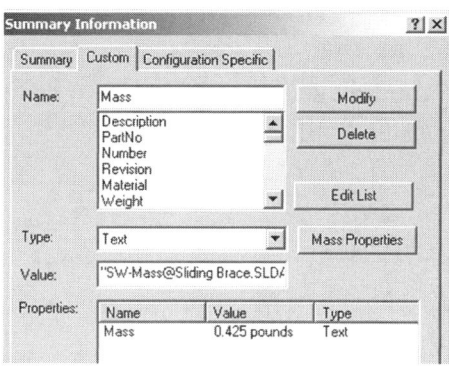

Fig. 16-30. Custom tab of the Summary Information window.

4. Specify the property Type, such as a text string or number. This is akin to formatting, and the property will be a text string by default.

5. Click in the section labeled Value and type in a value for the new property you are creating.

6. Click on the Add button to add the data to the file's properties.

7. Repeat steps 3 through 6 for each new property you wish to add.

8. Click on OK to close the Summary Information window.

If you refer back to figure 16-30, you will notice a button named Mass Properties. This button allows for creating a custom property linked to the mass properties of the model. This is what was done in the example used in this book. The entire Value field is truncated due to the text being longer than the space it is shown in, but in actuality it reads as follows.

"SW-Mass@Sliding Brace.SLDASM" pounds

What you see within quote marks is the code SolidWorks uses to link to the mass of the model. The word *pounds* was added as a suffix to the mass value. If you refer to figure 16-30 again, you will see that the property in its final form reads 0.425 pounds. Once again, it should be stressed that custom properties can be absolutely anything. You need not use the Mass Properties button. Additionally, the Edit List button allows for editing the default items found in the list of property names. This can be incorporated into a template and saved for future documents. With that said, how can we put this new custom property to good use?

Fig. 16-31. Link to Properties button.

The Link to Properties button, found in the Note panel (shown in figure 16-31), allows for linking to the properties associated with the file. If you need to refresh your memory on how to create notes, see the "Notes" section in Chapter 11. Although the Link to Properties button was described in Chapter 11, here the steps will be spelled out. How-To 16-4 takes you through the process of linking notes to custom properties.

How-To 16-4: Linking to Custom Properties

To link to a file's properties, perform the following steps. In the example used here, the custom property created in the previous How-To will be used in a note attached to the *Sliding Brace* assembly. Custom properties can be linked to notes in parts, assemblies, and drawings.

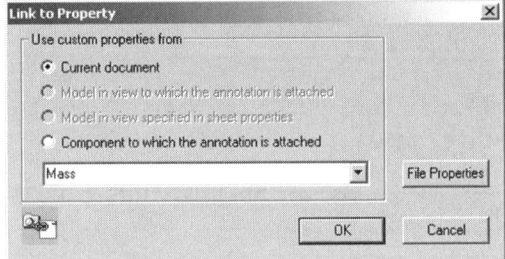

Fig. 16-32. Link to Property window.

1. Select Insert > Annotations > Note, or click on the Note icon.

2. Click where you would like the note to appear.

3. Click on the Link to Properties button. This will open the Link to Property window, shown in figure 16-32.

4. From the drop-down list, select the custom property you wish to insert into the note. You will be able to link to any of the file's properties added in the Summary Information window, along with other properties created automatically (prefixed by *SW*).

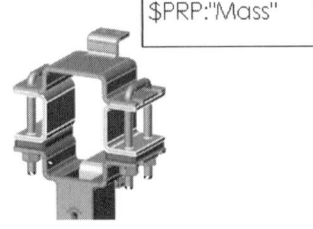

Fig. 16-33. Adding a custom property to a note.

5. Click on OK. SolidWorks places a code in the Note text box, shown in figure 16-33, which establishes a link back to the custom property.

6. Click on OK in the Note panel.

Where initially there was coding in the text box, now the note will contain text relative to the custom property it is referencing. You must not edit the code placed in the Note text box. To do so will break the link to the custom property. Figure 16-34 shows the final results of adding the note. Note that the text matches exactly the value of the original custom property shown in figure 16-30.

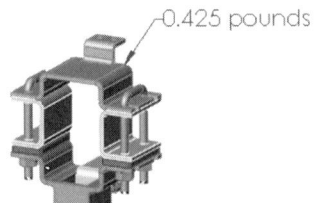

Fig. 16-34. Final note containing the custom property link.

Possibly the most important benefit of custom properties can now be shown. If the Summary Information window is opened and the custom property is modified, the data in the note will automatically update. Imagine a scenario in which a drawing template contains dozens of custom properties, such as for scale, number of sheets, drawing number, title, and so on. Some of these properties would update automatically, and others could be easily modified by the user. Instead of editing or adding notes to the drawing sheet format, the user could open the Summary Information window and make changes to the properties, whereas the changes would propagate to the drawing.

SolidWorks creates some predefined custom properties for you. These predefined properties are prefixed with *SW*. They can be found in the dropdown list in the Link to Property window. In particular, there are quite a few predefined properties that will prove extremely useful when creating drawing templates. These properties will automatically update, and can be used to show the number of sheets in a drawing, the current sheet, file name, sheet scale, and many other details.

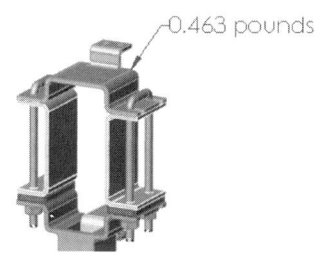

Fig. 16-35. The note automatically updates.

In the custom property example used in this book, the mass of the model was used rather than typing in a user-defined property. Therefore, if the mass of any of the components is changed, or if the configuration is changed to one containing more massive components, the note should automatically update. This is exactly what happens, an example of which is shown in figure 16-35.

Because not all custom properties automatically update on their own (such as with user-defined properties wherein a text string is entered for the value), you will need to know how to modify custom properties. How-To 16-5 takes you through the process of editing a file's custom properties.

How-To 16-5: Modifying Custom Properties

To edit a file's custom properties, perform the following steps.

1. Select Properties from the File menu.
2. Select the Custom tab.
3. Select the Property you want to modify.
4. Type in the new data in the Value field.
5. Click on the Modify button.
6. Click on OK when finished.

Because of the built-in associativity between SolidWorks documents, custom properties can be propagated to other files. In other words, custom properties associated with a part or assembly can filter through to a drawing. This is accomplished using the options found in the Link to Properties window (see "Text Format Panel" in Chapter 11). In short, the Model in view to which the annotation is attached or Model in view specified in sheet properties option could be selected, thereby allowing SolidWorks to link to properties in the model rather than in the drawing itself. After all, drawings are separate files, and therefore you may need to tell SolidWorks where to look for the properties so that it can establish the appropriate link.

When working with a drawing of an assembly, using the *Component to which the annotation is attached* option allows SolidWorks to access the properties of the component part file. This is important in that the properties of the drawing or assembly in the drawing view will likely have different properties containing different data than that of the part.

Configuration-specific Properties

When adding a custom property, it is possible to associate that property with a particular configuration. This is known as a configuration-specific custom property. To add a configuration-specific property, use the same procedure outlined in How-To 16-3, with one exception. Instead of clicking on the Custom tab in the Summary Information window, click on the Configuration Specific tab.

The Configuration Specific tab and Custom tab take you to identical windows. However, if a custom property has been added via the Configuration Specific tab, that property will only be available for the current configuration. Therefore, it is important to make the appropriate configuration active prior to adding the property.

Defining Custom Properties in Design Tables

Now that you have a better understanding of custom properties, let's get back to the main topic of this chapter. How do custom properties relate to design tables? The answer is that they can be defined within the design table. The heading used is $PRP. This would then be followed by what you want the name of the property to be. In all, the heading might look as follows.

```
$PRP@cost
```

Custom Properties

Fig. 16-36. A custom property in a design table.

Cost, in this case, is the name of the custom property. Below this heading, in the same column of the design table, would be the values of what the property should have. If cells are left blank, the property will contain a blank (empty) value for that configuration. An example of a design table with a custom property defined within it is shown in figure 16-36.

If you were to activate one of the configurations after closing the design table, and then access the files properties, you would see the property appear in the Configuration Specific tab. An example of this is shown in figure 16-37.

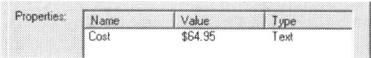

Fig. 16-37. The custom property appears in the Configuration Specific tab.

Associative Custom Properties

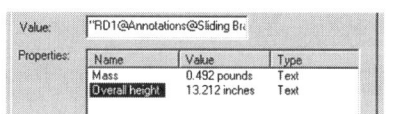

Fig. 16-38. Adding a dimension value as a custom property.

Associative custom properties are those linked to model dimensions. The custom property will automatically update when the model dimension changes. In other words, custom properties can link to and display dimension values. To create such a property, simply click on a dimension value when adding the custom property. SolidWorks will add a code to the custom property value, which translates to the dimension value itself. An example of this is shown in figure 16-38.

In the case of the example shown in figure 16-38, a reference dimension has been used. It was given the name *Overall height*. If desired, dimensions driven in a design table can be made to appear as custom properties. For example, dimensions in the design table might drive the size of a bracket. The dimensions might represent the length, height, and width of the bracket. The custom property names might be *Length, Height,* and *Width*, which would be sensible in the case of this example.

We have examined methods of inserting dimension names into a design table. We have also examined how those dimensions can be added to custom properties. However, to add associative dimensional custom properties via a design table, there is a bit more manual labor involved. To continue with the example of the bracket, let's see what the process would entail.

First, it would be necessary to establish the design table and insert the dimensions. This has been done and is shown in figure 16-39. The dimensions were renamed prior to creating the table. For the sake of simplicity, the configuration names are *Small*, *Medium*, and *Large*.

	A	B	C	D	E
1	Design Table for: bracket				
2		length@Sketch1	height@Sketch1	width@Baseplate	
3	Small	6	3	3	
4	Medium	8	3.5	3.5	
5	Large	10	4	4	
6					

Sheet1

Fig. 16-39. Preparing a design table.

Next, you would want to define some custom properties via the design table. We have recently examined this as well, so you should know the routine. Figure 16-40 shows the new additions to the design table, the new additions being in the form of the *$PRP* headings.

	E	F	G
plate	$PRP@Length	$PRP@Height	$PRP@Width
3			
3.5			
4			

Fig. 16-40. Defining the custom properties.

Now for the difficult part. The syntax used to define the property value is somewhat intimidating. If you can get past the weirdness of its appearance, you will be home free. The property value must take the form of the dimension name (e.g., *length@Sketch1*), followed by the configuration name (e.g., *Small*), followed by the name of the part (e.g., *Bracket.SLDPRT*). SolidWorks uses the "at" symbol (@) as a spacer, and thus the entire text string might look like as follows.

```
"length@Sketch1@@Small@Bracket.SLDPRT"
```

	E	
late	$PRP@Length	
3	"length@Sketch1@@Small@Bracket.SLDPRT"	"he
3.5	"length@Sketch1@@Medium@Bracket.SLDPRT"	"hei
4	"length@Sketch1@@Large@Bracket.SLDPRT"	"h

Fig. 16-41. Adding custom property values.

There are three things that are very important. First, the file extension must be in upper case. Second, use two (2) "at" symbols (@) after the dimension name. Third, make sure to include the quote marks. When the design table is finished, the portion containing the custom property information should look similar to that shown in figure 16-41. Note that only one column is being shown, due to space limitations.

Properties:

Name	Value	Type
Length	6.000	Text
Height	3.000	Text
Width	3.000	Text

Fig. 16-42. Configuration Specific tab complete with associative dimensional custom properties.

If you have done everything right, and you have somehow managed to avoid any typographical errors (a rarity the first time around), the configuration-specific property tab should now contain associative dimensional information that will update if those dimension values are altered in the design table. Figure 16-42 shows what you might see in the Configuration Specific tab of the file's properties.

Custom Properties and BOMs

In that BOMs can be customized (which you learned about in Chapter 12), we can now take custom properties to the next level and use them to populate a BOM. For this next example, the *Sliding Brace* assembly will be used once again. Cost has been added as a custom property in each of the components constituting the assembly. The cost will be the user-defined custom property we would like to have automatically populate the BOM.

bomtemp-cost.xls

	A	B	C	D	E	F	G
1	ITEM NO.	QTY.	PART NO.	DESCRIPTION	COST	TOTAL	$$END
2						$0.00	
3						$0.00	
4						$0.00	
5						$0.00	
6						$0.00	

Fig. 16-43. A customized BOM template.

Figure 16-43 shows an Excel spreadsheet that has been saved for use as a customized BOM template. It is worth noting that the TOTAL column utilizes formulas that will force Excel to do some of the work. Specifically, the TOTAL column will multiply the numeric value in the QTY column supplied by SolidWorks, and the value in the COST column added in by the user via custom properties.

ITEM NO.	QTY.	PART NO.	DESCRIPTION	COST	TOTAL
1	1	Slip bracket		$4.97	$4.97
2	1	Bracket with Tube		$16.26	$16.26
3	2	C-link		$1.37	$2.74
4	2	Insulator		$0.39	$0.78
5	1	Post		$2.53	$2.53
6	2	U-bolt assembly			
	1	U-bolt		$3.18	$3.18
	2	Nut		$0.42	$0.84
	2	Lock washer		$0.14	$0.28
7	1	Lynchpin		$1.79	$1.79

Fig. 16-44. The final BOM shown for the Sliding Brace assembly.

In our example, the BOM's TOTAL column has been formatted for currency. The same has been done for the COST column. The $0.00 values show up in the TOTAL column because that is the way Excel handles zero values when formulas have been used. If it would be desirable to not show a cost value if it were zero, an *If* statement could be used. Read through the Excel on-line documentation to gain an understanding of how this can be accomplished. The final BOM includes the *If* statement and in figure 16-44 is shown in a drawing, along with the *Sliding Brace* assembly.

Summary

Design tables are used to create more elaborate configurations where not only features but dimensions can be controlled. Entire family part tables can be created and modified through the use of a spreadsheet. You must use Microsoft Excel with SolidWorks to create a design table. It is helpful

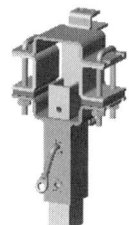

if you rename the dimensions that will be driven by the design table. This makes creating and editing the design table easier later on.

Generating a design table is done through a largely automated process. The Auto-create option allows for picking dimensions from a list to be added to a design table automatically in the proper format. Once generated, a design table can be saved as a file that can retain a link back to the model in which it was created.

Through optional settings accessible when defining a design table, tables can be made to update if a parameter (such as a dimension value) is changed in the model. New rows or columns can also be automatically added to the table if additional configurations or parameters are altered in the model. Alternatively, parameters can be locked so that any modifications must be made in the design table rather than in the model.

Assembly design tables are an excellent means of controlling various parameters of an assembly and of adding configurations. Assembly design tables differ from part design tables in that they cannot control component feature dimensions or part feature suppression. Assembly tables can control suppression of components and the dimensions associated with mating relationships. When controlling the suppression state of components, the proper terminology is suppressed or resolved.

An important aspect of assembly design tables is the ability to specify which configuration of a part or subassembly is being used in any particular assembly configuration. Component visibility can also be controlled. When a part is hidden, it is only hidden from view, but still exists in memory. When a part is suppressed, it is removed from memory, thereby disabling its respective mating relationships.

Use custom properties to create intelligent text in parts, assemblies, and especially drawings. Notes can be made to automatically update if certain model or file parameters change. Information in title blocks, such as sheet scale and drawing name, will update automatically if the notes containing that information are linked to custom properties. Users can also create their own custom properties that will propagate to BOMs, or define custom properties from within design tables.

Questions and Topics for Discussion

1. Name two parameters that can be controlled in a part design table that cannot be controlled in an assembly design table.

2. Dimensions from what type of objects can be controlled in an assembly design table?

3. What heading is used to control the suppression state of components in an assembly design table?

4. What are the controlling characters used to specify the suppression state of a component in an assembly design table?

5. What significance does the $PARTNUMBER heading have with regard to a BOM in an assembly drawing?

6. In your own words, describe the difference between linked and embedded objects within files.

7. Describe one important task you should perform prior to creating a design table.

8. Name at least three of the headings that can be used in an assembly design table, and describe what function they serve.

9. Describe a method of saving out the data from a previously inserted design table as a separate Excel spreadsheet.

10. When a design table is saved as a separate file, is it associated to the model from which it was created? Explain your answer.

11. How can a design table be shown in a drawing?

Optional Problem

Describe what you think would happen if a dimension or feature were renamed within SolidWorks after inserting a design table. Check your answer by opening an existing SolidWorks part and inserting a design table.

CHAPTER 17

Equations

OFTEN IT IS DESIRABLE TO PLACE MATHEMATICAL RULES ON DIMENSION VALUES. Equations allow this to take place. A simple example would be in the case of a bracket whose width should always be half its length. By establishing this rule through the process of equations, it becomes possible for the model's width to automatically update if the length is changed. Equations can be used within parts or assemblies.

As powerful as equations are, there is not a lot to them. In that there is quite a bit that can be accomplished with equations, a couple of examples of their usefulness are included to help you better understand how this functionality works.

This chapter builds on topics discussed previously, and therefore assumes that you understand fundamental concepts and know how to perform certain basic functions within SolidWorks. These functions and knowledge include such things as changing dimensions, renaming dimensions and features, accessing a dimension's properties, and what is meant by suppressing or unsuppressing objects.

Preparatory Work

Equations are similar to design tables in the sense that there is a bit of preparation that could take place prior to creating them. The preparatory work is essentially the same in both cases. Specifically, it is best to rename dimensions and features prior to creating equations. The reason for this is identical to that for design tables. Because dimensions will be used in equations, it makes sense to give dimensions (and related features) meaningful names. This makes it easier to understand what is taking place within the equation when it is viewed at a later time.

In actuality, SolidWorks is smart enough to know when a dimension has been renamed when equations are being used. That is, if a dimension is renamed, SolidWorks will change the name in the equation accordingly. However, to avoid potential problems, it is best to be cautious and take care of any renaming ahead of time.

Prior to creating an equation, it helps tremendously to physically jot down a rough draft of the equation. This is not necessarily true for very simple equations, but it sure does help when trying to work out something a bit more complex. Give yourself an idea of how the equation should be formed, and the chance of it turning out right the first time once it is defined in the software will improve.

Another extremely important aspect of equations relates to which dimensions will be solved for, and which dimensions will be changeable by the user. Any dimensions you want to be able to change should be on the right side of the equation's equal sign. The dimension value being solved for should be placed on the left-hand side of the equation's equal sign. In other words, driving dimensions appear on the right, and the driven dimension should appear on the left. An example follows.

```
"D1@BaseExtrusion" = "D2@Plate" - .25
```

In this equation, the *D2@Plate* dimension could be altered. Double clicking on the dimension value would result in the typical Modify window, which would allow you to enter a different value for the dimension. If you were to try this on the *D1@BaseExtrusion* dimension value, a window would appear informing you that the dimension is being driven in an equation and that its value cannot be changed. The window would appear similar to that shown in figure 17-1, though the dimension names would be different.

Fig. 17-1. A warning regarding changing a dimension driven by an equation.

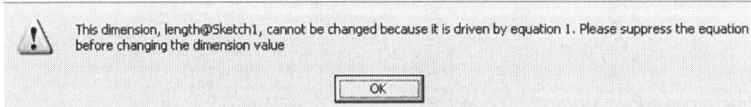

As is evident by the warning message, it is possible to suppress an equation in order to change the dimension being driven by it. This would only result in a temporary change, however, because as soon as the equation were unsuppressed, the dimension value would revert to whatever the equation dictates it should be. Suppressing equations is explored after an examination of equation basics.

The Equation Creation Process

The act of creating an equation is very similar to writing a sentence; start on the left and work your way to the right. Getting dimensions into the equation is just a matter of selecting them. To select a dimension, it must be shown on screen. Therefore, you have two choices: right-click on the *Annotations* object in FeatureManager and select Show Feature Dimensions or access dimensions a feature at a time by double clicking on the applicable features.

Accessing dimensions a feature at a time has the advantage of keeping the screen from becoming too cluttered. If it is possible to show all dimensions at once and still find the dimensions you are looking for, go ahead and do so, as this will make the task of getting the dimensions into the equation easier. In that you now know everything necessary to create and equation, let's try it. How-To 17-1 takes you through the process of creating an equation.

How-To 17-1: Creating an Equation

To create an equation, perform the following steps.

1. Select Equations from the Tools menu.

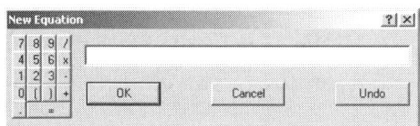

Fig. 17-2. New Equation window.

2. In the Equations window that appears, click on the Add button. This will open the New Equation window, shown in figure 17-2.

3. Click on the dimension you wish to drive in the equation. You may need to double click on the feature the dimension is associated with to gain access to the dimension.

✎ **NOTE:** *It is possible to drag the Equation and New Equation windows to different locations on the screen, which you may very well need to do in order to see the dimensions being placed into the new equation.*

4. Using either the keyboard or the New Equation window, add an equal sign to the equation.

5. Continue to click on dimensions, or add mathematical operators or numeric values, in order to finish writing the equation, depending on your requirements.

Chapter 17: Equations

6. Click on OK to close the New Equation window and create the equation.

7. Click on OK to close the Equation window.

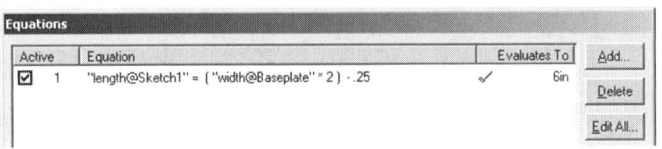

Fig. 17-3. A new equation being shown in the Equations window.

The Equations window will display any equations that have been created. This is shown in figure 17-3, with one equation listed in the upper portion of the Equations window. Any equations will also be displayed with what the equation evaluates to. In the case of the example shown in figure 17-3, this value happens to be 6 inches.

A green check in the Equations window means the equation is functioning normally. A red exclamation point indicates the equation cannot be solved for one reason or another and you may have to perform some troubleshooting. The check box at the far left of the equation allows for turning the equation on or off. Equations that are turned off (unchecked) are considered deactivated (or suppressed).

Once an equation has been created, the *Equations* folder will be listed in FeatureManager, as shown in figure 17-4. This folder is nothing more than a shortcut to certain operations involving equations. These operations, accessed via a right mouse button click on the *Equations* folder, consist of deleting, editing, and adding equations.

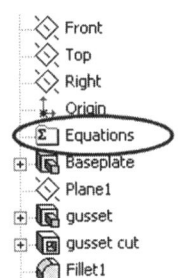

Fig. 17-4. Equations folder in FeatureManager.

✓ **TIP:** *Right-clicking on the* Equations *folder and selecting Delete Equation does not actually delete anything at that point in time, but rather takes you to the Equations window, where equations could be deleted. This serves as a nice shortcut.*

The steps listed in How-To 17-1 are generic. The steps you use may differ, depending on the number of dimensions you wish to incorporate into the equation and the mathematical operators used. One point that should remain constant is to place the dimension being solved for on the left-hand side of the equation by itself. This is standard, as there can be only one driven dimension per equation, which SolidWorks can then solve for.

The mathematical operators that can be used in equations are plus (+) and minus (–) for adding and subtracting, the asterisk (*) for multiplication, and the forward slash (/) for division. Additionally, parentheses can be used, and the carat symbol (^) can be used to signify exponents.

Equations in Action

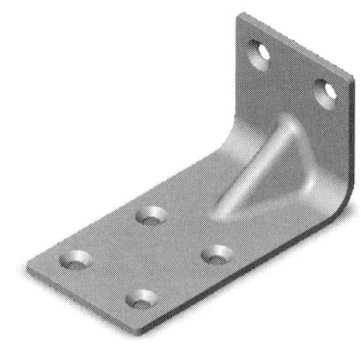

Fig. 17-5. Some equations will be used to drive this model.

To better understand equations, let's look at an example of how equations can help in modeling parts. Figure 17-5 shows the bracket used in the example.

It is necessary to first determine which dimensions will be used to drive the equations. In the case of the bracket, design requirements dictate that the width is a flexible dimension that will require certain values. The length of the bracket will be based on the width. Additionally, depending on the overall size of the bracket, the positioning and number of holes will change.

Once the general design requirements are known, we can begin to create a draft outline for how the equations should be written. Let's take it one step at a time. In that we know the length will be the driven dimension, it will need to be placed on the left side of the equation. For the sake of argument, let's also specify that the length should always be .25 inch less than twice the width. That means our equation should look as follows.

```
Length = (Width * 2) -.25
```

In the equation, the words *Length* and *Width* will be replaced with the actual dimension names, and the dimension names will need to be surrounded by quote marks, but SolidWorks will take care of that for us. The parentheses are important also, because we need to ensure that Solid-Works solves operations in the proper order. Parentheses are added by the user, so make sure to incorporate them in the equation, or the solution may not be what you think it should be. In the example shown, leaving the parentheses off would be okay because SolidWorks would still come up with the proper solution. However, it is a good idea to be in the habit of using them. If the process listed in How-To 17-1 is followed, the equation can now be added to the model. Once added, the equation itself appears as follows in the Equations window.

```
"length@Sketch1" = ( "width@Baseplate" * 2 ) - .25
```

The model is shown in figure 17-6, along with the two dimensions included in the equation. Note that the dimension values hold true to what they should be if they were to replace the dimension names in the equation. This is especially significant for the length value, in that it is the one that should update if the width value is altered.

Another aspect of our design requirement is that the holes in the model should change position in accordance with the overall size of the part. Currently, the countersunk holes reside at static positions relative to the sides of the model, as shown in figure 17-7.

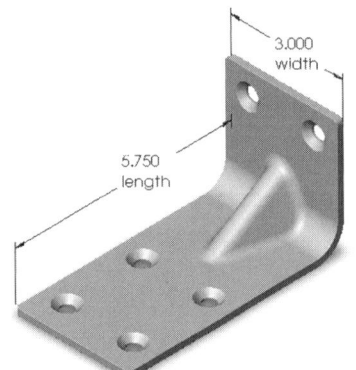

(LEFT): Fig. 17-6. The length dimension updates correctly.

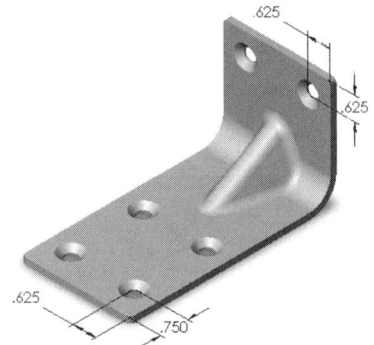

(RIGHT): Fig. 17-7. The edge offset dimensions of the holes.

The design requirements for this example will call for two things to happen. First, the edge offset dimensions should always be equal. Second, the edge offset value will change based on the width of the bracket. The first criterion will be met using a new command, Link Values, explored in the next section.

Link Values

Equations are a good tool to have for certain situations, but they can be overkill for simple applications. Because equations by their very nature create dimensions that are driven, they can sometimes reduce the amount of flexibility present in a model. For simple equalities, the Link Values command is the perfect choice.

Link Values allows for linking two or more similar dimensions so that their values are always the same. For example, two or more angular dimensions could be linked, but an angular dimension could not be linked to a diameter dimension. Link Values has an advantage over equations in that no driven states are created. When dimensions are linked, any of the linked values can be modified. There are no driving versus driven dimensions. How-To 17-2 takes you through the process of linking two or more similar dimensions.

How-To 17-2: Linking Dimension Values

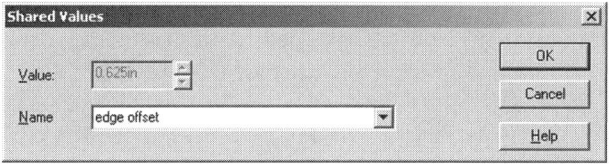

Fig. 17-8. Shared Values window.

To link two or more dimension values so that they are always equal, perform the following steps. Note that part feature dimensions cannot be linked in an assembly. Dimensions native to an assembly can be linked if they are similar.

1. Right-click on a dimension and select Link Values. The Shared Values window will open, shown in figure 17-8.

2. Type in a name that will be used to describe the linked dimensions. In the example used here (see figure 17-8), the name edge offset has been typed in.

3. Click on OK.

4. Right-click on another dimension, to be linked to the first, and select Link Values.

5. Either type in the same name used in step 2 or select the name from the drop-down list.

6. Click on OK.

7. Repeat steps 4 through 6 for every dimension to be linked to the original.

It is not really necessary to right-click on each dimension to be linked one at a time. If all dimensions to be linked are selected ahead of time, right-clicking on any one of them and supplying a name in the Shared Values window will link all values in one operation. The result is the same, which is to say that changing any one of the dimensions will also change any of the dimensions linked to it.

Unlinking dimensions can be done via the right mouse button as well. This is also an ideal way of determining if a dimension is linked in the first place. If you were to right-click on a dimension and see the option Unlink Value, the dimension must have previously been linked. Selecting Unlink Value removes the link, but only for that single dimension.

Linking dimensions changes the name of those dimensions. Changing the name of a dimension, as you have learned, can also be accomplished by accessing the dimension's properties. However, it should be mentioned that simply renaming dimensions so that they have the same name is not enough to link them. If dimensions have already been linked, changing one of the linked dimension names via its properties causes SolidWorks to rename all linked dimensions.

The Link Values command can only be accessed by right-clicking on a dimension value. The command does not exist anywhere else within the SolidWorks interface. For this reason, it is common to forget how to access the command. Always equate Link Values with the right mouse button, and you will find it easier to remember.

Now that it is understood how to link dimension values, the design requirements of the bracket can continue to be met. Let's assume that the four dimensions shown in figure 17-7 have been linked. That leaves us with one more task with regard to the hole locations: the edge offset values should increase if the part gets larger. That will require another equation.

The width of the bracket will range anywhere from 3 inches minimum to 5 inches maximum. If the edge offset value for the holes is set to 20% of the total width, that means the holes will be anywhere from .600 inch to 1 inch from the edge of the model. It should be noted that only the holes on the right side of the part (nearest the reader in the various illustrations) are controlled with dimensions. The holes on the left have been mirrored. The third hole on the right side of the part, which contains no locating dimensions (see figure 17-7), has been created via a linear pattern.

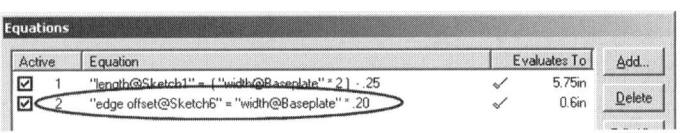

Fig. 17-9. Driving the edge offset dimensions.

The equation that needs to be created next will be an easy one. One of the linked dimensions will be driven by the width of the part multiplied by .20 (20%). The final equation, as it appears in the Equations window, is shown in figure 17-9.

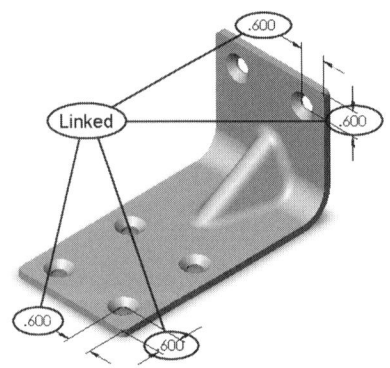

Fig. 17-10. The linked values have updated due to the new equation.

This situation works out wonderfully, because only one of the four linked dimensions needs to be driven in the equation. The rest of the linked dimensions update automatically if the width is changed. Linked dimensions are easier for SolidWorks to solve, and also less prone to errors from a user standpoint. Whenever possible, use Link Values instead of equations in order to make the model as efficient as possible. Figure 17-10 illustrates how all of the linked edge offset dimensions have changed to .600 inch, rather than their previous values, due to the new equation being added.

> **NOTE:** *A driven dimension cannot be linked. However, if dimensions are linked first, one of those linked values can then be added to an equation, thereby driving all of the linked dimensions via the equation.*

There is one equation that needs to be added, and that is the equation that will control the number of holes dependent on the size of the bracket. Initially, this may seem like a feat that may prove too complex for Solid-Works. Once you see how easily it can be done, you may be surprised.

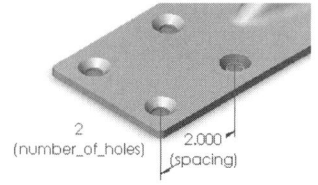

Fig. 17-11. The pattern dimensions.

Let's begin the same way the last two equation problems were tackled. Draft an equation first to understand what might be achieved. Currently, the spacing between the holes in the pattern is 2 inches, and there are a total of two (2) holes in the pattern (the holes on the other side of the part were mirrored). The pattern dimensions, shown in figure 17-11, have already been renamed to something very descriptive.

For this final problem, assume the spacing is going to remain a constant 2 inches, and the number of holes will change depending on the size of the bracket. If the width is between 3 and 3.5 inches, there should be two holes in the pattern. If the width is between 3.5 and 4.5, there should be three holes, and so on. The number of holes is being driven, so that value is placed on the left side of the equation, but what of the rest of the equation, and how can there be a fractional number of holes?

SolidWorks knows that when the left side of an equation contains a pattern instance dimension, that dimension must be rounded off. If the equation evaluates to 2.49, there will be two holes. If the equation evaluates to 2.5, there will be three holes. If the width of the bracket (currently 3 inches) has the value 1.01 subtracted from it, that leaves 1.99, which rounds to two holes. Let's look at some other possibilities, as follows.

```
3.5 inches - 1.01 = 2.49, which rounds to two holes
3.51 inches - 1.01 = 2.5, which rounds to three holes
```

This seems to do what we want, so let's write the equation as follows.

```
Number of holes = bracket width - 1.01
```

The final equation, once added to the bracket model, appears in the Equations window as shown in figure 17-12. There are now a variety of parameters in the bracket model. These will update automatically through changing a single value.

It is time to put all of our efforts to the test. The maximum value of the width dimension was originally specified as 5 inches. Changing the width to a value of 5 and rebuilding the model results in what you see in figure 17-13. It appears the efforts paid off!

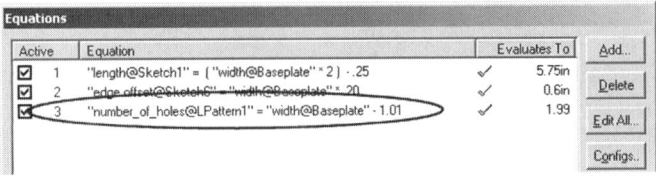

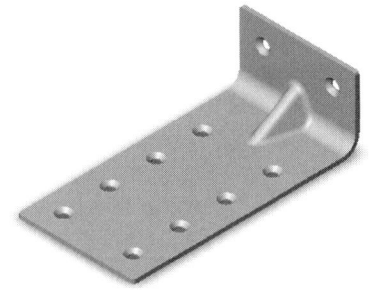

Fig. 17-12. Adding a third equation.

Fig. 17-13. Changing the width to 5 inches updates the rest of the model.

Equations Window Buttons

There is a column of buttons along the right side of the Equations window. These buttons are shown in figure 17-14. Most of the buttons are self-explanatory. OK and Cancel perform the usual functions, and Add opens the Add Equation window, as you have discovered.

To delete an equation, select it in the Equations window and click on the Delete button. Clicking on the Edit All button opens what amounts to a very rudimentary text editor. This is shown in figure 17-15. The window can be resized for convenience. If an equation needs to be rewritten, sometimes it is easier to simply delete it and create it again, rather than editing it, in that one little typo will cause the equation not to work.

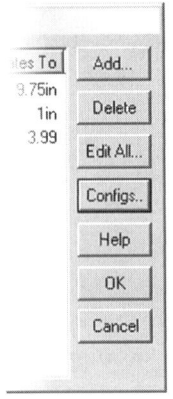

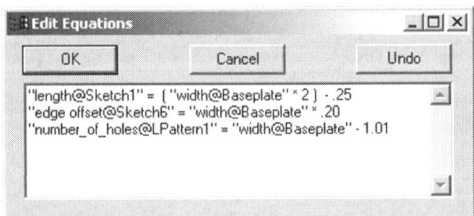

Fig. 17-14. Buttons in the Equations window.

Fig. 17-15. Edit Equations window.

When editing equations, feel free to add comments. Anything in an equation line that falls after a single quote is considered a comment. An example of a comment follows. Note the single quote.

```
"length@Sketch1" = ("width@Baseplate" * 2 ) - .25
'This text is considered a comment
```

The Configs button allows for specifying which configurations will be affected if you decide to change the suppression state of any of the equations. This utilizes the standard options, which should be familiar to you (see Chapter 10). You have the options This Configuration, All Configurations, and Specify Configurations.

Last, the Help button will provide you with instructions and tips for creating equations. The help also contains an excellent reference on supported functions. This includes trigonometric functions and others, such as calling out pi or absolute values in equations. Equation functions are outlined in Table 17-1.

Table 17-1: Equation Functions

Function	Name	Description
sin (α)	Sine	α is angle expressed in radians
cos (α)	Cosine	α is angle expressed in radians
tan (α)	Tangent	α is angle expressed in radians
atn (α)	Inverse tangent	α is angle expressed in radians
abs (α)	Absolute value	Returns the absolute value of α
exp (n)	Exponential	Returns e raised to the power of n
log (α)	Logarithmic	Returns the natural log of α to the base e
sqr (α)	Square root	Returns the square root of α
int (α)	Integer	Returns α as an integer
sgn (α)	Sign	Returns the sign of α
pi	pi	3.14159...

If you use any of the trigonometric functions, keep in mind that any angles need to be in radians. A radian is the angle described by traversing a distance around the perimeter of the circle equal to the radius of the circle. This works out to be approximately 57.3 degrees. It also means that because a circle's diameter times pi (π) equals its circumference (360°), then 2π * radians = 360. Solve for radians, and the formula reduces to radians = 180/π. Divide any degree values by 180 and multiply the result by pi, or just incorporate this information into any equation containing trigonometric functions and let SolidWorks do the math for you.

Changing Units in Models with Equations

If you have spent a lot of time on a model and have added equations to it, and then decide that the model's working units need to be changed, you are going to have a problem on your hands. However, if you think there is a chance the units will need to be changed sometime down the road, perhaps due to a customer or client overseas that may be working with the model, there is a precautionary measure that can be taken.

An equation used earlier in this chapter will be used again as an example. The equation reads as follows, and assumes the working units are in inches.

```
"D1@BaseExtrusion" = "D2@Plate" - .25
```

If it is assumed that *D2@Plate* equals 3 inches, the solution to the equation becomes 2.75. If the working units are switched to millimeters, this ruins the equation. Keep in mind that *D2@Plate* is a dimension value. Switching to millimeters means *D2@Plate* becomes three times the number of millimeters in an inch (approximately 25.4), or 76.2. This means the equation's solution becomes 75.95, which is a far cry from 2.75!

How can we keep this from happening? The solution is to create a sketch with a line in it. Dimension the line 1 inch. Rename the sketch to something like constant if you want. For the sake of argument, assume the dimension's full name is *D1@constant*. Hide the sketch, because it will not be used for anything else, and then plug the new dimension name into your equation. The equation might look as follows.

```
"D1@BaseExtrusion" = "D2@Plate" - (.25 * "D1@constant")
```

As long as the units are in inches, *D1@constant* will equal 1, but as soon as you switch to a different set of units, such as millimeters, *D1@constant* equals 25.4. The .25 value gets multiplied by *D1@constant*, and everything works out. Depending on the equation, you may have to apply the "constant" in a different manner, but that should not be too difficult to figure out. Incidentally, this works when starting out in any set of units, not just inches.

Common Equation Errors

The order in which equations are created is significant. Equations are solved in the order in which they appear in the Equations window. Equation order can be changed by copying and pasting via the Edit Equation window.

A circular reference can be created with equations and SolidWorks will not stop you from doing so. This is a nasty type of equation problem because it does not manifest itself as an error. There is no indication that the circular reference even exists, except that the model may behave strangely. Consider the following three equations.

```
"length@Base-Extrude" = "width@Sketch1" * 2
"height@Sketch1" = "length@Base-Extrude" / 2
"width@Sketch1" = "height@Sketch1" + 1
```

These three equations drive the size of a block. For the sake of discussion, assume the width starts out at 2 inches. That means the length will equal 4, the height will equal 2. But then the third equation rolls around and resets the width to 3, not 2.

This odd scenario causes two problems. The first main problem is that none of the equations can be modified because all three of the dimensions are being driven by an equation. The second issue is that the equations have the odd effect of causing the part to grow every time it is rebuilt. This is a circular reference, and is obviously not a desirable situation.

Another problem, more common than circular references, that arises when equations are used goes back to the matter of equation ordering. Let's take another set of three equations and break down what is occurring in each of them.

```
"A@Base-Extrude" = "B@Sketch1" * 2
"C@Sketch1" = "D@Sketch1" + .5
"D@Sketch1" = "B@Sketch1" + 1
```

If we understand what SolidWorks does when it solves these equations, we will gain an appreciation for the problem that arises. First, the software takes dimension B (shorthand names for the dimensions will suffice for this discussion) and multiplies it by 2 to solve for dimension A. It then attempts to take dimension D and add .5 to it to solve for C, but it cannot because D is also being driven by the third equation. This forces SolidWorks to place the second equation on "hold," in a manner of speaking. It then adds 1 to dimension B and solves for dimension D.

The result of this scenario is that the second equation is not solved for after the first rebuild. It takes a second rebuild to show the model geometry in its proper state, with all equations solved. The true problem is that there is no indication that a second rebuild is required. No green streetlight appears in FeatureManager to warn you a rebuild is needed. No warning of any kind is given.

This problem can be remedied by reordering the equations. If the second equation were solved last, the issue would go away and the model would need only a single rebuild to solve all three equations.

It was possible to create a circular reference with only three equations (or even just two, for that matter), and it was also possible to create a rebuild issue with only three equations. Just imagine what could happen with a dozen or more equations. If you do not pay attention to equation ordering, it might take multiple rebuilds to solve all of your equations and show the model correctly.

You should not be afraid of using equations as long as you are careful and as long as you follow one simple rule: simplicity breeds efficiency. In other words, use the easiest tool available to accomplish your design intent. If the Equal relation will suffice, use it. If that does not quite accomplish what you require, perhaps use Link Values instead. If Link Values falls short, use equations. Use the simplest tool that will get the job done, and your models will be less prone to errors and will rebuild much more quickly.

Summary

In this chapter, you have discovered the ability to link dimensions and create equations. Linking dimension values can be accomplished by right-clicking on a dimension value. Only similar dimension types can be linked, and linked dimensions must all be given the same name. Link Values only works for equalities (when two or more dimensions must be equal).

To set more complex relationships between dimension values, equations are required. The left-hand side of the equation can only contain one dimension, which will be the driven dimension. The right-hand side of the equation will contain the driving dimensions.

Even with the Add Equation window open, it is still possible to double click on features to access their dimensions. One click on a dimension value is all it takes to get that dimension placed into the equation.

Equations solve in the order in which they are listed in the Equations window. Use caution when creating equations so that circular references are not created. Also try not to create equations that drive dimensions that also appear as driving dimensions in earlier equations. This will cause your model to require multiple rebuilds in order to solve all equations.

Questions and Topics for Discussion

1. Which side of an equation contains the driving dimensions?
2. Is it possible to change the value of a dimension that appears on the left-hand side of an equation? Explain your answer.
3. What is the least number of equations required to create a circular reference? Give an example.
4. Can exponents be used in equations; and if so, how?
5. If two dimensions have had their names changed via their properties so that both dimensions are named the same, are they linked?
6. Where can the Link Values option be found? List all locations.

Optional Problem

Create a model with multiple equations so that with one dimensional change at least three different dimensional parameters on the model update. Be creative and design any type of model you wish.

Chapter 18

Advanced Assembly Modeling

THERE ARE SOME VERY IMPORTANT ASPECTS OF ASSEMBLY FILES that have not been covered to this point. This includes working with large assemblies and how to best optimize computer performance in such cases. This is an extremely important topic for anyone working with assemblies with more than a few hundred components.

In this chapter you will also explore ways in which an assembly can be restructured. Subassemblies can be created on-the-fly, and components can be organized into groups if the structure of the assembly needs modification. Advanced selection techniques when working with large assemblies is also be touched on, such as using volumetric envelopes with which to control component selection.

File references are looked at in depth. The topic of copying files and maintaining references is one of the most popular topics during a typical SolidWorks training session. External references to other files within an assembly can be very cumbersome when files need to be copied, moved, or renamed. The proper methods for performing all of these tasks are explored in this chapter.

System Performance

When working with large assemblies, or even complex part files, performance is always a factor. Not all of us (or all companies) can afford the most expensive, state-of-the-art computer systems with huge amounts of memory. Computer hardware has come down in price drastically, and will continue to do so. For the time being, though, some of us have to make due with what we have.

There are a number of ways in which part or assembly performance can be optimized, above and beyond putting in a requisition for faster hardware. In this section, you will take a look at some methods that can be used to speed things up a bit. Some of these methods are directly related to graphics card and video performance. Other options relate to settings that can be changed in the SolidWorks software.

Large Assembly Mode

The primary way in which performance can be increased when working with assemblies is known as large assembly mode. This mode of operation is highly customizable, and can be triggered when a certain user-definable threshold is crossed. The threshold is defined as the number of components you feel makes an assembly a "large" assembly. This value will differ, depending on the complexity of the individual components in an assembly or the computer's hardware.

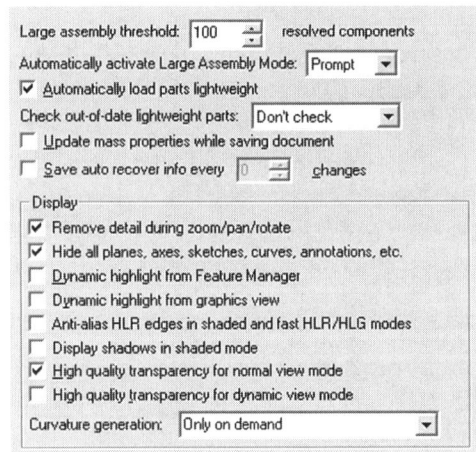

Fig. 18-1. A portion of the Large Assembly Mode section.

Large assembly mode is essentially a list of options that appear scattered throughout the system options. Certain options will change when large assembly mode becomes activated. There is a section titled Large Assembly Mode in System Options (Tools > Options), which is dedicated to this mode of operation. A portion of these settings is shown in figure 18-1.

A portion of the Large Assembly Mode section is devoted to drawings, as shown in figure 18-2. The settings contained therein pertain to drawings of assemblies whose component count is above the large assembly mode threshold.

Nearly all of the settings found in the Large Assembly Mode section are duplicated in other areas of System Options. When large assembly mode is activated, these options cannot be accessed within System Options unless it is via the Large Assembly Mode section. This can be a point of confusion for new users, and even some seasoned veterans. If certain options you feel

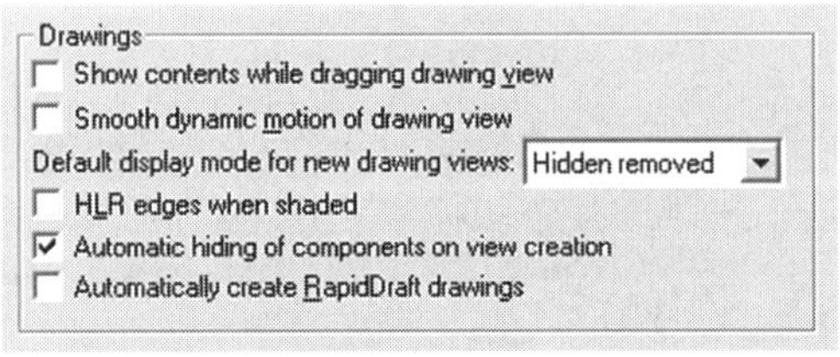

Fig. 18-2. Drawing settings for large assembly mode.

System Performance

should be accessible are grayed out, it is probably because large assembly mode has been triggered and is currently activated.

Large assembly mode can be triggered automatically, or turned on and off manually as well. How-To 18-1 takes you through the process of performing both of these tasks. It will also show you how to set the threshold level for large assembly mode.

How-To 18-1: Implementing Large Assembly Mode

To set the threshold for large assembly mode or to manually turn this mode of operation on or off, perform the following steps. Note that it is best to implement large assembly mode automatically rather than manually. This is due to lightweight components playing a large role in this mode of operation. If the threshold is triggered automatically, components will load lightweight. If large assembly mode is activated after an assembly is already loaded, the components will not revert to lightweight.

1. Select Options from the Tools menu.

2. Select the Large Assembly Mode section in the System Options tab.

3. Set the *Large assembly threshold* option to what you feel would be an appropriate value. 100 is a good value to start with.

4. Set Automatically activate Large Assembly Mode to either Prompt or Always, depending on your preference.

5. Check the option *Automatically load parts lightweight*. This will ensure lightweight mode is used anytime an assembly is opened that contains greater than the number of components specified for the threshold.

6. Click on OK to close the System Options window.

7. Optional: to manually activate or deactivate large assembly mode, check or uncheck Large Assembly Mode in the Tools menu.

If you were to open an assembly that contains more components than specified in the threshold setting, the assembly would load using lightweight components and large assembly mode would be activated. If an assembly was already open prior to enabling large assembly mode and you wish to activate it manually, use step 7. Alternatively, Large Assembly

Mode can be unchecked within the Tools menu as well, if you wish to turn it off manually.for

The Automatically activate Large Assembly Mode option is best left at Always. This keeps SolidWorks from prompting whether or not large assembly mode should be activated. After all, having large assembly mode activate automatically does not hurt anything, and it can always be turned off via the Tools menu.

It is best to leave the *Automatically load parts lightweight* option on, in that this setting alone has a more dramatic effect on assembly performance than anything else. Once again, it does not hurt to load the components lightweight, especially because they can be fully resolved easily enough if you choose to do so. See the section "Component Suppression States" in Chapter 12 to read more about lightweight components.

> **NOTE:** *If an assembly is saved with large assembly mode activated, this mode will be active when the assembly is next opened, whether or not the trigger threshold is reached.*

Performance Options

It is not the intention of this book to serve as a reference guide for every option in the System Options window. Therefore, the options in the Large Assembly Mode section will not be listed. The SolidWorks on-line help serves very well as a reference guide to the individual options found in the Large Assembly Mode section and System Options in general.

The Large Assembly Mode section itself can be used as a sort of listing for system options that will affect system performance. As a matter of fact, most of the options found in the Performance section of System Options are duplicated in the Large Assembly Mode section. There are some additional options or actions that can be taken to increase system performance that are not in the Large Assembly Mode section, and that is what this section is devoted to.

It should be noted that the options or actions discussed in the sections that follow will affect both part and assembly performance unless otherwise noted. Depending on your system hardware and your work habits, the following material may have a varying degree of impact on overall system performance.

General Transparency Settings

There are a few options that affect transparency in one form or another that can reduce graphics performance. Most 3D graphics cards contain

System Performance

enhancements that permit transparency to be accelerated, thereby diminishing any negative results of enabling options associated with transparency. You will have to experiment with these settings to see if they adversely affect system performance; in particular, performance regarding model rotation.

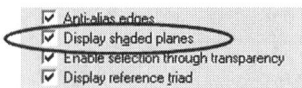

Fig. 18-3. Display shaded planes *option*.

The *Display shaded planes* option, shown in figure 18-3, allows planes to be displayed as translucent objects. This option is located in the Display/Selection section of System Options. When shaded planes are used, the fronts of planes appear green, and the backs of planes appear red. When shaded planes are not used, planes appear with gray borders only. Shaded planes can slow down model rotation on the screen, depending on the computer's graphics card.

Another transparency-related option is titled *Assembly transparency for in context edit*. This option is also located in the Display/Selection section of System Options. It only affects assemblies and only takes place when editing components in the context of an assembly (discussed in Chapter 12). Change this setting to *Opaque assembly* if you find transparency slows down graphics performance on your system to unacceptable levels.

Verification on Rebuild

The *Verification on rebuild* option is found in the Performance section of System Options. It is an often-misunderstood option. Turning it on does not mean SolidWorks will ask you for verification prior to performing a rebuild. Rather, it means that SolidWorks will perform extra error checking during a rebuild.

It is rarely necessary to turn *Verification on rebuild* on. It will increase the time it takes to perform rebuilds. It is suggested you only enable this option while working with problem geometry and suspect there may be errors in the model.

Image Quality

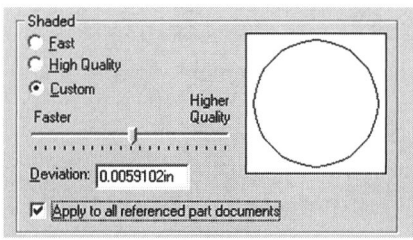

Fig. 18-4. Adjusting the display quality.

Here is a setting that can make a big difference in display performance and even load times for large assemblies. Under the Tools menu, click on Options and select the Document Properties tab. Select the section named Image Quality. There you will see slider bars for adjusting the image quality of both Shaded and Wireframe display. The Shaded image quality setting is shown in figure 18-4.

When saving a part in shaded mode, the tessellation quality of the part will make a difference in the size of the file. Tessellation has to do with the number of polygons used to create the shaded image. The higher the display quality, the more polygons used to generate the image. With regard to assemblies, it is possible to change the polygon count of every component in the assembly. To accomplish this task, the *Apply to all referenced part documents* option must be checked.

It stands to reason that if a file contains more polygons the file will be larger. This is not always significant when working with one part file. However, parts with many curved faces are affected to a greater degree when changing the display quality.

If every part in an assembly has the display quality turned up, the load time of the assembly can increase drastically. The speed at which the assembly can be manipulated on screen will be affected as well. Decreasing the display quality of a single part may reduce the file size by only 50 kilobytes, but an assembly containing 1,000 such components would subsequently have to load 50 million bytes less!

Bear in mind that if the *Apply to all referenced part documents* option is used, changing the shaded display quality setting will affect every component in the assembly at once. This shaded display quality setting will be saved with each part file used in the assembly.

Culling

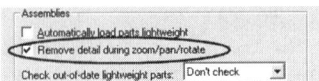

Fig. 18-5. Remove detail during zoom/pan/rotate *option.*

One option found in the Large Assembly Mode section and also on the Performance section is titled *Remove detail during zoom/pan/rotate*. This is otherwise known as culling, and will increase graphics performance drastically. Note that this option affects assemblies only. The option as it appears in the Performance section is shown in figure 18-5.

Checking the *Remove detail during zoom/pan/rotate* option results in faces of the assembly SolidWorks deems too insubstantial to be displayed during zoom, pan, or rotate operations. You may choose to override this option. Simply hold down the Alt key while zooming or rotating and all model faces will be shown. You may notice a big difference in graphics performance if you compare assembly rotation with and without the option enabled.

Add-ins

There are many add-in programs that can be run with SolidWorks. Examples of add-in programs are the PhotoWorks rendering package, the FeatureWorks feature recognition tool, and SolidWorks Toolbox, to name just a few. These add-in programs often add a great deal of functionality to the

System Performance

SolidWorks program, but they do not necessarily need to be turned on if they are not being used.

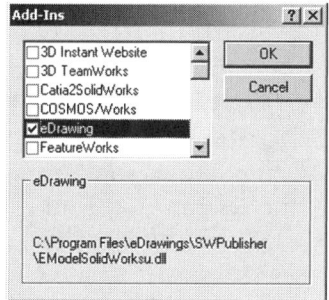

Fig. 18-6. Add-Ins window.

The Add-Ins window, shown in figure 18-6, is accessed by clicking on Add-Ins in the Tools menu. Add-in software should be disabled if it is not in use. This reduces the system resources required by the software. It just stands to reason that if less software is loaded or is running at the same time, more processor power will be left for SolidWorks to take advantage of. To make a long story short: if you are not using it, turn it off.

Suppression and Configurations

Suppressing features or components and part or assembly configurations are two topics that go hand in hand. These previously examined topics will not be reiterated here, but with regard to assembly performance you should be aware that component suppression and configurations can play a significant role.

When working with very large assemblies, it may not always be necessary to show every component in the assembly while working with the assembly file. If this is the case, create a configuration in which noncritical components can be suppressed. If you would rather have the advantage of being able to see certain components but do not want the memory overhead, make it a point to work with lightweight components as much as possible (discussed in Chapter 12).

With regard to configurations, they can be quite advantageous when used with large assemblies. Take the case of a part with many intricate or complex features. Such is the case with the *Knurled Knob* component, shown in figure 18-7. This part contains true helical threads, knurling, raised text, and a smattering of cosmetic fillets. Its file size is in the vicinity of 3.5 MB.

Adding a configuration and then suppressing features that do not need to be shown in the assembly result in the part shown in figure 18-8. The file size of the part shown in figure 18-8 is less than 1 MB, much smaller than that shown in figure 18-7. Perhaps more importantly, however, the part rebuilds much more quickly because the complex features have been suppressed.

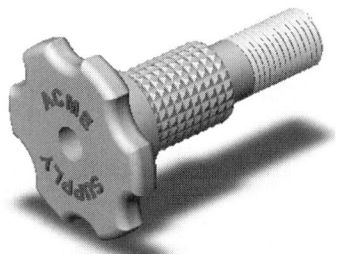

Fig. 18-7. Knurled Knob component.

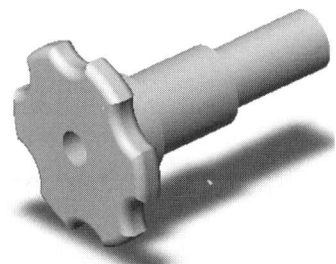

Fig. 18-8. The knob after suppressing noncritical features.

Creating simplified configurations of parts that will be inserted into large assemblies can have a cumulative effect. Assembly rebuild times will be significantly shortened, as will assembly load times. Graphics performance will also improve.

System Maintenance

There is a lot to be said for general system maintenance when it comes to increasing assembly performance in SolidWorks. Do not underestimate the power of defragmentation. Most Windows operating systems now contain a file defragmenter, which takes files on your hard drive and keeps them from being scattered over different physical locations. Instead, each file is stored in contiguous sections on the hard drive. This greatly reduces the time it takes for an assembly to locate and load individual part files.

Assembly Restructuring

In a perfect world, all subassemblies in an assembly would contain all components required, and no components would ever have to be moved from one subassembly to another. No top-level components would ever have to be moved into a subassembly. However, in an imperfect world we sometimes need to jog components around after the assembly has been built.

Restructuring components in an assembly could not be easier. The process is extremely easy to perform, and subassemblies can be created or dissolved on-the-fly. There are limits, however. For example, restructuring components may cause explode steps or assembly features to be deleted. In such a situation, a warning is issued, and you will always have the option to back out.

Incidentally, the terms *reorganize* and *restructure* are both used in the chapter. They both apply, and both terms have similar meaning. Restructuring pertains more to dissolving or creating assemblies, whereas reorganizing pertains more to moving components between assemblies. At certain times, one term may be used over another to more accurately reflect the action being taken.

When restructuring components in an assembly, breaking up a subassembly into top-level components offers the least resistance and is the least prone to problems. Therefore, let's tackle that capability first. The process is known as dissolving a subassembly, and is performed via the right mouse button. How-To 18-2 takes you through this process.

How-To 18-2: Dissolving Subassemblies

To dissolve a subassembly, perform the following steps.

Assembly Restructuring

1. In the assembly FeatureManager, right-click on the subassembly you wish to dissolve.

2. Select Dissolve Sub-assembly.

That is all there is to dissolving a subassembly. The actual assembly file will still exist on the hard drive, but the assembly will no longer be part of the overall top-level assembly. Instead, the separate components will be top-level components. Any mates in the subassembly's *Mates* folder will transfer to the top-level assembly's *Mates* folder.

One common problem you should be aware of is how the fixed attribute of a component will move with the component during restructuring. It is common in most assemblies to have one component that is fixed (locked) in position. During restructuring, this attribute is maintained. However, if one component in the top-level assembly is already fixed, you would not want an additional component moving up from a subassembly to also be fixed.

The problem is easily solved. Make it a point to mentally note which component is fixed in the subassembly prior to dissolving it. Once dissolved, right-click on the component and select Float, as opposed to Fixed. This will typically clear up any overdefining situations that arise from dissolving subassemblies.

Creating a subassembly on-the-fly is similar to dissolving a subassembly, only in reverse. When a subassembly is created, any mates belonging to the components being added to the subassembly will be transferred to the new subassembly's *Mates* folder. Additionally, the new subassembly must be given a name. This is because SolidWorks is actually creating a new assembly file during the process. There are a few methods by which a subassembly can be created within a top-level assembly. One method is presented in How-To 18-3.

How-To 18-3: Creating New Subassemblies

To create a subassembly from existing top-level components in an assembly, perform the following steps.

1. Select New Assembly from the Insert > Component menu.

2. If prompted, select a template to use for the new assembly, and then click on OK.

3. Type in a name for the new assembly.

4. Click on Save.

Following these steps, a new subassembly will be inserted at the bottom of FeatureManager. Then it is just a matter of inhabiting the new subassembly with components. This is done by dragging the desired components into the newly created subassembly. Place the cursor over a component in FeatureManager, hold down the left mouse button, and drag the component onto the new assembly's name. When you let go of the mouse, the component will transfer to the new subassembly.

The benefit of creating a new subassembly in the previously described manner is that the subassembly is created as an empty subassembly. It contains no components to start with, and you can decide at your convenience which components get added to the subassembly.

Another method of creating a subassembly is to use the right mouse button. By right-clicking over a component in FeatureManager, the option Form New Sub-assembly Here will appear. If this option is selected, SolidWorks will ask you to give the new assembly a name, just as with the previous How-To. The part that was right-clicked on will occupy the new assembly. You can then add other components to the new assembly as you see fit.

✓ *TIP: By control-selecting multiple components, a new subassembly can be created that contains all of those components.*

The Fixed attribute is something to watch out for when creating new subassemblies, just as it is when dissolving them. It may be necessary to either fix a component in the top-level assembly or to fix a component in the new subassembly, depending on your situation.

Reorganize Components Command

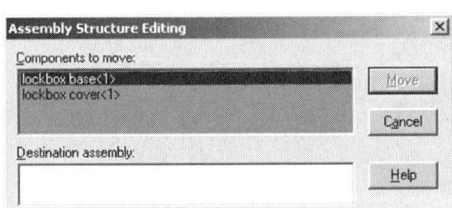

Fig. 18-9. Selecting components to be reorganized.

There is actually a command named Reorganize Components, located in the Tools menu. It opens the Assembly Structure Editing window, shown in figure 18-9. The command itself is very straightforward. Select the component or components to be moved, select a destination assembly, and then click on the Move button. Remember that the salmon-colored list box area is the one that is active.

Although the Reorganize Components command is simple enough, certain problems can arise when reorganizing. This is true whether the

Reorganize Components command is used or whether you simply drag and drop components into or out of different subassemblies.

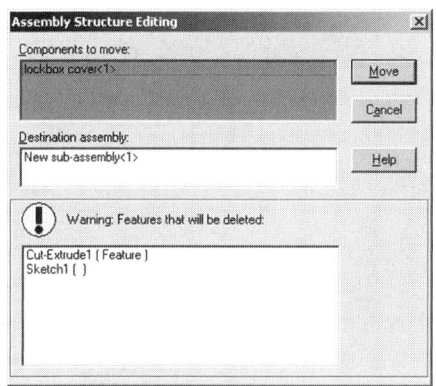

If a component is associated to an assembly feature, and you attempt to reorganize it, a warning message will appear. This is to warn you that if you proceed with reorganizing the component certain assembly features will need to be deleted for the restructuring to take place. An example of this warning window is shown in figure 18-10.

If it is decided to continue with the reorganizing process, any of the items listed in the window will be deleted. You will have to decide for yourself if the price is too high just to move a component to another assembly.

There are other factors that will make restructuring more difficult and prone to problems. External references often get in the way when trying to restructure assemblies. Equations can also have a tendency to fail, because component instance identification numbers often change during restructuring. Your safest bet is to perform any restructuring as early as possible to limit any unforeseen problems from cropping up.

Fig. 18-10. A reorganization warning.

Multi-user Environments and Copying Files

It may seem as though working in multi-user (networked) environments and copying files should be two separate topics, but in fact they are inexorably linked. Copying SolidWorks files in general can be a simple affair, which can be accomplished using Windows Explorer. However, when files contain external references or a multi-user environment is thrown into the mix, the situation becomes more complicated. This section addresses these complications.

Before continuing, let's first make sure some basic terms are fully understood. References (or referenced files) refers to files an assembly or drawing requires in order to be opened. For example, a drawing references a particular part file. External references refers to geometry in another model file that needs to be accessed in order to perform a proper rebuild of a part or assembly file. For example, a feature on a part may have an external reference to a feature on another part in an assembly.

Because copying files often requires different procedures, depending on whether or not a multi-user environment is being employed, we will examine the best way to work with files in two different scenarios. The first scenario involves working with files independently. The second scenario involves a networked workplace. The scenarios will start with a basic environment and get progressively complex. It is recommended that you read through the entire section to get the most thorough understanding of how SolidWorks handles copied files, referenced files, and external references.

Scenario 1: Working Independently

This first scenario assumes that all SolidWorks files of a particular project are being edited by a single individual. Others in the workgroup may also be using SolidWorks, but no more than one person will be working on a particular file.

It is always more efficient to work on files via a local hard drive than over a network. This helps cut down on network traffic and thereby speed up file access, file save times, and file performance. The general procedure would be to create the required parts, assemblies, and drawings on the user's workstation, and then back the files up by copying them to the server, either when completed or at the end of the day.

In that individuals are working on independent projects, there is no fear of overwriting someone else's work on the server. If an accident does occur, the files are duplicated on the user's workstation anyway.

This scenario does require that the user have a good understanding of how to copy files to the server using proper procedure. For the most part, the process of copying files can be broken down into the following three categories.

- Copying standalone files, such as parts
- Copying files with references, such as drawings or assemblies
- Copying files with external references, such as parts or assemblies

Because copying standalone files is not an issue, it will not be discussed here. It is a simple matter of using Windows Explorer to copy a file or files from one location to another, typically for purposes of backup. The other two possibilities are discussed in the to the sections that follow.

Copying Files with References

Assemblies that reference individual part files, and drawings that reference part or assembly files, are considered files with references. The referenced files are required in order to open the assembly or drawing. In other words, if for example an assembly containing 10 components were sent to a client, the client would not be able to open the assembly unless they were also sent the 10 part files corresponding to the components in the assembly.

Multi-user Environments and Copying Files

As long as the assembly contains no components with external references, and sometimes even if it does, the Copy Files function located in the Find References command works just fine. How-To 18-4 takes you through the process of copying a file that includes references.

How-To 18-4: Copying a File with References

To copy a drawing or assembly file and include the files that drawing or assembly references, perform the following steps.

1. Select Find References from the File menu.

2. Click on the Copy Files button in the Search Results window, shown in figure 18-11 (the image has been edited to conserve space).

Fig. 18-11. Search Results window.

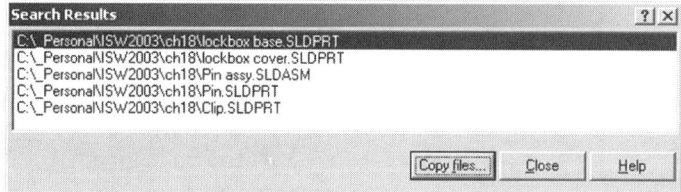

3. Specify whether or not you want to preserve the original directory structure. Clicking on No means that files scattered about can be saved in one location.

4. Browse for and specify a folder in which to save the copies of the referenced files.

5. Click on OK.

6. Click on Close.

This process will save a copy of the assembly or drawing and all files referenced by the assembly or drawing. Note that to open the copy you should close out of the current document first. Never try to open two files that have the same name at the same time.

If you click on Yes when asked to preserve the directory structure (step 3), the original directory structure of the assembly and its components will be recreated below the directory where the assembly's components are being copied. Be forewarned that SolidWorks has an easier time of finding referenced files if they are in the same directory as the assembly (or drawing).

Copying Files with External References

> **NOTE:** The warning messages shown throughout this chapter have been altered. Specifically, file names and paths have been replaced with generic text. Warning messages appearing on your screen will look slightly different.

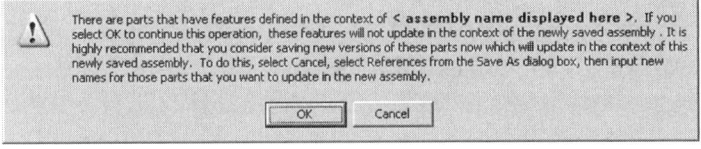

Fig. 18-12. Warning regarding external references.

This is where most people get into problems. If we use an assembly as an example, the problem will be easy to see. If an assembly containing a part with even so much as one seemingly innocuous external reference were opened, and a Save As attempted on the assembly, the warning shown in figure 18-12 would appear. The warning appears as soon you click on the Save button in the Save As window.

This message occurs for a reason most easily explained through example. Assume for a moment that *Part1* is referencing another component (an external reference, in other words) in *Assembly1*. Now imagine that a copy of *Assembly1* is going to be created. Using Save As, the name *Assembly2* is given to the new assembly. When the Save button is clicked on, the message shown in figure 18-12 is displayed. Clicking on OK would result in creating an assembly named *Assembly2* containing the same components as *Assembly1*. However, *Part1* in *Assembly2* now contains an out-of-context external reference. Why?

The answer is that the feature in *Part1* contains an external reference that was created in the context of *Assembly1*. This same feature no longer recognizes the name of the new assembly. Because the feature was not created in the context of *Assembly2*, it fails. This results not in error symbols appearing in FeatureManager but as geometry that no longer updates correctly. Any external reference symbols (->) are now displayed with question marks that never go away (->?).

The solution to this problem is given in the same message that warns you about the problem. Specifically, use the References button found in the Save As window. How-To 18-5 takes you through the process of copying assemblies with external references.

How-To 18-5: Copying Assemblies with External References

To copy an assembly file that contains components that include external references, perform the following steps.

1. In an assembly, select Save As from the File menu.

2. Tell SolidWorks where to save the copy, and type in a name. You do not have to check the Save As Copy option.

3. Click on the References button. This opens the Edit Referenced File Locations window, shown in figure 18-13.

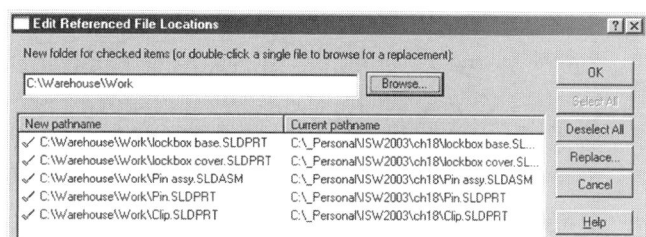

Fig. 18-13. Edit Referenced File Locations window.

4. At a bare minimum, select any component that contains an external reference. You can also use the Select All button to select every file in the list.

5. Optionally, rename any selected component under the *New pathname* heading. This is accomplished with a slow double click.

6. Click on the Browse button and decide where to place the copied components. By default, this will be the same location as the main assembly being copied, which is the typical choice.

7. Click on OK. Note that the *New pathname* column has updated for the selected components.

8. Click on OK to close the Edit Referenced File Locations window.

9. Click on Save.

At this point, as long as the Save As Copy option was not checked (see step 2), you will be editing the copy. If Save As Copy was checked, the files will be on your hard drive someplace and you will be editing the original. This option is explained in more detail in material to follow (see "Save As Copy Option").

To confirm that the external references are correct in the copied assembly, right-click on any of the components that contained an external reference and select List External References. You should find that the component references the new assembly, not the original. In other words, the external reference now refers to the copy. This is as it should be.

Copying Files in a Multi-user Environment

The procedures outlined in the previous material apply to multi-user environments as well. However, there are a few other situations that can arise with regard to read-only files. After all, you do not want to be writing over others' work. The following section explores this topic further.

Scenario 2: Working Over a Network

First and foremost it should be said that running files over the network requires the following.

- A server with the bandwidth to handle the workstations attached to it
- A fast network infrastructure with no bottlenecks, which includes but is not limited to network cards, hubs, and appropriate cabling

What constitutes a respectable server? That depends on a number of factors, including how many machines are attached to the server and the type of applications being run. This chapter will not delve into the hardware aspects of servers or the intricacies of network configurations, as those issues are outside the scope of this book. The network speed, however, is a little more cut and dry, and is an issue dealt with in material to follow.

Plain and simple, opening files over a network is not recommended unless your network can handle the load. This is not to say that you should not move files over the network. It just is not advisable to open files that do not reside on your local computer. This holds true for many applications and is not isolated to SolidWorks. However, because SolidWorks files often contain references to other files, the resultant network traffic can be very high.

Even most product data management (PDM) software will usually copy files to a local machine prior to opening them in SolidWorks. The PDM software is written this way in order to limit the performance bottlenecks inherent in networks. Internal computer data transfers are much quicker than any data transfers that occur over a network, especially considering how SolidWorks files are continually communicating with each other.

Any network running SolidWorks should be either a 100-Mb or 1-Gb network. A 10-Mbs network is not going to suffice for running SolidWorks in a multi-user environment. It will suffice for backing up files to the server at the end of the day, as long as everyone takes turns, but not for opening files over the network.

Opening Files Residing on a Server

As stated, it is not recommended that you open SolidWorks files (those that reside on a server) over a network. However, real-world circumstances dictate that under present situations, your company may be forced to work on files located on the server. Let's discover what happens in such a situation.

Probably one of the most common problems associated with multiple users working on the same project simultaneously is the issue of read-only file access. Using two hypothetical people (Ray and Frank), the situation plays out as follows. Ray opens a file named *Part1*. Frank opens an assembly that contains *Part1*. When the assembly gets to the point where it loads *Part1*, the file gets loaded without Frank ever knowing it is in use by Ray. As a matter of fact, Frank may never know *Part1* is being edited by another person unless he attempts to edit that particular component. When that happens, the message shown in figure 18-14 will appear.

Fig. 18-14. Write access warning.

So what do you do? Well, probably nothing. Do not attempt to edit a part someone else is already working on. On the other hand, if it is your responsibility to be completing this particular part, it would be a good idea to find out why someone else has your file open and is making changes to it. There are some legitimate reasons two different people may need to have access to the same part. The point of the matter is that you cannot have more than one person editing a part, no matter what the program. Communication is the key in this case.

In summary, the message in figure 18-14 is displayed when another user is editing a part contained within the assembly you have open and you attempt to edit that same part within the assembly. The message will also appear when another user has an assembly open that contains components also present in your assembly. The other user does not have to actually be editing the component in question. If you attempt to edit a component in your assembly that is present in theirs, the write access warning message in figure 18-14 will appear.

Opening a File Already In Use

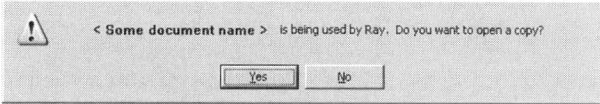

Fig. 18-15. The file is already in use.

Another common warning message occurs when trying to directly open a file already in use (open) by another. The same warning can occur if opening a part or assembly that is a component in another assembly already open by another person. The message that displays on screen is shown in figure 18-15. The situation is very straightforward and does not require much elaboration. Two people simply cannot be working on the same file at the same time.

Usually the wrong thing to do in this situation is to click on Yes. Do you really want to make a copy? Bear in mind that a copy will not be linked to the original, nor will it be referenced by any assembly the original file was in. However, it will be referenced if you save the copy with the same name as the original and in the same location, in which case you will overwrite any changes made to the original file. This would not be a good situation!

Nonetheless, let's say you do click on Yes and open a copy. Some changes are made, and now it is time to save the file. Figure 18-16 shows what you will see if you simply click on the Save icon.

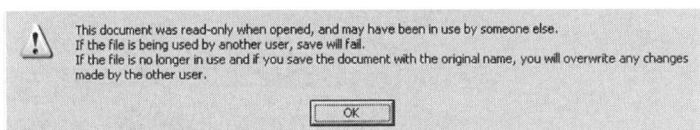

Fig. 18-16. A warning about overwriting a file.

This message is pretty clear because it tells it like it is. If you try to save the file with the same name and the original file is no longer being edited, the save operation will work, but the original will be overwritten with your copy. In this case, you run the risk of severely upsetting whoever modified the original, because you just destroyed all of their carefully thought-out modifications.

If you attempt to save the copy with the same name as the original and the original file is still open (in use), the save operation will fail and SolidWorks will inform you of that fact. Whether the original file is still in use or not, you will always be given one more warning whenever an attempt is made to overwrite another file. In either case, this is almost always a lose-lose situation.

When opening a copy, you would typically have no future intention of overwriting another person's work. In fact, the copy should be saved in some other project folder with a different name. After all, it is a copy and is probably going to get used somewhere else. When saving the copy, use the Save As command instead of Save and the warning message displayed in figure 18-16 will not be shown.

Opening Referenced Files Read-Only

You are in a multi-user environment, everyone is working on related projects, and no one knows from day to day what files they will be working on next. Sound familiar? To describe a situation a little more precisely:

User 1 will be working on an assembly, adding components to it and mating them into place. Other users will be making modifications to components already added to user 1's assembly. How can everyone be working on these interrelated files without worrying about a warning message cropping up every time a file is opened or edited?

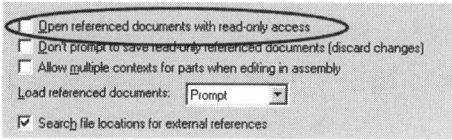

Fig. 18-17. Open referenced documents with read-only access *option*.

There is an option that should help. User 1 can turn on an option found in the External References section of System Options (Tools > Options) titled *Open referenced documents with read-only access*. This option is shown in figure 18-17. Even though the title of the System Options section is External References, the *Open referenced documents with read-only access* option does indeed refer to referenced files as well as externally referenced files.

➥ **NOTE:** *The* Load referenced documents *option in the External References section refers strictly to external references, such as those created using the Cavity command.*

The *Open referenced documents with read-only access* option can be useful in a number of ways. Because all referenced files in an assembly are loaded as read-only, no problems arise if another user attempts to open one of the components present in the open assembly. The user opening the component part file will see no warning messages and may not even be aware that another user has an assembly open that contains that file.

If you were working on the assembly and attempted to edit one of the components in the assembly, you would be faced with the message shown in figure 18-18. Clicking on Yes and accessing the file with write access means that you can then modify the component as you see fit. If another user happened to be working on that same component unbeknownst to you, the "You must obtain write access…" message, shown in figure 18-14, would have replaced the message shown here (in figure 18-18).

Fig. 18-18. Asking if you want to enable write access.

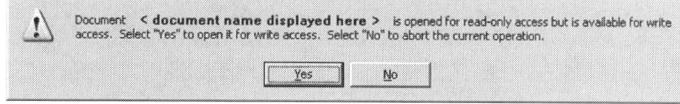

Reloading Read-Only Documents

Put yourself back in the shoes of the user with an open assembly where the referenced files have been loaded read-only. Mating components that might be under an editing process by someone else could cause conflicts. In this case, make it a point to reload any components currently being edited by other users. How-To 18-6 takes you through the process of reloading components loaded read-only.

HOW-TO 18-6: Reloading Components That Were Loaded Read-Only

To reload a component in an assembly that was originally opened as a read-only file, perform the following steps. Note that components are loaded read-only if you reply No to the warning message displayed (see figure 18-15).

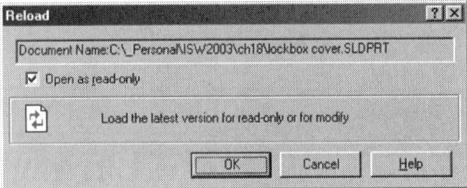

Fig. 18-19. Reload window.

1. Right-click on the component to be reloaded.

2. Select Reload. The Reload window, shown in figure 18-19, will appear.

3. Leave the *Open as read-only* option checked, and click on OK.

Any changes made to the component will now appear on your screen. Obviously, you will have to have some idea who is working on what part or assembly files in order to know which models to reload. This goes back to the issue of communication. SolidWorks will not tell you what files require reloading or who is working on what, unless an attempt is made to edit every component. Only a PDM package will give you this information in any useful form.

The *Open as read-only* option of the Reload window should be checked if the file in question was loaded read-only to begin with. You typically would not uncheck this option, because if you attempt to edit a read-only file, SolidWorks will ask if you want write access to the document. However, if components in an assembly were not initially loaded read-only, and someone needs to work on a component in your assembly, you could reload the component read-only so that the other person could then edit the part document. Meanwhile, you would be able to continue working within the assembly containing that part.

Editing Read-Only Documents

As mentioned in a previous section, referenced files opened as read-only can (under certain conditions) be given write access (see figure 18-18). It is then possible to edit the file and save it in the usual way. However, it is also possible to "trick" SolidWorks into letting you edit a part without getting write access to it first. The following is an example.

Multi-user Environments and Copying Files

User 1 has a file open we will call *Part1*. You open an assembly containing *Part1*. If you were to attempt to edit *Part1* by right-clicking on the component and selecting Edit Part, the "You must obtain write access..." warning, shown in figure 18-14, will appear. But let's say you double clicked on a feature belonging to *Part1* and made a dimensional change without gaining write access. Yes, it is possible. So what happens?

You would be able to perform a rebuild, see the changes take place, and never receive a warning. You would never have any idea the file was in use by another person until you attempted to save the assembly. First, you would see the familiar message (shown in figure 18-20) telling you that some of the referenced documents have been modified, and do you want to "Save the document and the referenced models now?" Well, of course you want to save your changes, unless you were just experimenting, which most people do not do on production documents anyway. Thus, you click on Yes.

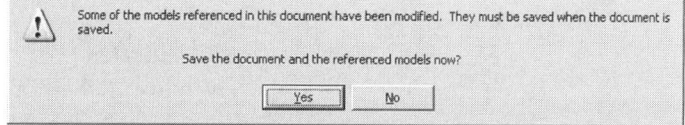

Fig. 18-20. Save the document and referenced models?

You are then presented with the warning message shown in figure 18-16. The warning message is referring to one of the components you changed without first gaining write access. One or more of the components you changed was apparently already opened by someone else, but you did not know it. Surprise! Well, there is not much you can do about it, and therefore you click on OK.

Lo and behold, the warning message displayed in figure 18-21 appears on screen. It is probably one you have seen before, though not yet discussed in this book. It is the message that typically appears after clicking on the Save As button in order to save a referenced file. It might be either an assembly or drawing that is referencing the file being saved. The reason it is appearing now is because SolidWorks knows it may not be possible to overwrite the original file you never gained write access to, and therefore is performing a Save As for you. You click on OK at this point.

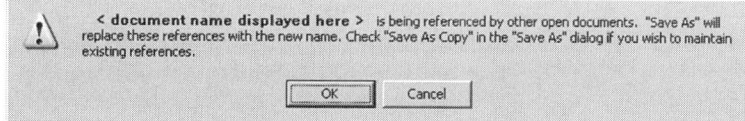

Fig. 18-21. The infamous Save As/Save As Copy warning message.

Up pops the Save As window, asking you for a file name. At this point you are back to deciding if you should overwrite the original file or not.

We have already decided that overwriting the original would be a bad idea (or altogether impossible). However, the assembly needs to be saved, and it appears your only option is to "Save the document and the referenced files now." In that you really do want to save the assembly, but not necessarily the read-only file you unknowingly modified, in this case you can give the file an indicative name (such as junk) and make absolutely certain you check the Save as copy option. The assembly will be saved, a copy of the read-only file will be saved with the name junk, and the assembly will continue to use the original read-only component and not the copy named junk.

How could you have avoided this confusing predicament? The answer is to have gained write access first. Avoid modifying component dimensions via double clicking without editing the component first. Additionally, avoid opening components in their own window without first editing the component to gain write access. You only need to obtain write access once for a component you wish to edit. Afterward, you can edit the component any way you see fit.

In summary, when the *Open referenced documents read-only* option is turned on, use caution when editing components in your assembly. Make it a point to right-click on the component to be edited and select Edit Part. This will inform you as to whether or not the part is open by another user. If a component of an assembly needs editing, obtain write access to all subassemblies in the hierarchy. This will ensure you have write access to subassemblies all the way down to the desired component level, and will reduce confusion when saving the modified file.

Special Note on Read-only Attributes

When files are opened read-only in SolidWorks, the Windows operating system does not see the document as read-only. Internal to the SolidWorks software, read-only is a term that dictates whether or not that file can be overwritten. A person with read-only access to a file cannot theoretically make changes to that file. Yet, under many circumstances a person with read-only access to a SolidWorks document can easily gain access to that document and change it any way he sees fit.

When a file is marked read-only by the operating system (not SolidWorks), there is an actual attribute that gets set on that file that allows only those with proper permission to edit the file. We will not go into depth regarding file permissions and operating system security, as these topics are outside the scope of this book. However, it is important to understand that operating system read-only file attributes and opening SolidWorks files as read-only are two different things.

The information presented in this section is really for those "in the know." System administrators need to understand how SolidWorks handles situations in which associated documents are accessed by multiple users. Be aware that opening read-only files in SolidWorks is more of a precautionary function and not a function of security.

Save As Copy Option

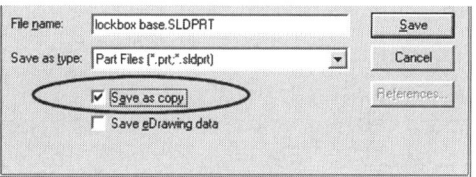

Fig. 18-22. Save As Copy option.

The Save As Copy option, shown in figure 18-22, has been mentioned a few times, but its meaning may still be unclear. Try this experiment. Open up any SolidWorks assembly, and then open up one of the assembly's components in its own window. With the component (part) window active, click on Save As in the File menu. The warning message shown previously (figure 18-20) will appear.

Let's use an example so that you may better understand this message. Assume the assembly you opened is named *Assembly1*. Assume the component you opened in its own window is called *Part1*. Plain and simple, if you check the Save As Copy option, *Assembly1* will continue to reference *Part1*. In other words, you will still see *Part1* listed in the assembly's FeatureManager.

If the Save As Copy option is left unchecked, you could save a copy of the part file (let's use the name *Part2*) and the assembly would begin using the copy. In other words, *Assembly1* will now list *Part2* in FeatureManager where previously it was listing *Part1*. *Part1* would still reside on the hard drive, but would no longer be a component in *Assembly1*.

This scenario indicates the more complex behavior of the Save As Copy option. Be aware that this behavior occurs only if the referencing assembly (or drawing) is open. If the referencing document is not open, there is no warning message, and the Save As Copy option does not exhibit the same behavior. Rather, the option reverts to its more primitive and mundane behavior of deciding whether or not you will be editing the copy or the original after saving the copy.

Most Windows programs will contain a Save As Copy option in one form or another, and they all function similarly. First and foremost, however, it should be stated that whether the Save As Copy option is used or not, a copy is still being saved. If the Save As Copy option is checked, a copy is placed on the hard drive and the original remains on screen. If not checked, the original is saved to the hard drive and the copy appears on screen.

Copying Drawings with Referenced Models

A common desire when creating design drawings is the capability to create a copy of a drawing along with a part (or assembly) that is in no way connected to the original drawing and part. This is often the case when minor changes have to be made to, for example, a part being sold to another customer and a second drawing is needed, but the SolidWorks user does not want to spend the time to detail a completely new drawing from scratch.

Creative use of the Save As Copy option can produce a copy of a drawing and the model it references, neither of which are linked back to the original model or drawing. Changes can then be made to the model that will update in the drawing, and a fair amount of tedious and time-consuming detail work can be avoided. How-To 18-7 takes you through the process of saving copies of drawings with models.

How-To 18-7: Save Copies of Drawings with Models

To save a copy of a detail drawing along with the model it references, perform the following steps. The copy of the drawing will reference the copy of the model, but neither the copied model nor the drawing will be related in any way to the original drawing or model.

1. Open both the drawing and the model you wish to create copies of.

2. With the Drawing window active, click on Save As in the File menu.

3. Type in a name for the new drawing you will be creating and decide where you want to place it. Make a mental note of this location.

4. Make sure Save As Copy is not checked, and then click on Save. You should now be looking at the copy, with the original drawing safely tucked away on the hard drive.

5. Leaving the new drawing open, activate the Model window. This can be done via the Window menu or by using the Windows standard Ctrl-Tab hot key combination.

6. Click on Save As in the File menu. The "Save As/Save As Copy" warning should appear.

7. Click on OK to close down the warning message.

8. Type in a name for the new part you will be creating and decide where you want to place it. The location should be the same place you placed the drawing in step 3.

9. Once again, make sure Save As Copy is not checked, and then click on Save. If the model is an assembly containing external references, you will need to use the References button to also save copies of the components containing the external references.

Once finished with the last step, activate the copy of the drawing (which should still be open). If you have followed the steps correctly, you will be able to click on Find References in the File menu and see that the drawing is now referencing the copied model file. Additionally, if the original drawing were opened and the Find References command employed, the references would point to the original model.

Search Order and Search Paths

To finish this section, let's examine one last important topic with regard to how SolidWorks searches for files. Search criteria play a very important role in how a company (or individual) implements directory structure. There are good ways in which to set up a directory structure, and then there are methods that can result in extremely inflexible file management options. Let's look at the recommended options first.

Every company is going to have its own file naming conventions and preferences with respect to the directory structure of where files are saved. When possible, one of the best ways to categorize a directory structure would be to use the project naming convention. In other words, create folders named after projects and save all documents relating to that project in that folder.

Project names might reflect the name of a customer, or perhaps a model number of a project, or some other criterion. What is important is that all assemblies, parts, and drawings reside in the same directory. With large assemblies, it might be prudent to break up the main project folder to include subdirectories for subassemblies and their related components and drawings.

The reason the project naming convention works so well has to do with how SolidWorks searches for referenced files. One of the very first places it searches is in the same directory as the file that was just opened. This means that if an assembly is opened, SolidWorks will look in the same folder for the components as well. The following list is the search order SolidWorks uses when it tries to find a referenced file.

1. Paths specified in the File Locations section of System Options, if the Search file locations for external references check box is selected (see material to follow).

2. The last path you specified to open a document.

3. The last path the system used to open a document (in the case of the system opening a referenced document last).

4. The path where the referenced document was located when the parent document was last saved.

5. The path where the referenced document was located when the parent document was last saved with the original disk drive designation.

Finally, if a referenced file still is not found, you are given the option to browse for it. Once the referenced file is found, all updated reference paths in the parent document are saved when the parent document is saved.

Most SolidWorks users do not use the option mentioned in the first search path. Therefore, SolidWorks reverts to the next search path, which is the last path specified to open a document. In other words, if the last file you opened was an assembly, SolidWorks will continue looking in the same folder the assembly was in for all components in the assembly.

If you do wish to set up a specific search path that SolidWorks will always use to look for referenced files, use the method outlined in Chapter 8 (see "Modifying Folder Location Paths"). Consider the implications, though. Unless all of your referenced files are in particular locations and all of the referencing files are someplace else, there is no advantage to setting up a particular search path.

Envelopes

An envelope, to put it simply, is a way of selecting components in an assembly. The envelope is a volumetric area that in actuality is really nothing more than a part. The area the envelope (or volume) defines can then be used to select components. For example, all components falling completely within the envelope can be selected.

What happens with the selected components is up to you. Perhaps the components need to be hidden or suppressed, or perhaps some other action needs to be carried out. You have a few options.

Where envelopes are most useful is when working with large assemblies. Envelopes simplify the selection process by making it so that each

Envelopes

component does not need to be separately selected, whether from the work area or FeatureManager. When a hundred components need to be selected, it is easy to see the benefit of using an envelope.

To use an envelope, one must first be created. An envelope uses the same extension as a part file. For that matter, an envelope is a part file, used for a special purpose. An existing part can be inserted as an envelope, or a new envelope can be created in much the same fashion as a new part can be added to an assembly. How-To 18-8 takes you through the first of these methods.

HOW-TO 18-8: Inserting a Part as an Envelope

To insert an existing part as an envelope, perform the following steps. It should be noted that parts used as envelopes are typically primitive geometric shapes created specifically for this function; they are not design components.

1. Select Insert > Envelope > From File.
2. Select the part file to be used as an envelope.
3. Click on Open.
4. Select a location to place the envelope.
5. Mate the envelope at the desired location.

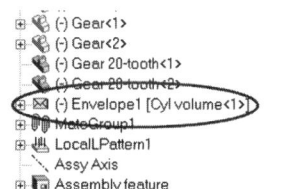

Fig. 18-23. An envelope in FeatureManager.

When a part is inserted as an envelope, it automatically has a translucent color. The envelope will also appear with a special symbol in FeatureManager. An example of an envelope in FeatureManager is shown in figure 18-23.

Note that the envelope is automatically given the name *Envelope1*. The name of the part file is shown in brackets, along with the ID number added by SolidWorks. This is displayed as *Cyl volume<1>* (short for cylindrical volume). The part file is nothing more than an extruded circle, and thus the act of creating a part file to use as an envelope can be a no-brainer. It simply has to be the right size for whatever it is that needs to be selected within the assembly. Size and position are most important when working with envelopes. Mate relationships allow for precisely controlling the position of the envelope.

Often it is easier to create an envelope within the context of the assembly. This is because the size of the volume can more easily be decided on when the rest of the assembly is being viewed at the same time. How-To 18-9 takes you through this second method.

Chapter 18: Advanced Assembly Modeling

HOW-TO 18-9: Inserting a New Envelope

To insert a new envelope when a part file containing the desired volumetric area does not already exist, perform the following steps.

1. Select Insert > Envelope > New.
2. Type in a name for the new part file.
3. Click on Save.
4. Select a plane or planar face to begin sketching on.

Fig. 18-24. An example of an envelope.

Once these steps have been completed, it will be necessary to create a sketch and subsequently create the feature that will define the volume of the envelope. Envelopes often contain only one simple feature, such as an extrusion, but this is by no means a requirement. An example of an envelope is shown in figure 18-24. It is the cylindrical part that encompasses the axle component and surrounding area.

If you have already read the section in Chapter 12 regarding inserting a new component into an assembly, you will recognize the procedure for inserting a new envelope. The process is exactly the same. However, the reason for carrying out the process is different. That reason, in the case of envelopes, is to more easily select components in the assembly. Read on to learn how the envelope function can be used to your advantage.

Envelope Selection Techniques

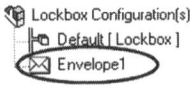

Fig. 18-25. An envelope displayed in Configuration-Manager.

There are a few simple methods that can be employed in order to select components with an envelope. This selection process will allow for selecting, for instance, all components that fall within the volume of the envelope. The entire process is quite simple.

Once an envelope has been created within your assembly, take a look at ConfigurationManager. You will notice the new envelope listed there. An example of this is shown in figure 18-25. Right-clicking on the envelope in ConfigurationManager is what gains access to the most basic selection techniques involving envelopes.

Advanced Selection Techniques

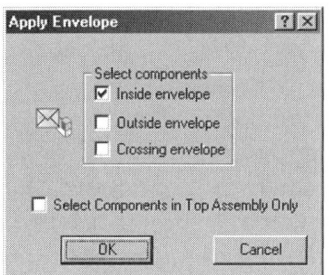

Fig. 18-26. Apply Envelope window.

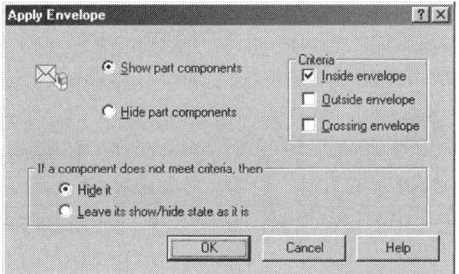

Fig. 18-27. Another variation of the Apply Envelope window.

One command found when right-clicking on the envelope in ConfigurationManager is titled, logically enough, *Select using envelope*. This opens the Apply Envelope window, shown in figure 18-26. Most of the options speak for themselves. The Select Components in Top Assembly Only option will ignore any components found in subassemblies.

The other command found in the right mouse button menu directly underneath *Select using envelope* is titled Show/hide using envelope. This opens another window with the same title of Apply Envelope, but with a slightly different purpose in mind. The window is shown in figure 18-27.

Where the *Select using envelope* command is used for selection purposes, Show/hide using envelope is used strictly for showing or hiding components. The interface needs no further elaboration, as it is self-explanatory. Simply check the desired criterion, specify the desired visibility setting, and click on OK.

Select using envelope can be used for functions other than hiding or showing components. Use it to select components for any reason, such as obtaining mass properties, suppressing components, or whatever the case. for the sections that follows explores more specific search criteria, which go beyond envelopes and includes custom properties.

Advanced Selection Techniques

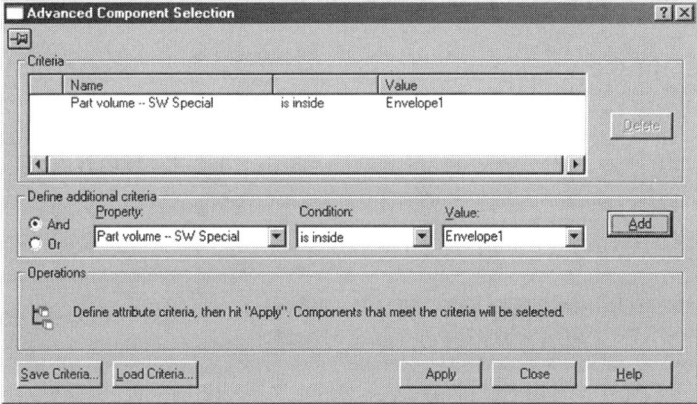

Fig. 18-28. Advanced Component Selection window.

Once an envelope has been inserted into an assembly, it can be used to select components. This is accomplished using the Advanced Select function, found under the Tools menu. An example of the Advanced Component Selection window is shown in figure 18-28.

There is a variety of selection criteria that can be added to the Advanced Component Selection window. For example, a property can be selected from the Property drop-down list, or a user-defined property can be entered. The property of an object directly relates to the properties of a file accessible through the File menu. As a matter of fact, the file's properties are exactly what are searched when using advanced selection techniques.

The area titled Condition allows for plugging in logical operators. For instance, if you are searching for a file that contains a property for *Date completed,* and the file being searched for must match a specific date, the condition should be set to is (exactly). The Value entry would be whatever date is being searched for. The value is something that would get typed in by the person performing the search.

SolidWorks has some special search properties it adds to the Property list. These special property options are *Part mass, Part volume, Configuration name,* and *Document name.* The document name would be the actual name of the file. Special properties can be used as search criteria without adding them to a file's custom properties. When a search is performed using any property other than one of the SolidWorks special properties, the file being searched for must contain relevant data in its properties. If not, the search will turn up nothing. How-To 18-10 takes you through the process of adding selection criteria.

How-To 18-10: Adding Selection Criteria

To add a selection criterion to the Criteria listing, perform the following steps. At least one criterion must be added.

1. Select a property from the Property list, or type in a name of a property to search for.
2. Select a condition from the Condition list.
3. Select or type in a value in the section named Value, if required.
4. Click on the Add button.
5. If additional criteria are required, select either the And or the Or option and then repeat steps 1 through 4.
6. Click on Apply.
7. Click on Close to accept the selection.

Selection criteria can be saved as a file once created. This same file can then be called up at a later time, as when performing another advanced selection in another assembly. When a selection criterion is saved as a file, a SolidWorks property criterion is created. This type of file has an *.sqy* file extension.

Using the advanced selection process will not highlight the icons of the components that get selected, as is usually the case when selecting files in FeatureManager. Nonetheless, the components will be selected. The selected components themselves will be highlighted in the usual manner if in shaded mode. You will notice that various functions on the Assembly toolbar, such as the Hide and Suppress icons, will be available. These icons are not available unless a component has been selected first.

After applying the selection criteria to the assembly and clicking on the Close button, you will want to carry out whatever command you had in mind. Hide, show, suppress, delete, check the mass properties of, or anything else you have learned how to accomplish to this point.

Summary

This chapter taught you how to use assemblies more efficiently. A number of settings were explored that should help you maximize computer power when working with large assemblies or complex part files. When using lightweight components, load times for assemblies can be significantly reduced. Components that have been loaded lightweight only load the data necessary to view the component. Because of this, assemblies must be saved in Shaded mode for them to take advantage of lightweight components.

Large assembly mode is a method of turning off certain resource-draining options when large assemblies are loaded. What is considered to be a large assembly is up to the user. By setting the large assembly threshold, you can determine the number of components an assembly can have before large assembly mode kicks in. Large assembly mode can be manually enabled or disabled through the Tools menu.

Subassemblies can be created on-the-fly while in an assembly. They can also be dissolved. Components can be dragged from one assembly to another within FeatureManager. There is a great deal of flexibility that exists, which allows for the reorganization or restructuring of components and subassemblies. Components with external references, however, often cannot be restructured without deleting the references. SolidWorks will always warn you if problems are encountered when performing any type of restructuring.

An important topic discussed in this chapter had to do with copying assemblies and components that exist within assemblies. As was discovered, there is a right way and a wrong way to copy files. If saving a copy of a referenced model, whether the model is referenced by an assembly or a drawing, make sure the Save As Copy option is checked if you want the assembly or drawing referencing it to continue to reference the original model.

When copying assemblies and components that exist in the assembly, the Save As command can be used. This is especially important when external references exist in the assembly. Use the References button inside the Save As window to copy any of the files containing external references. If this procedure is not followed, the files containing references may not update correctly.

Envelopes are nothing more than part files whose volume is used to select components. For example, if components fall within the envelope's volume, they will be selected. Other selection criteria can be performed by searching for selected file properties. Logical operators can be used to help define the criteria. Once the components have been selected, commands can be exercised on the selection set.

Questions and Topics for Discussion

1. Name three ways in which to speed up the performance of a sluggish part or assembly.

2. Describe what happens when an assembly is loaded "lightweight."

3. If an assembly were saved with its hidden lines displayed, could it be loaded lightweight? Explain.

4. What determines when large assembly mode is activated?

5. If a subassembly is dissolved, what happens to its mate relationships?

6. When copying an assembly and that assembly's components, and there are external references that exist within the assembly, how can all files be copied while maintaining proper external references between the copies?

7. What file extension does an envelope use?

8. If a part that exists as a component in an assembly is saved using the Save As command, and the referencing assembly is not open, does it matter if the Save As Copy option is checked?

CHAPTER 19

Advanced Feature Types

THERE ARE A NUMBER OF FEATURE TYPES worth mentioning that have not been discussed in previous material. These are features such as those created by the Dome or Shape command. There is also extended functionality related to topics previously discussed, such as the Fillet window. Areas to be covered include variable radius fillets and face blends, both of which are related to the Fillet command.

Domed Features

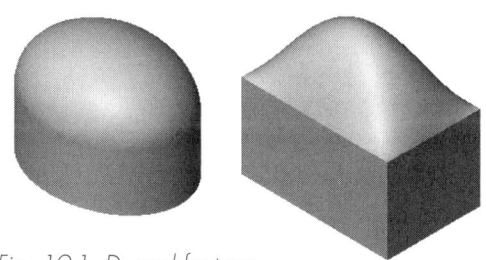

Fig. 19-1. Domed features.

Domed features are created with the Dome command. The mechanics behind the Dome command are very simple, and because a domed feature is an applied feature, no sketch is required. There is seemingly no limit to the geometric shapes that can be used to create a domed feature. The only stipulation is that the face is planar. Figure 19-1 shows two examples of domed features.

The simplicity of the Dome window, shown in figure 19-2, belies its complexity. Some very interesting and complex shapes can be created with the Dome command. How-To 19-1 takes you through the process of creating a domed feature.

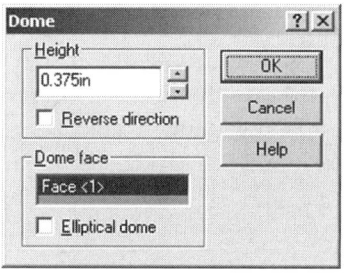

Fig. 19-2. Dome window.

How-To 19-1: Creating a Domed Feature

To create a domed feature, perform the following steps.

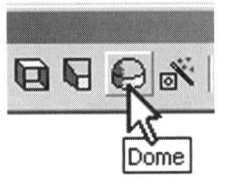

Fig. 19-3. Dome icon.

1. Select Dome from the Insert > Features menu, or click on the Dome icon, shown in figure 19-3. (The Dome icon can be added to the Features toolbar if it is not already present. See Chapter 23.)
2. Select a planar face to create the domed feature from.
3. Specify a height for the domed feature.
4. Click on the Reverse Direction option if necessary.
5. Click on OK to create the dome.

By default, the Dome command will add material to a part when creating the dome. However, clicking on the *Reverse direction* option will reverse the dome so that it points in the opposite direction. When this is done, the dome will remove material from the part. In essence, the Dome command then functions as a dome cut.

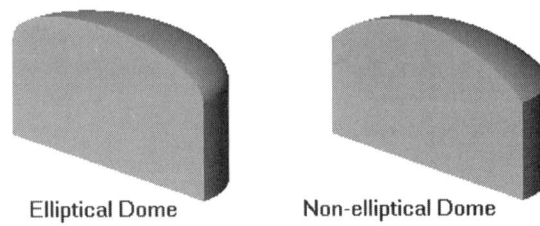

Fig. 19-4. An elliptical dome versus a nonelliptical dome.

If the face to be domed is circular or elliptical, the *Elliptical dome* option becomes available. An elliptical dome's tangency condition will be perpendicular to the plane being domed. The difference between the two dome types is more noticeable when the dome height is small. Figure 19-4 shows a dome on the left, where the *Elliptical dome* option has been checked. The dome on the right does not have the option checked. Look closely where the dome meets the rest of the part and you will see the difference. (Cross-sectional views have been used for clarity.)

Typically, many plastic containers have inverted dome shapes at their base. This is to help keep the base of the bottle from bowing outward from the weight of the content within, making it impossible for the bottle to stand upright on a flat surface. In Exercise 19-1 you will be asked to create a bottle using skills learned in chapters 6 and 9. Specifically, you will be asked to use a swept feature with guide curves, and to use the Split Line command. The exercise will then take you through the use of the Dome command.

EXERCISE 19-1: Using the Dome Command

In this exercise, you will build upon lessons learned in earlier chapters. Dimensions will not be given, aside from some basic overall design requirements. You will have free reign to create a bottle with a shape of your own choosing. The overall size requirements are as follows.

- The bottle must have an elliptical base measuring 1.5 x 3 inches.
- The neck of the bottle must be 1.25 inches in diameter.
- The bottle should be 4.5 inches tall to the base of the neck.

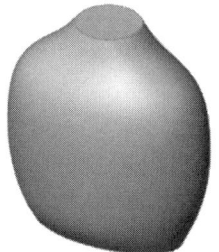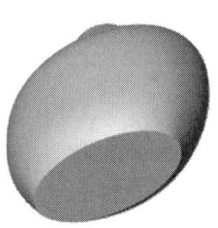

Fig. 19-5. Two views of a bottle. Yours may differ.

You could either use a lofted or swept feature to create the general shape of the bottle. The example shown in figure 19-5 uses a swept feature with two guide curves. Following figure 19-5 is an outline you could use to create a bottle similar to that shown. If you feel confident enough to embark on a different process, feel free, but try to maintain the overall size dimensions previously listed.

1. Create the guide curves. The guide curves will dictate the size of the base of the bottle, as well as the neck diameter. Splines were used to create the two guide curves used for the bottle shown in figure 19-5.

2. Create the path sketch. This should be the same height as the guide curves.

3. Create the sketch for the sweep profile. This will be an ellipse. Use the pierce relation to lock the ellipse to the guide curves. Use a pierce or coincident relation to center the ellipse on the path.

4. You should have four (4) sketches at this point, and can now complete the swept feature.

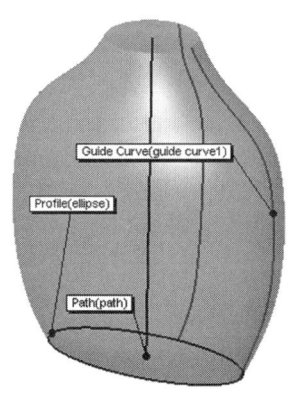

Fig. 19-6. Possible sketch geometry for the bottle.

To further help guide you along in the creation process of the bottle, examine figure 19-6 to see how the guide curves, path, and profile would appear. The bottle is shown in a translucent state, which is actually the preview shown when creating the sweep. The guide curve on the side of the bottle is nothing more than a sketch that describes the profile of the right side of the bottle as seen from the front. Likewise, the guide curve at the back of the bottle describes the profile as seen from the right side. Your guide curves will be different. How much different is up to you.

Once the overall shape of the bottle has been created, you will need to rotate the bottle so that you can see the base. This is where you will be concentrating your remaining efforts. You will be creating a domed feature at the bottle's base. A small indentation will be cut into the part. Because it is not desirable to dome the entire base (for aesthetic and manufacturing reasons), the face will be split and just a portion of it domed. To apply the split line operation to the bottle, continue the following steps.

5. Begin a sketch on the base of the bottle.

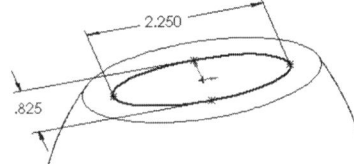

Fig. 19-7. Sketch for the split line.

6. Create the sketch shown in figure 19-7. (Hint: add a horizontal relation between two of the ellipses' quadrant points to help fully define it.)

7. Select Split Line from the Insert > Curve menu.

8. Specify Projection as the split type.

9. Select the bottom face of the bottle as the face to be split.

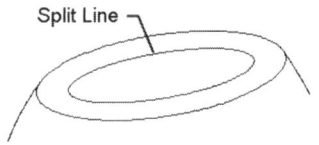

Fig. 19-8. Split line feature.

10. Check the *Single direction* option and make sure the preview arrow is pointing toward the part.

11. Click on OK to create the split line, shown in figure 19-8.

12. Now that a face exists that describes the area to be domed, select Dome from the Insert > Features menu.

13. Select the newly created face at the center of the bottle's base as the face to be domed.

Fig. 19-9. Completed dome feature.

14. Specify a height of .250 inch.

15. Check the *Reverse direction* option to create a cut with the domed feature.

16. Click on OK to create the dome. The completed feature is shown in figure 19-9.

17. To smooth things out, add a .500-inch-radius fillet to the edge of the dome cut.

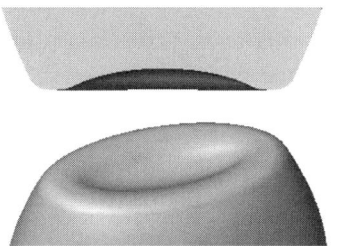

Fig. 19-10. The bottle's base with completed dome feature and added fillets.

18. Add a .25-inch-radius fillet to the outside edge of the base of the bottle.

19. Save your work.

Figure 19-10 shows the base of the bottle after adding the dome feature and fillets. The upper portion of the image shows a section view of the base, taken halfway through the bottle. Note the nice curvature, which should aid in strengthening the bottle and allow it to stand upright in a more stable position.

Shape Command

The Shape command has got to be the oddest of all commands in the SolidWorks arsenal. It can be used to create some truly strange geometry, and has slider bar controls for settings with interesting names, such as *Pressure* and *Stretch*. What the Shape command allows for is deforming existing faces on a part. The degree to which those faces are deformed is up to the SolidWorks user.

In its simplest form, the Shape command will allow for creating an effect of a container being pressurized, with its sides bowing outward. It can also be used to "suck in" the sides of a plastic part to give it an indented effect. Think of the Shape command as being able to deform a face as though it were a rubber membrane. This can add some realistic effects to certain types of parts, but is not something you would use every day.

An advanced application of the Shape command is to use curves to control the shape a face takes on. The control curves can be used to "draw out" areas of the face being deformed. The amount of control the curve (or curves) exerts on the deformed face can be adjusted.

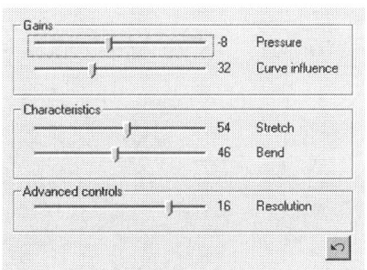

Fig. 19-11. Slider bar adjustments for the Shape command.

The slider bar adjustments (see figure 19-11) used when deforming a face require some time to act on the model. Once an adjustment is made, SolidWorks recalculates the new shape of the face. How long this takes depends on your computer and some of the settings within the Shape command. The Resolution setting will definitely increase the time it takes to calculate a deformed shape. Higher resolutions require more processor time.

The Shape command is best left to those that need to do a bit of free-form modeling. The features it creates are difficult to dimension in any meaningful way in a design drawing. As mentioned previously, though, it can be useful for creating interesting effects on plastic parts, such as squashed or inflated. How-To 19-2 takes you through the process of creating a shape feature.

How-To 19-2: Creating a Shape Feature

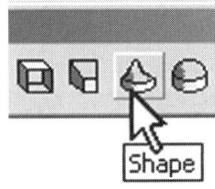

Fig. 19-12. Shape icon.

To deform a face on a model using the Shape command, perform the following steps.

1. Select Shape from the Insert > Features menu, or click on the Shape icon, shown in figure 19-12.

2. Select a face to be deformed. The face should appear in the *Face to shape* list box in the Shape Feature window, a portion of which is shown in figure 19-13.

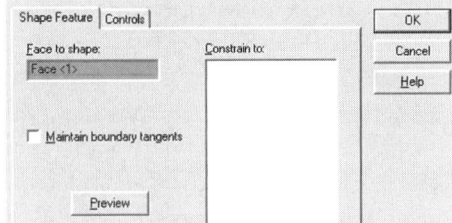

Fig. 19-13. Shape Feature window.

3. Select the *Maintain boundary tangents* option if the deformed shape should try to maintain tangency to the boundary of the selected face.

4. Optionally, click in the *Constrain to* list box and select one or more curves that will help constrain the shape of the face to be deformed.

5. Optionally, click on the Preview button, or simply select the Controls tab to see a preview, as well as the current control settings.

6. Use the slider controls to change settings for Pressure, Stretch, Bend, and Resolution.

7. If any curves were selected to constrain to in the Shape Feature tab, additionally specify a setting for the Curve Influence slider bar.

8. Click on OK to accept the shape shown in the preview.

The Shape command is one you should experiment with. It can be a frustrating command due to the shape wanting to update whenever a minute adjustment is made. It can also be frustrating when a slider is moved to a position that really blows the shape out of the water.

Shape Command

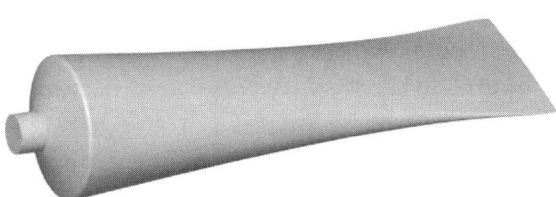

Fig. 19-14. The toothpaste tube.

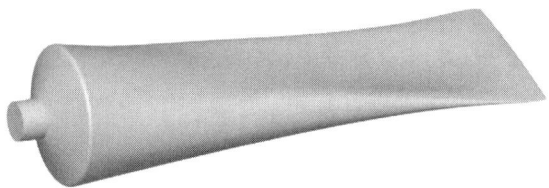

Fig. 19-15. After applying the Shape command to the tube.

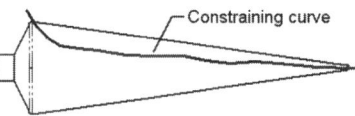

Fig. 19-16. A spline is used for deformation control.

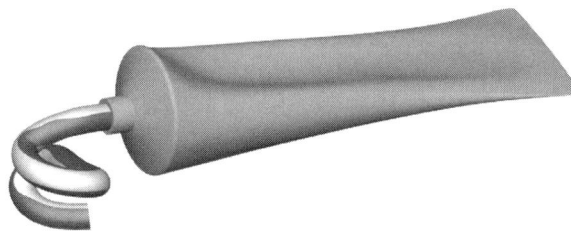

Fig. 19-17. Using the spline as a constraining curve and adding a swept feature.

This can almost be taken literally, because the shape can take on a big "splash" appearance, seemingly gaining a mind of its own. Luckily, there is an Undo icon built into the Control tab portion of the Command window. However, you will have to wait for the last adjustment to settle down before clicking on Undo. The following is a short illustrated example of what might be accomplished with the Shape command. Figure 19-14 shows a tube of toothpaste modeled with a lofted feature and variable radius fillets (discussed in material to follow).

Using the Shape command, the tube can be made to appear as though it is being squeezed, to some extent. This is shown in figure 19-15. The effect is not overly dramatic because the deformation is applied fairly equally over the entire top face. For a more dramatic squeeze effect, we need some way to make the larger end of the tube appear as though it is being squeezed more than the bottom end, which is already fairly flat to begin with.

To achieve better control over the shape the deformation takes, a curve will be used. In the case of the example being used here, the curve is a free-form spline that was sketched on the Front plane of the model. The spline is shown in figure 19-16, along with the toothpaste tube in wireframe display mode for reasons of clarity.

When creating the shape feature, the spline can be selected per step 4 of How-To 19-2. The spline curve will exert a force on the face selected to be deformed, thereby making the face conform to the curve. Additional curves could be selected to obtain a higher degree of control on the deformation. In the final image, figure 19-17, a swept feature was added (the toothpaste) to increase the squeeze effect even more. Better get a towel to clean up that toothpaste!

Advanced Filleting

There certainly is a heck of a lot more to the Fillet command than creating simple fillets and rounds. This book has only so far explored some of the more basic options available through the Fillet command. It would be an injustice to SolidWorks to ignore the more powerful functionality possible through this command. This introduces you to some advanced techniques and some previously unexplored territory regarding the use of the Fillet command.

Variable Radius Fillets

Typically, when creating a fillet or a round, edges are selected. This is not, however, always the case. If a face is selected, everything that comes into contact with that face will be filleted. If this is your design intent, selecting a face is certainly easier than selecting multiple edges. In the case of a variable radius fillet, edges must be selected. This is a requirement, and selecting faces is not allowed.

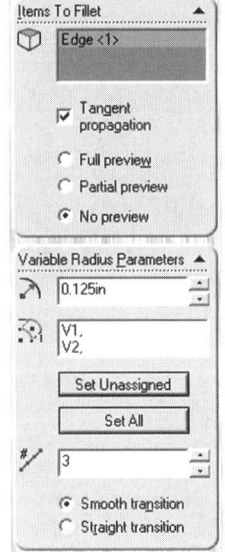

Edges must be selected for good reason. Selecting edges allows SolidWorks to populate the Vertex list box with vertex points (as the plural being vertices or vertexes) associated with the selected edges. It is not possible to click inside the Vertex list area and select vertex points. Only by selecting edges will the software place the appropriate vertices within the list box. Once there, the individual vertex points can be selected and assigned a radial value. What you would see when adding a variable radius fillet is shown in figure 19-18.

Variable radius fillets come in two varieties: smooth transition and straight transition. Examples of each of these transition types are shown in figure 19-19. The smooth transition will be parallel to the edge being filleted, where the edge begins and ends. A straight-transition variable radius fillet varies linearly from one radius to the next.

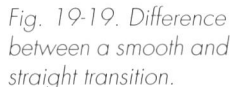

Fig. 19-18. Variable radius fillet panel.

Fig. 19-19. Difference between a smooth and straight transition.

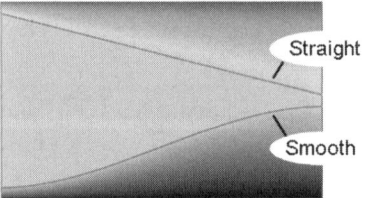

When creating a variable radius fillet, the prime concern is establishing radial values for each vertex point. There are two options that make

Advanced Filleting

this task easier, especially if there are numerous vertices listed. These options are in the form of buttons labeled Set Unassigned and Set All, which can be seen in figure 19-18.

By specifying a value for the radius and clicking on the Set All button, a radial value can easily be assigned to every vertex point in the list. Likewise, the Set Unassigned button will apply the specified radius to only those vertices that do not yet have a radius established. How-To 19-3 takes you through the process of creating a variable radius fillet.

HOW-TO 19-3: Creating a Variable Radius Fillet

To create a variable radius fillet, perform the following steps.

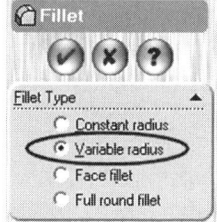

Fig. 19-20. Specifying Variable radius.

1. Select Fillet from the Insert > Features menu, or click on the Fillet icon.

2. For Fillet Type, select *Variable radius* from the list shown in figure 19-20.

3. Select the edges to be filleted.

4. Select a vertex from the Vertex list. The vertices will be labeled V1, V2, and so on. See figure 19-18 for an example of this listing.

5. Specify a radial value for the selected vertex.

6. Repeat steps 4 and 5 for the remaining vertices. Feel free to use the Set Unassigned or Set All buttons if it will make this process easier.

7. Select either Smooth or Straight transition.

8. Click on OK.

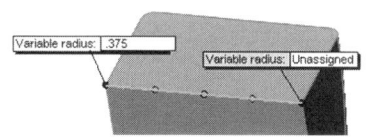

Fig. 19-21. Variable radius fillet callouts.

If you find it easier, use the callouts present on the screen to change the radial values for each vertex point. An example of these callouts is shown in figure 19-21. Click once inside a callout, type in a value, and then press Enter.

When a variable radius fillet needs to be edited, use techniques previously learned. When you double click on the fillet, all dimensions associated with the fillet will be displayed. That is, every dimension for each vertex point will appear on screen, where the dimensions can then be altered as required.

Utilizing Movable Fillet Points

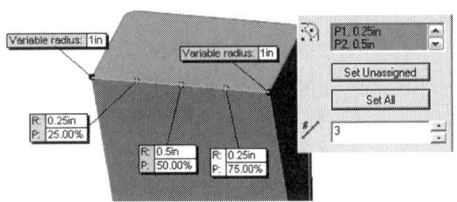

Fig. 19-22. Adding movable points to an edge.

If a single edge is selected that you wish to apply a variable radius fillet to, only two vertices appear in the vertex list. What happens if you want to dictate what the radius should be at certain positions along the edge? In a situation such as this, you would add movable points to the edge, which you could then supply radial values for. An example of what you would see on screen during this process is shown in figure 19-22. The inset shows a portion of the panel related to the movable points that can be added.

As can be seen in figure 19-22, there is a setting that dictates the number of points that will be added to the edge. This is what you would set first. Following that, select the points themselves, directly from the work area. Clicking on the points places them in the vertex list, where they are listed as P1, P2, and so on. You can then specify radial values for the points in the same manner values were applied to the vertices.

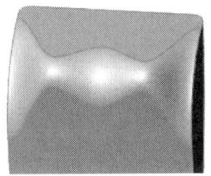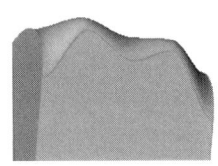

Fig. 19-23. Three movable points were added when defining this variable radius fillet.

To optionally specify a position for the movable points, enter a percentage value in the callout associated with the point. By default, the points will be evenly spaced along the edge. You will be able to change the position of the points after defining the fillet, but the number of points used is not something that can be altered once the fillet has been created. A variable radius fillet defined with three movable points is shown in figure 19-23. The fillet is shown from two angles so that you can better discern its shape.

Setback Fillets

Setback fillets are a special fillet type that can be applied to an area where filleted edges meet. The area where the fillets blend can be pulled back into the model, and a smoothly blended transition can still be maintained. This leaves the impression of a filleted corner that has been sanded down to give a nice rounded finish.

Advanced Filleting

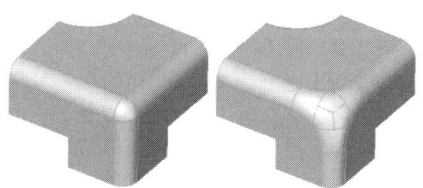

Figure 19-24 shows three edges, with a generic fillet on the left and a setback fillet on the right. The tangent edges have been highlighted so that the patchwork of the blend can be seen. How-To 19-4 takes you through the process of creating a setback fillet.

Fig. 19-24. A regular fillet versus a setback fillet.

HOW-TO 19-4: Creating a Setback Fillet

A setback fillet, by definition, should be created at a place where edges are coming together on the model. The setback area will be applied to a vertex point where the edges coincide. Keep this in mind when selecting edges where the setback fillet will be created. To create a setback fillet, perform the following steps.

1. Select Fillet from the Insert > Features menu, or click on the Fillet icon.

2. Leave the Fillet Type set to *Constant radius*.

3. Select the edges to be filleted. The edges should terminate at a common vertex.

4. Specify a radius for the fillets.

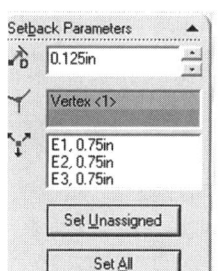

Fig. 19-25. Selecting a vertex point.

5. Click in the Setback Vertices list box, shown in figure 19-25, and select the vertex point where the edges meet. The edges common to the vertex point will be listed in the Setback Distances list box as E1, E2, and so on.

6. Select an edge from the Setback Distances list box and specify a distance (setback value) for the selected edge.

7. Repeat step 6 for every edge in the Setback Distances list box.

8. Click on OK.

When you double click on a setback fillet to access the feature's dimensions, all of the relative dimensions will appear, as is par for the course. There should be a dimension displayed for each of the setback distances, along with the radial dimension for the edges being filleted. If the *Multiple radius fillet* option is used (Chapter 6), dimensions will appear for each edge filleted. This falls in line with standard SolidWorks operating procedure.

Face Blends

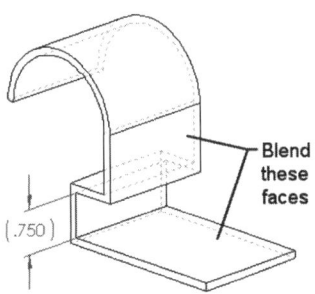

Fig. 19-26. Prior to creating a face blend fillet.

A face blend fillet can often be used where no other fillet type will work. Face blends can also be used in very difficult areas, where geometrically a fillet would be impossible to incorporate. Face blends can completely remove faces of existing geometry in order to add a fillet. Consider a simple example.

Figure 19-26 shows a simple part that serves as a good example of what a face blend fillet can accomplish. The object is to blend the two faces, shown in figure 19-26, thereby filling in the opening, which measures .750 inch.

Once the face blend fillet has been created, the part can be redesigned to create a somewhat inverted version of the original, with the opening on the other side. Figure 19-27 shows the same part after the face blend fillet was added.

Fig. 19-27. After adding the face blend fillet.

Note that when the fillet was added the software had to completely eliminate two faces in order to create the fillet. Two inner faces of the opening have been completely absorbed by the fillet. Generic fillets often will not allow for this type of behavior. That is why a face blend fillet can often prove a life saver when trying to complete a design in a situation in which a standard fillet just will not work. The finished part with an additional cut and fillet is shown in figure 19-28.

Fig. 19-28. Reworked sample part with the opening on the other side.

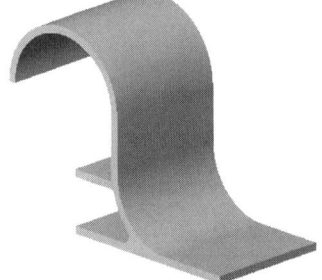

When creating a face blend fillet, two sets of faces must be selected that are to be blended. A radius must still be specified when creating a standard face blend. How-To 19-5 takes you through the process of creating a face blend fillet.

How-To 19-5: Creating a Face Blend Fillet

To create a face blend fillet, perform the following steps.

Advanced Filleting

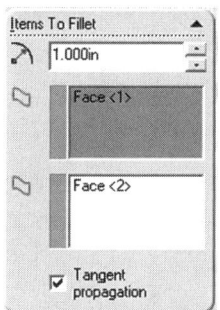

Fig. 19-29. Items selected in Items to Fillet panel for a face blend.

1. Select Fillet from the Insert > Features menu, or click on the Fillet icon.
2. For the Fillet type, select *Face fillet*.
3. Click in the list box labeled Face Set 1 and select the first set of faces to be filleted (see figure 19-29).
4. Click in the list box labeled Face Set 2 and select the second set of faces to be filleted.
5. Specify a value for the radius.
6. Click on OK.

During steps 3 and 4, faces are selected that the fillet will be blending between. Sometimes these faces meet along a common set of edges. If this is the case, one set of faces should reside on one side of the edges where the fillet will be defined, and the other set of faces should reside on the opposite side.

Sometimes, as in the example shown in figure 19-26, the faces do not meet at a common set of edges. Rather, there is a gap between the faces. Although only one face is in each face set in the example shown, this is not a limitation. Multiple faces can exist in each face set. If this is the case, make certain each set of faces resides on one side of the gap or the other. Accidentally placing a face in the wrong face set will cause the face fillet to fail.

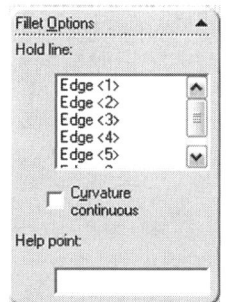

Fig. 19-30. Where hold lines are specified.

Using Hold Lines

An important aspect of face fillets is the ability to use what are known as hold lines. Hold lines control the fillet by establishing where the fillet will be tangent to the faces being filleted, thereby determining what the radius will be. For this reason, a radial value for the fillet does not need to be supplied. For that matter, the Radius setting is not even available.

When performing a face fillet, the hold lines can be supplied via the Fillet Options panel, shown in figure 19-30. After running through a couple of examples, you will have a much better idea of exactly what can be accomplished with hold lines when filleting.

Figure 19-31 shows an example of a free-form shape part that will have a face fillet applied to it. Note the edges that will be selected for the hold lines. The edges that are the intersection of the two face sets are the edges that will actually be filleted, but the hold lines will determine the radius of the fillet. The fillet will be tangent to the faces in Face Set 2, where the fillet

738   **Chapter 19: Advanced Feature Types**

meets the hold line. Incidentally, the part shown in figure 19-31 has an additional face that will be included in Face Set 2 (which could not be clearly shown in the illustration). It is the face on the left rear side of the part.

Figure 19-32 shows the same part with a face blend added. Because hold lines were used, the radius of the fillet was made to vary around the top outer perimeter of the part. As stated previously, the fillet is tangent to where the perimeter faces of the part used to be. Because of this, a part such as the one pictured would make a nice cover for some sort of container with vertical walls. The transition between the cover and container would be perfectly smooth.

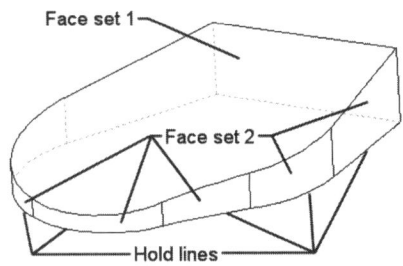

Fig. 19-31. Geometry that will be selected for a face blend with hold lines.

The *Help point* option is a very rarely used option. The only time a help point is needed is when an ambiguous situation arises while creating a face fillet. Sometimes it is not clear to the software where the face blend should take place. In other words, SolidWorks is not sure where to position the blend. When the *Help point* option is used, a point can be selected that is positioned near where the blend should occur. The faces closest to the help point are filleted. This is an option that most SolidWorks users need not concern themselves with.

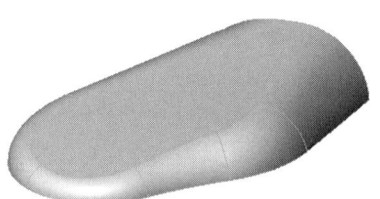

Fig. 19-32. After adding a face blend with hold lines.

Split Lines as Hold Lines

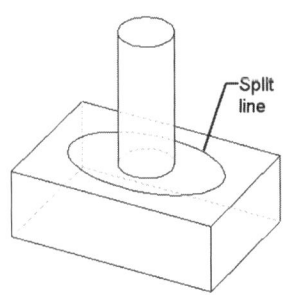

Fig. 19-33. Using a split line as a hold line.

A split line can also be used as a hold line when creating a face blend fillet. The results are very interesting. The fillet will wind up being tangent to the face being filleted at the hold line. This situation is similar to that discussed in the previous section. Examine figure 19-33, which shows a simple part with a split line. The split line was created using an ellipse.

When the face fillet is performed on the part, the cylindrical face is one face set, and the face within the boundaries of the elliptical split line is the other. The split line is selected as the hold line. Figure 19-34 shows the outcome of the face fillet. How-To 19-6 takes you through the process of creating a face blend with hold lines.

Fig. 19-34. Outcome of the face fillet using a split line as a hold line.

How-To 19-6: Creating a Face Fillet with Hold Lines

To create a face blend that uses hold lines to determine the radius of the fillet, perform the following steps.

1. Select Fillet from the Insert > Features menu, or click on the Fillet icon.
2. Select Face fillet as the Fillet type.
3. Click in the list box labeled Face Set 1 and select the first set of faces.
4. Click in the list box labeled Face Set 2 and select the second set of faces.
5. Click in the Hold Line list box and select the edges to be used for the hold lines. Edges created with the Split Line command are considered valid edges.
6. Click on OK to create the fillet.

It should be noted that hold lines can be used for both face sets. This situation only works under certain conditions. Use common sense is employing multiple hold lines. If the fillet is physically impossible to define, SolidWorks will obviously fail to create it.

Full Round Fillets

As the name implies, a full round fillet imparts a fully rounded face on an existing face of the model. This is similar to a face blend, both in physical aspects and in how the model faces are selected. Three face sets must be selected when creating a full round fillet.

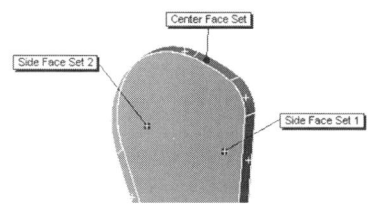

In figure 19-35, note the faces that have callouts attached. The model is of a plastic fan, and figure 19-35 shows one blade of the fan. Side Face Set 1 is the face on the back of the fan blade. Side Face Set 2 is the face on the front of the fan blade. Center Face Set represents all faces between the two side face sets.

Fig. 19-35. Creating a full round fillet.

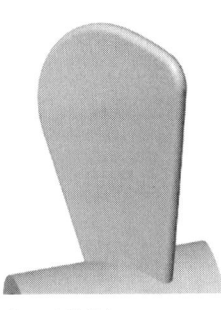

Fig. 19-36. Completed full round fillet.

Full round fillets do not have to be placed on material that has a uniform thickness. A radius value is not required, as the software calculates what the radius must be as defined by the thickness of the center faces. Figure 19-36 shows the outcome of creating a full round fillet on the fan blade.

Continuous Curvature Fillets

When performing a face blend fillet, the radius of the fillet is consistent. Its curvature does not change. This is true for generic edge fillets as well. If we examine the basic fillet shown in figure 19-37, we can see that the curvature is 0 on the flat faces, and that the curvature of the .5-inch-radius fillet is 2. Incidentally, curvature is defined as the reciprocal of the radius.

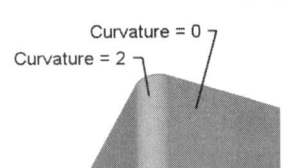

Fig. 19-37. Constant curvature on a generic fillet.

In certain cases, often for aesthetic reasons, it is desirable for the curvature to change gradually. In other words, the curvature should gently climb from a value of zero (a flat face) to the radius as dictated by the fillet's definition. This can be done through the use of the *Curvature continuous* option found in the Fillet Options panel. The option is only available when performing a face fillet. An example of a continuous curvature face fillet is shown in figure 19-38. The filleted area has been enlarged for clarity.

Fig. 19-38. A fillet with continuous curvature.

In SolidWorks terms, when a fillet is said to be "continuous," the meaning is that the curvature is continuously changing in a gradual manner, rather than abruptly. This fact can even be checked in a number of ways. The following section introduces you to two methods of examining face curvature.

Face Curvature

Fig. 19-39. Displaying curvature data.

When it becomes necessary to know what the curvature is of faces on a model, you have two choices. One is to use the Curvature option in the View > Display menu. This results in the model being displayed in a multi-color gradient, which indicates the degree of curvature present on the model. Moving the cursor over the model displays the curvature and radius of curvature at any given point. This is shown in figure 19-39, though the gradient colors are lost to the grayscale image. Areas of zero curvature are displayed as black.

Rolling Back FeatureManager

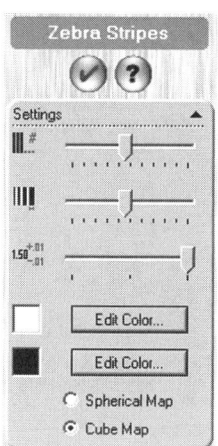

Fig. 19-40. Zebra Stripes panel.

Be forewarned that turning on curvature display can result in lengthy calculations. The math involved is extensive, and it may take some time to calculate the model's curvature. Once calculated, however, the data is saved with the model. Displaying curvature a second time will take little time unless the model has been edited. To control the colors associated with the curvature values, use the Curvature button found in the Colors section of Document Properties (Tools > Options).

A second display option with regard to curvature is known as zebra stripes. Use the View > Display menu once again, and select Zebra Stripes. The Zebra Stripes panel, shown in figure 19-40, will appear. Use the panel to control the number of stripes, the width and accuracy of the stripes, even the color of the stripes.

Zebra stripes can be cube mapped or spherically mapped on the model. Think of the model as having a reflective mirror coating, and there are black and white stripes painted on the ceiling. Those stripes can either be straight or circular.

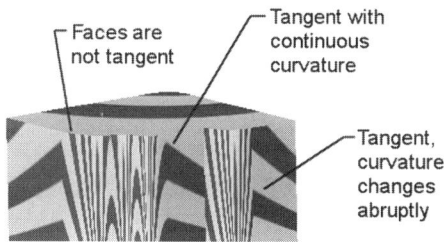

Fig. 19-41. Utilizing zebra stripes.

With zebra stripes, a number of factors regarding model faces can be determined. If the stripes do not continue unbroken across two faces, this indicates that the faces are not tangent. If the stripes continue unbroken, the faces are tangent. Furthermore, if the stripes bend gently across two faces, the curvature between those two faces is changing gradually. Figure 19-41 illustrates this.

It is difficult to properly show zebra stripes in a static image. Often, the model must be rotated in order to make a proper determination of the status of the curvature between two model faces.

Rolling Back FeatureManager

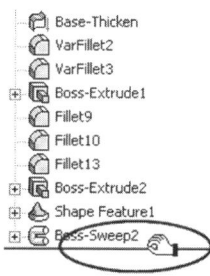

Fig. 19-42. Rollback bar.

The rollback bar provides a means of stepping back through time to see how a part was built. You will see the rollback bar at the bottom of FeatureManager. When the cursor is positioned directly over the rollback bar, a small hand appears, such as that shown in figure 19-42. The rollback bar can then be repositioned simply by dragging it with the left mouse button.

It is also possible to use the rollback bar to insert features at a point in time other than that represented by the bottom of the FeatureManager list. Wherever the rollback bar is located, none of the features below it are rebuilt by SolidWorks. Essentially, the program ignores them. If a feature

Chapter 19: Advanced Feature Types

is created, it gets added to FeatureManager at a position immediately above the rollback bar.

You will notice the rollback bar snapping between features in FeatureManager as it is being dragged. When you position the rollback bar, you will also notice that all features below the bar become grayed out. In figure 19-43, the same FeatureManager shown in figure 19-42 is shown again in a rolled-back state.

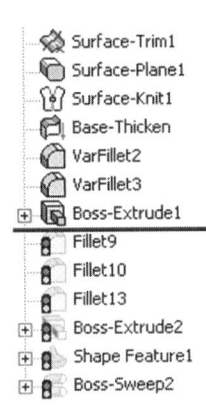

There will be a number of small green streetlights attached to the rolled-back features in FeatureManager. Whenever you see these small streetlight symbols, it means that the feature with the symbol attached requires a rebuild. Logically, because the features have been rolled back, they will eventually need to be rebuilt. This will happen automatically once you begin rolling the part forward.

Right-clicking on a feature in FeatureManager presents a set of rollback options. These options are Rollback, Roll to Previous, and Roll to End. If Rollback is selected, the model will be rolled back to the feature over which you right-clicked. Roll to Previous rolls back to the last position of the rollback bar. Roll to End refers to the last feature in the model.

Fig. 19-43. A rollback model.

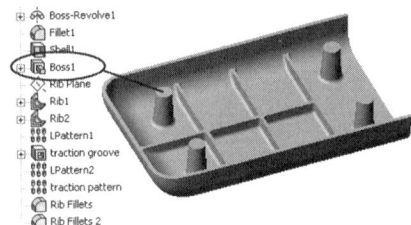

Fig. 19-44. Holes are needed in the shaver housing.

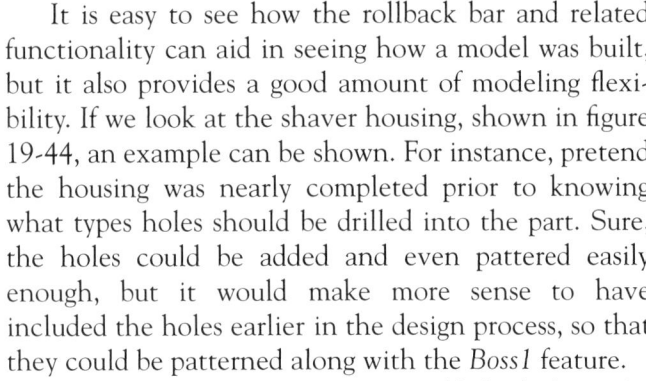

It is easy to see how the rollback bar and related functionality can aid in seeing how a model was built, but it also provides a good amount of modeling flexibility. If we look at the shaver housing, shown in figure 19-44, an example can be shown. For instance, pretend the housing was nearly completed prior to knowing what types holes should be drilled into the part. Sure, the holes could be added and even patterned easily enough, but it would make more sense to have included the holes earlier in the design process, so that they could be patterned along with the *Boss1* feature.

Using the rollback bar, we can add the hole at the appropriate position in FeatureManager. Figure 19-45 shows the same part, still in a rolled-back state, after adding a #4 countersunk hole. Note the location of the *CSK #4 Flat Head* feature.

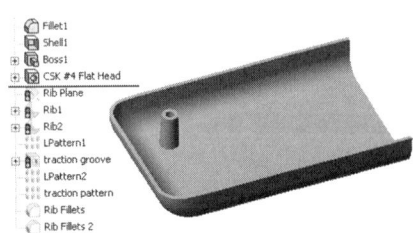

Fig. 19-45. After adding the hole.

As a side note, it is worth mentioning that when a new sketch is started while in a rolled-back state the sketch will temporarily be positioned at the bottom of FeatureManager. However, if the sketch is exited, or if it is used to build a feature (which is more likely), the sketch will reposition itself to its proper place just over the rollback bar.

After adding the hole, the model can be rolled forward once again. The definition for the linear pattern of the *Boss1* feature can be modified to include the new *CSK #4 Flat Head* feature. The hole gets patterned along with the boss, and everything is as it should be. An alternative to this process would have been to reorder the hole, rather than use the rollback bar. This is discussed in the following section.

Reordering

In the early chapters of this book, you became aware of the nature of FeatureManager, and the dependencies created therein. Think of FeatureManager as a chronological or historical list of events. Features created early in time are at the top of the list, and features created most recently will be at the bottom.

It is possible to change the place in time a feature was created. In other words, if you wanted a feature to exist earlier in the chronological history of events, you could position the feature higher up in the FeatureManager list. This is known as reordering, which works in either direction. You can reorder a feature up or down in the FeatureManager list. Parent/child relationships, discussed in Chapter 10, play a very important role in the reordering process.

It is not a requirement to check the parent/child relationships of a feature prior to reordering the feature, but if you run into problems, checking the parent/child relationships can help. Be aware that a child feature cannot be reordered before its parent. How-To 19-7 takes you through the process of reordering a feature.

How-To 19-7: Reordering a Feature

Reordering is nothing more than a simple drag-and-drop procedure. To reorder a feature, perform the following steps.

1. Using the left mouse button, select the feature to be reordered, keeping the mouse button held down.

2. Drag the feature to its new location and drop the feature on top of the feature you want it to appear after.

SolidWorks will rebuild the part automatically, with the features in the new order you specify. Did you notice the small symbol the cursor changed into while you were reordering? It is a small arrow pointing down and to the left. This is a reminder that the feature you are reordering will be placed immediately after whatever feature you drop it on top of.

Summary

This chapter taught you a number of techniques that can be used for some advanced feature types. Primarily, ways of deforming faces were discussed, as well as some advanced filleting techniques. You also learned how to analyze the curvature of faces.

Dome features allow for creating domed shapes on planar faces. A dome feature is an applied feature and does not require a sketch. If a circular or elliptical face is selected, an elliptical dome can be created, which means that the dome will be tangent to the selected plane's normal (a normal being a perpendicular vector).

The Shape command is used for creating pressurized effects on parts. A pressure value can be applied to a face of a model, giving it a look as though some force were pushing outward on the face. Negative pressures can be applied as well, giving the opposite effect of a force pushing inward on the face. In essence, the Shape command allows a face to be treated as a rubber membrane.

A number of advanced fillet types were covered in this chapter. Variable radius fillets were discussed, along with setback fillets, face fillets, and full round fillets. When creating a variable radius fillet, edges must be selected. Selecting edges populates the vertex list box with vertex points that can then be given radial values. Optionally, points can be added to an edge to gain more control over radial values at particular locations along that edge.

Face blend fillets can often be used to create a fillet where no other fillet type will work. A face blend fillet has the power to completely absorb other faces in the model, which makes it a very powerful filleting command. Hold lines can be used that will determine the radius of the faces being filleted together (blended). When hold lines are used, the fillet will be tangent to the faces adjacent to the hold line where the faces meet the hold line. Split lines can be used as hold lines, which can make it very easy to control the area where a fillet occurs.

Continuous curvature fillets can be created when a face fillet is performed. Fillets of this nature have a gradually changing curvature. Use the Curvature or Zebra Stripes display options to check curvature. Zebra stripes can also be used to determine tangency conditions between faces.

Questions and Topics for Discussion

1. What type of face must be selected when using the Dome command?

2. What is meant by the Dome command option Elliptical dome?

3. When using the Shape command, what slider bar setting will significantly increase the time it takes to calculate the shape feature?

4. Can vertex points be selected when creating a variable radius fillet? Explain your answer.

5. What are the two transition types for a variable radius fillet?

6. How many face sets are required for a full round fillet?

7. What is a continuous curvature fillet?

8. What are zebra stripes used for?

CHAPTER 20

Working with Surfaces

Fig. 20-1. Surface menu.

SURFACES WERE FIRST BRIEFLY MENTIONED IN Chapter 1, and in more detail in Chapter 13 (knit and radiated surfaces). SolidWorks does contain the ability to create surfaces, although its primary function is that of a solid modeler. The surface modeling capabilities of SolidWorks have evolved over the years, and can come in quite handy from time to time. The commands related to surface creation are found in the Insert > Surface menu, shown in figure 20-1.

It is interesting to note that the first four surface menu items seem extremely similar to the first four menu items in the Boss and Cut menus. Specifically, these four menu selections are for extruding, revolving, sweeping, and lofting. When you take a look at the PropertyManager panel used to create one of these surface entities, the situation gets even more interesting. Take, for example, the Surface-Extrude panel for creating an extruded surface, shown in figure 20-2.

This panel looks very much like the panel used when creating a basic solid feature. The only differences are that in the Surface-Extrude panel for a surface, the draft option is not available and there is no option for extruding as a solid or as a thin feature (thin features were covered in Chapter 7).

The same similarity between windows holds true for revolved, swept, and lofted surfaces. Suffice it to say that creating surfaces is no different from creating solids. It is just the geometry that is different, which raises the question of how surfaces are different from solids. If you do not already understand the differences between these two basic entity types, a quick review is in order.

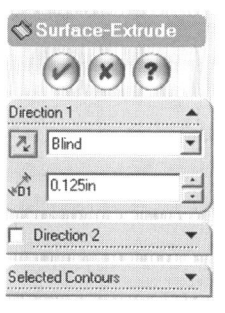

Fig. 20-2. Surface-Extrude panel for an extruded surface.

Surfaces Versus Solids

First, a surface has no mass from the standpoint of the computer. It is a theoretically perfect surface, with no thickness. A solid has mass. The computer software understands how all of the model faces are put together, and which side is in and which side is out. This is known as the topology of a model. Because the software (SolidWorks) knows all of the topological information and also knows what the density of the part is, it can calculate data such as volume, mass, centroid, and moments of inertia.

A solid model is essentially a set of surfaces (faces) with a little extra information thrown in that distinguishes it from a simple hollow set of surfaces. That extra information is enough for you to be able to extract information such as mass properties, and to perform operations such as interference checks in an assembly.

If solid modeling contains more information and has more downstream benefits, why even bother with surfaces at all? That is a very valid question, and the answer really comes down to one thing: geometry. There are certain shapes that are very difficult to create without some sort of surface manipulation. The end result may be a single solid model, but sometimes it takes surfacing commands to get there. Imagine trying to create a cast iron hook, for example.

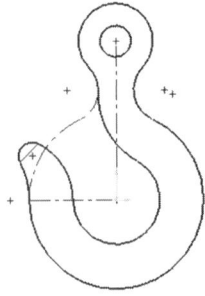

Fig. 20-3. How to model a cast iron hook?

Figure 20-3 shows an outline for a hook that we will model in this chapter. You have made it quite a way through this book, so considering what you have learned to this point, how would you model this hook? Revolved features might work, to a certain extent. Perhaps swept features would be a better choice. The point is, this is a difficult part to create from solid features. Instead, we will use surfaces.

Creating Surfaces

You have probably deduced by this time that creating surfaces is almost exactly the same as creating solids. You simply pick from a different menu and wind up with a different entity type. Normally, you will not have much reason to create a surface, as it just is not necessary for most models.

There may be times when surfaces will make a modeling job easier or may even be required to finish a job. When creating an extruded, revolved, swept, or lofted surface, the PropertyManager panels will be nearly the same as if you were creating a solid. How-To 20-1 presents an example of how to create a swept surface. The steps show the similarities between creating a swept solid and a swept surface. Because similarities exist between all solid/surface feature pairs, steps for creating all four main surface types will not be listed.

How-To 20-1: Creating a Swept Surface

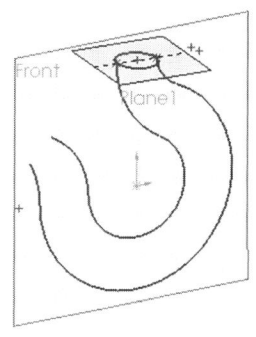

Fig. 20-4. Preparing to create a swept surface.

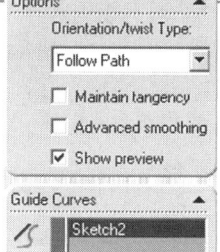

Fig. 20-5. Upper portion of the Surface Sweep panel.

To create a swept surface, perform the following steps. The hook will be used as an example. The main body of the hook will be modeled first. The sketch geometry used is shown in figure 20-4, and was converted from the original layout sketch shown in figure 20-3.

1. Create the required sketch geometry from which to create the surface. Use the same skills you have learned for creating any other sketched feature type.

2. Because this is a swept surface, exit out of the final sketch.

3. Select Insert > Surface > Sweep.

4. In the Surface-Sweep PropertyManager panel, select the appropriate sketches to represent the sweep profile and path. In our example, a guide curve is also used. A portion of the Surface-Sweep panel is shown in figure 20-5.

5. Click on OK to create the swept surface.

As you can see, creating surfaces does not require a lot of explanation. If you know how to create a swept solid, you can create a swept surface. As stated previously, these similarities exist for the other basic feature types as well; specifically, extruded, revolved, and lofted features. The only difference is where the commands are located.

The finished swept surface is shown in figure 20-6. The part is shown from two angles, along with the original underlying sketch geometry. Note that the sweep does not end at the appropriate location. The end of the hook needs to extend a short distance. In actuality, the top of the hook suffers from the same problem, but we will not worry about that yet.

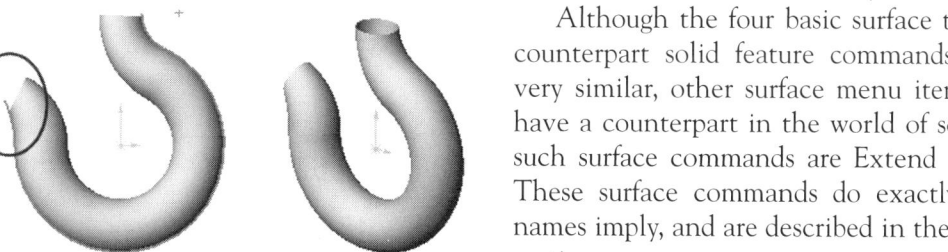

Fig. 20-6. The end of the hook falls short.

Although the four basic surface types have counterpart solid feature commands that are very similar, other surface menu items do not have a counterpart in the world of solids. Two such surface commands are Extend and Trim. These surface commands do exactly as their names imply, and are described in the following section.

Trimming and Extending Surfaces

Since the end of the hook does not extend quite as far as we would like it to, the Extend command will be used to draw it out a short distance. Alternatively, the path and guide curve could be extended as well, but using the Extend command gives us more options, as you will see.

Once the surface has been extended a sufficient amount, the end can be trimmed back accordingly. The original layout sketch will be used to establish were the trimming should occur. Finally, the Fill command will be used to round the end of the hook. First, however, let's extend the surface. How-To 20-2 takes you through the process of extending a surface.

HOW-TO 20-2: Extending a Surface

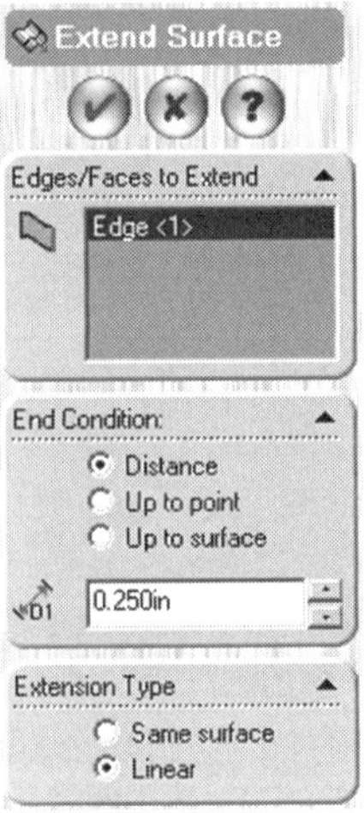

To extend a surface, perform the following steps.

1. Select Extend from the Insert > Surface menu.

2. Select an edge you wish to extend. Multiple edges can be selected.

3. In the Extend Surface panel, shown in figure 20-7, specify an end condition for the surface extension, or just enter an extension distance.

4. Specify whether the extension should use the *Same surface* or be Linear in nature. Figure 20-8 shows examples of each of these settings.

5. Click on OK to extend the surface.

Fig. 20-7. Extending a surface.

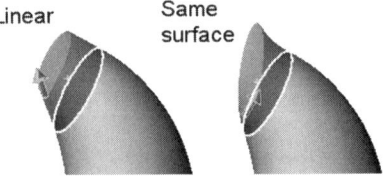

Fig. 20-8. Extended surfaces using Same surface *or* Linear.

In the case of the hook, the Linear option was used because the results are more desirable. Sometimes the *Same surface* option can cause the surface to start wrapping into itself, in a manner of speaking. There is usually a maximum limit to how far a surface can extend, and this is totally dependent on the surfaces being extended and extension option used. Often, the option used is dependent on aesthetics.

Trimming and Extending Surfaces

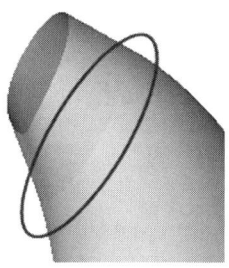

✓ **TIP:** *Dragging the preview arrow with the left mouse button can determine the extension length.*

Before moving on to the task of trimming, let's examine the newly extended surface a bit more closely. Note in figure 20-9 that the juncture between the original surface and the extended surface can be seen. The imperfection is not huge, but it is enough that you would not want it to appear in the finished product.

Fig. 20-9. An area of incongruity between two surfaces.

Deviation Analysis

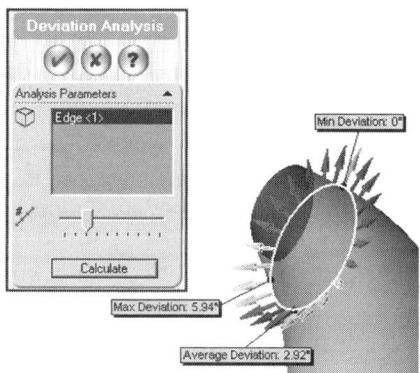

Fig. 20-10. Checking deviation between surfaces.

If we wanted to see what the actual deviation is between the two faces on the hook, a tool known as Deviation Analysis can be used. This command is found in the Tools menu. Once the Deviation Analysis panel appears, select the edges you wish to check and click on the Calculate button, shown in the inset in figure 20-10.

Once the Calculate button is pressed, multi-colored arrows will appear on the selected edge or edges. The effect is quite dramatic, actually, and cannot be appreciated in a grayscale image, though the arrows can be seen in figure 20-10. To increase or decrease the number of sample points where the deviation is measured, move the slider bar. It will be necessary to click on the Calculate bar again to redisplay the deviation arrows.

The callouts (also visible in figure 20-10) show what the average deviation is, along with the minimum and maximum deviation. In the case of the hook, the maximum deviation between the swept and extended surfaces is nearly 6 degrees. This is unacceptable, so what can be done in this situation? Fortunately, there is another surface tool that will help, explored in the next section.

Untrimming Surfaces

Extending a surface and untrimming a surface are very similar. Extending a surface actually creates a new surface, whereas untrimming a surface does not. The Untrim Surface panel, shown in figure 20-11, is accessed via the Insert > Surface menu.

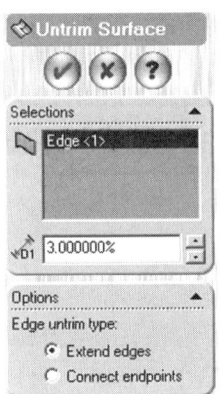

Untrimming a surface works much the same way extending one does. That is, an edge or edges to be untrimmed can be selected. In certain situations, the untrimmed edges can extend beyond their natural boundaries by a specified percentage. The options *Extend edges* and *Connect endpoints* are also available. The difference between these two options is shown in figure 20-12. The *Connect endpoints* option is used in the example on the right.

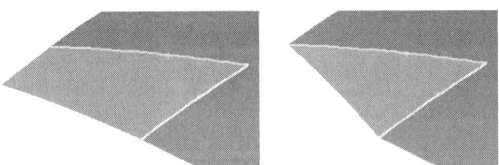

Fig. 20-12. Extend edges *versus* Connect endpoints.

Fig. 20-11. Untrim Surface panel.

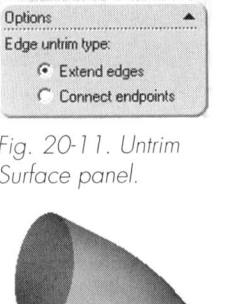

If the Untrim Surface command is applied to the end of the hook, the results are much cleaner than when using the Extend command. Figure 20-13 attests to this fact. Compare this to figure 20-9 and the difference in the outcome is clear to see. Additionally, there is no deviation between surfaces because there is only one surface.

Fig. 20-13. After untrimming the end of the hook.

Trimming Surfaces

Now that the end of the hook extends far enough beyond the original sketch geometry, it can be trimmed back to a precise position. Let's examine the Trim command in a bit more detail before performing the Trim operation on the hook.

Trimming a surface, for all intents and purposes, works similar to a cut-extrude command on a solid. Create a sketch that will describe the areas to be trimmed away, and then perform the trim. The sketch geometry will be projected onto the surface to be trimmed. Then it is just a matter of deciding on which pieces of the surface you wish to keep. The rest of the surface geometry is discarded. Figure 20-14 shows the Trim Surface panel.

Using the original layout sketch that contained the cross-section profile dimensions for the hook, a line can be converted that will dictate where the trim should occur. The resultant sketch would be an open profile, but either open or closed profiles are allowed. Figure 20-15 shows the trim operation being performed, as well as the final outcome.

Fig. 20-14. Trim Surface panel.

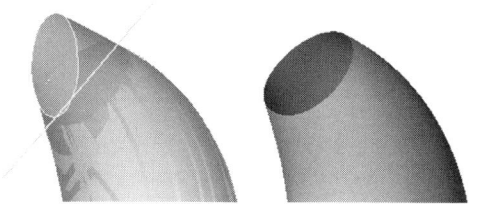

Fig. 20-15. Trimming the hook.

Another way in which the Trim command can be implemented is to trim a surface not with a sketch but with another surface. A mutual trim can also be performed, in which case multiple surfaces can be trimmed to each other. How-To 20-3 takes you through the process of performing a surface trim or mutual trim between surfaces.

How-To 20-3: Trimming Surfaces with Surfaces

Although sketch geometry can be used to trim a surface, as mentioned previously, this How-To provides the steps for trimming a surface to another surface. Optionally, a mutual trim of multiple surfaces can be accomplished. To trim a surface to another surface, perform the following steps.

1. Select Trim from the Insert > Surface menu.

2. Select *Trim tool* or *Mutual trim*, as necessary.

3. Select the surface or surfaces to be used as the cutting surfaces when performing the trim operation.

4. Click in the *Pieces to keep* list box and select the surfaces you wish to retain.

5. Click on OK to complete the operation.

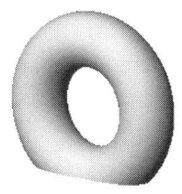
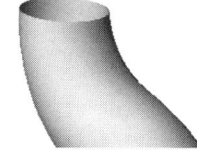

Fig. 20-16. Continuing to develop the hook.

A few new operations have been performed on the hook at this point. Specifically, a simple revolved surface has been added, and a bottom portion of the revolved surface has been trimmed away. The results of these operations are shown in figure 20-16. Only the upper portion of the hook is shown.

What needs to happen next is to create a surface between the upper and lower portions of the hook. This can be accomplished most easily through a loft operation. In that it has already been determined that creating a lofted surface is nearly identical to creating a regular lofted solid feature, it will not be necessary to go into the steps involved. However, there are a few details to be aware of that will make surface lofting easier.

Surface Lofting Techniques

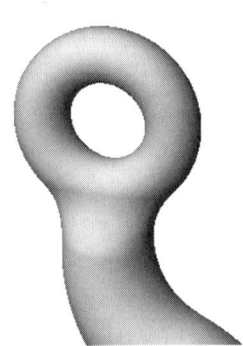

Fig. 20-17. After adding a lofted surface.

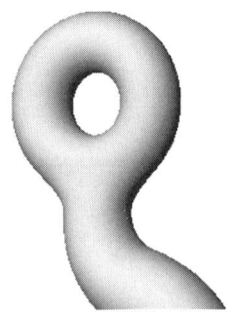

Fig. 20-18. After making some dimensional modifications to the hook's ring.

When lofting between surfaces, it is not necessary to create sketch geometry first. All that is required is to select the edges you wish to loft between. Occasionally, it is also possible to loft between entire faces. This makes surface lofting particularly easy.

The original rules of basic lofted features still apply. For example, if lofting between faces, each face should contain the same number of segments for the best results. Profile selection should be done via the work area, not FeatureManager. If lofting between edges, do not attempt to loft between a mixture of closed and open edges, as the software will refuse to create the loft. The resultant loft surface created on the hook is shown in figure 20-17.

It should be noted that the tangency condition was set to All (faces) for both the start and end of the loft. This helps to blend all faces, but the model still does not appear as smooth is it could be. This is due to the physical dimensions of the underlying layout sketch the model was based on. One of the advantages of a layout sketch is that the sketch's dimensions can be altered, thereby modifying the shape of the overall model. The dimensions for the size of the ring at the top of the hook were modified, and the results are shown in figure 20-18.

To finish up the hook, it will be necessary to use the Fill command (discussed in Chapter 13). This will allow for placing the rounded tip on the end of the hook. Finally, the hook can be turned into a solid using the Knit command. Knitting, also discussed in Chapter 13, is a prerequisite to turning surfaces into solids. The entire process, explored in the following section, involves the Thicken command as well.

Solidifying Surfaces

If surfaces were used to create a model, you would find that certain functions typical of solid geometry are not available. The mass properties of the model, for example, could not be determined. It would be advantageous to turn the surfaces into a solid. This is a two-part process that involves knitting the surfaces and then using the Thicken command.

Knitting is a simple matter, but it is a requirement if you wish to solidify a surface model. The model should form a completely closed area, or the Thicken command will not work the way you want it to.

Thicken can be used in two ways. In its simplest form, the command allows for adding thickness to a surface. This creates a solid from a surface,

Offset Surfaces

adding a thickness as specified by the user. This thickness can be applied to one side of the surface or the other, and there is an option for mid-plane as well. How-To 20-4 takes you through the process of using the Thicken command.

How-To 20-4: Thickening Surfaces

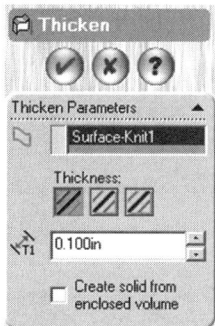

Fig. 20-19. Thicken panel.

To thicken a surface and turn it into a solid, perform the following steps. Also use these steps to solidify a surface model. Surface models must form completely enclosed boundary areas in order for the solidification process to work.

1. Select Thicken from the Insert > Boss/Base menu.
2. Select the surfaces to be thickened.
3. Specify a thickness for the surfaces in the Thicken panel, shown in figure 20-19.
4. Using the icons available on the Thicken panel, specify what direction the wall thickness should travel.
5. If the surface to be thickened is a knit surface, optionally select the *Create solid from enclosed volume* option. This will negate the need to perform steps 3 and 4.
6. Click on OK to complete the process.

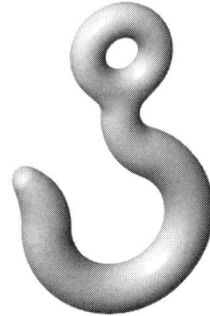

The *Create solid from enclosed volume* option will not even be present unless a knit surface was selected in step 2. This is true even if the selected surfaces form an enclosed boundary. They must first be knitted. Figure 20-20 shows the finished hook as a single solid. As far as appearance is concerned, it looks no different than if it had not been solidified, or even knitted, for that matter. It is the downstream benefits that make solidifying the surfaces a good idea.

Fig. 20-20. Completed hook as a solid model.

Offset Surfaces

Offset surfaces are very easy to create, but go a long way in creating some impressive geometry. The Offset Surface panel, shown in figure 20-21, is associated with this command. It has very few options and is straightforward and easy to use.

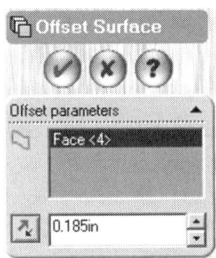

Fig. 20-21. Offset Surface panel.

Once the command is implemented, you would select a surface (or face) to be offset, specify the offset distance, and specify the direction for the offset through the use of the Flip Offset Direction button, which can be seen to the left of the offset distance parameter in figure 20-21. It is possible to offset more than one surface at a time.

The term *surface* is used loosely here. It is used to describe both a surface entity and an existing face, planar or nonplanar, of a solid model. Both are surfaces, but the term *face* is reserved to describe a surface that makes up a solid body.

There are two important details you will want to know about offset surfaces. One is that they can be used to cut solid geometry. This is crucial to being able to create features (or even text as a feature) that have uniform height over a nonplanar face. The second detail is that an offset value of zero can be used. This can also be advantageous when a cutting surface is needed and nothing else is available.

Planar Surfaces

Fig. 20-22. Planar Surface panel.

Like offset surfaces, planar surfaces are very easy to create. They are called *planar* because they are flat (if you are not already familiar with the term). These are defined by selecting entities that define the border of the surface. The objects must all be coplanar, meaning that all objects must reside on the same theoretical plane. Edges forming a closed profile from existing model geometry can be used, as well as a closed-profile sketch. The Planar Surface panel is shown in figure 20-22.

The steps for creating a planar surface need not be listed because it is just a matter of selecting the sketch or model edge geometry. Simply bear in mind that combinations of sketch entities and model edges cannot be used. If necessary, use the Convert Entities sketch tool to extract sketch entities from model edges if faced with the predicament of having a mix of geometry.

Manipulating Bodies

Bodies fall into two categories: solid bodies and surface bodies. Bodies can be moved, copied, rotated, mirrored, and patterned. Other commands typically carried out on features can also be carried out on bodies, such as scaling or filleting. The best part of this fact is that no new knowledge is required, for the most part. Commands previously learned can, in most cases, be used for surfaces. Take the Mirror command, for instance, whose PropertyManager panel is shown in figure 20-23. Note that there are list

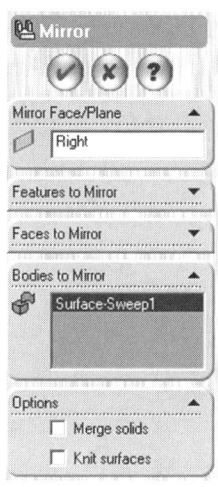

boxes for three different entity types. Specifically, these are features, faces, and bodies.

Depending on what it is that requires mirroring, select the appropriate list box. Features are typical solid features, faces are surfaces that make up a solid model, and bodies can be either solid bodies or surface bodies. Only one entity type can be mirrored at a time. This would hold true for other commands as well, such as patterning. In other words, you would not be able to pattern a model face and surface body at the same time.

✓ *TIP:* *If you are trying to add a fillet between surfaces, and a warning message appears regarding laminar edges, knit the surfaces prior to filleting them.*

Fig. 20-23. Mirroring a surface.

Moving, Copying, and Rotating Bodies

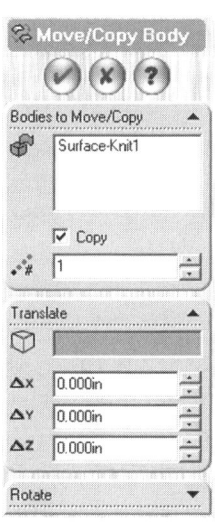

All three of the actions mentioned in the title of this section (moving, copying, and rotating) can be accomplished through one command. The command is titled Move/Copy Body, whose PropertyManager panel is shown in figure 20-24. It does not matter if the Move/Copy Body panel is accessed via the Insert > Surface menu or the Insert > Features menu. The same panel will be displayed in either case.

If the Move/Copy Body panel looks familiar, it is because a similar panel made an appearance earlier in the book. If you read the section in Chapter 13 on inserting a part into an existing part file, you would have seen a slimmed-down version of the Move/Copy Body panel. The panel was named Locate Part, which did not contain options to select other bodies or to copy selected bodies. In the Move/Copy Body panel, those options are present. How-To 20-5 takes you through the process of repositioning or copying bodies.

Fig. 20-24. Move/Copy Body panel.

How-To 20-5: Moving or Copying Bodies

To move (translate) or rotate solid or surface bodies, or to copy bodies, perform the following steps.

1. Select Insert > Surface > Move/Copy. Selecting Insert > Features > Move/Copy will display the same Move/Copy Body panel shown in figure 20-24.

2. Select the desired bodies, whether they are surfaces or solids. Bodies can be selected in any combination.

3. Select the Copy option if you wish to copy the bodies.

4. If creating copies, specify the quantity.

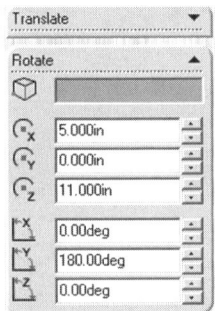

5. Expand either the Translate or Rotate panel, depending on the action you wish to perform on the bodies (use the small arrows to the right of the panel names).

6. If translating, specify the desired translation values for the *x*, *y*, and *z* axes.

7. If rotating, specify a point for the rotation origin, and then enter a rotation angle. By default, the rotation origin is the model's origin point. Changing the distance values in the Rotate panel (see figure 20-25) will alter this location.

Fig. 20-25. Rotating bodies.

8. Click on OK to finish translating or rotating the bodies.

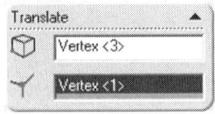

Fig. 20-26. Picking two vertices.

When moving bodies, it is not necessary to specify numeric distances. Geometry can be selected as well. For example, two vertices can be chosen that will represent the distance to move the body. This is illustrated in figure 20-26. Another option would be to select a linear edge, in which case a single distance value can be specified to indicate the distance the body will move along that edge.

If rotating bodies, geometry can also be used to indicate the rotation origin. This can be in the form of a single point, which would dictate the rotation origin, or a linear edge the body would rotate about. If an *x-y-z* location is given via the X, Y, and Z Rotation Origin parameter settings, a small point will be visible on screen that will visually clue you in as to where the point is located with respect to the bodies being rotated.

Regardless of whether bodies are being moved, rotated, or copied, a preview will always be visible in the work area. Use this to your advantage. It is easy to figure out what the software is doing by experimenting with the settings and watching what the preview is doing.

When finished with the Move/Copy Body command, a new feature will appear in FeatureManager. This new feature will be named *Body-Move/*

Copy, followed by the typical numeric value SolidWorks adds. Editing the definition of this feature allows for modifying the values specified for the translation or rotation, or for changing the number of copies created.

Cutting with Surfaces

Now that you understand how to create and manipulate surface geometry, it would be beneficial to understand some other functions that can be performed using surfaces. One such function would be to use a surface as a cutting tool. Let's bring the hook back for a curtain call in this next example.

Imagine that a series of ridges needs to exist on the ring of the hook. The ridges need to be at a uniform height of .060 inch, arranged radially around the top portion of the ring. They should also exist on both sides of the ring. The dimensions for one of the ridges are shown in figure 20-27.

There are a few problems faced in trying to create this feature. The problems can be overcome in a few different ways. One problem is determining the end condition of the extrusion when creating the first ridge. Using the Offset From Surface end condition does not work, because the ring surface is a single surface. SolidWorks cannot determine where the offset should be taken from because the surface exists both in front of and behind the sketch being extruded.

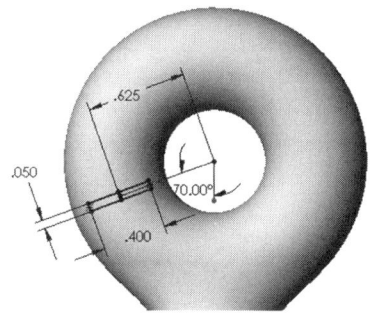

Fig. 20-27. Dimensions for one of the ridge features.

A solution to this dilemma would be to use the Split Line command to split the toroidal surface. A second solution would be to offset the surface, perform the extrusion, and then cut away what is not needed. A circular pattern will be needed as well, but that is immaterial to the problem faced.

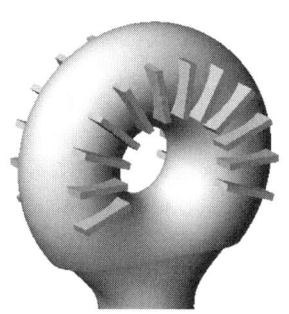

Fig. 20-28. The ridges prior to cutting them down to the proper height.

The offset surface must be created first. The reason for this is that the ridge changes the surface geometry when it is defined. Offset the ring's face the same amount as the height of the ridges. Next, the ridge can be extruded and patterned. For the extrusion end condition, the mid-plane type can be used. The extrusion distance can be anything, as long as the value is greater than the final ridge height. Figure 20-28 shows the ring with an offset surface, along with the extruded and patterned ridge.

Next, the cut needs to be made, so this is a good time to step through the process. How-To 20-6 outlines the steps required for cutting with a surface.

How-To 20-6: Cutting with a Surface

To cut with a surface, perform the following steps. It is assumed a surface has already been created and moved into position, if necessary.

1. Select Insert > Cut > With Surface.

2. Select the surface to cut with.

3. Paying attention to the preview arrow on screen, click on the Flip Cut icon if necessary. This icon, shown on the SurfaceCut panel in figure 20-29, determines what portion of the model will be discarded. The preview arrow on screen points toward this discarded geometry.

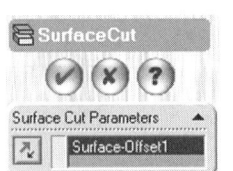

Fig. 20-29. SurfaceCut panel.

4. Click on OK to complete the cut.

Figure 20-30 shows the hook ring after performing the cut and adding some fillets. Note that the ridges are all a uniform height. This holds true for the ridges on either side of the hook ring.

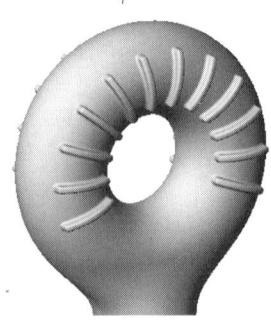

Fig. 20-30. Completed ridges on the hook ring.

Hiding and Showing Bodies

Similar to planes, sketch geometry, or other objects, bodies can be hidden or shown. This refers to both solid and surface bodies. The method used is the same in either case. Simply right-clicking on a body in FeatureManager gains access to an option to hide or show the body. The available option obviously depends on its current state of visibility.

Right-clicking on a feature in FeatureManager and selecting Hide Solid Body will hide the entire body associated with that feature. If it is your desire to hide a single feature, it must be suppressed. Review the section "Suppression States" in Chapter 10 if this is your wish.

An often more user-friendly way of hiding and showing bodies is to use the Hide/Show Bodies command. This command is accessible from the

Summary

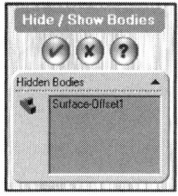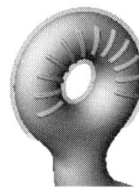

View menu. Once activated, the Hide/Show Bodies panel, shown in figure 20-31, appears.

The best part of using the Hide/Show Bodies command is that anything currently hidden appears as a transparent object. You can see exactly what is hidden and what is not. Knowing what is currently hidden is impossible to tell simply by looking at FeatureManager. Use the Hide/Show Bodies command and you can tell everything from a single glance.

Fig. 20-31. Hide/Show Bodies panel.

Summary

This chapter examined why surfaces can be important in the creation of certain models. Although it is usually advantageous to have a solid as a finished model, using surfaces along the way helps to obtain some shapes that might otherwise prove difficult or even impossible to achieve.

Basic surfaces can be created using surface extrude, revolve, sweep, or loft commands that are basically watered-down versions of their solid counterparts. Lofted surfaces can be created between existing surfaces without the need to first create a sketch. The need to select geometry from the work area with which to loft between is important. The primary reason for this is to match up vertices that should be connected during the loft operation.

Surfaces can be trimmed, or extended to other surfaces. When trimming, make sure to click on the surfaces you wish to retain. Untrimming a surface is a separate function, which closely resembles the Extend Surface command but whose end results differ in that an untrimmed surface does not create a new surface.

Extending a surface creates a new surface, which may deviate from the original. However, extending is sometimes necessary, as it allows for extending a surface edge in a linear fashion. Additionally, the same surface can be made to extend naturally. Use the Deviation Analysis command found in the Tools menu to see what the deviation is between adjacent faces.

To turn a surface into a solid, the surface can be thickened. To create a solid from a series of surfaces that form an enclosed boundary, the surfaces must first be knitted. Only then can they be knitted to form a solid representation of the model using the Thicken command.

Many commands that can be used on features can be used on surfaces. This includes patterning, scaling, and mirroring. Surfaces and solid bodies can also be moved, rotated, and copied. Surfaces can also be used as cutting tools to not just trim other surfaces but to cut solid geometry. If working with surfaces in conjunction with solid geometry, use the Hide/Show Bodies command found in the View menu to hide surfaces you are finished with.

Questions and Topics for Discussion

1. From a physical standpoint, what is the difference between surfaces and solids?
2. Do solid models contain surfaces? Explain you answer.
3. Describe the process used to turn a closed boundary set of surfaces into a solid.
4. Can a surface be offset a value of zero?
5. Is it possible to pattern surfaces in a circular fashion about a particular point in space without the aid of an axis, or even the circular pattern command? Explain your answer.
6. When cutting with a surface, what does the preview arrow point toward?
7. When using the Hide/Show Bodies command, how can you tell what bodies are being hidden?
8. Is it possible to use a mixture of sketch lines and model edges to create a planar surface?

Chapter 21

Engraved and Embossed Text

THIS CHAPTER DISCUSSES METHODS of adding engraved text to a part. The term engrave could be taken literally here, but that does not mean to say you could not use other methods to create text on a part in your area of manufacture. For instance, if your part contains text that has been stamped into the part, the methods learned in this chapter will still serve you well.

Other closely related scenarios, such as creating raised (embossed) text on a part, are covered as well. Creating raised or stamped text on a curved or irregular surface will also be looked into. You will be able to walk through the steps with the book to see exactly how to create these features. You will also learn how make text follow an arc or even a free-form curve.

Adding Text as a Feature

If text is to be used to create a feature, you must first add text to the sketch. This is in keeping with the normal mode of operation in SolidWorks: create a sketch, and then create a feature. The only difference is that in this case the sketch happens to consist of text. Figure 21-1 shows text added as a feature to a part.

Fig. 21-1. Embossed text on a model.

There is a distinction that should be made between adding text as a note and adding text as a feature. The methods (and reasons) used are very different. Notes are typically added to a drawing layout, but can be added to a part file as well. Notes are used to convey information about a part, to fill a title block, and so on. Details of the use of notes were presented in Chapter 11, which discusses detailed drawings.

When text is used for creating a feature, SolidWorks takes the text and turns it into a series of sketch entities. These entities might be lines, arcs, or splines. Any font loaded on your computer system can be used. Often, the font winds up as a series of splines, which results in quite a bit of computation required on the part of the computer. Although any font can be used to create sketch text, not all fonts will successfully translate into a solid feature. It really depends on the style and complexity of the font used.

If it becomes necessary to add text as a feature, and there is a lot of text to add, consider breaking the text up into separate features. The size of each text "chunk" will vary, depending on your system hardware and the font used. Before looking more closely at the various aspects of creating text features, let's first explore how to insert text into a sketch. How-To 21-1 takes your through the process of creating what SolidWorks terms sketch text.

How-To 21-1: Creating Sketch Text

This How-To represents the first phase in the process of creating a text feature. In the rest of the chapter, you will examine the process for creating the actual feature geometry. Start out just as you would create any other sketch. To add text to a sketch, perform the following steps.

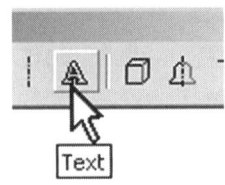

Fig. 21-2. Text icon.

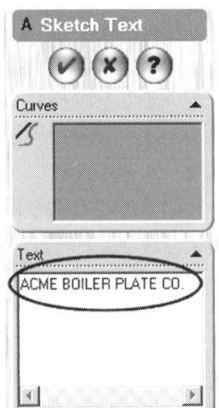

1. Select a plane or planar face on which to sketch.

2. Select Tools > Sketch Entity > Text, or click on the Text icon (shown in figure 21-2), found on the Sketch Tools toolbar. (Chapter 23 will show you how to add this icon to your toolbar if it does not already exist.)

3. Type the appropriate text into the Sketch Text panel, shown in figure 21-3. Multiple lines of text can be added, but your computer may have difficulty turning all of the text into a feature.

4. Pick a location for the text. This will be the lower left insertion point of the text string.

5. Click on OK when finished.

Fig. 21-3. Sketch Text panel.

Adding Text as a Feature

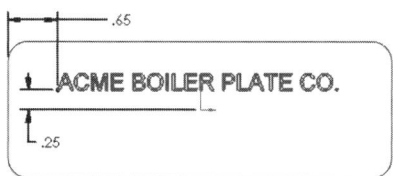

Fig. 21-4. Dimensions for the text location.

Normally you would want to position the text string at some specific location. It is possible to drag the text from its lower left insertion point to a new location, but this method is not accurate unless you are satisfied with eyeballing the text location. What works best is to place dimensions to the insertion point of the text from some other point or edge on the model. Figure 21-4 shows an example of this.

Sketch text does not always follow the same color code rules as other sketch geometry. Sometimes it will not turn black to show that it is fully defined. Do not spend too much time worrying about this. Text will align itself with the *x* axis of the origin point. It is possible to rotate text, or create backward text. You can accomplish this using the Modify Sketch command, discussed in Chapter 8. If you need to edit the text, either double click on it or right-click on the text and access its properties. You can do this only while editing the sketch that contains the sketch text.

Aligning Text on a Curve

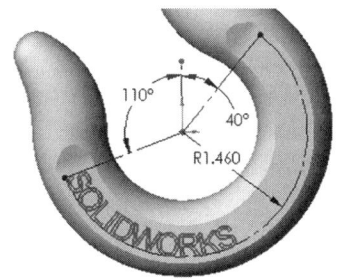

Fig. 21-5. Using a curve for sketch text.

Sketch text often benefits from construction geometry. This helps to align or dimension the text, or to force the text to follow a path. If a curve is selected and placed in the Sketch Text Curves list box, any text entered in the Sketch Text panel will follow that curve. An example of this is shown in figure 21-5.

In this example, the hook has been brought back again from Chapter 20. As you can see, a flat area has been machined from the hook (primarily a revolved cut in SolidWorks). Construction geometry in the form of two lines and an arc has been added and dimensioned. In this way, the positioning of the text can be controlled precisely. The spacing between the characters is examined in material to follow.

Text can follow any type of sketch entity, including splines. Text can also follow planar model edges. However, text is rarely positioned exactly on an edge of a part. More often, it would be desirable to use the Offset Entities sketch tool to offset an edge first. With a dimension controlling the offset distance, you can control exactly how far the text is from the edge by having the text follow the offset entity.

Sketch Text Options

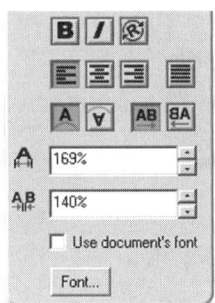

Fig. 21-6. Sketch Text panel options.

There are a few options in the Sketch Text panel that should be explained further. These options are in the lower portion of the Sketch Text panel, shown in figure 21-6. The sections that follow explore these options.

Use Document's Font

Uncheck the Use Document's Font option if you want to use a font other than the font you are currently using for your basic notes in the part file. This will activate the Font button. Clicking on the Font button will allow you to see a list of all fonts currently installed on your computer. This is the standard font window, containing the options Font Style (such as regular, bold, or italic), Font Height, and Effects (such as Strikeout or Underline).

Spacing and Width Factor

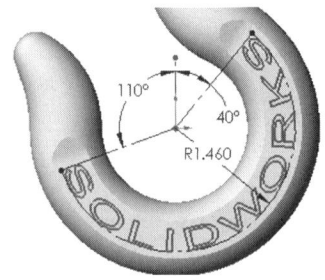

Fig. 21-7. Altering the spacing and width factor settings.

The Spacing and Width Factor settings control, respectively, the spacing between characters and the width of each character. Character spacing is also referred to as kerning. SolidWorks uses a percentage value to dictate the amount of kerning and character width. By altering these values, it is possible to space out a line of text so that it fills a particular area. The same effect can often be accomplished with full justification, but this can sometimes cause the distance between characters to be too great. It often comes down to tweaking settings until the text has the desired appearance.

Figure 21-7 shows the same SolidWorks text as that displayed in figure 21-5, except that the character spacing and width has been altered. In figure 21-7, the character width factor has been set to 196% and the spacing has been set to 120%. Note that you must uncheck the Use Document's Font option before the spacing and width factor settings become available.

Flip Vertical and Horizontal

Any text can be made to flip in a horizontal direction. This has the effect of creating text that reads backward (text that could be read if held up to a mirror). Only text made to follow a curve can be flipped vertically. This makes sense, in that it is not necessary to flip text vertically otherwise. Without a curve, flipping text horizontally and then rotating it 180 degrees has the same effect.

Adding Text as a Feature

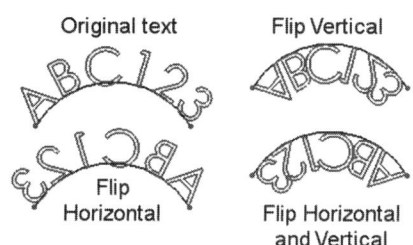

Fig. 21-8. Flipping sketch text.

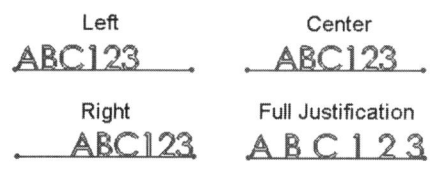

Fig. 21-9. Alignment conditions.

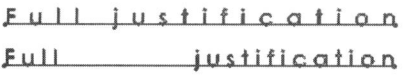

Fig. 21-10. Full justification of a line of text.

The need to flip text becomes more apparent when using curves to guide text. Some sample text is shown along a simple arc in figure 21-8. This shows you the various effects that can be achieved by flipping text.

Alignment and Justification

In terms of a sheet of paper, alignment refers to the positioning of the text horizontally. Text can be left, right, or center aligned. In other programs, this is often referred to as left, right, or center justification. Full justification means that the text is spaced out to fill the entire width of the page.

Because there is no "page" with regard to a SolidWorks model, there can be no alignment or justification unless there is something to gauge it by. That is, the alignment options are not available unless the sketch text is based on a curve. Figure 21-9 shows the three available alignment conditions and an example of full justification.

There is one other aspect of full justification you should be aware of. If you are used to working with publishing software, full justification will increase the spacing between words but leave the kerning (spacing) of the characters in each word alone. This is not the case in SolidWorks.

Figure 21-10 shows two examples of a line of sketch text with full justification. The upper example has a single space between the two words in the text string. As can be seen in the upper line of text, the kerning between characters has changed, making it difficult to distinguish each word. Admittedly, this is annoying behavior, but it has a simple enough workaround. To overcome this problem, spaces were added between the two words in the lower example, thereby making it easier to distinguish between the words.

Formatting and Rotating

Adding italics or bold formatting to text works much the same way as employing hypertext markup language (HTML) tags. However, if you have never created a web page, this fact would mean nothing. Without going into an explanation of HTML tags, let's look at the way formatting is applied to sketch text.

If there is a string of text you wish to apply formatting to, highlight the portion of the string (or the entire string) that requires the formatting.

Once that is done, click on the appropriate formatting icon, such as Bold, Italic, or Rotate. In the Text box, you will see that "tags" have been placed around the selected text. This allows for adding formatting to specific words in a string of text, an example of which is shown in figure 21-11.

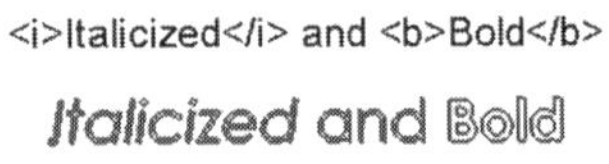

Figure 21-11 shows what would be seen in the Text box, and the end result. The tags themselves can even be typed in manually if you wish. The Rotate tag operates in the same manner as the Bold and Italic tags, but has an additional numeric component. The number associated with the Rotate tag dictates the angle the characters (not the text string) will be rotated. Positive or negative values can be used. Examples of Rotate formatting are shown in figure 21-12, along with the corresponding text that would appear in the Text box.

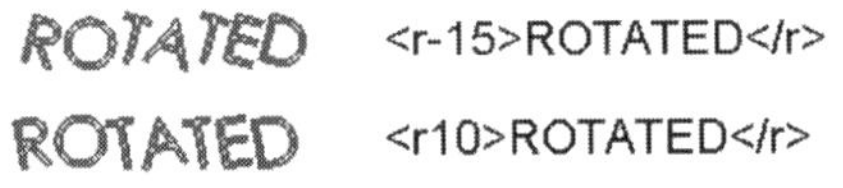

Italicized and Bold

Italicized and **Bold**

Fig. 21-11. Adding formatting to sketch text.

ROTATED   <r-15>ROTATED</r>
ROTATED   <r10>ROTATED</r>

Fig. 21-12. Rotating text.

It should be noted that rotating an entire string of text, rather than just individual characters, can be done with the Modify command, found in the Tools > Sketch Tools menu. The Modify command was discussed in Chapter 8.

Sketch Text as a Feature

The simplest type of text feature to create is one in which the text is on a simple planar surface. This is the text feature type you will learn how to create first. Both engraved and embossed text features can be created in the same fashion, with the only difference being that engraved features are created as a standard cut feature and embossed text is created as a boss.

Sketch text can be used to create revolved features, though actual applications requiring this would be limited. For that matter, sketch text can be used for swept or lofted features, but all of the basic rules you have already learned regarding these feature types must still be maintained. Maintaining these rules can be difficult with sketch text entities.

With regard to using sketch text for a boss or cut extrusion, all steps for creating the feature are identical to creating any other boss or cut extrusion. Exercise 21-1 takes you through the process of creating an identification plate for use on a larger assembly. Use the illustrations to help guide you through the process.

EXERCISE 21-1: Creating an Embossed Nameplate

Fig. 21-13. Nameplate created in this exercise.

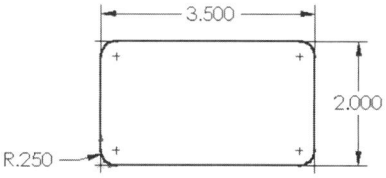

Fig. 21-14. Sketch for the base feature.

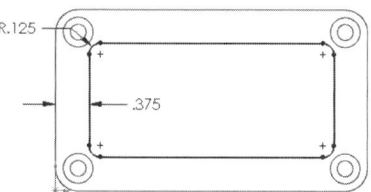

Fig. 21-15. Sketch for the recessed area.

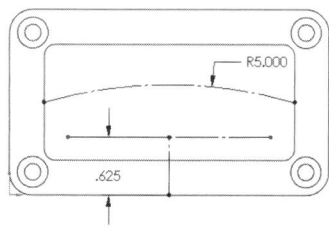

Fig. 21-16. Adding construction geometry.

Create a new part and save it with the name *Nameplate*. Set your units to inches. You will begin by creating the base plate itself. You will then add some bolt holes, create an indented area, and add the text feature. The completed nameplate is shown in figure 21-13. In that this exercise encompasses material you should be familiar with at this point, the details of every step are not provided. To create an embossed nameplate, perform the following steps.

1. On the Front plane, create the sketch for the base feature. The sketch, including dimensions, is shown in figure 21-14.

2. Extrude the sketch .125 inch.

3. Rename the base feature *Baseplate*.

4. Using the Hole wizard, create four (4) ANSI inch #8 flathead (82) machine screw holes, as shown in figure 21-15. (Hint: place the locating points concentric to the existing arc edges.)

5. Sketch on the front face of the *Baseplate* feature and create the sketch shown in figure 21-15. This will become the recessed area for the text. (Hint: use the Offset Entities and Sketch Fillet sketch tools.)

6. Cut the sketch .050 inch deep into the *Baseplate* feature.

7. Rename the cut-extrude feature *Recess*.

8. Sketch on the bottom of the *Recess* feature and add the construction geometry shown in figure 21-16. (Hint: use midpoint relations.)

9. Exit the sketch and rename the sketch *Text Placement*. Exiting the sketch allows you to reference the sketch for future geometry without necessarily having to see the dimensions.

10. Begin a new sketch on the bottom of the *Recess* feature once again.

11. Select Tools > Sketch Entity > Text.

12. Select the arc and add the text *ACME MACHINING*. Use a Century Gothic 18-point font and center alignment. (Hint: uncheck Use Document's Font.)

13. Click on OK to close the Sketch Text panel.

14. Select Tools > Sketch Entity > Text a second time. This is being done so that another string of text can be positioned on a different curve.

15. Select the horizontal line and add the text *Model XJ12-2003*. Use a Century Gothic 16-point font (slightly smaller than the text in step 12) and center alignment.

16. Click on OK to close the Sketch Text panel. Your sketch should look similar to that shown in figure 21-17.

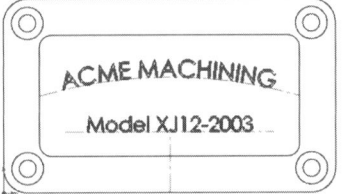

Fig. 21-17. After adding sketch text.

17. Extrude the sketch .050 inch.

18. Rename the boss extrusion feature *Text*.

19. Hide the *Text Placement* sketch.

20. Using the Edit Color icon, change the color of the *Nameplate* part, the *Recess* feature, and the *Text* feature to colors of your choosing. The reason for this is primarily to help the text stand out more prominently.

21. Save your work.

The *Nameplate* part you just finished creating should look similar to that shown in figure 21-13.

As you have just discovered, the entire process is pretty easy to accomplish, but the results can offer quite an impact. This exercise illustrated how to create what amounts to a fairly basic text feature. Not all are this easily created, so let's move on to more interesting ground.

Text Features on Curved Surfaces

Now that you have a better understanding of surface geometry (specifically, offset surfaces), you can take text features to the next level. Placing engraved text on a curved or irregularly shaped surface is fairly easy, but raised text is another matter. Because of this, engraved (such as stamped or machined) text will be tackled first.

There is really only one stipulation to placing engraved text onto a curved or irregularly shaped surface, and that is to use the Offset From Surface end condition type. By employing this end condition, it is possible to extrude the text into the solid geometry by a specific amount while retaining a uniform depth.

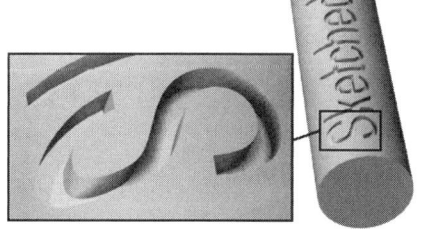

Fig. 21-18. Extruded cuts are always perpendicular to a sketch plane.

It is usually necessary to create a plane on which to sketch text. Because the extrusion will always be perpendicular to the plane, try to ensure that the plane is positioned at a point tangent to the area where the text feature is to be created on the surface in question. A common request is to be able to wrap a text feature around a cylinder. This is an extremely difficult task to accomplish. Figure 21-18 shows why.

Zooming in to the letter S in figure 21-18 clearly shows how the cut is not perpendicular to the curved surface. As stated previously, the cut is perpendicular to the sketch plane. This can manifest itself into some undesirable geometry if you are not careful. The geometry in figure 21-18 is an exaggerated situation, but it helps to illustrate what occurs when adding cuts to curved surfaces.

There is no easy way to ensure that engraved text cuts perpendicularly to every point on a surface where the text makes contact, but it is easy enough to keep the cut at a constant depth. As stated previously, use the Offset From Surface end condition. How-To 21-2 takes you through the process of creating engraved text on a curved surface.

How-To 21-2: Creating Engraved Text on a Curved Surface

To create engraved text on a curved surface, perform the following steps. The hotplate in figure 21-19 will be used as an example to help illustrate the process. An inset in figure 21-19 shows the cross section of the hotplate.

772   Chapter 21: Engraved and Embossed Text

Fig. 21-19. The hotplate where some engraved text will be added.

Fig. 21-20. Creating the sketch text.

Fig. 21-21. The engraved text.

1. Select a plane to sketch on and enter sketch mode in the usual fashion.

2. Create your sketch text (Tools > Sketch Entity > Text). Figure 21-20 shows an example of sketch text placed on three separate construction circles. This sketch contains a very large amount of text.

3. Select Insert > Cut > Extrude, or click on the Extruded Cut icon.

4. Specify Offset From Surface as the end condition type.

5. Select the surface you are adding the text feature to. This will be the surface you are offsetting the cut from.

6. Specify an Offset distance.

7. Flip the offset direction if necessary.

8. Click on OK to create the engraved text feature.

Figure 21-21 shows an example of the completed engraved text. One area has been enlarged to show the detail. Note that the text conforms to the surface of the model.

If you receive an error message during step 8 that states "Intended cut does not intersect the body," it is almost certainly because the offset direction is reversed. In other words, the text is being offset above the part instead of into the part. Check the Reverse Offset option and try clicking on OK again.

One other obstacle you may run into when performing this type of operation has to do with selecting (in step 5) the surface from which to offset. If the surface (or the face of the model, technically speaking) you are engraving the text in consists of multiple faces, it will be necessary to create a separate surface that can be selected as one. This was the case with the hotplate. The top of the hotplate consists of numerous faces.

Because multiple faces cannot be selected when using the Offset From Surface end condition, it was necessary to create a separate single surface that was a representation of the top of the hotplate. This sounds more difficult than it really is. By converting some entities from the original sketch used to create the hotplate, it was possible to create a simple revolved surface. This revolved surface is what was selected when indicating what surface to offset the text from.

Wrapping Text Around a Cylinder

It is generally desirable for the text feature, whether it is engraved or embossed text, to be perpendicular to the surface where the feature resides. As we have already discovered, that is not an easy request. Even though the text feature can be at a constant depth, making the cut perpendicular at every place it contacts the surface is next to impossible. But do not give up, as there is still hope.

Through inventive use of sheet metal functionality, it is possible to place text on a cylinder so that it completely wraps around the cylinder. The process basically involves creating a sheet metal part, unfolding it, adding the sketch text, creating a cut, and then folding it all back up. How-To 21-3 takes you through the process of creating text on a cylindrical feature. It is assumed that you understand how to create a sheet metal part, as discussed in Chapter 8.

How-To 21-3: Creating Text on a Cylindrical Feature

The following steps outline the procedure for creating text as a feature that wraps around cylindrical or even conical parts or features. The process is fairly extensive, and you must have an understanding of sheet metal commands in order to carry out this process. Numerous illustrations are included to help guide you through the process.

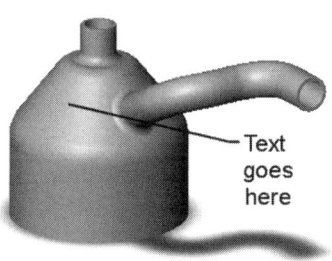

To make this How-To particularly interesting, a sample part created in Chapter 9 is used. The part is shown in figure 21-22. It is not a sheet metal part, though sheet metal functionality is required. Additionally, the text needs to be positioned on the conical face. If you are wondering how this can be done, let's put your curiosity to rest and get started.

Fig. 21-22. This part requires engraved text.

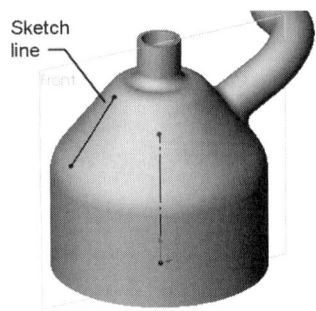

Fig. 21-23. Creating a new sketch.

Fig. 21-24. Creating a cut feature.

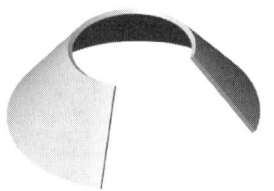

Fig. 21-25. Text Body part.

Fig. 21-26. Adding some sketch text to the unfolded part.

1. Create a sketch that will be used to cut away the area to be engraved. Later, this same area will be filled in by an engraved sheet metal "plate." In the sample part, a silhouette edge from the base feature was converted, using Convert Entities, to a sketch line in a new sketch (see figure 21-23).

2. Create a cut representing the area that will have the sheet metal plate fit into it at a later time. In the case of the sample part, the cut is a revolved feature .100 inch deep, with an included angle of 270 degrees. This feature is shown in figure 21-24.

3. Select the sketch you just used to create the cut, and press Ctrl-C to copy it to the Windows clipboard.

4. Create a new part that uses the same units as the part you wish to engrave.

5. Select the Front plane and press Ctrl-V to paste the sketch into the new part.

6. Recreate a feature that is the same volume as the cut made on the original part. In the sample part, this would be a boss-revolve feature .100 inch deep, with an included angle of 270 degrees. If you would like to edit the sketch first and define the locations of all sketch entities, that would be acceptable. However, do not reposition anything or this project will fail.

7. Save the new file with the name *Text Body*, or something similar. Figure 21-25 shows the *Text Body* part for this example.

8. Use the Insert > Sheet Metal > Bends command to turn the thin feature into a sheet metal part.

9. Add an Unfold feature in order to flatten out the sheet metal part so that some text can be cut into it.

10. Start a sketch on the face of the sheet metal part and add sketch text using the methods you have been taught so far in this chapter. Figure 21-26 shows an example of some sketch text added to the *Text Body* part.

11. Cut the text through the part. Use either the Through All end condition or the Link to Thickness option.

∞ **NOTE:** *If a font other than a stencil style font is used, the results will be unpredictable. SolidWorks may decide to not create the cut at all. This is because some letters create closed profiles (e.g., D and O). Because the closed profiles are interpreted as individual entities, this can result in a multiple-body sheet metal part, which is not allowed by SolidWorks. Stencil fonts, which leave gaps in characters (such as **0**), can often be found on the Internet.*

Fig. 21-27. After reforming (folding) the model.

12. Add a Fold feature to the part to form it back up. Believe it or not, SolidWorks actually will form the sheet metal part back to its original state with the text cut into it. This is shown in figure 21-27.

13. Save the part. You can also close it at this point in time.

14. Back in the original part, insert the *Text Body* part using the Insert > Part command.

15. Uncheck all Transfer options. Leave the Launch Move Dialog option checked, although you will probably not need to move the inserted part.

16. Click on OK to insert the *Text Body* part.

17. Ensure that the *Text Body* part is in the proper position, and then click on OK again.

18. Select Insert > Features > Combine.

19. Select the Add option and select the two solid bodies (the original model and the newly inserted *Text Body* part). The result of performing this operation in the sample model is shown in figure 21-28.

Fig. 21-28. After combining the two bodies.

20. If necessary, change the colors of some features to make the text stand out better.

And there you have it: engraved text the way it should be. Next, we will look at how to add embossed text to a nonplanar face. Embossing requires a new approach, which we will tackle in the next section.

Embossed Text on a Curved Surface

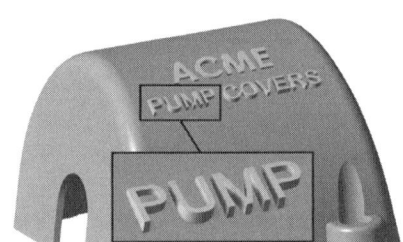

Fig. 21-29. The cover with improper embossed text.

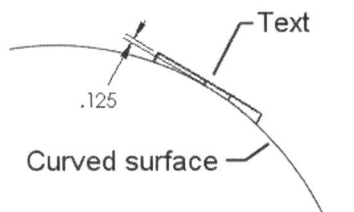

Fig. 21-30. Adding raised text to a curved surface.

The cover part from previous exercises will be used in the example that follows to show you how to add raised text to a curved surface. The cover is shown in figure 21-29. Some embossed text has already been added, but the resultant feature does not conform to the cover's surface. This is a typical situation when extruding. Where the extrusion begins, the feature is flat.

There are some special problems that arise when creating embossed text. The first problem is that of creating a plane to sketch on, the tools for which you have learned in Chapter 4. In particular, it may be necessary to create a plane tangent to a surface. The second problem is creating the extruded text at the required height. It is sometimes necessary to create the text feature larger than necessary with regard to the extrusion distance. The text then has to be cut down to size. An offset surface is usually used to perform the task of cutting the text.

A simple diagram can be used to illustrate the problem of creating the proper raised height for the text feature. Figure 21-30 shows some text that has been added to a curved surface, as seen from the side. The problem lies in the fact that the text height does not conform to the curvature of the model.

In figure 21-30, the .125-inch dimension is the desired height of the text. The problem should be obvious from the image. Due to the curvature of the surface, it is impossible to achieve the proper height for the text using nothing more than a simple extrusion. To achieve a uniform height, a different tactic is necessary.

Sometimes it is possible to create a plane on which to place the sketched text, and then extrude the text feature from underneath the surface of the model. The plane has to be essentially embedded within the material, so that when the text is extruded any excess is absorbed into the rest of the part. The extrusion would use an end condition of Offset From Surface, where the surface would be the outer curved surface of the part. This would be a direct way of controlling the height of the text feature. The only problem is that often the model is not thick enough to absorb the unwanted portion of the text feature.

Trying to extrude the sketch text from a plane above the curved surface while using the Offset From Surface end condition will not work either. This is because the offset is either going to be below the surface of the part (which does absolutely no good) or above the surface of the part, which creates a separate body (or in the case of extruded text, many separate bodies).

There is more than one way to create raised text on a curved surface, but there is one particular easily accomplished method that in this case will work well. This method utilizes an offset surface, and can usually be employed in what might otherwise be a difficult situation. How-To 21-4 takes you through the process of creating embossed text on a curved surface.

How-To 21-4: Creating Embossed Text on a Curved Surface

To create embossed text on a curved surface, perform the following steps.

1. Create an offset surface using Insert > Surface > Offset.

2. Offset the surface to the same height the embossed text is to be.

3. Create a sketch plane if necessary. Tangent planes work well, and the point of tangency should be near the area where the text feature will be placed. (Review How-To 4-14 if you need help with this.)

4. Start a new sketch and create the sketch text as desired. Add construction geometry and dimensions as necessary. Use the Modify Sketch command to rotate text.

5. Extrude the sketch. It is recommended that the extrusion be in two directions. Use Up To Next for the extrusion toward the part, and Blind to add some height to the text in the direction going away from the part. The precise height really is not important, as long as it is greater than the required final height.

6. Once the extrusion is complete, select Insert > Cut > With Surface.

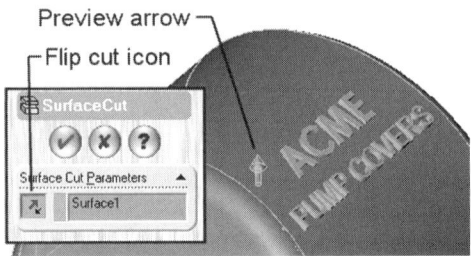

7. Select the offset surface you created in steps 1 and 2.

8. Make sure the preview arrow is pointing outward, away from the part. If it is not, click on the Flip Cut icon. This is shown in figure 21-31.

9. Click on OK to make the cut.

10. Hide the offset surface via the right mouse button.

Fig. 21-31. Cutting away the proper geometry.

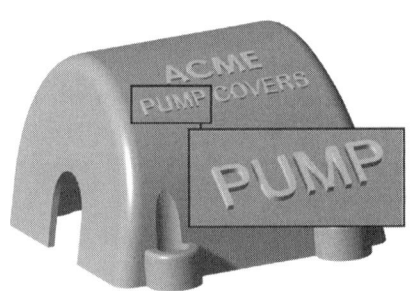

You can see how easy it is to cut with a surface. The height of the text can be changed by modifying the offset dimension associated with the offset surface. This assumes that the text was extruded far enough away from the part to begin with. Figure 21-32 shows the finished pump cover model with embossed text. Note that the text follows the curvature of the model.

Fig. 21-32. Embossed text at a consistent height.

Summary

In this chapter you discovered that adding text to a sketch is a simple matter of selecting the Text option from the Sketch Entity menu. It is possible to add sketch geometry or sketched text to a sketch, but not both. It has to be one or the other. Creating engraved text is quite straightforward with regard to flat, planar surfaces. Engraved text is nothing more than a cut extrusion, whereas raised text is a boss extrusion. It is possible to use any font on the computer when creating sketch text, but not all fonts are usable when it comes to creating features.

Creating engraved text on a curved surface requires the Offset From Surface end condition type. Creating embossed text on a curved surface often requires that first an offset surface be created, which can be used to cut the text feature down to the desired height. When this technique is used, it is possible to adjust the height of the text by adjusting the offset distance of the surface.

A part can be cut using the With Surface command, found in the Insert > Cut menu. This command is useful when creating raised text on a curved surface. The Surface Cut window is extremely simple, containing one option, which reverses the side to be cut away. Use the Surface Cut window to lop off unneeded feature geometry and to make an extruded text feature follow the contours of a curved surface.

Use the Modify Sketch command to manipulate sketched text. Sketched text can be rotated, mirrored (flipped), translated, and scaled. Many of these options will be unavailable if you have constrained the sketch with locating dimensions or constraints. If using a curve to guide the text, use the Flip Vertical and Flip Horizontal options to help position the text.

Because sketch text can follow a curve, use it to your advantage. Adding construction geometry to a sketch helps a great deal in controlling precise positioning of the text. Alignment and justification options help to further position the text.

Questions and Topics for Discussion

1. Describe the steps used to add text to a sketch.

2. Once text has been inserted into a sketch, how can an entire line of text be rotated?

3. Is it possible to rotate only a few characters in a line of sketch text?

4. How would you edit sketch text you have placed in a sketch?

5. What type of end condition must you use when creating an engraved text feature on an irregularly shaped or curved surface if you want the text feature to have a uniform depth?

6. What type of surface is usually needed when creating embossed text on a curved surface?

7. Would you be able to translate a sketch using the Modify Sketch command if the sketch were dimensioned to existing model geometry? Explain.

8. How can sheet metal functionality play an important role in creating engraved text?

9. Explain in your own words why it is necessary to turn sketch geometry into construction geometry if using sketch entities to guide your sketch text.

10. Are there any sketch entities that cannot be used as a guide curve for sketch text? Prove your answer.

Optional Problem

Create a sheet metal part that amounts to a simple rolled cylinder. Add some sketch text to the cylinder in its flattened state. You can also add a curve to the sketch containing the sketch text if you wish. Reform the model, and then add geometry to the rolled cylinder so that it appears like a solid cylinder (not a sheet metal part) with engraved text on it.

Note that unless SolidWorks changes its software error recognition flags, you will be required to use a stencil-type font to accomplish this exercise. In the past, any font was allowed, but resulted in an ambiguity that prompted the user to select which chunk of geometry was to be kept. As of this writing, if you attempt to cut completely through a sheet metal part using something other than a stencil font, a "Multiple sheet metal bodies are not supported" error will most likely appear (assuming the text contains at least one closed-profile character). You will need to use the following skills and commands in order to complete this exercise.

- Base Flange or Bends
- Fold and Unfold
- Sketch Text, Convert Entities, and other sketch tools
- Basic extruded cuts and bosses with various end conditions

Fig. 21-33. Optional problem exercise possible outcome.

If a stencil font is not available, you can still complete this exercise by not cutting all the way through the model. It will then be necessary to create extra bosses where necessary so that any islands within closed-profile letters are connected to the rest of the model. Finally, you will then need to create a second cut, which finishes cutting the modified text geometry completely through the model. An example of a part that could have been created for this exercise is shown in figure 21-33. A number of views are shown, as the text feature wraps around the part in a spiral fashion.

If you look closely at some of the letters (such as the lowercase e) in figure 21-33, you will see how small bosses were added to eliminate "islands." This action can be performed on all closed-profile letters in the same sketch at one time.

Chapter 22

Importing and Exporting Files

WHEN WORKING WITH COMPUTER-AIDED DESIGN or drafting software, there is almost always a need to import or export files. This is true whether you are in an area of manufacturing, surveying, mapping, architecture, or any other field. There are always customers, clients, or co-workers you will need to export files to or import files from.

SolidWorks has always been very strong in the area of translating files. There are numerous file translation options that allow for reading from or writing to almost any other CAD program. File translation is included with the SolidWorks program and is integrated in the command structure to make translating a very user-friendly process.

This chapter covers the processes involved in importing various file types and exporting SolidWorks documents. The various options available when importing or exporting files are explained in detail so that you will be able to make intelligent decisions regarding the translation options.

Importing Files

There is no special command for importing a file into SolidWorks. Simply use the File > Open window. Specify the type of file you want to import, and then select the file name to be imported. How-To 22-1 takes you through the process of importing a file into SolidWorks.

HOW-TO 22-1: Importing a File into SolidWorks

Although different files will have different import options associated with them, the process of importing a file is always the same. The various file types that can be imported and some of the available import options are discussed in material to follow. This How-To simply shows you how to open a nonnative file type in SolidWorks. To import a file into SolidWorks, perform the following steps.

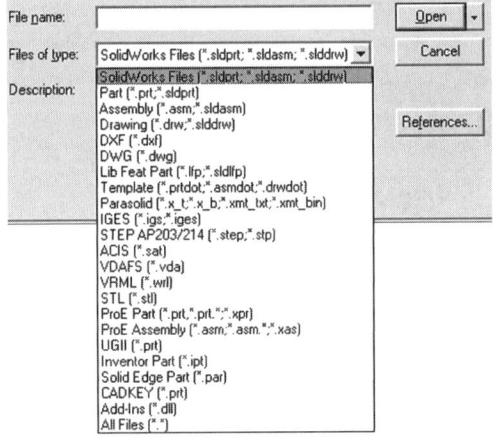

Fig. 22-1. Files of type *drop-down list.*

1. Select File > Open, or click on the Open icon. The Open window will appear.

2. From the *Files of type* drop-down list (shown in figure 22-1), select the type of file you want to import into SolidWorks.

3. Navigate to the appropriate drive or directory, if necessary, and select the file to be imported.

4. Click on Open to import the file.

After you click on the Open button, SolidWorks does the rest. The file will be translated and brought into a new SolidWorks document. The type of file you are importing determines which document type SolidWorks will translate the file into. For example, files with a *.dwg* or *.dxf* extension will be imported as drawings. An IGES file will be imported as a part or assembly.

Table 22-1 outlines all of the file types SolidWorks can read. It should be noted that the list of files recognized by the SolidWorks software is constantly expanding. Although every effort has been made to keep *Inside SolidWorks* as up to date as possible, there may be new file types supported by SolidWorks that are not listed in Table 22-1.

Table 22-1: File Formats Supported by SolidWorks

File Extension	Native Program, Modeling Kernel, or Standard	Description	Imports Into
DWG	AutoCAD	Drawing file	Drawing or part
DXF	AutoCAD	Drawing Exchange Format	Drawing or part
X_T, XMT_TXT	Parasolid kernel	Parasolid solid geometry text file	Part or assembly
X_B, XMT_BIN	Parasolid kernel	Parasolid solid geometry binary file	Part or assembly
IGS, IGES	Standard	Initial Graphics Exchange Specification	Part or assembly

Importing Files

File Extension	Native Program, Modeling Kernel, or Standard	Description	Imports Into
STP, STEP	Standard	International Standard for the Exchange of Product Data (AP203/214)	Part or assembly
SAT	ACIS kernel	ACIS solid geometry file	Part or assembly
VDA	Standard (German automobile industry)	VDAFS (Verband der Automobilindustrie Flaechenschnittstelle)	Part
WRL	Standard	Virtual Reality Modeling Language	Part or assembly
STL	Standard	Stereolithography file	Part
PRT, PRT.*, *.XPR	Pro/ENGINEER	Pro/ENGINEER part model file	Part
ASM, ASM.*, *.XAS	Pro/ENGINEER	Pro/ENGINEER assembly model file	Assembly
PRT	Unigraphics II	Unigraphics Parasolid model file	Part or assembly
IPT	Inventor	Autodesk Inventor part file	Part
PAR	SolidEdge	EDS SolidEdge part file	Part
PRT	Cadkey	Cadkey part file	Part

Most software programs will try to use a file extension that is unique to that particular application. For reasons of marketing and logic, however, this is not always the case. A perfect example happens to be the extension *.prt*. It is debatable which company first began using this extension. The *.prt* extension is used by both Unigraphics and Cadkey. It is also used by Pro/ENGINEER (Pro/E), but Pro/E files will typically have an additional numeric value following the *.prt* extension, so that it may look like *filename.prt.12*, or something similar.

SolidWorks used a *.prt* file extension in its earlier days. Assemblies and drawings used *.asm* and *.drw* extensions, respectively. Fortunately, Solid-Works Corporation realized that this was causing problems for some of their customers and switched to the six-character extensions you are now familiar with.

If you attempt to open a file type with the extension *.prt* without first specifying the type of file you are trying to open, the software will think you are trying to open an older SolidWorks file. The software program will not automatically detect the file type, and an error message will be issued. Of course, if the file actually is an old SolidWorks file, you need not specify the file type.

Certain file types have special considerations regarding how they import, and there are things you should know about importing certain types of files. There may be other functionality regarding the importation of files that requires further explanation. The following section describes aspects of individual file types that will help you make informed decisions about importing files into SolidWorks.

AutoCAD Drawing and DXF Files

There was a drawing file database change between AutoCAD R12 (Release 12) and R13, between R13 and R14, and again between R14 and 2000. There was no database change between AutoCAD 2000 and 2002. Releases R11 and R12 were compatible. What this means to the reader is that AutoCAD files from versions 12 through 2002 have different database structures and require different translators.

SolidWorks was nice enough to include read and write capabilities to all of the latest AutoCAD formats. This means that you can import and export AutoCAD DWG and DXF files from R11 (remember, R11 and R12 were compatible) right up through AutoCAD 2002. The export options for these files are discussed later in this chapter.

DWG and DXF files can be imported into either drawings or parts. When specifying a DWG or DXF file for importing, the DXF/DWG Import - Drawing Layer Mapping window, a portion of which is shown in figure 22-2, appears on your screen.

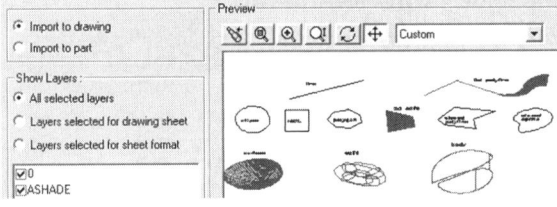

Fig. 22-2. DXF/DWG Import window.

It is recommended that you import files into a drawing rather than a part, even if the file is for a part whose geometry is to be used. If you import a DWG or DXF file into a drawing (rather than a part), layers can be turned on or off as needed. Therefore, only what you need to see is shown. Take advantage of the fact that AutoCAD layers are recognized by SolidWorks. Sketch geometry can then be copied and pasted into a part file if you wish to use that geometry in developing a solid model.

Because DWG and DXF files can consist of many types of geometry, you should understand the limitations in SolidWorks DWG and DXF translator capabilities. 2D geometry will, for the most part, come over fairly well into a SolidWorks drawing. There is often a problem with the appearance of dimensions and text, but this cannot be helped. Translating files is inherently difficult. SolidWorks does install AutoCAD fonts, which help relieve this problem, but AutoCAD text styles that have aspect ratios set to values other than 1 can be problematic.

AutoCAD DWG and DXF files can contain solid geometry. SolidWorks will not read solid geometry from either of these file types. Certain other entity types, such as polylines with width, do not translate well. If you must read in AutoCAD solids, use an IGES or SAT file type when exporting from AutoCAD. Most AutoCAD wireframe geometry will translate into a SolidWorks drawing, but it will be "flattened," losing its 3D aspect.

To import a DWG or DXF file as a drawing sheet format, use the Show Layers settings in combination with the layer listing. You can choose which layers are imported and whether each layer gets placed on the sheet or the sheet format, as indicated in figure 22-3. If importing to a part, you can only choose what layers are brought in, but remember that parts do not recognize layers, so everything will get piled into a single sketch.

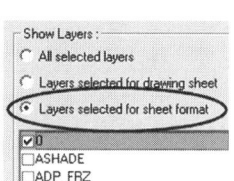

Fig. 22-3. Specifying which layers to place on the sheet or format.

If an AutoCAD file was imported with the intention of saving the data as a SolidWorks sheet format, you should save the newly imported geometry by using the SolidWorks File > Save Sheet Format command. Saving a format was discussed in Chapter 11.

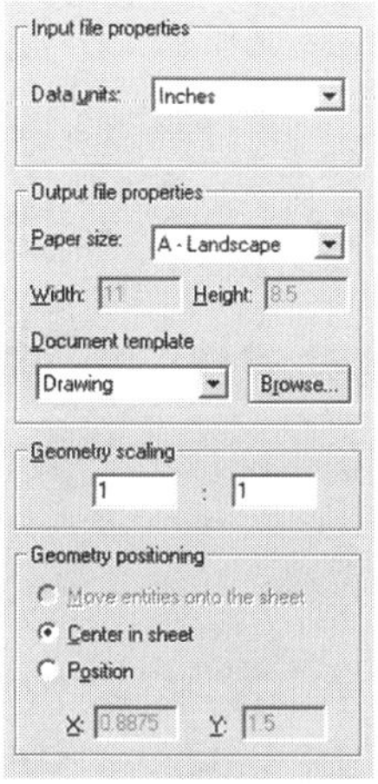

When importing AutoCAD files, there is a second window that appears after the one shown in figure 22-2. This second window is shown in part in figure 22-4, and is titled DXF/DWG Import – Document Settings. As can be determined from the image, it allows for establishing the drawing units, scale, and paper size. This is all very self-explanatory, and thus details regarding these options are unnecessary.

The Move entities onto the sheet option, grayed out in figure 22-4, is available only if SolidWorks detects that the x or y coordinate of the lower left-hand corner of the drawing being imported has an x-y location other than 0,0. If you are familiar with AutoCAD, this could mean that there is geometry to the left or below the User Coordinate System icon (as an example).

Fig. 22-4. DXF/ DWG Import – Document Settings window.

If you are going to be importing existing AutoCAD formats, it would be beneficial to first create some SolidWorks templates that contain no sheet format. As can be seen in figure 22-4, one of the options is to select what *Document template* you wish to use. Obviously, if you will be importing AutoCAD formats, you would not want to use a SolidWorks template that already contains one.

Parasolid Files

The Parasolid translator is probably one of the best methods for translating solid files into (or out of) SolidWorks. It is fast and efficient, and results in small files that can easily be moved from one location to the next. The Parasolid kernel is the solid modeling kernel used by SolidWorks. If you do need to import solid geometry, and the CAD program at the other end of the line can send a Parasolid file, you are better off using this method than any other.

SolidWorks can translate all Parasolid file versions up to and including version 14. Always use the highest version that both the sending and receiving machines can translate to ensure the highest possible quality and compatibility. Additionally, binary files are more efficient than text. Thus, if you have a choice, use binary.

IGES Files

It is not difficult to find a technical support person who receives more questions regarding IGES translation than any other topic. The reason for this is in large part the many IGES translation "flavors" available. There are always a number of settings or options that go along with IGES translation, and if those settings do not coincide with the requirements of the software on the receiving end, the translation does not work or is not complete.

When importing an IGES file, you do not have any choice but to hope the translation works. Hopefully the person at the sending end of the line knew what type of program you were running and tweaked her export settings correctly. If your goal is to import an IGES file as a solid model, the person sending you the file should make sure she sends you IGES trimmed surfaces. You will not want IGES wireframe, as there is very little you will be able to do with the resulting geometry.

IGES curves will translate into SolidWorks, but you will wind up with a bunch of 2D or 3D reference curves. This is normally not a desirable situation. On occasion, you may want to import a curve for use as a guide curve or sweep trajectory, but this is rare.

A situation that can develop when importing IGES trimmed surfaces from other programs is that SolidWorks refuses to knit the surfaces. Programs that create strictly surface (as opposed to solid) geometry will probably not translate well into SolidWorks. This has to do with how solid modelers differ from surface modelers.

If this is a troublesome topic for you, try to look at it this way. A solid can be broken down into surfaces, but surfaces do not necessarily have to form a solid. Solid geometry has faces (surfaces) that all meet edge to edge. Surface modelers can create surfaces that do not meet edge to edge. This is the crux of the problem. Some surface modelers do not have tolerances that allow a solid modeler to read the surface data and correctly interpolate it. Concessions must be made for one reason or another. When the solid modeler attempts to read the data, it cannot figure out where the surface edges meet. When this happens, the solid modeler cannot knit the surfaces and create the solid geometry.

SolidWorks will typically make numerous attempts to knit surfaces to form the solid if at first it does not succeed. Each time another attempt is made, the tolerance used to knit the surfaces is loosened to a less accurate level. If eventually the solid cannot be formed, a series of reference surfaces is created. At this point there is little you can do except request from the person who gave you the file that he try to clean up things at his end. Unfortunately, this is not always a politically feasible situation.

You do have some choices on your end as to how an IGES file is handled during importation. Figure 22-5 shows these options, which can be accessed by clicking on the Options button after selecting IGES as the type of file you wish to open. The Surface/solid entities option should be checked, assuming that is what is being imported, which is almost certainly the case. The General options shown in figure 22-5 will affect the import of SAT, STEP, and VDA files as well.

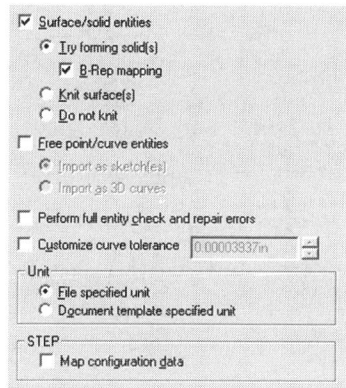

Fig. 22-5. General import options.

Importing surfaces without knitting them results in a bunch of reference surfaces that could then be manipulated using surface commands. This would obviously change the data so that it diverges from the original model, and because SolidWorks is primarily a solid (not surface) modeler, this option is not commonly used.

Knitting the surfaces results in one surface (usually), but not a solid. The most common option is to let the software attempt to knit the surfaces. Then you can try to create a solid from them. If importing small models or models with very small features, it may be necessary to check the *Custom curve tolerance* option and decrease the tolerance.

The *B-Rep mapping* option is okay to leave checked, as it will not hurt anything and may be faster. Boundary representation mapping is a bit different than knitting. Rather than knitting individual surfaces, the topology of the entire model is mapped out.

The *Perform full entity check and repair errors* option will force Solid-Works to perform extra checks, which will slow down the import process. It will also repair errors where possible, which can be a good thing when importing problem IGES files. However, the extra error checking and repairing may not be necessary.

In summary, when importing IGES files, leave *Surface/solid entities* checked, and use the *Try forming solid(s)* option. Turn on *B-Rep mapping*, because it is faster than the *Perform full entity check and repair errors* option (which you should turn off). It is almost never necessary to check *Free point/curve entities*, and it usually is not necessary to customize the tolerance. Use figure 22-5 for reference, which shows exactly what your typical settings should be.

STEP Files

The STEP file format has been constructed as a multi-part ISO standard (ISO 10303). It is a neutral file format constructed for the exchange of data that encompasses product development lifecycles rather than strictly CAD data. A number of application protocols have been developed, two of which are AP203 and AP214. These two protocols are supported by SolidWorks.

Its basic parts have been completed and published, but more aspects of the STEP format are still in development. There is only one option associated with STEP importation, which is the *Map configuration data* option. This option can be seen at the bottom of figure 22-5. Checking this option will result in any configuration data associated with the file being mapped out as custom properties. The Map configuration data option is in no way related to SolidWorks configurations.

Because STEP is not as widely accepted as IGES, and because it is not based on the Parasolid kernel, you should not use the STEP translator unless there is a valid reason to do so. If a customer specifically requests STEP, at least you can accommodate them.

ACIS SAT Files

The only thing noteworthy you should be aware of concerning SAT files is that there are many versions of ACIS SAT files and they are not all compatible. The ACIS solid modeling kernel (licensed from Spatial Technologies, Inc.) has many versions. Most modelers using this kernel have options for importing or exporting to various ACIS versions. SolidWorks can read and write many versions of ACIS files, up through and including version 7.

Virtual Reality Markup Language

VRML, as it is better known, was first thought to be a great new way to view the Internet, with 3D worlds taking the place of the flat 2D and multimedia web pages most of us see on the Web today. Huge 3D vistas and exciting worlds were imagined, in which "virtual reality" could be explored by daring individuals. The entire concept of virtual worlds was further glamorized by Hollywood and the media (Stephen King's *Lawnmower Man* being a perfect example).

Virtual reality on the Web was an interesting concept, and still is, but it has not come to fruition. The bandwidth needed to exchange files containing the polygons needed to describe the virtual worlds was not wide enough. Although web bandwidth is increasing, VRML is still not a very good solution. Viewing VRML models is often akin to viewing a construction paper model in a dark closet with a flashlight.

Some models lend themselves to VRML better than others. If your primary concern is to transmit 3D images of your solid files over the Internet, there are better ways, indeed. For instance, SolidWorks offers a free viewer that can be obtained from their web site at *http://www.solidworks.com*. Specifically, look into eDrawings and you will not be disappointed. There are other viewers for other file types available as well. Ask your local SolidWorks reseller for more information regarding viewers.

If your primary concern is to transfer solid files from one computer program to another, VRML files (having a *.wrl* extension) should be your last choice. VRML was not designed with file translation in mind. It is recommended that you do not use it for this purpose.

Stereolithography Files

Most individuals would want to export stereolithography (STL) files for reasons of rapid prototyping (discussed in the second half of this chapter). However, STL files can also be imported into SolidWorks.

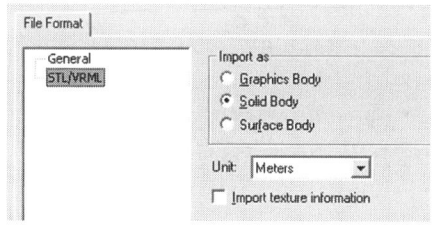

Fig. 22-6. STL import options.

When importing STL files, there are a few options available. These are shown in figure 22-6. Options that apply to STL files also apply to VRML files, but as was already mentioned, VRML is not the best choice for file translation.

Opening an STL file as a graphics body allows for viewing the file, but not much else. The imported body appears with a special icon in FeatureManager, as shown in figure 22-7. This is the quickest import option, so if the file needs no editing and you just want to see what it looks like, this is definitely the option to use.

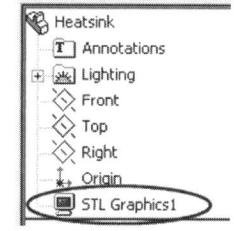

Fig. 22-7. STL graphics body.

Bringing in an STL file as a surface body (see figure 22-8) will result in a single knitted surface, but it will consist of hundreds, perhaps thousands, of polygons. This is the nature of STL files, which is to say they are polygonal representations of a solid model. You will be able to edit the surface file, but if editing is your intention, you are much better off importing as a solid body.

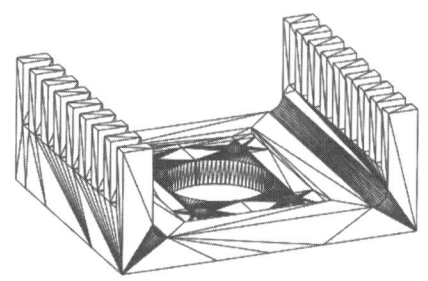

Fig. 22-8. An STL imported as a surface body.

Importing an STL file as a solid body will still result in a polygonal model, but at least it will be a solid. The solid can be edited just like any other solid model, but there will be a lot of separate faces to contend with.

Using STL as a file translation tool is not recommended. What STL is typically used for is rapid prototyping operations. The ability to import an STL file into SolidWorks is a benefit to those operating prototyping machines, because it means they now have the capability to make last-minute modifications to a model per a customer's request without the need for other file types. The other benefit is the ability to simply view the file. With all of this functionality built into SolidWorks, secondary programs are no longer necessary.

Proprietary Formats

SolidWorks can open up a whole range of file types, including proprietary file formats from other CAD programs. There are those in the manufacturing industry that have actually purchased SolidWorks mainly for its file translation capabilities. With so many options built into one program, SolidWorks pretty much covers the bases.

Because other software manufacturers are not sitting idly by, the programmers at SolidWorks Corporation have to stay on their toes. New versions of other CAD programs are constantly being released. Table 22-2 will help you better understand just what versions of software can be read by SolidWorks. Again, keep in mind that as newer versions of software are released by these other manufacturers, SolidWorks will undoubtedly continue to upgrade its translation capabilities.

Table 22-2: Proprietary Software Version Recognition

Software	Versions Recognized by SolidWorks
Pro/ENGINEER	17 through 2001, parts and assemblies
Unigraphics II	10 and higher (Parasolid data only)
AutoDesk Inventor	Up to and including version 5
EDS Solid Edge	All versions (Parasolid data only)
Cadkey	Surface and solid geometry, version 19 parts and assemblies

Software	Versions Recognized by SolidWorks
AutoDesk Mechanical Desktop	Versions 4.0, 5.0, and 6.0 (must have MDT installed on computer)
AutoCAD	R11 through 2002

Of particular merit is SolidWorks' capability of importing encrypted Pro/ENGINEER files. What is even better is that in many cases SolidWorks even recognizes feature data. This is a huge blessing for those who have struggled with the complexity of the Pro/ENGINEER software and wish to move to the ease of SolidWorks, while retaining the ability to open and modify existing data.

With regard to Mechanical Desktop (MDT) file translation, it is required that MDT be installed, though not necessarily running, in order to use the translation capabilities within SolidWorks. If MDT is installed, open the DWG or DXF files using the method described in How-To 22-1. Not all versions of MDT are supported on all operating systems. Check the SolidWorks help file for current information on this matter.

Dynamic Link Libraries

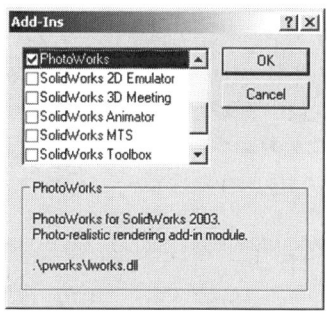

Fig. 22-9. Add-Ins window.

Those who dabble in programming will probably realize the benefit of being able to run a Dynamic Link Library (DLL) file from within SolidWorks. It is possible to run custom applications by running their DLL files directly from the File > Open window. This might be a custom routine for editing custom properties, for example. However, if a third-party add-on application has a listing in the Add-Ins window (shown in figure 22-9), it would be preferable to start the application from there.

Many third-party applications can easily be turned on or off through the Add-Ins window. The advantage of using this method is that the application will stay on or off until you go back into the Add-Ins window and change whether or not the application in question is being loaded. If the application has a check mark in front of it, it is being loaded. The Add-Ins window is accessed through the Tools menu.

Exporting Files

The list of file types that can be exported is nearly identical to those that can be imported, but with a few extra formats included that are primarily

792 Chapter 22: Importing and Exporting Files

viewing formats. The file types present in the *Save as type* list are dependent on whether you are exporting parts, assemblies, or drawings. The file types are shown in material to follow. First, however, you will learn how to export a SolidWorks file. How-To 22-2 takes you through the process, regardless of file type, with any additional comments or options discussed in material to follow.

How-To 22-2: Exporting to Other File Formats

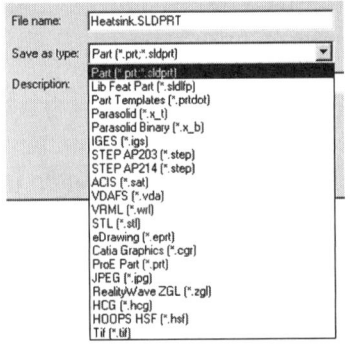

Fig. 22-10. Save as type *list for a part.*

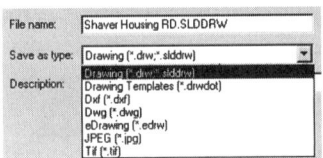

Fig. 22-11. Save as type *list for a drawing.*

To save a SolidWorks document as a different file format (in other words, to export a file), perform the following steps.

1. Select File > Save As.

2. Specify a file type from the *Save as type* drop-down list, shown in figure 22-10. (Note: this is the listing available when exporting a part file, not a drawing.)

3. Specify a name and location for the exported file, if necessary.

4. Click on Save.

If you are saving a part or assembly, you will see the drop-down list shown in figure 22-10, with a few exceptions. For instance, an assembly cannot be saved as a VDAFS file, as that format does not support assemblies. If you are saving a drawing, your export options will be limited to a much smaller list, as shown in figure 22-11.

You can see that the mechanics for saving the file are quite easy. However, when it comes to deciding on the file type or how that file type's export options should be set, the game becomes a good deal more complicated. For example, most file types that can be exported have their own set of options that can be tweaked by the user. In many cases, the default settings will work fine. In other cases, you may find it necessary to modify the optional export settings. When exporting a file, an Options button will appear on the Save As window, which will gain access to the Export Options window. Screen shots of this window's various sections are included throughout the next section for your reference.

The seemingly vast array of export options will not be as intimidating once you understand what many of the options accomplish. The following section is intended to make you feel more comfortable when deciding on an export option for a SolidWorks file. Whether you decide to read through the entire section or merely use it as reference, you should find it helpful.

Some export file types are available for 2D design drawings only, whereas others are available for part and assembly documents. This is noted throughout for the various export file types. Additionally, the most common file types and optional settings are recommended.

DWG and DXF Export Options

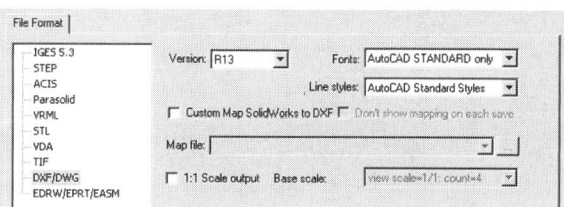

Fig. 22-12. DXF/DWG Export Options category.

The DWG and DXF file formats are available for 2D drawings only. When saving a drawing as one of these file types, click on the Options button in the Save As window to gain access to the optional settings, shown in figure 22-12. The export options for these two file types are explained in the following material.

Version

When presented with a choice for a version, you should always try to export the file at the highest version possible. Lower versions may not support as many objects as more recent versions. Make sure to check which version of AutoCAD the recipient of the exported drawing is using; otherwise, they may not be able to read the file.

Fonts and Line Styles

Your two choices for mapping fonts are *AutoCAD STANDARD only* and *TrueType fonts*, either of which are selectable from a drop-down list. The term *standard* refers to the font style in AutoCAD of the same name. If the *AutoCAD STANDARD only* option is used, all text in the SolidWorks drawing will be converted to the standard text style when brought into AutoCAD.

If TrueType is specified, the TrueType font used in SolidWorks will be used in AutoCAD if it is available on the computer importing the file. Otherwise, fonts will be mapped out according to AutoCAD's system variables. Consult your AutoCAD User's Guide for more information regarding font mapping and system variables.

When TrueType is used, text styles will be created automatically in AutoCAD. These styles use the TrueType font as the base font for each style required. This will happen automatically and will require no user intervention. The TrueType setting may not work correctly if exporting to AutoCAD version 12, because TrueType fonts were not available in that version. It is suggested that you select *AutoCAD STANDARD only* when exporting to R12.

Text will import into AutoCAD as blocks. If the blocks are exploded, the fonts appear as DTEXT entities. This is true no matter which version or file type (DXF or DWG) you are saving as. Consult the AutoCAD User's Guide for more information.

If SolidWorks Custom Styles is selected for the *Line styles* setting, SolidWorks line styles are retained. If AutoCAD Standard Styles is selected, SolidWorks line types are mapped out to the stock line styles used by AutoCAD. If saving as R2000 or later, line weights will be mapped out as well.

Custom Map SolidWorks to DXF

By checking the Custom Map SolidWorks to DXF option, you enable file mapping, and the Map file area is activated. Mapping allows you to specify layers for specific entity types. In other words, you can define layer names for specific SolidWorks entity types, such as text, dimensions, and sketch entities. You can also specify colors and line types for the layers you define. See the section "Creating a Map File," which follows, to see how this is done.

Map file Area

To create a custom map file, you must first save a drawing as a DWG or DXF file. The SolidWorks to DXF/DWG Mapping window will appear at that time, and you will then be able to define the map file and save it. Because it is possible to have more than one map file, you need to specify the map file you want to use in the Map file area, which defines the file that will be used whenever you save as a DXF or DWG file.

Don't show mapping on each save Option

The SolidWorks to DXF/DWG Mapping window can appear every time you use the Save As command and select DXF or DWG as the file type. If you do not want to have the SolidWorks to DXF/DWG Mapping window appear every time, check the *Don't show mapping on each save* option. Mapping will still take place, but the mapping window is not displayed.

Scale Settings

AutoCAD users typically design models at a scale of 1:1. This holds true for drawings of those models as well. Sheet formats, often inserted as blocks, are typically scaled up or down to fit the size of the model drawing views. Variables for controlling line type, text, and dimension scale are all changed to adjust for whatever the final print scale will be.

SolidWorks handles things differently. Parts and assemblies are always modeled 1:1. When creating a drawing, a sheet scale is set once and that is all the drafter need worry about. If a drawing with a sheet scale of 1:4 is exported as a DWG file, however, a decision must be made.

By not checking the *1:1 Scale output* option, you leave the sheet scale intact. This means the drawing will be opened in AutoCAD without the typical 1:1 scaling most AutoCAD users are accustomed to. On the other hand, checking *1:1 Scale output* allows for then choosing either the sheet or drawing view to base the 1:1 scale setting on.

If the *1:1 Scale output* option is enabled, select from the *Base scale* list to determine what view or views the 1:1 scale will be based on. The drawing is saved with a model geometry scale of 1:1 based on the views chosen from the *Base scale* list, and everything else is scaled accordingly. The sheet scale is not typically used in such a case.

Creating a Map File

Map files are not for everyone. They are used to specify, for example, what entity types get placed on which layers. Mapping can also override existing layer data, so that exported layers have a particular line type or color. It is important to understand that mapping can override the existing settings in a drawing. It is not necessary to use a map file to maintain layers, line types, and colors of your drawing during exporting.

If you decide to use mapping, it is probably because you would like certain layers to have certain properties in the exported file. This could be the case if sending a file to a customer accustomed to working with the typical black background of older versions of AutoCAD. It might be nice to have dark layer colors mapped out to lighter colors for the convenience of the person reading in the file.

Before you can define your first map file, first make sure you have the Custom Map SolidWorks to DXF option selected (described in the previous section). Once you have completed your SolidWorks drawing and have saved it as a DXF or DWG file, you will see the SolidWorks To DXF/DWG Mapping window, the upper portion of which is shown in figure 22-13. This window appears as soon as you click on the Save button when saving the drawing as a DXF or DWG file.

There are three tabs associated with the SolidWorks To DXF/DWG Mapping window: Define Layers, Map Entities, and Color Mapping. The

Define Layers tab is shown in figure 22-13. The Map Entities tab allows for assigning specific properties (such as color, line type, and layer) to various objects. These objects might be geometric tolerances, center marks, or any other object type in SolidWorks.

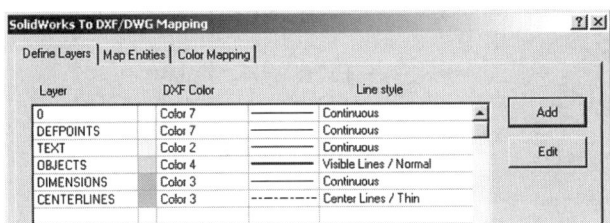

Fig. 22-13. SolidWorks To DXF/DWG Mapping window.

Finally, there is the Color Mapping tab, which allows for mapping SolidWorks colors to standard DXF colors. For example, you could choose to have the colors blue, red, and green map out to the color black in the exported file. Settings made in the Color Mapping tab will override settings made in the Map Entities tab, which in turn override settings in the Define Layers tab. Therefore, be careful.

There is an option at the bottom of the SolidWorks To DXF/DWG Mapping window that is labeled *Keep existing drawing layers for entities*. Checking this option will ensure that existing layer data is left intact, with only those objects not currently assigned to layers custom mapped. Be forewarned that not checking this option will cause all mapping data to overwrite existing layer data and entity attributes in the exported file.

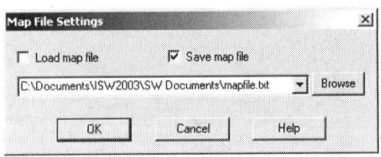

Fig. 22-14. Map File Settings window.

If you click on the Map File Settings button, you will bring up the window of the same name, shown in figure 22-14. To save a map file, uncheck *Load map file* and check *Save map file*. Map files are saved as simple text files, so they can even be edited in a basic text editor. To load an existing map file, check the *Load map file* option, and uncheck *Save map file*.

For first-time map file creators, make it a point to save the map file. Specify a name for the file, and a location. Use the Browse button to determine a location if that is easier for you. Give the file a descriptive name and a *.txt* extension. The map file will be a text file, but SolidWorks does not apply an extension automatically. This is an extremely rare exception, because file extensions are almost never added by the user. Typically that job is left to the operating system or software.

Once you have established a name and location for the map file in the Map File Settings window, click on the OK button. This will bring you back to the SolidWorks To DXF/DWG Mapping window. The following sections further describe the three tabs and how they work.

Define Layers Tab

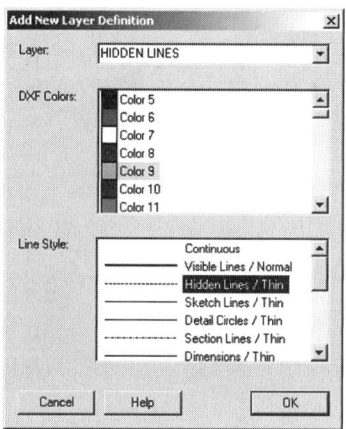

Fig. 22-15. Mapping colors and line types to a layer.

If you want to designate specific line types and colors for layers that will be created when you export a DXF or DWG file, this is where you would do it. First, click on the Add button, and then type in a name for the layer you want to create. Next, select a color and line type for the new layer, as shown in figure 22-15. Click on OK and the new layer will be created.

It would be logical to define, via the Define Layers tab, mapping layers that have the same name as those present in your drawing (for the purpose of assigning different properties to those layers during export), but that does not have to be the case. You could define new mapping layers, and then map entities to those layers using the Map Entities tab. This could all be done without ever having a single layer present in the original drawing. Layers defined in this tab will appear in the Entity Mapping tab, discussed in the following section.

✗ **WARNING:** *Be careful with the Reset All button, because it will wipe out your settings and you will have to start over.*

Map Entities Tab

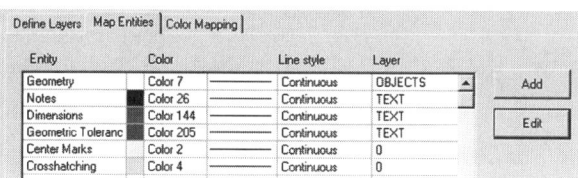

Fig. 22-16. Map Entities tab.

The upper portion of the Map Entities tab is shown in figure 22-16. As you can see, some entities have already been mapped out. If entities are mapped to a particular layer, the color and line style settings specified in the Map Entities tab will take precedence over what is specified in the Define Layers tab.

Clicking on the Add or Edit button brings up the Add New Entity Mapping window, shown in figure 22-17. The Entity list at the top of the window contains a wide-ranging list of objects, from blocks to weld symbols. At the bottom of the window is a list of layers that can be selected. The list will contain all of the layers previously defined in the Define Layers tab.

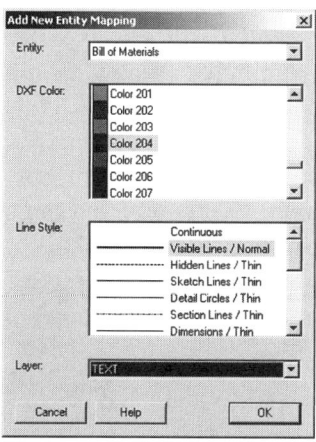

Fig. 22-17. Add New Entity Mapping window.

Color Mapping Tab

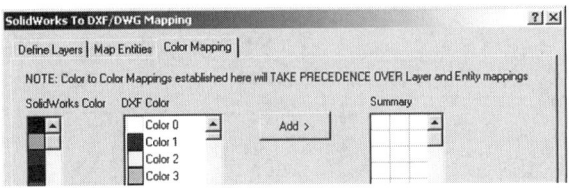

Fig. 22-18. Color Mapping tab.

The Color Mapping tab is shown in figure 22-18. In their default state, all SolidWorks colors are mapped out to the same colors in AutoCAD. You may very well decide to leave well enough alone when it comes to mapping out colors. If you wish to change how the colors are mapped out, first select a color from the SolidWorks Color list on the left, and then from the DXF Color list on the right select the color the first color will be mapped to. Clicking on the Add button maps the selected colors and places the map data in the Summary list for your reference.

Parasolid Export Options

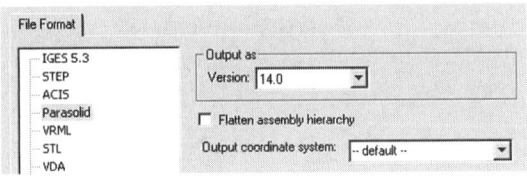

Fig. 22-19. Parasolid export options.

There are few options when it comes to saving a part or assembly as a Parasolid file, as can be seen from figure 22-19. This is good, because fewer choices mean fewer places to make mistakes. Any solid modeler using the Parasolid kernel (such as Solid Edge and Unigraphics) should be able to read a Parasolid file. Because SolidWorks also uses the Parasolid kernel, this is the translator most highly recommended.

Always try to specify the highest version supported by the target system. Check *Flatten assembly hierarchy* if you wish all subassembly components to become top-level components in the exported file. Specifying a user-defined coordinate system is also an option for defining the exported geometry relative to something other than an assembly's default coordinate system, but this is rarely necessary.

When selecting Parasolid from the *Save as type* drop-down list in the Save As window, always go with Parasolid Binary unless you have a good reason not to. Binary files will be smaller than text files, which makes for easier transfer of data, and will load faster during import. Of course, you must also weigh the fact of whether or not the person you are exporting the file to can read it. If a software program can read a Parasolid text file, it should be able to read a Parasolid binary file.

IGES Export Options

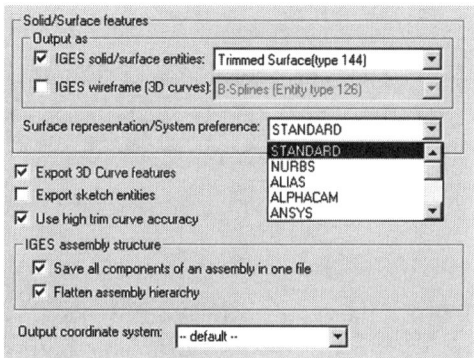

Fig. 22-20. IGES export preferences.

IGES is one of the most frequently used file export options and one of the most common neutral file formats used for file translation. Both parts and assemblies can be exported as IGES files. There are many types of IGES entities. Not all software programs can read all IGES entity types. There are numerous preset selections for specific programs in a drop-down listing, shown in figure 22-20. Specifically, this listing is labeled *Surface representation/ System preference*.

If the target system application is listed in this drop-down list, you should be all set. Programs or named settings include STANDARD (the default setting), NURBS, ANSYS, COSMOS, MasterCAM, SurfCAM, SmartCAM, and TekSoft, to name a few. If the target application is not listed, use the default STANDARD setting. The following sections describe the various options present when exporting IGES files.

Trimmed Surfaces Versus Manifold Objects

Always use the Trimmed Surfaces option if you want to export surfaces that can be imported as solid geometry and are not sure if the target system can read manifold solid B-rep objects. Trimmed surfaces can be imported and knitted into a solid in just about any solid modeling program that accepts IGES files. Not all programs can read in manifold solid B-rep objects, so check the documentation for the target system software prior to using this setting. (Dassault Systèmes' CATIA software and EDS's I-deas software can read manifold solid B-rep object data.)

When IGES solid/surface entities are being exported, select an IGES "flavor" from the *Surface representation/System preference* drop-down list. If the target system software is not represented, stick with the STANDARD setting. Chances are the receiving party will be able to read the file without any problems.

IGES wireframe (3D curves) Option

Select the 3D Curves option only if you want to export 3D wireframe geometry. Wireframe geometry can be thought of as a bunch of wires glued end to end. They can be curved, and they can exist in 3D space, but they are not solid geometry. If you select this option, the target system will not be able to convert the data to solid (or even surface) geometry automatically. It is not suggested you use this option unless the target system owner specifically requests wireframe geometry.

Export 3D Curve features Option

Turning on this option allows for features such as composite curves or helical curves to be exported. Typically you would not want to include curves when exporting trimmed surfaces. It just is not necessary, and additional curves may make it difficult to distinguish between the various IGES surface entities once imported into the target system.

Export Sketch Entities

Checking the *Export sketch entities* option will enable SolidWorks to export all 2D and 3D sketch objects, along with all other IGES surfaces being exported. This option is turned off by default, and you will more than likely want to leave it that way under normal circumstances. Similar to the *Export 3D Curve features* option, it just is not necessary to export sketch data.

Use high trim curve accuracy Option

Checking the *Use high trim curve accuracy* option will result in the exported IGES file being more accurate. If file size is not an issue, which it usually is not these days, leave this setting checked. A higher trim curve accuracy can reduce possible import problems on the target system. If, however, you are trying to keep the size of the file as small as possible, uncheck this option.

IGES Assembly Structure Options

If the *Save all components of an assembly in one file* option is checked, an assembly will be saved as a single file as opposed to separate files. In other words, SolidWorks exports an assembly as separate IGES files for each component in the assembly when this option is disabled. It should be noted that checking *Save all components of an assembly in one file* does not turn the assembly into a single, contiguous solid file. It simply exports the assembly as a single file containing multiple components, rather than separate files for each component.

When the *Flatten assembly hierarchy* option is checked, any subassembly components in the original assembly will become top-level components in the exported assembly. The assembly hierarchy will have no depth, as all components will be top-level components. This is known as "flattening"; hence the name *Flatten assembly hierarchy*.

Output coordinate system Option

Exported geometry is defined mathematically and must reference a coordinate system. Most of the time, the default coordinate system is fine. Occasionally, it may be necessary to define a coordinate system and require that the exported model reference the user-defined coordinate system instead. This might occur, for example, if the target application requires the geometry to reside in positive 3D space. In most cases, however, this setting can be ignored.

STEP Export Options

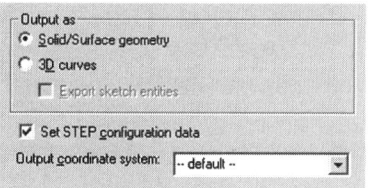

Fig. 22-21. STEP export options.

There are not many options when creating a STEP file. There is the option for selecting a user-defined coordinate system, which was explained in the previous section, and there are options for exporting solid and surface data, or 3D curves, as can be seen in figure 22-21. Always export solid and surface data whenever possible, as this data is the most complete. Not much can be accomplished on the target system with curve data.

The only time the *Set STEP configuration data* option will be available is when you are exporting STEP Application Protocol 203 (AP203). AP214 does not support configuration data. By checking this option, you gain the ability to enter data relative to the part or assembly you are saving. Data regarding the design or about yourself or your company can be added as well. The STEP Configuration Data for Export window will appear when saving the file, a portion of which is shown in figure 22-22.

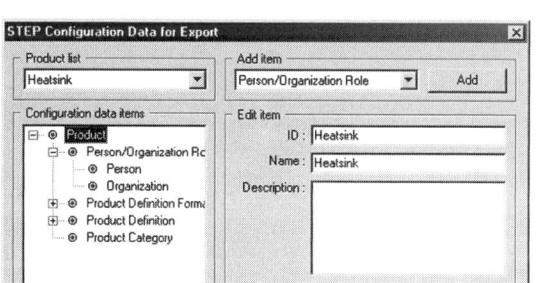

Fig. 22-22. STEP Configuration Data for Export window.

It certainly is not mandatory to enter STEP configuration data. It is there if you want to use it, but can be ignored if the configuration data is not something that concerns you. STEP configuration data is in no way associated with SolidWorks configurations. Checking the *Set STEP configuration data* option will not affect the model being saved.

ACIS Export Options

For those using Autodesk products such as AutoCAD and Mechanical Desktop, be aware that these products often used old versions of the ACIS kernel. Depending on the version of the Autodesk product being used, it may be necessary to export an ACIS file using a very early version of the kernel. The options associated with ACIS export are shown in figure 22-23.

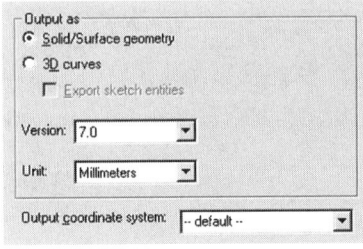

Fig. 22-23. ACIS export options.

ACIS export versions can be anywhere from version 1.6 to the most recent versions in use by CAD systems. If you are exporting to AutoCAD or Mechanical Desktop, be aware of which ACIS file type is being exported. AutoCAD releases 13 or 14 can read ACIS versions 1.5 or 1.6 only. AutoCAD 2000 can only read up to ACIS version 4.0. Mechanical Desktop, on the other hand, can read higher versions of ACIS, but make sure to check what version of Mechanical Desktop the person on the target system is using, and what versions of ACIS she can import.

If exporting ACIS files, specify the units you want to export into. This is not necessarily the units you are working in, but the units you want to be used when the file is imported to another program. Your options are millimeters, centimeters, meters, inches, and feet.

Additional ACIS output options are the same as for STEP export. If you have any questions about these options, see the section "STEP Export Options."

VDAFS Export Options

The only option when exporting a VDAFS file is to use a different coordinate system other than the default. This option was explained previously, in the "IGES Export Options" section.

VRML Export Options

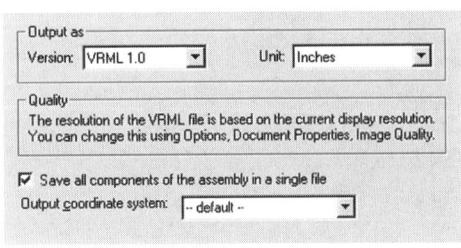

Fig. 22-24. VRML export options.

When exporting as VRML, it is most likely that you will be using the exported geometry for reasons of web site development. It has already been stated that VRML should not be used for reasons of file data transfer between two systems. There are much better choices available for accomplishing that task, such as Parasolid or IGES.

VRML export options, shown in figure 22-24, are limited. Choices exist for version (VRML 1.0 being the most recent as of this writing) and units.

If the *Save all components of the assembly in a single file* option is not checked, SolidWorks will create a directory with the same name as your assembly and then create a separate file (with a *.wrl* extension) for every component in the assembly. Otherwise, all assembly data is saved in one file. The option is irrelevant if exporting a part file.

Be aware that image quality settings will affect the size of the VRML file by changing how many polygons are used to represent the VRML model. Adjust display quality via the Image Quality section of Document Properties (Tools > Options). The Output coordinate system option has been explained previously, in the "IGES Export Options" section.

Stereolithography Export Options

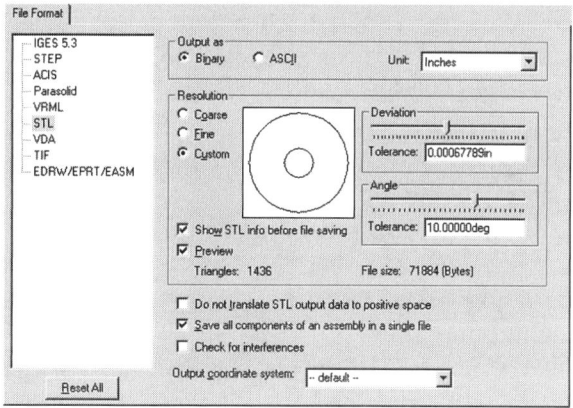

Fig. 22-25. Exporting STL files presents many choices.

Stereolithography (STL) files, polygonal representations of parts or assemblies, are used for rapid prototyping. After specifying that you wish to save a model as an STL file, click on the Options button. You will then be able to make adjustments that control the size of the polygons that make up the STL file. Export options for STL files (shown in figure 22-25) are a bit more involved than other file types.

STL file export is an important topic, and the rapid prototyping business is a growing industry. Changes made in the STL export settings will directly affect how accurate the STL file is. Accuracy and other settings are discussed in material to follow. Refer to figure 22-25 when reading the descriptions of the various STL options.

Output Format and Units

The output format can be either binary or ASCII. Most rapid prototyping machines can read binary files, and you should always choose binary files if you have a choice. ASCII files are text based and are much larger than binary. You should be safe creating a binary file unless your rapid prototyping service specifically requests ASCII. As far as the units are concerned, it would be most common to use the same units as the model being exported. However, if the model were modeled in mils or angstroms (as examples), to minimize problems with the rapid prototyping process you would want to select a more common unit of measurement.

Resolution

The Resolution section has settings that control the overall accuracy of the STL file. Parts with flat surfaces do not require as high a quality setting as parts with small, curved surfaces. If the quality is set too low for parts with curved surfaces, the curved surface geometry of the prototype may require a good deal of sanding or rework, and small curved areas (such as holes) may not form correctly. Both Coarse and Fine have their quality settings preset. Selecting Custom allows you to use the slider-bar settings Deviation and Angle Tolerance (see figure 22-25).

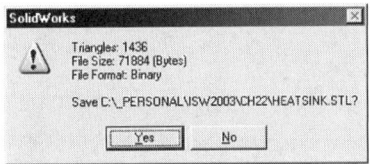

Fig. 22-26. Showing the STL information.

Adjusting the Deviation or Angle Tolerance setting too high will result in a very large file. It might take some trial and error before finding the best settings for your particular style of parts. It is better to err on the side of accuracy and wind up with a large STL file. Use the *Show STL info before file saving* option to preview the file size just prior to saving the file. The window shown in figure 22-26 will appear immediately after clicking on the Save button, and you will still have a chance to back out of the save process and readjust the tolerance quality settings.

Perhaps a better choice would be to check the Preview option. This will give triangle and file size information in the STL Export Options window. This makes it easier to experiment with the settings and see what the results are right away. If the file is complex and the tolerance settings are set too high, however, you may be in for a wait.

If you do check the Preview option, it will probably be necessary to move the Save As window and the Export Options window so that they are not blocking the model (prior to clicking on the Options button). Otherwise, it will not be possible to view the preview. Changing the Deviation and Angle tolerance settings will not update the preview automatically. It will be necessary to uncheck and then recheck the Preview option. The triangle and file size values will update, though, as soon as the tolerance settings are altered. See figure 22-8 for an idea of what the STL preview would look like.

Do not translate STL output data to positive space Option

Due to the nature of rapid prototyping machines, the polygon files must exist in the positive quadrant of an *xyz* Cartesian coordinate system. This is a technicality that you as a SolidWorks user should not have to worry about. Most STL machines can move the geometry if it does not already exist in positive space. However, you might as well let SolidWorks move the geometry for you if the geometry needs to be moved. Your rapid prototyping service will have one less detail to worry about, and it takes no effort on your part. Just make sure you leave this option unchecked.

Save all components of an assembly in a single file Option

Typically, STL files are exported one part at a time. You can choose to export an entire assembly as an STL file if you prefer. If you decide to do this, you can pack all of the assembly components into one STL file by checking this option. If you do not check this option, each component will be written to an individual file, using the name of the component as the file name.

Check for interferences Option

This option is only available if the previous option to save an assembly as a single file is checked. Some rapid prototyping systems do not take very kindly to interference. You should probably have already checked your assembly for interference anyway, in which case it would not be necessary to check it again.

The *Output coordinate system* option was discussed earlier. See the section "IGES Export Options" for more information regarding this topic.

Exporting Proprietary Formats

Not all of the proprietary file formats recognized by SolidWorks for importation are also available when saving (exporting) files. For example, Inventor, Solid Edge, Mechanical Desktop, Unigraphics, and Cadkey formats are not available for exportation. Use one of the neutral file formats discussed previously if exporting files for use in these applications.

Pro/ENGINEER parts and assemblies can be exported directly from SolidWorks. This results in a native Pro/E file that can be read in directly using Pro/E. SolidWorks currently saves Pro/E files as nonencrypted Pro/E version 20 parts or assemblies.

Other export options exist, but these are not available as import options. These additional export options only pertain to graphics files or streaming web media. They would not be useful to those looking to transfer solid model data to other systems. Table 22-3 outlines these proprietary export file formats and what they are used for.

> **NOTE:** *TIF and JGP image formats, discussed in the next section, are not considered proprietary.*

Table 22-3: Proprietary Graphic Export File Formats

File Extension	Name or Related Application	Description
.cgr	CATIA	CATIA graphics file
.hcg	CATIA CATWeb	Highly compressed graphics file
.hsf	HOOPS	Streaming Internet graphics data
.zgl	RealityWave	Streaming Internet graphics data

Saving Image Files

Images of SolidWorks documents can be captured from the screen in a couple of ways. One often-neglected method is to simply press the Print Screen button on your keyboard to capture a screen image. After performing this action and starting any basic photo editing software, you will find that pasting in whatever is in the Windows clipboard results in an image of whatever was on the screen when the Print Screen button was clicked on.

There are other more formal methods for saving images of SolidWorks documents. Two types of image formats are supported, which are JPEG and TIFF. The JPEG format (pronounced "jay-peg") derived its name from the committee that wrote the standard, the Joint Photographic Experts Group. JPEG files utilize a *.jpg* file extension.

JPEG files are often related to web page creation, as they are compressed files that transfer well via the Internet. JPEG images are recognized by every major Internet web browser on the planet, and most e-mail programs. It is a very common graphics format, which makes it a great choice for sharing graphics data.

TIFF Files

TIFF files (rhymes with cliff) derive their acronym from Tagged Image File Format, and utilize a *.tif* file extension. One advantage of TIFF files is that they are platform independent, meaning that they can be viewed and edited on a variety of platforms. These include Windows and Macintosh operating systems, as well as UNIX systems. TIFF files also provide advanced options for various compression schemes (including no compression), and color models, including RGB (red-green-blue) and CMYK (cyan-magenta-yellow-black). TIFF is the format many graphics professionals use. As a matter of fact, TIFF images were used throughout Inside SolidWorks 2003.

Image files can be exported from any SolidWorks document: parts, drawings, or assemblies. Use the Save As command to export a JPEG or TIFF file using the same technique you would use to export any other file type. When saving a JPEG image, there are no export options. What you see in the work area of your SolidWorks screen is exactly what the image will look like. This holds true for any document type, including drawings. Therefore, if the text is not legible in the drawing when saving a JPEG image, it will not be legible in the exported image. TIFF images, explored in the following section, are a different matter.

TIFF Export Options

The main reason for saving files as TIFFs is because this file type produces a high-resolution image, required for archival or publishing purposes. If it is your desire to create an image that can be sent to a customer for reasons of adding notes or redlining a drawing, image files work particularly well for this purpose. Anyone with photo editing software can open and edit a TIFF file.

Reasons for exporting an image of a drawing are often different from those for exporting images of parts or assemblies. Images of drawings are often used to create electronic hard copy for the purpose of archiving. Images of parts or assemblies are saved as high-resolution images for reproduction in a brochure, perhaps for reasons of advertising or company literature. Unfortunately, saving high-resolution images of parts or assemblies is not possible unless additional software is purchased.

> **NOTE:** *For photorealistic images, you might want to look into the Photo-Works product available from SolidWorks Corporation. PhotoWorks, which offers high-resolution output, can make your SolidWorks models nearly indistinguishable from photographs of real-world objects.*

Regardless of your reason for wanting to save a high-quality image of a SolidWorks document, the following section will help familiarize you with the options available. TIFF export options are accessed by clicking on the Options button in the Save As window, once you have specified TIFF as the file type to save. The TIFF export options are shown in figure 22-27.

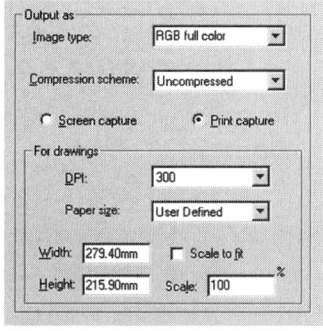

Fig. 22-27. TIFF export options.

Image Type

There are only two options for image type. Always select RGB full color for parts or assemblies. You may also want to use RGB full color with drawings that contain shaded views or colored lines or layers. Select *Black & white (bilevel)* for monochrome images. Do not try to use the *Black & white (bilevel)* setting for anything other than drawings. Unless you are trying to save an image of your part or assembly in a wireframe display mode, you will wind up with a black blob for an image.

Keep in mind that bilevel is strictly black or white. That is, there are no 256 shades of gray. Therefore, if you try to use the black-and-white setting for a shaded part or assembly, or on drawings with shaded views, the results will not be desirable.

Compression Scheme

There are actually more than a few types of compression available for TIFF files, but SolidWorks offers just two. This is quite satisfactory for most purposes. Packbits is the compression you should use if saving a color image. Use Group 4 Fax if saving a black-and-white image. Use Uncompressed (no compression) to ensure the highest quality and compatibility.

The Uncompressed setting offers the highest possible quality, although Packbits is a lossless compression scheme. This means that if you wish to use Packbits compression, the image should suffer no ill effects in the slightest degree. When the file is viewed by a TIFF image viewer, the file will be uncompressed into an exact duplicate of the original.

Screen Capture Versus Print Capture

If you are saving a TIFF image of a part or assembly, your only option will be Screen Capture. This means that you cannot export the file to a higher resolution even if you wanted to. For this reason, it is best to turn on anti-aliasing and zoom in to the part or assembly as far as possible while keeping all portions of the model on screen. This will make for the best possible image, which can then be downsampled if necessary. It is usually not an issue to reduce an image in size and retain quality, but it is impossible to increase the size of an image without losing quality.

Anti-aliasing is a method of removing jagged lines from graphic images. Intermediate colors are interpolated between the background and the model color in order to give the appearance of very smooth model edges. To turn anti-aliasing on, select the *Anti-alias edges* option, found in the Display/Selection section of System Options (Tools > Options). This setting is shown in figure 22-28.

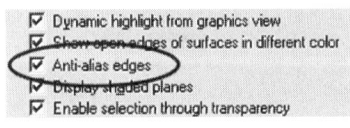

Fig. 22-28. Turning on anti-aliasing.

If you are saving a TIFF image of a drawing, make sure to use Print Capture. If set to Screen Capture, your drawing will be saved with the same resolution as that displayed on the computer screen, with no regard for scale or dots-per-inch (dpi) settings. This will make for a very poor print if you decide to try to print the image at a later time. About the only benefit of using Screen Capture with drawings is that it is quick.

When using Print Capture, you can adjust many aspects of the saved drawing file. The quality will be much higher than if you were to perform a basic screen capture. The settings described in the following material are only available if Print Capture is selected, and as noted previously, Print Capture is only available for drawing files.

DPI

DPI stands for dots per inch. Your options are 50, 100, 200, 300, or 600. The higher the DPI setting, the greater the clarity of the image and its resolution. However, file size will increase as the DPI setting is increased. Use some experimentation to determine what is right for you. You may want to start with a DPI setting of 100. If the file size is not too large for you to work with, and the processing time for the image is reasonable, you can gradually increase the DPI setting to a higher level.

If creating an RGB full color image type with a large sheet size and high DPI setting, the processing time for the image may be quite long. Depending on your computer speed, this could take quite a few minutes. Additionally, file sizes can be extreme. As an example, a full-color uncompressed TIFF image of a C-size drawing at a DPI setting of 300 will be right around 100 MB in size. Be reasonable when adjusting the settings for your TIFF file.

Paper Size and Scale

This speaks for itself. All of the typical sheet sizes are available. You would typically want to use a sheet size to match the sheet size of your drawing. If necessary, you can also specify a setting under User Defined, which will make the Width and Height settings available to you. The unit used for the Width and Height options is always millimeters, but you can type in inches as long as the "in" suffix is attached to the end of the numerical values.

If scale is important, such as if people will be taking measurements from the print, you will want to leave the *Scale to fit* option unchecked. If, on the other hand, you simply want to fit your drawing on a particular sheet size and do not care about scale, place a check in the *Scale to fit* option.

If the *Scale to fit* option is not checked, the Scale area will become available so that you can specify a scale. Type in an appropriate value to scale your drawing to the sheet size you want to save the image to. This setting is set to 100% by default. Fifty percent, of course, would be half-scale, and 200% would be 2:1 scale. Enter whatever percentage value you require.

eDrawings

Electronic drawings, or eDrawings, are an excellent means of communicating with others that do not have SolidWorks. Drawings can be animated, displaying all of a drawing's views and dimensions, much to the benefit of the individual on the receiving end. The model can be rotated and examined by the user, who does not need any other software to accomplish this. Best of all, eDrawings are free.

The eDrawings publishing software is all that is necessary to create an eDrawing. This is an add-on utility that can be obtained by anyone from SolidWorks' web site at *http://www.solidworks.com/edrawings/*. You must have a copy of SolidWorks or AutoCAD (see web site for currently supported version information) to publish eDrawings. Once created, however, the published eDrawing can be viewed by anyone with a computer running Windows.

eDrawings can also be created for parts and assemblies, and are not limited strictly to drawings. None of the feature data is included in an eDrawing. The only data sent via an eDrawing is the data necessary to view the model. This makes eDrawings small and easy to send via e-mail. eDrawings, when created, can embed the viewer directly into the eDrawing itself. In this way, the file can be self-contained.

If you have never had the pleasure of checking out an eDrawing, you may wonder how it is possible to animate a 2D drawing. When an eDrawing is animated, the model is shown in a shaded state as the view transitions on the screen. For example, the front view of the model may zoom in to a detail view if one were present in the drawing. The view pauses on the detail view for just a moment, and then transitions to the next view. The person viewing the eDrawing can pause the animation or zoom and pan the model at will.

If you find it necessary to communicate your drawings to others that do not have the good fortune to own a copy of SolidWorks, consider the eDrawings publishing software. The price is certainly right, and it makes for an excellent collaboration and communication tool. The eDrawing software is included on the SolidWorks installation disk.

Summary

In this chapter you learned that the mechanics behind importing or exporting files are quite easy. The settings behind file translation can be somewhat confusing. When importing files, you are dependent on the person sending you the file. You hope that the person on the sending end is making the appropriate adjustments so that you will be able to read the file. You have learned what types of files SolidWorks can read, so you can now relate this information to the sender and receive a readable file. Use the File > Open command to import a file. Make sure you specify the file type you are importing.

When exporting a file, you would use the Save As command and specify the file type. Exporting a file requires that you are knowledgeable of the applicable settings for specific file types. To make changes to export settings, click on the Options button, which will appear in the Save As window after you have specified what type of file you are exporting.

This chapter is intended to help serve as a guide to sending and receiving files. It should help you make the correct decisions regarding export settings when exporting a file to another party. There are always inherent difficulties when translating files, and no translator is perfect. Always communicate with the party you are translating files to or from, as an awareness of the software on the other end of the line will aid in the translation process.

Questions and Topics for Discussion

1. Describe the steps for importing a file into SolidWorks.
2. Describe the steps for exporting a file from SolidWorks.
3. What type of 2D files can SolidWorks import into a drawing?
4. Can SolidWorks import an STL file?
5. What does STL stand for, and what type of process is this file type generally used for?
6. What two file types can mapping be used for when exporting drawings?
7. Name three file types you can use to export an assembly.
8. What is the solid modeling kernel used by SolidWorks?

9. If exporting a solid model to Unigraphics, what format would you use?

10. Name four proprietary file formats SolidWorks can read.

Chapter 23

Customizing SolidWorks

THERE ARE MANY WAYS YOU CAN CUSTOMIZE SOLIDWORKS. The Customize window is only the beginning, as customization really goes beyond that. For instance, one could also modify what tabs appear in the New SolidWorks Document window when beginning a new part, assembly, or drawing.

The Feature palette can be highly customized to contain an array of parts, assemblies, library features, or forming tools. Folders containing these objects can be created with user-defined names. Even lighting conditions can be modified to change how models look on screen. All of these topics are covered in this chapter.

Through the Customize window, it is possible to make minor modifications to how the toolbars function, change which icons the toolbars contain, modify the pull-down menus, and create keyboard shortcuts. The Customize window is separated into four tabs that reflect these areas of customizability: Toolbars, Commands, Menus, and Keyboard. The functionality of each of these tabs is described in the sections that follow.

✗ **WARNING:** *Before reading further, a note of caution: If you are very new to the SolidWorks program, be careful when customizing the interface. It is best to know the software in its default, out-of-the-box state before making changes to the interface. It is possible that you can make changes that will impede your productivity when using the software. Always use a good deal of discretion when customizing SolidWorks.*

Customize Window

To access the Customize window, you must have a SolidWorks document open. The document can be a part, drawing, or assembly. If you do not have a document open, the Customize menu item will be grayed out. Select Customize from the Tools menu to open the Customize window. Customize is also available from the View > Toolbars menu. The various tabs available are detailed in the following section.

Toolbars

Of the four customize tabs, the Toolbars tab is the safest to use. Figure 23-1 shows a portion of the Toolbars tab. You cannot get into much trouble here. Turn various toolbars on or off by clicking on the toolbar's name to toggle the check marks on or off. If you see a check mark, the toolbar is visible.

The lower portion of this tab has three options. The first option, *Large icons*, will use toolbars that have a larger set of icons. The toolbars are not different; the icons are just larger. The second switch, *Show tooltips*, turns the toolbar icon tips on or off. Tooltips are those small yellow boxes that appear when you hold the mouse motionless over an icon. A sample of a tooltip is shown in figure 23-2. Once you get familiar with the software, you may decide to turn the tooltips off.

The third option, *Auto-activate sketch toolbars*, is the setting that makes the various sketch-related toolbars automatically appear when sketching in a part or assembly document. Drawing documents are not affected, as there is no such thing as "sketch mode" when in a drawing. You are basically always in sketch mode when working in a drawing.

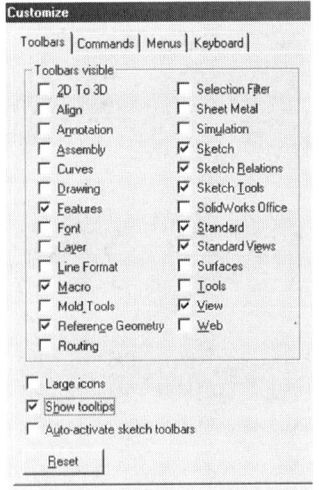

Fig. 23-1. Toolbars tab.

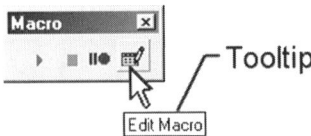

Fig. 23-2. Example of a tooltip.

Setting up toolbars to remain somewhat consistent while opening up various SolidWorks document types was first discussed in Chapter 11. Up to Chapter 11, part files were the main topic of discussion, so maintaining toolbar positioning was not an issue. If you are having problems regarding toolbars moving around, see How-To 11-2.

The Reset button will remove any changes made to the Toolbars tab. This will work only for the current session of modifications made. This means that if you make a change and click on OK to exit the window you will not be able to use the Reset button. It will remove only those modifications made since the Customization window was open.

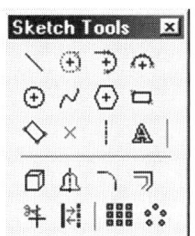

Fig. 23-3. After resizing a toolbar.

Resize toolbars by dragging a toolbar's border. This can be done at any time and does not require that you be in the Customize window. Toolbars can be made into different shapes and sizes (as long as the shape is rectangular) by dragging their edges. In figure 23-3, the Sketch Tools toolbar has been resized and is floating, rather than docked to one side of the screen.

∞ *NOTE: Turning any of the toolbars on or off can be accomplished by right-clicking over any docked toolbar.*

Commands

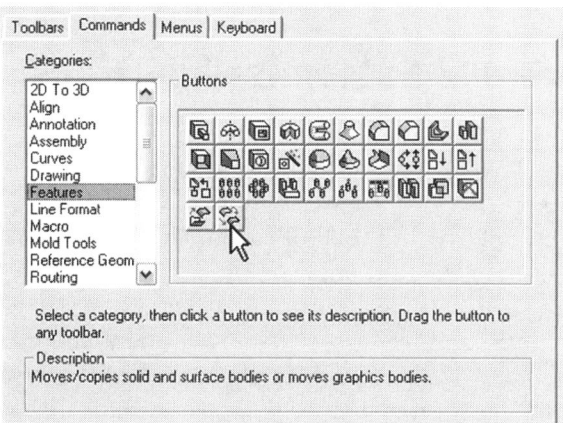

Fig. 23-4. Commands tab.

The Commands tab, shown in figure 23-4, is probably the most popular of the four customize tabs. Be careful that you do not accidentally delete icons from toolbars when using this tab. (Actually, you can accidentally delete icons whenever the Customize window is open. You do not have to be in the Commands tab.)

The name of this tab does not mean to imply that you can change how commands work. You can, however, change the commands present on certain toolbars. It is possible to add, remove, or relocate icons to different positions on most of the toolbars. To do this, hold the left mouse button down on an icon and drag it to its new location. Again, this location can be a new position on the same toolbar, or on an alternate toolbar.

✗ **WARNING:** *Before you do much toolbar customization, see the section "Resetting Toolbars and Customized Settings," which follows this section.*

As of this writing, SolidWorks does not give you the capability of creating your own toolbar, but in essence you have that functionality anyway. For example, you could take a toolbar you do not often use and add all of the icons you would like to see present on that toolbar. Not all toolbars can be customized, however, and therefore all icons for all toolbars will not be accessible.

To add an icon to a toolbar, first access the toolbar category from the Categories list. The icons for that particular toolbar are displayed in the Buttons area to the right (see figure 23-4). To add an icon to a toolbar, drag the icon from the Buttons area to the toolbar located elsewhere on the screen. Needless to say, the toolbar you are customizing must be visible.

If you are not sure what a particular icon is for, click on it in the Buttons area. A description of that icon will be given in the Description section at the bottom of the Commands tab. For example, the description for the Move/Copy Bodies icon is shown in figure 23-4.

Icons you may wish to add to the Features toolbar are the Suppress, Unsuppress, and Unsuppress with Dependents icons. These icons are typically not present on the Features toolbar after initially installing the SolidWorks software. The Sketch Tools toolbar is another toolbar that often needs customizing. Certain individuals use certain sketch tools more than others, so make your job easier and customize the toolbar (or the toolbars in general) to suit your needs.

Resetting Toolbars and Customized Settings

It is important that you understand there is no easy way to reset your toolbars to their default layout if your customization endeavors get out of hand. Once icons have been added or removed from toolbars, that is where they will stay. The Reset button in the Toolbars tab of the Customize window will not reverse the toolbar customization process.

Most alterations made to the interface through the Customize window are not easily reversible. In the simple case of making some toolbar changes and then changing your mind and wishing you had left things alone, the detrimental effects are minimal. In the worst scenario, your menus, toolbars, and keyboard shortcuts have gotten so fouled up that SolidWorks has been rendered useless. In a situation such as this, there is only one solution. You will need to uninstall the SolidWorks software and then reinstall it.

Updating the current installation by reinstalling SolidWorks over the existing version will not work because SolidWorks does you the courtesy of maintaining your software registry settings. This is as it should be. When installing a new release of SolidWorks, you would not want the software to reset all of your finely tuned customized settings. For this reason, uninstalling and reinstalling the program is the only answer.

To uninstall SolidWorks, access the Windows control panel and select the Add/Remove Programs function. This function may be named differently, depending on your operating system. You will find a listing for SolidWorks there, and you will then be able to uninstall the software. You will then need to reinstall the software, which includes supplying serial number and authorization code information. If you are not familiar with these tasks, have your system administrator or SolidWorks reseller help you with the process. Do not delete the SolidWorks directory manually, as this will almost certainly cause problems further down the road.

Menus

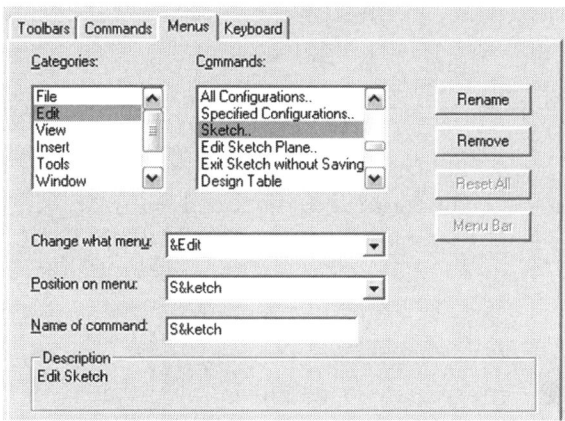

Fig. 23-5. Menus tab.

The Menus tab, shown in figure 23-5, is the area in which you can wreak the most havoc. It is strongly suggested that new users steer clear of this area. A lot of thought has gone into the layout of the menu structure to make it intuitive and easy to use. There are few reasons you would want to change it.

You have the capability of adding items to menus, deleting menu items, and repositioning and renaming menu items. The buttons in the Menus tab will change depending on what menu item has been selected. For example, the Rename button will change to an Add or Add Below button in certain cases.

Not all menu items can be added, and not all menu items can be renamed or deleted. This is to help prevent users from completely messing up the menu structure. Imagine what would happen if the menus were poorly customized and then the Customize menu option were deleted! Actually, you can delete the Customize item from the Tools menu, but not the View > Toolbars menu. Be careful! How-To 23-1 takes you through the process of adding an item to a menu.

HOW-TO 23-1: Adding An Item to a Menu

To add an item to a menu, perform the following steps.

1. Select the desired category from the Categories list.

2. Select the desired command from the Commands list. This will be the command you want to insert (add) into a menu.

3. From the *Change what menu* drop-down list, select the menu you want to add the command to.

4. Determine a position for the command in the *Position on menu* drop-down list. You have for the options At Top, At Bottom, and Auto, in which case SolidWorks automatically determines where the new item goes using its own internal criteria. If you select an item from the Position on menu list, the new command will be inserted below that item.

5. Type a name for the command into the *Name of command* area. Typically, you would accept the default value for the name. The ampersand (&) symbol determines the hot key for the command. The letter following the ampersand symbol will be the hot key, and will be underlined in the menu. Normal Windows conventions dictate that three dots (...) following a menu item, indicating that a window will open if the menu item is chosen.

6. Click on the Add button to add the item to the specified menu. If you selected an existing menu item in the *Position on menu* list, the Add button will read Add Below.

7. Click on OK to accept your changes.

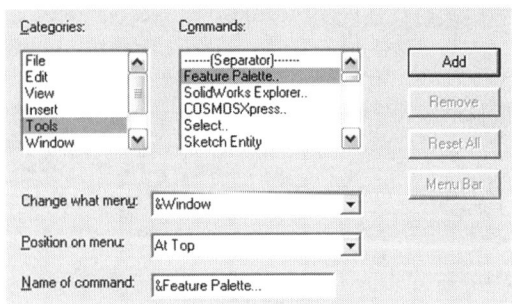

Fig. 23-6. Placing the Feature Palette menu item in the Window menu.

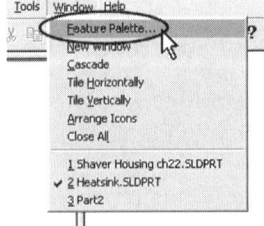

Fig. 23-7. Results of adding the Feature Palette menu item to the Window menu.

In figure 23-6, the Feature Palette item is being added to the Window menu. It is being positioned at the top of the menu. The name of the item is &Feature Palette. As mentioned in step 5 of the previous How-To, when naming items the ampersand symbol designates which letter in the menu item will be the hot key.

When designating a hot key, make sure there is not already a hot key using the same letter in the menu where the new menu item is being added. Figure 23-7 shows the results of adding the Feature Palette menu item to the Window menu. Note in particular that the letter F in Feature Palette is underlined. This is due to the ampersand in the name of the menu item.

For those not familiar with hot keys, it is just an alternative to using mouse picks. For instance, the Window menu can be accessed by holding down the Alt key and pressing the W key. The Feature Palette item could then be selected by pressing the F key. Because the menu has already been accessed, it is not necessary to hold down the Alt key when pressing F.

Assuming you have gotten this far, you are undoubtedly modifying your SolidWorks menus. Considering this fact, you would benefit from knowing how to remove menu items from the default SolidWorks menus. How-to 23-2 takes you through the process of deleting a menu item.

How-To 23-2: Deleting a Menu Item

If you have been experimenting and now find it is time to remove a menu item, or if you just deem it necessary, perform the following steps to remove an item from a menu.

1. Select the desired category from the Categories list. This must be the category where the original menu item was located, and not where the user-added menu item exists.

2. From the Commands list, select the menu item to be deleted. The Remove button can be pressed at this time to remove the selected menu item, but if the item to be deleted is a user-added menu item, continue.

3. From the *Change what menu* list, select the menu from which the menu item is to be deleted.

4. From the *Position on menu* list, select the item to be deleted.

5. Click on the Remove button.

6. Click on OK to accept your changes.

Sometimes it is difficult to remove a menu item that was added. Solid-Works will sometimes remove the original menu item instead, unless you are careful to follow the previously listed steps precisely. Even so, sometimes the original menu item is deleted. If this happens, repeat the process of deleting the added menu item from where it does not belong, and then add the original command back in at its original location.

Of course, you should have some experience with SolidWorks menus in order to know where the original menu item should be. If necessary, talk to another individual that has access to the SolidWorks software. It will never be impossible to add a menu item, because all menu items, even if deleted from the menus, will still appear in the Commands listing. How-To 23-3 takes you through the process of renaming a menu item.

HOW-TO 23-3: Renaming a Menu Item

As previously stated, you can also rename menu items. The steps that follow take you through the process of renaming a menu item.

1. Select the desired category from the Categories list.
2. From the *Change what menu* list, select the menu containing the item.
3. If the item is a top-level command, select it from the Commands list. Otherwise, from the *Position on menu* list, select the menu item to be renamed.
4. Type in a new value for the menu item's name in the *Name of command* area.
5. Click on the Rename button.
6. Click on OK to exit the window.

Not all menu items can be renamed, including those menu items added by the user. Top-level menus, such as File and Insert, cannot be renamed, not that you should want to anyway. Remember, customizing the menu structure could prove dangerous. New users beware, and tread lightly. There is no easy way to reestablish the default menu structure short of uninstalling and reinstalling the software.

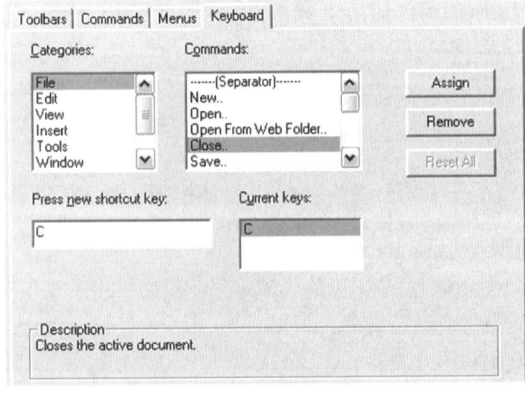

Keyboard

A fairly safe way of customizing SolidWorks is to set up your own keyboard shortcuts. If there is one command you use often, assign a keyboard shortcut to it so that you can easily call up that command. You use the Keyboard tab, shown in figure 23-8, for this purpose.

Fig. 23-8. Keyboard tab.

The Commands section will reflect whatever document type happens to be open. Therefore, if you wish to add a keyboard shortcut for use with an assembly, make sure an assembly document is open. This also holds true for parts and drawings. How-To 23-4 takes you through the process of adding a keyboard shortcut.

How-To 23-4: Adding a Keyboard Shortcut

To add a keyboard shortcut, perform the following steps.

1. Select a menu category from the Categories list.

2. From the Commands list, select a menu item to assign a keyboard shortcut to.

3. Depress a key on the keyboard that will be used as the shortcut key. You can use key combinations, such as Ctrl + A, or some other suitable key combination. If using a single key, such as the V key, you may need to depress the key more than once.

4. Click on the Assign button.

5. Click on OK when finished.

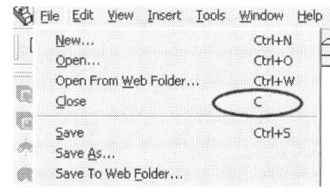

Fig. 23-9. Keyboard shortcuts are displayed in menus.

To remove a keyboard shortcut, complete the first two steps listed previously and click on the Remove button. Duplicating keyboard shortcuts used to be an issue, but this is no longer the case, as a warning will be issued if an attempt is made to add a keyboard shortcut already in use. The software simply will not allow it. After adding a keyboard shortcut, the shortcut key will be listed to the right of the menu item. An example of this is shown in figure 23-9. Note, for example, the letter C, which has been assigned to the Close command.

Transferring Hot Keys

A very common question with regard to hot keys is how to transfer hot key settings to another computer. This is a fairly simple operation, and involves a file that is created whenever a user begins customizing hot key settings. The file will use the file extension of *.cus* and will have the name of whatever the user's log-in name is for the computer worked on. For example, if the name you log onto your computer with is *Thompson*, the name of the file with your hot key settings will be *thompson.cus*.

The CUS file will reside in the user directory under the SolidWorks main program directory. To transfer the CUS hot key file from one machine to another, simply copy it to the target machine's *SolidWorks/user* directory. The only catch is that the CUS file must have the same name as your user log-in name on the machine it is transferred to, and therefore you may need to rename it.

Macros

In the world of computer software, a macro can be described as a collection of commands that can be recorded and then played back at will to simplify a task. In many programs, a macro is a collection of key strokes that can be saved and then run back to repeat a process that would otherwise be tedious and time consuming. If a macro does not save you time and effort, it is not worth creating in the first place.

In the case of SolidWorks, macros are a collection of SolidWorks API (Application Program Interface) calls. In other words, they are lines of computer code that talk to the SolidWorks program and ask it to do certain things. SolidWorks macros do not remember key strokes. If they did, it would be easy for the majority of SolidWorks users to create their own macros. Because this is not the case, macros have been reserved for those who have some basic understanding of the SolidWorks API and how it works.

There are a number of resources an individual can use to better understand the SolidWorks API. The best method most easily within your reach is to use the Help menu in SolidWorks to access SolidWorks API Help Topics. This is an excellent place to start. For more information regarding the SolidWorks API, contact the reseller where you purchased the software.

Because learning and implementing the SolidWorks API is a specialized field within itself, that area of expertise is outside the scope of this book. You will need to have experience in Visual Basic or C/C++ programming languages in order to fully understand SolidWorks macros. Assuming this is the case, *Inside SolidWorks 2003* can at least show you how to create macros.

One way of picking up quite a bit of information regarding the API is to record a macro and then edit it to better understand the internal language SolidWorks is using to carry out commands. By piecing this information together, you might be surprised how much you can pick up. The Macro toolbar is shown in figure 23-10. It has been modified a bit, so yours will look different.

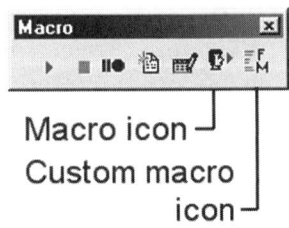

Fig. 23-10. Macro toolbar.

In the toolbar shown, the first five icons are the most important. These icons are, from left to right, Run Macro, Stop Macro, Record/Pause Macro, New Macro, and Edit Macro. The additional two icons shown were added to run custom macros that were previously recorded. Customizing the Macro toolbar is covered in upcoming material. How-To 23-5 takes you through the process of recording a macro.

How-To 23-5: Recording a Macro

To record a macro, perform the following steps.

1. Make sure you are at a point in time where you want the macro to start recording. For example, this might be prior to creating a new part, prior to starting a new sketch, and so on. This will make editing the macro easier when the time comes.

2. Click on the Record/Pause Macro button.

3. Carry out any tasks to be recorded. Perform the tasks in the correct sequence, because the order will be reflected in the macro.

4. Click on the Stop Macro icon when finished recording.

5. Type in a name for the macro.

6. Click on Save to save the macro.

At any point during the recording phase, click on the Record/Pause Macro icon to pause the recording process if necessary. Click on the same icon to start recording once again. These actions can be carried out in the previously listed step 3. Once recording has ended and the macro is saved, you will notice that the macro is saved with an *.swp* extension.

Older versions of SolidWorks created text files that had an *.swb* extension (SolidWorks Basic file). These were text files, unlike modern-day macro files. To edit a macro file, you must open the file in an editor capable of reading it. SolidWorks includes the Visual Basic for Applications (VBA) editor for this purpose. How-to 23-6 takes you through the process of editing a macro.

How-To 23-6: Editing a Macro

Once the macro has been recorded, you will almost certainly need to edit it in some way. To edit a macro, perform the following steps.

1. Click on the Edit Macro icon.
2. Select the macro to be edited.
3. Click on Open. This will open the VBA editor, shown in figure 23-11.

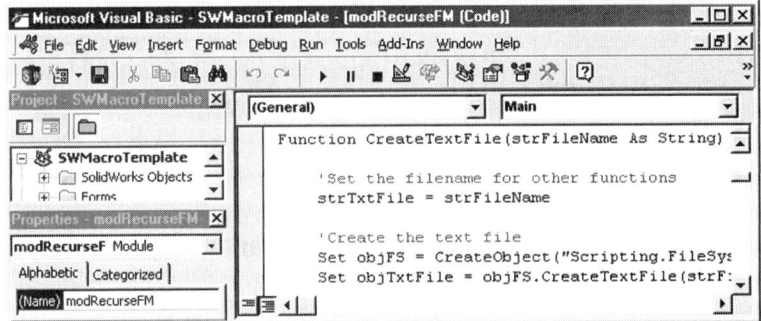

Fig. 23-11. VBA editor.

4. Edit the macro as necessary.
5. Select File > Close and Return to SolidWorks to save the macro when finished.

If an old macro with an .*swb* file extension is opened, it will automatically be saved with an .*swp* file extension. As mentioned earlier, *Inside SolidWorks 2003* will not go into the finer aspects of the SolidWorks API, nor will we delve into how to use the VBA editor, as these topics are outside the scope of this book. The VBA editor is not a part of SolidWorks per se, but is licensed by SolidWorks from Microsoft Corporation and installed during a SolidWorks software installation.

For the sake of argument, let's assume you now have a fully functional macro designed to perform some sort of function that will save time and increase productivity. Alternatively, perhaps it does not increase productivity at all, but just does something really cool. In any event, now you need to run it. That is the easy part. How-To 23-7 takes you through the process of running a macro.

How-To 23-7: Running a Macro

To run a macro, perform the following steps.

1. Click on the Run Macro icon.
2. Select the macro to be run.
3. Click on Open.

Assuming the macro being run is a functional macro without any glitches, the macro will run. You may have to interact with the macro in some way, depending on the nature of the macro. After all, a macro is really just a miniature program. It may include dialog boxes or request information from the user.

If the macro does not work as intended, it is back to the drawing board, or at least the VBA (macro) editor. In regard to How-To 23-8, let's assume the macro works. The Commands tab of the Customize window can be used to assign a macro to an icon. How-To 23-8 takes you through the process of assigning a macro to an icon.

How-To 23-8: Assigning a Macro to an Icon

Macros can only be assigned to certain icons reserved just for that purpose. Those icons are found in the Commands tab of the Customize window, shown in figure 23-12. Actually, there is only one icon, but it can be used over and over. Custom bitmaps (images) can also be assigned to macro icons. To create a macro icon and assign a macro to it, perform the following steps.

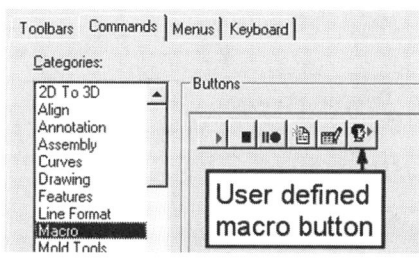

Fig. 23-12. Icon used for customizing the Macro toolbar.

1. Select Customize from the Tools menu.
2. Click on the Commands tab.
3. Drag the icon shown in figure 23-12 to the Macro toolbar (or any other toolbar of your choice). The Customize Macro Button window will appear as shown in figure 23-13.

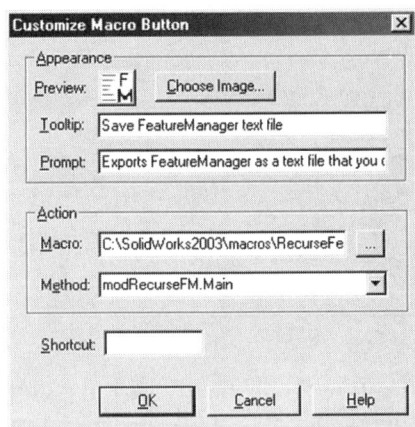

Fig. 23-13. Customize Macro Button window.

4. Optionally, select a bitmap file to use as the image on the icon (macro button). Images must be in the Windows Bitmap file format with a *.bmp* file extension.

5. Optionally add a tooltip. This is the yellow box that appears when holding the cursor over the icon.

6. Optionally add a prompt. This is the text that appears in the status bar when the cursor is held over the icon.

7. Select a macro to run when the icon is clicked. Use the button adjacent to the Macro listing to browse for a macro, if necessary.

8. Optionally select a method to run from your macro.

9. Optionally establish a hot key to run your macro.

10. Click on OK to finish creating your custom macro button.

Clicking on the corresponding icon on the macro toolbar (or whatever other toolbar you may have added the icon to) will now run the macro. If you added a hot key when defining the new macro button, you will also be able to use that hot key to run the macro.

It is not necessary to assign a custom bitmap to a macro icon, but macro icons will be difficult to tell apart if you do not. Some icons are included with the software. Look in the *SolidWorks\data\user macro icons* folder for predefined icons, or feel free to create your own.

⇨ **NOTE:** *This is just a note for those that have used previous versions of SolidWorks. Assigning hot keys was done through the Macro tab of the Customize window. The Macro tab no longer exists. All of the previous functionality associated with the Macro tab can now be implemented via the Commands tab. This includes adding hot keys. Review How-To 23-8 for the information you are looking for.*

Feature Palette

In Chapter 8 you learned a good deal about how the Feature palette worked and how palette-forming tools and library features could be created. In addition to creating files of these types, it is also possible to customize the folders that contain these files. The subcategories found in the Feature palette are highly customizable. Basically, if you understand how to create directories (folders) in Windows, you can customize the Feature palette.

There are four main folders in the Feature palette. This fact you also learned in Chapter 8. There can be more folders shown, but that would be a byproduct of telling SolidWorks where to look for specific palette objects. For example, a number of directories could be specified as directories containing palette features. There would then be a palette features folder listed for every referenced directory location.

No more than the four main types of palette folders are available at this time, nor can any other main folder categories be created, as this is a function of the software. The four main categories are Palette Parts, Palette Assemblies, Palette Features (library features), and Palette Forming Tools. However, below these four main categories there can exist any number of branches. Figure 23-14 shows an example of a portion of Windows Explorer and the directory structure containing these "branches."

Fig. 23-14. Windows Explorer showing Palette Part branches.

As stated previously, all it takes to create additional folders in the Feature palette is the know-how necessary to create folders in Windows. Note the folders under *Palette Parts* in figure 23-14. The subdirectories are named *Hardware, Inserts, Knobs, Pins,* and *Transformers*. Now compare that listing with the folders shown in the Palette Parts listing of the Feature palette, shown in figure 23-15. As you can see, the names match precisely.

Fig. 23-15. Folders in Feature palette match those in Windows Explorer.

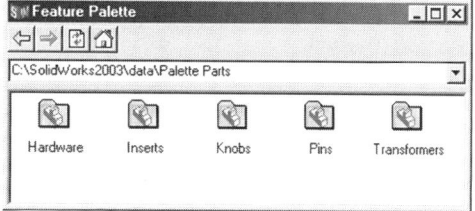

If a folder is created or if items are added to folders (such as a part file to the *Palette Parts* folder) and the Feature palette is open, the Feature palette will need to be reloaded. The Reload icon located on the Feature Palette window is specifically designed for this task. Clicking on it will reload any new folders or palette items added to the Feature palette, and the new items will then be displayed in the Feature Palette window.

Another aspect of customizing the Feature palette has to do with adding search paths of where SolidWorks looks for various palette items. This task was covered in Chapter 8 (see How-To 8-16). In essence, adding location paths allows SolidWorks to look in multiple locations for palette items. Figure 23-16 shows an example of this. Two extra paths were added, which point to C:*Purchased Parts* and C:*Purchased Assemblies*.

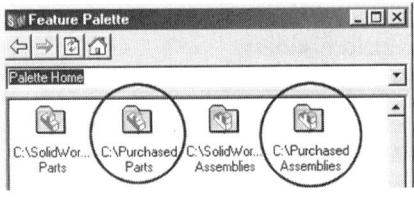

Of course, paths do not necessarily have to reside on your local hard drive. They can point to anywhere, including over the network. Folder names can be anything you wish, as long as it falls within the naming conventions established by the Windows operating system.

Fig. 23-16. After adding paths for the Feature palette.

New SolidWorks Document Tabs

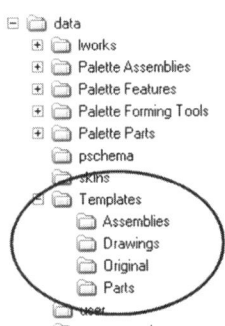

Adding new tabs to the New SolidWorks Document window is very similar to adding folders to the Feature palette. The directory names are different, but the theory is the same. There is a folder named *Templates* in the *SolidWorks\data* directory. This is shown in figure 23-17. Note that four new folders have been added under the *Templates* folder.

Once the new folders have been created via Windows Explorer, template files can be moved into these folders. Likewise, new templates can be created and saved at these new locations. Once that has been done, tabs will appear in the New SolidWorks Document window, a portion of which is shown in figure 23-18. The tabs will have the same names as the folders created below the *Templates* folder, as you can see from the image.

Fig. 23-17. Folder structure under Templates.

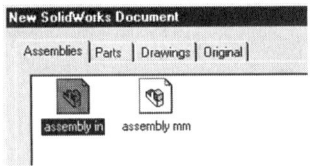

Fig. 23-18. New tabs have been added.

✏ **NOTE:** *If at least one template file is not added to a new folder, the folder will not appear as a tab in the New SolidWorks Document window.*

There is another way to get a tab to appear in the New SolidWorks Document window, which has to do with adding a new file location where SolidWorks will search for templates. To accomplish this task, see How-To 8-16.

Gradient or Image Backgrounds

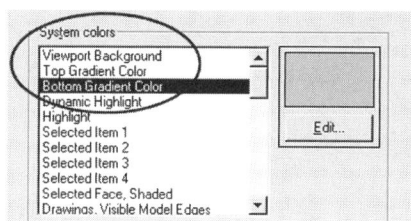

Fig. 23-19. Changing the work area background.

SolidWorks' default background is white. If you are tired of this background color, it is possible to change it. The background can even be made to appear as a color gradient. The settings for changing these colors, located in the Colors section of System Options (Tools > Options), are shown in figure 23-19.

Many CAD users migrating to SolidWorks from other programs often ask about changing the background color. The common request is to change the color to black, but it is highly recommended that black or dark backgrounds be avoided. Changing the background to a dark color makes it difficult to see the various color codes SolidWorks employs.

If you must change the background color, a soft pastel color or light eggshell is recommended. If the gradient function is used, try to avoid using dark colors for both top and bottom gradient colors. You may notice a loss in performance when rotating models on the screen when using gradient backgrounds. It really depends on the graphics card and computer you are running SolidWorks on. Experiment to see if a gradient background degrades display performance on your particular computer. How-To 23-9 takes you through the process of establishing a gradient background, or changing the background color in general.

How-To 23-9: Altering Background Colors

To change the background color of a part or assembly, or to specify a gradient background, perform the following steps.

1. Select Options from the Tools menu.
2. Select the Colors section in the System Options tab.
3. In the listing titled *System colors*, select Viewport Background > Top Gradient Color or Bottom Gradient Color, depending on what it is you wish to change.
4. Click on the Edit button and select the desired color.
5. Click on OK when finished selecting a color.
6. Repeat steps 3 through 5 for other color settings as necessary.

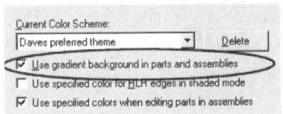

Fig. 23-20. Enabling a gradient background.

7. Make sure the option *Use gradient background in parts and assemblies* is checked. This option is shown in figure 23-20.

8. Click on OK.

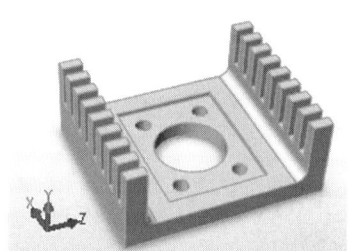

In figure 23-21, the color for the Top Gradient Color option was set to white, and the Bottom Gradient Color option was set to a fairly light shade of a grayish-blue color. This makes for a very eye-pleasing background. It removes the harshness of a pure white background while adding an illusion of depth to the models on screen.

Fig. 23-21. After adding a gradient background to the work area.

Using Background Images

Although it is possible to use background images for parts or assemblies, it is not what one might call a "productivity enhancer." There are not many technical reasons it would be necessary to insert an image behind a model. If looking for a backdrop to a rendered image, there are much better alternatives that are present in the PhotoWorks add-on software.

Background images are often more well suited to Windows desktop backgrounds, but if you really want to insert an image into your SolidWorks model file, the process is an easy one. Select Picture from the Insert menu, and then select the picture you wish to use as your SolidWorks background. The picture will not be used for every part or assembly, but only the one in which the image was inserted.

SolidWorks will only accept TIFF files (.*tif* file extension) as the file format used for a background image. The image will scale with the size of the window, and you will have no control over this behavior. Background images are embedded objects that are saved with the file. In other words, the original image will no longer be needed in order to display the SolidWorks document with the image background.

Figure 23-22 shows an example of using an image as a background to a part file. The image in this case is a scenic background, complete with Ponderosa pines and a country trail winding through the hills. There is something inher-

Fig. 23-22. Using a background image in a part file.

Gradient or Image Backgrounds

ently odd about seeing a SolidWorks model floating above the landscape. Call it an acquired taste.

If you have added a background image and have decided you want to turn the image off, uncheck Picture in the View > Display menu. Checking the same option brings the picture back. To replace the picture being used with another, select Replace from the View > Modify > Picture menu. To simply delete the image, select Delete from the View > Modify > Picture menu.

Manager Backgrounds and Schemes

Not only can the background be changed for the work area, it can be changed for all of the "managers." This includes FeatureManager and PropertyManager. ConfigurationManager will take on whatever background is specified for FeatureManager, but the background for PropertyManager can be controlled independently. How-To 23-10 takes you through the process for changing the manager backgrounds and saving these settings as a theme.

How-To 23-10: Controlling Manager Backgrounds

FeatureManager can have different colors (albeit a limited selection) assigned to its background, and PropertyManager can have its appearance changed through the use of bitmap images. Once these backgrounds have been established, it is then possible to save the settings as a theme. To accomplish these tasks, perform the following steps.

1. Select Options from the Tools menu.
2. Select the Colors section of the System Options tab.
3. Select a color for FeatureManager from the FeatureManager Color drop-down list, as shown in figure 23-23.
4. Select a background to use for PropertyManager from the PropertyManager Skin drop-down list, also shown in figure 23-23.
5. Click on the Save As Scheme button if you wish to save the changes made to the manager background settings. Note that saving a scheme can also save changes made to system colors.

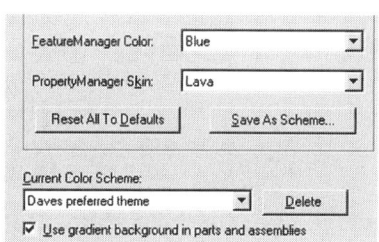

Fig. 23-23. Setting backgrounds for the managers.

6. Type in a name for the scheme to be saved, and then click on OK.

7. Click on OK again to close out of the System Options window.

If you would rather have PropertyManager use the same background as FeatureManager, select None from the PropertyManager Skin drop-down list. To force FeatureManager to use the same color background as specified for the Windows operating system, select Windows from the FeatureManager Color drop-down list.

PropertyManager skins can be created by those with artistic inclinations. Image files can be created and used as backgrounds (skins) for PropertyManager, but they must be in Windows Bitmap format (*.bmp* file extension). Skins should be placed in the *data\skins* folder under the *SolidWorks* main installation folder. Color depth is typically 256 colors, but can be 16 million with no ill effects. Tiled images work best, but are not required.

PropertyManager buttons can also be customized. It is not terribly difficult to take a look inside the *SolidWorks\data\skins* folder to see what is already available, and then to adapt some of the existing button bitmap files to suit your needs. The button backgrounds should have an RGB value of 255-0-255 in order to have a transparent background.

If you do not understand what the preceding two paragraphs are talking about, you should not attempt to create your own PropertyManager skins or custom buttons without performing further research into the matter. This type of customization is not for everyone. If you feel up to the task, the SolidWorks Help file has additional information that will help. Otherwise, further elaboration on this topic is outside the scope of this book.

Lighting

When speaking of customizing SolidWorks, the things that come to mind are menus, toolbars, and the interface in general. Maybe the lighting characteristics of a part or assembly really do not fit into this train of thought, but it seems like a good enough place to talk about this subject. Changing the lighting characteristics of a model is, after all, a form of customizing the appearance of the model.

There are a number of ways in which the lighting can be controlled. Individual light sources can be created and adapted to any part. Point light sources can be created, spot lights can be used to showcase a model, and directional lights can be made that simulate light sources infinitely far away.

Characteristics of a shaded model can also be adjusted. How a model reacts when a light source interacts with it can be controlled to some extent, such as how much light passes through the model (transparency) or the reflective properties of the model (specularity). All of these topics are covered in this section. First, however, you will explore the various characteristics that can be attributed to a model and to some light sources.

Advanced Shading Properties

You have already learned one method of controlling the color of a part. One such method is to use the Edit Color icon, discussed in Chapter 8. Another method is to access the Colors section of Document Properties (Tools > Options). This section is different than the Colors section of System Options, but has the same name, and is accessed via the same menu picks. Colors changed in System Options affect all documents globally, and change different settings than those in the Colors section of Document Properties.

In the Colors section of Document Properties, additional options that go beyond the simple ability to control the color of the part will become available. These properties are accessed through the Advanced button, shown in figure 23-24.

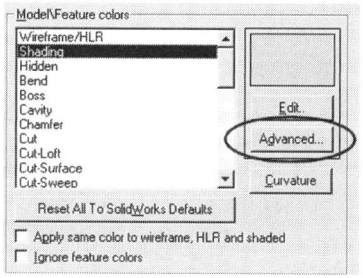

Fig. 23-24. Advanced button in the Colors section of Document Properties.

Clicking on the Advanced button opens up the Advanced Properties window, containing slider bars that can be adjusted to control properties of the shaded model. The window used to control the lighting characteristics is shown in figure 23-25. The six characteristics that can be modified are not necessarily self-explanatory. Therefore, some further elaboration is in order. The sections that follow describe these various advanced properties.

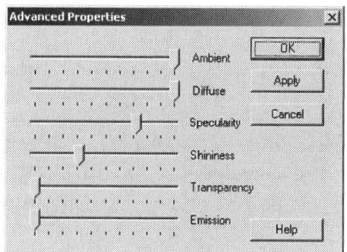

Fig. 23-25. Modifying the advanced properties of a shaded part.

Ambient

Ambient light is the light scattered throughout the room. Any light reflected from other objects that adds to the overall brightness of the scene is considered ambient light.

Diffuse

When light hits an object, a certain amount of light is reflected by the surface of the object equally in all directions. This is known as diffuse light.

Specularity

A surface's specularity is its ability to reflect light. Increasing the Specularity setting will increase the amount of light shining from a "hot spot" directly toward the user. This is similar to the shininess property (see following section), but where specularity changes the brightness of a hot spot shininess will change its size.

Shininess

If a part is very shiny, it will reflect light in a more concentrated beam. Increasing the Shininess setting will make the reflective hot spot smaller. Decreasing the Shininess setting all the way has the effect of making the part bright and washed-out looking because the reflective hot spot is no longer isolated to a narrow beam pointing toward the user.

Transparency

Transparency is the one setting that everybody understands. Technically, it is the amount of light that can pass through an object. If the Transparency setting is turned up all the way, the part is invisible because it is passing 100% of the light being directed toward it.

Emission

Light being emitted from a part is essentially light being generated and projected outward from a surface. This setting is the least commonly used of all six settings. The Emission setting has its place in the grand scheme of things, but is not typically used on a day-to-day basis for most parts.

Light Sources

There are three light sources that can be added by the user. They are directional lights, spot lights, and point light sources. This is in addition to the ambient light source that exists in every part and assembly. The user can control certain aspects of the default ambient light source, but cannot create additional ambient light sources. There would be no point in it.

All lights are controlled through a special folder in FeatureManager named Lighting, shown in figure 23-26. The *Lighting* folder contains two default light sources: Ambient and Directional1. New sources can be added via the *Lighting* folder, and existing light sources can be modified.

Be forewarned that experimenting with lighting sources can be very entertaining and may decrease productivity! However, they can also be

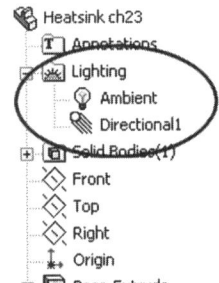

Fig. 23-26. Lighting folder in FeatureManager.

Lighting

used to create some handsome effects if you are looking to spruce up a model. The following sections show you how to create and modify the various light sources. Following this, you will learn more about the individual light sources themselves. How-To 23-11 takes you through the process of adding and editing a light source.

How-To 23-11: Adding and Editing a Light Source

To add a light source, perform the following steps.

1. Right-click on the *Lighting* folder.

2. Select Add Directional Light, Add Point Light, or Add Spot Light, as required.

3. The new light source will appear in the *Lighting* folder. Right-click on the new light source and select Properties.

4. Modify the light's properties as required. These properties are explained in detail in material to follow.

5. Click on OK to accept the changes.

After selecting the light source to be added, that particular light source automatically gets added to the *Lighting* folder. You do not have to modify the new light sources properties, but you will almost certainly want to. The difficult part is understanding all of the options available for the various light sources. This is covered in the following material, which starts with the easiest of all the light sources, the ambient light source.

Ambient

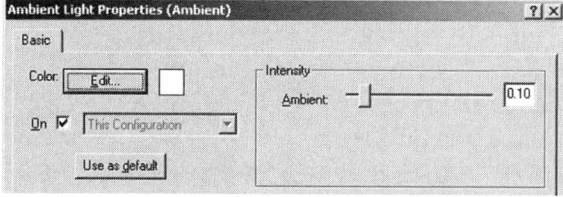

Fig. 23-27. Ambient Light Properties window with its default settings.

The one and only ambient light source is the one provided by SolidWorks. When the properties of the ambient light source are accessed, the Ambient Light Properties window appears. The upper portion of this window is shown in figure 23-27. There are not many adjustments that can be made, so this will be a short topic.

Adjusting the Intensity slider bar will increase or decrease the ambient light surrounding the model. The default setting is .10, which is a very good level for this setting. However, the setting can be adjusted anywhere from 0 to 1. Use the Edit button to control the color of the ambient light source. The On checkbox can be disabled to turn off the light source. If more than one configuration is present, a light source can be turned on or off in the configuration of your choice.

The default values for the ambient light source do not usually need adjusting. Sometimes, though, when enough additional light sources are added, the ambient source can be disabled completely (turned off). With the correct lighting, the model will still be presentable without ambient light.

The *Use as default* button is actually a holdover from earlier versions of the software. It is supposed to establish default lighting settings, but can safely be ignored. You will probably see this button disappear in a future release of the software. If it is your intention to change the default characteristics of light sources, incorporate the desired changes into your templates as necessary.

Directional

Directional lighting is also added as a default light source. Unlike ambient light sources, directional light sources can be added as required. As the name would imply, the direction of the light source can be altered to shine from any position surrounding the model.

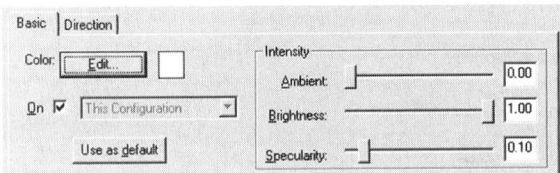

Fig. 23-28. Directional Light Properties window with its default settings.

Directional light sources are collimated, meaning that the light rays are parallel, which makes directional lights act as if they were from a very distant light source, such as the sun. The target point of a directional light source (where the light points) is always at an x-y-z location of 0,0,0. The first tab of a directional light source (titled Basic) is shown in figure 23-28.

Most of the basic properties of a directional light are identical to that of an ambient light source, but with the addition of slider bars for Brightness and Specularity settings. The Brightness setting controls the intensity of the directional light, whereas the Specularity setting controls the reflective properties of the light source. It would seem that specularity should be a property of the model, and in truth it is. The specularity of the model (see "Advanced Shading Properties") can be used in conjunction with the specularity setting of individual light sources to produce an increased effect.

Because the specularity of individual light sources (directional, point, and spot) can be controlled, it is important to pay attention to the settings

Lighting

being used for specularity for any of these light sources in the model. Otherwise, certain areas of the model will be more reflective than others. This is sometimes desirable, but not usually.

Typically one part is made of one material and exhibits the same amount of reflectiveness over the entire part. If all of the light sources have the same specularity setting, all of the reflections (hot spots) will be consistent and will have the same intensity. This will give the part a much more realistic appearance. There is an exception to this rule, however. When using colored light sources other than white, the hot spots may appear dimmer, and the specularity may need to be increased to compensate for this.

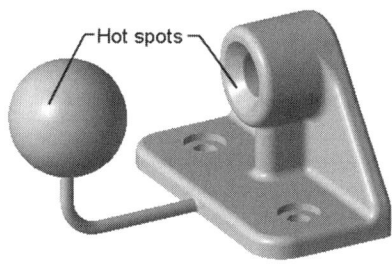

Fig. 23-29. A sample part with ambient and directional lighting. Note the reflective hot spots.

In figure 23-29, only the default ambient and directional light sources are being used. The settings for the directional light source have been modified slightly to produce a somewhat more plastic-looking part. Note the shiny areas called out by the annotation. These are the hot spots referred to on occasion in this chapter. They are the reflections caused by the light source shining from the part toward the user. The intensity of these reflections is what is controlled by the Specularity setting.

When modifying the properties of a directional light source, a graphical representation of the light source is displayed on screen. This is shown in figure 23-30. The beam's direction is also displayed. Some detail is lost in the black-and-white images, but one can still make out the second hot spot on the spherical shape of the model. This is caused by the second directional light source that is being modified in the image. Additionally, the graphical representation of the light source (or "lamp," as one might call it) takes on the color of the light being used, indicating the source's color.

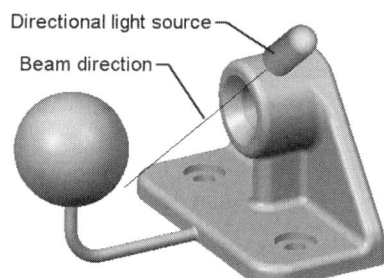

Fig. 23-30. Modifying a second directional light source.

When the second directional light shown in figure 23-30 was created, the light source added a fair amount of additional light to the model. By modifying the ambient light source, the overall brightness could easily be adjusted to accommodate the additional light source. The additional light source being used in figure 23-30 was given a rose color, as opposed to the default white light source. This adds some interesting highlights and tones to the model.

The Position tab of the Directional Light Properties window controls the direction from which the light source is emanating. This is accomplished via the slider bars (figure 23-31), or by plugging in a value in degrees for the latitude and longitude. Move the window out of the way if you have to, and then move the slider bars. You will see the "lamp" move around appropriately, and the lighting on the part will dynamically update.

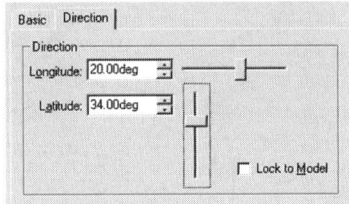

Fig. 23-31. Use the slider bars for controlling the light's position.

Lock to Model

It is sometimes difficult to understand what the Lock to Model option accomplishes until the model is rotated. When Lock to Model is checked, the light source will remain in a constant position with respect to the model. The bright, reflective areas on the model will not change. In essence, rotating the model is akin to the viewer walking around a model within a studio in which light sources have been positioned on tripods. The camera is moving, but the model is not.

When Lock to Model is not checked, the light source is still stationary, but now so is the camera. When the model is rotated, it is as if someone really is rotating the model and the viewer is standing in one spot looking directly at the model.

Whether or not the Lock to Model option is checked is dependent on your personal preferences and what effects you are trying to achieve. In the case of a light source meant to represent a light bulb in a lamp, the light should be locked to the model. If setting up light sources as though you were showcasing a model in a studio, leave Lock to Model off. With Lock to Model off, you can be the museum curator, and the model is your showpiece.

Point

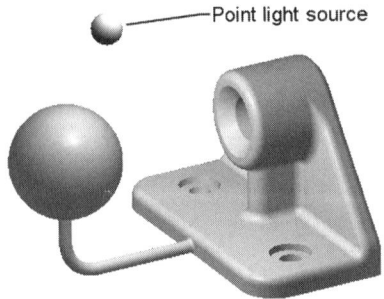

Fig. 23-32. A point light source.

Point lights act as very small pinpoint light sources that shine in all directions simultaneously. The location of the point light can be altered, but not its direction, because there is no specific direction in which the light is shining. The basic properties of a point light source are nearly identical to those of a directional light source.

A point light source is graphically represented as a sphere. This serves to remind the user that light is being emitted from the point light in all directions simultaneously. An example of a point light source is shown in figure 23-32. This graphical representation will only be shown when accessing the properties of the light source.

Lighting

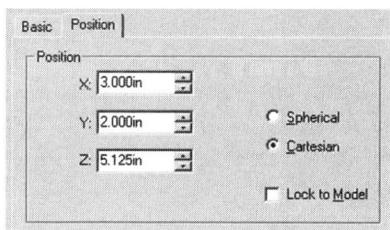

Fig. 23-33. Modifying the point light position using the Cartesian coordinates option.

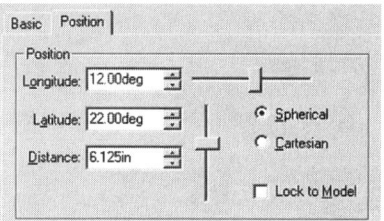

Fig. 23-34. Modifying the point light position using spherical coordinates.

Fig. 23-35. The point light source is inside the sphere.

The Position tab of the Point Light Properties window is used to locate the light source. In the case of a point light, this can be done via the Cartesian or the Spherical coordinates option. Depending on which option is used, the window will change to show the appropriate settings. Figure 23-33 shows the Position tab when using the Cartesian coordinates option.

Looking at figure 23-34, we can see what the Position tab appears like when employing spherical coordinates. The only difference between using spherical or Cartesian coordinates is how the position of the light source is specified. Slider bars available when using spherical coordinates can be more user friendly if precise *x-y-z* positioning is not critical. The option used will not affect the light source in any other way. The Lock to Model option is always present, and has the same meaning it did for directional lights (see the section "Lock to Model").

In the example shown in figure 23-35, the point light source was placed directly at the center of the spherical feature. Note that the shading of the model seems to reinforce this fact. The sphere itself, however, does not appear to be radiating any light. This is due to the way light is handled in a SolidWorks model.

Surfaces that face the point light source will be brighter, as though light is being shone upon them. This is true even for faces behind other objects (see figure 23-36). Plain and simple, if faces point toward the light source, the faces will be lit. If faces point away from the light source, they will be dark. Other than this, there are no shadows displayed in SolidWorks. This is depicted in figure 23-36.

- **NOTE:** *Regarding shadows, the Display Shadows in Shaded Mode option (View > Display) is in no way associated with light sources. To obtain photorealistic style shadows, a program such as PhotoWorks is necessary. Contact a SolidWorks reseller for additional information.*

Fig. 23-36. Light shines through objects in SolidWorks, and no shadows are displayed.

How did the point light get positioned precisely in the center of the sphere? The task was easier than you might think. The Measure tool (found in the Tools menu) was used to find the x-y-z coordinates of the center of the arc used in the sketch for the spherical feature. Although the Measure tool has not been mentioned yet in this text, it is very easy to use and should not pose a challenge.

Once the x-y-z coordinates of the center of the sphere are noted, they can be used to position the point light source. Cartesian coordinates rather than spherical coordinates were used when positioning the light source. Additionally, the Lock to Model option was enabled to lock the point light source to the model, thereby keeping it centered within the sphere even if the model were rotated.

Spot

Spot light sources are extremely useful because they offer the unique capability of being able to have not just their position set but their target as well. Another useful function of spot lights is the ability to alter the cone angle at which the light beam is being projected. The cone can be narrowed to display a tight beam of light that shines in any direction the user desires. Spot lights (and the other light sources) are represented in FeatureManager as shown in figure 23-37. The image has been enlarged for clarity.

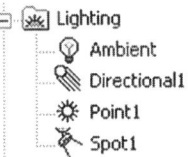

Fig. 23-37. Light sources as shown in FeatureManager.

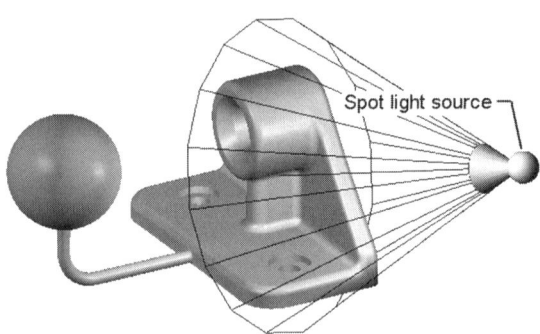

Fig. 23-38. Graphical representation of a spot light while editing its properties.

Spot lights have the nicest graphical representation of all light sources. The light source appears as a sphere, similar to a point light, and the cone of the spot light's beam is depicted as well. The beam of the spot light will fall within the area encompassed by the cone. Any geometry that falls outside the cone will not be illuminated, at least not by the spot light source in question. Figure 23-38 shows what you might see while editing the properties of a spot light.

The Basic tab of the Spot Light Properties window is exactly the same as that for a point light. The default settings are also exactly the same. The Position tab, shown in figure 23-39, is different, however.

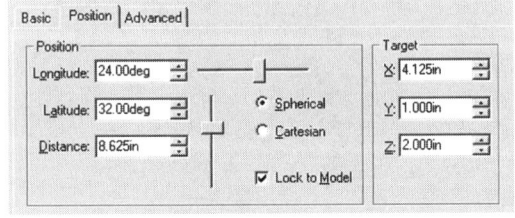

Fig. 23-39. Position tab of the Spot Light Properties window.

Lighting

The Position tab, with the exception of the additional area labeled Target, is identical to the Position tab in the Point Light Properties window. There is the ability to specify the location of the light source via either spherical or Cartesian coordinates, and there is the Lock to Model option, described previously. The Target, relative to the model's origin point, can be used to adjust the x-y-z location on which the light source is pointing. The ability to modify both the light source's location and the beam direction can make for some interesting effects.

In figure 23-40, a spot light has been positioned at the center of the hole in the cylindrical boss. The beam of the spot light has been modified to point directly at the spherical feature. The cone angle in the example has been reduced to only 10 degrees. (The default value for the cone angle is 45 degrees.)

Fig. 23-40. Positioning a spot light and directing its beam.

The location of the spot light in the previous example was made by adding a reference point in a sketch. The reference point is just a point entity that was constrained to be concentric to the circular edge of the hole. Once this was accomplished, the x-y-z location of the point could then be found using the Measure tool. The x-y-z coordinates can then be supplied to the Position tab of the spot light's properties. Make sure the Cartesian coordinates option is specified, along with the Lock to Model option. The light source must be locked to the model in this case so that it will not fall out of position if the model is rotated.

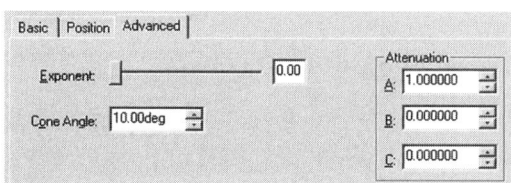

Fig. 23-41. Advanced tab of the Spot Light Properties window.

There is a third tab in the Spot Light Properties window. This third tab, Advanced, is shown in figure 23-41. With the Advanced tab, the cone angle of the light's beam can be adjusted and the "fall-off" aspect of the light can also be adjusted. The Attenuation setting is another way of controlling the brightness of the light source.

The cone angle does not need further elaboration, particularly because it is represented graphically on screen while editing the properties of the spot light. It is very easy to see where the light will shine. The Attenuation setting, as stated previously, is just another way of adjusting the brightness of the light source. Decreasing the value in the box labeled A will increase the brightness. Adding values to the box labeled B or C will serve to decrease the brightness level of the light. Valid values can be anywhere from 0 to 1.

Basically, the A setting can be thought of as a course adjustment. The B and C settings are medium and fine brightness level adjustments. This is the easiest way of looking at these settings. The actual formula used to calculate the brightness level is as follows, where D is the distance. It is much easier to simply adjust the brightness level in the Basic tab of the light source's properties.

```
Attenuation = 1 / A + (B * D) + (C * D2)
```

The Exponent setting is what controls the light level's fall-off (also known as roll-off). To put it simply, it is how sharp the edges of the light's beam are. The default value of zero makes for a very sharp beam. In figure 23-42, a spot light is being projected onto a "screen." The spot light source is positioned inside the hole in the cylindrical feature.

Fig. 23-42. A spot light with Exponent set to zero.

Note that the circle of light is choppy-looking, not defining a true circle. This has to do with how the shading is handled internally. The polygons used to make up the model either fall within the light's beam or outside it. If a larger number of polygons is used, the circle of light will appear less choppy, but only to a certain extent. This setting is controlled in the Image Quality section of Document Properties (Tools > Options).

Increasing the value of the Exponent setting will result in a circle that is smoother. The line of demarcation between light and shadow is not as sharply defined. The light falls off gradually as it moves away from the beam's axis. Figure 23-43 shows an example of this. In the image, the Exponent value has been set to .50. Because increasing the Exponent setting also decreases the size of the area of brightness, the cone angle was also increased a bit to compensate.

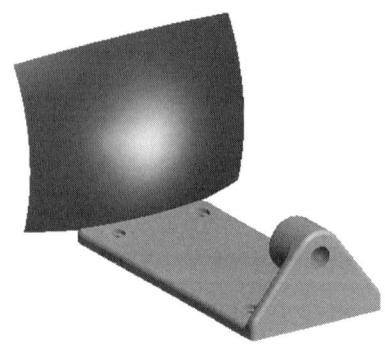

Fig. 23-43. After increasing the Exponent value.

As is evident in figure 23-43, the area between shadow and light now blends much more smoothly. This works well for getting rid of the "jaggies" when it is necessary to focus the beam on a small area. One last point is that the "screen" in the image is curved. This is necessary in order to force the software to use more polygons to create the screen.

If a simple flat face were used, SolidWorks would create one or two polygons, as opposed to dozens or hundreds. The screen would either appear bright or dark, with nothing in between. It would not be possible to focus the beam of light to form a circle in such a situation, even with the Exponent value set to 1, its highest setting.

Summary

In this chapter, a number of methods of customizing the SolidWorks software were explored. One such method was to use the actual Customize window. This allows for customizing a number of aspects of the software. It can also be used for turning toolbars on or off.

Icons can be moved to or from toolbars using the Command tab of the Customize window. Icons can be added to toolbars you may find useful that are not otherwise present on a toolbar. SolidWorks does not stop a person from placing icons on toolbars where they do not really belong, so some care should be taken when customizing the toolbars.

Menus can be customized to a large extent, but this is an area in which you should tread very carefully. It is best to get to know the software well before attempting to customize menus. Most users will never find it necessary to customize the menu structure at all.

Hot keys can be established using the Keyboard tab of the Customize window. This allows for adding shortcut keys to make certain tasks more easily accessible. The software will not allow you to add a shortcut key that has already been assigned to another task. Macros can be created in SolidWorks, but they require some basic understanding of software programming, and a knowledge of the SolidWorks API.

The Feature palette can be customized to the extent that different folders can be added to it. These folders correspond to the various categories one would like to use to store the parts and library feature part files. Any name can be given to these folders. Additionally, SolidWorks can be made to look at various file locations when searching for palette items. This is accomplished by adding file locations in the System Options window. Template file locations can be specified in the same manner.

An easy way to add template file tabs to the New SolidWorks Document window is to add new directories (folders) under the *Template* directory. As long as at least one template file exists in the new directory, the directory name will appear as a tab in the New SolidWorks Document window.

Changing the background color of the work area can be accomplished quite easily, but try to avoid using dark colors, as they will make it difficult to distinguish some color codes used by SolidWorks. Background colors can be chosen for FeatureManager, and bitmap files can be used as backgrounds for PropertyManager. Various schemes that employ these different backgrounds can also be saved by the user.

A number of light sources can be created in a part or assembly file. These include directional, point, and spot light sources. This can increase the realistic look of the model. The shading characteristics of the model and how it reacts to light sources can also be controlled. These include characteristics such as transparency and shininess. All in all, adding and modifying light sources can enhance the look of your model, and can be fun. Experimenting with light sources can be one of the more entertaining aspects of SolidWorks.

Questions and Topics for Discussion

1. Is it possible to add Drawing toolbar icons to the Assembly toolbar? Explain.
2. Is it possible to add the same icon to a toolbar more than once?
3. Is it possible to add the same hot key to more than one command?
4. How can customized keyboard shortcuts be transferred from one computer to another?
5. Describe one method of adding custom folders to the Feature palette.
6. Describe one method of adding a new tab to the New SolidWorks Document window.
7. Where are the commands (or options) located for establishing a gradient color background in the work area?
8. Can FeatureManager be made to use the same color background as that used by the operating system? Explain your answer.
9. How can PropertyManager and FeatureManager be made to use the same color background?
10. Name the three types of light sources that can be created in SolidWorks.
11. What are two advanced characteristics of a shaded model that can be changed to make the model appear more shiny?
12. Does SolidWorks lighting calculate shadows? Explain your answer.
13. Which light source allows for controlling both the position of the light source and the conical angle of the beam?

Index

A
Add Corner-Trim option, 294–295
Add Relations command, 49, 131, 345
Add Relations panel, 50
Align to seed option, 266
Align with End Faces option, 348–349
aligning
   Aligned Section View icon, 436
   Area Hatch/Fill window, 439
   changing crosshatch, 437–440
   conditions, 512
   creating an aligned section view, 436
   improper alignment, 538
   notes, 472
   Part option, 439
   Remove crosshatch option, 439
ambient light source, 835–836
angle mate, 510
Angle option, grayed out, 276
angled rectangles, 66
angle-distance chamfers, 193
annotations. *See* notes.
application-specific requirements, 390
applied features, 109
applied features vs. sketched, 21–22
Apply to all option, 473
arcs
   3 point, 64–65
   centerpoint, 65
   dimensioning, 160–161
   displaying centerpoints, 80
   tangent, 60, 64
area hatch, adding, 440
Area Hatch/Fill window, 439
arrows
   Arrows/Leaders panel, 472–473
   dimension arrows, 157, 159
   in the Dimension Property Manager, 159–160
   section line arrows, 434
Arrows/Leaders panel, 472–473

assemblies
   balloons, 572–575
   Bill of Materials, 559–566
   collision detection, 545–547
   component editing, 539–544
   component patterns, 519–523
   component replacement, 533–535
   component suppression states, 524–528
   definition, 497–498
   dimensions in design tables, 659–660
   drawings, 554–559
   dynamic clearance, 547–548
   exploded views, 567–572
   features, 551–554
   hiding components, 523–524
   inserting components, 499–502
   interference checking, 544–545
   mate troubleshooting and repair, 536–539
   mating relationships, 505–517
   moving and rotating components, 502–504
   moving components in specific directions, 504
   palette parts and palette assemblies, 517–519
   pattern types, 554
   performance, 390
   physical dynamics, 548–549
   reloading components, 535–536
   saving as part files, 623–626
   section views, 555–556
   simulation, 549–551
   starting a new assembly, 498–499
   working with subassemblies, 528–533
assembly configurations
   adding, 629–630
   bill of material properties, 632–633
   color, 631–632
   component configuration, 633–639

   in drawings, 639–641
   hiding component models, 632
   properties, 630–633
   showing alternate components, 639
   suppressing component models, 632
   suppressing features and mates, 632
assembly drawings
   creating, 554
   hiding components, 556–557
   section views, 555–556
   show hidden edges, 557–558
Assembly FeatureManager, 498
assembly features
   creating, 553
   description, 551–552
   editing, 552–553
   patterns, 554
Assembly Instance Color window, 631–632
assembly modeling
   copying drawings with models, 704–708, 716–717
   creating subassemblies, 701–702
   dissolving subassemblies, 700–701
   editing read-only documents, 712–714
   envelopes, 718–723
   multi-user environments, 704–718
   opening files, 709–710
   overwriting files, 710
   performance, 694–700
   read-only attributes, 714–715
   reloading read-only components, 712
   reloading read-only documents, 711–712
   reorganizing components, 702–703
   restructuring, 700–702
   saving as a copy, 715
   search order, 717–718
   search paths, 717–718
   selecting components. *See* envelopes.

Assembly transparency for in context edit, 697
associativity, 617–619, 673–674
At Angle button, 137
Auto fillet option, 285
auto relief, 291–292
Autodimension Sketch icon, 91–92
Autodimension Sketch panel, 91–92
autodimensioning, 91–92
Auto-show PropertyManager option, 27
auto-transitioning, 60
Auxiliary View icon, 428
auxiliary views, 428
Axis icon, 254

B

balloons, 572–575
base features, 110
Base Flange command, 287–288
Base Flange feature, 292
base flanges, creating, 288
Base Flange/Tab icon, 302
Base-Loft panel, 206
bend allowance, 289–291, 298–299, 334
Bend Allowance option, 289
Bend from Virtual Sharp option, 297
bend tables, 290–291
bending sheet metal. *See* sheet metal.
bend-outside flange position, 297–298
Bends icon, 333
Bends panel, 333
bill of material properties, 632–633
Bill of Materials. *See* BOM.
Bill of Materials Properties window options
 adding new items, 562
 note font, 560
 recreation of a SolidWorks BOM, 562–563
 setting an anchor point, 561
 show assemblies and parts in an indented list option, 561
 show parts only option, 560
 show top level subassemblies and parts only option, 560
 summary information title, 560
 use table anchor point option, 561
blank design tables, 649–650
Blank option, 649
blending closed profiles, 203
Blind end condition, 111, 119–120
blocks, 484–488
BOM (Bill of Materials)
 adding custom properties, 563–564
 and custom properties, 675
 editing, 563–564
 inserting, 559–560
 mapping from design tables, 651–653, 663
 overriding a part number, 566
 properties, 565–566
border panel on notes, 475
Break command, 307–308
Break Corner command, 307–308
Break Corner icon, 308
Break Corner panel, 308
break gap, 446
break line extension, 446
breaking corners, 307–308
broken views, 445–446
Broken-out Section icon, 441
Broken-out Section panel, 441
broken-out section views, 441

C

callouts, 87–88
Cap Ends option, 287
castings
 applied features, 109
 cavities, 581–582, 591–592, 605–606
 cores, 591–592, 607
 creating features, 109-114
 creating planes, 116–147
 defining the sketch, 108
 derived parts, 587–589
 determine the best profile, 108
 determining best profile, 108
 dies, 580–581
 editing techniques. *See* editing.
 fully define sketch, 108
 InPlace mates, 601–602
 inserting parts, 582–587, 601
 knitting surfaces, 603–605
 lofted surfaces, 598–600
 mate errors, 591
 Mirroring Sketch Geometry, 116
 mirroring sketch geometry, 129–134
 mold assembly, 600
 molds, 585–586
 Multiple Bodies, 116
 multiple bodies, 134–137
 open-profile cutting, 586–587
 Pivot Arm, 112–116
 pivot arm project, 112–115, 122–126, 132–134
 radiated surfaces, 596–597
 referencing configurations, 585
 scaling, 592–593
 select the appropriate plane, 108
 shutoff surfaces, 593–596
 splitting parts, 589–591
 stretched features, 109
 types of features, 109
cavities, 581–582, 591–592, 605–606
Cavity command, 581–582
center marks, 463–465
Centerline icon, 71
Centerline option, 210–211
centerlines
 adding diameter dimensions, 154
 creating features around centerline, 109
 creating revolved features, 150
 definition, 71
 deleting, 130
 rules crossing/touching, 151–152
 using more than one centerline, 150–151
Centerpoint Arc icon, 65
centerpoint arcs, 65
centerpoint ellipses, 69–70
centerpoints, 128
centroids, 260, 385
Chamfer panel, 193
chamfers, 193–195
Check Sketch For Feature function, 82–83
child feature, 395–396
Circle icon, 65
Circular Pattern icon, 251
Circular Pattern panel, 249
circular patterns, 250–252
circular references, 690–691
Circular Sketch Step and Repeat command, 277–278
Circular Sketch Step and Repeat icon, 277
Clockwise option, 355
Closed Corner command, 308–309
Closed Corner icon, 309
Closed Corner panel, 309
Closed hems, 306–307
closed-profile thin features, 286–287
closing corners, 308–309
coincident mate, 505–507
coincident symbol, 58
Collect All Bends function, 314–315
collision detection, 545–546
Collision Detection option, 546
color
 assembly configurations, 631–632
 modifying part color, 328–329

settings of draft analysis, 229
specifying in design tables, 651–652
Color Advanced Properties, 544
color coding
   black sketches, 43
   blue sketches, 43
   brown sketches, 43
   open areas, 43
   planes, 33–35
   red open areas, 323
   red sketches, 43
   sketch planes, 43–44
   table of codes, 43
   text features, 765
   yellow sketches, 43
Color Mapping tab, 798
Combine panel, 136
comments, 651–652, 663, 688
component models, suppressing, 632
component suppression states, 524–528
components
   configuration, 633–639
   configuration, specifying, 662–663
   definition, 497–498
   derived pattern, 522–523
   description, 530–531
   design tables, 660–662
   disassociating joined, 623
   editing, 540–544
   first component, 500
   hiding, 523–524
   inserting, 499–502
   joining, 619–623
   moving, 502–504
   names, 530
   opening, 540
   patterns, 519–523
   read-only, reloading, 712
   reloading, 535–536, 712
   reorganizing, 702–703
      replacement, 533–536
   resolving, 527–528
   rotating, 502–504
   selecting. *See* envelopes.
   showing alternates, 639
   showing/hiding, 662
   suppressing, 524–661
   transparency, 543–544
Composite Curve command, 359, 598
Composite Curve icon, 360
Composite Curve panel, 360
composite curves
   Composite Curve command, 359
   Composite Curve icon, 360

Composite Curve panel, 360
   creating, 360–363
concentric mates, 505–507
ConfigurationManager, 391–395
ConfigurationManager symbols, 655
configurations, 391–392, 648–651
configuration-specific custom properties, 672
Constrain angle between axes option, 275
constraining corners, 164–165
constraints. *See also* geometric relations.
   adding, 110
   callouts, 87–88
   definition, 86–87
   deleting, 88–89
   description, 49–50
   midpoint, 127
   table of, 51–52
context-sensitive menu, 31–32
Contour Select tool, 81–82
Convert Entities command, 168
Convert Entities icon, 168
Coordinate System icon, 261
coordinate systems, 261–263
Copy icon, 238
copying
   bodies, 757–759
   configurations, 394–395
   drawings with models, 716–717
   exploded views, 570–571
   features. *See* patterns.
   Files, 704–708
   and pasting features, 238–240
   between views, 458
   to Windows clipboard, 394
cores, 591–592, 607
Corner treatment option, 294
corners, sheet metal
   breaking, 307–308
   closing, 308–309
   relief cuts, 294
   trim, 294–295
Cosmetic Thread icon, 477
Cosmetic Thread option, 270–271
Cosmetic Thread window, 478
   cosmetic threads, 476–478
Counterclockwise option, 355
Create solid surface from enclosed volume option, 755
cropped views, 446–447
Curvature continuous option, 740
Curve Driven Pattern icon, 265
Curve Driven Pattern panel, 265

Curve Through Free Points command, 378–379
Curve Through Reference Point command, 36–379
curve-driven patterns, 264–267
curves through points
   curve files, 379–380
   Curve Through Free Points command, 36–379
   Curve Through Reference Point command, 378
   picking points on the screen, 378
   reference points, 378
   through free points, 378
   typing in xyz coordinates, 378
Custom Bend Allowance option, 298–299
custom properties. *See also* properties.
   associative, 673–674
   and bills of materials, 675
   configuration-specific, 672
   creating, 668–670
   description, 667–668
   in design tables, 672–673
   linking to, 670–671
   modifying, 671–672
Custom Relief Type option, 298–299
Customize window, 814–822
   Commands tab, 815–816
   Keyboard tab, 820—821
   Menu tab, 817–820
   toolbars, 814–815, 816
customizing SolidWorks, 813–844
   adding tabs to New SolidWorks Document window, 828
   background colors, 829–830
   background images, 830–831
   BOM template, 564–565
   controlling Manager backgrounds 831–832
   Customize window, 814–822
   Feature palette, 826–828
   hot keys, transferring, 821–822
   keyboard, 820–821
   lighting, 832–842
   macros, 822–826
   menus, 817–820
   resetting toolbars and settings, 816–817
   shading properties, 834
Cut icon, 238
cutting with surfaces, 759–760

D

dangling dimensions, 44, 238–239
datum features, 481–482
Datum Features Symbol icon, 481–482
Datum Target icon, 482
datum targets, 482
Define Layers tab, 797
definitions, editing, 118–119
deformations. *See* forming tools.
deforming faces, 729–731
Delete icon, 59
density, defining for a part, 383–384
Derived Component Part command, 587–589
derived component pattern, 522–523
derived configurations, 651
derived parts, 587–589
design drawings
    adding sheets, 420–422
    annotations, 470–488
    of assemblies, 554–559
    assembly configurations, 639–641
    blocks, 484–488
    center marks, 463–465
    creating. *See* new drawings.
    dimensioning, 452–463
    exploded views, 571
    file associatively, 408
    layers, 465–469
    line formatting, 469–470
    properties of multiple objects, 469
    RapidDraft files, 488–491
    section views, 432
    tabulated, 491–492
    templates, 416–422
design history. *See* rolling back.
design table files, linking to, 654–655
Design Table panel, 647
design tables
    adding rows and columns, 653–654
    assembly dimensions, 659–660
    blank, 649
    comments and notes, 653, 663
    component configuration, specifying, 662–663
    configuration control, 651
    configuration names, 648–649
    ConfigurationManager symbols, 655
    creating, 648–650
    custom properties, 668–675
    deleting configurations, 657–658
    derived configurations, 651
    descriptive dimension names, 647
    dimension names, 646–647
    edit control, 653
    editing, 656–657
    file extension names, 674
    formulas, 648
    headings for parts, 650–653
    legacy tables, 657
    linking to design table files, 654–655
    linking vs. embedding, 645–646
    mapping to bill of materials, 651–652, 663
    mate dimensions, 659–660
    model color, 651–652
    nested configurations, 652
    parameter insertion failure, 657
    part vs. assembly, 658–659
    renaming dimensions, 647, 667
    sample table, 664–666
    saving, 654
    showing dimension names, 647–648
    showing in drawings, 666–667
    showing/hiding components, 662
    showing/hiding dimensions, 648
    subassembly expansion, 663–664
    suppressing, 650–653, 660–661
    updating from model changes, 653
Detail View icon, 442
Detail View panel, 443
detail views, 442–444
Deviation Analysis tool, 751
diameter dimension, creating, 154
dimension arrows, 157, 159
Dimension icon, 89
Dimension Properties window, 156
Dimension Property Manager. *See also* dimension arrows.
    adding text to dimension values, 157, 161–162
    adding tolerances, 157
    changing decimal place precision, 157
    changing dimension line's arrow position, 157
    dimension and tolerance settings, 158
    modifying tolerance values, 157
    symbols in the dimension values, 157
dimensioning
    annotations, 453
    automatic, 91–92
    dimension favorites, adding, 459–461
    extension lines, 462–463
    inserting, 452–456
    Model Items icon, 452
    moving, 457–459
    reference dimensions, turning off parentheses, 461–462
    reference geometry, 455–456
    to a tangency point, 160–162
dimensions
    adding, 89–91, 110
    dangling, 44
    diameter dimensions from centerline, 161–162
    driven by equations, 680
    dual dimension display, 162
    favorites, adding, 459–461
    linking values, 684–688
    modifying, 119–120
    modifying text size, 158
    names, 646–647
    properties, 155–156
    renaming, 327–328, 647, 667
    restricting access, 326–327
    settings, 158
    showing/hiding, 648
direction light source, 836–838
Direction Vector option, 214
Display arc centerpoints option, 80
Display entity points, 80
Display shaded planes option, 697
Display/Delete Relations command, 88–89, 602
displaying arc centerpoints, 80
displaying planes, 32–35
dissolving features, 318–319
distance mate, 509–510
distance-distance parameters, 193–194
documents, modifying properties, 39–40
Dome command, 727–729
domed features, 727–729
Dowel Pin Symbol icon, 484
draft
    example, 112–113
    face propagation options, 222–223
    linear vs. natural, 224–225
    Neutral Plane method, 220
    neutral planes, 220—221
    Parting Line method, 220
    parting lines, 218–220
    reference for ribs, 226
    Rib command, 224
    Rib tool, 223–224
    ribs to model, 223–224
    step draft, 221–222
draft analysis
    adding features to shaver housing, 229–232
    analyzing draft, 227
    draft analysis color settings, 229
    Draft Analysis tool, 226
    face classification, 227–228
    face classification designations, 228
    Face classification option, 227–228

Index

Draft Analysis tool, 226
drawing manager, 456
drawing templates. *See* new drawings.
drawings. *See* design drawings.
drilled holes, 198
dual dimensioning, 162–163
dual processors, 14–15
duplicating features. *See* patterns.
dynamic clearance, 547–548
Dynamic Clearance option, 548

E

ECNs (Engineering Change Notices), 1
Edge Flange command, 295–297
Edge Flange icon, 296
edge flanges
   creating, 296–297
   custom bend allowance, 298–299
   custom relief type, 298–299
   editing the profile, 299
   position offset, 298
   positions, 297–298
   virtual sharp, 297
Edge-Flange panel, 296
edges
   fillets and rounds, 183, 189–190
   flattening sheet metal, 189–190, 293–294
   hiding, 448–449, 557–558
   Keep Edge, 189–190
   model edges, 169
   rounded. *See* fillets and rounds.
   show hidden edges, 557–558
   showing hidden edges, 557–558
   simplifying, 293–294
   simplifying edges, 293–294
   swept features, 346–347
   Tangent edge display option, 449
   tangent edges, 449–451
   view appearance, 448–449
Edit Color icon, 329
Edit Color window, 329
edit control, design tables, 653
Edit Crop option, 447
Edit Part icon, 542
editing. *See also* modifying.
   Blind End condition, 119–120
   changing a dimension value, 119–120
   components, 539–544
   cutting with an open profile, 121
   definitions, 118
   design tables, 656–657
   dimensions, 119–120

editing a definition, 118
exploded views, 569
feature definitions, 118–119
features, 326–328
in-context components, 541–542
knowing what you are editing, 47–48
modifying dimensions, 119–120
part colors, 328–329
read-only documents, 712–714
rebuilding a model, 117
regenerating a model, 118
renaming features, 120
sketches, 116
step-and-repeat patterns, 276
table-driven patterns, 263–264
Through All end condition, 125
through all end condition, 125
Up to Surface end condition, 121
Ellipse icon, 68–69
embedding design tables vs. linking, 645–646
embossing text features, 769–770, 776–778. *See also* forming tools.
empty views, 447–448
end conditions, 111, 400
Engineering Change Notices (ECNs), 1
engraving text features, 771–773
envelopes, 718–723
Equal spacing option, 251–252
equations
   changing working units, 689–690
   circular references, 690–691
   comments, 688
   common errors, 690–692
   creating, 681–683
   Equations window buttons, 688–689
   functions, table of, 689
   Link Values command, 684–688
   linking dimension values, 684–688
   linking driven dimensions, 687
   modeling parts, 683–684
   ordering, 690–692
   preparing for, 679–680
Equations window buttons, 688–689
error messages. *See* messages.
error symbology, 44–45, 536
exiting commands, 61
expanding an exploded view, 570
exploded views
   collapsing, 570
   copying, 570–571
   in drawings, 571–572
   editing, 569

   expanding, 570
   exploding an assembly, 567–569
exponentiation, 196
exporting files
   ACIS files, 801–802
   DWG files, 793–795
   DXF files, 793–795
   eDrawings, 810
   IGES files, 799–801
   map files, 795–798
   to other file formats, 792–793
   parasolid files, 798
   proprietary formats, 805–806
   STEP files, 801
   stereolithography files, 803–805
   TIFF files, 806–810
   VDAFS files, 802
   VRML (Virtual Reality Markup Language), 802–803
Extend icon, 167
extending
   entities, 167–168
   lines, 160–161
   surfaces, 750–754
external references, 271–273
Extrude command, 109
Extruded Cut icon, 127
extrusions, 111

F

face blend fillets, 736–739
face classification designations, 228
Face classification option, 227–228
face curvature fillets, 740–741
face propagation options, 222–223
faces
   Align with End Faces option, 348–349
   classification, 227–228
   converting, 170
   deforming, 729–731
   filleting, 186–187
   holes, 200
   Merge faces option, 293
   mirroring, 268–269
   model face, 170
   projecting a sketch, 376–377
   removing, 322–323
   selecting, 110
   selecting a view, 37
   sheet metal, merging, 293
   Sketch onto Face(s) option, 377
   tangency conditions, 214–215

Feature palette
  creating, 320–322
  creating features, 319–320
  customizing, 826–828
  dissolving features, 318–319
  editing items, 326–328
  file paths, 325
  inserting features, 316–318
  modifying paths, 325–326
  open areas, 323
  removing faces, 322–323
  renaming dimensions, 327–328
  restricting dimension access, 326–327
  simplifying features, 320
  structure, 324
Feature Properties window, 398
Feature Scope panel, 135
feature-based modeling, 3–4
FeatureManager. *See also*
     PropertyManager.
  description, 25–26, 392
  Normal To function, 37
  Orientation window, 35–36
  pane control, 29–30
  plane display, 32–35
  PropertyManager, 26–29
  renaming objects, 30–31
  reordering, 743
  rolling back, 741–743
  separator bar, moving, 26
  Standard Views toolbar, 36–37
  view orientation, 35–37
  What's Wrong? function, 45
  work area, 26
features
  adding dimensions and constraints,
     110
  applied, 109
  assemblies, 551–554
  base, 110
  blind end condition, 111
  create sketch, 110
  create the feature, 110
  creating, 109–116, 319–320
  dissolving, 318–319
  editing, 326–328
  editing feature definition, 118–119
  end condition types, 111
  enter sketch mode, 110
  exceeding allowed geometry, 81–82
  extruded, 111
  inserting, 316–318
  lofting (blending), 109
  open areas, 323

  overview, 20–21
  removing faces, 322–323
  renaming dimensions, 327–328
  restricting dimension access, 326–327
  revolving, 109
  select a plane or planar face, 110
  simplifying, 320
  sketched vs. applied, 21–22
  suppressing, 524, 632
  sweeping, 109
  terminating the extrusion, 111
  tracking. *See* FeatureManager.
  types of, 109
files, manipulating, 704–711
Fill command, 594–596
Fillet command, 183–188
fillet options
  Keep Edge, 189–190
  Keep features, 188–189
  Keep Surface, 190
  overflow control, 189
  round corners, 188
filleting
  continuous curvature, 740
  face blends, 736–739
  face curvature, 740–741
  full round, 739–740
  hold lines, 737–738
  movable points, 734
  setback fillets, 734–735
  smooth transition, 732
  split lines as hold lines, 738–739
  straight transition, 732
  variable radius fillets, 732–734
fillets and rounds
  adding a fillet as a feature, 184–185
  applying to edges, 183
  control-selecting, 184
  faces, 186–187
  features, 187–188
  as features, 183–188
  Fillet command, 183–188
  fillet options, 188–190
  multiple radius fillets, adding, 185–186
  previews, 186
  as sketches, 183
  tangent propagation, 185
Fix projected length option, 305
Fixed for either direction option, 275
Fixed size weld symbols option, 480
Flat Pattern panel, 293
Flat surface icon, 594
FlatBend1 feature, 335–336

Flat-Pattern1 feature, 293–295, 313
Flat-Sketch1 feature, 335
Flatten-Bends1 feature, 334
Flattened icon, 293
flattening bends, 310–311
flattening sheet metal. *See* sheet metal.
Flip Radiate Direction option, 597
Flip side to cut option, 587
flipping, 332, 766–767. *See also*
     mirroring.
Fold command, 313–315
Fold icon, 314
follow 1st and 2nd guide curves, 352–353
Follow path option, 350–352
fonts
  document notes, 560
  settings, 450
  size, modifying, 158
  text features, 766
  for text features, 766
  Use Document Fonts option, 766
formatting text features, 767–768
forming tools
  creating, 320–322
  creating library features, 319–320
  description, 315–316
  dissolving palette features, 318–319
  inserting, 316–318
  open areas, 323
  removing faces, 322–323
  simplifying library features, 320
formulas, design tables, 648
Full outline option, 444
full round fillets, 739–740
functions, table of, 689

G

gaps, unintended, 80
GCS (geometric characteristics symbol),
     480
geometric characteristics symbol (GCS),
     480
geometric relations, 86–89, 402–403.
     See also constraints.
Geometric Tolerance icon, 480
Geometric Tolerance Options window,
     481
geometric tolerances
  adding, 480
  adding geometric tolerances, 480
  attaching symbol to dimension, 481
  GCS (geometric characteristics
     symbol), 480

Geometric Tolerance icon, 480
Geometric Tolerance Options
    window, 481
MC (material condition), 480
geometry
    dragging, 61–62
    exceeding allowed, 81–82
    overlapping, 79–80
    self-intersecting, 79
Geometry pattern option, 252–253, 267
geometry patterns, 252–253
graphics cards, 16
Grid/Snap settings, 84–85
groove patterns, 255–257
guide curves
    description, 206–207
    follow 1st and 2nd, 352–353
    sweeping, 353–354

H

hardware requirements, 15–17
headings, design tables, 650–653
helical curves
    Clockwise option, 355
    Counterclockwise option, 355
    creating, 355–357
    definition, 354–355
    Helix icon, 356
    Reverse Direction option, 355, 357
    spirals, 357
Helix icon, 356
helixes. *See* helical curves.
help. *See also* color coding; system
    feedback.
    command execution, 27
    error symbology, 44–45
    overdefined geometry, 44–45
    PropertyManager, 27
    system feedback, 48–49
    What's Wrong? function, 45
Hem icon, 306
Hem panel, 307
hems, 306–307
Hide command, 32–33
Hide component models option, 632
hide model, 531
Hide/Show Bodies panel, 761
Hide/Show Component icon, 523
hiding. *See also* showing; suppressing.
    annotations, 478
    assembly components, 523–524
    bodies, 760–761
    component models, 632

components, 523–524, 555–557, 662
cosmetic threads, 478
design tables, 662
dimensions, 458–459, 648
edges, 448–449, 557–558
planes, 33
vs. suppressing, 524
toolbars, 23
view borders, 424–425
history of design. *See* rolling back.
hold lines, 737–738
Hole Callout command, 475–476
Hole Callout icon, 475
hole callouts, 475–476
Hole wizard
    adding favorites, 200–201
    creating holes, 198–199
    holes on nonplanar faces, 200
    legacy tab, 200
    parameter settings, 200
    Simple Hole command, 201–202
    simple holes, 201
holes
    counterbored, 198
    countersunk, 198
    creating with Hole Wizard, 198–199
    drilled, 198
    on nonplanar faces, 200
    simple, 201
Home icon, 325
horizontal alignment, 426
hot keys, 98, 821–822
hypertext links on notes, 473
hyphen (-), configuration names, 647

I

image quality, and performance, 697–698
Import items into views option, 455
importing files
    ACIS SAT files, 788
    AutoCAD files, 784–785
    DLLs (Dynamic Link Libraries), 791
    DWG files, 784–785
    DXF files, 784–785
    file extensions, 783–784
    formats supported, 782–783
    generic procedure, 781–784
    IGES files, 786–787
    Parasolid files, 785–786
    proprietary formats, 790–791
    STEP files, 788
    stereolithography files, 789–790

VRML (Virtual Reality Markup
    Language), 788–789
improper alignment, 538
Include Hidden Bodies/Components
    option, 384
in-context components, editing, 541–542
Inference Detection tool, 544
InPlace mates, 601–602
Insert Part panel, 583
inserting
    3 standard views, 422–424
    blocks, 486–487
    components, 499–502
    components with SmartMates, 515–516
    cosmetic threads, 477–479
    dimensioning, 452–456
    drag and drop from file, 423
    FeatureManager symbology, 501–502
    fixed components vs. floating, 500–501
    Insert Block icon, 486–487
    Insert Block panel, 487
    from Insert Menu, 423
    instance identification numbers
        (instance ID numbers), 501–502
    methods, 499–500
    Move Component icon, 503
    moving components, 502–504
    Rotate Component icon, 503–504
    rotating components, 503–504
    views, 422
    from Window Explorer, 423
instance ID number, 530
interference checking/detection, 544–546

J

Jog icon, 305
Jog panel, 305
jogs, 304–306
joining components, 619–623
justifying text features, 767

K

Keep Edge, 189–190
Keep features, 188
Keep features option, 188–189
Keep normal constant option, 351
Keep Surface, 190
keyboard, customizing, 820–821
keyboard shortcuts. *See* hot keys.

K-factor, 289
Knit Surface icon, 603
Knit Surface panel, 603
knitting surfaces, 603–605, 754–755

L

lances. *See* forming tools.
large assembly mode, 694–696
layers
    creating, 465–467
    deleting a layer, 467
    Layer Properties icon, 466
    Layer Window, 466
    making a layer current, 467
    moving objects between layers, 467–469
    turning layers on and off, 467
leaders, flipping, 159
legacy parts, 337–338
Legacy tab, 200
lighting, 833–842
    light sources, 834–842
    shading, 833–834
lightweight components, 524–527
Line Format toolbar, 469–470
Line icon, 56
linear pattern, 554
Linear Pattern icon, 242
linear patterns, 241–249
Linear Sketch Step and Repeat command, 274–276
Linear Sketch Step and Repeat icon, 274
Linear Sketch Step and Repeat window, 274
Linear Spring panel, 550–551
linear vs. natural, 224–225
lines, 56–58
lines, split. *See* split lines.
Link to file option, 654–655
link to thickness, 311
Link to thickness option, 311
Link Values command, 684–688
linking
    to custom properties, 670–671
    design tables vs. embedding, 645
    dimension values, 684–688
    driven dimensions, 687
    properties on notes, 474–475
Load referenced documents option, 711
Locate Part panel, 584
Loft command, 109, 203
loft options, 211–212

Lofted Bend command, 310
Lofted Bend icon, 310
lofted bends, 310–311
Lofted Bends PropertyManager, 310
lofted parts
    advanced smoothing, 211
    Base-Loft panel, 206
    blending closed profiles, 203
    Centerline option, 210–211
    close loft, 211
    creating a lofted feature, 205–206
    direction, 210
    guide curves, 206–207
    Loft command, 203
    maintain tangency, 211
    Merge Result, 212
    pierce constraint, 207–210
    selecting profiles, 206
    show preview, 212
    Split Curve command, 204–205
    surfaces, 598–600, 754
    swept features, 203
    tangency conditions, 212–215
    twisting, 206
Lofted Surface, 599
Lofted Surface icon, 599
logos, adding, 415
louvers. *See* forming tools.

M

macros, 822-826
    assigning macro to an icon, 825–826
    editing a macro, 824
    Macro toolbar, 822
    recording a macro, 823
    running a macro, 825
Map Entities tab, 797
mass properties, 383–385
Mass Properties command, 380
Mate panel, 507–508
Mate Reference panel, 517
Mated Entities panel, 534–535
material condition (MC), 480
material-inside flange position, 297–298
material-outside flange position, 297–298
mates
    angle, 510
    coincident, 505–507, 509
    concentric, 505–508
    dimensions, 659–660
    distance, 509–510
    errors, 591

    InPlace, 601–602
    options, 507–512
    parallel, 510–511
    perpendicular, 510–511
    previewing, 512
    relationships, 507
    suppressing, 632
    symmetric, 511
    tangent, 49–509
    troubleshooting and repair, 536–539
mates, InPlace, 601–602
Mates folder, 513
mathematical operations, 196
mating
    references, 516–517
    relationships, 505–519
    surfaces, weld beads, 615–616
MC (material condition), 480
Measurement Options Window, 382
measurement settings, 382
memory requirements, 15–16
menus, customizing 817–820
Merge faces option, 293
Merge Result option, 134, 212, 350
mid plane end condition type, 112–113
midpoint constraint, 127
Mirror Feature panel, 267
Mirror icon, 130, 268
mirror images, 269–271
Mirror Part command, 269–271
mirroring. *See also* Modify Sketch command.
    cutting with an open profile, 129-136
    deleting centerline, 130
    dynamic mirroring, 130
    faces and bodies, 268–269
    feature geometry, 267–268
    flipping, 332
Miter Flange icon, 300
Miter Flange panel, 301
miter flanges, 300–302
model document path, 531
model edges, 169
model face, 170
modeling parts with equations, 683–684
model rebuild/regenerate, 117–118
Modify Sketch command, 329–332
Modify Sketch icon, 330
Modify Sketch window, 329
modifying. *See* editing.
mold assembly, 600
molded parts
    adding draft, 218–223
    chamfers, 193–197

draft analysis, 226–232
fillets and rounds, 182–190
hole wizard, 198–202
lofted parts, 202–215
Rib tool, 223–226
selection techniques, 190–192
Shell command, 178–182
shell command, 178–182
silhouette method, 217–218
split lines, 215–217
thin-walled parts, 178
molds, 585–586
monitors, 16–17. *See also* screens.
mouse buttons
   context-sensitive menu, 31–32
   left, 31
   middle, 32, 98
   right, 31–32
   selecting entities, 31
   zooming, 32
mouse wheel, 32
Move and Rotate panel, 504
Move Component command, 504
Move/Copy Body panel, 757–759
moving
   bodies, 757–759
   components, 502–504
   hypertext links on notes, 474
Multi-jog Leader icon, 482–484
Multiple Bodies, 134–137
multiple bodies
   adding bodies together, 136
   Combine command, 136
   creating, 134–136
   in joined parts, 623
multiple radius fillets, 185–186
multi-thickness shell, 179
multi-user environments, 704–718

N

named views, 429–430
naming/renaming, 30–31, 120, 327–328, 647, 667
Nested configuration option, 404
nested configurations, 652
nested profiles, 80–81
networks, working over, 708–718
Neutral Plane method, 220
neutral planes, 220—221
new assembly, starting, 498–499
new drawings. *See also* drawing templates.
   adding a company logo, 415
   aligned sections, 436–440

Auxiliary View icon, 428
broken views, 444–446
Broken-out Section icon, 441
Broken-out Section panel, 441
broken-out section views, 441
creating, 441
creating a sheet format, 413
creating named views, 429–430
cropped views, 446–447
   detail views, 442–444
drawing interface, 410–413
drawing sheet formats, 413
empty views and creating tables, 447–448
hiding/showing view borders, 424–425
inserting views, 422–424
modifying view alignment, 426
New icon, 409
Project View icon, 427
projected views, 427–428
relative to model view method, 430–431
Relative View icon, 430
Save Sheet Format window, 416
saving a sheet format, 415–416
section views
   activating views, 431
   creating a section view, 433–434
   flipping section line arrows, 434
   Partial section option, 434
   section line positioning, 435
   Section View icon, 433
   Section View panel, 434
   section view properties, 434–435
starting a new drawing, 409
New icon, 23, 498
New SolidWorks Document window, 828
New View icon, 93–94
non-parametric modelers, 8
Normal cut option, 311–312
normal cuts, 312
Normal To function, 37
Normal To icon, 37
normal to profile, 213–214
Note icon, 471
notes
   adding, 471–472
   aligning, 472
   Apply to all option, 473
   arrow "smart" mode, 473
   Arrows/Leaders panel, 472–473
   border panel, 475

callouts, 87–88
cursor as small hand, 474
cursor changes, 474
design tables, 651, 663
hypertext links, 473
linking properties, 474–475
moving a hypertext link, 474
Note icon, 471
small yellow star, 473
vs. text features, 763–764
Text Format panel, 473–475
underlined blue text, 473

O

obround relief, 292
Offset curve option, 266
offset entities, 170–171, 298, 301–302
Offset Entities command, 170–171
Offset option, 298
Offset Surface panel, 756
offset surfaces, 755–756
On Surface plane creation, 144–146
open areas, 323
Open for read-only option, 712
Open hems, 306–307
Open referenced documents... option, 711
open-profile cutting, 586–587
ordinate dimensions, 258–259
orientation and twist control, 350–351
Orientation window, 35–36
Orientation/twist Type option, 350
origin point, 46–47
overdefined entities, 44–45, 537
overflow control, 189
overlapping geometry, 79–80
overwriting files, 710

P

palette assemblies, 517–519
palette parts, 517–519
Pan icon, 97
panning, 97, 98
Parabola icon, 70
parabolas, 70
parallel lines, 58
parallel mate, 510–511
Parallelogram icon, 66–67
parallelograms, 66–67
parameter insertion failure, 657
parametric modeling/modelers, 4, 8–9
parent/child relationship, 395–396
part crosshatch, 437–438
part history. *See* rolling back.

Index

Part option, 439
Partial section option, 434
Parting Line method, 220
parting lines, 113, 218–220
parts
   ConfigurationManager, 391–394, 399
   dimensional configurations, 399
   nested configurations, 404–405
   other configurable objects, 399–404
   suppression states, 395–399
Paste icon, 238
Pattern Seed Only option, 243
patterned components, deleting, 523
patterns
   axes, 253–255
   centroids, 260
   chaotic, 258–260
   circular, 250–252
   control-drag technique, 240–241
   copying and pasting features, 238–240
   curve-driven, 264–267
   dangling dimensions, 238–239
   deleting instances, 245–246
   exercises, 246–249, 255–257
   external references, 271–273
   geometry, 252–253
   irregular, 258–260
   linear, 241–243
   mirror images, 269–271
   mirroring, 267–269
   ordinate dimensions, 258–259
   reference points, 260
   seeding, 243
   sketch-driven, 258–260
   skipping instances, 245–246
   step-and-repeat, 274–278
   symmetrical parts, 269
   table-driven, 260–264
   vary sketch, 243–245
   on x- and y-axis coordinates. *See* table-driven patterns.
patterns of components, 519–523
Per Standard option, 443
performance add-in programs, 698–699
pierce constraint, 207–210
Pin position option, 444
plan view, 37
Planar Surface panel, 756
planar surfaces, 756
Plane icon, 137
planes. *See also* sketch planes.
   color coding, 33–35
   creating, 138–149
   displaying, 32–35
   hiding/showing, 33
   naming, 30–31
   neutral, 220—221
   new planes, 137–138
   plan view, 37
   selecting a planar face, 110
   selecting a plane, 110
   selecting a view, 37
   selecting appropriate plane, 108
   starting a plane, 138
   transparency, 34–35
Point command, 199
Point icon, 71
point light source, 838–840
point of tangency, 146
points, 70–71, 145–146
Polygon icon, 67–68
polygons, 67–68
Previous View icon, 96
previous views, 96
Process-Bends1 feature, 335
Profile option, 443
profiles
   closing, 80
   cutting with an open profile, 121
   determining the best, 108
   nested, 344
   single, 344
   swept, 359
   valid, 344
Project View icon, 427
Projected Curve command, 375
projected curves
   creating a projected curve, 377
   Projected Curve command, 375, 377
   projecting a sketch, 375–377
   Projection Icon, 377
   Sketch onto Face(s) option, 377
   Sketch onto Sketch option, 377
projected views, 427–428
Projection Icon, 377
properties. *See also* PropertyManager.
   assembly configurations, 630–633
   cosmetic threads, 476–478
   custom, 668–675
   document, modifying, 39–40
   of newly inserted items, 393–394
PropertyManager. *See also* FeatureManager.
   accessing item properties, 28–29
   automatic display, 27
   changing item properties, 28–29
   help, 27
   opening, 26–27
   in place of FeatureManager, 27
   purpose of, 27
punching. *See* forming tools.

R

Radiate Surface command, 596–597
Radiate Surface panel, 597
radiated surfaces, 596–597
RapidDraft files
   creating RapidDraft drawings, 489–490
   Load Model option, 491
   properties, 488–489
   synchronizing model geometry, 490–491
read-only
   attributes, 714–715
   components, reloading, 712
   documents, editing, 712–714
   documents, reloading, 711–712
   files, opening, 710–711
Rebuild icon, 116
rebuilding a model, 117
rectangles, 66
rectangular relief, 292
reference dimensions, turning off parentheses, 461–462
Reference Geometry toolbar, 137
Reference point option, 260
reference points, 260, 378
referenced configuration, 532
references, missing, 537
referencing configurations, 585
refresh rate, 16–17
regenerating a model, 118
relations, adding, 49–50
relief, 291–294, 298–299
reloading
   components, 535–536
   read-only components, 712
   read-only documents, 711–712
Remove crosshatch option, 439
Remove detail during pan/zoom/rotate option, 698
renaming
   dimensions, 327–328, 647, 667
   features, 120
   objects, 30–31
Reorder Bends window, 336–337
reordering FeatureManager, 743
Reorganize Components command, 702–703
repeating features. *See* patterns.
Replace command, 533

Index

Replace panel, 534
replacing components, 533–536
Reset All button, 797
Reset Standard Views icon, 93, 95–96
resetting views, 95–97
resolving components, 527–528
Reverse Direction option, 355, 357
Revolve, 109
Revolve command, 109, 149
Revolved Boss/Bass icon, 154
revolved features
    adding diameter dimensions from centerlines mistake, 154
    creating a centerline, 150
    editing, 172–174
    Revolve panel, 152
    rules, 151–152
revolving an open profile, 285–286
Rib command, 224
Rib tool, 223–226
ribs, adding, 223–224
Rip command, 338–340
Rip features, 338–340
Rip icon, 339
Rip panel, 339
rolled bends, 336–337
Rolled hems, 306–307
rolling back, 741–743
Rotary motor panel, 550
Rotate Component command, 504
rotate hot key, 98
Rotate View icon, 97
rotating. *See also* Modify Sketch command.
    bodies, 757–759
    components, 502–504
    geometry, with the mouse, 97
    text features, 767–768
    views, 97
round corners, 188
RoundBend1 feature, 335
rounded inside edge. *See* fillets and rounds.
rounds. *See* fillets and rounds.
rules, revolved features, 151–152

S

Save As Copy options, 715
saving
    assemblies as part files, 623–626
    block files, 487
    copies of palette assemblies, 518
    as a copy, 715
    design tables, 654
    relationship copies, 518
    sheet formats, 415–416
    template files, 40–41
    toolbar layouts, 410–412
Scale command panel, 593
Scale hatch pattern option, 444
Scale icon, 593
scaling. *See also* Modify Sketch command.
    cores and cavities, 592–593
    sketches, 332
search order, 717–718
search paths, 717–718
section line arrows, 434
section line positioning, 435
section properties
    how they work, 381
    Mass Properties command, 380
    Measurement Options Window, 382
    measurement settings, 382
    Section Properties command, 380
    system defaults, 382
Section Properties command, 380
Section View icon, 433
Section View panel, 434
seeding patterns, 243
Select icon, 61
Select Loop option, 191
Select Other command, 192
Select Tangent, 191
selecting
    appropriate plane, 108
    entities, 31
    loop, 191
    midpoint, 190–191
    planar faces, 110
    planes, 55–56, 110
    something behind a transparent object, 543
    tangency, 191
    transparent object, 543
selection techniques
    select chain, 191
    select loop, 191
    Select Loop option, 191
    select midpoint, 190–191
    select other, 192
    Select Other command, 192
    select tangency, 191
    Select Tangent, 191
    transparency, 543
self-intersecting
    geometry, 79
    path, 345–346
    sweep, 345–346
separator bar, moving, 26
set origin on curve, 143
setback fillets, 734–735
Shape command, 729–731
SharpBend1 feature, 335
Sharp-Sketch1 feature, 334
sheet metal. *See also* Feature palette.
    bending
        auto relief, 291–292
        bend allowance, 289–291, 298–299, 334
        bend angle, 335–336
        bend direction, 335
        bend radius, 334, 335
        bend tables, 290–291
        flattening bends, 310–311
        inserting bends, 333–336
        jogs, 304–306
        K-factor, 289
        lofted bends, 310–311
        placing bends, 334
        processing bends, 335
        relief types, 292, 298–299
        reordering bends, 336
        ripping the metal, 291–292
        rolled bends, 336–337
        selecting all bends, 314
        simplifying edges, 293–294
        sketched bends, 303–304
        unrolling bends, 337
    converting from other CAD programs, 332–338
    corners
        breaking, 307–308
        closing, 308–309
        relief cuts, 294
        trim, 294–295
        cutting, 312–315
    edge flanges, 296–299
    flattening
        adding features while flattened, 313–314
        bends, 310–311
        corner relief cuts, 294
        corner trim, 294–295
        description, 293
        folding, 313–315
        merging faces, 293
        selecting all bends, 314
        simplifying bend edges, 293–294
        unfolding, 313–315
    forming tools
        creating, 320–322
        creating library features, 319–320
        description, 315–316
        dissolving palette features, 318–319

inserting, 316–318
open areas, 323
removing faces, 322–323
simplifying library features, 320
hems, 306–307
legacy parts, 337–338
miter flanges, 300–302
modifying part color, 328–329
from shelled parts, 338–340
tabs, 302
thin-feature parts, 283–285
Sheet Metal toolbar, 287
sheet setup, 417
Sheet-Metal1 feature, 288–292
Sheet-Metal1 panel, 289
Shell command, 178–182
shell feature panel, 179
shell operation, 179
shelled parts, 178, 338–340
shortcuts, keyboard, 821. *See also* hot keys.
Show command, 32–33
Show dimension names option, 647–648
Show Feature Dimensions option, 648
show preview, 212
showing. *See also* hiding.
alternate components, 639
alternates, 639
arc centerpoints option, 80
bodies, 760–761
components, 662
design table in drawings, 666–667
design tables, 662
dimension names, 647–648
dimensions, 648
exploded views, 571
hidden edges, 557–558
planes, 33
preview, 349–350
toolbars, 23
view borders, 424–425
shutoff surfaces, 593–596
silhouette method, 217–218
Silhouette option, 217
Simple Hole command, 201–202
simple holes, 201
Simplify bends option, 293–294
simplifying edges, 293–294
simplifying features, 320
simulation, 549–551
Simulation toolbar, 550
Single command per pick option, 62
Sketch Chamfer icon, 165
sketch chamfers, creating, 165–166

Sketch Driven Pattern panel, 259
Sketch Extend command, 167
sketch fillets, 163–165
sketch geometry, 170, 276
Sketch icon, 42
sketch mode, entering, 42–43, 110
Sketch onto Face(s) option, 377
Sketch onto Sketch option, 377
sketch planes. *See also* planes.
color coding, 43–44
on a curve, 356
entering sketch mode, 42–43
indicators, 85–86
selecting, 41–42
Sketch Relations toolbar, 25
Sketch toolbar, 24
sketch toolbar auto-activation, 412–413
Sketch Tools toolbar, 25
Sketch Trim command, 167, 752–753
sketch-driven patterns, 258–260, 554
Sketched Bend command, 303–304
Sketched Bend icon, 303–304
Sketched Bend panel, 304
sketched features vs. applied, 21–22
sketches
creating base feature sketch, 110
creating features, 109
defining, 108
editing, 116
exiting accidentally, 115
knowing you are in one, 47–48
mirroring, 130–134
reentering, 115–116
rotating, 329–330
translating, 331–332
sketching
3 point arcs, 64–65
adding relations, 49–50
angled rectangles, 66
autodimensioning, 91–92
auto-transitioning, 60
bends, 303–304
callouts, 87–88
centerlines, 71
centerpoint arcs, 65
centerpoint ellipses, 69–70
checking validity, 82–83
circles, 65
click-click method, 57–58
click-drag method, 57
closing profiles, 80
command persistence, 62
constraints, 49–52
deleting entities, 58–59

deleting geometric relations, 88–89
dimensioning, 89–92
displaying arc centerpoints, 80
displaying entity points, 80
dragging geometry, 61–62
ellipses, 68–69
exceeding allowed geometry, 81–82
exiting commands, 61
full definition, 83
geometric relations, 86–89
Grid/Snap settings, 84–85
guidelines, 79–83
keeping it simple, 83
lines, 56–58
nested profiles, 80–81
new views, 94
overlapping geometry, 79–80
overview, 19–20
panning, 97
parabolas, 70
parallel lines, 58
parallelograms, 66–67
points, 70–71
polygons, 67–68
previous views, 96
rectangles, 66
resetting views, 95–97
rotating views, 97
select mode, 60
selecting a plane, 55–56
self-intersecting geometry, 79
sketch plane indicators, 85–86
snap behavior, 85–86
specific contours, 81–82
splines, 73–79
stopping short, 90
system feedback, 48–49
system views, 93
tangent arcs, 60, 64
unintended gaps, 80
units of measure, 84
updating views, 94–95
user-defined views, 94
view orientation, 92–99
watching the cursor, 48–49
zooming, 96–99
SmartMates icon, 514–515
smooth transition fillets, 732
smoothing, 349
snap behavior, 85–86
software requirements, 12–14. *See also* hardware requirements.
solidifying surfaces, 754–755
solids vs. surfaces, 748

Index

SolidWorks Basic, 13–14
SolidWorks web site, 13
solve as rigid or flexible property, 532
    spacing, 251–252, 275
Spacing option, grayed out, 276
specific contours, 81–82
spirals, 357
splines
    adding control points, 73–74
    construction entities, 78–79
    creating, 72
    deleting control points, 74
    inflection points, 77
    inspecting sketch curvature, 77
    minimum radius, 77
    moving frames, 76
    nonproportional, 72–73
    proportional, 73
    simplifying, 74–75
    tangency conditions, 75–76
Split Assembly panel, 591
Split command, 589–591
Split command panel, 590
Split Curve command, 204–205
Split Line command, 759–760
split lines
    creating a split line, 216–217
    as hold lines, 738–739
    Parting Line command, 215
    Split Line command, 215
splitting parts, 589–591
spot lighting, 840–842
spreadsheets. See design tables.
springs, 358–359
Standard toolbar, 24
Standard Views toolbar, 36–37
Start/End Offset option, 301–302
step draft, 221–222
step-and-repeat patterns
    angle control, 275
    changing number of instances, 276
    Circular Sketch Step and Repeat
        command, 277–278
    dimension values, 276
    editing, 276
    Linear Sketch Step and Repeat
        command, 274–276
    previewing, 275
    sketch geometry pattern vs. feature,
        276
    spacing, 275
Stop at collision option, 546–547
straight transition fillets, 732
stretched features, 109

stretching. See deforming.
subassemblies
    component, 528–529
    component properties, 529–533
    creating, 701–702
    dissolving, 700–701
    expanding, 663–664
Suppress component models option, 632
Suppress features and mates option, 632
Suppress features option, 393
Suppress icon, 526
suppressing
    assembly features, 661
    components, 524–528, 660–661
    configurations, performance, 699–700
    vs. deleting, 395
    features, 524, 650–653
    mates, 661
states, 395–398, 532
Surface Fill command, 594–596
Surface Fill panel, 594
Surface Finish icon, 481
SurfaceCut panel, 760
Surface-Extrude panel, 747
surfaces
    bodies, 757–761
    creating, 748–749
    cutting with, 759–760
    Deviation Analysis tool, 751
    extending, 750–754
    knitting, 754–755
    lofting, 754
    offsets, 755–756
    planar, 756
    solidifying, 754–755
    vs. solids, 748
    sweeping, 749
    trimming, 750–754
    untrimming, 751–752
Sweep, 109
Sweep command, 109, 348
sweep cut, material remaining, 348
Sweep icon, 346
sweep options, smoothing, 349
Sweep Panel, 346–347
sweep path, existing edges, 346–347
sweeping surfaces, 749
swept features
    Add Relations command, 345
    Align with End Faces option, 348–349
    creating a sweep, 346–347
    definition, 344
    orientation and twist control, 350–353

    remaining material after sweep cut,
        348
    rules for creating, 345
    self-intersecting path, 345–346
    self-intersecting sweep, 345–346
    Sweep command, 348
    Sweep icon, 346
        sweep options, 349–350
    Sweep Panel, 346–347
    sweep path, existing edges, 346–347
    sweeping with guide curves, 353–354
    Tangent propagation option, 348
    valid profiles, 344
    valid sketch geometry, 345–346
symmetric mate, 511
symmetrical parts, 269
symmetrical relationships, adding, 131–132
system defaults, 382
system feedback, 48–49
system maintenance, performance, 700
system requirements. See hardware
        requirements; software
        requirements.
system views, 93

T

Tab command, 302
Table Driven Pattern window, 263
table-driven patterns, 261–263, 554
tables, creating, 447–448
tabs, 302
    tangency conditions, 213–215
tangency point dimensioning, 161–162
Tangent Arc icon, 64
tangent arcs, 60, 64
Tangent edge display option, 449
tangent edges, 449–451
tangent mate, 49–509
tangent propagation, 185
Tangent propagation option, 185, 348
tear relief, 292
template files, saving, 40–41
templates
    drawing, 416–420
    file extensions, 38–39
    modifying document properties, 39–40
    saving a template file, 40–41
    text features, 763–778
text features 763–780
    adding text to dimension values, 157, 161–162
    color coding text features, 765

embossing text features, 769–770, 776–778. *See also* forming tools.
engraving text features, 771–773
fonts, 766. See also formatting tools.
formatting text features, 767–768
justifying text features, 767
modifying text size, 158
notes vs. text features, 763–764
rotating text features, 767–768
templates and text features, 763–778
underlined blue text, 473
Text Format panel, 473–475
Thicken command, 754–755
Thicken panel, 755
Thin Feature option, 284, 287
thin-feature parts, 178. *See also* sheet metal.
   Auto fillet option, 285
   base flanges, creating, 288
   closed-profile thin features, 286–287
   description, 283–285
   revolving an open profile, 285–286
thin-walled parts, 178, 364–367
3 Point Arc icon, 64–65
3 point arcs, 64–65
3 standard views, 422–424
3 view drawings option, 419
3D sketcher
   3D Sketch command, 369
   3D Sketch icon, 368
   3D Sketch Relations, 371
   3D Sketch tool, 368
   Add Relations icon, 370
   creating a 3D sketch, 371–374
   drawing splines in 3D, 371
Through All end condition, 125
through free points, 378
Through Reference Point command, 378
tolerances
   adding, 157
   basic, 157
   bilateral, 157
   geometric, 480
   modifying font size, 158
   setting, 158
   symmetric, 157
toolbars
   customizing, 814–815
   listing, 23–24
   resetting, 816–817
   Sketch, 24

   Sketch Relations, 25
   Sketch Tools, 25
   Standard, 24
   turning on/off, 23
   View toolbar, 24
tooltips, 814
Transform curve option, 266
translating sketches, 331–332. *See also* Modify Sketch command.
transparency, 34–35, 542–543, 696–697
trim corners, 294–295
Trim icon, 166
Trim side bends option, 297
Trim Surface panel, 752
trimming, 166–167, 750–754
   turned parts, 155–170
twist control, 350–351

U

underdefined sketches, color code, 44
undoing. *See* rolling back.
Unfold command, 313–315
Unfold icon, 314
Unfold panel, 314
units of measure, 84, 689–690
unrolling bends, 337
Unsuppress icon, 526
Untrim Surface command, 752
Untrim Surface panel, 752
untrimming surfaces, 751–752
Up To Surface end condition, 121
Update and Delete buttons, 201
update holders, 618
Update Standard Views icon, 93–94
updating views, 94–95
Use Document Fonts option, 766
user interface, 22–25. *See also* FeatureManager; PropertyManager.

V

valid sketch geometry, 345–346
validity checking, 82–83
Variable radius fillet panel, 732
variable radius fillets, 732–734
Vary Sketch option, 243–245
vary sketch patterns, 243–245
Verification on rebuild option, 697
vertex chamfer, 194
view crosshatch, 438–440
view orientation, 35–37, 92–99

View Orientation icon, 36
View toolbar, 24, 448
viewing mates with PropertyManager, 514
views
   3 standard, 422–424
   aligned section, 436
   auxiliary, 428
   broken, 445–446
   broken-out section, 441
   copying between, 458
   cropped, 446–447
   detail, 442–444
   exploded, 567–571
   hiding individual edges, 448–449
   inserting, 422
   named, 429–430
   projected, 427–428
   section, 431
   selecting, 37
   tangent edges, 449–451
virtual sharps, 165
visibility option, 531

W

weld bead files, 618–619
weld beads, 611–619
Weld Symbol icon, 479
Weld Symbol Properties window, 479
weld symbols
   Fixed size weld symbols option, 480
   Weld Symbol icon, 479
   Weld Symbol Properties window, 479
welding assemblies
   disassociating joined components, 623
   joining components, 619–623
   multiple bodies in joined parts, 623
   saving assemblies as part files, 623–626
work area, 26

Z

Zebra Stripes panel, 740
Zoom In/Out icon, 96
Zoom To Area icon, 96
Zoom To Fit icon, 96
Zoom To Selection icon, 97
zooming, 32, 96–99